Pin assignments for the RS232 EIA-574 and EIA-561 protocols.

D0780898

DB25 Pin	RS232 Name	DB9 Pin	EIA-574 Name	RJ45 Pin	EIA-561 Name	Signal	Description			
1						FG	Frame Ground/Shield			
2	BA	3	103	6	103	TxD	Transmit Data	-12	out	in
3	BB	2	104	5	104	RxD	Receive Data	-12	in	out
4	CA	7	105/133	8	105/133	TSR	Request to Send	$+12$	out	in
5	CB	8	106	7	106	CTS	Clear to Send	$+12$	in	out
6	CC	6	107			DSR	Data Set Ready	$+12$	in	out
7	AB	5	102	4	102	SG	Signal Ground			
8	CF	1	109	2	109	DCD	Data Carrier Detect	$+12$	in	out
9							Positive Test Voltage			
10							Negative Test Voltage			
11							Not Assigned			
12						sDCD	Secondary DCD	$+12$	in	out
13						sCTS	Secondary CTS	$+12$	in	out
14						sTxD	Secondary TxD	-12	out	in
15	DB					TxC	Transmit Clk (DCE)		in	out
16						sRxD	Secondary RxD	-12	in	out
17	DD					RxC	Receive Clock		in	out
18	LL						Local Loopback			
19						sRTS	Secondary RTS	$+12$	out	in
20	CD	4	108	3	108	DTR	Data Terminal Rdy	$+12$	out	in
21	RL					SQ	Signal Quality	$+12$	in	out
22	CE	9	125	1	125	RI	Ring Indicator	$+12$	in	out
23						SEL	Speed Selector DTE		in	out
24	DA					TCK	Speed Selector DCE		out	in
25	TM					TM	Test Mode	$+12$	in	out

General specification of various types of capacitor components[2, 3]

Type	Range	Tolerance	Temperature coef	Leakage	Frequencies
Polystyrene	10pF to 2.7μF	±0.5	Excellent	10 G•	0 to 10^{10} Hz
Polypropylene	100pF to 50μF	Excellent	Good	Excellent	
Teflon	1000pF to 2μF	Excellent	Best	Best	
Mica	1pF to 0.1μp	±1 to ±20%		1000 M•	10^3 to 10^{10} Hz
Ceramic	1pF to 0.01μF	±5 to ±20%	Poor	1000 M•	10^3 to 10^{10} Hz
Paper (oil-soaked)	1000pF to 50μF	±10 to ±20%		100 M•	100 to 10^8 Hz
Mylar (polyester)	5000pF to 10μF	±20%	Poor	10 G•	10^3 to 10^{10} Hz
Tantalum	0.1μF to 220μF	±10%	Poor		
Electrolytic	0.47μF to 0.01F	±20%	Ghastly	1 M•	10 to 10^4 Hz

[2] Modified from Wolf and Smith, *Student Reference Manual,* Prentice Hall, pg. 302, 1990.

[3] From Horowitz and Hill, *The Art of Electronics,* Cambridge University Press, pg. 22, 1989.

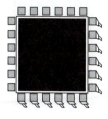

Embedded Microcomputer Systems

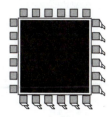

Embedded Microcomputer Systems

Real Time Interfacing

Jonathan W. Valvano

University of Texas at Austin

Brooks/Cole
Thomson Learning™

Australia • Canada • Mexico • Singapore • Spain • United Kingdom • United States

Sponsoring Editor: *Heather Woods*
Marketing Team: *Nathan Wilbur, Christina De Veto,*
  Samantha Cabaluna
Editorial Assistant: *Shelley Gesicki*
Production Coordinator: *Kelsey McGee*
Production Service: *Carlisle Communications, Inc./*
  Terry Routley
Manuscript Editor: *Alice Goehring*

Media Editor: *Marlene Thom*
Permissions Editor: *Mary Kay Hancharick*
Interior Design: *John Edeen*
Cover Design: *Roy Neuhaus*
Typesetting: *Carlisle Communications, Inc.*
Cover Printing, Printing and Binding: *R. R. Donnelley*
  and Sons, Crawfordsville

For more information, contact:
BROOKS/COLE
511 Forest Lodge Road
Pacific Grove, CA 93950 USA
www.brookscole.com

For permission to use material from this work, contact us by
Web: www.thomsonrights.com
fax: 1-800-730-2215
phone: 1-800-730-2214

Printed in United States of America

10 9 8 7 6 5 4 3 2 1

Library of Congress Cataloging-in-Publication Data

Valvano Jonathan W.
Embedded Microcomputer Systems:
Real Time Interfacing/Jonathan W. Valvano.
        p.   cm.

  ISBN 0 534-36642-2
  1. Embedded computer systems.   2. Microprocessors.   3. Computer interfaces.   I. Title
TK7895.E42V35   1999
004.16—dc21   99-38316

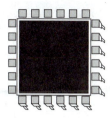

Preface

Embedded computer systems, which are electronic systems that include a microcomputer to perform a specific dedicated application, are ubiquitous. Every week millions of tiny computer chips come pouring out of factories like Motorola and Mitsubishi and find their way into our everyday products. Our global economy, food production, transportation system, military defense, communication systems, and even quality of life depend on the efficiency and effectiveness of these embedded systems. As electrical and computer engineers we play a major role in all phases of this effort: planning, design, analysis, manufacturing, and marketing.

This book is unique in several ways. Like any good textbook, it strives to expose underlying concepts that can be learned today and applied later in practice. The difference lies in the details. You will find that this book is rich with detailed case studies that illustrate the basic concepts. After all, engineers do not simply develop theories but rather continue all the way to an actual device. During my years of teaching I have found that the combination of concepts and examples is an effective method of educating student engineers. Even as a practicing engineer, I continue to study actual working examples whenever I am faced with learning new concepts.

Also unique to this book is its simulator, called *Test Execute and Simulate* (TExaS). This simulator, like all good applications, has an easy learning curve. It provides a self-contained approach to writing and testing microcomputer hardware and software. It differs from other simulators in two aspects. If enabled, the simulator shows you activity internal to the chip, like the address/data bus, the instruction register, and the effective address register. In this way the application is designed for the educational objectives of understanding how a computer works. On the other end of the spectrum, you have the ability to connect such external hardware devices as switches, keyboards, LEDs, LCDs, serial port devices, motors and analog circuits. Logic probes, voltmeters, oscilloscopes, and logic analyzers are used to observe the external hardware. The external devices together with the microcomputer allow us to learn about embedded systems. The simulator supports many of the I/O port functions of the microcomputers, like interrupts, serial port, output compare, input capture, key wake up, timer overflow, and the ADC. You will find the simulator on the CD that accompanies this book. Get the CD out now, run the *Readme.exe,* install TExaS, and follow the tutorial example. In particular, double-click the **tut.uc** file in the MC6812 subdirectory and run the tutorial on the simulator. If you are still having fun, then run the other four tutorials: **tut2.*** shows the simple serial I/O functions, **tut3.*** is an ADC data acquisition example, **tut4.*** shows interrupting serial I/O functions, and **tut5.*** is an interrupting square-wave generator. Although many programs are included in the

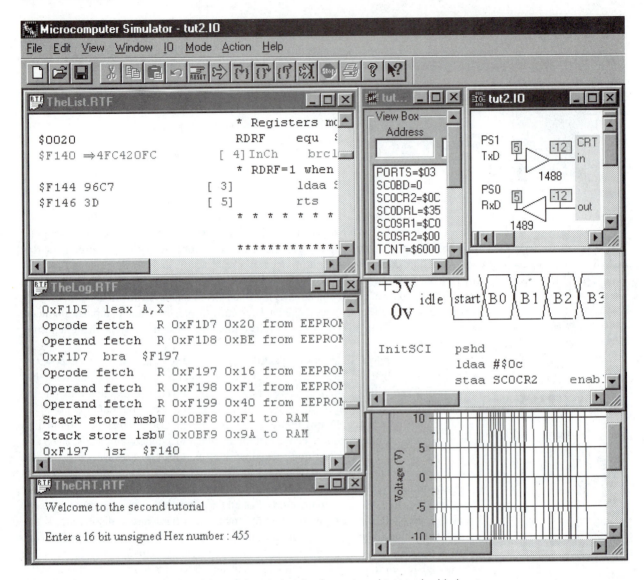

Figure P.1 The TExaS simulator can be used to design, implement, and test embedded systems.

application, these five give a broad overview of the capabilities of the simulator. The screen shot in Figure P.1 was obtained when running the **tut2.rtf** example. Notice these features in the figure: (1) address/data bus activity, (2) embedded figures in the source code, (3) external hardware, (4) voltmeters and logic probes, and (5) an oscilloscope.

P.1 General Objectives

This book constitutes an in-depth treatment of the design of embedded microcomputer systems. It includes the hardware aspects of interfacing, advanced software topics like interrupts, and a systems approach to typical embedded applications. The book is unique to other microcomputer books because of its in-depth treatment of both the software and hardware issues important for real-time embedded applications. The specific objectives of the book include the understanding of:

1. Advanced architecture
   - Timing diagrams
   - Memory and I/O device interfacing to the address/data bus
2. Interfacing external devices to the computer
   - Switches and keyboards
   - LED and LCD displays
   - DC and stepper motors
   - Amplifiers, analog filters, DAC and ADC
   - Synchronous and asynchronous serial ports
3. Advanced programming
   - Debugging of real time embedded systems
   - Interrupts and real time events
   - Signal generation and timing measurements
   - Threads and semaphores
4. Embedded applications
   - Data acquisition systems with digital filters
   - Linear and fuzzy logic control systems
   - Simple communication networks

P.2　Prerequisites

This book is intended for a junior- or senior-level course in microcomputer programming and interfacing. We assume the student has knowledge of:

- Microcomputer programming
- Digital logic (multiplexers, Karnaugh maps, tristate logic)
- Data structures in C (queues, stacks, linked lists)
- Test equipment like multimeters and oscilloscopes
- Discrete analog electric circuits (resistors, capacitors, inductors, and transistors)

Although this book focuses on how to design embedded systems, extensive tutorials on the CD help students with the important issues of how to program in assembly language and how to program in C. A section later in this preface contains more information about what is on the CD.

P.3　Structure of the Book

We will use `this font` whenever displaying assembly language and `this font` for high-level code written in C. Usually, we present assembly language followed by C software for each case study. For the most part, software that is encapsulated in a box has equivalent examples for all four microprocessors. Because of the limited features of the 6805 and 6808, not all examples are available for these two computers. The program number in the caption will assist you in finding the software on the CD. For example, the assembly Program P.1 will be named on the CD as file **Preface.asm** in the corresponding MC6805, MC6808,

```
; MC68HC705J1A
PORTA equ   $0000
      org   $0300 Programs in EPROM
Start lda   PORTA input Port A
      sta   $00C0 save in RAM
      bra   Start
      org   $07FE
      fdb   Start beginning address
```

```
; MC68HC11A8
PORTA equ   $1000
      org   $E000    Programs in ROM
Start ldaa  PORTA    input Port A
      staa  0        save in RAM
      bra   Start
      org   $FFFE
      fdb   Start    beginning address
```

```
; MC68HC708XL36
PORTA equ   $0000
      org   $6E00 Programs in EPROM
Start lda   PORTA input Port A
      sta   $00C0 save in RAM
      bra   Start
      org   $FFFE
      fdb   Start beginning address
```

```
; MC68HC812A4
PORTA equ   $0000
      org   $F000    Programs in EEPROM
Start ldaa  PORTA    input Port A
      staa  $0800    save in RAM
      bra   Start
      org   $FFFE
      fdb   Start    beginning address
```

Program P.1 Program examples that read from input Port A and save the value in memory.

MC6811, or MC6812 subdirectory. All programs in an entire chapter are grouped together into one file. Each chapter has eight files, grouped by microcomputer (MC6805, MC6808, MC6811, or MC6812) and by language (assembly and C).

In assembly language Program P.1, execution begins at the line labeled `Start`. The first assembly instruction (`ldaa PORTA`) reads the current value on input port A (an input port has external pins so that you can connect signals to the microcomputer). The second assembly instruction (`staa`) writes this value into memory. The last assembly instruction (`bra Start`) causes a program branch so that the three instructions (`ldaa staa bra`) are repeated indefinitely. Assembly code can be developed using the editor/assembler included in the TExaS simulator.

When presenting C code, sometimes the implementation for the different computers is so identical that only one version is given, but usually there are important differences. In most situations C code is given just for the MC68HC11 and MC68HC12. The C code in this book was written, compiled, and tested using the ImageCraft compilers (ICC11 for the 6811 and ICC12 for the 6812). We have been very successful transporting these software examples to the Hiware compiler (6805, 6808, 6811, or 6812), with only minor syntax modifications required. The major differences between the ImageCraft™ and HIWARE compilers are the syntax for embedded assembly and the defining interrupt handlers. We have tried using various free SmallC compilers, but the software in this book is too complex to be compiled using SmallC. The biggest limitation of SmallC is its lack of support for data structures. In other words, substantial editing would be required to use SmallC in a course based on this book. Like the assembly language programs, the various C implementations are shown in box format like Program P.2. The C Program P.2 will be named on the CD as file **Preface.c** in the corresponding MC6811 or MC6812 subdirectory.

We assume you have access to the Motorola programming reference guides for your particular microcomputer that show the details of each assembly instruction. You should also have the microcomputer technical reference manuals that detail the I/O ports you will be using. For example, if you are using input capture to measure period on the Motorola 6808, you will find basic principles and examples in this book, but you need to access the 6808 Technical Summary for a complete list of all the details. In other words, we intend

```
// MC68HC11A8
#define PORTA *(char volatile*) (0x1000)
char data;
void main(void){
   while(1){
     data=PORTA;}
}
#pragma abs_address:0xfffe
void (*reset_vector[]) () = { _start };
#pragma end_abs_address
```

```
// MC68HC812A4/MC68HC912B32
#define PORTA *(char volatie*) (0x0000)
char data;
void main (void){
   while(1){
     data=PORTA;}
}
#pragma abs_address:0xfffe
void (*reset_vector[]) () = { _start };
#pragma end_abs_address
```

Program P.2 Program examples that read from input Port A and save the value in memory.

that you use this book along with the manuals from Motorola. Although these reference manuals are available as pdf documents on the CD, it might be better to order physical documents from Motorola's literature center or to download the latest version from the Motorola web site.

Although software development is a critical aspect of embedded system design, we did not intend for this book to serve as an introduction to C programming. Consequently, you will find it convenient to have a C programming reference available. Fortunately, located on the CD that accompanies this book is a reference called *Developing Embedded Software in C using ICC11/ICC12/HIWARE,* written as an HTML document. Although this reference is not as complete a programming guide as some books on C, it is specific for writing embedded software for the 6811 and/or 6812.

In this book we will discuss programming style and develop debugging strategies specific for embedded real-time systems from both an assembly language and C perspective. Because of the nature of single-chip computers (they are very slow and have very, very little RAM memory when compared to today's desktop microcomputers), most of the available C compilers for the single-chip microcomputers we will be using in this book do not support objects, double integers, or floating-point. We feel floating-point should be used only in situations where the range of values spans many orders of magnitude or where the range of values is unknown at software design time. Our experience is that numbers used in embedded systems usually have a narrow and known range, so integer mathematics is sufficient. On the other hand, interest is growing for applying some of the object-oriented approaches of C++ to embedded system design. A few illustrations of object-oriented design can be found in this book.

In the same fashion, this book is not intended as an introduction to assembly language programming. Unfortunately, some students may be using this book without a formal course in assembly language, or they have had assembly language training on a different machine. We strongly believe that a working knowledge of assembly is necessary even when virtually all of our software is written in C. Specifically, we believe you must know enough assembly language to be able to follow the assembly listing files generated by our compiler. This understanding of assembly language is vital when debugging, writing interrupt handlers, calculating real-time events, and considering reentrancy. Consequently, detailed information on assembly language programming is included on the CD. In particular you will find microcomputer data sheets, interactive on-line help as part of the TExaS simulator, and an introduction to assembly language programming as an HTML document.

The electrical components used in this book span a wide range. Examples include the 2764 PROM, 1N914 diode, 2N2222 transistor, 7406 open-collector TTL driver, 74LS74

LSTTL flip flop, 74HC573 high-speed CMOS transparent latch, 3140 op amp, 75451 interface driver, ULN2074 driver, IRF540 MOSFET, 6N139 optocoupler, MC1488 RS232 driver, Dallas Semiconductor DS1620 temperature controller, and various ADCs. It is unrealistic for each student to build a personal library that contains the data books for all these devices. On the other hand, it is appropriate for the company or university to establish a reference library that can be accessed by all the students. The circuit diagrams in this book usually include chip numbers and component values, but not pin numbers or circuit board layout information. Consequently, armed with the appropriate data sheets, most circuits can be readily built.

P.4   How to Teach a Course Based on This Book

The first step in the design of any course is to create a list of educational objectives, along with an understanding of the topics taught in previous classes and the topics needed as prerequisites to subsequent classes. Some important topics like computer architecture and modular software design may be included in multiple courses. Armed with this list, the instructor (or department committee) searches for a book that covers most of the topics in an effective manner. It is unrealistic to teach the entire contents of this book in a single one-semester class, but the rationale in writing this book was to cover a wide range of possible classes. There are two approaches to selecting a subset of material from this book to cover. The first approach is to pick a microcomputer and programming language. For example, one could teach just assembly language on the 6811 or just C programming on the 6812. In this situation, one simply skips the other cases.

The other approach to selecting the appropriate subset is to pick and choose topics. For example, a junior-level laboratory class might introduce the student to microcomputer interfacing. This class might focus on interfacing techniques and may cover Chapters 1 to 4, 6 to 8 and a little bit of Chapters 11 to 13. For these students, the remaining parts of the book become a resource to them for projects later in school or on the job. Another possibility is a senior-level project laboratory class. The objectives of this class might focus on the systems aspect of real time embedded systems. In this situation, some microcomputer programming has been previously taught, so this course might cover the advanced interfacing techniques in Chapters 5, 9, 10, and 15 and the applications in Chapters 12 to 14. For these students, the first half of the book becomes a review, allowing them to properly integrate previous learned concepts to solve complex embedded system applications.

In most departments analog circuit design (for example, op amps and analog filters) is taught in separate classes, so Chapter 11 will be a review chapter. Specific and detailed information about analog circuit design was included in this book to emphasize the system integration issues when designing embedded systems. In other words, developing embedded systems does not rely solely on the tools of computer and software engineering but rather involves all of electrical, computer, and software engineering.

The next important decision to make is the organization of the student laboratory. As engineering educators we appreciate the importance of practical "hands on" experiences in the educational process. On the other hand, space, staff, and money constraints force all of us to compromise, doing the best we can. Consequently, we present three laboratory possibilities that range considerably in cost. Indeed, you may wish to mix two or more approaches into a hybrid simulated/physical laboratory configuration. We do believe that the role of simulation is becoming increasingly important as the race for technological superiority is run with shorter and shorter design cycle times. On the other hand, we should expose our students to all the phases of engineering design, including problem specification, conceptualization, simulation, construction, and analysis.

In the first laboratory configuration, we use the traditional approach to an interfacing laboratory. A physical microcomputer development board is made available for each laboratory group of two students. There are numerous possibilities here. Motorola itself makes evaluation boards for its microcomputers. The breadth of possibilities from commercial vendors can be seen at the back of various trade magazines like *Circuit Cellar INC.* and *Electronics Now.* In addition to the microcomputer board, each group will need a power supply, a prototyping area to build external circuits, and the external I/O devices themselves. A number of shared development/debugging stations will also have to be configured. It is on these dedicated PC-compatible computers that the assembler or compiler is installed. If you develop software in assembly, then the TExaS simulator can be used to edit and assemble software. TExaS will create the standard S19 object code records for downloading. An early version of ICC11 can be used to create 6811 EVB code. In most cases, when programming in C, a current version of a C cross-compiler is greatly preferable. As mentioned earlier, the specific C examples in this book were developed with ImageCraft's ICC11 and ICC12 (ImageCraft, 706 Colorado Ave. Suite 10-88, Palo Alto, CA 94303 *http://www.imagecraft.com*). On the other hand, the HIWARE compiler is also a good but expensive alternative. Test equipment like an oscilloscope, a digital multimeter, and a signal generator are required at each station. Expensive equipment like logic analyzers and printers can be shared. Some mail-order companies sell used or surplus electronics that can be configured into laboratory experiments (for possibilities, see my web site *http://www.ece.utexas.edu/~valvano/book.HTML*). Many laboratory assignments are available using this traditional configuration. For universities that adopt this book, you will be allowed to download these assignments in Microsoft Word format, then rewrite, print out, and distribute to your students laboratory assignments based on these example laboratory assignments. Because of the detailed and specific nature of the laboratory setup, rewriting will certainly be necessary. The Adapt812 board from Technological Arts and the ImageCraft ICC12 cross-compiler is the specific configuration presented in the example laboratory assignments, but the assignments are appropriate for most microcomputer development boards based on the 6805, 6808, 6811, or 6812.

The second laboratory configuration is based entirely on the TExaS simulator. Each book comes with a CD that allows the student to install the application on a single computer. Students, for the most part, work off campus and come to a teaching assistant station for help or laboratory grading. In this configuration you can either develop software in assembly using the TExaS assembler or develop C programs using the demonstration version of ICC11 or HIWARE. The simulator itself becomes the platform on which the laboratory assignments are developed and tested. As mentioned earlier, this freeware version of ICC11 only can be used to create 6811 EVB code. Fortunately, the simulator supports the 6811 EVB architecture. For examples of this ICC11/TExaS combination, run some of the dem-onstration examples in the ICC11 subdirectory of the TExaS application. The demonstration version of HIWARE is also very limited, but it can be used to write very short programs. Laboratory assignments are also available using the simulator. Again, for universities that adopt this book, you will be allowed to download these assignments in Microsoft Word format, then rewrite, print out, and distribute to your students laboratory assignments based on these example laboratory assignments.

The third configuration performs simulation of 6812 C programs. In this laboratory setup, a standard PC-compatible laboratory room can be used. TExaS and a professional 6812 cross-compiler (like ImageCraft ICC12 or HIWARE) are installed, and the students perform laboratory assignments in this central facility. Other than these two pieces of software, no additional setup costs are required. The available laboratory assignments mentioned in both the preceding paragraphs could be adapted for this configuration. The advantage of this approach is that C programming is performed on the newest microcomputer technology (for example, 6812) with only a modest expense.

The exercises at the end of each chapter can be used to supplement the laboratory assignments. In actuality, these exercises, for the most part, were collected from old quizzes and final examinations. Consequently, these exercises address the fundamental educational objectives of the chapter, without the overwhelming complexity of a regular laboratory assignment.

P.5 What's on the CD?

The Readme.exe is a 15-minute introductory tutorial about TExaS. This document does not need to be copied to your hard drive; you can simply watch the movie from the CD itself.

TExaS is a complete editor, assembler, and simulator for the 6805, 6808, 6811, and 6812 microcomputers. It simulates external hardware, I/O ports, interrupts, memory, and program execution. It is intended as a learning tool for embedded systems. This software is not freeware, but the purchase of the book entitles the owner to install one copy of the program. The **Texas** directory contains the installer, which you must do before using the application. Once installed, TExaS creates these nine subdirectories:

- MC6805   that contains 6805 assembly examples
- MC6808   that contains 6808 assembly examples
- MC6811   that contains 6811 assembly examples
- MC6812   that contains 6812 assembly examples
- Hiware08   that contains 6808 C examples using the demo compiler
- Hiware11   that contains 6811 C examples using the demo compiler
- Hiware12   that contains 6812 C examples using the demo compiler
- ICC11   that contains 6811 C examples using the freeware compiler
- ICC12   that contains 6812 C examples (not a freeware compiler)

The subdirectory ICC11 also contains the ImageCraft Freeware compiler. You can run the existing ICC12 examples, but to edit and recompile you will need the commercial C compiler.

The **PDF** directory contains many data sheets in Adobe pdf format. This information does not need to be copied to your hard drive; you can simply read the data sheets from the CD itself. In particular there are data sheets for microcomputers, digital logic, memory chips, op amps, ADCs, DACs, timer chips, and interface chips. If you wish to download more current data sheets, or search for other devices, we have a page of web links on the site *http://www.ece.utexas.edu/~valvano/book.HTML.*

The **example** directory contains software from the book. For example, all the assembly language programs from Chapter 1 can be found in the file "Chap.1.asm." Similarly, all the C language programs from Chapter 1 can be found in the file "Chap1.c." The versions for the various microcomputers are located in the corresponding subdirectories MC6805, MC6808, MC6811, and MC6812. Not all programs have versions for each microcomputer.

The **assembly** directory contains an HTML document describing how to program in assembly for embedded systems using the TExaS application. This document does not need to be copied to your hard drive; you can simply read the HTML document from the CD itself. (Note also that the TExaS application itself contains a lot of information about assembly language development as part of its on-line help).

The **embed** directory contains an HTML document describing how to program in C for embedded systems. This document does not need to be copied to your hard drive; you can simply read the HTML document from the CD itself.

The **lab** directory contains ICC11 and ICC12 software that could be used in a laboratory setting.

The **HIWARE** directory contains a demonstration version of the HIWARE 6805/6808/6811/6812 C/C++ compiler. This limited version can be used to develop small programs that are less than 10 functions and less than 1000 bytes of object code. The HIWARE application must be installed before it can be used.

P.6 Acknowledgments

Many shared experiences contributed to the development of this book. First, I would like to acknowledge the many excellent teaching assistants I have had the pleasure of working with. Some of these hardworking, underpaid warriors include Pankaj Bishnoi, Rajeev Sethia, Adson da Rocha, Bao Hua, Raj Randeri, Santosh Jodh, Naresh Bhavaraju, Ashutosh Kulkarni, Bryan Stiles, V. Krishnamurthy, Paul Johnson, Craig Kochis, Sean Askew, George Panayi, Jeehyun Kim, Vikram Godbole, Andres Zambrano, and Ann Meyer. I dreamed of writing this book the first time I taught microcomputer interfacing on the old Motorola MC6809 in 1981. Over the intervening years my teaching assistants have contributed greatly to the contents of this book, particularly to its laboratory assignments. In a similar manner, my students have recharged my energy each semester with their enthusiasm, dedication, and quest for knowledge.

Second, I appreciate the patience and expertise of my fellow faculty members at the University of Texas at Austin. From a personal perspective, Dr. John Pearce provided much needed encouragement and support throughout my career. Also, Dr. John Cogdell and Dr. Francis Bostick helped me with analog circuit design, and Dr. Baxter Womack and Dr. Robert Flake provided good information about control systems. The book and accompanying software include many finite-state machines derived from the digital logic examples explained to me by Dr. Charles Roth. I continue to appreciate the encouragement and support of Dr. G. Jack Lipovski. An outside observer might conclude that Dr. Jack and I enjoy taking opposite sides of every issue. In actuality this friendly competition makes us organize our otherwise erratic thoughts, and in the process everything we do is the better for it.

Last, I appreciate the valuable lessons of character and commitment taught to me by my grandparents and parents. Most significantly, I acknowledge the love, patience, and support of my entire family, especially my children, Ben, Dan, and Liz.

Jonathan W. Valvano

Good luck!

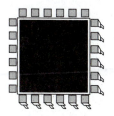

Contents

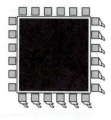

Embedded Microcomputer Systems

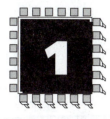

1 Microcomputer-Based Systems

Chapter 1 objectives are to introduce the basic architecture of:

❏ Motorola MC68HC05
❏ Motorola MC68HC08
❏ Motorola MC68HC11
❏ Motorola MC68HC12

The overall objective of this book is to design microcomputer-based systems. In this first chapter, we begin with a discussion of the applications, and then we introduce the architectures of the specific microcomputers that we will be using. It is in this chapter that we develop a feel for the range of possible applications and solutions. For more information concerning a microcomputer, refer to the respective Motorola manual. Data sheets for most of the devices we will use can be found in portable document file (PDF) format on the compact disc (CD) that accompanies this book.

If you are studying just one or two of these microcomputers, you can skip the sections that apply to the other microcomputers. If your overall goal is to develop assembly language software, then the C code can serve to clarify the software algorithms. If your overall goal is to develop C code, then I strongly advise you to learn a little assembly language so that you can understand the machine code that your compiler generates. From this understanding, you can evaluate, debug, and optimize your system.

1.1 Embedded Computer Systems

1.1.1 Applications
The computer engineer has many microcomputer choices when designing real-time embedded applications. The Motorola 8-bit microcontrollers are used for embedded applications. In this book we will develop computer engineering designs using Motorola microcontrollers. We will develop both the design methodology and specific solutions for the various architectures. The particular 8-bit Motorola microcontrollers we will use include:

- Motorola 6805
- Motorola 6808
- Motorola 6811
- Motorola 6812

In this book we will sometimes refer to devices such as the Motorola MC68HC11 simply as the 6811. The different versions of the microcomputers contain various amounts of memory and input/output (I/O) devices. More information about the various programming environments can be found in this chapter.

An *embedded computer system* is an electronic system that includes a microcomputer like the Motorola 6805, 6808, 6811, or 6812 that is configured to perform a specific dedicated application. The software that controls the system is programmed or fixed into ROM and is not accessible to the user of the device. Software for embedded systems typically solves only a limited range of problems. The microcomputer is embedded or hidden inside the device. A typical automobile now contains an average of ten microcontrollers. In fact, upscale homes may contain as many as 150 microcontrollers, and the average consumer now interacts with microcontrollers up to 300 times a day. General areas that employ embedded microcomputers encompass every field of engineering (Figure 1.1):

- Communications
- Automotive
- Military
- Medical
- Consumer
- Machine control

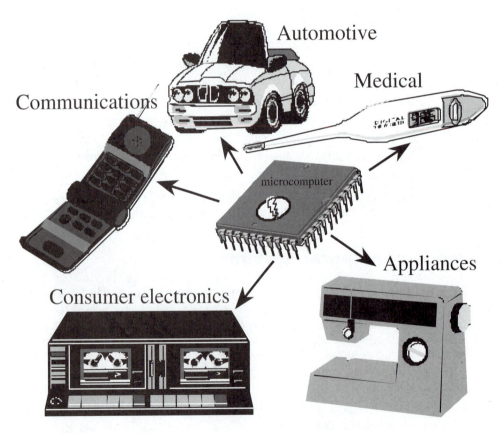

Figure 1.1
Examples of embedded computer systems.

Table 1.1 lists typical embedded microcomputer applications and the function performed by the embedded microcomputer. Each microcomputer accepts inputs, performs calculations, and generates outputs. The other general concept involved in most of these systems is that they run in *real time*. In a real-time computer system, we can put an upper

Table 1.1

Examples of typical microcomputer applications.

Consumer:	Function Performed By the Microcomputer
Washing machine	Controls the water and spin cycles
Exercise equipment	Measures speed, distance, calories, heart rate, logs workouts
Remote controls	Accepts key touches and sends infrared (IR) pulses to base system
Clocks and watches	Maintains the time, alarm, and display
Games and toys	Entertains the user, joystick input, video output
Audio/video	Interacts with the operator and enhances performance
Communication:	
Telephone answering machines	Plays outgoing message, saves and organizes messages
Telephone system	Interactive switching and information retrieval
Cellular phones and pagers	Key pad inputs, sound I/O, and communicates with central station
ATM machines	Provides both security and banking convenience
Automotive:	
Automatic braking	Optimizes stopping on slippery surfaces
Noise cancellation	Improves sound quality by removing background noise
Theft deterrent devices	Keyless entry, alarm systems
Electronic ignition	Controls sparks and fuel injectors
Power windows and seats	Remembers preferred settings for each driver
Instrumentation	Collects and provides the driver with necessary information
Military:	
Smart weapons	Recognizes friendly targets
Missile guidance systems	Directs ordinance at the desired target
Global positioning systems	Determines where you are on the planet
Industrial:	
Setback thermostats	Adjusts day/night thresholds, thus saving energy
Traffic control systems	Senses car positions and controls traffic lights
Robot systems	Input from sensors, controls the motors
Bar code readers and writers	Input from readers, output to writers for inventory control and shipping
Automatic sprinklers	Used in farming to control the wetness of the soil
Medical:	
Apnea monitors	Detects breathing and alarms if the baby stops breathing
Cardiac monitors	Measures heart functions
Renal monitors	Measures kidney functions
Drug delivery	Administers proper doses
Cancer treatments	Controls doses of radiation, drugs, or heat
Pacemakers	Helps the heart beat regularly
Prosthetic devices	Increases mobility for the handicapped
Dialysis machines	Performs functions normally done by the kidney

bound on the time required to perform the input/calculation/output sequence. Because of the real-time nature of these systems, we will study the rich set of features built into these microcontrollers to handle all aspects of time. An *interface* is defined as the hardware and software that combine to allow the computer to communicate with the external hardware. We must also learn how to interface a wide range of inputs and outputs that can exist in either digital or analog form.

In contrast, a *general-purpose computer system* typically has a keyboard, disk, and graphics display and can be programmed for a wide variety of purposes (Figure 1.2). Typical general-purpose applications include word processing, electronic mail, business accounting, scientific computing, and data base systems. The user of a general-purpose computer does have access to the software that controls the machine. In other words, the user decides which operating system to run and which applications to launch. Because the general-purpose computer has a removable disk or network interface, new programs can be added easily to the system. The most common type of general-purpose computer is the personal computer—for example, the Pentium-based IBM PC compatible and the PowerPC-based Macintosh. Computers more powerful than the personal computer can be grouped in the workstation ($10,000 to $50,000 range) or supercomputer (more than $50,000) categories. These computers often employ multiple processors and have much more memory than the typical personal computer. The workstations and supercomputers are used for handling large amounts of information (business applications) or performing large calculations (scientific research). We will not cover the general-purpose computer, although many of the basic principles of embedded computers do apply to all types of systems.

Figure 1.2
Examples of general-purpose computer systems.

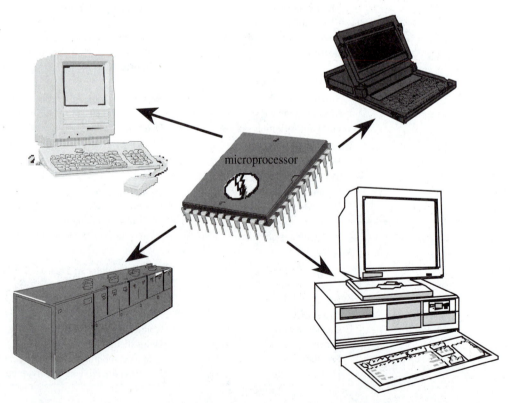

1.1.2
Software Issues

In an embedded system, the software is usually programmed into read-only memory (ROM) or electrically erasable programmable ROM (EEPROM) and therefore fixed. Even so, *software maintenance* (e.g., verification of proper operation, updates, fixing bugs, adding features, extending to new applications, end user configurations) is still extremely important. We will develop techniques that facilitate this important aspect of system design.

In the next chapter, we will study software development in detail, but for now we introduce simple instructions in order to understand the architecture of each computer. The TExaS simulator that accompanies this book can be used to visualize program execution by activating the *CycleView* mode. The following notation will be used in describing the action of specific assembly language instructions.

w	any signed 8-bit −128 to +127	or any unsigned 8-bit 0 to 255
n	any signed 8-bit −128 to +127	
u	any unsigned 8-bit 0 to 255	
W	any signed 16-bit −32768 to +32767	or any unsigned 16-bit 0 to 65535
N	any signed 8-bit −32768 to +32767	
U	any unsigned 8-bit 0 to 65535	
[addr]	the 8-bit contents read at addr	
{addr}	the 16-bit contents read at addr "big endian"	
[addr]=	write 8-bit data to addr	
{addr}=	write 16-bit data to addr big endian	

Assembly language instructions have four fields. The label field is optional and starts in the first column, and it is used to identify the position in memory of the current instruction. You choose a unique name for each label. The operation code (opcode) field specifies the microcomputer command to execute. There are many similarities in the opcodes among the four computers studied in this book. The operand field specifies where to find the data to execute the instruction. We will see opcodes that have 0, 1, 2, or 3 operands. The comment field is also optional and is ignored by the computer, but it allows you to describe the software, making it easier to understand. Good programmers add comments to explain what the instruction does—for example,

```
here        ldaa  100      RegA = [100]
 ↑           ↑     ↑         ↑
Label      Opcode Operand  Comment
```

As we will learn in Chapter 2, it is much better to add comments to explain how or (even better) why we do the action. But for now we want to learn what the instruction is doing, so comments in this chapter will describe what the instruction does. The assembly language instructions (like the above example) are translated into machine instructions. The `ldaa 100` instruction is translated into 2 bytes of machine code:

Object code	Instruction	Comment	
$A6 $64	lda 100	RegA=[$0064]	for the 6805 and 6808
$86 $64	ldaa 100	RegA=[$0064]	for the 6811 and 6812

Note that the numbers that start with $ (e.g., $64) are specified in hexadecimal notation. In C code, we start hexadecimal numbers with 0x (e.g., 0x64). Intel assembly language adds an "H" at the end to specify hexadecimal notation (e.g., 64H).

1.1.3
Memory-Mapped
Architecture

A computer system combines a processor, random-access memory (RAM), ROM, input devices, and output devices. In a system with *memory-mapped I/O,* the I/O devices are connected to the processor in a manner similar to memory. I/O devices are assigned addresses, and the software accesses I/O using reads and writes to the specific I/O address. The software inputs from an input device by using the same instructions as it would if it were reading from memory. Similarly, the software outputs from an output device by using the same instructions as it would if it were writing to memory (Figure 1.3). The *bus* contains address,

Figure 1.3
A memory-mapped
computer system.

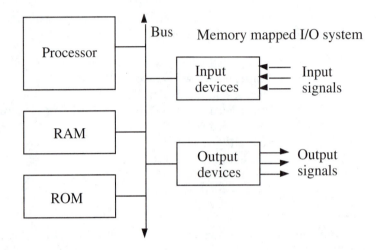

data, and control information that provides data transfer between the various modules in the system. The address specifies which module (input, output, RAM, or ROM) will communicate with the processor. The data contains the information that is being transferred. Control signals specify the direction of the transfer. We call a complete data transfer a *bus cycle.* In a simple computer system, like the 6805/6811/6812, only two types of transfers are allowed (Table 1.2). In this simple system, the processor always controls the address (where to access), the direction (read or write) (Figures 1.4, 1.5), and the control (when to access).

Table 1.2
Simple computers
generate two types of
cycles.

Type	Address Driven By	Data Driven By	Transfer
Read cycle	Processor	RAM, ROM, or input	Data copied to processor
Write cycle	Processor	Processor	Data copied to output or RAM

Figure 1.4
A read cycle copies data
from RAM, ROM, or
input device into the
processor.

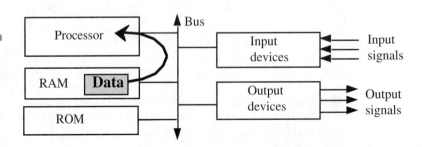

Figure 1.5
A write cycle copies data from the processor into RAM or an output device.

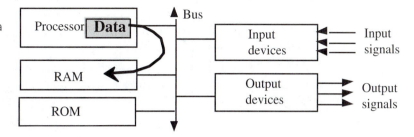

You see that if we wish to transfer data from an input device into RAM, we must first transfer it from input to the processor, then from the processor into RAM. In more complex systems like the Motorola 6808, Motorola 68340, and IBM PC, we will be able to transfer data directly from input to RAM or RAM to output using *direct memory access* (DMA) (Figures 1.6, 1.7). The *bandwidth* of an I/O device is the number of bytes per second that can be transferred. Because DMA is faster, we will use this method to interface high-bandwidth devices like disks and networks (See Chapter 10).

Figure 1.6
A DMA read cycle copies data from RAM, ROM, or input device into an output device.

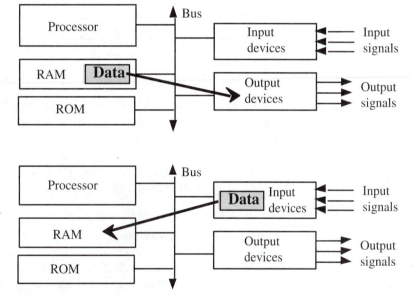

Figure 1.7
A DMA write cycle copies data from the input device into RAM or an output device.

In a computer system with *isolated I/O,* the control bus signals that activate the I/O are separate from those that activate the memory devices (Figure 1.8). These systems have a separate address space and instructions to access the I/O devices. The Intel x86 family has four control bus signals: MEMR, MEMW, IOR, and IOW. The advantages of isolated I/O are that software does not inadvertently access I/O when it thinks it is accessing memory, and the I/O interfaces are simpler. On the IBM PC, there are 32 memory address lines, but only 11 of those 32 lines are used to access I/O devices. Rather than use the regular memory access instructions, the Intel x86 computer uses special IN and OUT instructions to access the I/O devices.

The Motorola microcomputers implement memory-mapped I/O; therefore each I/O device is assigned an address. To input from a port, we simply read from its address. For

Figure 1.8
An isolated I/O computer system.

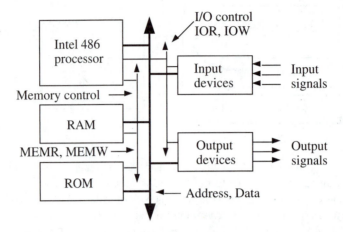

example, if we wish to bring into Register A the current values on Port B (assuming Port B is located at $0001), we could execute

```
ldaa $0001    RegA=[$0001] copy to values of PortB into RegA
```

On an isolated I/O system like the Intel 486, if we wish to bring into Register AL the current values of an input port that is located at $10, we could

```
in AL,$10¹    Copy the values of Port $10 into Register AL
```

To output to a port, we simply write to its address. For example, if we wish to set the output bits of Port A (on the 6811 Port A is located at $1000) to the value in Register A, we could

```
staa $0000    [$0000]=RegA copy the values of RegA out to Port A
```

On an isolated I/O system like the Intel 486, if we wish to send the value in Register AL to an output port that is located at $11, we could

```
out AL,$11    copy the values of Register AL out to Port $11
```

In this book we will study a wide range of on chip peripherals supported by the Motorola microcomputers. The synchronous Serial Peripheral Interface (SPI) and asynchronous Serial Communications Interface (SCI) have a wide range of software-selectable baud rates and are optimized to minimize central processing unit (CPU) overhead. The SPI also enables synchronous communication between the microcontroller and peripheral devices such as:

Shift registers
Liquid crystal display (LCD) drivers
Analog-to-digital converters (ADCs)
Other microprocessors

The timer features on the Motorola microcomputers include:

Fixed periodic rate interrupts
Computer Operating Properly (COP) protection against software failures

---

[1]Hexadecimal numbers on the Intel assemblers are usually specified as 10H.

Pulse accumulator for external event counting or gated time accumulation

Pulse-width-modulation (PWM) outputs

Event counter system for advanced timing operations

Input capture used for period and pulse width measurement

Output compare used for generating signals and frequency measurement

Analog-to-digital (ADC) systems are available with 8- and 10-bit resolution. The ADC system is software-programmable to provide single or continuous conversion modes.

1.2 MC68HC05 Architecture

Like most microcomputers, the 6805 is available in a wide range of memory and I/O configurations. In this book we choose to study the MC68HC705J1A because of its low cost and simple packaging. We also include the MC68HC05C4 because of the large volume of applications currently using the MC68HC05C family.

1.2.1 MC68HC705J1A

The 20-pin MC68HC705J1A microcomputer is available in plastic dual inline package (PDIP) or small outline integrated circuit (SOIC). To make it run, we add a crystal to pins 1, 2 and power to pins 9, 10 (see Figure 1.9). The MC68HC705J1A can operate at either +5 V (with a maximum crystal frequency of 4.2 MHz) or +3.3 V (with a maximum crystal frequency of 2 MHz). The microcomputer executes with clock frequency of half the crystal frequency. With a 2-MHz crystal, the 6805 cycle time is 1 μs. This means it can read/write memory every 1 μs. Each 6805 instruction takes a fixed number of memory cycles to execute. Using this clock frequency, we can determine exactly how long each instruction takes to execute. The two bypass capacitors across power (V_{DD}) and ground (V_{SS}) are used to reject a wide range of noise on the power line. A large polarized tantalum capacitor (10 μF) rejects low-frequency noise, and the small nonpolarized (0.01 μF) ceramic capacitor rejects high-frequency noise. The RESET line can be connected to a reset button or a power-on-reset circuit. The IRQ line is an external active low-interrupt request. More

Figure 1.9
The MC68HC705J1A
has 14 I/O pins.

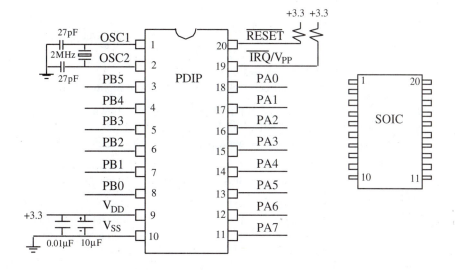

information about interrupts will be presented in Chapter 4. The general features of the MC68HC705J1A are:

- 1240 bytes of erasable programmable ROM (EPROM)
- 64 bytes of RAM
- 15-stage multifunction timer
- 14 bidirectional I/O lines

The address map of the MC68HC705J1A is shown in Table 1.3. The MC68HC05J1A is similar to the MC68HC705J1A except that the EPROM is replaced with factory program- mable ROM. Typically we develop the prototype using the EPROM part and switch over to the ROM part for large production. In particular, we will study the MC68HC705J1A I/O devices listed in Table 1.4. Recall that the memory clock period is twice the crystal clock period. For example, if the MC68HC705 is driven with a 2-MHz crystal, then the memory clock period is 1 μs, and TCR is incremented every 4 μs.

Table 1.3
The MC68HC705J1A has about 1K of EPROM and 64 bytes of RAM.

Address (hex)	Size	Device	Contents
0000 to 001F	32	I/O	
00C0 to 00FF	64	RAM	Variables and stack
0300 to 07CF	1232	EPROM	Programs and fixed constants
07F0 to 07FF	16	EPROM	Vectors

Table 1.4
The MC68HC705J1A has two external I/O ports and an internal counter.

I/O device	Address	Function
PORTA	$0000	8-bit bidirectional I/O port
PORTB	$0001	6-bit bidirectional I/O port
DDRA	$0004	Specifies whether each bit of PORTA is input (0) or output (1)
DDRB	$0005	Specifies whether each bit of PORTB is input (0) or output (1)
TCR	$0009	8-bit up counter incremented every memory clock period

1.2.2 MC68HC05C4

The general features of the MC68HC705C4/MC68HC05C4 are:

- 4160 bytes of EPROM/ROM
- 176 bytes of RAM
- 16-bit timer interface module (input capture/output compare)
- Synchronous SPI module
- Synchronous SCI module
- 24 bidirectional I/O lines and seven input-only lines

The address map of the MC68HC705C4 is shown in Table 1.5. The MC68HC05C4 is simi- lar to the MC68HC705C4 except that the EPROM is replaced with factory-programmable ROM. In particular, we will study the MC68HC05C4 I/O devices listed in Table 1.6.

Table 1.5
The MC68HC705C4 has 4K of EPROM and 176 bytes of RAM.

Address (hex)	Size	Device	Contents
0000 to 001F	32	I/O	
0050 to 00FF	176	RAM	Variables and stack
0100 to 10FF	4096	EPROM	Programs and fixed constants
1FF0 to 1FFF	16	EPROM	Vectors

Table 1.6
The MC68HC05C4 has four external I/O ports and an internal counter.

I/O device	Address	Function
PORTA	$0000	8-bit bidirectional I/O port
PORTB	$0001	8-bit bidirectional I/O port
PORTC	$0002	8-bit bidirectional I/O port
PORTD	$0003	7-bit fixed input port (shared with SPI, SCI)
DDRA	$0004	Specifies whether each bit is input (0) or output (1)
DDRB	$0005	Specifies whether each bit is input (0) or output (1)
DDRC	$0006	Specifies whether each bit is input (0) or output (1)
TSC	$0012	Timer status and control
TCNT	$0018	16-bit up counter incremented every clock period

1.2.3 MC68HC05 Registers

The 6805 registers are depicted in Figure 1.10. Typically Register A contains data (numbers), while Register X contains addresses (pointers). Register SP+1 points to the top element of the stack. Register PC points to the current instruction. For compatibility with other members of the Motorola microcomputer family, we can think of SP and PC as 16-bit registers, but in actuality the SP on the 6805 is only 6 bits and the PC is only 11 bits. The condition code bits are shown in Figure 1.11.

Figure 1.10
The 6805 has five registers.

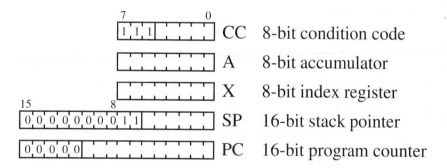

CC	8-bit condition code
A	8-bit accumulator
X	8-bit index register
SP	16-bit stack pointer
PC	16-bit program counter

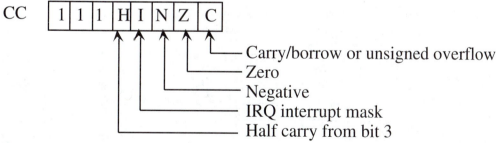

Figure 1.11
The 6805 condition code register has 5 bits.

The H bit is used for binary-coded decimal (BCD) addition (see the `add adc bhcs` and `bhcc` instructions). When I=0, IRQ interrupts are enabled. N, Z, C bits signify the status of the previous arithmetic-logic unit (ALU) operation. Notice that the 6805 does not have a signed overflow bit (V bit) like the 6808, 6811, and 6812. Although the 6805 implements the half carry bit, it does not have the `daa` instruction like the 6808, 6811, and 6812, so BCD addition is more awkward on the 6805.

The bidirectional I/O pins on the MC68HC705J1A (Port A bits 7–0 and Port B bits 5–0) have direction registers that specify whether the pin is an input or output. For example, the direction register for Port A is located at $0004. If we wish to make Port A bits 7, 6, 5, 4 outputs and bits 3, 2, 1, 0 inputs, we could execute

```
lda #$F0      RegA=$F0    0 means input, 1 means output
sta $0004     [$0004]=RegA
```

If we wish to make all six pins of Port B bits inputs, we could execute

```
clr $0005     [$0005]=0
```

1.2.4 MC68HC05 Addressing Modes

The 6805 instruction set has six addressing modes:

1. Inherent addressing mode has no operand field. Examples:

Machine code	Opcode	Operand	Comment	Cycles
$42	mul		Reg X:A = RegA • RegX (unsigned)	11
$5C	incx		RegX = RegX + 1	3
$9A	cli		Condition Code I bit = 0	2
$9C	rsp		RegSP = $00FF	2
$83	swi		Software interrupt	10
$97	tax		RegX = RegA	2

2. Immediate addressing mode uses a fixed constant (Figure 1.12). The data themselves are included in the machine code. Examples:

Machine code	Opcode	Operand	Comment	Cycles
$AE**C0**[2]	ldx	#$C0	Initialize index register	2
$A9**00**	adc	#0	RegA = RegA+C (Add carry bit to RegA)	2
$A8**80**	eor	#$80	Invert bit 7 of RegA	2
$AA**02**	ora	#2	Set bit 1 of RegA	2

Common Error: It is illegal to use the immediate addressing mode with instructions that store data into memory (e.g., `sta` and `stx`).

Figure 1.12
Example of the 6805
immediate addressing
mode.

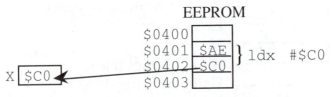

3. Direct page addressing mode uses an 8-bit address to access from addresses 0 to $00FF (Figure 1.13). These addresses include the 6805 I/O ports and RAM. Notice that some instructions can be used with both direct page and extended addressing (e.g., `adc add and bit cmp cpx eor jmp jsr lda ldx ora sbc sta stx sub`), while others can be used with direct but not extended mode. The < operator forces direct page addressing. Examples:

---

[2]**Boldface** type represents the operand portion of the machine code.

Machine code	Opcode	Operand	Comment	Cycles
$B6**32**	lda	50	RegA = [$0050]	3
$BD**64**	jsr	100	Jump subroutine at location 100	5
$BF**34**	stx	<$1234	Will store RegX at $0034	4
$17**0A**	bclr	3,10	Clear bit 3 of location 10	2
$08**05**rr	brset	4,5,FUN	Go to FUN if bit 4 of location 5 is 1	5
$05**07**rr	brclr	2,7,OK	Go to OK if bit 2 of location 7 is 0	5

For an explanation of "rr," see relative addressing made in item 6.

Figure 1.13
Example of the 6805 direct page addressing mode.

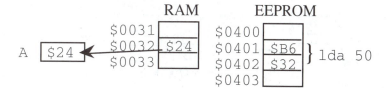

4. Extended addressing mode uses a 16-bit address to access all memory and I/O devices (Figure 1.14). Notice that the instructions that implement bit manipulations or perform read-modify-write operations cannot be used with extended addressing (e.g., bclr brclr brset bset asl asr clr com dec inc lsl lsr neg rol ror tst). The > operator forces extended addressing. Examples:

Machine code	Opcode	Operand	Comment	Cycles
$C6**0301**	lda	$0301	Read memory at $0301 and put into RegA	4
$CB**0401**	add	$0401	RegA = RegA + [$0401]	4
$CE**0000**	ldx	>0	Force 16-bit addressing	4

Figure 1.14
Example of the 6805 extended addressing mode.

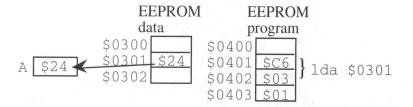

5. Indexed addressing mode uses one of three possibilities: no offset, an 8-bit uns; offset, or a 16-bit offset with RegX. Typically, indexed mode is used in two ster RegX is initialized to point to memory or an I/O port. Then, indexed mode ? is used to access the information. Indexed mode is useful when address; structures. With the 6808, 6811, and 6812, indexed mode will allow acc tion on the stack. We will place local variables on the stack, but unfortv ables can not be created with the 6805. Register X is not modified dressing mode. In each case, the 16-bit sum of RegX and the 0-, 8 used as a pointer. With 8-bit index mode, the 8-bit fixed constant is For example if RegX were $D0, and the 8-bit constant were $C0, .

address would be $00D0+$00C0, or $0190. With 16-bit index mode, negative constants are possible. The first three examples use 0-, 8-, and 16-bit constants, respectively. In the first example (Figure 1.15), $F7 is the `sta 0,X` instruction, and there is no extra machine code to represent the operand:

Machine code	Opcode	Operand	Comment	Cycles
$F7	sta	0,X	[X] = RegA	4

Assuming Register X=$23, the instruction `sta 0,X` will store a copy of the value in Register A at $23, leaving Register X unchanged. The effective address (EA) is $0023.

Figure 1.15
Example of the 6805 indexed addressing mode.

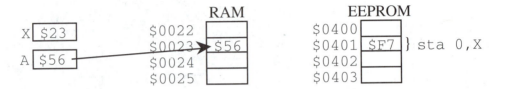

In the next example (Figure 1.16) $E7 is the `sta` instruction and **$04** is the machine code representing the operand:

Machine code	Opcode	Operand	Comment	Cycles
$E7**04**	sta	4,X	[X+4] = RegA	5

Assuming Register X=$23, the instruction `sta 4,X` will store a copy of the value in Register A at $27, leaving Register X unchanged. The effective address (EA) is $0023+4=$0027.

Figure 1.16
A second example of 6805 indexed addressing mode.

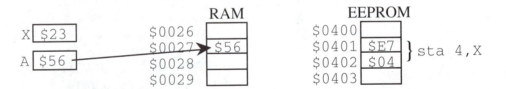

In the third example (Figure 1.17), $D7 is the `sta` instruction and **$FFFC** is the machine code representing the operand:

Machine code	Opcode	Operand	Comment	Cycles
$D7**FFFC**	sta	-4,X	[X+$FFFC] = RegA	6

Assuming Register X=$23, the instruction `sta -4,X` will store a copy of the value in Register A at $1F, leaving Register X unchanged. The effective address (EA) is $0023-4=$001F.

Figure 1.17
third example of
5 indexed
essing mode.

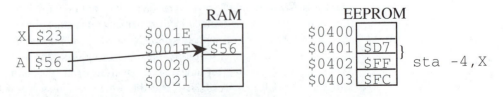

More indexed mode examples:

Machine code	Opcode	Operand	Comment	Cycles
$AED0	ldx	#$D0	RegX points to RAM variables	2
$73	com	,X	Invert all bits of location $00D0	5
$6304	com	4,X	Invert all bits of location $00D4	6
$AE03	ldx	#3	RegX is the index into array at $0600	2
$D60600	lda	$0600,X	Read byte #3 of the array (4th one)	4

6. Program counter (PC) relative addressing mode is used for the branch and branch-to-subroutine instructions. Stored in the instruction is not the absolute address of where to branch but the 8-bit signed offset relative distance to the current PC value. When the branch address is being calculated, the PC already points to the next instruction. The address of the next instruction is (location of instruction) + (number of bytes in the machine code). In the following examples assume the instruction is located at address $0480. The "rr" operand field is calculated as

$$\mathbf{rr} = \text{(location of destination address)} - \text{(location of instruction)} - \text{(number of bytes in the machine code)}$$
$$= \$0440 - \$0480 - 2 = -\$42 = \$BE$$
$$= \$0500 - \$0480 - 2 = \$7E$$
$$= \$0489 - \$0480 - 3 = \$06$$

Machine code	Opcode	Operand	Comment	Cycles
$A105	cmp	#5	Compare RegA to 5	2
$23BE	bls	$0440	go to $0440 if RegA•5 (unsigned)	3
$20BE	bra	$0440	go to location $0440	3
$AD7E	bsr	$0500	branch to subroutine at $0500	6
$080506	brset	4,5,$0489	go to $489 if bit 4 of location 5 is 1	5

Common Error: Since not every instruction supports every addressing mode, it would be a mistake to use an addressing mode not available for that instruction.

1.3 MC68HC08 Architecture

1.3.1 MC68HC708XL36 Introduction

The 6808 is upward-compatible with the 6805 family and provides a number of advantages over its predecessor. It too is available in a wide range of memory and I/O configurations. In this book, we choose to study the MC68HC708XL36 because it is one of the first 6808 devices made and has a wide range of I/O devices including DMA capabilities. The MC68HC708XL36 microcomputer is available in a 56-pin plastic shrink dual inline package (SDIP) or in a 64-pin plastic quad flat pack (QFP) (Figure 1.18 on p. 16).

Ports G and H are available only in the QFP package. One of the unique aspects of the 6808 family is its customer-specific integrated circuit (CSIC). This means that if you buy enough of them, Motorola will create a 6808 microcomputer just for you with the exact memory and I/O devices to satisfy your requirements. To make it run, we add a crystal to OSC1 and OSC2, power to V_{DD}, and ground to V_{SS}.

The MC68HC708XL36 can operate at either +5 or +3.3 V. The microcomputer executes with a clock frequency of half the crystal frequency. With a 16-MHz crystal, the 6808 cycle time is 125 ns. This means that it can read/write memory every 125 ns. Each 6808 instruction takes a fixed number of memory cycles to execute. Using this clock frequency, we

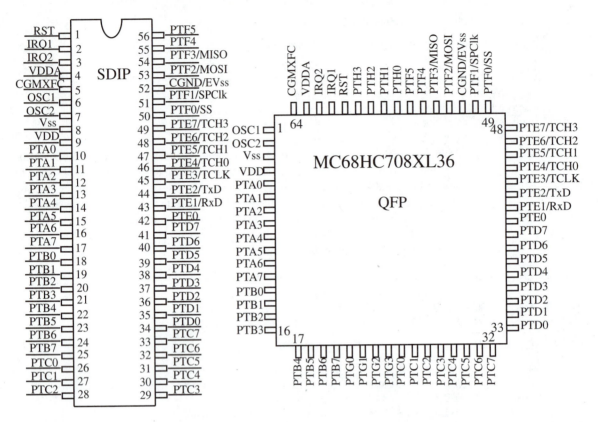

Figure 1.18
The MC68HC708XL36 comes in a SDIP or a QFP package.

can determine exactly how long each instruction takes to execute. The two bypass capacitors across power (V_{DD}) and ground (V_{SS}) are used to reject a wide range of noise on the power line. A large polarized tantalum capacitor (10 μF) rejects low-frequency noise, and the small nonpolarized (0.01 μF) ceramic capacitor rejects high-frequency noise. The RST line can be connected to a reset button or a power-on-reset circuit. The IRQ lines are external active low-interrupt requests. More information about interrupts will be presented in Chapter 4. The general features of the MC68HC708XL36 are:

- 36K bytes of EPROM
- 1024 bytes of RAM
- 16-bit, four-channel Timer Interface module
- Synchronous SPI module
- Asynchronous SCI module
- Three-channel DMA module
- 54 bidirectional I/O lines

1.3.2
MC68HC708XL36
Memory Map

The address map of the MC68HC708XL36 is shown in Table 1.7. The MC68HC08XL36 is similar to the MC68HC708XL36 except the EPROM is replaced with factory-programmable ROM. Typically we develop the prototype using the EPROM part and switch over to the ROM part for large production. In particular, we will study the MC68HC708XL36 I/O

Table 1.7
The MC68HC708XL36 has 36K of EPROM and 1 kbyte of RAM.

Address (hex)	Size	Device	Contents
0000 to 004F	80	I/O	
0050 to 044F	1024	RAM	Variables and stack
6E00 to FDFF	36K	EPROM	Programs and fixed constants
FFE0 to FFFF	32	EPROM	Vectors

devices listed in Table 1.8. Recall that the memory clock period is twice the crystal clock period. For example, if the MC68HC708 is driven with a 16-MHz crystal, then the memory clock period is 125 ns. With the default value of TSC, TCNT is incremented every 125 ns. Actually, eight possible configurations can be programmed into the TSC register to specify the rate at which TCNT increments.

Table 1.8
The MC68HC708XL36 has eight external I/O ports and an internal counter.

I/O device	Address	Function
PORTA	$0000	8-bit bidirectional I/O port
PORTB	$0001	8-bit bidirectional I/O port
PORTC	$0002	8-bit bidirectional I/O port
PORTD	$0003	8-bit bidirectional I/O port
DDRA	$0004	Specifies whether each bit of PORTA is input (0) or output (1)
DDRB	$0005	Specifies whether each bit of PORTB is input (0) or output (1)
DDRC	$0006	Specifies whether each bit of PORTC is input (0) or output (1)
DDRD	$0007	Specifies whether each bit of PORTD is input (0) or output (1)
PORTE	$0008	8-bit bidirectional I/O port
PORTF	$0009	6-bit bidirectional I/O port
PORTG	$000A	4-bit bidirectional I/O port
PORTH	$000B	4-bit bidirectional I/O port
DDRE	$000C	Specifies whether each bit of PORTE is input (0) or output (1)
DDRF	$000D	Specifies whether each bit of PORTF is input (0) or output (1)
DDRG	$000E	Specifies whether each bit of PORTG is input (0) or output (1)
DDRH	$000F	Specifies whether each bit of PORTH is input (0) or output (1)
TSC	$0021	Timer status and control
TCNT	$0022	16-bit up counter incremented every memory clock period

1.3.3
MC68HC08
Registers

Both the 6808 assembly source codes and machine object codes are fully upward-compatible with the 6805 family. We can see that the 6808 adds the H register and increases the size of the SP and PC (Figure 1.19). The CC register adds the signed overflow bit, V. Typically Register A contains data (numbers), while Register H together with Register X (H:X) contains addresses (pointers). Register SP+1 points to the top element of

Figure 1.19
The 6808 also has five registers.

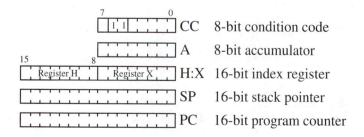

Figure 1.20
The 6808 condition
code includes an
overflow bit.

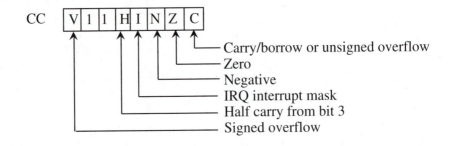

the stack. Register PC points to the current instruction. The condition code bits are shown in Figure 1.20.

The H bit is used for BCD addition (see the `add adc daa bhcs` and `bhcc` instructions). When I=0, IRQ interrupts are enabled. Software sets the I bit to 1 when it wishes to disable interrupts. V, N, Z, C bits signify the status of the previous ALU operation.

The 6808 implements memory-mapped I/O, and therefore each I/O device is assigned an address. To input from a parallel port, we simply read from its address. For example, if we wish to bring into Register A the current values on Port B (which is located at $0001), we could execute

```
lda $0001   RegA=[$0001]
```

The bidirectional I/O pins on the MC68HC708XL36 have direction registers that specify whether the pin is an input or output. For example, the direction register for Port A is located at $0004. If we wish to make the Port A bits 7, 6, 5, 4 outputs and the Port A bits 3, 2, 1, 0 inputs, we could execute

```
lda #$F0      RegA=$F0          0 means input, 1 means output
sta $0004     [$0004]=RegA
```

If we wish to make all six pins of Port B bits inputs, we could execute

```
clr $0005     [$0005]=0
```

**1.3.4
MC68HC08
Addressing Modes**

The 6808 instruction set has eight addressing modes:

1. Inherent addressing mode has no operand field. Compare the number of cycles required to execute on the 6808 versus the same instruction on the 6805. Most are the same, but a few are faster. The biggest speed improvement of the 6808 over the 6805 comes from the faster clock. Examples:

Machine code	Opcode	Operand	Comment	Cycles
$42	mul		Reg X:A = RegA • RegX (unsigned)	5
$8B	pshh		Push RegH on the stack	2
$9A	cli		Condition Code I bit = 0	2
$52	div		RegA=(RegH:X)/RegX and H=remainder	7
$83	swi		Software interrupt	9
$95	txs		RegH:X = (RegSP+1)	2

2. Immediate addressing mode uses a fixed constant (Figure 1.21). The data themselves are included in the machine code. Examples:

Machine code	Opcode	Operand	Comment	Cycles
$45**C000**[3]	ldhx	#$C000	Initialize index register H:X	3
$A9**00**	adc	#0	RegA = RegA+C (Add carry bit to RegA)	2
$A8**80**	eor	#$80	Invert bit 7 of RegA	2
$AA**02**	ora	#2	Set bit 1 of RegA	2

Figure 1.21
Example of the 6808 immediate addressing mode.

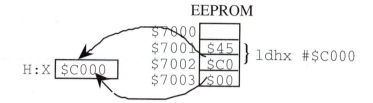

3. Direct page addressing mode uses an 8-bit address to access from addresses 0 to $00FF (Figure 1.22). These addresses include the 6808 I/O ports and RAM. Notice that some instructions can be used with both direct page and extended addressing (e.g., adc add and bit cmp cpx eor jmp jsr lda ldx ora sbc sta stx sub), while others can be used with direct, but not extended, mode. The < operator forces direct page addressing. Examples:

Machine code	Opcode	Operand	Comment	Cycles
$B6**32**	lda	50	RegA = [$0050]	3
$BD**64**	jsr	100	Jump subroutine at location 100	4
$35**34**	sthx	<$1234	{$0034} = RegH:X	4
$17**0A**	bclr	3,10	Clear bit 3 of location 10	2
$08**05**rr	brset	4,5,FUN	Go to FUN if bit 4 of location 5 is 1	5
$05**07**rr	brclr	2,7,OK	Go to OK if bit 2 of location 7 is 0	5

For an explanation of "rr," see relative addressing mode in item 8.

Figure 1.22
Example of the 6808 direct page addressing mode.

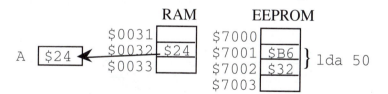

4. Extended addressing mode uses a 16-bit address to access all memory and I/O devices (Figure 1.23). Notice that the instructions that implement bit manipulations or perform read-modify-write operations cannot be used with extended addressing (e.g., bclr brclr brset bset asl asr clr com dec inc lsl lsr neg rol ror tst). The > operator forces extended addressing. Examples:

Machine code	Opcode	Operand	Comment	Cycles
$C6**0301**	lda	$0301	RegA = [$0301]	4
$CB**0461**	add	$0461	RegA = RegA + [$0461]	4
$CE**0000**	ldx	>0	Force 16-bit addressing	4

---

[3]**Boldface** type represents the operand portion of the machine code.

Figure 1.23
Example of the 6808 extended addressing mode.

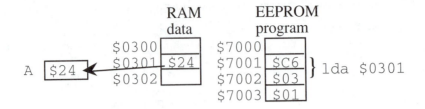

5. Indexed addressing mode uses either no offset, an 8-bit unsigned offset, or a 16-bit offset with RegH:X (Figure 1.24). Typically, indexed mode is used in two steps. First RegH:X is initialized to point to memory or an I/O port. Then, indexed mode addressing is used to access the information. If none of the HC08 instructions that modify RegH is used (e.g., `aix ldhx div pulh tsx` or using the X+ addressing mode), then RegH will remain zero, and the HC05 code will run on the HC08. Indexed mode is useful when addressing the data structures. Indexed mode will allow access to information on the stack. Register X is not modified by these indexed addressing modes. In each case, the 16-bit sum of RegH:X and the 0-, 8-, or 16-bit constant is used as a pointer. With 8-bit index mode, the 8-bit fixed constant is considered unsigned. For example, if Reg H:X were $1BD0 and the 8-bit constant were $C0, then the effective address would be $1BD0+$00C0, or $1C90. With 16-bit index mode, negative constants are possible. The first three examples use 0-, 8-, and 16-bit constants, respectively. In the first example, $F7 is the `sta 0,X` instruction, and there is no extra machine code to represent the operand:

Machine code	Opcode	Operand	Comment	Cycles
$F7	sta	0,X	[X] = RegA	2

Assuming Register H:X=$0323, the instruction `sta 0,X` will store a copy of the value in Register A at $0323, leaving Register H:X unchanged. The effective address (EA) is $0323 (Figure 1.24).

Figure 1.24
Example of the 6808 IDX indexed addressing mode.

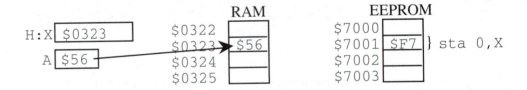

In the next example, $E7 is the `sta` instruction and **$04** is the machine code representing the operand:

Machine code	Opcode	Operand	Comment	Cycles
$E7**04**	sta	4,X	[X+4] = RegA	3

Assuming Register H:X=$0323, the instruction `sta 4,X` will store a copy of the value in Register A at $0327, leaving Register H:X unchanged. The effective address (EA) is $0323+4=$0327 (Figure 1.25).

Figure 1.25
Example of the
6808 IDX1 indexed
addressing mode.

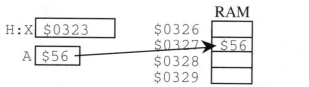

In the next example, $D7 is the sta instruction and **$FFFC** is the machine code representing the operand:

Machine code	Opcode	Operand	Comment	Cycles
$D7**FFFC**	sta	-4,X	[X+$FFFC] = RegA	4

Assuming Register H:X=$0323, the instruction sta -4, X will store a copy of the value in Register A at $031F, leaving Register H:X unchanged. The effective address (EA) is $0323−4=$031F (Figure 1.26).

Figure 1.26
Example of the 6808
IDX2 indexed
addressing mode.

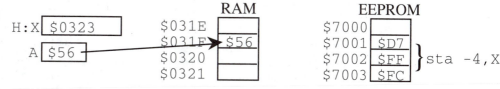

More indexed mode examples:

Machine code	Opcode	Operand	Comment	Cycles
$45**0050**	ldhx	#$0050	RegH:X points to RAM variables	3
$73	com	,X	Invert all bits of location $0050	3
$63**04**	com	4,X	Invert all bits of location $0054	4
$45**0003**	ldhx	#3	RegX is the index into array at $9000	2
$D6**0600**	lda	$9000,X	Read byte #3 of the array (4th one)	4

6. Stack pointer addressing mode uses either an 8-bit unsigned offset or a 16-bit offset with RegSP. The stack pointer is usually initialized once at the beginning of the program. Stack pointer indexed mode is useful for passing parameters and accessing local variables. In each case, the 16-bit sum of RegSP and the 8- or 16-bit constant is used as a pointer. With 8-bit index mode, the 8-bit fixed constant is considered unsigned. For example, if RegSP were $03FF and the 8-bit constant were $C0, then the effective address would be $03FF+$00C0, or $04BF. With 16-bit index mode, negative constants are possible (although negative indexing with the stack is an illegal stack operation). This addressing mode simplifies the object code generated by a C compiler. Examples:

Machine code	Opcode	Operand	Comment	Cycles
$9EE6**04**	lda	4,SP	RegA = [SP+4]	4
$9EDB**0102**	add	$102,SP	RegA = RegA + [SP+$0102]	5

7. Postincrement index addressing mode uses either no offset or an 8-bit unsigned offset with RegH:X. This mode is similar to indexed mode, except H:X is incremented after the data are accessed. A typical application is the memory-to-memory move. Only a few instructions support this addressing mode. In the first example, 8-bit data are read from location $0018 and stored at the address pointed to by H:X, then H:X is incremented.

Machine code	Opcode	Operand	Comment	Cycles
$5E18	mov	$18,X+	[H:X]=[$0018], H:X = H:X + 1	4

In the second example, 8-bit data are read from the address pointed to by H:X and stored at location $0050, then H:X is incremented.

Machine code	Opcode	Operand	Comment	Cycles
$7E50	mov	,X+,$50	[$0050]=[H:X], H:X = H:X + 1	4

In the third example, H:X is initialized to $0050. RegA contains a search code. This program will search the array, starting at $0060, for the first value in the array that does not match RegA. The symbol "*" refers to the address of the current instruction.

Machine code	Opcode	Operand	Comment	Cycles
$450050	ldhx	#$0050	RegH:X points to RAM variables	3
$6110FD	cbeq	$10,X+,*	go to * if RegA==[H:X], H:X = H:X + 1	5

8. PC relative addressing mode is used for the branch and branch to subroutine instructions. Stored in the instruction is not the absolute address of where to branch but the 8-bit signed offset relative distance to the current PC value. When the branch address is being calculated, the PC already points to the next instruction. The address of the next instruction is (location of instruction) + (number of bytes in the machine code). In the following examples assume the instruction is located at address $0480. The "rr" operand field is calculated as

$$\begin{aligned} \mathbf{rr} &= \text{(location of destination address)} - \text{(location of instruction)} - \text{(number of bytes in the machine code)} \\ &= \$0440 - \$0480 - 2 = -\$42 = \$BE \\ &= \$0500 - \$0480 - 2 = \$7E \\ &= \$0489 - \$0480 - 3 = \$06 \end{aligned}$$

Machine code	Opcode	Operand	Comment	Cycles
$A105	cmp	#5	Compare RegA to 5	2
$93BE	ble	$0440	go to $0440 if RegA·5 (signed)	3
$AD7E	bsr	$0500	branch to subroutine at $0500	6
$080506	brset	4,5,$0489	go to $489 if bit 4 of location 5 is 1	5

1.4 MC68HC11 Architecture

1.4.1 MC68HC11 Family

The MC68HC11 CPU is optimized for low-power consumption and high-performance operation at bus frequencies up to 4 MHz. The 6811 uses either two separate 8-bit accumulators (A, B) or one combined 16-bit accumulator (D). There are two 16-index registers (X, Y). Like all the Motorola microcomputers, the 6811 has powerful bit-manipulation instructions. The 6811 supports 16-bit add/subtract, 16×16 integer divide, 16×16 fractional divide, and 8×8 unsigned multiply. Although not as convenient as the 6812, the 6811 instruction set does lend itself to C compiler implementations, primarily because it handles 16-bit manipulations well.

In many applications, the MC68HC11 provides a single-chip solution with mask-programmed ROM or user-programmable EPROM. All family derivatives are also expandable for the incorporation of external memory in the design. A four-channel (DMA) unit on some devices permits fast data transfer between two blocks of memory (including externally mapped memory in expanded mode), between registers, or between registers and memory. Within the MC68HC11 family, there are nine major series of microcontroller units. The following are examples of the features the MC68HC11 can offer as shown through specific devices within each series:

A series The 68HC11A8 is the basic model, featuring 8 kbytes of ROM, 256 bytes of RAM, 512 bytes of EEPROM, a nominal bus speed of 3 MHz, and an eight-channel 8-bit ADC.

D series The 68HC11D3 chip with 4 kbytes of ROM offers an economical alternative for applications when advanced 8-bit performance is required with fewer peripherals and less memory.

E series The 68HC11E9 family comes in a wide range of I/O capabilities. It combines EEPROM and EPROM on a single chip. The E series offers multiple memory sizes in a pin-compatible package.

F series High-speed expanded systems required the development of the 68HC11Fl. This particular chip series stands out with its extra I/O ports, an increase in static RAM (1 kbyte), chip selects, and a 4-MHz non-multiplexed bus.

G series The 68HC11G5 offers 10-bit A/D resolution. This series also includes the most sophisticated timer systems in the family.

K series The 68HC11K4 offers high speed, large memories, a memory management unit, and PWM along with standard I/O ports.

L series The 68HC11L6 (or L6, for short) is a high-speed, low-power chip with a multiplexed bus capable of operation up to 3 MHz. The high-performance design of the L6 is based on the 68HC11E9. It includes 16 kbytes of ROM plus an additional bidirectional port. Its fully static design allows operation at frequencies down to 0 Hz.

M series These enhanced, high-performance microcontrollers are derived from the 68HC11K4 and include large memory modules, a 16-bit math coprocessor, and four channels of DMA.

P series The 68HC11P2 offers a power-saving programmable phase-locked loop (PLL) based clock circuit along with many I/O pins, large memory, and three SCI ports.

The 68HC11's with EPROM (68HC711) and EEPROM (68HC811) can be used during product development or in applications that require a small number of devices to be built. The memory maps shown in this section are the default values and can be relocated by writing to the INIT and CONFIG registers.

**1.4.1.1
MC68HC11A8 Single-
Chip Mode**

The 6811 can operate in one of four modes. We will focus on expanded mode, which we can use during development, and single-chip mode, which we embed into our final product.[4] In the *single-chip operating mode,* the MC68HC11A8 has 8 kbytes of internal

---

[4]The other two modes are bootstrap mode, which can be used to load programs into RAM, and test mode, which is used by Motorola to verify that the chip is operational.

Figure 1.27
ADAPT-11 Single Chip
6811 Module from
Technological Arts used
in a motor controller.

ROM, 512 bytes of EEPROM, and 256 bytes of RAM (Figure 1.27). Other versions of the 6811 have differing amounts of these types of memory. In single-chip mode, the 6811 implements a complete microcomputer, where all its I/O ports are available. This mode is used for the final product with the application software programmed into the ROM. During hardware and software development, typically we operate the 6811 in expanded mode on a development system like the Motorola Evaluation Board (EVB) so that the edit/assemble/load/run software development cycle is short. Once the system is debugged, the program can be burned into ROM, and the 6811 single-chip computer can be embedded into the system. A block diagram of a 6811 is shown in Figure 1.28.

Figure 1.28
Block diagram of the
Motorola
MC68HC11A8.

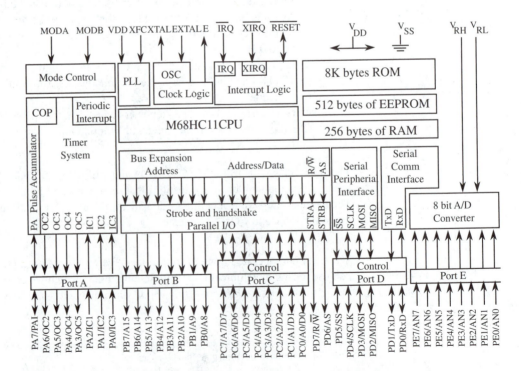

The address map of a single-chip MC68HC11A8 is shown in Table 1.9. Table 1.10 describes the five ports on the MC68HC11A8.

Table 1.9
The MC68HC11A8 has 8K of ROM and 256 bytes of RAM.

Address (hex)	Size	Device	Contents
0000 to 00FF	256	RAM	Variables and stack
1000 to 103F	64	I/O	
B600 to B7FF	512	EEPROM	Constants unique to each system
E000 to FFFF	8192	ROM	Programs and fixed constants

Table 1.10
The MC68HC11A8 has five external I/O ports.

Port	Input Pins	Output Pins	Bidirectional Pins	Shared Functions
Port A	3	4	1	Timer
Port B	—	8	—	High-order address
Port C	—	—	8	Low-order address and data bus
Port D	—	—	8	SCI and SPI
Port E	8	—	—	ADC

The MC68HC11A8 comes in both a 48-pin dual inline package (DIP) and a 52-pin plastic leaded chip carrier (PLCC) package (Figure 1.29).

Figure 1.29
A single-chip
MC68HC11A8 system.

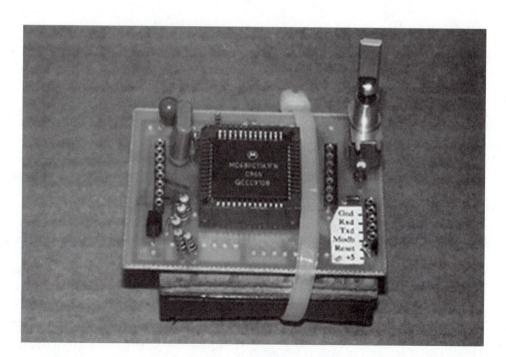

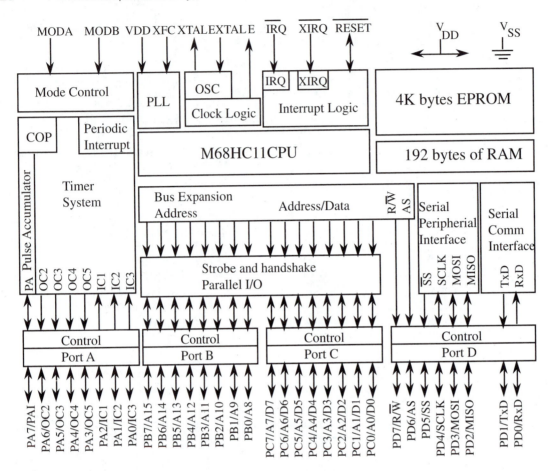

Figure 1.30
Block diagram of the MC68HC711D3.

1.4.1.2
MC68HC711D3 Single-
Chip Mode

Compare the block diagram of the D3 computer (Figure 1.30) with the other 6811 products, and notice that it has fewer I/O devices but is available in a smaller, less expensive package. The on-board 192-byte RAM is initially located at $0040 after reset, but it can be placed at any other 4-kilobyte boundary ($x040) by writing an appropriate value to the INIT register. The 64-byte register block originates at $0000 after reset, but it can be placed at any other 4-kbyte boundary ($x000) after reset by writing an appropriate value to the INIT register. There is no EEPROM or ADC on this version of the 6811. The 4-kbyte EPROM is located at $C000 through $FFFF.

The MC68HC711D3 has four 8-bit I/O ports: A, B, C, and D (Table 1.11). In the 40-pin package, Port A bits 4 and 6 are not connected to pins. In single-chip and bootstrap modes, all ports are parallel I/O data ports. In expanded-multiplexed and test modes, Ports B and C and lines AS and R/W are a memory expansion bus, with Port B serving as the high-order address bus, Port C as the multiplexed address and data bus, AS as the demultiplexing signal, and R/W as the data bus direction control.

The EPROM-based MC68HC711D3 is available in a 40-pin PDIP or a 44-pin PLCC.

The address map of a single-chip MC68HC711D3, one-time programmable (OTP) or ultraviolet (UV) EPROM, is shown in Table 1.12.

Table 1.11
The MC68HC711D3 has four external I/O ports.

Port	Input Pins	Output Pins	Bidirectional Pins	Shared Functions
Port A	3	3	2	Timer
Port B	—	—	8	High-order address
Port C	—	—	8	Low-order address and data bus
Port D	—	—	8	SCI, SPI, AS, and R/W

Table 1.12
The MC68HC711D3 has 4K of EPROM and 192 bytes of RAM.

Address (hex)	Size	Device	Contents
0000 to 003F	64	I/O	
0040 to 00FF	192	RAM	Variables and stack
F000 to FFFF	4K	EPROM	Programs and fixed constants

1.4.1.3
MC68HC711EA9 Single-Chip Mode

The E series 6811 provides more memory. The on-board 512-byte RAM is initially located at $0000 after reset, but it can be placed at any other 4-kbyte boundary ($x000) by writing an appropriate value to the INIT register. The 64-byte register block originates at $1000 after reset, but it can be placed at any other 4-kbyte boundary ($x000) after reset by writing an appropriate value to the INIT register. The 12-kbyte EPROM is located at $C000 through $FFFF.

The EPROM-based MC68HC711EA9 is available in a windowed ceramic leaded chip carrier (CLCC). A OTP version of the MC68HC711EA9 is available by ordering the device in a nonwindowed package. The address map of a single-chip MC68HC711EA9, which has OTP ROM, is shown in Table 1.13. Table 1.14 describes the five ports on the MC68HC711EA9.

Table 1.13
The MC68HC711EA9 has 12K of EPROM and 512 bytes of RAM.

Address (hex)	Size	Device	Contents
0000 to 01FF	512	RAM	Variables and stack
1000 to 103F	64	I/O	
B600 to B7FF	512	EEPROM	Constants unique to each system
C000 to FFFF	12K	EPROM	Programs and fixed constants

Table 1.14
The MC68HC711EA9 has five external I/O ports.

Port	Input Pins	Output Pins	Bidirectional Pins	Shared Functions
Port A	—	—	8	Timer
Port B	—	—	8	High-order address
Port C	—	—	8	Low-order address and data bus
Port D	—	—	2	SCI
Port E	8	—	—	ADC

1.4.2
MC68HC11
Expanded Mode

1.4.2.1
Expanded Mode with an
A, D, or E Series

During the hardware/software development, we can use an A, D, or E series 6811 in the expanded or microprocessor mode because the edit/compile/download time is short and we can utilize a debugger. In expanded mode, 6811 ports B and C are used for the address and data bus. The Peripheral Recovery Unit (PRU, 6824) is added to reconstruct I/O ports B and C. External ROM and RAM are used, and the 6811 acts like a microprocessor.

The address map of a Motorola 6811 EVB development board is shown in Table 1.15.

Table 1.15
The 6811 EVB board has 8K of PROM, 16K of PROM/RAM, and 256 bytes of RAM.

Address (hex)	Device	Description
0000 to 0035	RAM	Internal to 6811, can used by you
0036 to 00FF	RAM	Internal to 6811, BUFFALO variables, interrupt vectors
1000 to 103F	I/O	Internal I/O of the 6811/6824 system
4000	Switch	Output a 1 to this location to use the HOST Serial Port; 0 to use PD0 for other purposes
6000 to 7FFF	8K RAM/PROM	You put your software here
9800 to 9801	6850	Universal Asynchronous Receiver/Transmitter
B600 to B7FF	512 bytes EEPROM	Internal to 6811, electrically erasable PROM
C000 to DFFF	8K RAM/PROM	You put your software here
E000 to FFFF	8K PROM	BUFFALO

Adding a debugger in PROM like Bit User's Fast Friendly Aid to Logical Operation (BUFFALO) provides features like download, breakpoints, single stepping, viewing memory, and I/O. The Motorola 6811 EVB board has three 8-kbyte PROM/RAM sockets and a 6850 serial port (Figure 1.31).

Figure 1.31
Block diagram of a 6811 EVB system.

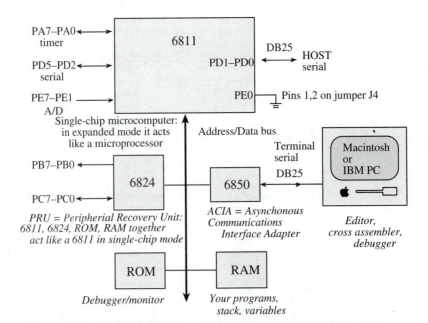

Figure 1.32
ADAPT-11C75DX
Expanded Mode 6811
Module from
Technological Arts.

You can replace one or both of the 8-kbyte RAMs on the EVB with 2764A 8-kbyte by 8-bit PROMs. To make a stand-alone system using the EVB, you first develop the software using two RAM chips in the 8-kilobyte sockets. You segment your software, placing programs/fixed constants in one 8-kbyte block and variables/stack in the other 8-kbyte block. Once the program is debugged, you can burn the program/ fixed constants into an 8-kbyte PROM. Next, you use BUFFALO to program a jump to the starting address of your software at $B600, $B601, and $B602 in EEPROM. If the jumper J4 is set to pins 2, 3, then the 6811 will jump to $B600 on a reset or power-on. In this way, your system will operate without a Macintosh or personal computer (PC) to download each time you power up the system. Expanded mode systems, like the one shown in Figure 1.32, have much more memory than the single chip microcomputers.

1.4.2.2
MC68HC11F Series

The MC68HC11F1 is a high-performance member of the M68HC11 family of microcontroller units (MCUs). High-speed expanded systems required the development of this chip with its extra I/O ports, an increase in static RAM (1 kbyte), internal chip-select functions, and a non-multiplexed bus that reduces the need for external interface logic. The timer, serial I/O, and ADC enable functions similar to those found in the MC68HC11E9. The MC68HC11FC0 is a low-cost, high-speed derivative of the MC68HC11F1. It does not have EEPROM or an ADC. The MC68HC11FC0 can operate at bus speeds as high as 6 MHz. The address map of a single-chip MC68HC11F1 is shown in Table 1.16.

Table 1.16
The MC68HC11F1 has
512 bytes of EEPROM
and 1 kbytes of RAM.

Address (hex)	Size	Device	Contents
0000 to 00FF	1024	RAM	Variables and stack
1000 to 105F	96	I/O	
FE00 to FFFF	512	EEPROM	Constants unique to each system

The 16-bit address bus can access 64 kbytes of memory. Because the MC68HC11F1 and MC68HC11FC0 are intended to operate principally in expanded mode, there is no internal ROM, and the address bus is non-multiplexed. Both devices include 1 kbyte of static RAM, a 96-byte control register block, and 256 bytes of bootstrap ROM. The MC68HC11F1 also includes 512 bytes of EEPROM. RAM and registers can be remapped on both the MC68HC11F1 and the MC68HC11FC0. On both the MC68HC11F1 and the MC68HC11FC0, out-of-reset RAM resides at $0000 to $03FF and registers reside at $1000 to $105F. On the MC68HC11F1, RAM and registers can both be remapped to any 4-kbyte boundary. On the MC68HC11FC0, RAM can be remapped to any 1-kbyte boundary, and registers can be remapped to any 4-kbyte boundary in the first 16 kbytes of address space. RAM and control register locations are defined by the INIT register, which can be written only once within the first 64 E clock cycles after a reset in normal modes. It becomes a read-only register thereafter. If RAM and the control register block are mapped to the same boundary, the register block has priority of the first 96 bytes. The MC68HC11F1 has 512 bytes of EEPROM. A nonvolatile EEPROM-based configuration register (CONFIG) controls whether the EEPROM is present or absent and determines its position in the memory map. In single-chip and bootstrap modes the EEPROM is positioned at $FE00–$FFFF. In expanded and special test modes, the EEPROM can be repositioned to any 4-kbyte boundary ($xE00–$xFFF). Table 1.17 shows the seven ports on the MC68HC11F1.

Table 1.17
The MC68HC11F1 has seven external I/O ports.

Port	Input Pins	Output Pins	Bidirectional Pins	Shared Functions
Port A	3	4	1	Timer
Port B	—	—	8	High-order address
Port C	—	—	8	Data bus
Port D	—	—	6	SCI, SPI
Port E	8	—	—	ADC
Port F	—	—	8	Low-order address
Port G	—	—	8	Chip selects

1.4.3 MC68HC11 Registers

The 6811 registers are depicted in Figure 1.33. Registers A and B concatenated together form a 16-bit accumulator, RegD, with RegA containing the most significant byte. Typically Registers A and B contain data (numbers) while Registers X and Y contain addresses (pointers). Register SP+1 points to the top element of the stack. Register PC points to the current instruction.

Figure 1.33
The 6811 has six registers.

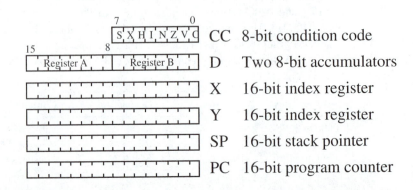

CC	8-bit condition code
D	Two 8-bit accumulators
X	16-bit index register
Y	16-bit index register
SP	16-bit stack pointer
PC	16-bit program counter

The condition code bits are shown in Figure 1.34.

Figure 1.34
The 6811 has eight
condition code bits.

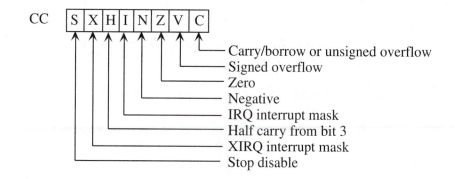

When S=1, the stop instruction is disabled. When X=0, XIRQ interrupts are allowed. Once X is set to zero, the software cannot set it back to 1. The H bit is used for BCD addition (see the `adda` and `daa` instructions). When I=0, IRQ interrupts are enabled. N, Z, V, and C bits signify the status of the previous ALU operation.

The bidirectional I/O pins on the 6811 (PortA bit 7, PortC, and PortD) have direction registers that specify whether the pin is an input or output. For example, the direction register for PortC is located at $1007. If we wish to make PortC bits 7, 6, 5, 4 outputs and bits 3, 2, 1, 0 inputs, we use

```
ldaa #$F0    0 means input, 1 means output
staa $1007
```

1.4.4
MC68HC11
Addressing Modes

The 6811 instruction set has five addressing modes.

1. Inherent addressing mode has no operand field. Examples:

Machine code	Opcode	Operand	Comment	Cycles
$3D	mul		RegD = RegA • RegB (unsigned)	10
$02	idiv		RegX=RegD/RegX, RegD=remainder unsigned	41
$03	fdiv		RegX=(65536 • RegD)/RegX, RegD=remainder (unsigned)	41
$1808	iny		RegY = RegY + 1	4
$3F	swi		Software interrupt	14
$1B	aba		RegA = RegA + RegB	2
$0E	cli		Condition Code I bit = 0	2
$3A	abx		RegX = RegX + RegB (unsigned)	3
$30	tsx		RegX = RegSP + 1	3

2. Immediate addressing mode uses a fixed constant (Figure 1.35). The data are included in the machine code. Examples:

Machine code	Opcode	Operand	Comment	Cycles
$8E**00FF**[5]	lds	#$00FF	Initialize stack	3
$89**00**	adca	#0	RegA = RegA+C (Add carry bit to RegA)	2
$C8**80**	eorb	#$80	Invert bit 7 of RegB	2
$8A**02**	oraa	#2	Set bit 1 of RegA	2

---

[5]**Boldface** type represents the operand portion of the machine code.

Figure 1.35
Example of the 6811
immediate addressing
mode.

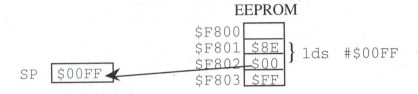

Observation: With immediate mode addressing, the information is stored in the machine code.

Common Error: It is illegal to use the immediate addressing mode with instructions that store data into memory (e.g., staa stab std stx sty and sts).

3. Direct page addressing mode uses an 8-bit address to access from addresses 0 to $00FF (Figure 1.36). These addresses include the 6811 single-chip RAM. Notice that some

Figure 1.36
Example of the 6811
direct page addressing
mode.

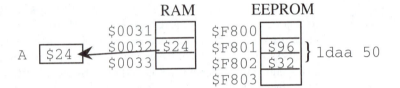

instructions can be used with both direct page and extended addressing (e.g., adca adcb adda addb addd anda andb bita bitb cmpa cmpb cpx cpy eora eorb jsr ldaa ldab ldd lds ldx ldy oraa orab sbca sbcb staa stab stx sts sty suba subb subd), while others can be used with extended mode (e.g., asl asr clr com dec inc jmp lsl lsr neg rol ror tst) but not direct mode. The < operator forces direct page addressing. Examples:

Machine code	Opcode	Operand	Comment	Cycles
$9632	ldaa	50	RegA = [$0050]	3
$9D64	jsr	100	Jump subroutine at location 100	5
$DF34	stx	<$1234	{$0034} = RegX	4
$150A03	bclr	10 3	Clear bits 1,0 of location 10	6
$140A81	bset	10 $81	Set bits 7,0 of location 10	6
$120518rr	brset	5 $18 FUN	Go to FUN if both bits 4,3 of location 5 = 1	6
$13070Frr	brclr	7 $0F OK	Go to OK if bits 3-0 of location 7 all are 0	6

For an explanation of "rr," see relative addressing mode in item 6.

Observation: With direct and extended mode addressing a fixed pointer to the information is stored in the machine code. The data may change dynamically, but their location is fixed.

4. Extended addressing mode uses a 16-bit address to access all memory and I/O devices (Figures 1.37, 1.38). Notice that the bit manipulation instructions which can be used with

Figure 1.37
Example of the 6811
extended addressing
mode.

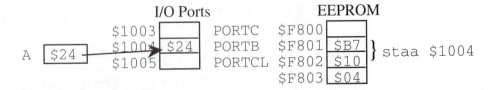

Figure 1.38
Another example of the 6811 extended addressing mode.

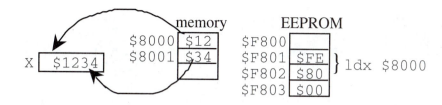

direct and indexed addressing cannot be used with extended addressing (e.g., `bclr brclr brset bset`). The > operator forces extended addressing. Examples:

Machine code	Opcode	Operand	Comment	Cycles
$B71004	staa	$1004	[$1004]=RegA, Output RegA to Port B	4
$FE8000	ldx	$8000	RegX = {$8000}	5
$BF9000	sts	$9000	{$9000} = RegSP	5
$B60000	ldaa	>0	Force 16-bit addressing	4

Common Error: It is wrong to assume that the < and > operators affect the amount of data transferred. The < and > operators will affect the addressing mode, which is how the address is represented.

Observation: Accesses to Registers A, B, and CC transfer 8 bits, while accesses to Registers D, X, Y, SP, and PC transfer 16 bits regardless of the addressing mode.

5. Indexed addressing mode uses an 8-bit unsigned offset with either RegX or RegY. The 8-bit unsigned offset is included even if the offset is zero. Notice that instructions that use RegX take less object code and execute faster than those using RegY do. Indexed mode is useful when addressing the 6811 I/O ports located from $1000 to $103F. Indexed mode is also useful when addressing the data structures and information on the stack. In each case, the 16-bit register is used as a pointer (index) and is unmodified by the instruction. In the first example, $A7 is the `staa` instruction and $04 is the machine code representing the operand:

Machine code	Opcode	Operand	Comment	Cycles
$A704	staa	4,X	[X+4] = RegA	3

Assuming Register X=$0023, the instruction `staa 4,X` will store a copy of the value in Register A at $0027, leaving Register X unchanged. The effective address (EA) is $0023+4=$0027 (Figure 1.39).
More indexed mode examples:

Machine code	Opcode	Operand	Comment	Cycles
$CE1000	ldx	#$1000	RegX points to internal I/O block	3
$6304	com	4,X	Invert all bits of PortB ($1004)	6
$1E3080rr	brset	$30,X,$80,DONE	Branch to DONE if $1030 bit 7 set	7
$1D0313	bclr	3,X,%00010011	Clear Port C($1003) bits 4,1,0	7
$18CE100E	ldy	#$100E	RegY points to TCNT	4
$18EC00	ldd	,Y	Read TCNT ($100E) into RegD	6

Observation: With indexed mode addressing the pointer to the information is calculated at run time, so the data and their location may change dynamically.

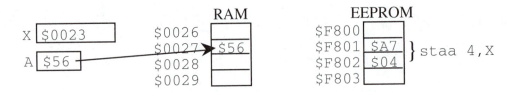

Figure 1.39
Example of the 6811 indexed addressing mode.

> *Observation:* 6811 instructions involving Register Y require an extra byte of machine code.

6. PC relative addressing mode is used for the branch and branch-to-subroutine instructions. Stored in the instruction is not the absolute address of where to branch but the 8-bit signed offset relative distance to the current PC value. When the branch address is being calculated the PC already points to the next instruction. The address of the next instruction is (location of instruction) + (number of bytes in the machine code). In the following examples assume each branch instruction is located at address $F880. The "rr" operand field is calculated as

$$\mathbf{rr} = (\text{location of destination address}) - (\text{location of instruction}) - (\text{number of bytes in the machine code})$$
$$= \$F840 - \$F880 - 2 = -\$42 = \$BE$$
$$= \$F8C8 - \$F880 - 2 = \$46$$
$$= \$F889 - \$F880 - 4 = \$05$$

Machine code	Opcode	Operand	Comment	Cycles
$8105	cmpa	#5	Compare RegA to 5	2
$25**BE**	blo	$F840	go to $F840 if RegA <5 (unsigned)	3
$20**BE**	bra	$F840	go to location $F840	3
$8D**46**	bsr	$F8C8	branch to subroutine at $F8C8	6
$1205**10**05	brset	5,$10,$F889	go to $F889 if bit 4 of location 5 is 1	6

> *Common Error:* Since not every instruction supports every addressing mode, it would be a mistake to use an addressing mode not available for that instruction.

> *Common Error:* You cannot access the I/O ports of the 6811 (locations $1000–$107F) with the `bset bclr brset brclr` using extended addressing mode. You must bring the I/O address into the X or Y index register and access the I/O ports using indexed addressing mode.

> *Observation:* The number of cycles required to execute 6805, 6808, and 6811 is fixed for each instruction.

> *Observation:* When the 6811 operates in expanded mode with external memory, the memory cycle time (time to fetch data) is exactly the same for all data accesses. This fact makes the task of counting cycles (predetermining how long a program will take to execute) difficult but possible.

**1.4.5
Erasing and
Programming the
MC68HC11
EEPROM**

The 6811 has 512 bytes of EEPROM. Since it is nonvolatile, information stored here will remain even when power is interrupted. This is a convenient place to store calibration constants or mode configurations. The EEPROM can also be used to store a small amount of information in a data acquisition system. The idea is to place in EEPROM information that is different from system to system or that you wish to change in the near future. The code in Program 1.1 can be used to erase and program the 6811 EEPROM.

Program 1.1
6811 software that programs its internal EEPROM.

```
PPROG   = 0x103B
; ************erases all of 6811 EEPROM***************
; returns RegA=FF if successful
EraseAll:
      ldab #0x06        ; Bulk erase mode
      stab PPROG        ; ERASE=1, ELAT=1, EEPRGM=0
      stab 0xB600       ; write to any EEPROM address
      incb              ; ERASE=1, ELAT=1, EEPRGM=1
      stab PPROG
      bsr  Wait10ms
      clr  PPROG
      rts
; *********program 1 byte*********************
; RegX points into $B600-$B7FF
; RegA has data, returns actual data written
Prog: ldab #0x02        ; Byte program mode
      stab PPROG        ; ERASE=0, ELAT=1, EEPRGM=0
      staa 0,x          ; write to the EEPROM address
      incb              ; ERASE=0, ELAT=1, EEPRGM=1
      stab PPROG
      bsr  Wait10ms
      clr  PPROG
      ldaa 0,x          ; actual value in EEPROM
      rts
; **********Wait for 10 ms*****************
Wait10ms: pshx
      ldx  #3334        ; E clock=2MHz
wloop: dex              ; 6 cycles*3334*0.5us/cycle=10ms
      bne  wloop
      pulx
      rts
```

1.5 MC68HC12 Architecture

1.5.1
MC68HC12 Family

The MC68HC12 is a highly integrated, general-purpose family of microcomputers with a 16-bit microcontroller architecture specifically designed for low-power consumption. It is source-code-compatible with the popular 68HC11 8-bit microcontroller family, but some of the I/O ports operate slightly differently. Features include:

Low-power consumption and low-voltage operation at full bus speed
Single-wire Background Debug mode for nonintrusive in-circuit programming and debugging
High-level language optimization
Flash EEPROM and byte-erasable EEPROM integrated on a single device
Fuzzy logic instructions
Enhanced arithmetic instructions over the 68HC11

1.5.1.1
MC68HC812A4 Single-Chip Mode

The MC68HC812A4 MCU is a 16-bit device composed of standard on-chip peripheral modules connected by an intermodule bus. Modules include a 16-bit CPU12, a Lite integration module (LIM), two asynchronous SCIs (SCI0 and SCI1), an SPI, a timer and pulse accumulation module, an 8-bit ADC, a 1-kbyte RAM, a 4-kbyte EEPROM, and memory expansion logic with chip selects, key wakeup ports, and a PLL. A block diagram of the MC68HC812A4 is shown in Figure 1.40.

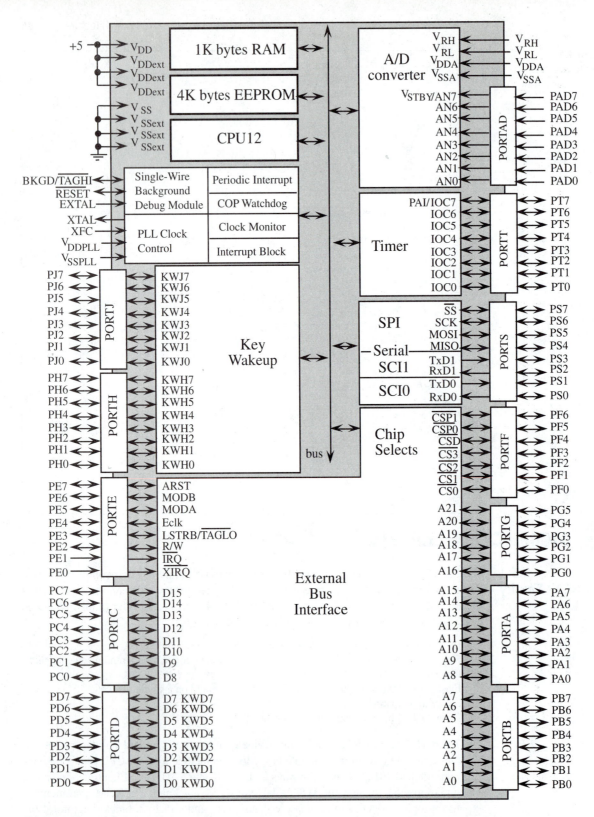

Figure 1.40
Block diagram of a Motorola MC68HC812A4.

BKGD	MODB	MODA	Mode	Port A	Port B	Port C	Port D
0	0	0	Special single chip	I/O	I/O	I/O	I/O
0	0	1	Special expanded narrow	A15-8	A7-A0	D7-D0	I/O
0	1	0	Special peripheral	A15-8	A7-A0	D15-D8	D7-D0
0	1	1	Special expanded wide	A15-8	A7-A0	D15-D8	D7-D0
1	0	0	Normal single chip	I/O	I/O	I/O	I/O
1	0	1	Normal expanded narrow	A15-8	A7-A0	D7-D0	I/O
1	1	1	Normal expanded wide	A15-8	A7-A0	D15-D8	D7-D0

Table 1.18   The MC68HC812A4 has seven operating modes.

The 68HC812A4 can operate in one of seven modes as specified by the three input signals BKGD, MODA, and MODB (Table 1.18). In single-chip mode, the 6812 contains the four major building blocks required to make a complete computer system: processor, I/O, RAM, and (P)ROM. In this mode, all ports are available for I/O. In expanded mode some of the ports are used for the address and data bus. A prototyping system based on this chip is shown in Figure 1.41.

We use expanded mode during development because it provides a fast edit/assemble/ download cycle, and we use single-chip mode to embed the microcomputer into our final product. Once the system is debugged, the program can be burned into ROM or programmed into PROM. In the *single-chip operating mode,* the MC68HC812A4 has 4096 bytes of EEPROM and 1024 bytes of RAM. Other versions of the 6812 have differing amounts of these types of memory. In single-chip mode, the 6812 implements a complete microcomputer, where all its

Figure 1.41   ADAPT-812 MC68HC812A4 Module from Technological Arts.

I/O ports are available. This mode is used for the final product, with the application software programmed into the ROM. The address space of the I/O devices and the RAM are mappable to any 2-kbyte space. The EEPROM is mappable to any 4-kbyte space. The standard address map of a single-chip MC68HC812A4 is shown in Table 1.19.

The 6812 implements memory-mapped I/O; therefore each I/O device is assigned an address. Some of the MC68HC812A4 addresses are included in Table 1.20. Refer to your 6812 reference for a complete list.

Table 1.19
The MC68HC812A4 has 4K of EEPROM and 1 kbytes of RAM.

Address (hex)	Size	Device	Contents
0000 to 01FF	512	I/O	
0800 to 0BFF	1024	RAM	Variables and stack
F000 to FFFF	4096	EEPROM	Programs and fixed constants

Table 1.20
The MC68HC812A4 has 12 external I/O ports and an internal timer.

Port	Address	Usage	Alternate Usage
PORTA	$0000	General-purpose I/O port	Address A15–A8 in expanded mode
PORTB	$0001	General-purpose I/O port	Address A7–A0 in expanded mode
DDRA	$0002	Data direction register	
DDRB	$0003	Data direction register	
PORTC	$0004	General-purpose I/O port	Data D15–D8 in expanded wide mode
PORTD	$0005	General-purpose I/O port	Data D7–D0 in expanded wide mode
DDRC	$0006	Data direction register	
DDRD	$0007	Data direction register	
PORTE	$0008	General-purpose I/O port	Mode and bus control
DDRE	$0009	Data direction register	
PORTH	$0024	General-purpose I/O port	Key wakeup, interrupt
DDRH	$0025	Data direction register	
PORTJ	$0028	General-purpose I/O port	Key wakeup, interrupt
DDRJ	$0029	Data direction register	
PORTF	$0030	General-purpose I/O port	Chip select logic
PORTG	$0031	General-purpose I/O port	Address A21–A16 Memory expansion
DDRF	$0032	Data direction register	
DDRG	$0033	Data direction register	
PORTAD	$006F	Analog input	General-purpose input port
TCNT	$0084	16-bit timer	
PORTT	$00AE	Timer I/O port	General-purpose I/O port
DDRT	$00AF	Data direction register	
SC0BD	$00C0	16-bit baud rate register	
SC0CR1	$00C2	Serial port control register	
SC0CR2	$00C3	Serial port control register	
SC0SR1	$00C4	Serial port status register	
SC0SR2	$00C5	Serial port status register	
SC0DRH	$00C6	High byte of Serial port data	
SC0DRL	$00C7	Serial port data register	
PORTS	$00D6	Serial I/O port	General-purpose I/O port
DDRS	$00D7	Data direction register	

1.5.1.2
MC68HC912B32 Single-
Chip Mode

The MC68HC912B32 MCU is a 16-bit device composed of standard on-chip peripherals including a 16-bit CPU12, 32-kbyte flash EEPROM, 1-kbyte RAM, 768-byte EEPROM, an asynchronous SCI, a SPI, an 8-channel timer and 16-bit pulse accumulator, an 8-bit ADC, a four-channel PWM, and a J1850-compatible byte data link communications (BDLC) module. The MC68HC912B32 has full 16-bit data paths throughout; however, the multiplexed external bus can operate in an 8-bit narrow mode, so single 8-bit-wide memory can be interfaced for lower-cost systems. Figure 1.42 is a block diagram of the MC68HC912B32. Table 1.21 shows the address map of the MC68HC912B32.

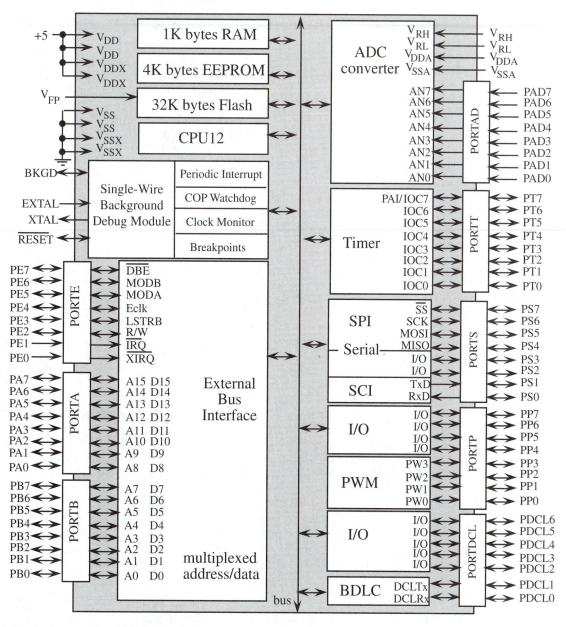

Figure 1.42
Block diagram of a Motorola MC68HC912B32.

Table 1.21
The MC68HC912B32 has 32K of flash EEPROM and 1 kbytes of RAM.

Address (hex)	Size	Device	Contents
0000 to 01FF	512	I/O	
0800 to 0BFF	1028	RAM	Variables and stack
0D00 to 0FFF	768	EEPROM	Programs and fixed constants
8000 to FFFF	32768	EEPROM	Programs and fixed constants

BKGD	MODB	MODA	Mode	Port A	Port B
0	0	0	Special single chip	I/O	I/O
0	0	1	Special expanded narrow	Addr[15:8] / Data[7:0]	Addr[7:0]
0	1	0	Special peripheral	Addr[15:8] / Data[15:8]	Addr]7:0] / Data[7:0]
0	1	1	Special expanded wide	Addr[15:8] / Data[15:8]	A7–A0
1	0	0	Normal single chip	I/O	I/O
1	0	1	Normal expanded narrow	Addr[15:8] / Data[7:0]	Addr[7:0]
1	1	1	Normal expanded wide	Addr[15:8] / Data[15:8]	Addr[7:0] / Data[7:0]

Table 1.22
The MC68HC912B32 has seven operating modes.

Table 1.23
The MC68HC912B32 has eight external I/O ports and an internal timer.

Port	Address	Usage	Alternate Usage
PORTA	$0000	General-purpose I/O port	Address/Data 15–8 in expanded mode
PORTB	$0001	General-purpose I/O port	Address/Data 7–0 in expanded mode
DDRA	$0002	Data direction register	
DDRB	$0003	Data direction register	
PORTE	$0008	General-purpose I/O port	Mode and bus control
DDRE	$0009	Data direction register	
PORTP	$0056	General-purpose I/O port	PWM
DDRP	$0057	Data direction register	
PORTAD	$006F	Analog input	General-purpose input port
TCNT	$0084	16-bit timer	
PORTT	$00AE	Timer I/O port	General-purpose I/O port
DDRT	$00AF	Data direction register	
SC0BD	$00C0	16-bit baud rate register	
SC0CR1	$00C2	Serial port control register	
SC0CR2	$00C3	Serial port control register	
SC0SR1	$00C4	Serial port status register	
SC0SR2	$00C5	Serial port status register	
SC0DRH	$00C6	High byte of serial port data	
SC0DRK	$00C7	Serial port data register	
PORTS	$00D6	Serial I/O port	General-purpose I/O port
DDRS	$00D7	Data direction register	
PORTDLC	$00FE	BDLC	General-purpose I/O port
DDRDLC	$00FF	Data direction register	

Like the MC68HC812A4, the MC68HC912B32 operates in one of seven modes as specified by the three input signals BKGD, MODA, and MODB (Table 1.22, 1.23). In single-chip mode, the 6812 contains the four major building blocks required to make a complete computer system: processor, I/O, RAM, and (P)ROM. In this mode, all ports are available for I/O. In expanded mode some of the ports are used for the address and data bus.

1.5.2 MC68HC12 Expanded Mode

In expanded mode, MC68HC812A4 Ports A, B, C, and D are used for the address and data bus. Port E contains bus control signals. Ports F and G can be optionally used for chip selects and extended address, respectively (Figure 1.43) In this mode, we may have up to 4 Mbytes of program memory and 1 Mbyte of RAM memory. The chip selects provide a mechanism to build the expanded mode computer with few external components. (In expanded mode, MC68HC912B32 Ports A and B are used for the address and data bus. Port E contains bus control signals.) In this mode, external PROM and RAM are used, and the 6812 acts like a microprocessor. The MC68HC912B32 does not support expanded mode memory, so external RAM and ROM are restricted to 64 kbytes. Wide expanded mode creates a 16-bit data bus, while narrow expanded mode has only an 8-bit data bus. A 16-bit bus can transfer data to/from memory twice as fast as 8 bits, but the 8-bit bus is simpler to use when connecting external 8-bit devices. Since both the MC68HC812A4 and MC68HC912B32 include a background debug module, it is not usually necessary to include a debugger/monitor program in the PROM, like we needed on the 6811 expanded mode systems.

Figure 1.43
Bus interface for an expanded mode MC68HC812A4 system.

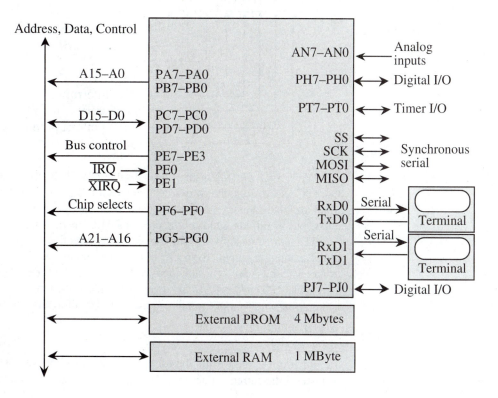

Figure 1.44
The 6812 has six registers.

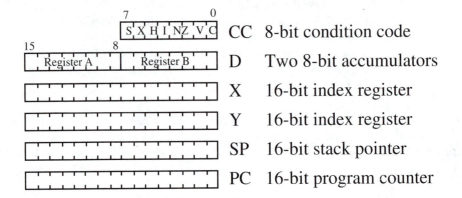

1.5.3 MC68HC12 Registers

The 6812 registers are the same as the 6811; this provides a simple upgrade path from the 6811 to the 6812 (Figure 1.44). Registers A and B concatenated together form a 16-bit accumulator, RegD, with RegA containing the most significant byte. Typically Registers A and B contain data (numbers), while Registers X and Y contain addresses (pointers). Different from the 6811, where SP+1 points to the top of the stack, the 6812 Register SP itself points to the top element of the stack. Register PC points to the current instruction.

The condition code bits of the 6812 (Figure 1.45) are also the same as the 6811.

Figure 1.45
The 6812 has eight condition code bits.

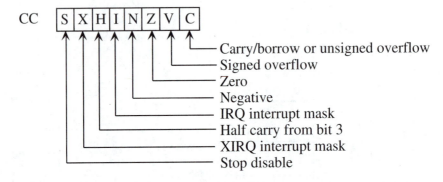

When S=1, the stop instruction is disabled. When X=0, XIRQ interrupts are allowed. Once X is set to zero, the software cannot set it back to 1. The H bit is used for BCD addition (see the `adda` and `daa` instructions). When I=0, IRQ interrupts are enabled. N, Z, V, and C bits signify the status of the previous ALU operation.

The bidirectional I/O pins on the 6812 have direction registers that specify whether the pin is an input or output. To specify port pins as inputs, we clear their corresponding direction register bits to 0. Typically there is one direction bit for each port bidirectional I/O pin. Port E bits 1, 0 and Port AD are always inputs, and the remaining ports can be configured as inputs or outputs. For example, to make Port A all inputs, we clear DDRA ($0002).

```
ldaa #$00
staa $0002    make Port A all input
```

To input from a port, we simply read from its address. For example, if we wish to bring into Register B the current values on Port A (which is located at $000), we could execute

```
ldab $0000
```

To specify port pins as outputs, we set their corresponding direction register bits to 1. For example, to make Port B all outputs, we set DDRB ($0003) to $FF.

```
ldaa #$FF
staa $0003    make Port B all output
```

To output to a port, we simply write to its address. For example, if we wish to set the output bits of Port B (which is located at $0001) to the value in Register A, we could execute

```
staa $0001
```

We can have some pins input and some output on the same port. For example, the direction register for Port C is located at $0006. If we wish to make Port C bits 7, 6, 5, 4 outputs and bits 3, 2, 1, 0 inputs, we execute

```
ldaa #$F0    0 means input, 1 means output
staa $0006
```

An output to Port C affects only the output pins—for example,

```
staa $0004    affects only output pins 7,6,5,4
```

whereas an input from Port C returns the current values of both input pins and output pins—for example,

```
ldaa $0004    current value of all 8 pins
```

1.5.4
MC68HC12
Addressing Modes

The 6812 instruction set has 15 addressing modes.

1. Inherent addressing mode has no operand field. Examples:

Machine code	Opcode	Operand	Comment	Cycles
$12	mul		RegD = RegA • RegB (unsigned)	3
$1814	edivs		RegY = Y:D/X and RegD = remainder (32 by 16-bit signed)	12
$13	emul		RegY:D = Y•D (unsigned)	3
$02	iny		RegY = RegY + 1	1
$3F	swi		Software interrupt	9
$1806	aba		RegA = RegA + RegB	2
$183A	rev		fuzzy logic rule evaluation	3•n
$B775	tsx		RegX = RegSP (same as tfr SP,X)	1

2. Immediate addressing mode uses a fixed constant (Figure 1.46). The data themselves are included in the machine code. Compare the number of cycles required to execute on the 6812 versus the same instruction on the 6811. Many of the same instructions are faster because the 6812 fetches 16 bits per memory cycle. Examples:

Machine code	Opcode	Operand	Comment	Cycles
$CF**0C00**[6]	lds	#$0C00	Initialize stack	2
$89**00**	adca	#0	Add carry bit to RegA	1
$C8**80**	eorb	#$80	Invert bit 7 of RegB	1
$8A**02**	oraa	#2	Set bit 1 of RegA	1

---

[6]**Boldface** type represents the operand portion of the machine code.

Figure 1.46
Example of the 6812
immediate addressing
mode.

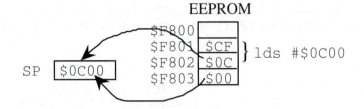

EEPROM

```
       $F800
       $F801  $CF
       $F802  $0C   } lds #$0C00
       $F803  $00
SP  $0C00
```

3. Direct page addressing mode uses an 8-bit address to access I/O devices 0 to $00FF (Figure 1.47). Notice that some instructions which can be used with extended and indexed addressing cannot be used with direct page addressing (e.g., asl asr clr com dec inc jmp lsl lsr neg rol ror tst). Examples:

Machine code	Opcode	Operand	Comment	Cycles
$96**01**	ldaa	$01	Read location $01, PortB	3
$4D**08**30	bclr	8,$30	Clear bits 5,4 of location 8, Port E	4
$4C**AF**81	bset	$AF,$81	Set bits 7,0 of location $00AF (DDRT)	4
$4E**05**18rr	brset	5,$18,FUN	Go to FUN if both bits 4,3 of Port D are 1	4
$4F**07**0Frr	brclr	7,$0F,OK	Go to OK if bits 3-0 of location 7 all are 0	4

Figure 1.47
Example of the 6812
direct page addressing
mode.

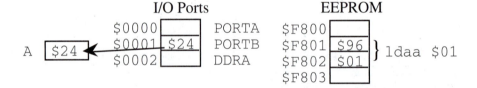

I/O Ports

```
      $0000              PORTA
      $0001  $24         PORTB
      $0002              DDRA
A  $24
```

EEPROM

```
$F800
$F801  $96
$F802  $01   } ldaa $01
$F803
```

4. Extended addressing mode uses a 16-bit address to access all memory and I/O devices (Figures 1.48, 1.49). We use the same syntax for direct and extended addressing, and the assembler will automatically create direct address if possible. Examples:

Machine code	Opcode	Operand	Comment	Cycles
$FE**0800**	ldx	$0800	RegX = {$0800}	3
$7B**0901**	stab	$0901	[$0901] = RegB	3
$B6**0000**	ldaa	>0	Force 16-bit addressing	3

Figure 1.48
Example of the 6812
extended addressing
mode.

RAM

```
       $0800  $12
       $0801  $34
X  $1234
```

EEPROM

```
$F800
$F801  $FE
$F802  $08   } ldx $0800
$F803  $00
```

Figure 1.49
Another example of the
6812 extended
addressing mode.

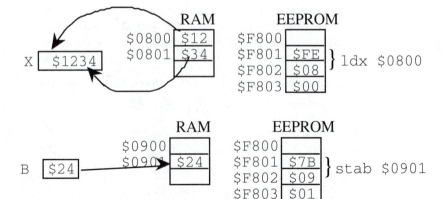

RAM

```
      $0900
      $0901  $24
B  $24
```

EEPROM

```
$F800
$F801  $7B
$F802  $09   } stab $0901
$F803  $01
```

5. Indexed addressing mode uses a fixed offset with the 16-bit registers: X, Y, SP, or PC. On the 6811, instructions that use Register Y take more memory and run slower than the equivalent instruction using Register X. The 6812 eliminates this speed and memory cost of using RegY. In addition, the 6812 indexing modes can be used with the SP and the PC. The offset can be 5 bits (−16 to +15), 9 bits (−256 to +127), or 16 bits. Five-bit (−16 to +15) index mode requires one machine byte to encode the operand. In the first example that uses 5-bit indexed mode, $6A is the `staa` instruction and $**5C** is the index mode operand. Tables A.3 and A.4 of the Motorola CPU12 Reference Manual show the machine codes for the indexed instructions.

Machine code	Opcode	Operand	Comment	Cycles
$6A**5C**	staa	-4,Y	[Y-4] = RegA	2

Assuming Register Y=$0823, the instruction `staa -4,Y` will store a copy of the value in Register A at $081F, leaving Register Y unchanged. The effective address (EA) is $0823−4=$081F (Figure 1.50).

Figure 1.50
Example of the 6812 indexed addressing mode.

Nine-bit (−256 to +255) indexed mode requires two machine bytes to encode the operand.

Machine code	Opcode	Operand	Comment	Cycles
$6A**E840**	staa	$40,Y	[Y+$40] = RegA	3

Assuming Register Y=$0823, the instruction `staa $40,Y` will store a copy of the value in Register A at $0863, leaving Register Y unchanged. The effective address (EA) is $0823+$40=$0863 (Figure 1.51).

Sixteen-bit indexed mode requires three machine bytes to encode the operand.

Machine code	Opcode	Operand	Comment	Cycles
$6A**EA0200**	staa	$200,Y	[Y+$200] = RegA	3

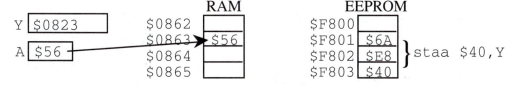

Figure 1.51
Another example of the 6812 indexed addressing mode.

Assuming Register Y=$0823, the instruction `staa $200,Y` will store a copy of the value in Register A at $0A23, leaving Register Y unchanged. The effective address (EA) is $0823+$200=$0A23 (Figure 1.52). Due to the properties of 16-bit addition, the 16-bit offset can be interpreted either as unsigned (0 to 65535) or signed (−32768 to +32767). Indexed mode is useful when addressing the data structures and information on the stack. In each case, the 16-bit register used as a pointer (index) is not modified by the instruction. More examples:

Machine code	Opcode	Operand	Comment	Cycles
$CE0000	ldx	#$0000	RegX points to internal I/O	2
$6101	com	1,X	Invert all bits of PortB=$0001	3
$0EE02880rr	brset	$28,X,$80,DONE	go to DONE if $0028 bit 7 set	6
$0D0413	bclr	4,X,%00010011	Clear Port C($0004) bits 4,1,0	4
$CD0084	ldy	#$0084	RegY points to TCNT	2
$EC40	ldd	,Y	Read TCNT ($100E) into RegD	3

Common Error: SP relative indexed addressing with a negative constant is usually defined as an illegal stack access.

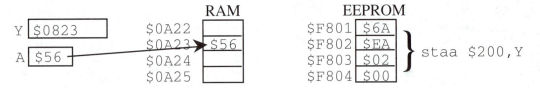

Figure 1.52
A third example of the 6812 indexed addressing mode.

6. Auto pre/post decrement/increment indexed addressing uses the 16-bit registers X, Y, or SP. The PC cannot be used with these index modes which modify the index register. In each case, the 16-bit register used as a pointer (index) is modified either before (pre-) or after (post-) the memory access. These modes are useful when addressing the data structures. The 6812 allows the programmer to specify the amount added to (subtracted from) the index register from 1 to 8. In each case assume RegY is initially 2345.

Postincrement examples:

```
staa 1,Y+    Store a copy of the value in Reg A at 2345, then Reg Y=2346
staa 4,Y+    Store a copy of the value in Reg A at 2345, then Reg Y=2349
```

Preincrement examples:

```
staa 1,+Y    Reg Y=2346, then store a copy of the value in Reg A at 2346
staa 4,+Y    Reg Y=2349, then store a copy of the value in Reg A at 2349
```

Postdecrement examples:

```
staa 1,Y-    Store a copy of the value in Reg A at 2345, then Reg Y=2344
staa 4,Y-    Store a copy of the value in Reg A at 2345, then Reg Y=2341
```

Predecrement examples:

```
staa 1,-Y    Reg Y=2344, then store a copy of the value in Reg A at 2344
staa 4,-Y    Reg Y=2341, then store a copy of the value in Reg A at 2341
```

Observation: Usually we would add/subtract 1 when accessing an 8-bit value and add/subtract 2 when accessing a 16-bit value.

Common Error: The improper use of these index modes with the SP can result in an illegal stack access or unbalanced stack.

7. Accumulator offset indexed addressing uses two registers. The offset is located in one of the accumulators A, B or D, and the index (memory address) uses the 16-bit registers X, Y, SP, or PC. In each case, the accumulator used for the offset and the index register used as a pointer (index) are not modified by the instruction. Examples:

```
ldab  #4
ldy   #2345
staa  B,Y    Store a copy of the value in Reg A at 2349 (B & Y unchanged)
```

8. Indexed indirect addressing mode uses a fixed offset with the 16-bit registers X, Y, SP, or PC (Figure 1.53). The fixed offset is always 16 bits. The fixed 16-bit value is added to the index register (X, Y, SP, or PC) and used to fetch a second 16-bit big endian address from memory. The load or store is performed at this second address. Indexed indirect mode is useful when data structures contain pointers. In each case, the 16-bit index register and the memory pointer are not modified by the instruction. Examples:

```
ldy   #2345
staa  [-4,Y]    fetch 16-bit address from 2341, store 56 at 1234
```

Figure 1.53
Another example of the 6812 indexed addressing mode.

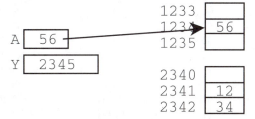

9. Accumulator D offset indexed indirect addressing uses two registers (Figure 1.54). The offset is located in accumulator D, and the index (memory address) is in one of the 16-bit registers X, Y, SP, or PC. The value in D is added to the index register (X, Y, SP, or PC) and used to fetch a second 16-bit big endian address from memory. The load or store is performed at this second address. This mode is also useful when data structures contain

Figure 1.54
Example of the 6812 accumulator offset indexed indirect addressing mode.

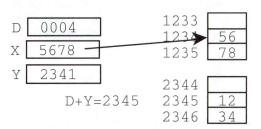

pointers. In each case, accumulator D and the index register used as a pointer (index) are not modified by the instruction. Examples:

```
ldd    #4
ldy    #2341
stx    [D,Y]    Store a copy of the value in Reg X at 1234 (D & Y unchanged)
```

In Tables 1.24 and 1.25, rr can specify X, Y, SP, or PC.

Table 1.24
The MC68HC812A4 indexed register code.

rr	Register
00	X
01	Y
10	SP
11	PC

Table 1.25
Postbyte values for the 6812 indexed addressing modes.

Postbyte(xb)	Syntax	Mode	Explanations
rr000000	,r	IDX	5-bit constant, n=0
rr00nnnn	n,r	IDX	5-bit constant, n from 0 to +15
rr01nnnn	−n,r	IDX	5-bit constant, n from −16 to −1
rr100nnn	n,+r	IDX	preincrement, n from 1 to 8
rr101nnn	n,−r	IDX	predecrement, n from 1 to 8
rr110nnn	n,r+	IDX	postincrement, n from 1 to 8
rr111nnn	n,r−	IDX	postdecrement, n from 1 to 8
111rr100	A,r	IDX	RegA accumulator offset
111rr101	B,r	IDX	RegB accumulator offset
111rr110	D,r	IDX	RegD accumulator offset
111rr000 ff	n,r	IDX1	9-bit constant, n 16 to 255
111rr001 ff	−n,r	IDX1	9-bit constant, n −256 to −16
111rr010 ffee	n,r	IDX2	16-bit constant, any 16-bit n
111rr111	[D,r]	[D,IDX]	RegD offset, indirect
111rr011 ffee	[n,r]	[IDX2]	16-bit constant, indirect

The expression ff refers to the 8-bit offset needed for IDX1 modes, and ffee refers to the 16-bit offset needed for IDX2 modes.

10. PC relative addressing mode is used for the branch and branch-to-subroutine instructions. Stored in the instruction is not the absolute address of where to branch but the 8-bit signed offset relative distance to the current PC value. When the branch address is being calculated, the PC already points to the next instruction. The address of the next instruction is (location of instruction) + (number of bytes in the machine code). In the following examples assume each branch instruction is located at address \$F880. The "rr" operand field is calculated as

$$\mathbf{rr} = \text{(location of destination address)} - \text{(location of instruction)} - \text{(number of bytes in the machine code)}$$
$$= \$F840 - \$F880 - 2 = -\$42 = \$BE$$
$$= \$F8C8 - \$F880 - 2 = \$46$$
$$= \$F889 - \$F880 - 4 = \$05$$

Machine code	Opcode	Operand	Comment	Cycles
$8105	cmpa	#5	Compare RegA to 5	1
$25**BE**	blo	$F840	go to $F840 if RegA<5 (unsigned)	3/1
$20**BE**	bra	$F840	go to location $F840	3
$07**46**	bsr	$F8C8	branch to subroutine at $F8C8	4
$4E05**1005**	brset	5,$10,$F889	go to $F889 if bit 4 of location 5 is 1	4

The 6812 also supports 16-bit relative branches. The "rr" operand field is calculated using the same equation, except now **rr** is a 16-bit number.

$$\mathbf{rr} = \text{(location of destination address)} - \text{(location of instruction)} - \text{(number of bytes in the machine code)}$$
$$= \$FAC8 - \$F880 - 4 = \$0244$$

Machine code	Opcode	Operand	Comment	Cycles
$8105	cmpa	#5	Compare RegA to 5	1
$1825**0244**	lbls	$FAC8	go to $FAC8 if RegA<5 (unsigned)	4/3

Common Error: Since not every instruction supports every addressing mode, it would be a mistake to use an addressing mode not available for that instruction.

Observation: The number of cycles required to execute 6805, 6808, and 6811 is fixed for each instruction.

Observation: Some of the conditional branch instructions on the 6812 require a different number of cycles to execute depending on whether or not the branch is taken. This fact complicates the task of counting cycles (predetermining how long a program will take to execute).

Observation: When the 6812 operates in expanded mode with external memory, the memory cycle time (time to fetch data) may be slower for data accesses in external memory. This fact makes the task of counting cycles impossible.

1.5.5 Erasing and Programming the MC68HC12 EEPROM

The MC68HC812A4 (or A4, for short) has 4 kbytes of EEPROM. The MC68HC912B32 (or B32, for short) has 768 bytes of regular EEPROM and 32 kbytes of flash EEPROM. Since it is nonvolatile, information stored here will remain even when power is interrupted. The flash EEPROM of the B32 is intended for program storage. The EEPROM on a single-chip A4 is also meant for program storage. But the 768-byte EEPROM on the B32 and the 4-kbyte EEPROM on an expanded-mode A4 is a convenient place to store calibration constants or mode configurations. This EEPROM can also be used to store a small amount of information in a data acquisition system. The idea is to place in EEPROM information that is different from system to system or that you wish to change in the near future. To wait 10 ms we will need to active TCNT so that it increments at 1MHz (Program 1.2).

```
TSCR = $0086              #define TSCR *(unsigned char volatile*)(0x0086)
TMSK2 = $008D             #define TMSK2 *(unsigned char volatile*)(0x008D)
   movb #$80,TSCR             TSCR = 0x80;  // enable TCNT
   movb #$33,TMSK2            TMSK2= 0x33;  // 1 µs clock
```

Program 1.2
6812 software to enable the timer system.

The code in Program 1.3 can be used to erase and program the 6812 EEPROM. In C these same operations could be defined as shown in Program 1.4.

Program 1.3
6812 assembly software
that programs its
internal EEPROM.

```
TCNT    = $0084
EEPROT  = $00F1
EEPROG  = $00F3
EraseAll = $06      ; bulk erase all EEPROM
EraseWord= $16      ; erase an aligned 16-bit word
EraseRow = $0E      ; erase 32 bytes in one row
ProgWord = $02      ; prgram an aligned 16-bit word
; *********program EEPROM*************************
; RegX points into aligned address to change
;   (RegX needs to point into EEPROM for erase functions)
; RegY has data, returns actual data at that address
; RegA has command $06,$16,$0E,$02
Prog:   ldab EEPROT       ; save previous block protect
        pshb
        clr  EEPROT       ; allow changes
        staa EEPROG       ; BULKP=0, ELAT=1, EEPRGM=0
        sty  0,x          ; latches address, data
        inca              ; sets EEPRGM=1
        staa EEPROG
        ldd  TCNT         ; 10000 counts at 1µs each
        addd #10000       ; EndT=RegD is TCNT value 10ms from now
wloop:  cpd  TCNT         ; EndT-TCNT
        bpl  wloop        ; wait for TCNT to pass EndT
        clr  EEPROG
        pulb
        stab EEPROT       ; restore previous block protect
        ldy  0,x          ; actual value in EEPROM
        rts
```

Program 1.4
6812 C software that
programs its internal
EEPROM.

```
#define EraseAll  0x06
#define EraseWord 0x16
#define EraseRow  0x0E
#define ProgWord  0x02
int Program(char command, volatile int *address, int data){
char oldEEPROT; int Endt;
    oldEEPROT=EEPROT;   // read and save
    EEPROT=0;
    EEPROG=command;     // BULKP=0, ELAT=1, EEPRGM=0
    EEPROG++;           // sets EEPRGM=1
    (*address)=data;    // latch address, data
    Endt=TCNT+10000;    // TCNT value 10 ms from now
    while(EndT-TCNT>0); // wait until TCNT passes EndT
    EEPROG=0;           // normal mode
    EEPROT=oldEEPROT;   // restore
    return (*address);} // value at that address
```

The MC68HC912B32 has 32 kbytes of flash EEPROM. Since it is nonvolatile, information stored here will remain even when power is interrupted. Information in the flash EEPROM will last for 10 years. Motorola claims the microcomputer can support 100 erasure/program cycles. Flash programming is different from regular EEPROM in a couple of ways. First, it needs to have $+12$ V applied to the V_{FP} pin to erase and/or program. Second, it does not support word or row erase. Rather, the entire memory must be erased. Under normal operation, though, the 2 kbytes from \$F800 to \$FFFF is protected from erasure/programming and will contain the bootstrap loader. To sometimes wait 10 μs and sometimes wait 110 ms, we will need to change how often TCNT increments. The following code can be used to erase and program the MC68HC912B32 flash EEPROM.

The first part of the code (Program 1.5) contains constants and helper functions.

Program 1.5
Helper functions used to program MC68HC91B32 flash EEPROM.

```
void wait(int time){ int Endt;  // works for time<32678
    Endt=TCNT+time;      // TCNT value time cycles from now
    while(EndT-TCNT>0); // wait until TCNT passes EndT
}
int IsFlashErased(void){int *pt; // returns true if flash erased
    for(pt=(int*)0x8000;pt<(int*)0xf800;pt++){
        if(*pt!=0xFFFF) return 0;  // not erased
    }
    return 1;} // erased
```

The next function in Program 1.6 will erase the flash.

Program 1.6
Software that erases the MC68HC91B32 flash EEPROM.

```
#define tEPULSE 27500     // erase pulse 110000µs/4µs
#define tVERASE   250     // verify erase 1000µs/4µs
#define nEP 5             // maximum erase pulses
int EraseFlash(void){  // returns true if successful
unsigned int numErases=0; // number of erases attempted
int bErased=0;            // set to true when flash erased
    if(!(FEECTL&0x08)) return 0;   // SVFP return if no +12
    TSCR = 0x80;      // enable TCNT
    TMSK2= 0x35;      // 4µs clock
    FEECTL=0x06;      // ERAS and LAT
    *(int*)0x8000=0; // write anywhere to latch address
    while((!bErased)&&(numErases<nEP)){
      FEECTL=0x07;   // ERAS LAT and ENPE enable erase
      wait(tEPULSE); // wait 110 ms
      FEECTL=0x06;   // turn off ENPE
      wait(tVERASE); // wait 1 ms
      numErases++;
      bErased=IsFlashErased();
    }
    if(bErased){      // if successful after numErases tries
      do{
        FEECTL=0x07;   // ERAS LAT and ENPE enable erase
        wait(tEPULSE); // wait 110 ms
        FEECTL=0x06;   // turn off ENPE
        wait(tVERASE); // wait 1 ms
```

continued on p. 52

Program 1.6
Software that programs the MC68HC91B32 flash EEPROM.

```
continued from p. 51
        }while (--numErases);
    }
    bErased=IsFlashErased();
    FEECTL=0x00;  // turn off
return bErased;}
```

The last function in Program 1.7 will program the 16-bit data at the specified address.

Program 1.7
Software that programs the MC68HC91B32 flash EEPROM.

```
#define tPPULSE 200    // program pulse 25µs/125ns
#define tVPROG 80      // verify program 10µs/125ns
#define nPP 50         // maximum program pulses
int ProgramFlash(volatile int *address, int data){
unsigned int numProg=0; // number of programs attempted
int bProg=0;            // set to true when flash programmed
    if(!(FEECTL&0x08)) return 0;   // SVFP return if no +12
    TSCR = 0x80;  // enable TCNT
    TMSK2= 0x30;  // 125ns clock
    FEECTL=0x02;  // LAT
    *address=data; // address and data latched
    while((!bProg)&&(numProg<nPP)){
      FEECTL=0x03;   // LAT and ENPE enable erase
      wait(tPPULSE); // wait 25µs
      FEECTL=0x02;   // turn off ENPE
      wait(tVPROG);  // wait 10µs
      numProg++;
      bProg=(*address==data);
    }
    if(bProg){      // if successful after numProg tries
      do{
          FEECTL=0x03; // LAT and ENPE enable erase
          wait(tPPULSE); // wait 25µs
          FEECTL=0x02;   // turn off ENPE
          wait(tVPROG);  // wait 10µs
      } while (--numProg);
    }
    bProg=(*address==data);
    FEECTL=0x00;  // turn off
    return bProg;}
```

1.6 Digital Logic and Open Collector

Normal digital logic has two states. There are four currents of interest when analyzing if the inputs of the next stage are loading the output. I_{IH} and I_{IL} are the currents required of an input when high and low, respectively. Similarly, I_{OH} and I_{OL} are the maximum currents available at the output when high and low. For the output to properly drive all the inputs of the next stage, the maximum available output current must be larger than the sum of all the required input currents for both the high and low conditions (Figure 1.55).

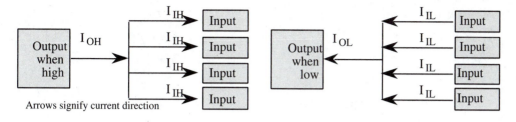

Arrows signify current direction

Figure 1.55
Sometimes one output must drive multiple inputs.

When we design circuits using devices all from a single logic family, we can define fan-out as the maximum number of inputs that one output can drive. For transistor-transistor logic (TTL) we can calculate fan-out from the input and output currents:[7]

$$\text{fan-out} = \text{minimum } (I_{OH}/I_{IH},\ I_{OL}/I_{IL})$$

For circuits that mix devices from one family with another, we must look individually at the input and output currents, voltages, and capacitive loads. Table 1.26 shows typical current values for the various digital logic families.

Observation: For TTL devices the low currents are much larger than the high currents.

Table 1.26
The input and output currents of various digital logic families and microcomputers.

Family	Example	I_{OH}	I_{OL}	I_{IH}	I_{IL}	Fan-out
Standard TTL	7404	0.4 mA	16 mA	40 μA	1.6 mA	10
Schottky TTL	74S04	1 mA	20 mA	50 μA	2 mA	10
Low-power Schottky TTL	74LS04	0.4 mA	4 mA	20 μA	0.4 mA	10
High-speed complementary metal-oxide semiconductor (CMOS)	74HC04	4 mA	4 mA	1 μA	1 μA	
Motorola microcomputer	68HC705J1A	0.2 mA	0.4 mA	10 μA	10 μA	
	PA7-4	0.2 mA	5 mA	10 μA	10 μA	
Motorola microcomputer	68HC708	2 mA	1.6 mA	1 μA	1 μA	
Motorola microcomputer	68HC11A8	0.8 mA	1.6 mA	1 μA	1 μA	
Motorola microcomputer	68HC812A4	0.8 mA	1.6 mA	1 μA	1 μA	
Intel microcomputer	87C51 P0	7 mA	3.2 mA	10 μA	10 μA	
	87C51 P1, P2, P3	60 μA	1.6 mA		50 μA	

Figure 1.56 compares the input and output voltages for many of the digital logic families. The four parameters that affect our choice of logic families are:

■ Power supply current
■ Speed
■ Output drive, I_{OL}, I_{OH}
■ Noise immunity

---

[7]For CMOS, fanout is determined by capacitive loading of chips and wiring.

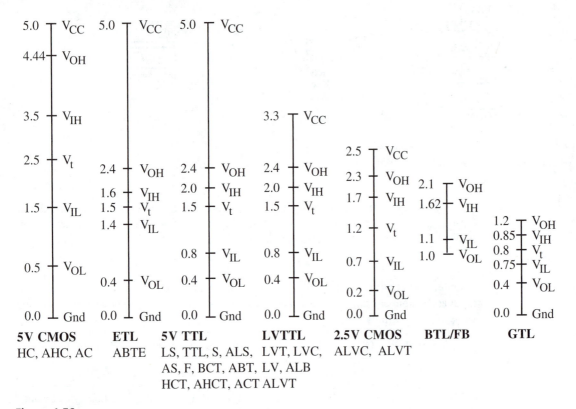

Figure 1.56
Voltage thresholds for various digital logic families.

Tables 1.27 and 1.28 illustrate the wide range of devices available for digital logic design. Not all logic families have the same choice of logic functions. The 74HC245 chip has eight bidirectional tristate drivers (Figure 1.57). The drivers are active when the gate $\bar{G}$ input is low, and the direction is controlled by the DIR input. Devices like the 245 are used in microcomputer systems because of the large output current and bidirectional tristate outputs.

The voltage values presented in Table 1.27 refer to the classifications presented in Figure 1.56.

Family Technology	V_{IL} V_{IH}	V_{OL} V_{OH}	$I_{OL,}$, mA	I_{OH}, mA	I_{CC}, µA	t_{pd}, ns
LVT: Low-voltage BiCMOS	LVTTL	LVTTL	64	−32	190	3.5
ALVC: Advanced low-voltage CMOS	LVTTL	LVTTL	24	−24	40	3.0
LVC: Low-voltage CMOS	LVTTL	LVTTL	24	−24	10	4.0
ALB: Advanced low-voltage BiCMOS	LVTTL	LVTTL	25	−25	800	2.0
AC: Advanced CMOS	CMOS	CMOS	12	−12	20	8.5
AHC: Advanced high-speed CMOS	CMOS	CMOS	4	−4	20	11.9
LV: Low-voltage CMOS	LVTTL	LVTTL	8	−8	20	14

Table 1.27
Comparison of the output drive, power supply current, and speed of various 3.3-V logic 245 gates.

Family Technology	V_{IL} V_{IH}	V_{OL} V_{OH}	I_{OL}, ma	I_{OH}, ma	I_{CC}, ma	t_{pd}, ns
AHC: Advanced high-speed CMOS	CMOS	CMOS	8	−8	0.04	7.5
AHCT: Advanced high-speed CMOS	TTL	CMOS	8	−8	0.04	7.7
AC: Advanced CMOS	CMOS	CMOS	24	−24	0.04	6.5
ACT: Advanced CMOS	TTL	CMOS	24	−24	0.04	8.0
HC: High-speed CMOS logic	CMOS	CMOS	6	−6	0.08	21
HCT: High-speed CMOS logic	TTL	CMOS	6	−6	0.08	30
ABT: Advanced BiCMOS	TTL	TTL	64	−32	0.25	3.5
74F: Fast logic	TTL	TTL	64	−15	120	6.0
BCT: BiCMOS	TTL	TTL	64	−15	90	6.6
AS: Advanced Schottky logic	TTL	TTL	64	−15	143	7.5
ALS: Advanced low-power Schottky	TTL	TTL	24	−15	58	10
LS: Low power Schottky logic	TTL	TTL	24	−15	95	12
S: Schottky logic	TTL	TTL	64	−15	180	9
TTL: Transistor-transistor logic	TTL	TTL	16	−0.4	22	22

Table 1.28
Comparison of the output drive, power supply current, and speed of various 5-V logic 245 gates.

Figure 1.57
Block diagram of a 245 tristate driver.

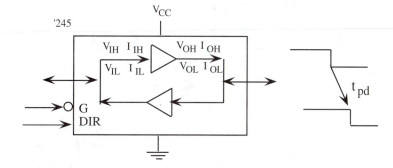

The 74LS04 is a low-power Schottky NOT gate (Figure 1.58). The output is *high* when there is an active transistor Q4 driving the output to +5 V. The output is *low* when there is an active transistor Q5 driving the output to 0.

The 74HC04 is a high-speed CMOS NOT gate (Figure 1.59). The output is *high* when there is an active transistor Q1 driving the output to +5 V. The output is *low* when there is an active transistor Q2 driving the output to 0. Most high-speed CMOS microcomputer outputs behave like the Q1/Q2 "push/pull" transistor pair. These output ports are not inverting. That is, when you write a 1 to an output port, then the output voltage goes high, and when you write a 0 to an output port, then the output voltage goes low.

Open-collector or open-drain logic outputs also have two states. The 7405 is a TTL open-collector NOT gate (Figure 1.60). The output is *open* when there is no active transistor driving the output. In other words, when the input is low, the output floats. This "not driven" condition is called the *open-collector state*. The output is low when there is an active transistor Q3 driving the output to 0.

The 74HC05 is a high-speed CMOS open-collector NOT gate (Figure 1.61). The output is *open* when there is no active transistor driving the output. In other words, when

Figure 1.58
Transistor
implementation of a
low-power Schottky
NOT gate.

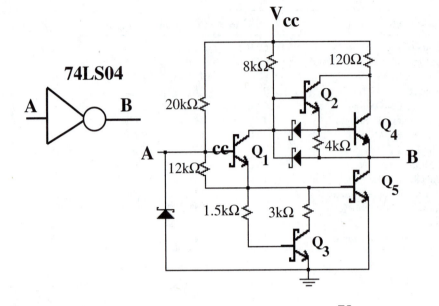

Figure 1.59
Transistor
implementation of a
high-speed CMOS NOT
gate.

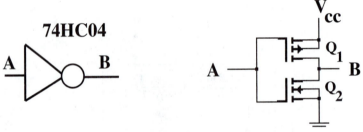

Figure 1.60
Transistor
implementation of a
regular TTL open-
collector NOT gate.

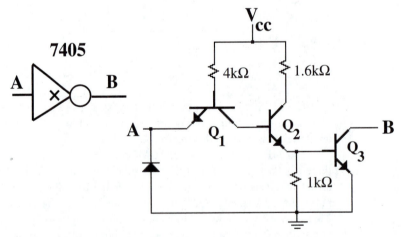

Figure 1.61
Transistor implementation
of a high-speed CMOS
open-collector NOT gate.

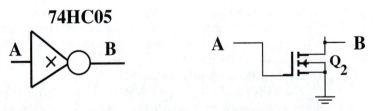

the input is low, the output floats. The output is low when there is an active transistor Q2 driving the output to 0.

Because of the multiple uses of open-collector gates, many microcomputers can implement open-collector logic. All the ports of the Intel 8051 are inherently open collector. In the 6811, when the CWOM bit in the PIOC is set to 1, the outputs of Port C behave like the outputs of the 74HC05 open-collector gate. Also in the 6811, the DWOM bit in the SPCR specifies whether Port D output pins are regular or open drain. The 6812 serial ports also support open-collector outputs.

Observation: The 7405 and 74HC05 are inverting open-collector examples, but most microcomputer output ports are not inverting. That is, when you write a 1 to an open-collector output port, the output floats, and when you write a 0 to an open-collector output port, the output voltage goes low.

You can use an open-collector gate with a +5V pull-up resistor to create digital logic functions. For example, two 7405 open-collector NOT gates with the two outputs tied together with a 1k pull-up resistor to +5V will act like a NOR gate (the output will be high only if both outputs are floating—that is, if both inputs are 0). I do not recommend using an open collector this way, because the output high current is not very good, which in turn will slow down the rise time during a low to high output transition. On the other hand, this example is included because it illustrates an effective approach to combining multiple interrupt request lines to a common interrupt line on the computer (Figure 1.62). In this configuration consider **A** and **B** to be positive logic interrupt requests from two I/O devices and **Out** to be a negative logic request line into the computer. An interrupt is requested (**Out**=0) whenever one or more I/O devices are requesting (**A**=1 or **B**=1). This is a good approach because we don't have to know the total number of I/O devices a priori. Additional interrupting I/O devices can be added without redesigning the existing devices. More about interrupt request lines will be covered in Chapter 4.

How do we select the value of the pull-up resistor? In general the smaller the resistor, the larger the I_{OH} it will be able to supply when the output is high. On the other hand, a larger resistor does not waste as much I_{OL} current when the output is low. One way to calculate the value of this pull-up resistor is to first determine the required output high voltage V_{out} and output high current I_{out}. To supply a current of at least I_{out} at a voltage above V_{out}, the resistor must be less than

$$R \leqslant (+5-V_{out})/I_{out}$$

Figure 1.62
Open-collector
implementation of a
NOR function.

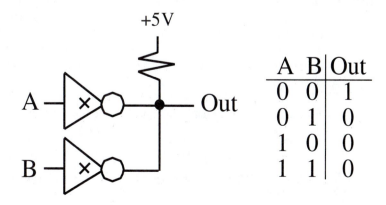

A	B	Out
0	0	1
0	1	0
1	0	0
1	1	0

As an example we will calculate the resistor value for the situation where the circuit needs to drive five regular TTL loads. We see from Figure 1.63 that V_{out} must be above V_{IH} (2 V) for the TTL inputs to sense a high logic level. We can add a safety factor and set V_{out} at 3V. For the high output to drive all five TTL inputs, I_{out} must be more than 5 I_{IH}. From Table 1.28, we see that I_{IH} is 40 μA, so I_{out} should be larger than 5•40 μA, or 0.2 mA. For this situation the resistor must be less than 10 kΩ.

Another good application of open-collector logic exists with a network using a bidirectional communication bus, where outputs are mixed together. This approach works when only one output is active at a time. Since the output ports are readable, a computer can read data from the bus. The ability to transfer in both directions but in only one direction at a time is called half-duplex (Figure 1.63). Recall that open-collector logic has two states: low (0 V) and off (floating). The computer makes its output inactive by using the floating state. We will present the details of open-collector logic to implement a half-duplex communication network in Chapter 7.

Figure 1.63
Open-collector implementation of a half-duplex communication network.

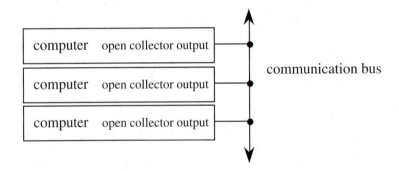

1.7 Initializing and Accessing I/O Ports

1.7.1 Basic Concept of an I/O Port Address and Direction Register

On most embedded microcomputers, the I/O ports are memory-mapped. This means that the software accesses an I/O port simply by reading from or writing to the appropriate address. Usually, an input port is read-only. This means that a read cycle access from the port address returns the current values existing on the inputs. In particular, the tristate driver (triangle-shaped circuit) will drive the input signals onto the data bus during a read cycle from the port address. A write cycle access to an input port usually produces no effect.[8] To make our software more readable we usually include symbolic definitions for the I/O ports. The following assembly lines define some of the ports for the 6805, 6808, 6811, and 6812 microcomputers. For a complete list of I/O ports refer to the respective data sheets.

The assembly code in Program 1.8 specifies some of the I/O port definitions for the MC68HC705J1A. A full list of I/O ports can be found in the MC68HC705J1A reference manual.

---

[8] An input port on the 8051 is created by writing a 1 to the port address; therefore writing a 0 to the port will switch the input to an output.

```
PORTA equ  $0000   8-bit bidirectional I/O port
PORTB equ  $0001   6-bit bidirectional I/O port
DDRA  equ  $0004   specifies whether each bit is input(0) or output(1)
DDRB  equ  $0005   specifies whether each bit is input(0) or output(1)
TCR   equ  $0009   8-bit up counter incremented every memory clock
```

Program 1.8
Assembly definitions of some of the MC68HC705J1A I/O ports.

The assembly code in Program 1.9 specifies some of the I/O port definitions for the MC68HC05C4 and MC68HC05C8. A full list of I/O ports can be found in the MC68HC05C reference manual.

```
PORTA equ  $0000   8-bit bidirectional I/O port
PORTB equ  $0001   8-bit bidirectional I/O port
PORTC equ  $0002   8-bit bidirectional I/O port
PORTD equ  $0003   7-bit fixed input port (shared with SPI, SCI)
DDRA  equ  $0004   specifies whether each bit is input(0) or output(1)
DDRB  equ  $0005   specifies whether each bit is input(0) or output(1)
DDRC  equ  $0006   specifies whether each bit is input(0) or output(1)
TSC   equ  $0012   Timer status and control
TCNT  equ  $0018   16-bit up counter incremented every clock period
```

Program 1.9
Assembly definitions of some of the MC68HC05C I/O ports.

The assembly code in Program 1.10 specifies some of the I/O port definitions for the MC68HC708XL36. A full list of I/O ports can be found in the MC68HC708XL36 reference manual.

```
PORTA equ  $0000   8-bit bidirectional I/O port
PORTB equ  $0001   8-bit bidirectional I/O port
PORTC equ  $0002   8-bit bidirectional I/O port
PORTD equ  $0003   8-bit bidirectional I/O port
DDRA  equ  $0004   specifies whether each bit is input(0) or output(1)
DDRB  equ  $0005   specifies whether each bit is input(0) or output(1)
DDRC  equ  $0006   specifies whether each bit is input(0) or output(1)
DDRD  equ  $0007   specifies whether each bit is input(0) or output(1)
PORTE equ  $0008   8-bit bidirectional I/O port
PORTF equ  $0009   6-bit bidirectional I/O port
PORTG equ  $000A   4-bit bidirectional I/O port
PORTH equ  $000B   4-bit bidirectional I/O port
DDRE  equ  $000C   specifies whether each bit is input(0) or output(1)
DDRF  equ  $000D   specifies whether each bit is input(0) or output(1)
DDRG  equ  $000E   specifies whether each bit is input(0) or output(1)
DDRH  equ  $000F   specifies whether each bit is input(0) or output(1)
TSC   equ  $0021   Timer status and control
TCNT  equ  $0022   16-bit up counter incremented every clock period
```

Program 1.10
Assembly definitions of some of the MC68HC708XL36 I/O ports.

The assembly code in Program 1.11 specifies some of the I/O port definitions for the MC68HC11. A full list of I/O ports can be found in the MC68HC11 reference manual.

```
PORTA equ  $1000  PA7 input/output, PA6-PA3 outputs, PA2-PA0 inputs
PACTL equ  $1026  Bit 7(DDRA7) specifies whether PA7 is input or output
PORTB equ  $1004  PB7-PB0 are all read able outputs
PORTC equ  $1003  PC7-PC0 can be input or output
DDRC  equ  $1007  Direction register for Port C
PORTD equ  $1008  PD5-PD0 can be input or output
DDRD  equ  $1009  Direction register for Port D
PORTE equ  $100A  PE7-PE0 are all inputs
PIOC  equ  $1002  Parallel I/O Control Register
```

Program 1.11 Assembly definitions of some of the MC68HC11 I/O ports.

The C code in Program 1.12 defines these same MC68HC11 I/O port addresses. The `volatile` qualifier means the contents at that address can change by means other than explicit software action of the current module. It is used to prevent the compiler from optimizing a C program that accesses I/O ports. A full list of I/O ports can be found in the file HC11.H on the CD accompanying this book.

Program 1.12
C definitions of some of the MC68HC11 I/O ports.

```
#define PORTA *(unsigned char volatile *)(0x1000)
#define PACTL *(unsigned char volatile *)(0x1026)
#define PORTB *(unsigned char volatile *)(0x1004)
#define PORTC *(unsigned char volatile *)(0x1003)
#define DDRC  *(unsigned char volatile *)(0x1007)
#define PORTD *(unsigned char volatile *)(0x1008)
#define DDRD  *(unsigned char volatile *)(0x1009)
#define PORTE *(unsigned char volatile *)(0x100A)
#define PIOC  *(unsigned char volatile *)(0x1002)
#define TCNT  *(unsigned short volatile *)(0x100E)
```

The assembly code in Program 1.13 specifies some of the I/O port definitions for the MC68HC812A4. A full list of I/O ports can be found in the MC68HC812A4 reference manual.

```
PORTA equ  $0000  General Purpose Input/Output Port
PORTB equ  $0001  General Purpose Input/Output Port
DDRA  equ  $0002  Data Direction Register
DDRB  equ  $0003  Data Direction Register
PORTC equ  $0004  General Purpose Input/Output Port
PORTD equ  $0005  General Purpose Input/Output Port
DDRC  equ  $0006  Data Direction Register
DDRD  equ  $0007  Data Direction Register
PORTE equ  $0008  General Purpose Input/Output Port
DDRE  equ  $0009  Data Direction Register
```

```
PORTH equ  $0024  General Purpose I/O Port, Key wakeup, interrupt
DDRH  equ  $0025  Data Direction Register
PORTJ equ  $0028  General Purpose I/O Port, Key wakeup, interrupt
DDRJ  equ  $0029  Data Direction Register
PORTF equ  $0030  General Purpose I/O Port, Chip select logic
PORTG equ  $0031  General Purpose I/O Port
DDRF  equ  $0032  Data Direction Register
DDRG  equ  $0033  Data Direction Register
PORTAD equ $006F  Analog input   General Purpose Input Port
TCNT  equ  $0084  16-bit Timer
PORTT equ  $00AE  Timer Input/Output Port General Purpose I/O Port
DDRT  equ  $00AF  Data Direction Register
SC0BD equ  $00C0  16-bit Baud rate Register
SC0CR1 equ $00C2  Serial Port Control Register
SC0CR2 equ $00C3  Serial Port Control Register
SC0SR1 equ $00C4  Serial Port Status Register
SC0SR2 equ $00C5  Serial Port Status Register
SC0DRH equ $00C6  high byte of Serial Port Data
SC0DRL equ $00C7  Serial Port Data Register
PORTS equ  $00D6  Serial Input/Output Port General Purpose I/O port
DDRS  equ  $00D7  Data Direction Register
```

Program 1.13
Assembly definitions of some of the MC68HC812A4 I/O ports.

HC12.H is a C header file that defines all the I/O ports for both the MC68HC812A4 and the MC68HC912B32. This file can also be found on the accompanying CD. The syntax of this file is the same as Program 1.12.

A simple fixed input port behaves similar to the circuit shown in Figure 1.64. The digital values existing on the input pins are copied into the microcomputer when the software executes a read from the port address.

A latched input port behaves similar to the circuit shown in Figure 1.65. The digital values existing on the input pins are copied into an internal latch on an appropriate edge of the external control signal. At a later time the data are transferred to the microcomputer when the software executes a read from the latch address. Notice that this latched input port also supports the regular input function. In other words, the software has the option of reading the port address to get information directly from the input port pins or from the latch address to get information that existed at the time of the previous active edge of the external control signal.

Figure 1.64
A read-only input port allows the software to read external digital signals.

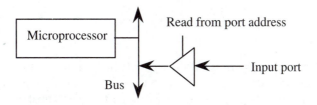

Figure 1.65
A latched input port allows the software to read external digital signals that are captured via the external control signal.

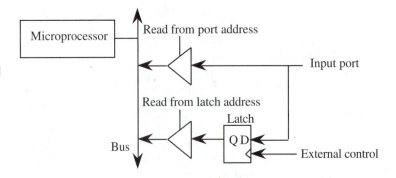

Often the latched input port allows the software to select the strategic edge to affect the latch function. In other words, the software specifies whether the rise or fall of an external control signal latches the input data. It is important to remember that when the software reads from the latch address, it obtains the values of the input signals occurring at the time of the active edge of the external control signal. In this way, the external device can provide the data at the input, issue an edge on the control signal latching into the computer, and then the software can process the data at a later time without requiring the external device to maintain the valid data at the input. One of the limitations of this particular configuration is that it does not include a way for the software to signal back (acknowledge) to the external hardware that the previous data have been read by the software. In Chapter 3 we will develop more sophisticated handshaking protocols that provide for two-way synchronization.

While an input device usually just involves the software reading the port, an output port can participate in both the read and write cycles very much like a regular memory. Figure 1.66 describes a readable output port. For an 8-bit output port there will be eight D flip-flops to hold the values on the output pins. A write cycle to the port address will affect the values on the output pins. In particular, the microcomputer places information on the data bus and that information is clocked into the D flip-flops. Since it is a readable output, a read cycle access from the port address returns the current values existing on the port pins. A write-only output port does not allow software to read the current values.

Figure 1.66
A readable output port allows the software to generate external digital signals.

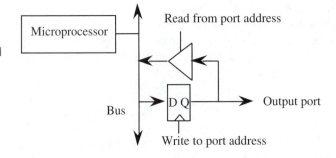

The primary purpose of the example in Program 1.14 is to illustrate how software interacts with its I/O ports. Its secondary purpose is to implement the equivalent of an 8-bit digital NOT gate. In particular, the software will read the digital information from Port A and write the complement of the data out to Port B. The first line of the func-

```
; Program to implement a NOT gate
; Port A are the 8 digital inputs and Port B are the 8 digital outputs
; Software continuously repeats the following
; 1) Read value from Port A
; 2) Calculate the logical complement
; 3) Write the result out to Port B
NotGate ldaa PORTA    Read from Port A into Register A
        coma          Logical complement
        staa PORTB    Write from Register A to Port B
        bra  NotGate
```

Program 1.14
Assembly software that reads from Port A and writes to Port B.

tion, `ldaa PORTA`, will copy the digital values from Port A into Register A. The expression "copy" is used because this operation does not directly affect the status of these input pins. Therefore, after the first instruction is executed, there are indeed two copies of the data, one in Register A and the other on the input pins themselves. Recall that registers are high-speed storage devices that reside inside the processor. Registers do not have specific memory addresses like RAM and ROM data but rather have specific machine instructions that operate with the register. For example, the `ldaa` instruction brings data into Register A. In a similar manner, the `staa` instruction copies data from Register A out to a memory or port location. To perform arithmetic and logical operations we usually first bring the data into registers, perform the operations, then store the results. The second line of this subroutine, `coma`, will perform a logical complement of the value in Register A. The third line of this subroutine, `staa PORTB`, will copy the current value in Register A out to Port B. The last line, `bra NotGate`, causes the program execution to branch to the place `NotGate`. The program cycles through the four instructions continuously.

We can implement this system in C as well (Program 1.15). For most compilers we can include a definitions file that allows us to write software using the symbolic names of the I/O ports. The definition files for the 6811 and 6812 (HC11.H and HC12.H) are included on the accompanying CD. HC11.H and HC12.H can be used with either the ImageCraft compiler or HiWare compiler.

Program 1.15
C software that reads from Port A and writes to Port B.

```c
// Program to implement a NOT gate
// Port A are the 8 digital inputs and Port B are the 8 digital outputs
// Software continuously repeats the following
// 1) Read value from Port A
// 2) Calculate the logical complement
// 3) Write the result out to Port B
void main(void){ unsigned char data;
  while(1){
    data = PORTA;   // Read from Port A
    data = ~data;   // Logical complement
    PORTB = data;   // Write from to Port B
}}
```

Some ports are fixed as inputs or outputs, while others can be specified by the software to be inputs or outputs. Motorola uses the concept of a direction register to determine whether a pin is an input (direction register bit is 0) or an output (direction register bit is 1) (Figure 1.67). We define a *ritual* as a program executed during start-up that initializes hardware and software. If the direction bit is 0, the port behaves like a simple input, and if the direction bit is 1, it becomes a readable output port.

Figure 1.67
A bidirectional port can be configured as a read-only input port or a readable output port.

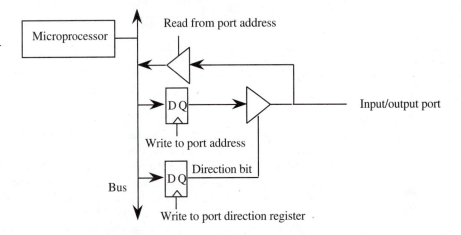

Intel uses open-collector outputs to determine whether a pin is an input or an output (Figure 1.68). Port P0 is open collector. If the software writes a 1 to the port, it behaves like a simple input port. If Port P0 is used as an output, you need to add external pull-up resistors. On the Intel 8051, Ports P1, P2, and P3 are open collector with internal pull-up (Figure 1.69). If the software writes a 1 to the port, it behaves like a simple input port with the

Figure 1.68
A bidirectional port is sometimes implemented with an open-collector readable output port.

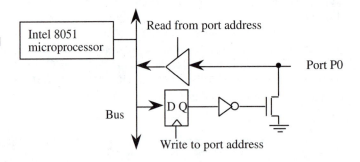

Figure 1.69
Some of the bidirectional ports on the Intel 8051 have internal pull-up resistors.

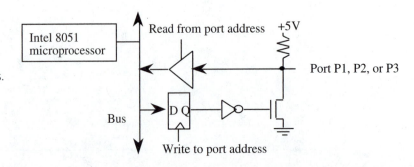

pull-up to +5 V. If the software writes a 0 to the port, the output will be pulled low, and if it writes a 1, the output will be pulled up by the internal resistor.

Common Error: An output port that is created with open-collector outputs with pull-up may not have enough I_{OH} to drive its external circuits.

Common Error: Many program errors can be traced to confusion between I/O ports and regular memory. For example, you should not write to an input port, and sometimes we cannot read from an output port.

Observation: If a port pin is configured as a readable output, but external loading causes the pin voltage to be different from the value written by the software, then a read from the port will return the current voltage level at the pin and not the value last written by the software. This fact can be used by the software to detect excess loading by the external circuits.

The 6811, 6812, and 8051 can provide address and data bus signals for connecting external memory and I/O devices. In this expanded mode, some of the I/O port signals are used for the address/data bus and thus are not available for I/O.

**1.7.2
Simple I/O
Software**

The first example is a single 8-bit I/O port that can be used as either read-only inputs or readable outputs. The hardware configuration for this example is shown in Figure 1.70. The first step to access a bidirectional I/O port is to set the direction register. The direction register specifies bit for bit whether the corresponding pins are input (0) or output (1). To make all pins input, you should clear the direction register as shown in Program 1.16, or in C you could execute as shown in Program 1.17.

Figure 1.70
Examples of bidirectional ports.

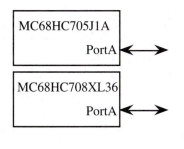
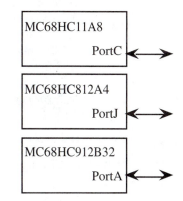

Program 1.16
Assembly software that initializes an I/O port to input.

```
; MC68HC705J1A              ; MC68HC11A8
    clr   DDRA                  clr   DDRC

; MC68HC708XL36/MC68HC912B32  ; MC68HC812A4
    clr   DDRA                  clr   DDRJ
```

Program 1.17
C software that initializes an I/O port to input.

```
// MC68HC705J1A             // MC68HC11A8
    DDRA=0x00;                  DDRC=0x00;

// MC68HC708XL36/MC68HC912B32 // MC68HC812A4
    DDRA=0x00;                  DDRJ=0x00;
```

The software in Program 1.18 reads a regular input and saves the information into a global variable called Happiness.

In C (Program 1.19), an I/O port name (defined in a manner similar to Program 1.12) on the left-hand side of an assignment operator is implemented with an input port read access. In actuality a read input port access is produced whenever the I/O port name exists anywhere except on the right-hand side of an assignment operator (e.g., if while for switch, etc.).

Program 1.18
Assembly software that reads from an I/O port input.

```
; MC68HC705J1A              ; MC68HC11A8
    lda   PORTA                 ldaa  PORTC
    sta   Happiness             staa  Happiness

; MC68HC708XL36/MC68HC912B32 ; MC68HC812A4
    lda   PORTA                 ldaa  PORTJ
    sta   Happiness             staa  Happiness
```

Program 1.19
C software that reads from an I/O port input.

```
// MC68HC705J1A              // MC68HC11A8
   Happiness=PORTA;             Happiness=PORTC;

// MC68HC708XL36/MC68HC912B32 // MC68HC812A4
   Happiness=PORTA;             Happiness=PORTJ;
```

During the execution of the lda(a) instruction, the computer will perform a read bus cycle to the address of the port. It is during this exact time that the current value of the input is transferred into the computer. If a pin is programmed as an input, a write to that port has no effect on that pin.

If a pin is programmed as an output, a write to that port will set/clear that pin, and a read from that port will return the value that we had previously written. To make pins 7–4 input and pins 3–0 output, you should set the direction register to $0F (Program 1.20), or in C you could execute as shown in Program 1.21.

Program 1.20
Assembly software that initializes an I/O port to output.

```
; MC68HC705J1A              ; MC68HC11A8
    lda   #$0F                  ldaa  #$0F
    sta   DDRA                  staa  DDRC

; MC68HC708XL36/MC68HC912B32 ; MC68HC812A4
    lda   #$0F                  ldaa  #$0F
    sta   DDRA                  staa  DDRJ
```

Program 1.21
C software that initializes an I/O port to output.

```
// MC68HC705J1A              // MC68HC11A8
   DDRA=0x0F;                   DDRC=0x0F;

// MC68HC708XL36/MC68HC912B32 // MC68HC812A4
   DDRA=0x0F;                   DDRJ=0x0F;
```

The software in Program 1.22 will set the outputs (pins 3–0) high but have no effect on the inputs (pins 7–4), or in C you could execute as in Program 1.23.

Program 1.22
Assembly software that outputs an I/O port output.

```
; MC68HC705J1A              ; MC68HC11A8
    lda  #$FF                   ldaa #$FF
    sta  PORTA                  staa PORTC
```
```
; MC68HC708XL36/MC68HC912B32  ; MC68HC812A4
    lda  #$FF                   ldaa #$FF
    sta  PORTA                  staa PORTJ
```

Program 1.23
C software that outputs an I/O port output.

```
// MC68HC705J1A              // MC68HC11A8
    PORTA=0xFF;                  PORTC=0xFF;
```
```
// MC68HC708XL36/MC68HC912B32  // MC68HC812A4
    PORTA=0xFF;                  PORTJ=0xFF;
```

We could write general subroutines to access the parallel port. The advantage to developing subroutines for parallel I/O access is that changes can be easily made. When you develop a layered approach to the I/O ports, you are creating an I/O abstraction, or a hardware abstraction layer (HAL). The separation of the policy (how parameters are passed to/from `Init Set Read`) from the implementation (`staa PORTA ldaa PORTJ`) makes it easy to debug and maintain. For example, if you had a system implemented on the 6811 using Port C and wished to redesign it on the 6812 using Port J, then it should be a simple manner of changing the three subroutines. These assembly language subroutines in Program 1.24 pass the data in and out using Register A, or in C you could execute as in Program 1.25. When `Set` is called, only the output pins are affected. On the other hand, the `Read` subroutine reads from the port and gets the current value of all pins, both inputs and outputs.

Program 1.24
Assembly subroutines that provide initialization, read, and write access an I/O port.

```
; MC68HC705J1A                ; MC68HC11A8
Init sta   DDRA              Init staa DDRC
     rts                          rts
Set  sta   PORTA             Set  staa PORTC
     rts                          rts
Read lda   PORTA             Read ldaa PORTC
     rts                          rts
```
```
; MC68HC708XL36/MC68HC912B32  ; MC68HC812A4
Init sta   DDRA              Init staa DDRJ
     rts                          rts
Set  sta   PORTA             Set  staa PORTJ
     rts                          rts
Read lda   PORTA             Read ldaa PORTJ
     rts                          rts
```

Program 1.25
C functions that provide initialization, read, and write access an I/O port.

```
// MC68HC705J1A
void Init(unsigned char value){
    DDRA=value;}
void Set(unsigned char value){
    PORTA=value;}
unsigned char Read(void){
    return(PORTA);}
```
```
// MC68HC11A8
void Init(unsigned char value){
    DDRC=value;}
void Set(unsigned char value){
    PORTC=value;}
unsigned char Read(void){
    return(PORTC);}
```
```
// MC68HC708XL36/MC68HC912B32
void Init(unsigned char value){
    DDRA=value;}
void Set(unsigned char value){
    PORTA=value;}
unsigned char Read(void){
    return(PORTA);}
```
```
// MC68HC812A4
void Init(unsigned char value){
    DDRJ=value;}
void Set(unsigned char value){
    PORTJ=value;}
unsigned char Read(void){
    return(PORTJ);}
```

1.7.3
I/O Example

In the previous NOT gate implementation, shown in Program 1.14, we skipped over the important issue of direction registers. In this example, we will create a four-input–four-output NOT gate using an embedded microcomputer. The four input signals will be connected to an input port of the microcomputer, and the four output signals will be connected to an output port. Some pins of Port C are used as input, and some pins are used as output (Figure 1.71).

Figure 1.71
External inputs are connected to PC3–PC0, and PC7–PC4 are the outputs.

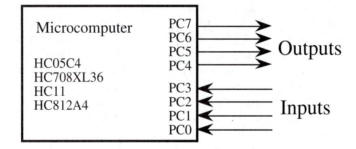

The first software step is to write an initialization function that specifies Port C bits PC7–PC4 will be outputs and that Port C bits PC3–PC0 will be inputs (Program 1.26). This function, which we usually execute once at the start of our program, is called a ritual. The direction register, DDRC, specifies whether each pin is an output (1) or an input (0).

Program 1.26
Assembly subroutines that initialize the I/O port.

```
; MC68HC05C4
PORTC equ $0002 ;I/O port
DDRC  equ $0006
init  lda #$F0   ;PC7-PC3 out
      sta DDRC   ;PC3-PC0 in
      rts
```
```
; MC68HC11A8
PORTC equ  $1003 ;I/O port
DDRC  equ  $1007
init  ldaa #$F0   ;PC7-PC3 out
      staa DDRC   ;PC3-PC0 in
      rts
```
```
; MC68HC708XL36
PORTC equ $0002 ;I/O port
DDRC  equ $0006
init  mov #$F0,DDRC
      rts
```
```
; MC68HC812A4
PORTC equ  $0004 ;I/O port
DDRC  equ  $0006
init  movb #$F0,DDRC
      rts
```

The main program will input from Port C, perform the logical complement, shift the data into the proper position, then output it back to Port C. The input operation performed on the output pins will return the previous value written, while the output operation to the input pins has no effect. In each case, the stack pointer is initialized to a RAM location (Program 1.27).

Program 1.27
Assembly programs that implement the 4-bit NOT gate.

```
; MC68HC05C4                        ; MC68HC11A8
     org  $0100 ;ROM                     org  $E000 ;ROM
main rsp        ;SP=$00FF          main lds #$00FF ;SP=$00FF
     bsr  init  ;ritual                  bsr  init  ;ritual
loop lda  PORTC ;input             loop ldaa PORTC ;input
     coma       ;logical not            coma       ;logical not
     lsla       ;shift                  lsla       ;shift
     lsla                               lsla
     lsla                               lsla
     lsla                               lsla
     sta  PORTC ;output                 staa PORTC ;output
     bra  loop  ;repeat                 bra  loop  ;repeat
     org  $1FFE                         org  $FFFE
     fdb  main  ;reset vector           fdb  main  ;reset vector
```

```
; MC68HC708XL36                     ; MC68HC812A4
     org  $6E00 ;EEPROM                  org  $F000 ;EEPROM
main rsp        ;SP=$00FF          main lds #$0C00 ;SP=$00FF
     bsr  init  ;ritual                  bsr  init  ;ritual
loop lda  PORTC ;input             loop ldaa PORTC ;input
     coma       ;logical not            coma       ;logical not
     nsa        ;swap 7-4, 3-0          lsla       ;shift
     sta  PORTC ;output                 lsla
     bra  loop  ;repeat                 lsla
     org  $FFFE                         lsla
     fdb  main  ;reset vector           staa PORTC ;output
                                        bra  loop  ;repeat
                                        org  $FFFE
                                        fdb  main  ;reset vector
```

We could implement these operations in C. In these C programs we will assume that the compiler will handle the segmentation (placing variables in RAM, programs in ROM/EEPROM), initializing the stack and setting the reset vector (Program 1.28).

Program 1.28
C programs that implement the 4-bit NOT gate.

```c
// HC05C4, HC708XL36, HC11A8, HC812A4
void init(void){
    DDRC=0xF0;} // PC7-PC4 are outputs, PC3-PC0 are inputs
void main(void){ unsigned char data;
    init();      // call ritual once
    while(1){
        data=PORTC;       // input
        data=(~data)<<4;  // complement and shift
        PORTC=data;}}     // output
```

Notice how different the assembly solutions are for the various microcomputers, while the software system can be implemented on all four microcomputers with the same C code. Portability is a measure of how easy it is to convert software that runs on one machine to run on another machine.

Observation: C code is more portable than assembly language.

1.8   Choosing a Microcomputer

This book focuses on the four Motorola microcomputers, but the computer engineer is often faced with the task of selecting the microcomputer for the project. One reason Motorola products were selected is that year after year, Motorola continues to be the leading supplier of microcontrollers. In 1997 Motorola had 17.4% of the 4/8/16+ bit microcontroller market, Hitachi had 15.6%, and NEC had 13.2%. Table 1.29 breaks the 8-bit market share down by architecture.

Table 1.29
1997 market share in dollars of 8-bit microcontrollers.

Architecture	Manufacturer	Dollars (millions)
8051	Intel, Philips, Siemens, Dallas Semiconductor	1027
HC05	Motorola	864
HC11	Motorola	643
H8	Hitachi	505
78K	NEC	497

Table 1.30 illustrates that the microcontroller field is currently dominated by 8-bit devices. The 6812 is classified as a 16-bit device because it maintains 16-bit internal data paths.

Table 1.30
1997 market share in dollars.

Category	Dollars (millions)
4-bit microcontrollers	1305
8-bit microcontrollers	5556
16+ bit microcontrollers	2547

As of 1997, Motorola had shipped over 2 billion 68HC05 8-bit microcontrollers.[9] This cumulative number on a unit level is more than all microprocessors from all other chip vendors. As of 1998, Motorola had shipped over 400 million 68HC11 8-bit microcontrollers. Table 1.31 lists the manufacturers and their web sites.

Often only those devices for which the engineers have hardware and software experience are considered. Fortunately, this blind approach often yields an effective and efficient product, because many of the computers overlap in their cost and performance. In other words, if a microcomputer that you are familiar with can implement the desired functions for the project, then it is often efficient to bypass that more perfect piece of hardware in favor of a faster development time. On the other hand, sometimes we wish

---

[9]Motorola press release of April 27, 1997.

Company	Products	Web site
Motorola	HC05 HC08 HC11 HC12 HC16 683xx 68K MCORE Coldfire PowerPC	http://www.motorola.com/
Hitachi	H8	http://www.hitachi.com/
NEC	78K	http://www.nec.com/
Intel	8051 80251 8096 80296	http://www.intel.com/
Mitsubishi	740 7600 7700 M16C	http://www.mitsubishichips.com/index.htm
Philips	8051	http://www.philips.com/home.html
Siemens	C500 C166 Tricore	http://www.siemens.de/en/
Microchip	PIC12 PIC12 PIC16PIC17	http://www.microchip.com/

Table 1.31 Web sites of companies that make microcontrollers.

to evaluate all potential candidates. Sometimes, it may be cost-effective to hire or train the engineering personnel so that they are proficient in a wide spectrum of potential computer devices. Many factors should be considered when selecting an embedded microcomputer.

- Labor costs include training, development, and testing
- Material costs include parts and supplies
- Manufacturing costs depend on the number and complexity of the components
- Maintenance costs involve revisions to fix bugs and perform upgrades
- ROM size must be big enough to hold instructions and fixed data
- RAM size must be big enough to hold locals, parameters, and global variables
- EEPROM to hold nonvolatile fixed constants that are field-configurable
- Speed must be fast enough to execute the software in real time
- I/O bandwidth affects how fast the computer can input/output data
- 8-, 16-, or 32-bit data size should match most of the data to be processed
- Numerical operations like multiply, divide, signed, floating point
- Special functions like multiply/accumulate, fuzzy logic, complex numbers
- Enough parallel ports for all the input/output digital signals
- Enough serial ports to interface with other computers or I/O devices
- Timer functions to generate signals, measure frequency, measure period
- PWM for the output signals in many control applications
- ADC that is used to convert analog inputs to digital numbers
- Package size and environmental issues affect many embedded systems
- Second source availability
- Availability of high-level language cross-compilers, simulators, emulators
- Power requirements, because many systems will be battery-operated

When considering speed it is best to compare time to execute a benchmark program similar to your specific application, rather than just comparing bus frequency. One of the difficulties is that the microcomputer selection depends on the speed and size of the software, but the software cannot be written without the computer. Given this uncertainty, it is best to select a family of devices with a range of execution speeds and memory configurations. In this way a prototype system with large amounts of memory and peripherals can be purchased for software and hardware development, and once the design is in its final stages, the specific version of the computer can be selected now knowing the memory and speed requirements for the project. In conclusion, while this book focuses on the four Motorola microcomputers, it is expected that once the study of this book is completed, the reader will be equipped with the knowledge to select the proper microcomputer and complete the software design.

1.9 Glossary

bidirectional Digital signals that can be either input or output.

BUFFALO Bit User's Fast Friendly Aid to Logical Operation, an interactive monitor that runs on the target microcomputer, a 6811 EVB.

burn The process of programming a ROM, PROM, or EEPROM.

direct An addressing mode where the data or address value for the instruction is located in memory at address $0000 to $00FF.

direction register Specifies whether a bidirectional I/O pin is an input or an output. We set a direction register bit to 0 (or 1) to specify the corresponding I/O pin to be input (or output).

download The process of transferring object code from the Macintosh or personal computer to the target microcomputer.

dummy PC A computer bus cycle that fetches data pointed to by the PC, but the data are not used.

dummy SP A computer bus cycle that fetches data pointed to by the SP, but the data are not used.

EEPROM Electrically erasable programmable read-only memory that is nonvolatile and easy to reprogram.

embedded computer system A system that performs a specific dedicated operation where the computer is hidden or embedded inside the machine.

EPROM Same as PROM. Electrically programmable read-only memory that is nonvolatile and requires external devices to erase and reprogram. It is usually erased using UV light.

erase The process of clearing the information in a PROM or EEPROM. The information bits are usually all set to logic 1.

EVB Evaluation Board, a Motorola product used to develop microcomputer systems.

expanded mode The mode where some of the I/O ports are used to create an external data bus (control, address, data), allowing external memory to be connected.

extended An addressing mode where the data or address value for the instruction is located anywhere in memory.

fan-out The number of inputs that a single output can drive if the devices are all in the same logic family.

general-purpose computer system A system like the IBM PC or Macintosh with a keyboard, disk, and display that can be programmed for a wide variety of purposes.

hold time (t_h) When latching data into a device with a rising or falling edge of a clock, the hold time is the time after the active edge of the clock that the data must continue to be valid.

I_{IH} Input current when the signal is high.

I_{IL} Input current when the signal is low.

immediate An addressing mode where the operand is a fixed data or address value.

indexed An addressing mode where the data or address value for the instruction is located in memory pointed to by an index register.

inherent An addressing mode where there is no operand or where the operand is implied (not explicitly stated).

interrupt A mechanism where the external hardware can cause a special software routine (handler) to be executed.

I_{OH} Output current when the signal is high.

I_{OL} Output current when the signal is low.

I/O device A computer component capable of bringing information from the external environment into the computer (input device) or sending data out from the computer to the external environment (output device).

I/O port A hardware device that connects the computer with external components.

IRQ An interrupt mechanism on the 6805, 6808, 6811 and 6812.

latched input port An input port where the signals are latched (saved) on an edge of an associated strobe signal.

memory A computer component capable of storing and recalling information.

memory-mapped I/O A system where I/O devices exist at specific addresses.

microcomputer (μC) An electronic device capable of performing I/O functions containing a microprocessor, memory, and I/O devices.

microcontroller A single-chip microcomputer such as the Motorola 6811, Motorola 6816, Intel 8051, Intel 8096, PIC16, or the Texas Instruments TMS370.

nonvolatile A condition where information is not lost when power is removed. When power is restored, then the information is in the state that occurred when the power was removed.

null cycle A computer bus cycle that fetches data at address \$FFFF, but the data are not used.

open collector A digital logic output that has two states, low and off.

operating system (OS) System software for managing computer resources and facilitating common functions such as I/O, memory management, and file system.

parallel port A port where all signals are available simultaneously. In this book the parallel ports are 8 bits wide.

PC relative An addressing mode where the effective address is calculated by its position relative to the current value of the program counter.

personal computer system A small general-purpose computer system having a price low enough for individual people to afford and used for personal tasks.

port External pins through which the microcomputer can perform I/O.

PROM Same as EPROM. Programmable read-only memory that is nonvolatile and requires external devices to erase and reprogram. It is usually erased using UV light.

RAM Random-access memory, a type of memory where is the information can be stored and retrieved easily and quickly. Since it is volatile, the information is lost when power is removed.

real-time computer system A system where time-critical operations occur when needed.

ritual Software, usually executed once at the beginning of the program, that defines the operational modes of the I/O ports.

ROM Read-only memory, a type of memory where the information is programmed into the device once but can be accessed quickly. It is low cost, must be purchased in high volume, and can be programmed only once.

serial port An I/O port where the bits are input or output one at a time.

setup time (t_{su}) When latching data into a device with a rising or falling edge of a clock, the setup time is the time before the active edge of the clock that the data must be valid.

software maintenance Process of verifying, changing, correcting, enhancing, and extending software.

stack Last-in–first-out data structure used to temporarily save information.

tristate The state of a tristate logic output when off or not driven.
tristate logic A digital logic device that has three output states: low, high, and off.
volatile A condition where information is lost when power is removed.
workstation A powerful general-purpose computer system having a price in the $10,000 to $50,000 range and used for handling large amounts of data and performing many calculations.
XIRQ A high-priority interrupt mechanism available on the 6811 and 6812.

1.10   Exercises

1.1 Consider the single-chip microcomputer (6805, 6808, 6811, or 6812) with memory-mapped I/O. Assume there is no DMA on your microcomputer. For this problem, we specify four classes of devices: *processor, RAM, ROM,* and *I/O.* .

a) List the devices that can drive the address bus (1) during a CPU read cycle; (2) during a CPU write cycle.

b) List the devices that can drive the data bus (1) during a CPU read cycle; (2) during a CPU write cycle.

c) List the devices that can drive the R/W line (1) during a CPU read cycle; (2) during a CPU write cycle.

d) List the devices that can receive the information from data bus (1) during a CPU read cycle; (2) during a CPU write cycle.

1.2 Two same instructions were executed on three different machines. The following data were collected on a MC68HC05J1A using an in-circuit emulator:

R/W	Address	Data		R/W	Address	Data
1	$0400	$49		1	$0402	$DF
1	$FFFF	$00		1	$FFFF	$00
1	$FFFF	$00		1	$FFFF	$00
1	$0401	$E7		0	$00F0	$9B

The following data were collected on a MC68HC08XL36 using an in-circuit emulator:

R/W	Address	Data
1	$7000	$49
1	$7001	$E7
1	$7002	$DF
1	$FFFF	$00
0	$00F0	$9B

The following data were collected on an expanded mode 6811 using a logic analyzer:

R/W	Address	Data
1	$E100	$49
1	$E101	$A7
1	$E101	$A7
1	$E102	$DF
1	$FFFF	$00
0	$00F0	$9B

a) Assuming the first cycle represents the start of a new instruction, list the instructions that were executed.

b) What is the value of RegX?

c) List two differences and two similarities between these three computers.

Figure 1.72 6811 architecture diagram used in Exercise 1.3.

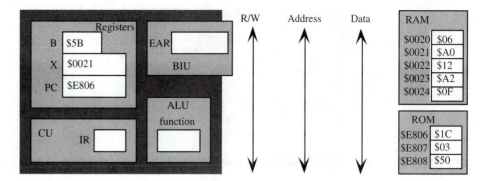

1.3 Figure 1.72 depicts the initial state of the 6811 system.
  a) What instruction will be executed next? Show both the opcode and operand(s).
  b) List below all the changes in the 6811 registers, except the CC register, and/or 6811 memory that occur due to the execution of this instruction. Show only the changes. In other words, if a register or memory is not modified by this instruction, do not include it below.

1.4 You will describe in detail the execution of the branch-to-subroutine instruction. Initially on the 6805, we have

```
SP = $00FD    $0500 $AD $30    bsr function
PC = $0500
```

Initially on the 6808, we have

```
SP = $00FD    $9000 $AD $30    bsr function
PC = $9000
```

Initially on the 6811, we have

```
SP = $00FD    $F000 $8D $30    bsr function
PC = $F000
```

Initially on the 6812, we have

```
SP = $0BFD    $F800 $07 $30    bsr function
PC = $F800
```

  a) What is the absolute address of `function`? That is, where in memory is `function` located?
  b) What is the return address? That is, where will the computer resume execution after the subroutine `function` finishes?
  c) How many cycles does it take to execute?
  d) Show the R/W, address, and data bus for each cycle. For the 6812, simply list the operations that must occur to execute the instruction.

1.5 You will describe in detail the execution of this one 6811 instruction. Initially

```
B = $5B
X = $0001    $F000 $E8 02    eorb 2,x
PC = $F000
```

At the end of each of its four cycles, fill in all boxes in Figure 1.73. In particular, determine the values of the R/W (1 is read, and 0 is write), Address, Data, IR, EAR, and ALU operation. If there is no ALU function, write *none*. If a register (B, X, PC) or memory

Figure 1.73
6811 architecture
diagram used in Exercise
1.5.

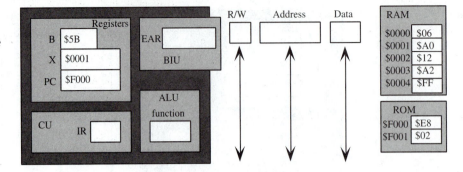

location changes, cross out the old value and put in the new value. Do *not* worry about the CC register.

a) Update after first cycle.   **c)** Update after third cycle.
b) Update after second cycle.   **d)** Update after fourth and last cycle.

1.6 Hand-execute the ONE 6811 instruction located at $E100. Assume the initial values:

Register	Value before	Memory address	Contents (all other memory has a value of zero)
PC	$E100	$0010	$60
RegA	$47	$0011	$61
RegB	$5C	$0012	$62
RegX	$0010	$0020	$70
RegY	$0020	$0021	$71
RegS	$00FE	$0022	$72
		$E100	$18
		$E101	$68
		$E102	$01

a) Disassemble the instruction at $E100. That is, what instruction is located at $E100? Part of the instruction may exist in the bytes that follow: $E101, $E102, etc. The *MC68HC11 Programming Reference Guide* has an opcode/instruction cross-reference table.
b) Show the values of the registers after the instruction is executed in the above table.
c) Show the R/W line (Read is 1, and Write is 0), the memory address bus (16-bit unsigned hex), and data bus (8-bit unsigned hex) while this one instruction is executed.

1.7 The original values of some registers/memory locations are specified for the 6811 or 6812.

C=1	RegA=$79	RegB=$AC	RegX=$2000	RegY=$4000
[$0030]=$A5	[$0031]=$B6	[$3000]=$C7	[$3001]=$D8	[$2000]=$00
[$2001]=$11	[$2002]=$22	[$2003]=$33	[$2004]=$44	[$2005]=$55
[$2006]=$66	[$4000]=$40	[$4001]=$41	[$4002]=$42	[$4003]=$43

a) Specify the addressing mode, the effective address, the opcode, the operand, and the number of cycles it takes to execute the following instructions.

b) Assume each instruction is at memory location 0. Start over with PC=0 each time, and hand-execute each instruction one at a time. In other words, it is not a program in which you execute one instruction after another. After each instruction is executed, indicate the registers and/or memory locations that get affected by the instructions. Give the new contents of the relevant registers and/or memory locations.

```
ldy     #$40
ldx     $30
ldy     $2002
ldy     2,X
ldy     2,Y
adca    #$30
adca    $30
aba
blo     6
psha
```

1.8 Hand-execute the following 6811/6812 program. After each instruction is executed, illustrate the condition of the stack and the values of the registers after each of the instructions. The original condition is RegA=$55, RegB=$66, RegX=$A5B6, and RegY=$A1B2. Assume that the stack is initially empty, with the RegSP=$00FF (for the 6811) or $0C00 for the 6812.

```
psha
pshx
pshy
pulb
pula
pulx
pula
```

1.9 a) What is a *direction register?*
  b) Why does the microcomputer have direction registers?

1.10 You can make the 6811 Port C generate open-collector outputs by setting the appropriate bit in the direction register, DDRC, to 1 and setting the CWOM bit to 1. This open-collector mode can be found in some of the ports of the Intel 8051 microcontroller family, but it is missing in the other 8-bit Motorola single-chip microcomputers 6805, 6808, and 6812.

Motorola claims you can easily upgrade 6811 systems to the more powerful 6812. You have just graduated from Prestigious University, and the first task at your new high paying job at *We Are Nerds INC* is to upgrade an existing *WAN INC* 6811 system to the 6812. Seems easy, so you begin by reading all about the 6812. The 6812 has more instructions and addressing modes, but luckily all existing 6811 assembly instructions and addressing modes are still available on the 6812. The 6812 has more parallel ports, more serial ports, more input captures, and more output compares. In particular, the existing system uses Port C, and the 6812 has a parallel port C, PORTC, and a direction register, DDRC. So far so good, but all of a sudden, BAM, it hits you: Looking at you square in the face is a big problem. The existing system uses open-collector output mode with the 6811 bit CWOM set, but the 6812 has no open-collector modes on any of its parallel port outputs. So now you, the young engineer from PU, must solve your first engineering problem.

Lucky for you, the previous engineers who worked on the problem were also from PU. They implemented the low-level access to PORTC as a device driver and organized the software with a layered approach. This layered system will allow you to replace the low (hardware access) level without having to modify the upper levels. In particular, here are the existing low-level routines in both assembly and C language.

```
* Initialize Parallel Port
* Input: Reg B specifies which parallel port bits will be input(0) or output(1)
* Outputs: none
Pinit: ldaa PIOC
       oraa #$20 ; set CWOM so outputs are open collector
       staa PIOC
       stab DDRC ; 1 means open collector output, 0 means input
       ldaa #$FF
       bsr Pout  ; any output pins are initialized to HiZ
       rts
* Set output (only output pins are effected)
* Input: Reg B specifies new output values
* Outputs: none
Pout: stab PORTC ; modifies output pins only
      rts
* Get input
* Input: none
* Outputs: Reg B returned with current values of both inputs and outputs
Pin: ldab PORTC
     rts
```

```
void Pinit(unsigned char direction){
    PIOC |= 0x20;      // set CWOM so outputs are open collector
    DDRC = direction; // 1 means open collector output, 0 means input
    Pout(0xFF);}       // any output pins are initialized to HiZ
void Pout(unsigned char data){
    PORTC = data;}    // modifies output pins only
unsigned char Pin(void){
    return PORTC;}
```

Your specific task is to rewrite the three device driver routines to run on the 6812 either in assembly *or* in C. You may use any 6811 assembly language instruction and addressing mode. You may use any C language operation normally available on a single-chip microcomputer. Refer to the 6812 parallel port using the symbol PORTC and to the direction register using the symbol DDRC. A global variable will be required.

Test your answer with this example

Assume the four input signals (C7–C4) are set by external hardware to C7=0, C6=1, C5=0, C4=0. Assume the four output signals (C3–C0) have +5-V pull-up resistors.

```
void main(void){ unsigned char info;
    Pinit(0x0F);  // C7-C4 inputs(HiZ), C3-C0 outputs are HiZ
    info=Pin();   // info set to 0x4F (because of pullups)
    Pout(0x00);   // C7-C4 still inputs(HiZ), but C3-C0 are low
    info=Pin();   // info set to 0x40
    Pout(0x05);   // C7-C4,C2,C0 are HiZ, but C3,C1 are low
    info=Pin(); } // info set to 0x45
```

For the remaining exercises, substitute 6805, 6808, 6811, or 6812 for *YourComputer* as appropriate.

1.11 How many general registers are there in *YourComputer*? List them.

1.12 What is the CC register in *YourComputer*?

1.13 Write down all the instruction mnemonics for *YourComputer* that you can use to add two 8-bit numbers.

1.14 Write down the instruction mnemonics for *YourComputer* that you can use to move data between processor registers and memory (both from registers to memory, and vice versa.)

1.15 Write down the instruction mnemonics for *YourComputer* that you can use to move data between processor registers and I/O ports (both from registers to I/O, and vice versa.)

1.16 What are the bidirectional ports on *YourComputer?*

1.17 What are the unidirectional ports on *YourComputer* (if any)?

1.18 How do you control the direction of the bidirectional ports on *YourComputer?*

1.19 List all the I/O ports on *YourComputer,* and give their addresses.

1.20 What instruction gives the 2s complement of a number?

1.21 What instruction gives the 1s complement of a number?

1.22 Describe the action of the following instructions on *YourComputer.* The starting conditions are RegA=$CA; RegB=$43 (if you are studying the 6811 or 6812); C=0. Start over with the initial conditions for each instruction.

Instruction
```
1. lsla
2. lsra
3. asra
4. asla
5. rora
6. rola
7. lsrd  (if you are studying the 6811 or 6812)
8. asld  (if you are studying the 6811 or 6812)
```

1.23 Choose from the following possibilities the most accurate description of the V bit in the CC register that occurs after an addition or subtraction. Choose from the following possibilities the most accurate description of the C bit in the CC register that occurs after an addition or subtraction. (Recall that the 6805 does not have a V bit.)

The bit is set on a signed overflow.
The bit is set on an unsigned overflow.
The bit is set on a signed underflow.
The bit is set on an unsigned underflow.
The bit is set on either a signed overflow or a signed underflow.
The bit is set on either an unsigned overflow or an unsigned underflow.

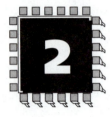

2 Software Development

Chapter 2 objectives are to:

- ❏ Discuss memory allocation on an embedded system, comparing it to segmentation on the Intel x86
- ❏ Develop techniques for writing modular or structured software
- ❏ Introduce layered software organization
- ❏ Present a software model for device drivers
- ❏ Define the concept of threads
- ❏ Describe effective techniques for software debugging

The ultimate success of an embedded system project depends on both its software and its hardware. Computer scientists pride themselves in their ability to develop quality software. Similarly, electrical engineers are well-trained in the processes to design both digital and analog electronics. Manufacturers, in an attempt to get designers to use their products, provide application notes for their hardware devices. The main objective of this book is to combine effective design processes with practical software techniques to develop quality embedded systems. As the size and especially the complexity of the software increase, the software development changes from simple "coding" to "software engineering," and the required skills also vary along this spectrum. These software skills include modular design, layered architecture, abstraction, and verification. Real-time embedded systems are usually on the small end of the size scale, but nevertheless these systems can be quite complex. Therefore the above-mentioned skills are essential for developing embedded systems. This chapter on software development is placed early in the book because writing good software is an art that must be developed and cannot be added on at the end of a project. Good software combined with average hardware will always outperform average software on good hardware. In this chapter we will outline various techniques for developing quality software, then apply these techniques throughout the remainder of the book.

2.1 Quality Programming

Software development is similar to other engineering tasks. We can choose to follow well-defined procedures during the development and evaluation phases, or we can meander in a haphazard way and produce code that is hard to test and harder to change. The ultimate goal of the system is to satisfy the stated objectives such as accuracy, stability, and I/O relationships. Nevertheless, it is appropriate to separately evaluate the individual components of the system. Therefore in this section we will evaluate the quality of our software. There are two categories of performance criteria with which we evaluate the "goodness" of our software. Quantitative criteria include dynamic efficiency (speed of execution), static efficiency (ROM and RAM program size), and accuracy of the results. Qualitative criteria center around ease of software maintenance. Another qualitative way to evaluate software is ease of understanding. If your software is easy to understand, then it will be:

Easy to debug (fix mistakes)
Easy to verify (prove correctness)
Easy to maintain (add features)

Common Error: Programmers who sacrifice clarity in favor of execution speed often develop software that runs fast but doesn't work and can't be changed.

Golden Rule of Software Development
Write software for others as you wish they would write for you.

2.1.1 Quantitative Performance Measurements

To evaluate our software quality, we need performance measures. The simplest approaches to this issue are quantitative measurements. *Dynamic efficiency* is a measure of how fast the program executes. It is measured in seconds or CPU cycles. *Static efficiency* is the number of memory bytes required. Since most embedded computer systems have both RAM and ROM, we specify memory requirement in global variables, stack space, fixed constants, and program objcct code. The global variables plus maximum stack size must be less than the available RAM. Similarly, the fixed constants plus program size must be less than the ROM or PROM size. We can judge our software system according to whether or not it satisfies given constraints, like software development costs, memory available, and timetable.

2.1.2 Qualitative Performance Measurements

Qualitative performance measurements include those parameters to which we cannot assign a direct numerical value. Often in life the most important questions are the easiest to ask but the hardest to answer. Such is the case with software quality. So we ask the following qualitative questions: Can we prove our software works? Is our software easy to understand? Is our software easy to change? Since there is no single approach to writing the best software, we can only hope to present some techniques that you may wish to integrate into your own software style. In fact, we will devote most of this chapter to the important issue of developing quality software. In particular, we will study self-documented code, abstraction, modularity, and layered software. These parameters indeed play a profound effect on the bottom-line financial success of our projects. Although quite real, because there often is not an immediate and direct relationship between a software's quality and profit, we may be tempted to dismiss its importance.

To get a benchmark on how good a programmer you are, we challenge you to two tests. In the first test, find a major piece of software that you have written over 12 months ago, then see if you can still understand it enough to make minor changes in its behavior. The second test is to exchange with a peer a major piece of software that you have both recently written (but not written together), then in the same manner see if you can make minor changes to each other's software.

> **Observation:** You can tell that you are a good programmer if (1) you can understand your own code 12 months later and (2) others can make changes to your code.

2.2 Memory Allocation

Memory allocation is the decision of where in memory we put the various pieces of our software. The memory on a PC-compatible computer is physically configured as a simple linear array. In other words, if you have 64 Mbytes of RAM, then this memory exists as a continuous linear object, with no fundamental difference in the behavior of one memory cell to the next. Although the memory itself forces no structure in the way it is used, the Intel x86 processors have implemented a memory access scheme that requires the programmer to separate in memory three segments: machine codes, global variables, and local variables. The term *x86* refers to any Intel processor from the 8086 through the current Pentiums. Actually there can be more than three segments, but three are enough to illustrate our point. The mechanism to access these segments is called *segmentation*. Figure 2.1 shows a simple view of the memory allocation on the Intel x86 family.

In particular, when the Pentium fetches a machine code, it uses two registers. The code segment (CS) register points to the beginning of the CS, and the instruction pointer (IP) register contains the offset within this segment of the opcode to fetch. Similarly, when the

Figure 2.1
The Intel x86 uses segmented memory allocation.

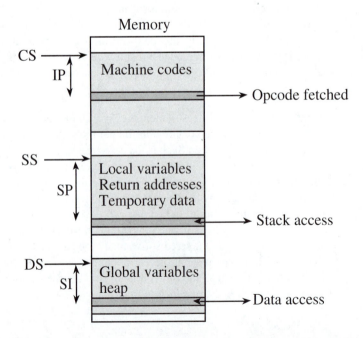

Pentium accesses a global variable, it uses two different registers. The data segment (DS) register points to the beginning of the DS, and a data pointer (source index [SI], destination index [DI]), contains the offset within this segment of global variable. Finally, when the Pentium accesses a local variable, it uses a stack segment (SS) register and either the stack pointer (SP) or the base pointer (BP). The SS register points to the beginning of the SS, and a SP or a BP contains the offset within this segment of local variable. Segmentation forces the programmer to allocate in memory information that has similar properties. In other words, all the machine codes are placed in one group, the global variables are in another group, and the stack is in a third group. This allocation scheme provides for protection so that a stack overflow, a stack underflow, or an illegal stack pointer do not modify machine codes.

We will allocate memory on our embedded system in a fashion similar to segmentation, but for a different reason. Because different types of memory on an embedded computer behave in different fashions, it makes sense to group together in memory information that has similar properties or usage. Typical examples of this grouping include global variables, the heap, local variables, fixed constants, and machine instructions. Figure 2.2 shows a typical memory allocation scheme for an embedded system.

Figure 2.2
We place variables in RAM and programs in ROM on an embedded system.

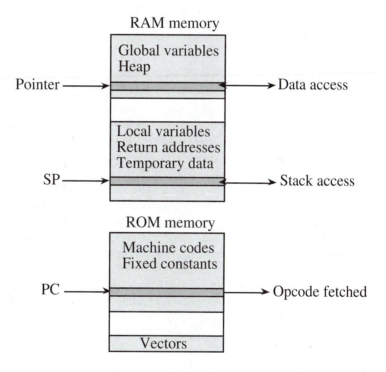

Global variables are permanently allocated and usually accessible by more than one program. We must use global variables for information that must be permanently available or for information that is to be shared by more than one module. We will see many applications later in this book of the first-in–first-out (FIFO) queue, a global data structure that is shared by more than one module. Some software systems use a heap to dynamically allocate and release memory. This information can be shared or not shared, depending on which modules have pointers to the data. The heap is efficient in situations where storage is needed for only a limited amount of time. Local variables are usually allocated on the stack at the beginning of the subroutine/function, used within the subroutine/function, and deallocated at the end of the subroutine/function. Local variables are not shared with other modules. Fixed constants do not change and include information such as numbers, strings, sounds, and pictures. Just like the heap the fixed constants can be shared or not shared, depending on which modules have pointers to the data. As we saw in Chapter 1, the assembler or compiler translates our software into machine instructions (opcodes and operands) that when executed perform the intended operations. For single-chip microcomputers, there are three types of memory (Table 2.1). The RAM contains temporary information that is lost when the power is shut off (i.e., volatile.) This means that all variables allocated in RAM must be explicitly initialized at run time by the software. Some C compilers initialize all RAM-based global variables to zero, and others do not. It is good software development practice to set globals to the desired initial value explicitly. As we saw in Chapter 1, many Motorola microcomputers have a little bit of EEPROM. The ROM is a low-cost nonvolatile storage that can be programmed only once.

Table 2.1

There are different types of memory in an embedded microcomputer.

Memory	When Power is Removed	Ability To Read/Write
RAM	Volatile	Random and fast access
EEPROM	Nonvolatile	Easily erased and reprogrammed
ROM	Nonvolatile	Programmed once

In an embedded application, we usually put global variables, the heap, and local variables in RAM because these types of information can change during execution. When software is to be executed on a regular computer, the machine instructions are usually read from a mass storage device (like a disk) and loaded into memory. Because the embedded system usually has no mass storage device, the machine instructions and fixed constants must be stored in nonvolatile memory. If both EEPROM and ROM are on our microcomputer, we put some fixed constants in EEPROM and some in ROM. If it is information that we may wish to change in the future, we could put it in EEPROM. Examples include language-specific strings, calibration constants, finite-state machines (FSMs), and system identification numbers. This allows us to make minor modifications to the system by reprogramming the EEPROM without throwing the chip away. If our project involves producing a small number of devices, then the program can be placed in EPROM or EEPROM. For a project with a large volume it will be cost-effective to place the machine instructions in ROM.

The assembly language program of Program 2.1 is a simple illustration of how we allocate various sections of our software using the org pseudo-operation. The program outputs to a port the sequence 0, 5, 10, 15. . . . The global variable is placed at the start of the RAM, the stack is initialized to the top of RAM (and grows down), and the program is placed in ROM. The constants are placed in the 6811 EEPROM.

In C, this memory allocation is usually handled as compiler-specific setup parameters. Figure 2.3 illustrates how the ImageCraft ICC12 compiler places the global variables (Data section) at the start of the RAM, the stack at the top of RAM, and the program (Text section) at the start of EEPROM for the MC68HC812A4.

Program 2.1
Memory allocation places variables in RAM and programs in ROM.

```
; MC68HC05J1A
        org $00C0 ;RAM
cnt     rmb 1      ;global

        org $0300 ;ROM
const fcb 5 ;amount to add

init  lda #$FF
      sta DDRA ;outputs
      clr cnt
      rts


main  rsp       ;sp=>RAM
      bsr init
loop  lda cnt
      sta PORTA ;output
      add const
      sta cnt
      bra loop

      org $07FE ;ROM

      fdb main ;reset vector
```

```
; MC68HC11A8
        org $0000 ;RAM
cnt     rmb 1      ;global

        org $B600 ;EEPROM
const fcb 5  ;amount to add

        org  $E000 ;ROM
init  ldaa #$FF
      staa DDRC ;outputs
      clr  cnt
      rts

main  lds  #$00FF ;sp=>RAM
      bsr  init
loop  ldaa cnt
      staa PORTC ;output
      adda const
      staa cnt
      bra  loop

      org $FFFE ;ROM

      fdb main ;reset vector
```

```
; MC68HC08XL36
        org $0050 ;RAM
cnt     rmb 1      ;global

        org $6E00 ;ROM
const fcb 5  ;amount to add

init  mov #$FF,DDRA ;outputs
      clr  cnt  ;
      rts
main  rsp       ;sp=>RAM
      bsr init
loop  lda cnt
      sta PORTA ;output
      add const
      sta cnt
      bra loop

      org $FFFE ;ROM
      fdb main ;reset vector
```

```
; MC68HC812A4
        org $0800 ;RAM
cnt     rmb 1      ;global

        org $F000 ;EEPROM
const fcb 5  ;amount to add

init  movb #$FF,DDRA ;outputs
      clr  cnt
      rts
main  lds  #$0C00 ;sp=>RAM
      bsr  init
loop  ldaa cnt
      staa PORTA ;output
      adda const
      staa cnt
      bra  loop

      org $FFFE ;EEPROM
      fdb main  ;reset vector
```

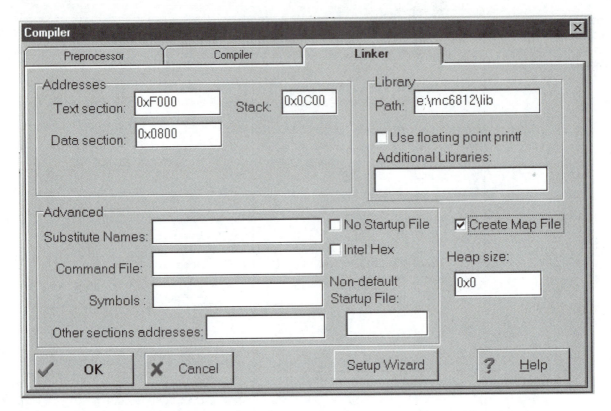

Figure 2.3
ImageCraft ICC12 Compiler Options dialog is used to specify the memory allocation.

2.3 Self-Documenting Code

The goal of this section is to present ideas concerning software documentation in general and writing comments in particular. Maintaining software is the process of fixing bugs, adding new features, optimizing for speed or memory size, porting to new computer hardware, and configuring the software system for new situations. Maintenance is the *most important* phase of software development. Documentation should therefore assist software maintenance. In many situations the software is not static but is continuously undergoing changes. Because of this liquidity, we believe that flowchart and software manuals are not good mechanisms for documenting programs because it is difficult to keep these types of documentation up to date when modifications are made. Therefore, the term *documentation* in this book refers almost exclusively to comments that are included in the software itself. There are two types of readers of our comments. Our client is someone who will use our software, incorporating it into a larger system. Client comments focus on the *policies* of the software. What are the possible valid inputs? What are the resulting outputs? What are the error conditions? The other reader of our comments is a colleague, who is someone charged with software maintenance. Colleague comments focus on the *mechanisms* of the software. How does it work? What algorithms are used?

Observation: The simulator, TExaS, that accompanies this book is one of the few regular software development applications that allows you to add color drawings into the comment fields.

As software developers, our goal is to produce code that not only solves our current problem but can serve as the basis of our future problems. To reuse software we must leave our code in a condition such that future programmers (including ourselves) can easily understand its purpose, constraints, and implementation. Documentation is not something tacked onto software after it is done, but rather it is a discipline built into it at each stage of the development. Writing comments as we develop the software forces us to think about what the software is doing and more importantly why we are doing it. Therefore, we should carefully develop a programming style that provides appropriate comments. A comment that tells us why we perform certain functions is more informative than comments that tell us what the functions are. We should assume the reader of our comment already knows the syntax of the language. The following are examples of bad comments because they provide no additional information:

```
X=X+4;         /* add 4 to X */
Flag=0;        /* set Flag=0 */
```

Common error: A comment that simply restates the operation does not add to the overall understanding.

Common error: Putting a comment on every line of software often hides the important information.

Good comments assist us now while we are debugging and will assist us later when we are modifying the software, adding new features, or using the code in a different context. The following are examples of good comments because they explain why the function is being executed:

```
X=X+4;    /* 4 is added to correct for the offset (mV) in the transducer */
Flag=0;   /* means no key has been typed */
```

When a variable is defined, we should add comments to explain how the variable is used. If the variable has units, then it is appropriate to include them in the comments. It may be relevant to specify the minimum and maximum values. A typical value and what it means often will clarify the usage of the variable. For example:

```
int SetPoint;
/* The desired temperature for the temperature control system
   16-bit signed temperature with a resolution of 0.5C, a range of -55C to +125C
   a value of 25 means 12.5C, a value of -25 means -12.5C */
```

When a constant is used, we could add comments to explain what the constant means. If the number has units, then it is appropriate to include them in the comments. For example:

```
V=999;   /* 999mV is the maximum possible voltage */
err=1;   /* error code of 1 means out of range */
```

When a subroutine or function is defined, there will be two types of comments. The first type is directed to the user of the routine (client). These comments explain how the function is to be used, how to pass parameters, what sort of errors might happen, and how the results are returned. If we are writing in C language, then these comments should be included in the *.h file along with the function prototypes. If we are writing in assembly language, then these comments should be included at the beginning of the subroutine. If the parameters have units, then it is appropriate to include them in the comments. Just like a variable, it may be relevant to specify the minimum and maximum values for the I/O parameters. Typical I/O values and what they mean often will clarify the usage of the function. Frequently we give entire software examples showing how the functions could be used. The second type of comments is directed to the programmer responsible for debugging and software maintenance (colleague). These comments explain how the function works. Generally we separate these comments from the ones intended for the user of the function. This separation is the first of many examples in this book of the concept "separation of policy from mechanism." The policy is what the function does, and the mechanism is how it works. Specifically, we place this second type of comments within the body of the function. If we are writing in C, then these comments should be included in the *.c file along with the function implementation.

Self-documenting code is software written in a simple and obvious way such that its purpose and function are self-apparent. Descriptive names for variables, constants, and functions will go a long way to clarify their usage. To write wonderful code like this, we first must formulate the problem, organizing it into clear, well-defined subproblems. How we break a complex problem into small parts goes a long way toward making the software self-documenting. The concepts of abstraction, modularity, and layered software, all presented later in this chapter, address this important issue of software organization.

Observation: The purpose of a comment is to assist in debugging and maintenance.

We should use careful indenting, and descriptive names for variables, functions, labels, and I/O ports. Liberal use of C language #define and assembly language equ provide explanation of software function without cost of execution speed or memory requirements. A disciplined approach to programming is to develop patterns of writing that you consistently follow. Software developers are not like short story writers. When writing software it is OK to use the same *function outline* over and over again. In the programs of this chapter, notice the following assembly language style issues:

1. Begins and ends with a line of *s
2. States the purpose of the function
3. Gives the I/O parameters, what they mean, and how they are passed
4. Different phases (submodules) of the code delineated by a line of -'s

Observation: It is better to write clear and simple software that is easy to understand without comments than to write complex software that requires a lot of extra explanation to understand.

#define statements, if used properly, can clarify our software and make our software easy to change. Notice in Program 2.2, that if one changes the value of size, then a bug would occur in initialize.

Program 2.2
An inappropriate use of #define.

```
#define size 10
short data[size];
void initialize(void){ short j
    for(j=0;j<10;j++)
        data[j]=0;
};
```

It is proper to use `size` in all places that refer to the size of the data array (Program 2.3).

Program 2.3
An appropriate use of
`#define`.

```
#define size 10
short data[size];
void initialize(void){ short j
    for(j=0;j<size;j++)
        data[j]=0;
};
```

Common error: A programmer may employ a `#define` or `equ` for the sole purpose of making the software easier to read, and a software bug may occur if you change the value of the constant.

Software documentation is an important communication tool between software developers. It also provides invaluable information for software that will be modified in the future. The approach to good documentation is to provide information that enhances understanding, use, and modification. It is good practice to tailor documentation specifically to the intended reader. For example, we might give different information to a user of our module (programmer writing code that calls our module) than to a developer (programmer responsible for testing and upgrading). Clearly state in the comments:

> Purpose of the module
> Input parameters
>> How passed (call by value, call by reference)
>> Appropriate range (does the module assume the input is within range?)
>> Format (8 bit/16 bit, signed/unsigned, etc.)
> Output parameters
>> how passed (return by value, return by reference)
>> format (8 bit/16 bit, signed/unsigned, etc.)
> Example inputs and outputs if appropriate
> Error conditions
> Example calling sequence
> Local variables and their significance

2.4 Abstraction

2.4.1 Definitions

Software abstraction is when we can define a complex problem with a set of basic abstract principles. If we can construct our software system using these building blocks, then we have a better understanding of the problem, because we can separate what we are doing from the details of how we are getting it done. This separation also makes it easier to optimize. It provides for a proof of correct function and simplifies both extensions and customization. A good example of abstraction is the *finite-state machine* (FSM) implementation. The abstract principles of FSM development are the inputs, outputs, states, and state transitions. If we can take a complex problem and map it into a FSM model, then we can solve it with simple FSM software tools. Our FSM software implementation will be easy to understand, debug, and modify. Other examples of software abstraction include *proportional integral derivative* (PID) digital controllers, fuzzy logic digital controllers, neural networks, and linear systems of differential equations (e.g., PSPICE). In each case, the problem is mapped into a well-defined model with a set of abstract yet powerful rules. Then, the software solution is a matter of implementing the rules of the model.

Linked lists are lists or nodes where one or more of the entries are a pointer (link) to other nodes of similar structure. We can have statically allocated fixed-size linked lists that are defined at assemble or compile time and exist throughout the life of the software. On the other hand, we implement dynamically allocated variable-size linked lists that are constructed at run time and can grow and shrink in size. We will use a data structure similar to a linked list, called a *linked structure,* to build a FSM controller. Linked structures are very flexible and provide a mechanism to implement abstraction.

The FSM controller is a good example of the concept of program abstraction. A well-defined model or framework is used to solve our problem (implemented with a linked structure). The three advantages of abstraction are (1) it can be faster to develop because a lot of the building blocks preexist, (2) it is easier to debug (prove correct) because it separates conceptual issues from implementation, and (3) it is easier to change. An important factor when implementing FSMs using linked structures is that there should be a clear and one-to-one mapping between the FSM and the linked structure; that is, there should be one structure for each state.

Program 2.4
6811 C implementation
of a Mealy FSM.

```
const struct State
{      unsigned char Time;      /* Time to wait in each state */
       unsigned char Out[2];    /* Output if input=0,1 */
       const struct State *Next[2]; /* Next state if input=0,1 */
};
typedef const struct State StateType;
#define SA &fsm[0]
#define SB &fsm[1]
#define SC &fsm[2]
#define SD &fsm[3]
StateType fsm[4]={
        {100,{0,0},{SB,SD}},
        {100,{0,8},{SC,SA}},
        { 15,{0,0},{SB,SD}},
        { 15,{8,8},{SC,SD}}
};
void Wait(unsigned int delay){ int Endt;
    Endt=TCNT+delay;              /* Time (125ns cycles) to wait */
    while((Endt-(int)TCNT)>0);    /* wait */
};
void main(void){ StatePtr *Pt;  /* Current State */
        unsigned char Input;
    Pt=SA;                     /* Initial State */
    DDRC=0x08;                 /* PortC bit3 is output */
    while(1){;                 /* PortC bit7 is input */
        Wait(Pt->Time);        /* Time to wait in this state */
        Input=PORTC>>7;        /* Input=0 or 1 */
        PORTC=Pt->Out[Input];  /* Perform output for this state */
        Pt=Pt->Next[Input];    /* Move to the next state */
    }};
```

Figure 2.4
Mealy FSM.

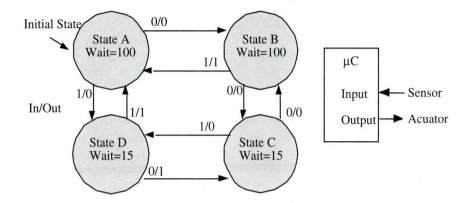

2.4.2
Mealy FSM

The outputs of a Mealy FSM are a function of both the input and the current state. To illustrate the usage of a linked structure, a simple FSM (Figure 2.4) will be implemented (Program 2.4). For example if we are in StateC, we will wait 15 time units, and if the input is 1, then we will make the output 0 and go to StateD. In this example, the delay function is implemented using the internal timer (more information on the timer can be found in Chapter 6). A 6812 implementation would require setting TSCR—for example, TSCR=0×80;

Again proper memory allocation is required if we wish to implement a stand-alone or embedded system. The `org` pseudo-operations are used to place the FSM data structure in EEPROM and the assembly language program in ROM of a single chip 6811. This allows us to make minor modifications to the FSM (add/delete states, change I/O values) by changing the linked structure without modifying the assembly language controller. In this way small modifications/upgrades/options to the FSM can be made by reprogramming the EEPROM without throwing the chip away (Program 2.5).

Program 2.5
6811 assembly implementation of a Mealy FSM.

```
        org  $B600  Put in EEPROM so it can be changed
* Finite State Machine
Time   equ  0      Index for time to wait in this state
Out0   equ  1      Index for output pattern if input=0
Out1   equ  2      Index for output pattern if input=1
Next0  equ  3      Index for next state if input=0
Next1  equ  5      Index for next state if input=1
IS     fdb  SA     Initial state
SA     fcb  100    Time to wait
       fcb  0,0    Outputs for inputs 0,1
       fdb  SB     Next state if Input=0
       fdb  SD     Next state if Input=1

SB     fcb  100    Time to wait
       fcb  0,8    Outputs for inputs 0,1
       fdb  SC     Next state if Input=0
       fdb  SA     Next state if Input=1
```

continued on p. 92

Program 2.5
6811 assembly
implementation of a
Mealy FSM.

continued from p. 91

```
SC      fcb   15        Time to wait
        fcb   0,0       Outputs for inputs 0,1
        fdb   SB        Next state if Input=0
        fdb   SD        Next state if Input=1

SD      fcb   15        Time to wait
        fcb   8,8       Outputs for inputs 0,1
        fdb   SC        Next state if Input=0
        fdb   SA        Next state if Input=1
        org   $E000        Place assembly program in ROM
* 6811 program
* Initialization of 6811 and linked structure
GO      lds   #$00FF    Initialize stack
        ldaa  #$08      PC3=output, rest are input
        staa  $1007     Set DDRC
        ldx   IS        Reg X => current state
*Linked structure interpreter
LL      ldaa  Time,X    Time to wait
        bsr   WAIT      Reg A is call by value
        ldaa  $1003
        bita  #$80      Test input PC7
        bpl   is0       Go to is0 if Input=0
is1     ldaa  Out1,X    Get desired output from linked structure
        staa  $1003     Set PC3=output
        ldx   Next1,X   Input is 1
        bra   LL
is0     ldaa  Out0,X    Get desired output from linked structure
        staa  $1003     Set PC3=output
        ldx   Next0,X   Input is 0
        bra   LL        Infinite loop
        org   $FFFE
        fdb   GO        reset vector
```

Figure 2.5
Moore FSM.

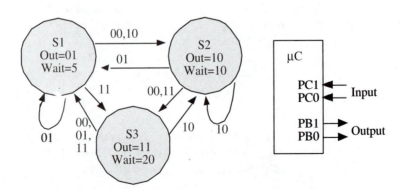

```
/* PortC bits 1,0 are input, Port B bits 1,0 are output */
const struct State {
        unsigned char Out;          /* Output to Port B */
        unsigned int Time;          /* Time in 100 µsec to wait in this state */
        const struct State *Next[4]; /* Next state if input=0,1,2,3 */
};
typedef const struct State StateType;
#define SA &fsm[0]
#define SB &fsm[1]
#define SC &fsm[2]
StateType fsm[3]={
        {0x01, 4000,{SB,SA,SB,SC}},  /* SA out=1, wait= 500usec, next states */
        {0x02, 8000,{SC,SA,SB,SC}},  /* SB out=2, wait=1000usec, next states */
        {0x03,16000,{SA,SA,SB,SA}}   /* SC out=3, wait=2000usec, next states */
};
void Wait(unsigned int delay){ int Endt;
    Endt=TCNT+delay;                /* Time (125ns cycles) to wait */
    while((Endt-(int)TCNT)>0);   /* wait */
};
void main(void){ StatePtr *Pt;  /* Current State */
  unsigned char Input;
  Pt=SA;                    /* Initial State */
  DDRB=0xFF;                /* Make Port B outputs  */
  DDRC=0x00;                /* Make Port C inputs */
  TSCR=0x80;                /* Enable TCNT, default rate 8 MHz */
  while(1){
    PORTB=Pt->Out;      /* Perform output for this state */
    Wait(Pt->Time);     /* Time to wait in this state */
    Input=PORTC&0x03;   /* Input=0,1,2,or 3 */
    Pt=Pt->Next[Input]; /* Move to next state */
  }
};
```

Program 2.6
6812 C implementation of a Moore FSM.

2.4.3
Moore FSM

The outputs of a Moore FSM are only a function of the current state. The simple Moore FSM, shown in Figure 2.5, is implemented in Program 2.6. For example, if we are in State S2, we will set the output to %10 and wait 10 time units. Then, if the input is %11, we will go to State S3. The Moore FSM is shown in Figure 2.5. The times are in 100 µs (e.g., 5 means 500 µs). A 6811 implementation would not require setting the DDRB and TSCR— for example, remove DDRB=0xFF; TSCR=0x80;

2.5 Modular Software Development

In this section we introduce the concept of modular programming and demonstrate that it is an effective way to organize our software projects. There are three reasons for forming modules. Functional abstraction allows us to reuse a software module from multiple locations. Complexity abstraction allows us to divide a highly complex system into smaller, less complicated components. The third reason is portability. If we create modules for the I/O devices, then we can isolate the rest of the system from the hardware details. This approach will be presented later in Section 2.6 on layered software.

2.5.1
Modules

The key to completing any complex task is to break it down into manageable subtasks. Modular programming is a style of software development that divides the software problem into distinct and independent modules. The parts are as small as possible, yet relatively independent. Complex systems designed in a modular fashion are easier to debug because each module can be tested separately. Industry experts estimate that 50 to 90% of software development cost is spent in maintenance. All five aspects of software maintenance:

- Correcting mistakes
- Adding new features
- Optimizing for execution speed or program size
- Porting to new computers or operating systems
- Reconfiguring the software to solve a similar related program

are simplified by organizing the software system into modules. The approach is particularly useful when a task is large enough to require several programmers (Figure 2.6).

Figure 2.6
Block diagram of a software module.

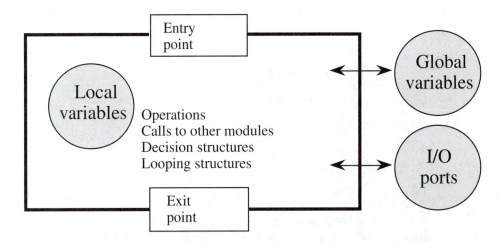

A *program module* is a self-contained software task with clear entry and exit points. We make the distinction between a module and the assembly language subroutine or C language function. A module can be a collection of subroutines or functions that in its entirety performs a well-defined set of tasks. The collection of serial port I/O functions presented in Section 2.7.2 can be considered one module. A collection of 32-bit math operations is another example of a module. The main program and subroutines are obvious modules, but code fragments that perform a well-defined task also can be written as modules. Modular programming involves both the specification of the individual modules and the connection scheme whereby the modules are connected together to form the software system. While the module may be called from many locations throughout the software, there should be only one *entry point*.

In assembly language, the entry point is used to call the subroutine. In this example, the input parameter (e.g., global variable ss) is first pushed on the stack by the instruction movb ss,1,-sp. Next, the instruction jsr sqrt calls the subroutine. After the subroutine returns, the output parameter is in Register B. The input parameter is no longer needed, so ins is executed to discard it. Finally, the output parameter is stored in the global variable tt by the instruction stab tt.

```
movb ss,1,-sp    ;push parameter on the stack (binary fixed point)
jsr  sqrt        ;subroutine call to the module "sqrt"
ins
stab tt          ;save result
```

The stack contents are shown in Figure 2.7 as the calling routine pushes the input parameter ss on the stack, calls the subroutine (Figure 2.8) and then the subroutine allocates three local variables.

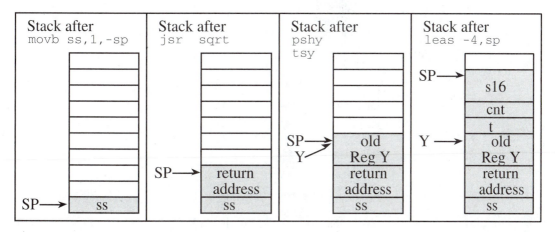

Figure 2.7
6812 stack contents before and after the function call.

Common error: In many situations the input parameters have a restricted range. It would be inefficient for both the module and the calling routine to check for valid input. On the other hand, an error may occur if neither one checks for valid input.

The entry point for a C module (Figure 2.9) is also the name of the function, and it is also used to call the function [e.g., tt=sqrt(ss);].

The main module is the entry point of the entire program. The Motorola microcomputers employ a reset vector to specify where to begin execution after a reset. For Motorola microcomputers, the reset vector location is usually the last 2 bytes of the program ROM (EEPROM) space:

Computer	ROM (hex)	Reset vector (hex)
MC68HC705J1	0300 to 07FF	07FE to 07FF
MC68HC05C8	0100 to 1FFF	1FFE to 1FFF
MC68HC708XL36	6E00 to FFFF	FFFE to FFFF
MC68HC11A8	E000 to FFFF	FFFE to FFFF
MC68HC812A4	F000 to FFFF	FFFE to FFFF
MC68HC912B32	8000 to FFFF	FFFE to FFFF

Figure 2.8
Example of a 6812
assembly language
module.

Entry point

Input parameter

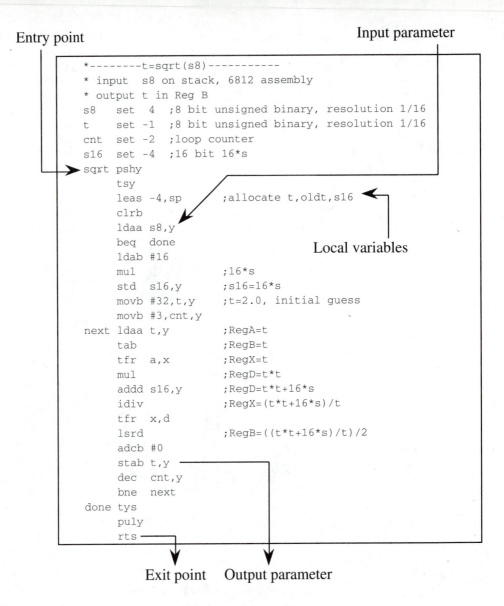

```
*--------t=sqrt(s8)-----------
* input   s8 on stack, 6812 assembly
* output  t in Reg B
s8    set  4  ;8 bit unsigned binary, resolution 1/16
t     set -1  ;8 bit unsigned binary, resolution 1/16
cnt   set -2  ;loop counter
s16   set -4  ;16 bit 16*s
sqrt pshy
     tsy
     leas -4,sp          ;allocate t,oldt,s16
     clrb
     ldaa s8,y
     beq  done
     ldab #16
     mul                 ;16*s
     std  s16,y          ;s16=16*s
     movb #32,t,y        ;t=2.0, initial guess
     movb #3,cnt,y
next ldaa t,y            ;RegA=t
     tab                 ;RegB=t
     tfr  a,x            ;RegX=t
     mul                 ;RegD=t*t
     addd s16,y          ;RegD=t*t+16*s
     idiv                ;RegX=(t*t+16*s)/t
     tfr  x,d
     lsrd                ;RegB=((t*t+16*s)/t)/2
     adcb #0
     stab t,y
     dec  cnt,y
     bne  next
done tys
     puly
     rts
```

Local variables

Exit point   Output parameter

Figure 2.9
Example of a C
language module.

Entry point

Input parameter

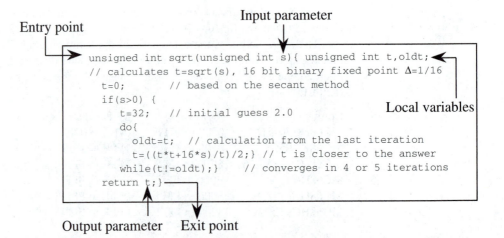

```
unsigned int sqrt(unsigned int s){ unsigned int t,oldt;
// calculates t=sqrt(s), 16 bit binary fixed point Δ=1/16
   t=0;        // based on the secant method
   if(s>0) {
     t=32;    // initial guess 2.0
     do{
       oldt=t;  // calculation from the last iteration
        t=((t*t+16*s)/t)/2;} // t is closer to the answer
     while(t!=oldt);}    // converges in 4 or 5 iterations
   return t;}
```

Local variables

Output parameter   Exit point

The following assembly language software defines the reset vector:

```
org $FFFE
fdb main
```

The following ICC11/ICC12 C code defines a reset vector:

```
#pragma abs_address:0xfffe
void (*reset_vector[])() =
    {
    _start      /* RESET */
    };
#pragma end_abs_address
```

An *exit point* is the ending point of a program module. The exit point of a subroutine is used to return to the calling routine (e.g., `rts`). We need to be careful about exit points. It is important that the stack be properly balanced at all exit points. Similarly if the subroutine returns parameters, then all exit points should return parameters in an acceptable format.

> **Common error:** It is an error if all the exit points of an assembly subroutine do not balance the stack and return parameters in the same way.

If the main program has an exit point, it either stops the program or returns to the debugger. The *input parameters* (or arguments) are pieces of data passed from the calling routine into the module during execution. The *output parameter* (or argument) is information returned from the module back to the calling routine after the module has completed its task.

It is easy to return multiple parameters in assembly language. If just a few parameters need to be returned, we can use the registers. In this simple example, the numbers 1, 2, 3, 4 are to be returned:

```
module: ldaa #1
        ldab #2
        ldx  #3
        ldy  #4
        rts        ; returns four parameters in 4 registers
********calling sequence******
        jsr  module
* Reg A,B,X,Y have four results
```

If many parameters are needed, then the stack can be used. Space for the output parameters is allocated by the calling routine, and the module stores the results into those stack locations.

```
data1 equ 2
data2 equ 3
data3 equ 4
data4 equ 5
module movb #1,data1,sp ;first parameter onto stack
       movb #2,data2,sp ;second parameter onto stack
       movb #3,data3,sp ;third parameter onto stack
       movb #4,data4,sp ;fourth parameter onto stack
       rts
```

continued on p. 98

continued from p. 97

```
********calling sequence******
        leas  -4,sp  ;allocate space for results
        jsr   module
        pula          ;first parameter from stack
        staa  first
        pula          ;second parameter from stack
        staa  second
        pula          ;third parameter from stack
        staa  third
        pula          ;fourth parameter from stack
        staa  fourth
```

There are two approaches for returning multiple parameters in C. In the first approach, we pass a pointer to the module and return the parameters through the pointer.

```
void module(int *pt){
  (*pt)=1;   *(pt+1)=2;  *(pt+2)=3;  *(pt+3)=4;}
void main(void){ int data[4];
    module(&data);}
```

In the second approach we create a structure, and return the parameters within the structure:

```
struct data{ int first,second,third,fourth;};
typedef struct data dataType;
dataType module(void){ dataType myData;
  myData.first=1;  myData.second=2;  myData.third=3;  myData.fourth=4;
 return myData;}
void main(void){ dataType theData;
    theData=module();}
```

2.5.2
Dividing a
Software Task
into Modules

The overall goal of modular programming is to enhance clarity. The smaller the task, the easier it will be to understand. *Coupling* is defined as the influence one module's behavior has on another module. To make modules more independent we strive to minimize coupling. Obvious and appropriate examples of coupling are the I/O parameters explicitly passed from one module to another. On the other hand, information stored in shared global variables can be quite difficult to track. In a similar way shared accesses to I/O ports can also introduce unnecessary complexity. Global variables cause coupling between modules that complicates the debugging process, because now the modules may not be able to be separately tested. On the other hand, we must use global variables to pass information into and out of an interrupt service routine and from one call to an interrupt service routine to the next call. Another problem specific to embedded systems is the need for fast execution, coupled with the limited support for local variables. The 6805 instruction set has no support for local variables, so all 6805 variables are global. On the 6811 it is possible, but inefficient, to implement local variables on the stack. Consequently, many 6811 programmers opt for the less elegant, yet faster approach of global variables. When passing information through global variables is required, it is better to use a well-defined abstract technique like a FIFO queue. We assign a logically complete task to each module. The module is logically complete when it can be separated from the rest of the system and placed into another application. The interfaces are extremely important. The interfaces determine the policies of our modules. In other words, the interfaces define the operations of our software system. The interfaces also represent the coupling between modules. In general we wish to minimize the amount of information passing between the modules yet maximize the number of modules. Of the following three objectives when dividing a software project into subtasks, only the first one really matters:

- Make the software project easier to understand
- Increase the number of modules
- Decrease the interdependency (minimize coupling)

We can develop and connect modules in a hierarchical manner: Construct new modules by combining existing modules. In Figure 2.10 the modules of the software project are organized into a calling graph. An arrow points from the calling routine to the module it calls. The I/O ports are organized into groups (e.g., all the serial port I/O registers are in one group). This graph allows us to see the organization of the project. Figure 2.10 shows a calling graph for the simple example of outputting decimal numbers via the serial port. To make simpler calling graphs on large projects we can combine multiple related subroutines into

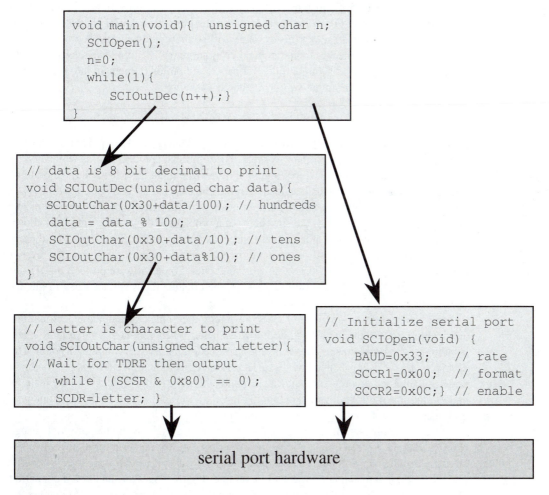

Figure 2.10
A simple calling graph.

a single module. The main program is at the top, and the I/O ports are at the bottom. In a hierarchical system the modules are organized both in a horizontal fashion (grouped together by function) and in a vertical fashion (decisions on overall policies at the top and implementation details at the bottom). Since one of the advantages of breaking a large software project into subtasks is concurrent development, it makes sense to consider concurrency when dividing the tasks. In other words, the modules should be partitioned in such a way that multiple programmers can develop the subtasks as independently as possible. On the other hand, careful and constant supervision is required as modules are connected together and tested.

Observation: Recursive functions do not fall into this hierarchical calling structure.

Observation: If module A calls module B and module B calls module A, then you have created a special situation that must account for these mutual calls.

In Figure 2.10 `SCIOpen` will initialize the serial port interface, establishing the baud rate and protocol format. `SCIOutChar` transmits a single 8-bit byte using the serial port. `SCIOutDec` accepts an 8-bit unsigned byte and calls `SCIOutChar` three times to transmit the three ASCII characters. For example, if n = 145, then the three ASCII characters will be $31, $34, and $35. The details of this example will be presented later in Chapter 7.

There are two approaches to hierarchical programming. The top-down approach starts with a general overview, like an outline of a paper, and builds refinement into subsequent layers. A top-down programmer was once quoted as saying:

"Write no software until every detail is specified"

It provides a better global approach to the problem. Managers like top-down because it gives them tighter control over their workers. The top-down approach works well when an existing operational system is being upgraded or rewritten. On the other hand, the bottom-up approach starts with the smallest detail, building up the system "one brick at a time." The bottom-up approach provides a realistic appreciation of the problem because we often cannot appreciate the difficulty or the simplicity of a problem until we have tried it. It allows programmers to start coding immediately and gives programmers more input into the design. For example, a low-level programmer may be able to point out features that are not possible and suggest other features that are even better. Some software projects are flawed from their conception. With bottom-up design, the obvious flaws surface early in the development cycle.

We believe bottom-up is better when designing a complex system and specifications are open-ended. On the other hand, top-down is better when you have a very clear understanding of the problem specifications and the constraints of your computer system. The best software we ever produced was actually written twice. The first pass was programmed bottom-up and served only to provide a clear understanding of the problem and the features and limitations of our hardware. We literally threw all the source code in the trash and programmed the second pass in a top-down manner.

One of the biggest mistakes beginning programmers make is the inappropriate usage of I/O calls (e.g., screen output and keyboard input). Our explanation for their foolish behavior is that they haven't had the experience yet of trying to reuse software they have written for one project in another project. For example, assume you wrote and tested a function that found the median of three numbers. The goal of this first program was to display the result, so you solved it as shown in Program 2.7.

```
unsigned int Median (unsigned int u1, unsigned int u2, unsigned int u3) { unsigned int result;
  printf("The inputs are %d, %d, %d.\n",u1,u2,u3);
  if(u1>u2)
    if(u2>u3)    result=u2;     // u1>u2,u2>u3         u1>u2>u3
      else
        if(u1>u3) result=u3;    // u1>u2,u3>u2,u1>u3 u1>u3>u2
        else      result=u1;    // u1>u2,u3>u2,u3>u1 u3>u1>u2
  else
    if(u3>u2)    result=u2;     // u2>u1,u3>u2         u3>u2>u1
      else
        if(u1>u3) result=u3;    // u2>u1,u2>u3,u1>u3 u2>u1>u3
        else      result=u1;    // u2>u1,u2>u3,u3>u1 u2>u3>u1
  printf("The median is %d.\n",result);
  return(result);}
```

Program 2.7 A median function that is not very portable.

Software portability of this function is diminished because it is littered with printf's. To use this function in another situation, you will almost certainly have to remove the printf statements. In general we avoid interactive I/O at the lowest levels of the hierarchy; rather we return data and flags and let the higher-level program do the interactive I/O. Often we add keyboard input and screen output calls when testing our software. It is important to remove the I/O that is not directly necessary as part of the module function. This allows you to reuse these functions in situations where screen output is not available or appropriate. Obviously screen output is allowed if that is the purpose of the routine.

Common Error: Performing unnecessary I/O in a subroutine makes it harder to reuse at a later time.

From a formal perspective, I/O devices are considered as global because I/O devices reside permanently at fixed addresses. From a syntactic viewpoint any module has access to any I/O device. To reduce the complexity of the system we will restrict the number of modules that actually do access the I/O device. It will be important to clarify which modules have access to I/O devices and when they are allowed to access it. When more than one module accesses an I/O device, it is important to develop ways to arbitrate (which module goes first if two or more want to access simultaneously) or synchronize (make a second module wait until the first is finished). These arbitration issues will be presented in detail in both Chapters 4 and 5.

Information hiding is similar to minimizing coupling. It is better to separate the mechanisms of software from its policies. We should separate what the function does (the relationship between its inputs and outputs) from how it does it. It is good to hide certain inner workings of a module, and simply interface with the other modules through the well-defined I/O parameters. For example, we could implement a FIFO by maintaining the current byte count in a global variable, CNT. A good module will hide how CNT is implemented from its users. If the user wants to know how many bytes are in the FIFO, it calls one of the FIFO routines that returns the count. A badly written module will not hide CNT from its users. The user simply accesses the global variable CNT. If we update the FIFO routines, making them faster or better, we might have to update all the programs that access CNT, too. The object-oriented programming environments provide well-defined mechanisms to support information hiding. This separation of policies from mechanisms is discussed further in Section 2.6 on layered software.

The *keep it simple stupid* approach tries to generalize the problem so that it fits an abstract model. Unfortunately, the person who defines the software specifications may not understand the implications and alternatives. As software developers, we always ask ourselves these questions:

"How important is this feature?"
"What if it worked this different way?"

Sometimes we can restate the problem to allow for a simpler (and possibly more powerful) solution.

2.5.3 Rules for Developing Modular Software in Assembly Language

The objective of this section is to present modular design rules and illustrate these rules with assembly language examples. This set of rules is meant to guide, not control. In other words, the rules serve as general guidelines rather than as fundamental law.

The single entry point is at the top. In assembly language, we place a single entry point at the first line of the code. This guarantees that registers will be saved and local variables will be properly allocated on the stack. By default, C functions have a single entry point. Placing the entry point at the top provides a visual marker for the beginning of the subroutine.

The single exit point is at the bottom. We prefer to use a single exit point as the last line of the subroutine. Many good programmers employ multiple exit points for efficiency reasons. In general, we must guarantee that the registers, stack, and return parameters are at a similar and consistent state for each exit point. In particular we must deallocate local variables properly. If you do employ multiple exit points, then we suggest you develop a means to visually delineate where one subroutine ends and the next one starts. You could use one line of comments to signify the start of a subroutine and a different line of comments to show the end of it, as in Program 2.8.

Program 2.8 An assembly subroutine that uses comments to delineate its beginning and end.

```
;**************Abs**********************
; Input: RegA is signed 8 bit
; Output: Reg A is absolute value 0 to 127
Abs: tsta      ; already positive?
     bpl  ok
     nega
ok   rts
* ------------end of Abs -----------------
```

Observation: Having the first and last lines of a module be the entry and exit points makes it easier to debug, because it will be easy to place debugging instruments (like breakpoints).

Common error: If you place a debugging breakpoint on the last rts of a subroutine with multiple exit points, then sometimes the subroutine will return without generating the break.

Write structured programs. A structured program is one that adheres to a strict list of program structures. When we program in C (with the exception of goto, which, by the way, you should never use), we are forced to write structured programs because of the syntax of the language. One technique for writing structured assembly language is to adhere to the same strict list of program structures available in C. In other words, restrict the assembly language branching to configurations that mimic the software behavior of `if`, `if-else`, `do-while`, `while`, `for`, and `switch`. Assembly language examples of these control structures are included with the TExaS simulator in the files UIF.RTF, SIF.RTF, WHILE.RTF, and FOR.RTF. Structured programs are much easier to debug, because execution proceeds only through a limited number of well-defined pathways. When we reuse existing assembly branching structures, then our debugging can focus more on the overall function and less on how the details are implemented.

The registers must be saved. When working on a software team, it is important to establish a rule whether or not subroutines will save/restore registers. Establishing this convention is especially important when a mixture of assembly and C is being used or if the software project remains active for long periods of time. It is safest to save and restore registers that are modified (most programmers do not save/restore the CCR). Exceptions to this rule can be made for those portions of the code where speed is most critical.

Common error: If the calling routine expects a subroutine to save/restore registers and it doesn't, then information will be lost.

Observation: If the calling routine does not expect a subroutine to save/restore registers and it does, then the system executes a little slower and the object code is a little bigger than it could be.

Common error: When a mixture of C and assembly language programs is integrated, then an error may occur when the compiler is upgraded because there may be a change if registers are saved/restored or in how parameters are passed.

Use high-level languages whenever possible. It may seem odd to have a rule about high-level languages in a section about assembly language programming. In general we should use high-level languages when memory space and execution speed are less important than portability and maintenance. When execution speed is important, you could write the first version in a high-level language, run a profiler (which will tell you which parts of your program are executed the most), then optimize the critical sections by writing them in assembly language. If a C language implementation just doesn't run fast enough, you could consider a more powerful compiler or a faster microcomputer.

Minimize conditional branching. Every time software makes a conditional branch, there are two possible outcomes that must be tested (branch or not branch). For example, assume we wish to add two 16-bit numbers $u3 = u2 + u1$. A conditional branch could be avoided by solving the problem in another way, as shown in Program 2.9.

Program 2.9
Sometimes we can remove a conditional branch and simplify the program.

```
; 6805/6808                      ; 6811/6812
; uses conditional branch        ; uses conditional branch
add16a lda  u1+1  ;lsb           add16a ldaa u1+1  ;lsb
       add  u2+1                         adda u2+1
       sta  u3+1                         staa u3+1
       lda  u1    ;msb                   ldaa u1    ;msb
       bcc  noc                          bcc  noc
       inca       ;carry                 inca       ;carry
```

continued on p. 104

Program 2.9
Sometimes we can
remove a conditional
branch and simplify the
program.

continued from p. 103
```
noc      add  u2
         sta  u3
         rts
; no conditional branch
add16b lda  u1+1  ;lsb
         add  u2+1
         sta  u3+1
         lda  u1     ;msb
         adc  u2
         sta  u3
         rts
```

```
noc      adda u2
         staa u3
         rts
; no conditional branch
add16b ldaa u1+1  ;lsb
         adda u2+1
         staa u3+1
         ldaa u1     ;msb
         adca u2
         staa u3
         rts
```

Observation: Software can be made easier to understand by reworking the approach to reduce the number of conditional branches.

2.6   Layered Software Systems

As the size and complexity of our software systems increase, we learn to anticipate the changes that our software must undergo in the future. In particular, we can expect to redesign our system to run on newer and more powerful hardware platforms. A similar expectation is that better algorithms may become available. The objective of this section is to use a layered software approach to facilitate these types of changes.

We can use the calling graph defined in Section 2.5 to visualize software layers. A module in a layer can call a module within the same layer or a module in a layer below it. Some layered systems restrict the calls only to modules within the same layer or to a module in the most adjacent layer below it. If we place all the modules that access the I/O hardware in the bottommost layer, we can call this layer a *hardware abstraction layer*. Each layer of modules only calls modules of the same or lower levels, but not modules of higher level. Usually the top layer consists of the main program. In a multithreaded environment (e.g., Unix, Windows NT) there can be multiple main programs at the topmost level, but for now assume there is only one main program. The arrows in Figure 2.11 point from the calling module to the module it calls.

To develop a layered software system we begin with a modular system. The main advantage of layered software is the ability to separate the modules into groups or layers such that one layer may be replaced without affecting the other layers. For example, you could change which ports the printer is connected to by modifying the low level without any changes to the middle or high levels. Figure 2.11 depicts a layered implementation of a printer interface. In a similar way, you could replace the IEEE488 printer with a serial printer by replacing the bottom two layers. If we were to employ buffering and/or data compression to enhance communication bandwidth, then these algorithms would be added to the middle level. A layered system should allow you to change the implementation of one layer without requiring redesign of the other layers.

A *gate* is used to call from a higher- to a lower-level routine. Another name for this gate is application program interface (API). Unfortunately the 6805, 6806, 6811, and 6812 each have only one SWI instruction (the Intel x86 and Motorola 680x0 both have 256 software

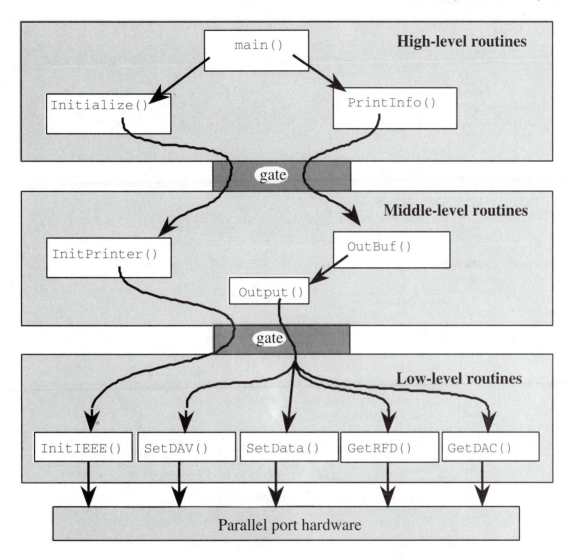

Figure 2.11
A layered approach to interfacing an IEEE488 parallel port printer.

interrupts). If you wished to simulate multiple software interrupts, you could perform an indirect jsr to specific fixed addresses (e.g., one of the unused interrupt vectors $FF80). In this example, to call a *low-level* routine from within a *middle-level* program, you could set RegA equal to the function type and RegB equal to the input data if needed. Then you will jump subroutine indirect through location $FF80. RegB is also used to return any results. On the other hand, to call a *middle-level* routine from a *high-level* routine, you could set RegA equal to the function type, and RegB contains the data or RegX points to the data. Then execute the swi instruction. An error code is returned in RegB. To illustrate one possible implementation of this layering, Program 2.10 shows a skeleton of Figure 2.11.

```
***************************High level***************************
            org   $F000      ; High level routines start at $F000
main:       lds   #$0C00
            bsr   Initialize ; simple call to a function in the same layer
loop:       bsr   PrintInfo  ; simple call to a function in the same layer
            bra   loop
Initialize: ldaa  #1         ; function code for InitPrinter
            swi              ; call to middle level
            rts
            org $FFFE        ; high level vector (reset)
            fdb   main
**************************Middle level***************************
            org   $F400      ; Middle level routines start at $F400
swihandler: cmpa  #1         ; function code for InitPrinter
            bne   notIP
            bsr   InitPrinter
            bra   swidone
notIP:      cmpa  #2         ; function code for PrintInfo
            bne   notPI
            bsr   PrintInfo
            bra   swidone
notPI:                       ; ****error
swidone:    rti
InitPrinter: ldaa #1         ; function code for InitIEEE
            ldx   $FF80       ; vector address into the lower level
            jsr   0,x         ; call lower level
            rts
            org   $FFF6      ; middle level vector (SWI)
            fdb   swihandler
***************************Lower level***************************
            org   $F800      ; Lower level routines start at $F800
lowhandler: cmpa  #1         ; function code for InitIEEE
            bne   notInit
            bsr   InitIEEE
            bra   lowdone
notInit:    cmpa  #2     ; function code for SetDAV
            bne   notDAV
            bsr   SetDAV
            bra   lowdone
notDAV:           ; rest of the functions
lowdone:    rts
InitIEEE:         ; lower level implementation, access to hardware
            rts
            org   $FF80      ; Lower level vector
            fdb   lowhandler
```

Program 2.10 Assembly implementation of a three-layer software system.

Notice that the three layers can be assembled and loaded separately. The gates provide a mechanism to link between the layers. The following example uses simulated software interrupts to implement the gate. Because the size of the software on an embedded system is small, it is possible and appropriate to implement a layered system using standard function calls by simply assembling/compiling all software together. A minor advantage of the above implementation is that the three individual layers can be assembled/compiled and loaded separately.

The following rules apply to layered software systems:

1. A module may make a simple call to other modules in the same layer.
2. A module may make a call to a lower-level module only by using the gate.
3. A module may not directly access any function or variable in another layer (without going through the gate).
4. A module may not call a higher-level routine.
5. A module may not modify the vector address of another level's handler(s). For example:
   The software interrupt handler address, memory contents at $FFF6, is specified by the middle level. The low-handler address, memory contents at $FF80, is specified by the low level.
6. (Optional) A module may not call farther down than the immediately adjacent lower level.
7. (Optional) All I/O hardware access is grouped in the lowest level.
8. (Optional) All user interface I/O (InString, OutString, OutDec, etc.) is grouped in the highest level unless it is the purpose of the module itself to do such I/O.

The purpose of rule 6 is to allow modifications at the low layer to not affect operation at the highest layer. On the other hand, for efficiency reasons you may wish to allow module calls farther down than the immediately adjacent lower layer. To get the full advantage of layered software, it is critical to design functionally complete interfaces between the layers. The interface should support all current functions as well as provide for future expansions.

2.7 Device Drivers

2.7.1 Basic Concept of Device Drivers

A device driver consists of the software routines that provide the functionality of an I/O device. The driver consists of the interface routines that the operating system or software developer's program calls to perform I/O operations as well as the low-level routines that configure the I/O device and perform the actual I/O. The issue of the separation of policy from mechanism is very important in device driver design. The policies of a driver include the list of functions and the overall expected results. In particular the policies can be summarized by the interface routines that the operating system or software developer can call to access the device. The mechanisms of the device driver includes the specific hardware and low-level software that actually performs the I/O. As an example, consider the variety of mass storage devices that are available. Floppy disk, RAM disks, integrated device electronics (IDE) hard drive, Small Computer Systems Interface (SCSI) hard drive, tape, and even a network can be used to save and recall data files. A simple mass storage system might have the following C level interface functions, as explained in the following prototypes (in each case the functions return 0 if successful and an error code if the operation fails):

```
int CreateFile(char *FileName);  // create a new empty file with name Filename
int OpenFile(char *FileName);    // open a file for reading, and rewind to the beginning
int WopenFile(char *FileName);   // open a file for writing, and rewind to the beginning
int WriteFile(char *FileName, char data); // save data into file
int ReadFile(char *FileName, char *data); // read data from file
int PosFile(char *FileName, unsigned int position); // move file pointer to position
```

Building a HAL is the same idea as separation of policy from mechanism. A diagram of this layered concept was shown in Figure 2.11. In the above file example, a HAL would treat all the potential mass storage devices through the same software interface. Another example of this abstraction is the way some computers treat pictures on the video screen and pictures printed on the printer. With the abstraction layer, the software developer's program draws lines and colors by passing the data in a standard format to the device driver, and the operating system redirects the information to the video graphics board or color laserwriter as appropriate. This layered approach allows one to mix and match hardware and software components but does suffer some overhead and inefficiency.

Low-level *device drivers* normally exist in the basic I/O system (BIOS) ROM and have direct access to the hardware. They provide the interface between the hardware and the rest of the software. Good low-level device drivers allow:

1. New hardware to be installed
2. New algorithms to be implemented
   a. Synchronization with gadfly, interrupts, or DMA
   b. Error detection and recovery methods
   c. Enhancements like automatic data compression
3. Higher-level features to be built on top of the low level
   a. Operating system features like blocking semaphores
   b. Additional features like function keys

and still maintain the same software interface. In larger systems like the Workstation and IBM PC, where the low-level I/O software is compiled and burned in ROM separate from the code that will call it, it makes sense to implement the device drivers as software interrupts (SWIs) and specify the calling sequence in assembly language. We define the "client programmer" as the software developer who will use the device driver. In embedded systems like we use, it is okay to provide device.H and device.C files that the client programmers can compile with their application. In a commercial setting, you may be able to deliver to the client only the device.H together with the object file, device.O.[1] *Linking* is the process of resolving addresses to code and programs that have been complied separately. In this way, the routines can be called from any program without requiring complicated linking. In other words, when the device driver is implemented with a TRAP, the linking is simple. In our embedded system, the compiler will perform the linking. The device driver software is grouped into four categories. Protected items can only be directly accessed by the device driver itself, and public items can be accessed by the software developer's program.

---

[1]The object code files of the ImageCraft and Hiwave compilers have the *.O extension.

The concept of a device driver can be illustrated with the following prototype of a keyboard device driver.

1. Data structures: global (protected)

`OpenFlag` Boolean that is true if the keyboard port is open

This global is initially false and is set to true by `KeyOpen`. It is then set to false by `KeyClose`. It exists as static storage (or dynamically created at bootstrap time—that is, when loaded into memory). It can be accessed only by internal device driver routines and not by the client programmer.

`Fifo` queue, with functions `InitFifo`, `PutFifo`, `GetFifo`

This FIFO queue can exist statically or can be dynamically allocated by `KeyOpen`. A heap is required to implement dynamic allocation. The FIFO provides the data linkage between the keyboard interrupt handler and the function `KeyIn`. FIFO queues will be discussed in detail later in Chapter 4.

2. Initialization routines (public, called by the client once in the beginning)

`KeyOpen` Initialize the keyboard port

This function will set the `OpenFlag` to true. The purpose of this function is to initialize the hardware and the FIFO queue. It will return an error code if unsuccessful. Typical reasons the initialization might fail include the hardware is nonexistent, the driver is already open, the heap is out of memory, the hardware has failed, and the client passed an illegal parameter. Typical parameters are listed below:

> Input parameters (FIFO size)
> Return value (error code)
> Typical calling sequence: `if(!KeyOpen(100)) error();`

`KeyClose` Release the keyboard port

This function will set the `OpenFlag` to false, and release the allocated memory of FIFO queues back to the heap. There are no input parameters, but it will return an error code if not previously open.

> Return value (error code)
> Typical calling sequence: `if(!KeyClose()) error();`

3. Regular I/O calls (public, called by client to perform I/O)

`KeyIn` Input an ASCII character from the keyboard port

This routine tries to get a byte from the FIFO (calls `GetFifo`). It will return the data if successful and will return an error code if unsuccessful. There are two reasonable approaches to handling a FIFO empty situation. One good implementation is to return an error code if the FIFO is empty and let the client decide what to do. This approach is illustrated in the typical calling sequence shown below. Another approach would be to wait for key input. If you do implement this second approach, then it makes sense to provide an additional function, like `KeyStatus` shown below, that the client can call before calling `KeyIn` to see if there are data in the FIFO. Typical situations that result in an unsuccessful `KeyIn` include device not open, FIFO empty, and hardware failure (probably not applicable here).

This function illustrates the difference between a return value (in this case the error code) and an output parameter (in this case a call by reference passing of the data). There are no input parameters:

> Output parameter (data)
> Return value (error code)
> Typical calling sequence: `while(!KeyIn(&data)) process();`

`KeyStatus` Returns the status of the keyboard port (checks FIFO to see if data is waiting)

This function returns a true if a call to `KeyIn` would return with a key and a false if a call to `KeyIn` would not return right away but rather would wait. There may be other error conditions as well, such as device not open, FIFO full errors that resulted in lost data, and hardware failure (probably not applicable here).

> Typical calling sequence: `if(KeyStatus()) KeyIn(&data);`

4. Support software (protected) There is one interrupt service handler: `KeyHan`

An interrupt occurs whenever new data are available (the operator has typed a key). The handler will scan, debounce, deal with one or two key rollover, and put the ASCII code into FIFO. It can only be accessed by internal device driver routines and not by the client programmer.

Notice that this keyboard example implements a layered approach. The low-level functions provide the mechanisms and are protected (hidden) from the client programmer. The high-level functions provide the policies and are accessible (public) to the client. When the device driver software is separated into `device.H` and `device.C` files, you need to pay careful attention as to how many details you place in the `device.H` file. A good device driver separates the policy (overall operation, how it is called, what it returns, what it does, etc.) from the implementation (access to hardware, how it works, etc.). In general, you place the policies in the `device.H` file (to be read by the client) and the implementations in the `device.C` file (to be read by you and your coworkers). Think of it this way: If you were to write commercial software that you wished to sell for profit and you delivered the `device.H` file and its compiled object file `device.O`, how little information could you place in the `device.H` file and still have the software system be fully functional? In object-oriented terms the policies will be public, and the implementations will be private.

> *Observation:* A layered approach to I/O programming makes it easier for you to upgrade to newer technology.

> *Observation:* A layered approach to I/O programming allows you to do concurrent development.

**2.7.2
Serial
Communications
Interface Device
Driver**

In this section we will develop a simple device driver using the SCI. This serial port allows the microcomputer to communicate with devices such as other computers, printers, input sensors, and LCD displays. (Figure 2.12). The details of this interface will be presented in Chapter 7. Here in this chapter we will focus on the software aspects of the device driver.

Figure 2.12
Hardware interface
implementing an
asynchronous RS232
channel.

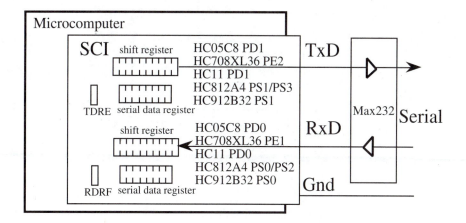

The initialization routine turns on the SCI receiver and transmitter and establishes the 1200 bits/s baud rate. The format used is 1 start bit, 8 data bits (M=0), no parity, and 1 stop bit. This software assumes that the E clock for the MC68HC05 and MC68HC11 is 2 MHz, while the MC68HC708 and MC68HC812 operate at 8 MHz. Again the details of these initializations will be presented in Chapter 7. The initialization rituals are included here for completeness. For now, don't worry about the details of the SCI init functions, except to know that this code must be executed before the SCI can be used (Program 2.11).

Program 2.11 Assembly
implementation of SCI
initialization.

```
; MC68HC05C8
init lda #$33   ;1200 baud
     sta BAUD
     lda #$00   ;mode
     sta SCCR1
     lda #$0C   ;tie=rie=0,
     sta SCCR2 ;te=re=1
     rts
```

```
; MC68HC11A8
init  ldaa #$33   ;1200 baud
      staa BAUD
      ldaa #$00   ;mode
      staa SCCR1
      ldaa #$0C   ;tie=rie=0,
      staa SCCR2 ;te=re=1
      rts
```

```
; MC68HC708XL36
init lda #$34 ;1200 baud
     sta SCBR
     lda #$00 ;mode
     sta SCC1
     lda #$0C ;tie=rie=0,
     sta SCC2 ;te=re=1
     lda #$00 ;no DMA
     sta SCC3 ;no interrupts
     rts
```

```
; MC68HC812A4
init  ldd  #417   ;1200 baud
      std  SC0BD
      ldaa #$00   ;mode
      staa SC0CR1
      ldaa #$0C  ;tie=rie=0,
      staa SC0CR2 ;te=re=1
      rts
```

The input routine waits in a loop until receive data register full (RDRF) is set, then reads the data register, SCDR. The function returns the new ASCII character by value in Register A (Program 2.12).

Program 2.12
Assembly implementation of SCI input.

```
; MC68HC05C8                      ; MC68HC11A8
; Return RegA = character    InChar ldaa  SCSR    ;status
InChar brclr 5,SCSR,InChar          bita #$20    ;rdrf?
       lda SCDR    ;SCI data         beq  InChar
       rts                           ldaa SCDR    ;SCI data
                                     rts
```

```
; MC68HC708XL36                   ; MC68HC812A4
InChar brclr 5,SCS1,InChar    InChar brclr SC0SR1,#$20,InChar
       lda SCDR    ;SCI data          ldaa SC0DRL ;SCI data
       rts                            rts
```

The output routine first waits in a loop until transmit data register empty (TDRE) is set, then writes data to the data register, SCDR. The function accepts the new ASCII character as a call by value in Register A (Program 2.13). As we will see later in Section 3.3, this is an efficient way to implement an output gadfly loop.

Program 2.13
Assembly implementation of SCI output.

```
; MC68HC05C8                      ; MC68HC11A8
; RegA is data to output     OutChar ldab  SCSR    ;status
OutChar brclr 7,SCSR,OutChar         bitb #$80    ;tdre?
        sta SCDR ;output             beq  OutChar
        rts                          staa SCDR    ;output
                                     rts
```

```
; MC68HC708XL36                   ; MC68HC812A4
OutChar brclr 7,SCS1,OutChar    OutChar brclr SC0SR1,#$80,OutChar
        sta SCDR    ;output             staa  SC0DRL ;sci data
        rts                             rts
```

These routines for the 6811 and 6812 in C are shown in Program 2.14.

```
// 68HC11A8 SCI routines          // 68HC812A4 SCI routines
void init(void) {                 void init(void) {
    BAUD=0x33;  // 1200 baud          SC0BD=417;    // 1200 baud
    SCCR1=0x00; // 8data, 1stop       SC0CR1=0x00;  // 8data, 1stop
    SCCR2=0x0C;}// enable gadfly      SC0CR2=0x0C;} // enable gadfly
#define RDRF 0x20                 #define RDRF 0x20
#define TDRE 0x80                 #define TDRE 0x80
#define TC 0x40                   #define TC 0x40
unsigned char insci(void){        unsigned char insci(void){
    while ((SCSR & RDRF) == 0);       while ((SC0SR1 & RDRF) == 0);
    return(SCDR); }                   return(SC0DRL); }
/* letter is character to print, */  /* letter is character to print, */
void OutChar(unsigned char letter){  void OutChar(unsigned char letter){
/* Wait for TDRE then output */   /* Wait for TDRE then output */
    while ((SCSR & TDRE) == 0);       while ((SC0SR1 & TDRE) == 0);
    SCDR=letter; }                    SC0DRL=letter; }
```

Program 2.14   C implementation of SCI basic I/O.

This next function inputs an 8-bit unsigned integer (0 to 255) using InChar. The input data is echoed using OutChar. A global variable is used in the calculation (Program 2.15).

```
; MC68HC05C8 or MC68HC708XL36        ; MC68HC11A8 or MC68HC812A4
; Input a byte from the SCI          ; Input a byte from the SCI
; Inputs:  none                      ; Inputs:  none
; Outputs: Reg X 0 to 255            ; Outputs: Reg B 0 to 255
;          C=1 if error              ;          C=1 if error
DIGIT   rmb 1       ;global          DIGIT   rmb  1        ;global
InUDec  clrx        ;N=0             InUDec  clrb          ;N=0
InUDloop bsr InChar ;Next input      InUDloop bsr  InChar  ;Next input
        bsr OutChar ;Echo                    bsr  OutChar ;Echo
        cmp #13     ;done if cr               cmpa #13     ;done if cr
        beq InUDret ;with C=0                 beq  InUDret ;with C=0
        cmp #'0                               cmpa #'0
        blo InUDerr ;error?                   blo  InUDerr ;error?
        cmp #'9                               cmpa #'9
        bhi InUDerr ;error?                   bhi  InUDerr ;error?
        and #$0F    ;0-9 digit                anda #$0F    ;0-9 digit
        sta DIGIT                             staa DIGIT
        lda #10                               ldaa #10
        mul                                   mul
        tstx        ;overflow?                tsta         ;overflow?
        bne InUDerr                           bne  InUDerr
        add DIGIT ;N=10*N+DIGIT               addb DIGIT ;N=10*N+DIGIT
        tax                                   bra  InUDloop
        bra  InUDloop                InUDerr ldaa #'?
InUDerr ldaa #'?                             bar  OutChar
        bsr OutChar                          clrb
        clrx                                 sec          ;error flag
        sec         ;error flag      InUDret rts
InUDret rts
```

Program 2.15 Assembly implementation of input decimal number.

The function in Program 2.16 outputs a null-terminated string using OutChar.

```
; MC68HC05C8 or MC68HC708XL36        ; MC68HC11A8 or MC68HC812A4
; Output a string to the SCI         ; Output a string to the SCI
; Inputs: X points to string         ; Inputs:  Reg X points to string
;         String ends with 0         ;          String ends with 0
; Outputs:none                       ; Outputs: none
OutString lda 0,X                    OutString ldaa 0,X
        beq OSdone  ;0 at end                 beq  OSdone  ;0 at end
        bsr OutChar                           bsr  OutChar
        incx                                  inx
        bra OutString                         bra  OutString
OSdone    rts                        OSdone    rts
```

Program 2.16 Assembly implementation of SCI output string.

The function `OutUDec` takes an 8-bit number and prints it as an unsigned decimal using `OutChar`. The 6808 subroutine could have been optimized using the `div` instruction (Program 2.17).

```
; MC68HC05C8 or MC68HC708XL36
; Output unsigned byte to SCI
; Inputs:  Reg A= 0 to 255,
;    print as 3 digit ascii
; Outputs: none
OutUDec lda #100
        bsr div8 ;A=num/100,
        add #'0  ;A=100's ascii
        bsr OutCh
        lda  #10
        bsr div8 ;A=num/10,
        add #'0  ;A=tens ascii
        bsr OutCh
        txa
        add #'0  ;A=ones ascii
        bsr OutCh
        rts
; Inputs: RegA, RegX
; Outputs: A=X/A quotient
;        X=X%A   remainder
div8    clr quot
        tsta
        beq divdone
        sta divisor
        txa       ;a=dividend
divloop cmp divisor
        blo divdone
        inc quot
        sub divisor
        bra divloop
divdone tax       ;remainder
        lda quot ; quotient
        rts
quot    rmb  1
divisor rmb  1
```

```
; MC68HC11A8 or MC68HC812A4
; Output unsigned byte to the SCI
; Inputs:  Reg B= 0 to 255,
;    print as 3 digit ascii
; Outputs: none
OutUDec  clra    ;Reg D=number
         ldx  #100
         idiv    ;X=num/100,
         xgdx    ;B=100s digit
         tba
         adda #'0 ;A=100's ascii
         bsr  OutCh
         xgdx    ;D=num
         ldx  #10
         idiv    ;X=num/10,
         xgdx    ;B=tens digit
         tba
         adda #'0 ;A=tens ascii
         bsr  OutCh
         xgdx    ;D=num
         tba
         adda #'0 ;A=ones ascii
         bsr  OutCh
         rts
```

Program 2.17 Assembly implemen-tation of SCI output decimal number.

A full set of SCI I/O routines can be found as tutorial 2 (files tut2.*) on the TExaS simulator. Assembly implementations exist for the 6805, 6808, 6811, and 6812, and C programs exist for the 6808, 6811, and 6812.

2.8 Object-Oriented Interfacing

2.8.1
Encapsulated
Objects Using
Standard C

Object-oriented software development in C++ involves three fundamental issues: encapsulation, polymorphism, and inheritance. *Encapsulation* is the grouping of functions and variables into a single class. C++ provides the mechanisms to implement modular software as described in this section. *Polymorphism* is the ability to reuse function names so that the exact operation depends on which class is being operated. *Inheritance* allows you to derive one class upon a previous class, reusing code and extending its functionality. For embedded systems, encapsulation is much more important than polymorphism and inheritance.

The example in this subsection will show you how to write C code that incorporates most of the important issues of encapsulation. This example also illustrates the top-down approach and includes three modules: the LCD interface, the COP functions, and some timer routines (Program 2.18). Notice that function names are chosen to reflect the module in which they are defined. If you are a C++ programmer, consider the similarities between this C function call `LCDclear()` and a C++ LCD class and a call to a member function `LCD.clear()`. The *.H files contain function declarations and the *.C files contain the implementations.

Program 2.18 Main program with three modules.

```
#include "HC12.H"
#include "LCD12.H"
#include "COP12.H"
#include "Timer.H"
void main(void){ char letter; int n=0;
    COPinit(); // Enable TOF interrupt to make COP happy
    LCDinit();
    TimerInit()
    LCDString("Adapt812 LCD");
    TimerMsWait(1000);
    LCDclear();
    letter='a'-1;
    while(1){
        if (letter=='z')
            letter='a';
        else
            letter++;
        LCDputchar(letter);
        TimerMsWait(250);
        if(++n==16){
            n=0;
            LCDclear();
}}}
#include "LCD12.C"
#include "COP12.C"
#include "Timer.C"
#include "VECTORS.C"
```

For every function definition, ICC11 and ICC12 generate an assembler directive declaring the function's name to be public. This means that every C function is a potential entry point and so can be accessed externally. One way to create private/public functions is to control which functions have declarations. Now let us look inside the Timer.H and Timer.C files. To implement private and public functions we place the function declarations of the public functions in the Timer.H file (Program 2.19).

The implementations of all functions are included in the Timer.C file. We can apply this same approach to private and public global variables. Notice that in this case the

Program 2.19 Timer.H header file has public functions.

```
void TimerInit(void);
void TimerMsWait(unsigned int time);
```

global variable, TimerClock, is private and cannot be accessed by software outside the Timer.C file (Program 2.20).

Program 2.20 Timer.C implementation file defines all functions.

```
unsigned int TimerClock; // private global
void TimerInit(void){ // public function
  TSCR |=0x80; // TEN(enable)
  TMSK2=0xA2;  // TOI arm, TPU(pullup) timer/4 (500ns)
  TimerClock=2000; // 2000 counts per ms
}
void TimerWait(unsigned int time){ // private function
  TC5=TCNT+TimerClock;  // 1.00ms wait
  TFLG1 = 0x20;         // clear C5F
  while((TFLG1&0x20)==0){};}
void TimerMsWait(unsigned int time){ // public function
  for(;time>0;time--)
    TimerWait(TimerClock); // 1.00ms wait
}
```

2.8.2 Object-Oriented Interfacing Using C++

The three characteristics of object-oriented programming are encapsulation, polymorphism, and inheritance. We defined these terms in Section 2.8.1 and discussed how to encapsulate modules using standard C. In this section we will introduce the concept of object-oriented interfacing using C++ by discussing the software environment on the IBM PC compatible. Then, we will show an example of how C++ might provide support to improve the portability of our embedded system software.

Since its inception C++ has steadily replaced C as the preferred programming environment for developing both system programs and user applications for the *WinTel* platform (Windows operating system on an Intel microprocessor). The reasons for this software evolution (revolution?) stem from the inherent advantages of C++. Some of the fundamental difficulties for developing software for the *WinTel* platform are:

1. We need a common user interface on top of a multitude of similar but not identical computers.
2. The hardware platform makes a fundamental advancement every 6 months.
3. Many hardware and software companies act in concert to produce a product.
4. The newer software must run on the older computers.
5. The older software must run on the newer computers.
6. The hardware/software configurations may change at run time.

Because of these constraints, a layered software model was adopted so that changes in one aspect of the system could be made without having to reengineer the entire system. Objects in C++ allow the programmer to use hardware and software modules without complete knowledge of how they work. The member functions provide a clean yet powerful mechanism to implement the interface between the software modules. Especially at the hardware interface level, classes provide a mechanism for abstraction. In other words, the HAL is a set of C++ objects that define basic input/output operations.

There are some similarities but many differences between the *WinTel* and embedded platforms. If we examine the same six constraints for an embedded microcomputer, we see that only the first constraint is similar. In other words, we are interested in making embedded system software run on multiple microcomputers (code reuse). Software is portable if it is easy to convert it to run on another platform. The other five constraints for the most part do not exist in the embedded system development environment:

2. Embedded microcomputers have a much longer lifetime than a x86 micro-processor.

3. Usually a single company develops the hardware and software.

4., 5. Hardware and software are upgraded together.

6. Configurations are usually well-defined at compile time.

2.8.3 Portability Using Standard C and C++

Even though assembly and C are currently the primary software development approaches for embedded systems, it is appropriate to consider software development C++. As a case study, we will address the issue of portability using C and C++. First, we will show an enhanced C software implementation of the Moore FSM first presented in Section 2.4.3 and Program 2.6. To make this program more portable using C, we create #define macros for those parameters that are likely to change (port assignments and clock speed) (Program 2.21).

```
/* 6812 PortC bits 1,0 are input, Port B bits 1,0 are output */
#define OutPort (*(unsigned char  volatile *)(0x0001))
#define OutDDR  (*(unsigned char  volatile *)(0x0003))
#define InPort  (*(unsigned char  volatile *)(0x0004))
#define InDDR   (*(unsigned char  volatile *)(0x0006))
/* rate is the number of cycles/100usec */
#define rate 800
const struct State{
        unsigned char Out;          /* Output values */
        unsigned int Time;          /* Time in 100 µsec to wait in this state */
        const struct State *Next[4]; /* Next state if input=0,1,2,3 */
};
typedef const struct State StateType;
#define SA &fsm[0]
#define SB &fsm[1]
#define SC &fsm[2]
StateType fsm[3]={
  {0x01,5*rate,{SB,SA,SB,SC}},   /* SA out=1, wait= 500usec, next states */
  {0x02,10*rate,{SC,SA,SB,SC}},  /* SB out=2, wait=1000usec, next states */
  {0x03,20*rate,{SA,SA,SB,SA}}   /* SC out=3, wait=2000usec, next states */
};
void Wait(unsigned int delay){ int Endt;
    Endt=TCNT+delay;              /* Time (125ns cycles) to wait */
    while((Endt-(int)TCNT)>0);    /* wait */
};
void main(void){ StateType *Pt;  unsigned char Input;
  Pt=SA;                  /* Initial State */
  OutDDR=0xFF;            /* Make Output port outputs  */
  InDDR=0x00;            /* Make Input port inputs  */
  TSCR=0x80;             /* Enable TCNT, default rate 8 MHz */
  while(1){
    OutPort=Pt->Out;
    Wait(Pt->Time);       /* Time to wait in this state  */
    Input=InPort&0x03;    /* Input=0,1,2,or 3 */
    Pt=Pt->Next[Input];
  }
};
```

Program 2.21 Enhanced C implementation of a Mealy FSM.

In C++ we will define a class to describe a generic I/O port. The attributes of the port include its address, the existence and address of its data direction register, and its type T. Program 2.22 uses the same FSM and Wait function as Program 2.21.

```
// a DDRAddress of 1 means fixed output port with no DDR
// a DDRAddress of 0 means fixed input port with no DDR
template <class T> class port{
  protected :
      unsigned char *PortAddress;    // pointer to data
      unsigned char *DDRAddress;     // pointer to data direction register
   public : port(unsigned short ThePortAddress, unsigned short TheDDRAddress){
      PortAddress = (unsigned char *)ThePortAddress; // initialize pointer to I/O port
      DDRAddress  = (unsigned char *)TheDDRAddress;} // initialize pointer to DDR
virtual short Initialize(unsigned char  data){
      if((int)DDRAddress==1)   // fixed output port
          return(data==0xFF);  // OK if initializing all bits to output
      if(DDRAddress==0)        // fixed input port
          return(data==0);     // OK if initializing all bits to input
      (*DDRAddress) = data;    // configure direction register
      return 1;}    // successful
virtual void put(unsigned char  data){
      if((int)DDRAddress==0) return;  // fixed input
      if((*DDRAddress)==0) return;    // all input
      (*PortAddress) = data;}         // output data to port
virtual unsigned char get(void){
      return (*PortAddress);} // input data from port
}
// 6812 PortC bits 1,0 are input, Port B bits 1,0 are output
port<unsigned char> OutPort(0x0001,0x0003);
port<unsigned char> InPort(0x0004,0x0006);

void main(void){ StateType *Pt;  unsigned char Input;
  Pt=SA;                   // Initial State
  OutPort.Initialize(0xFF);        // Make Output port outputs
  InPort.Initialize(0x00);         // Make Input port inputs
  TSCR=0x80;               // Enable TCNT, default rate 8 MHz
  while(1){
    OutPort.put(Pt->Out);
    Wait(Pt->Time);        // Time to wait in this state
    Input=InPort.get()&0x03;  // Input=0,1,2,or 3
    Pt=Pt->Next[Input];
  }
}
```

Program 2.22 C++ implementation of a Mealy FSM.

To configure this software for the 6811, we change the way the I/O ports are blessed.

```
// 6811 PortC bits 1,0 are input, Port B bits 1,0 are output
port<unsigned char> OutPort(0x1004,1);       // fixed output port
port<unsigned char> InPort(0x0003,0x0007);  // bidirectional port
```

To make this system truly portable we would have to create an object for the timer functions as well. You can derive classes from this class and override (enhance) the I/O functions. Additional member functions can also be added to enhance this C++ object, as in Program 2.23.

Program 2.23
Additional member functions for the I/O port class.

```
T operator = (T data){
      put(data);                 // output to port
      return data;}              // returns data itself
operator T () (T data){
      return get();}             // returns port data
virtual T operator |= (T data){
      put(data |= get());        // read modify write port access
      return data;}              // returns new data
virtual T operator &= (T data){
      put(data &= get());        // read modify write port access
      return data;}              // returns new data
virtual T operator ^= (T data){
      put(data ^= get());        // read modify write port access
      return data;}              // returns new data
```

Observation: The issues of software clarity, portability, and modularity are important no matter which programming language you are using.

2.9 Threads

2.9.1 Single-Threaded Execution

Software (e.g., program, code, module, procedure, function, subroutine) is a list of instructions for the computer to execute. A thread, on the other hand, is defined as the path of action of software as it executes. The expression "thread" comes from the analogy shown in Figure 2.13. This simple program prints the 8-bit numbers 000 001 002. . . . If we connect the statements of our executing program with a continuous line (the thread), we can visualize the dynamic behavior of our software.

The execution of the main program is called the *foreground thread*. In most embedded applications, the foreground thread executes a loop that never ends. We will learn later that this thread can be broken (execution suspended, then restarted) by interrupts and DMA.

2.9.2 Multithreading and Reentrancy

With interrupts we can create multiple threads. Some threads will be created statically, meaning they exist throughout the life of the software, while others will be created and destroyed dynamically. There will usually be one foreground thread running the main program, as in Figure 2.13. In addition to this foreground thread, each interrupt source has its own background thread, which is started whenever the interrupt is requested. Figure 2.14 shows a software system with one foreground thread and two background threads. The "key" thread is invoked whenever a key is touched on the keyboard, and the "time" thread is invoked every 1 ms in a periodic fashion. Because there is but one processor, the currently running thread must be suspended to execute another thread. In Figure 2.14 the suspension of the main program is illustrated by the two breaks in the foreground thread. When a key is touched, the main program is suspended, and a keyhandler thread is created with

Figure 2.13
Illustration of the
definition of a thread.

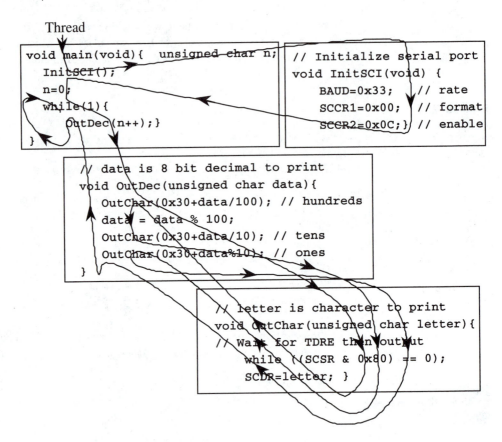

Figure 2.14
Interrupts allow us to
have multiple
background threads.

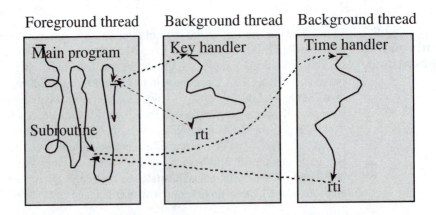

an "empty" stack and uninitialized registers. When the `keyhandler` is done, it executes RTI to relinquish control back to the main program. The original stack and registers of the main program will be restored to the state before the interrupt. In a similar way, when the 1-ms timer occurs, the main program is suspended again, and a `timehandler` thread is created with its own "empty" stack and uninitialized registers. We can think of each thread as having its own registers and its own stack area. In Chapter 4, we will discuss in detail this approach to multithreaded programming. In Chapter 5, we will implement a preemptive thread scheduler that will allow our software to have multiple foreground threads.

A program segment is reentrant if it can be concurrently executed by two (or more) threads. In Figure 2.14 we can conceive of the situation where the main program starts executing a subroutine, is interrupted, and the background thread calls that same subroutine. For two threads to share a subroutine, the subroutine must be reentrant. To implement reentrant software, place local variables on the stack and avoid storing into I/O devices and global memory variables. The issue of reentrancy will be covered in detail later in Chapter 4.

2.10 Recursion

A recursive program is one that calls itself. When we draw a calling graph like the one in Figure 2.10, a circle is formed. Although many algorithms can be defined using recursion, it does require special care when implementing. In particular, recursive subroutines must be reentrant. For some sorting, computational, and database functions, recursion affords a more elegant solution. Recursive algorithms are often easy to prove correct and use less permanent memory but require more temporary stack space and execute slower than nonrecursive algorithms. For example, consider the recursive implementation of the function OutUDec, which will output a 16-bit unsigned decimal using the serial port, as shown in Program 2.24.

Program 2.24
Recursion is defined as a function calling itself.

```
//-----------------------Start of OutUDec-----------------------------------
// Output a 16 bit number in unsigned decimal format
// Variable format 1 to 5 digits with no space before or after
// This function uses recursion to convert a decimal number of
//    unspecified length as an ASCII string
void OutUDec(unsigned int number){
    if (number>=10){
        OutUDec(number/10);
        OutUDec(number%10);
    }
    else
        OutChar(number+'0');
}
```

2.11 Debugging Strategies

All programmers are faced with the need to debug and verify the correctness of their software. In this section we will study hardware-level probes like the logic analyzer and in-circuit emulator (ICE); software-level tools like simulators, monitors, and profilers; and manual tools like inspection and print statements.

2.11.1
Debugging Tools

Microcomputer-related problems often require the use of specialized equipment to debug the system hardware and software. Two very useful tools are the logic analyzer and the ICE. A logic analyzer (Figure 2.15) is essentially a multiple-channel digital storage scope with many ways to trigger. As a troubleshooting aid, it allows the experimenter to observe numerous digital signals at various points in time and thus make decisions based upon such ob-

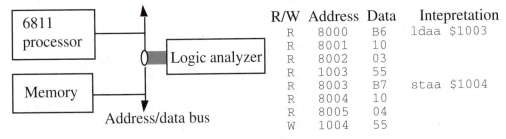

Figure 2.15   A logic analyzer and example output.

servations. Typically, the logic analyzer is attached to the address/data bus in a passive manner, which allows the user to view the real-time execution of the software. One problem with logic analyzers is the massive amount of information that it generates. To use an analyzer effectively one must learn proper triggering mechanisms to capture data at appropriate times, eliminating the need to sift through volumes of output. With less expensive logic analyzers, the user must interpret the R/W, address, and data information by hand. This can be quite tedious, especially when the original program is written in a high-level language. More expensive logic analyzers can disassemble the output and show the assembly level instructions. This type of output is also difficult to interpret because the comments, labels, and program structure are not integrated into the assembly instructions. With the advent of today's microprocessor technology (in particular the cache, multiple instruction queues, branch predictions, and internal buses) it is very difficult to interpret software activity simply by observing read/write cycles to external memory. Even though the Motorola 8-bit microcomputer architectures are quite simple (having none of those above fancy features, they are therefore a good candidate for a logic analyzer), this discussion will focus on debugging techniques that can be applied to most computer architectures. The 6812 does have an instruction queue, so the address/data bus activity precedes program execution.

An ICE is a hardware debugging tool that recreates the I/O signals of the processor chip (Figure 2.16). To use an emulator, we remove the processor chip and insert the emulator

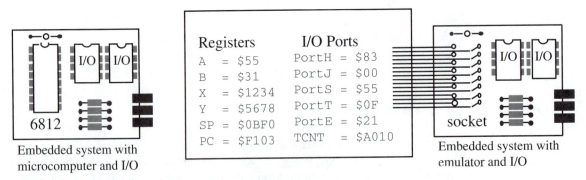

Figure 2.16   In-circuit emulator and example output.

cable into the chip socket. In most cases, the emulator/computer system operates at full speed. The ICE allows the programmer to observe and modify internal registers of the processor. An ICE is often integrated into a personal computer so that its editor, hard drive, and printer are available for the debugging process.

Most software-based debuggers implement a breakpoint by replacing the existing instruction with a software trap (6805/6808/6811/6812 instruction `swi`). This procedure cannot be performed when the software is programmed in ROM. To debug this type of system we can use another class of emulator called the ROM emulator (Figure 2.17). This debugging tool replaces the ROM with cable connects to a dual-port RAM within the emulator. While the software is running, it fetches information from the emulator RAM just as if it were the ROM. While the software is halted, you can modify its contents.

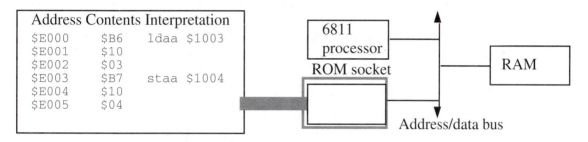

Figure 2.17   In-circuit ROM emulator and example output.

The only disadvantage of the ICE is its cost. To provide some of the benefits of this high-priced debugging equipment, the MC68HC812A4 and MC68HC912B32 both have a *background debug module* (BDM). The BDM hardware exists on the microcomputer chip itself and communicates with the debugging computer via a dedicated two- or three-wire serial interface. Although not as flexible as an ICE, the BDM can provide the ability to observe software execution in real time, the ability to set breakpoints, the ability to stop the computer, and the ability to read and write registers, I/O ports, and memory. Although both the MC68HC812A4 and the MC68HC912B32 have BDMs, only the B32 version supports hardware breakpoints. The registers can be observed only when the computer is halted, but the memory and I/O ports are accessible while the program is executing.

**2.11.2
Debugging
Theory**

Research in the area of program monitoring and debugging has not kept pace with developments in other areas of computer programming. This area is comparatively deficient in the structure and unity now common to programming languages. The area is compartmentalized because of specialized tools. For example, run-time profile generators and execution monitors are tools that have many commands and functions in common with "debuggers," but in practice, because of tool specialization, a user needs to use two tools: a monitor to examine time behavior and a debugger for functional behavior. This area is also fragmented because of a multiplicity of terms. Terms such as program testing, diagnostics, performance debugging, functional debugging, tracing, profiling, instrumentation, visualization, optimization, verification, performance measurement, and execution measurement have specialized meanings, but they are also used interchangeably, and they often describe overlapping functions. For example, the terms profiling, tracing, performance measurement, or execution measurement may be used to describe the process of examining a program from a time viewpoint. But, tracing is also a term that may be used to describe the process of monitoring a program state or history for functional errors or to describe the process of stepping through a program with a debugger. Usage of these terms among researchers and users vary.

Furthermore, the meaning and scope of the term debugging itself is not clear. We hold the view that the goal of debugging is to maintain and improve software, and the role of a debugger is to support this endeavor. We define the debugging process as testing, stabilizing, localizing, and correcting errors. And in our opinion, although testing, stabilizing, and localizing errors are important and essential to debugging, they are auxiliary processes: The primary goal of debugging is to remedy faults or to correct errors in a program.

Although a variety of program monitoring and debugging tools are available today, in practice it is found that an overwhelming majority of users either still prefer or rely mainly upon "rough and ready" manual methods for locating and correcting program errors. These methods include desk checking, dumps, and print statements, with print statements being one of the most popular manual methods. Manual methods are useful because they are readily available, and they are relatively simple to use. But the usefulness of manual methods is limited: They tend to be highly intrusive, and they do not provide adequate control over repeatability, event selection, or event isolation. A real-time system, where software execution timing is critical, usually cannot be debugged with simple print statements, because the print statement itself will require too much time to execute.

We define a debugging instrument as software code that is added to the program for the purpose of debugging. A print statement is a common example of an instrument. Using the editor, we add print statements to our code that either verify proper operation or illustrate the programming errors. A key to writing good debugging instruments is to provide for a mechanism to reliably and efficiently remove them all when the debugging is done. Consider the following mechanisms as you develop your own unique debugging style.

- Place all print statements in a unique column (e.g., first column) so that the only code that exists in this column must be a debugging instrument.
- Define all debugging instruments as functions that all have a specific pattern in their names. In this way, the find/replace mechanism of the editor can be used to find all the calls to the instruments.
- Define the instruments so that they test a run-time global flag. When this flag is turned off, the instruments perform no function. Notice that this method leaves a permanent copy of the debugging code in the final system, causing it to suffer a run-time overhead, but the debugging code can be activated dynamically without recompiling. Many commercial software applications utilize this method because it simplifies "on-site" customer support.
- Use conditional compilation (or conditional assembly) to turn on and off the instruments when the software is compiled. When the compiler supports this feature, it can provide both performance and effectiveness.

The emergence of concurrent languages and the increasing use of embedded real-time systems place further demands on debuggers. The complexities introduced by the interaction of multiple events or time-dependent processes are much more difficult to debug than errors associated with sequential programs. The behavior of non-real-time sequential programs is reproducible: For a given set of inputs their outputs remain the same. In the case of concurrent or real-time programs this does not hold true. Control over repeatability, event selection, and event isolation is even more important for concurrent or real-time environments.

2.11.3
Functional
Debugging

Functional debugging involves the verification of I/O parameters. It is a static process where inputs are supplied, the system is run, and the outputs are compared against the expected results. We will present seven methods of functional debugging.

2.11.3.1
Single Stepping or Trace

Many debuggers allow you to set the program counter to a specific address then execute one instruction at a time. To single-step a program with the Kevin Ross BDM/Adapt812 combination, use the following commands:

```
reset
er pc f000
t
t
```

The first line (`reset`) enables the 6812 background debug module. The second line (`er pc f000`) sets the program counter to $F000. The subsequent lines execute a single assembly instruction of the user program before returning control back to the debugger. The Hiware debugger allows you to execute single assembly instructions or single C level instructions. The TExaS simulator provides Step, Few, StepOver, and StepOut commands. *Step* is the usual execute one assembly instruction. *Few* will execute some instructions and stop (you can set how many "some" is). *StepOver* will execute one assembly instruction, unless that instruction is a subroutine call, in which case the simulator will execute the entire subroutine and stop at the instruction following the subroutine call. *StepOut* assumes the execution has already entered a subroutine and will finish execution of the subroutine and stop at the instruction following the subroutine call.

2.11.3.2
Breakpoints Without
Filtering

The first step of debugging is to *stabilize* the system with the bug. In the debugging context, we stabilize the problem by creating a test routine that fixes (or stabilizes) all the inputs. In this way, we can reproduce the exact inputs over and over again. Once stabilized, if we modify the program, we are sure that the change in our outputs is a function of the modification we made in our software and not due to a change in the input parameters. A **breakpoint** is a mechanism to tag places in our software, which when executed will cause the software to stop.

A crude way to set breakpoints on the Kevin Ross BDM/Adapt812 combination is to define

```
#define bkpt asm(" bgnd");
```

and insert `bkpt` at the places you wish to break. The software must be recompiled and reloaded after every breakpoint is added or removed. In addition, the software must be started from the BDM debugger using the commands

```
reset
g f000
```

After a breakpoint, registers and memory can be viewed or changed. To single-step you can execute `t`. To restart execution after the breakpoint, type `g`. To remove a breakpoint, you have to edit, compile, and load.

2.11.3.3
Conditional Breakpoints

One of the problems with breakpoints is that sometimes we have to observe many breakpoints before the error occurs. One way to deal with this problem is the conditional breakpoint. Add a global variable called `count` and initialize it to zero in the ritual. Add the following conditional breakpoint to the appropriate location. And run the system again (you can change the 32 to match the situation that causes the error).

```
if(++count==32)
       bkpt
```

Notice that the breakpoint occurs only on the 32nd time the break is encountered. Any appropriate condition can be substituted.

**2.11.3.4
Instrumentation: Print
Statements**

The use of print statements is a popular and effective means for functional debugging. The difficulty with print statements in embedded systems is that a standard "printer" may not be available. Another problem with printing is that most embedded systems involve time-dependent interactions with their external environment. The print statement itself may be so slow that the debugging process itself causes the system to fail. The print statement is *intrusive*. Therefore this section will focus on debugging methods that do not rely on the availability of a printer.

**2.11.3.5
Instrumentation: Dump
into Array Without
Filtering**

One of the difficulties with print statements are that they can significantly slow down the execution speed in real-time systems. Many times the bandwidth of the print functions cannot keep pace with the existing system. For example, our system may wish to call a function 1000 times a second (or every 1 ms). If we add print statements to it that require 50 ms to perform, the presence of the print statements will significantly affect the system operation. In this situation, the print statements would be considered extremely intrusive. Another problem with print statements occurs when the system is using the same output hardware for its normal operation, as is required to perform the print function. In this situation, debugger output and normal system output are intertwined.

To solve both these situations, we can add a debugger instrument that dumps strategic information into an array at run time. We can then observe the contents of the array at a later time. One of the advantages of dumping is that the 6812 BDM module allows you to visualize memory even when the program is running. So this technique will be quite useful in systems with a BDM (the 68332 has one too).

**2.11.3.6
Instrumentation: Dump
into Array with Filtering**

One problem with dumps is that they can generate a tremendous amount of information. If you suspect a certain situation is causing the error, you can add a filter to the instrument. A filter is a software/hardware condition that must be true to place data into the array. In this situation, if we suspect the error occurs when the pointer nears the end of the buffer, we could add a filter that saves in the array only when the pointer is above a certain value.

**2.11.3.7
Monitor Using
Fast Displays**

Another tool that works well for real-time applications is the monitor. A monitor is an independent output process, somewhat similar to the print statement, but one that executes much faster and thus is much less intrusive. The LCD display can be an effective monitor for small amounts of information. Small LCDs can display up to 16 characters, while the larger ones can hold four lines by 40 characters. The hardware/software interface for such a display will be presented in Chapter 8. You can place one or more LEDs on individual otherwise unused output bits. Software toggles these LEDs to let you know what parts of the program are running. A LED is an example of a Boolean monitor.

**2.11.4
Performance
Debugging**

Performance debugging involves the verification of timing behavior of our system. It is a dynamic process where the system is run, and the dynamic behaviors of the I/Os are compared against the expected results. We will present two methods of performance debugging, then apply the techniques to measure execution speed.

2.11.4.1
Instrumentation
Measuring with an
Independent Counter,
TCNT

There is a 16-bit counter, called TCNT, which is incremented every E clock. There is a prescaler that can be placed between the E clock and the TCNT counter. It automatically rolls over when it gets to $FFFF. If we are sure the execution speed of our function is less than 65,535 counts, we can use this timer to collect timing information with only a modest amount of intrusiveness.

2.11.4.2
Instrumentation Output
Port

Another method to measure real-time execution involves an output port and an oscilloscope. Connect a microcomputer output bit to your scope. Add a debugging instrument that sets/clears these output bits at strategic places. Remember to set the port's direction register to 1.

2.11.4.3
Measurement of
Dynamic Efficiency

There are three ways to measure dynamic efficiency of our software. To illustrate these three methods, we will consider measuring the execution time of the sqrt function presented earlier as Programs 2.3 and 2.4. The first method is to count bus cycles using the assembly listing (Program 2.25). This approach is appropriate only for very short programs and becomes difficult for long programs with many conditional branch instructions. Often this is a very tedious process, but luckily the TExaS assembler will look up and keep a running count of the number of cycles. The assembly pseudo-operation `org *` will reset the cycle counter, shown between the parentheses. A portion of the assembly output is presented in Program 2.25. Notice that the total cycle count for a 6812

```
$F019                                    org   *     ;reset cycle counter
$F019 35              [ 2](   0)sqrt pshy
$F01A B776            [ 1](   2)      tsy
$F01C 1B9C            [ 2](   3)      leas  -4,sp       ;allocate t,oldt,s16
$F01E C7              [ 1](   5)      clrb
$F01F A644            [ 3](   6)      ldaa  s8,y
$F021 2723            [ 3](   9)      beq   done
$F023 C610            [ 1](  12)      ldab  #16
$F025 12              [ 3](  13)      mul               ;16*s
$F026 6C5C            [ 2](  16)      std   s16,y       ;s16=16*s
$F028 18085F20        [ 4](  18)      movb  #32,t,y     ;t=2.0, initial guess
$F02C 18085E03        [ 4](  22)      movb  #3,cnt,y
$F030 A65F            [ 3](  26)next ldaa  t,y          ;RegA=t
$F032 180E            [ 2](  29)      tab               ;RegB=t
$F034 B705            [ 1](  31)      tfr   a,x         ;RegX=t
$F036 12              [ 3](  32)      mul               ;RegD=t*t
$F037 E35C            [ 3](  35)      addd  s16,y       ;RegD=t*t+16*s
$F039 1810            [12](  38)      idiv              ;RegX=(t*t+16*s)/t
$F03B B754            [ 1](  50)      tfr   x,d
$F03D 49              [ 1](  51)      lsrd              ;RegB=((t*t+16*s)/t)/2
$F03E C900            [ 1](  52)      adcb  #0
$F040 6B5F            [ 2](  53)      stab  t,y
$F042 635E            [ 3](  55)      dec   cnt,y
$F044 26EA            [ 3](  58)      bne   next
$F046 B767            [ 1](  61)done tys
$F048 31              [ 3](  62)      puly
$F049 3D              [ 5](  65)      rts
$F04A 183E            [16](  70)      stop
```

Program 2.25 Assembly listing from TExaS of the sqrt subroutine.

implementation is 70 cycles. At 8 MHz, 70 cycles is 8.75 μs. Because the loop (between next and bne next) is executed exactly three times, the actual time will be 140 cycles, or 17.5 μs. For most programs it is actually very difficult to get an accurate time measurement using this technique.

The second method uses an internal timer called TCNT. Most Motorola microcomputers have this 16-bit internal register that is incremented at the bus frequency. In Chapter 6 we will study the timer features in detail, but for now assume we have a 16-bit unsigned counter that is automatically incremented every bus cycle. If we are sure the function will complete in a time less than 65,535 bus cycles, then the internal timer can be used to measure execution speed empirically. The assembly language call to the function is modified so that TCNT is read before and after the subroutine call. The elapsed time is the difference. Since the execution speed may be dependent on the input data, it is often wise to measure the execution speed for a wide range of input parameters. There is a slight overhead in the measurement process itself. To be more accurate you could measure this overhead and subtract it from your measurements. Notice that in Program 2.26 the total time including parameter passing is measured.

Program 2.26 Empirical measurement of dynamic efficiency in assembly language.

```
before  rmb 2       ; TCNT value before the call
elasped rmb 2        ; number of cycles required to execute sqrt
    movw TCNT,before
    movb ss,1,-sp ; push parameter on the stack (binary fixed point)
    jsr  sqrt     ; subroutine call to the module "sqrt"
    ins
    stab tt       ; save result
    ldd  TCNT     ; TCNT value after the call
    subd before
    std  elasped  ; execute time in cycles
```

This same technique can also be used in C language programs (Program 2.27).

Program 2.27 Empirical measurement of dynamic efficiency in C language.

```
unsigned short before,elasped;
void main(void){
    ss=100;
    before=TCNT;
    tt=sqrt(ss);
    elasped=TCNT-before;
}
```

The third technique can be used in situations where TCNT is unavailable or where the execution time might be larger than 65,535 counts. In this empirical technique we attach an unused output pin to an oscilloscope or to a logic analyzer. We will set the pin high before the call to the function and set the pin low after the function call. In this way a pulse is created on the digital output with a duration equal to the execution time of the function. We assume Port B is available and that bit 7 is connected to the scope. By placing the function call in a loop, the scope can be triggered. With a storage scope or logic analyzer, the function need be called only once. Program 2.28 shows the assembly language measurement. Program 2.29 shows the same technique in C language.

Program 2.28 Another empirical measurement of dynamic efficiency in assembly language.

```
          org  $0800
ss        rmb  1
tt        rmb  1
          org  $F000
main lds #$0C00
          movb #$FF,DDRB ; PB7 is connected to the scope
          movb #100,ss
loop bset PORTB,#$80 ; set PB7 high
          movb ss,1,-sp ; push parameter on the stack (binary fixed point)
          jsr  sqrt     ; subroutine call to the module "sqrt"
          ins
          stab tt        ; save result
          bclr PORTB,#$80 ; clear PB7 low
          bra  loop      ; repeat to trigger the scope
```

Program 2.29 Another empirical measurement of dynamic efficiency in C language.

```
void main(void){
   DDRB=0xFF;  // PB7 is connected to a scope
   ss=100;
   while(1){
      PORTB |= 0x80;   // set PB7 high
      tt=sqrt(ss);
      PORTB &= ~0x80; // clear PB7 low
   }
}
```

2.11.5 Profiling

Profiling is similar to performance debugging because both involve dynamic behavior. Profiling is a debugging process that collects the time history of strategic variables. For example, if we could collect the time-dependent behavior of the program counter, then we could see the execute patterns of our software. We can profile the execution of a multiple-thread software system to detect reentrant activity.

2.11.5.1 Profiling Using a Software Dump to Study Execution Pattern

In this section, we will discuss software instruments that study the execution pattern of our software. To collect information concerning execution we will add a debugging instrument that saves the time and location in an array (like a dump). By observing this data we can determine both a time profile (when) and an execution profile (where) of the software execution (Program 2.30).

Program 2.30 A time/position profile dumping into a data array.

```
unsigned short time[100];
unsigned short place[100];
unsigned short n;
void profile(unsigned short p){
  time[n]=TCNT; // record current time
  place[n]=p;
  n++;
}
unsigned short sqrt(unsigned short s){ unsigned short t,oldt;
profile(0);
  t=0;        // based on the secant method
  if(s>0) {
```

continued on p. 130

Program 2.30 A time/position profile dumping into a data array.

continued from p. 129

```
profile(1);
    t=32;    // initial guess 2.0
    do{
profile(2);
      oldt=t;  // calculation from the last iteration
      t=((t*t+16*s)/t)/2;} // t is closer to the answer
    while(t!=oldt);}    // converges in 4 or 5 iterations
profile(3);
  return t;}
```

Since the debugging instrument is implemented as a function, we could read the return address off the stack (the place from which it was called). This is a good approach when we are adding/subtracting many debugging instruments. A symbol table (Table 2.2) shows the mapping between the physical address (PC data) and the logical address (name of the function). The symbol prog2_30.8 refers to program file prog2_30 line 8.

Table 2.2
Symbol table for Program 2.30 generated by ICC12.

Addr	Global
F000	
F03B	
F03F	
F04F	
F05E	prog2_30.8
F067	prog2_30.9
F06D	_sqrt
F073	prog2_30.11
F079	prog2_30.12
F07E	prog2_30.13
F085	prog2_30.14
F08B	prog2_30.15
F090	prog2_30.17
F096	prog2_30.18
F09A	prog2_30.19
F0B8	prog2_30.20
F0BE	prog2_30.21
F0C4	prog2_30.22
F0D2	prog2_30.24
F0D2	_main
0800	_n
0802	_place
08CA	_time

2.11.5.2
Profiling Using an Output Port

In this section, we will discuss a hardware/software combination to visualize program activity. Our debugging instrument will set output port bits. We will place these instruments at strategic places in the software. If we are using a regular oscilloscope, then we must stabilize the system so that the function is called over and over. We connect the output pins to a scope or logic analyzer and observe the program activity (Program 2.31).

Program 2.31 A
time/position profile
using two output bits.

```
unsigned int sqrt(unsigned int s){ unsigned int t,oldt;
PORTB=0;
  t=0;        // based on the secant method
  if(s>0) {
PORTB=1;
    t=32;    // initial guess 2.0
    do{
PORTB=2;
      oldt=t;  // calculation from the last iteration
      t=((t*t+16*s)/t)/2;} // t is closer to the answer
    while(t!=oldt);}    // converges in 4 or 5 iterations
PORTB=3;
  return t;}
```

2.11.5.3
Thread Profile

When more than one program (multiple threads) is running, you could use the previous technique to visualize the thread that is currently active (the one running). For each thread, we assign an output pin. The debugging instrument would set the corresponding bit high when the thread starts and clear the bit when the thread stops. We would then connect the output pins to a multiple-channel scope to visualize in real time the thread that is currently running. For an example of this type of profile, run one of the thread.* examples included with the TExaS simulator and observe the logic analyzer.

2.12 Glossary

assembler System software that converts an assembly language program (human readable format) into object code (machine readable format).

break or trap A break or a trap is an instrument that halts the processor. With BUFFALO, the SWI instruction will stop your program and jump into the debugger. Therefore, a break essentially halts all processes but may allow specific and essential processes, such as clock, display, or keyboard processes, to continue. The condition of being in this state is also referred to as a break.

breakpoint The place where a break is inserted, the time when a break is encountered, or the time period when a break is active. With the BUFFALO monitor we can add four assembly level breakpoints using the BR command. It is implemented by replacing the existing opcode with a SWI instruction.

compiler System software that converts a high-level language program (human readable format) into object code (machine readable format).

cross assembler An assembler that runs on one computer but creates object code for a different computer.

cross compiler A compiler that runs on one computer but creates object code for a different computer.

desk checking or dry run We perform a desk check (or dry run) by determining in advance, either by analytical algorithm or explicit calculations, the expected outputs of strategic intermediate stages and final results for a typical inputs. We then run our program and compare the actual outputs with this template of expected results.

emulator An ICE is an expensive debugging hardware tool that mimics the processor pin outs. To debug with a 6811 emulator, you would remove the 6811 processor chip and attach the emulator cable into the 6811 processor socket. The emulator would sense the processor input signals and recreate the processor outputs signals on the socket as if a 6811 chip were actually there running at 2 MHz. Inside the emulator you have internal read/write access to the registers and processor state. Most emulators allow you to visualize/record strategic information in real time without halting the program execution. You can also remove ROM chips and insert the connector of a ROM emulator. This type of emulator is less expensive, and it allows you to debug ROM-based software systems.

EPROM programmer   System hardware/software that burns the object code into the microcomputer's EPROM.

filter   In the debugging context, a filter is a Boolean function or conditional test used to make run-time decisions. For example, if we print information only if two variables x, y are equal, then the conditional (x==y) is a filter. Filters can involve hardware status as well. For example, if we halt when the serial port has an overrun error, then (SCSR&0x08) is the filter, and if(SCSR&0x08)asm(" swi"); would be the entire instrument.

functional debugging   The process of detecting, locating, or correcting functional and logical errors in a program, and the process of instrumenting a program for such purposes is called functional debugging or often simply debugging.

instrument   An instrument is the code injected into a program for debugging or profiling. This code is usually extraneous to the normal function of a program and may be temporary or permanent. Instruments injected during interactive sessions are considered to be temporary, because these instruments can be removed simply by terminating a session. Instruments injected in source code are considered to be permanent, because removal requires editing and recompiling the source. An example of a temporary instrument is the SWI instruction added by the BUFFALO monitor when you add a breakpoint with the BR command. This temporary instrument can be removed dynamically by clearing breakpoints with the BR command. A print statement added to your source code is an example of a permanent instrument, because removal requires editing and recompiling.

instrumentation   Instrumentation is the process of injecting or inserting an instrument.

loader   System software that places the object code into the microcomputer's memory. If the object code is stored in EPROM, the loader is also called a EPROM programmer.

monitor or **debugger window**   A monitor is a debugger feature that allows us to passively view strategic software parameters during the real-time execution of our program. An effective monitor is one that has minimal effect on the performance of the system. When debugging software on a windows-based machine, we can often set up a debugger window that displays the current value of certain software variables.

nonintrusive/intrusive   Nonintrusiveness is the characteristic or quality of a debugger that allows the software/hardware system to operate normally as if the debugger did not exist. Intrusiveness is used as a measure of the degree of perturbation caused in program performance by an instrument. For example, a print statement added to your source code and single stepping are very intrusive because they significantly affect the real-time interaction of the hardware and software. When a program interacts with real-time events, the performance is significantly altered. On the other hand, an instrument that outputs a strategic variable on the LEDs (that requires just 5 μs to execute) is much less intrusive. A logic analyzer that passively monitors the address and data is completely nonintrusive. An ICE is also nonintrusive because the software input/output relationships will be the same with and without the debugging tool.

noninvasive/invasive   Noninvasiveness is the characteristic or quality of a debugger that makes the order of invocation immaterial. The debugger and the user program coexist in the same global environment. On the other hand, an invasive debugger requires the user program to execute within an environment defined by the debugger. The debugger is invoked first, and the program is then loaded either by the debugger or by the user from within the debugger. Invasiveness is also a measure of the degree of source code modification to debug or monitor a program. The BUFFALO monitor is invasive because it exists first and then your program is loaded on top of it. This program development environment is very invasive because the 6811 in expanded mode with BUFFALO is very different from a single-chip embedded computer. An ICE is noninvasive because it can coexist (be added or deleted) from our system without changing the way our system runs.

object code Programs in machine readable format created by the compiler or assembler. The S19 records are examples of object code.

performance debugging or profiling The process of acquiring or modifying timing characteristics and execution patterns of a program and the process of instrumenting a program for such purposes is called performance debugging or profiling.

private Can be accessed only by software modules in that local group.

public Can be accessed by any software module.

scan Any instrument used to produce a side effect without causing a break (halt) is a scan. Therefore, a scan may be used to gather data passively or to modify functions of a program. Examples include software added to your source code that simply outputs or modifies a global variable without halting.

scan point A scan point is similar to a breakpoint but used in the context of a scan.

simulator A simulator is a software application that simulates or mimics the operation of a processor or computer system. Most simulators recreate only simple I/O ports and often do not effectively duplicate the real-time interactions of the software/hardware interface. On the other hand, they do provide a simple and interactive mechanism to test software. Simulators are especially useful when learning a new language, because they provide more control and access to the simulated machine than one normally has with real hardware.

source code Programs in human readable format created with an editor.

stabilize The process of stabilizing a software system involves specifying all its inputs. When a system is stabilized, the output results are consistently repeatable. Stabilizing a system with multiple real-time events, like input devices and time-dependent conditions, can be difficult to accomplish. It often involves replacing input hardware with sequential reads from an array or disk file.

thread The execution of software.

time profile and execution profile Time profile refers to the timing characteristic of a program, and execution profile refers to the execution pattern of a program.

2.13 Exercises

2.1 Write a subroutine, called FUZZY, that performs the I/O function shown in Figure 2.18. In and Th are inputs and Out is the result. All parameters are 8-bit unsigned integers. A typical

Figure 2.18 Fuzzy logic membership function.

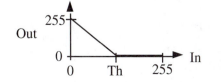

If In≥Th then Out=0

If In<Th then Out = $\dfrac{(255 \cdot (Th-In))}{Th}$

calling shows that the two inputs, In and Th, are passed on the stack and the return parameter, Out, is returned in RegB.

```
ldaa #150 value for Th
psha      Th pushed on the stack
ldaa #90  value for In
psha      In on the stack
jsr FUZZY your function
ins
ins       pop off Th and In
* Reg B = (255*(150-90))/150 = 102
```

You **must** use at least one local variable. Comments will be graded.

2.2 Consider the reasons why one chooses which technique to create a variable.
   a) List three reasons why one would implement a variable using a register.
   b) List three reasons why one would implement a variable on the stack and access it using RegX indexed mode addressing.
   c) List three reasons why one would implement a variable in RAM and access it using direct or extended mode addressing.

2.3 Write a subroutine that performs the following C function explicitly. Draw lines in your assembly code that delineate the implementation for each line of C explicitly. The input parameter, channel, is passed by value on the stack. You must implement the local variable, result, on the stack. The output parameter is returned in RegD. Notice that Adr1 and Adr2 are 8-bit unsigned, while result is 16-bit unsigned.

```
AtoD(channel) unsigned char channel; { unsigned int result;
    Adctrl=channel;                    /* Start A/D              */
    while ((Adstat& 0x80) == 0){};     /* Wait for A/D to finish */
    result=Adr1;                       /* Read first A/D result  */
    result=result1Adr2;                /* Combine two A/D results */
    return(result); }
```

An example calling sequence might be

```
        ldab #5       ; A/D channel 5
        pshb          ; parameter on stack (not in Reg B)
        jsr AtoD
        ins           ; discard input parameter
    * Reg D has result
```

2.4 List three factors that we can use to evaluate the "goodness" of a program.

2.5 Consider the following simple C program. To solve this problem you can read Chapter 6 of the HTML document in the embed folder on the accompanying CD or simply type it in to your compiler and observe the assembly listing file it creates.

```
int x1;
const int x3=1000;
int add3(int z1, int z2, int z3){ int y1;
    y=z1+z2+z3;
/* ***** answer the questions when the program is here ***** */
    return(y);}
void main(void){ int y2; static int x2;
    x1=1000;
    x2=1000;
    y=add3(x1,x2,x3);
```

a) Where in memory are x1, x2, x3, y1, y2, z1, z2, z3, add3, and main allocated? For each object simply specify global RAM, stack RAM, or EEPROM. Answer the questions when the scope of execution is just after the addition in add3.

b) Draw a stack picture at the point just after the addition in add3. Show in your picture the three parameters z1, z2, z3 and the two local variables y1, y2. Show where Registers X and SP are pointing.

3 Interfacing Methods

Chapter 3 objectives are to:

❑ Introduce basic performance measures for I/O interfacing
❑ Outline various interfacing approaches
❑ Interface simple I/O devices using blind cycle synchronization
❑ Discuss the basic concepts of gadfly synchronization
❑ Describe general approach to I/O interface design
❑ Present the basic hardware/software for parallel port interfaces
❑ Introduce the general concept of a handshake interface, then present many
examples

One factor that makes an embedded system different from a regular computer is the special I/O devices we attach to our embedded system. While the entire book addresses the design and analysis of embedded systems, this chapter serves as an introduction to the critical task of I/O interfacing. Interfacing includes both the physical connections of the hardware devices and the software routines that affect information exchange. The chapter begins with performance measures to evaluate the effectiveness of our system (latency, bandwidth, priority). As engineers we are not asked simply to design and build devices, but we also are required to evaluate our products. Latency and bandwidth are two quantitative performance parameters we can measure on our real-time embedded system. Next, the basic approaches to I/O interfacing are presented (blind cycle, gadfly, interrupts, periodic polling, and DMA). Although a complete understanding of interrupts and DMA won't come until you complete Chapters 4, 6, 7, and 10, the discussion in this chapter will point to situations that require these more powerful interfacing methods. The rest of the chapter presents simple examples to illustrate the blind cycle and gadfly approaches to interfacing.

3.1 Introduction

3.1.1
Performance
Measures

Latency is the time between when the I/O device needs service and when service is initiated. Latency includes hardware delays in the digital gates plus computer hardware delays. Latency also includes software delays. For an input device, software latency (or software response time) is the time between new input data ready and the software reading the data. For an output device, latency is the delay from output device idle and the software giving the device new data to output. In this book, we will also have periodic events. For example, in our data acquisition systems, we wish to invoke the ADC at a fixed time interval. In this way we can collect a sequence of digital values that represent the continuous analog signal. Software latency in this case is the time between when the ADC is supposed to be started and when it is actually started. The microcomputer-based control system also employs periodic software processing. Similar to the data acquisition system, the latency in a control system is the time between when the control software is supposed to be run and when it is actually run. A *real-time* system is one that can guarantee a worst-case latency. In other words, there is an upper bound on the software response time. *Throughput* or *bandwidth* is the maximum data flow (bytes per second) that can be processed by the system. Sometimes the bandwidth is limited by the I/O device, while other times it is limited by computer software. Bandwidth can be reported as an overall average or a short-term maximum. *Priority* determines the order of service when two or more requests are made simultaneously. Priority also determines if a high-priority request should be allowed to suspend a low-priority request that is currently being processed. We may also wish to implement equal priority so that no one device can monopolize the computer. In some computer literature, the term *soft real time* is used to describe a system that supports priority.

3.1.2
Synchronizing the
Software with the
State of the I/O

One can think of the hardware as being in one of three states. The *idle* state occurs when the device is disabled or inactive. No I/O occurs in the idle state. When active (not idle), the hardware toggles between the *busy* and *done* states. For an input device, a status flag is set when new input data are available (Figure 3.1). The busy-to-done state transition will cause a gadfly loop (polling loop) to complete. Once the software recognizes that the input

Figure 3.1
The input device sets a flag when it has new data.

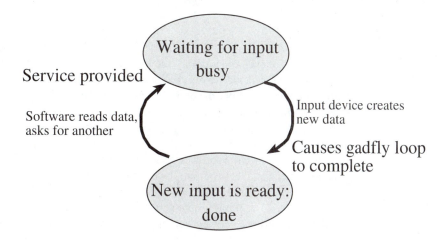

device has new data, it will read the data and ask the input device to create more data. These hardware state transitions are illustrated in Figures 3.1 and 3.3. It is the *busy-to-done* state transition that signals to the computer that service is required. When the hardware is in the done state, the I/O transaction is complete. Often the simple process of reading the data will clear the flag and request another input. Later in this chapter we will present examples of this type of interface.

The problem with I/O devices is that they are usually much slower than software execution. Therefore, we need synchronization, which is the process of the hardware and software waiting for each other in a manner such that data are properly transmitted. A way to visualize this synchronization is to draw a state versus time plot of the activities of the hardware and software (Figure 3.2). For an input device, the software begins by waiting for new input. When the input device is busy, it is in the process of creating new input. When the input device is done, new data are available. When the input device makes the transition from busy to done, it releases the software to go forward. In a similar way, when the software accepts the input, it can release the input device hardware. The arrows from one graph to the other represent the synchronizing events. In this example, the time for the software to read and process the data is less than the time for the input device to create new input. This situation is called *I/O bound*. If the input device were faster than the software, a situation called *CPU bound,* then the software waiting time would be zero. From Figure 3.2 we can see that the bandwidth depends on both the hardware and the software.

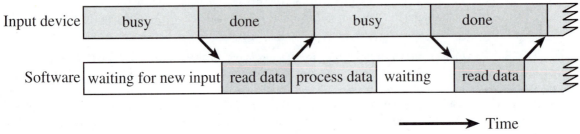

Figure 3.2
The software must wait for the input device to be ready.

This configuration is also labeled as *unbuffered* because the hardware and software must wait for each other during the transmission of each piece of data. A buffered system allows the input device to run continuously, filling a buffer as fast as it can. In the same way, the software can empty the buffer whenever it is ready and whenever data are in the buffer. We will implement a buffered interface in Chapter 4 using interrupts.

For an output device, a status flag is set when the output is idle and ready to accept more data (Figure 3.3). The busy-to-done state transition causes a gadfly loop (polling loop) to complete. Once the software recognizes that the output is idle, it gives the output device another piece of data to output. It will be important to make sure the software clears the flag each time new output is started.

Figure 3.4 contains a state versus time plot of the activities of the output device hardware and software. For an output device, the software begins by generating data, then

Figure 3.3
The output device sets a
flag when it has finished
outputting the last data.

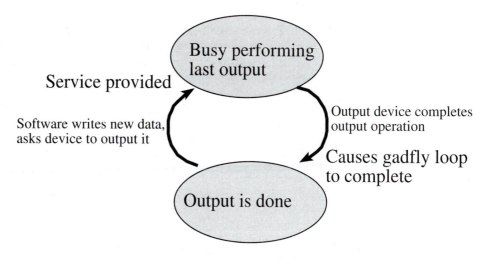

Figure 3.4
The software must wait for the output device to finish the previous operation.

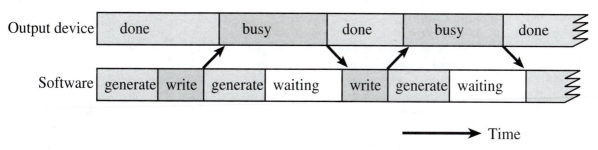

sending them to the output device. When the output device is busy, it is processing the data. Normally when the software writes data to an output port, that only starts the output process. The time it takes an output device to process data is usually longer than the software execution time. When the output device is done, it is ready for new data. When the output device makes the transition from busy to done, it releases the software to go forward. In a similar way, when the software writes data to the output, it releases the output device hardware. The output interface illustrated in Figure 3.4 is also I/O bound because the time for the output device to process data is longer than the time for the software to generate and write it.

This output interface is also unbuffered, because when the hardware is done, it will wait for the software, and after the software generates data, it waits for the hardware. A buffered system would allow the software to run continuously, filling a buffer as fast as it wishes. In the same way, the hardware can empty the buffer whenever it is ready and whenever there are data in the buffer. We will implement a buffered interface in Chapter 4 using interrupts.

The purpose of our interface is to allow the microprocessor to interact with its external I/O device. There are five mechanisms to synchronize the microprocessor with the I/O device. Each mechanism synchronizes the I/O data transfer to the busy-to-done transition. The methods are discussed in the following paragraphs.

Blind cycle is a method whereby the software simply waits a fixed amount of time and assumes the I/O will complete after that fixed delay. For an input device, the software triggers (starts) the external input hardware, waits a specified time, then reads data from device. For an output device, the software writes data to the output device, triggers (starts) the device, then waits a specified time. We call this method *blind* because there is no status information about the I/O device reported to the computer software. This method will be used in situations where the I/O speed is short and predictable.

Gadfly or *busy waiting* is a software loop that checks the I/O status waiting for the done state. For an input device, the software waits until the input device has new data, then reads them from the input device. For an output device, the software writes data, triggers the output device, then waits until the device is finished. Another approach to output device interfacing is for the software to wait until the output device has finished the previous output, write data, then trigger the device. We will discuss these two approaches to output device interfacing later in the chapter. Gadfly synchronization will be used in situations where the software system is relatively simple and real-time response is not important.

An *interrupt* uses hardware to cause special software execution. With an input device, the hardware will request an interrupt when input device has new data. The software interrupt service will read the data from the input device and save them in a global structure. With an output device, the hardware will request an interrupt when the output device is idle. The software interrupt service will get data from a global structure, then write them to the device. Sometimes we configure the hardware timer to request interrupts on a periodic basis. The software interrupt service will perform a special function. A data acquisition system needs to read the ADC at a regular rate. Details of data acquisition systems can be found in Chapters 11 and 12. The Motorola microcomputers will execute special software when it tries to execute an illegal instruction. Other computers can be configured to request an interrupt on an access to an illegal address or a divide by zero.[1] Interrupt synchronization will be used in situations where the software system is fairly complicated or when real-time response is important.

Periodic polling uses a clock interrupt to periodically check the I/O status. With an input device, a ready flag is set when the input device has new data. At the next periodic interrupt, the software will read the data and save in global structure. With an output device, a ready flag is set when the output device is idle. At the next periodic interrupt, the software will get data from a global structure and write them. Periodic polling will be used in situations that require interrupts, but the I/O device does not support interrupt requests.

Direct memory access is an interfacing approach that transfers data directly to/from memory. With an input device, the hardware will request a DMA transfer when input device has new data. Without the software's knowledge or permission, the DMA controller will read from the input device and save in memory. With an output device, the hardware will request a DMA transfer when the output device is idle. The DMA controller will get data from memory, then write to the device. Sometimes we configure the hardware timer to request DMA transfers on a periodic basis. DMA can be used to implement a high-speed data acquisition system. Details of DMA can be found in Chapter 10. DMA synchronization will be used in situations where bandwidth and latency are important.

---

[1]The Motorola microcomputers do not provide for a divide-by-zero trap, but most computers do.

**3.1.3
Variety of
Available I/O
Ports**

Microcomputers perform digital I/O using their ports. In this chapter we will focus on the input and output of digital signals. We will present a detailed description of analog interfacing later in Chapter 11. Microcomputers have a variety of configurations, only a few of which are listed in Table 3.1. There are 18 different 6805 families (B, BD, C, D, E, F, J, JB, JJ, JP, K, KJ, L, P, T, V, X, and V). The 6808 has five families (AB, AS, AZ, MP, and XL), the 6811 has nine families (A, D, E, F, G, K, L, M, and P) and the 6812 has two families (A and B). Except for the 6812, there are many derivatives in each family. Many of the port pins can be used for alternative functions other than parallel I/O.

Table 3.1
The number of I/O ports and alternative function.

MC68HC	Port Pins	Alternative Functions
705J1A	14	
708XL36	54	Serial, time measurement, signal generation
11A8	40	Serial, time measurement, signal generation, A/D, address/data bus
812A4	93	Serial, time measurement, signal generation, A/D, background debug, chip select, address/data bus
912B32	64	Serial, time measurement, signal generation, A/D, background debug, address/data bus
Intel 8751	32	Asynchronous serial, time measurement, address/data bus

It is good practice to use the same technology of the microprocessor for the design of the I/O interface. When faced with the problem of designing an I/O interface to our microcomputer, we have the choice of using sophisticated devices made with large-scale integrated circuits [e.g., metal-oxide semiconductor (MOS), large-scale integration (LSI), very large scale integration (VLSI)] or simple devices made from small-scale integrated devices like standard TTL (e.g., 7400), low-power Schottky TTL (e.g., 74LS00) and high-speed CMOS (e.g., 74HC00). Table 3.2 shows many differences between large- and small-scale integrated circuits.

Table 3.2
Comparison between integrated technology used to make microcomputers versus standard digital logic.

Large-scale Integrated Circuits	Small-scale Integrated Circuits
Sophisticated functions:	Simple function:
Multiply, sqrt, sort, merge	And, or, move
10^6 to 10^9 transistors/chip	10^1 to 10^3 transistors/chip
Single-chip operation	100s of chips for the same function
$5 to $50	$1000
40-pin DIP or 52-pin QFP	Large PC board
2-MHz clock speed	100-MHz clock speed
Low power (Icc)	High power (Icc)

The interfacing issues (speed, voltage levels, complexity) are matched when both the processor and I/O are designed from similar technologies. To produce a marketable LSI I/O device, one must:

■ Increase the market by making the device flexible, able to perform many functions, and use only a fraction of its power
■ Increase flexibility by making the device programmable and by decreasing pins while increasing function
■ Increase yield by including redundancy, using a modular design, including on-chip diagnostics, making the device a standard size, and decreasing pins

Table 3.3 clearly shows us that LSI technology is appropriate for designing I/O devices.

Table 3.3
Advantages and disadvantages of using LSI technology to design I/O devices.

Advantages	Disadvantages
Shorter design time	Increased software complexity
Increased performance	Need to write a software ritual
Reduced size	Added LSI design costs
Fewer bugs	
Easier to maintain	
Easier to modify	
Increased flexibility	
Lower power	

To be successful each computer family needs a set of high-performance I/O devices. In single-chip microcomputer systems like the 6805, 6808, 6811, and 6812, most of these I/O devices are built-in. Similarly, Cyrix has an integrated microprocessor chip, Gx86, that implements the x86 microprocessor and the associated I/O ports (video controller, sound blaster, peripheral component interconnect (PCI) controller, etc.). We call the Gx86 an integrated microprocessor instead of a single-chip microcomputer because external memory is required to complete the system. Motorola has a line of integrated microprocessors built around the CPU32 (68000-like) processor. Examples include the 68332, 68333, and 68340.

In the early days of microcomputer interfacing before single-chip microcomputers, design engineers would first evaluate the needs of their project. They would select a basic microprocessor (like the 6800, 6809, or 68000) that could handle the software tasks. Then they added external RAM, PROM, and I/O devices to build the microcomputer system. Table 3.4 lists some of these "older style" I/O devices available from Motorola. Adding devices is a very expensive and complex process but provides for a wide range of possibilities.

The current trend in the microcomputer industry is customer-specific integrated circuits (CSICs). A similar term for this development process is application-specific integrated circuits (ASICs). With these approaches, the design engineers (customer) first evaluate the needs of their project. In many ways this new development process is similar to the "older way" of design, but now the design engineers work more closely with the microcomputer

Chip	Acronym	Name	Functions
6821	PIA	Peripheral interface adapter	Parallel I/O and interrupts
6824	PRU	Peripheral recovery unit	6811/6824 work like a microcomputer
6840	PTM	Programmable timer module	Timing
6850	ACIA	Asynchronous communication interface adapter	Serial interface
6844	DMAC	Direct memory access controller	High-speed interface
6843	FDC	Floppy disk controller	Storage
14469	AART	Addressable asynchronous receiver/transmitter	Remote serial interface
6860	Modem	Digital modem interface	Telephone communication
6862	Modem	Phase-encoded modem	High-speed remote
68488	GPIA	General-purpose interface adapter	IEEE 488 bus

Table 3.4
Motorola external I/O devices.

manufacturer. The design engineers together with the microcomputer manufacturer make a list of features the microcomputer requires. For example:

CPU type	CPU05, CPU08, CPU11, CPU12, CPU16, CPU32[2]
Memory	RAM, EEPROM, EPROM, Flash, ROM
ADC	8 or 10 bits
Timer	PWM, input capture, output compare, etc.
DMA	Number of channels
Parallel ports	Key wakeup, pull-up, pull-down, reduced drive for low-power applications
Serial interface	Asynchronous (SCI), synchronous (SPI)

The manufacturer then designs and manufactures a microcomputer specific for that project.

3.2   Blind Cycle Counting Synchronization

Blind cycle counting is appropriate when the I/O delay is fixed and known. This type of synchronization is blind because it provides no feedback from the I/O back to the computer.

3.2.1
Blind Cycle Printer Interface

For example, consider a printer that can print 10 characters every second. With blind cycle counting synchronization, there is no printer status signal from the printer telling the computer when the last character output is complete. A simple software interface would be to output the character, then wait 100 ms for it to finish (Figure 3.5).

Figure 3.5
A simple printer interface

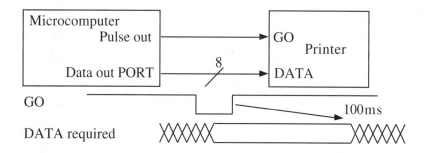

The subroutine that outputs one character follows these steps: (1) The software places the character to be printed on the output part, (2) the software issues a GO pulse (set GO to high, clear GO to low), and (3) the software waits the 100 ms for the character to be printed (Program 3.1).

Program 3.1 A software function that outputs to a simple printer.

```
void Output (unsigned char LETTER) { unsigned short cnt;
    PORT=LETTER;        /* sets Port outputs */
    Pulse();            /* pulses GO */
    for(cnt=0,cnt<10000,cnt++);    /* Wait for 100 ms */
}
```

---

[2]The CPU within the 6805, 6808, 6811, 6812, 6816, 683xx microcontrollers.

The advantage of blind cycle counting is that it is simple and predictable. It does not have the chance of hanging up (i.e., never returning). Unfortunately, there are several disadvantages of the blind cycle counting technique. If the output rate is variable (like a "carriage return," "tab," "graphics," or "formfeed"), then this technique is awkward. If the input rate is unknown (like a keyboard), this technique is inappropriate. The time delay is wasted. If the delay time is long (as it is in the above example), then this technique is dynamically inefficient. This wait time could be used to perform other useful functions. It does not allow for error checking or special conditions.

**3.2.2
Blind Cycle ADC
Interface**

Nevertheless, blind cycle counting can be appropriate for simple high-speed interfaces. An ADC has an analog input (e.g., $0 \leq In \leq +5V$) and converts this analog signal into digital form (e.g., $0 \leq data \leq 255$). Consider the following example of a high-speed 8-bit ADC interface (Figure 3.6). A positive logic pulse, GO, starts the ADC conversion. The result, DATA, is available 5 μs later. There are no error conditions to consider in this problem.

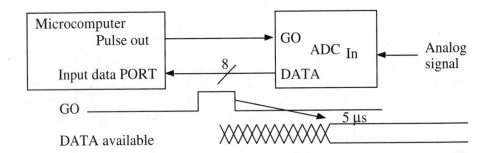

Figure 3.6 A simple A/D interface.

To perform one ADC conversion the subroutine does the following steps (Program 3.2). First, the software starts the ADC conversion by sending a GO pulse (set GO to high, clear GO to low). Second, the software waits for the ADC to convert the analog input signal into digital form (only takes 5 μs). Last, the software inputs the 8-bit result. It is a "blind" interface because there is no ADC status signal telling the software when the conversion is complete.

Program 3.2 A software function that inputs from an ADC.

```
unsigned char Input(void); { int dummy;
    Pulse();                    /* pulses GO     */
    dummy=1000;                 /* Wait for 5us  */
    return(PORT);               /* Read ADC result */
}
```

3.3 Gadfly or Busy Waiting Synchronization

To synchronize the software with the I/O device, the microcomputer usually must be able to recognize the busy-to-done transition. With *gadfly* or *busy waiting* synchronization, the software checks a status bit in the I/O device and loops back un-

Figure 3.7
A software flowchart for gadfly input.

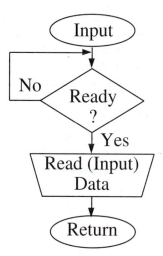

til the device is ready. The gadfly loop must precede the data transfer for an input device (Figure 3.7).

Two steps are involved when the software interfaces with hardware to perform an output function. One step is for the software to output the new data to an output port. This step usually executes in a short amount of time because it involves just a few instructions, with no backward jumps. The other software step is a gadfly loop that executes until the output device is ready. The time in this step is usually long compared to the other operations (I/O-bound situation). These two steps can be performed in either order as long as that order is consistently maintained, and we assume the device is initially ready. Polling before the output allows the computer to perform additional tasks while the output is occurring. Therefore, polling before the output will have a higher bandwidth than polling after the output. On the other hand, polling after the output allows the computer to know exactly when the output has been completed (Figure 3.8).

To illustrate the differences between polling before and after the write-data operation, consider a system with three printers. Each printer can print a character in 1 ms. In other

Figure 3.8 Two software flowcharts for gadfly output.

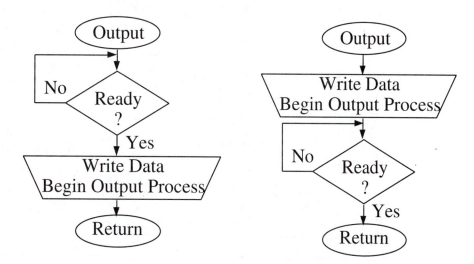

words, a printer will be ready 1 ms after the write-data operation. We will also assume all three printers are initially ready. Since the execution speed of the microcomputer is fast compared to the 1 ms it takes to print a character, we will neglect the software execution time (I/O bound). In the gadfly-before-output system, all three outputs are started together and will operate concurrently. In the gadfly-after-output system, the software waits for the output on printer 1 to finish before starting the output on printer 2. In this system, the three outputs are performed sequentially—that is, about three times slower than the first case (Figure 3.9).

Time(ms)	Gadfly before output	Gadfly after output
0	Start 1,2,3	Start1
From 0 to 1	Wait for 1	Wait for 1
1	Start 1,2,3	Start 2
From 1 to 2	Wait for 1	Wait for 2
2	Start 1,2,3	Start 3
From 2 to 3	Wait for 1	Wait for 3
3	Start 1,2,3	Start1
From 3 to 4	Wait for 1	Wait for 1
4	Start 1,2,3	Start 2
From 4 to 5	Wait for 1	Wait for 2

Figure 3.9　Two software flowcharts for multiple gadfly outputs.

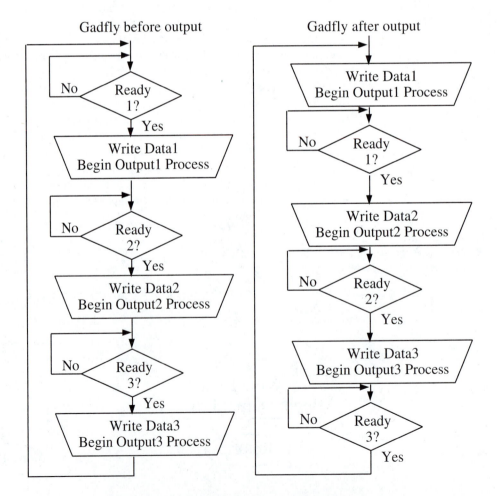

Figure 3.10
A software flowchart for multiple gadfly inputs and outputs.

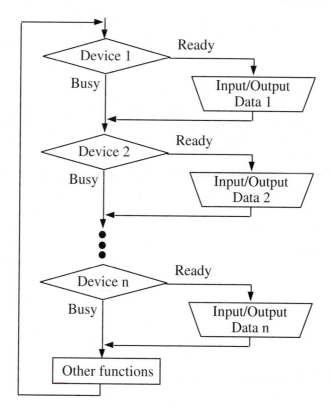

Performance Tip: Whenever we can establish concurrent I/O operations, we can expect an improvement in the overall system bandwidth.

To implement gadfly synchronization with multiple I/O devices, simply poll them in sequence and perform service as required. Figure 3.10 implements a fixed priority and does not allow high-priority devices to suspend the service of lower-priority devices. Therefore the software response time to high-priority devices will be poor.

Interrupts, introduced later in the book, provide an efficient mechanism to reduce the software latency of high-priority devices (like power failure or temperature overflow).

3.4 Parallel I/O Interface Examples

A parallel interface encodes the information as separate binary bits on individual port pins, and the information is available simultaneously. For example, if we are transmitting ASCII characters, then the ASCII code is represented by digital voltages on seven (or eight) digital signals. We begin with parallel I/O devices because they are widely used and simple to understand.

The examples in this section can be run on all four of the computers presented in this book. The differences involve the port names and port addresses. Some of the 6811 ports are fixed direction, and therefore they provide less flexibility but don't require setting the direction register (e.g., 6811 Port B is a fixed output port). The 6811 also has built-in handshaking hardware that uses STRA and STRB.

3.4.1
Blind Cycle Printer
Interface

Consider the printer that can print 100 characters every second. We use the pulse output on GO to start the printer. It is a *blind cycle* interface because no printer status signal reports to the software when the character has been printed. This simple software interface outputs the character, then waits 10 ms for it to finish (Figure 3.11). The timing diagram of Figure 3.12 says the input data must be stable at the time of the rising edge of GO and must remain stable for at least 10 ms. After 10 ms, another output can be performed.

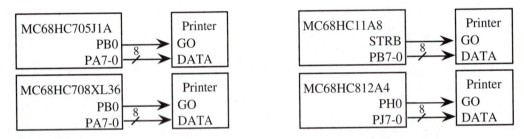

Figure 3.11
Hardware interface between a printer and the microcomputer.

Figure 3.12
Timing diagram for
simple parallel printer.

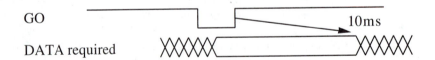

The 6805, 6808, and 6812 use simple output operations to implement the software interface. The 6811 ritual software sets Port B,C mode to no handshake, and STRB is a negative logic pulse output whenever the software writes to Port B. The subroutine outputs one character by sending a GO (STRB) pulse, then waits for the character to be printed. It is assumed that the cycle time is 0.5 μs. The 10-ms wait counter is calculated as the number of cycles to wait divided by the number of cycles in the loop (Table 3.5).

Table 3.5
Data used to calculate
the number of loops
required to wait 10 ms.

Microcomputer	E Period	Cycles to Wait	Cycles in Loop	Counter
General	p	$n = 10$ ms/p	L	n/L
68705	500 ns	20,000	4	5000
68708	125 ns	80,000	4	20,000
6811	500 ns	20,000	7	2857
6812	125 ns	80,000	5	16,000

The 68705 loop counter of 20 causes the wait loop to be executed 256 * 20, or 5120 times. The 68708 loop counter of 79 causes the wait loop to be executed 256 * 79, or 20,224 times. Only the inner two instructions of the 68705 and 68708 loops are considered because the other two are executed only once out of every 256 times. The assembly language subroutines pass the data in and out using Register A (Program 3.3), or in C you could execute Program 3.4.

Program 3.3 Assembly language routines to initialize and output to a printer.

```
; MC68HC705J1A
; PortA is DATA, PB0=GO
Init    lda   #$FF ;Ports A,B
        sta   DDRA ;are
        sta   DDRB ;outputs
        rts
Out     sta   PORTA ;set data
        lda   #0
        sta   PORTB ;GO=0
        lda   #1
        sta   PORTB ;GO=1
        lda   #20
        ldx   #0     ;5120
OLoop   decx         ;[1]
        bne   OLoop ;[3]
        deca
        bne   Oloop ;wait
        rts
```

```
; MC68HC11A8
; PortB is DATA, STRB=GO
Init    clr   PIOC
        rts
Out     staa  PORTB
        ldd   #2857
OLoop   subd  #1      ;[4]
        bne   OLoop   ;[3]
        rts
```

```
; MC68HC708XL36/MC68HC912B32
; PortA is DATA, PB0=GO
Init    lda   #$FF
        sta   DDRA
        lda   #$01
        sta   DDRA
        sta   PORTB ;GO=1
        rts
Out     sta   PORTA ;set data
        lda   #0
        sta   PORTB ;GO=0
        lda   #1
        sta   PORTB ;GO=1
        lda   #79
        ldx   #0     ;20224
OLoop   decx         ;[1]
        bne   OLoop ;[3]
        deca
        bne   Oloop ; wait
        rts
```

```
; MC68HC812A4
; PortJ is DATA, PH0=GO
Init    ldaa #$FF
        staa DDRJ ;outputs
        ldaa #$01
        staa DDRH
        staa PORTH ;GO=1
        rts
Out     staa PORTJ ;set data
        ldaa #0
        staa PORTH ;GO=0
        ldaa #1
        staa PORTH ;GO=1
        ldd  #16000
OLoop   subd #1      ; [2]
        bne  OLoop ; [3]
        rts
```

Program 3.4
C language routines to initialize and output to a printer.

```
// MC68HC705J1A
void Init(void){
    DDRA=0xFF; // outputs
    DDRB=0x01;
    PORTB=1;} // GO=1
void Out(unsigned char value){
unsigned int n;
    PORTA=value;
    PORTB=0;  // GO=0
    PORTB=1;  // GO=1
    for(n=0;n<40000;n++);}
```

```
// MC68HC11A8
void Init(void){
    PIOC=0x00;}
void Out(unsigned char value){
unsigned int n;
    PORTB=value;
    for(n=0;n<28571;n++);}
```

```
// MC68HC708XL36/MC68HC912B32
void Init(void){
    DDRA=0xFF; // outputs
    DDRB=0x01;
    PORTB=1;} // GO=1
void Out(unsigned char value){
unsigned int n;
    PORTA=value;
    PORTB=0;  // GO=0
    PORTB=1;  // GO=1
    for(n=0;n<40000;n++);}
```

```
// MC68HC812A4
void Init(void){
    DDRJ=0xFF; // outputs
    DDRH=0x01;
    PORTH=1;} // GO=1
void Out(unsigned char value){
unsigned int n;
    PORTJ=value;
    PORTH=0;  // GO=0
    PORTH=1;  // GO=1
    for(n=0;n<40000;n++);}
```

The for loop constants should be calculated by counting the number of cycles in the loop. The next section will introduce a better mechanism to affect time delays.

**3.4.2
Accurate Time
Delays**

Counting cycles is a tedious exercise and must be recalculated when:

- Software algorithm changes
- Computer speed changes (e.g., 1 MHz to 2 MHz)
- Computer is upgraded (e.g., 6811 to 6812)
- Compiler is upgraded (e.g., compiler version upgrade)

Counting cycles is inaccurate when background interrupts are occurring. Some of these inaccuracies and inconveniences can be avoided by using the built-in clock of the computer. All microcomputers have at least a basic mechanism to measure time. In this section, we will develop some simple means to delay a fixed amount of time. Since the hardware clock continues during interrupt processing, the time delay will still be accurate. The computers each have an unsigned integer counter that gets incremented at a fixed rate. Most of the Motorola microcomputers allow for a programmable clock rate (Table 3.6). The program-

Table 3.6
Motorola microcomputers have internal counters.

Microcomputer	Counter	Address	Size	Increment Rate
68HC705J1A	TCR	$0009	8 bits	Every 4 memory cycles
68708	TCNT	$0022	16 bits	Programmable 1 to 64 memory cycles[3]
6811	TCNT	$100E	16 bits	Programmable 1 to 16 memory cycles
6812	TCNT	$0084	16 bits	Programmable 1 to 32 memory cycles

[3]Also has a mode where an external clock can be used to increment the counter.

mable clock rate on the 6812 can be adjusted to match the timing of a previous 6811 system or a previous 6812 system with a slower clock.

One method of creating a time delay is to read the counter and add a constant. This sum represents the counter value when the delay is over. The software waits until the counter reaches this value. This method takes advantage of the fact that the counter operation and the addition instruction operate in a similar manner with regard to overflow from $FFFF back to $0000. The assembly language subroutine (Program 3.5) has an input parameter containing the number of cycles to wait (Register D). To create a delay with the 68705, one could use the TOF flag. The TOF flag on the 68705 gets set every 256 counts of the timer (TCR).

Program 3.5
MC68HC705J1A assembly language routine to create an accurate time delay.

```
; 68705
; 256*(RegA) is the number of cycles to wait
Wait    tax         RegX is a loop counter
Loop    lda   #$08  TOFR bit
        sta   TSCR  clear TOF
Loop2   tst   TSCR  wait for TOF
        bpl   Loop2 set every 256 counts
        decx
        bne   Loop
        rts
```

The easiest and most flexible way to create a delay on the 6805, 6808, 6811, and 6812, would be to use the output compare feature that will be covered in detail in Chapter 6. But for now, we can generate a flexible delay program by (1) calculating the time at the end of the delay, then (2) waiting for TCNT to get there (Program 3.6).

Program 3.6 6811 or 6812 assembly language routine to create an accurate time delay.

```
; 6811 or 6812
; Reg D is the number of cycles to wait
;        can range from 25 to 32767
Wait    addd TCNT    time at the end of delay
Loop    cpd  TCNT    wait for Endt-TCNT>0
        bpl  Loop
        rts
```

When we implement this delay function in C, it is essential to define `Endt` as a signed integer. Because both `Endt` and `(short)TCNT` are signed, the compiler will then use the `bpl` conditional branch for the `while` loop as desired (Program 3.7).

Program 3.7 6811 or 6812 C language routine to create an accurate time delay.

```
// 6811 or 6812, numCycles can range from 25 to 32767
void Wait(short numCycles){
short EndT;  // TCNT at the end of the delay
   EndT=TCNT+numCycles;
   while(EndT-(short)TCNT>0);} // wait until TCNT passes EndT
```

3.4.3
Blind Cycle ADC
Interface

Consider again the example of the high-speed 8-bit ADC. A positive logic pulse, GO, starts the A/D conversion. The result, DATA, is available 5 μs later. The interface is "blind" because there is no feedback from the ADC back to the software (Figure 3.13).

To perform an ADC conversion, the software issues a GO pulse, waits 5 μs, and reads the result (Figure 3.14).

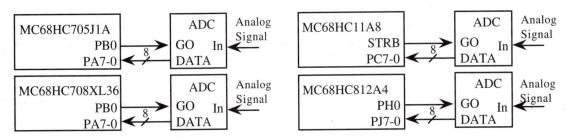

Figure 3.13
Hardware interface between an ADC and the microcomputer.

Figure 3.14
Timing diagram for an ADC interface.

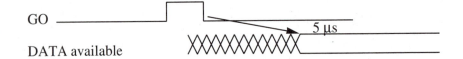

The ritual software sets the DATA port to inputs and the GO pin to an output. On the 6811 the STRB is configured as a positive logic pulse output on write Port B. The subroutine, In, starts the ADC conversion by sending a GO pulse, waits for the ADC to convert the analog input signal into digital form, then inputs the 8-bit result into Register A. It is assumed that the cycle time for the 68705 and 6811 is 0.5 μs (10 cycles is 5 μs) and the cycle time for the 68708 and 6812 is 0.125 μs (40 cycles is 5 μs) (Program 3.8), or in C you could execute Program 3.9. The software delay should be calculated by counting cycles in the listing file generated by the compiler.

Program 3.8 Assembly language routines to initialize and read from an ADC.

```
; MC68HC705J1A
; PortA is DATA, PB0=GO
Init   clr   DDRA
       lda   #$FF
       sta   DDRB
       clr   PORTB  ;GO=0
       rts
In     lda   #1
       sta   PORTB  ;GO=1
       clr   PORTB  ;GO=0
       mul          ;[11]
       lda   PORTA
       rts
```

```
; MC68HC11A8
; PortC is DATA, STRB=GO
Init   ldaa  #$01   ;INVB=1
       staa  PIOC
       clr   DDRC
       rts
In     staa  PORTB  ;GO pulse
       mul          ;[10]
       ldaa  PORTC
       rts
```

```
; MC68HC708XL36/MC68HC912B32        ; MC68HC812A4
; PortA is DATA, PB0=GO             ; PortJ is DATA, PH0=GO
Init   clr   DDRA                   Init   clr   DDRJ
       lda   #$FF                          ldaa  #$FF
       sta   DDRB                          staa  DDRH
       clr   PORTB ;GO=0                   clr   PORTH ;GO=0
       rts                                 rts
In     ldaa  #1                     In     ldaa  #1
       staa  PORTB ;GO=1                   staa  PORTH ;GO=1
       clr   PORTB ;GO=0                   clr   PORTH ;GO=0
       lda   #10                           ldaa  #10'
loop   deca        ; [1]           loop   deca        ; [1]
       bne   loop  ; [3]                   bne   loop  ; [3]
       lda   PORTA                         lda   PORTJ
       rts                                 rts
```

Program 3.9
C language routines to initialize and read from an ADC.

```
// MC68HC705J1A                     // MC68HC11A8
void Init(void){                    void Init(void){
    DDRA=0x00; // PortA DATA            PIOC=0x01;  // GO=STRB
    DDRB=0x01; // PB0 GO                DDRC=0;}    // PORTC is data
    PORTB=0;} // GO=0               unsigned char In(void){ int n;
unsigned char In(void){ int n;         PORTB=value; // GO pulse
    PORTB=1;  // GO=1                   for(n=0;n<1;n++);
    PORTB=0;  // GO=0                   return(PORTC);}
    for(n=0;n<1;n++);
    return(PORTA);}
```

```
// MC68HC708XL36                    // MC68HC812A4
void Init(void){                    void Init(void){
    DDRA=0x00; // PortA DATA            DDRJ=0x00; // PortJ DATA
    DDRB=0x01; // PB0 GO                DDRH=0x01; // PH0 GO
    PORTB=0;} // GO=1                   PORTH=0;}  // GO=0
unsigned char In(void){int n;      unsigned char In(void){int n;
    PORTB=1;  // GO=1                   PORTH=1;  // GO=1
    PORTB=0;  // GO=0                   PORTH=0;  // GO=0
    for(n=0;n<8;n++);                   for(n=0;n<8;n++);
    return(PORTA);}                     return(PORTJ);}
```

3.4.4 Gadfly Keyboard Interface Using Latched Input

The 6811 is the only computer in the group with true latched inputs. The other three implementations differ in that the data read is the value that exists when the read data port instruction is executed. On the 6811, the data read from PORTCL is the value that exists at the time of the rising edge of STROBE. On the 68705, 68708, and 6812, the rise of STROBE triggers the software wait to complete, and the data are read (Figure 3.15).

When the user types a key on this keyboard, the 7-bit ASCII code becomes available on the DATA , followed by a rise in the signal STROBE (Figure 3.16). The data remain available until the next key is typed. On the 6811, the simple latched input mode of Port C (HNDS bit in PIOC=0) is used to capture the input data on the rise of STROBE (STRA) and to set the flag STAF. On the 6811, latched input means the edge of a control signal (e.g., rise of STRA) clocks the input data into a register (e.g., Port CL). We use gadfly

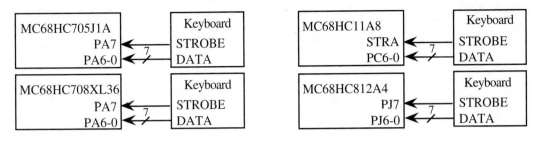

Figure 3.15
Hardware interface between a simple keyboard and the microcomputer.

Figure 3.16
Timing diagram for
simple parallel keyboard.

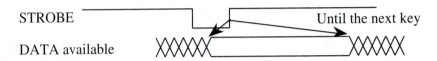

Program 3.10
Assembly language
routines to initialize and
read from a keyboard.

```
; MC68HC705J1A
; PA6-0 is DATA, PA7=STROBE
Init  clr   DDRA
      rts
In    brset 7,PORTA,In
Loop  brclr 7,PORTA,Loop
      lda   PORTA
      and   #$7F
      rts
```

```
; MC68HC11A8
; PortC=DATA, STRA=STROBE
Init ldaa #$02  ;EGA=1
     staa PIOC
     ldaa #$80  ; PC7=output
     staa DDRC
     ldaa PIOC  ;STAF=0
     ldaa PORTCL
     clr  PORTC ;PC7=0
     rts
In   ldaa PIOC ; wait for STAF
     bpl  In
     ldaa PORTCL
     rts
```

```
; MC68HC708XL36/MC68HC912B32
; PA6-0 is DATA, PA7=STROBE
Init  clr   DDRA
      rts
In    brset 7,PORTA,In
Loop  brclr 7,PORTA,Loop
      lda   PORTA
      and   #$7F
      rts
```

```
; MC68HC812A4
; PJ6-0 is DATA, PJ7=STROBE
Init clr  DDRJ
     ldaa #$80  ;rise on PJ7
     staa KPOLJ ;key wakeup
     staa KWIFJ ;clear flag
     rts
In   brclr KWIFJ,$80,In
     ldaa #$80
     staa KWIFJ ;clear flag
     ldaa PORTJ
     anda #$7F
```

synchronization that is a software loop to wait for STAF to be set. In this interface STAF is set when new data are available. The two-step sequence, read parallel I/O control (PIOC) register with STAF set followed by read Port CL, will clear the STAF flag (Program 3.10).

In the 68705, and 68708[4] implementations, we first wait for `STROBE=0` (`brset 7,` `PORTA,In`), then wait for `STROBE=1` (`brclr 7,PORTA,Loop`). In the 6811, the STRA mode is used to wait for the rise of `STROBE`. The 6812 implementation uses the key wakeup feature. The KPOLJ register determines if the rise (1) or fall (0) edge on the Port J inputs will set the corresponding bit in the KWIFJ register. The 6812 `Init` routine sets KPOLJ bit 7 to 1, signifying that the rise of `STROBE` will set bit 7 in the KWIFJ register. The 6812 `In` routine first waits for KWIFJ bit 7 to be set, then clears the KWIFJ bit 7 (writing a 1 to this register clears the bit). Notice the difference between the `brset`, `brclr` instructions on the 6805/8 versus those same instructions on the 6811/12. Because of the gadfly loop, no software delays need to be calculated. Program 3.11 shows the same algorithms implemented in C.

Program 3.11
C language routines to initialize and read from a keyboard.

```
// MC68HC705J1A
void Init(void){ // PA7=STROBE
    DDRA=0x00;}  // PA6-0 DATA
unsigned char In(void){
    while(PORTA&0x80);
// wait for PA7=0
    while((PORTA&0x80)==0);
// wait for PA7=1
    return(PORTA&0x7F);}
```

```
// MC68HC11A8
void Init(void){
// PC6-0 is DATA
unsigned char dummy;
    PIOC=0x02;    // EGA=1
    DDRC=0x80;    // STRA=STROBE
    PORTC=0x00;   // PC7=0
    dummy=PIOC;  dummy=PORTCL;}
unsigned char In(void){
    while ((PIOC & STAF) == 0);
    return(PORTCL); }
```

```
// MC68HC708XL36/MC68HC912B32
void Init(void){ // PA7=STROBE
    DDRA=0x00;}  // PA6-0 DATA
unsigned char In(void){
    while(PORTA&0x80);
// wait for PA7=0
    while((PORTA&0x80)==0);
// wait for PA7=1
    return(PORTA&0x7F);}
```

```
// MC68HC812A4
void Init(void){ // PJ7=STROBE
    DDRJ=0x00;    // PJ6-0 DATA
    KPOLJ=0x80;   // rise on PJ7
    KWIFJ=0x80;} // clear flag
unsigned char In(void){
    while((KWIFJ&0x80)==0); // wait
    KWIFJ=0x80;   // clear flag
    return(PORTJ&0x7F);}
```

Common Error: CMOS inputs have a very large input impedance and a very small input current. If a CMOS input port is left unconnected, then it may begin to toggle, causing a significant power drain.

Observation: A gadfly loop like the one implemented in Programs 3.10 and 3.11 will hang (crash) the computer if the input device is broken and no `STROBE` pulse ever comes.

Performance Tip: Unused CMOS inputs should be connected to +5 or to ground to prevent power loss.

---

[4]The 68708 has a keyboard interrupt mechanism that could have been used.

Performance Tip: Unused CMOS port pins could be programmed as outputs to prevent power loss.

Observation: Some pins on the 6812 can be configured with pull-ups to +5 or pull-downs to ground.

3.4.5
Gadfly ADC
Interface Using
Simple Input

To perform an ADC conversion, the software first issues a GO pulse. For the 6805/6808/6812 implementations, the ADC GO pulse is generated by writing a 1, then a 0 to the port where GO is connected. A gadfly loop waits for the rise of DONE. A simple input of the data (not latched) is used to capture the ADC input data (Figure 3.17).

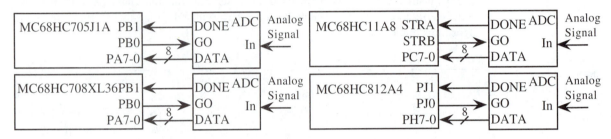

Figure 3.17
Hardware interface between an ADC and the microcomputer.

For the 6811 interface, the ADC GO pulse is generated by writing to Port B (HNDS=0 in the PIOC), and the simple latched input mode of Port C is used to capture the ADC input data on the rise of DONE (STRA) and to set the flag STAF. The rising edge of DONE (STRA) clocks the input data into Port CL. We use gadfly synchronization to wait for STAF to be set, meaning ADC conversion done. The two-step sequence, read PIOC with STAF set followed by read Port CL, will clear the STAF flag.

The timing diagram of Figure 3.18 says that the rise of GO will initialize the ADC, making DONE go low. The fall of GO will start the ADC conversion process. After 25 μs, the 8-bit digital result first becomes available on the DATA lines, followed by the rise of DONE. DONE will remain high, and the DATA will remain valid until the next ADC conversion operation.

Figure 3.18
Timing diagram for an ADC.

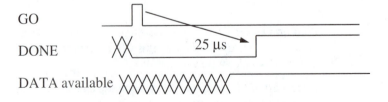

In the 68705, 68708, and 6812 implementations, we explicitly make GO high, then low, and first wait for DONE=1. In the 6811, the handshake mode automatically performs the interlocked communication. The 6812 implementation uses the key wakeup feature. The KPOLJ register determines if the rise (1) or fall (0) edge on the Port J inputs will set the corresponding bit in the KWIFJ register. The 6812 Init routine sets KPOLJ bit 1 to

1, signifying that the rise of DONE will set bit 1 in the KWIFJ register. The 6812 In routine first clears the KWIFJ bit 1 (writing a 1 to this register clears the bit), issues the pulse on GO, then waits for KWIFJ bit 1 to be set (Program 3.12).

Program 3.12
Assembly language routines to initialize and read from an ADC.

```
; MC68HC705J1A                    ; MC68HC11A8
; PortA=DATA PB1=DONE PB0=GO      ; PortC=DATA STRA=DONE STRB=GO
Init lda  #$01  ;PB1 input        Init  ldaa #$03  ;EGA=1, INVB=1
     sta  DDRB  ;PB0 output             staa PIOC
     clr  DDRA  ;PA inputs              ldaa #$00  ;PC=input
     clr  PORTB ;GO=0                   staa DDRC
     rts                                ldaa PIOC  ;STAF=0
                                        ldaa PORTCL
In   lda  #1                            rts
     sta  PORTB ;GO=1             In    staa PORTB ;GO pulse
     clr  PORTB ;GO=0             loop  ldaa PIOC  ;wait for STAF
;wait for rise of DONE                  bita #$80
Loop brclr 1 PORTB Loop                 beq  loop
     lda  PORTA                         ldaa PORTCL
     rts                                rts
```
```
; MC68HC708XL36/MC68HC912B32      ; MC68HC812A4
; PortA=DATA PB1=DONE PB0=GO      ; PortH=DATA PJ1=DONE PJ0=GO
Init lda  #$01  ;PB1 input        Init ldaa #$01  ;PJ1 input
     sta  DDRB  ;PB0 output            staa DDRJ  ;PJ0 output
     clr  DDRA  ;PA inputs             ldaa #$02  ;rise on PJ1
     clr  PORTB ;GO=0                  staa KPOLJ ;key wakeup
     rts                               clr  DDRH  ;PA inputs
                                       clr  PORTJ ;GO=0
                                       rts
In   lda  #1                      In   ldaa #$02
     sta  PORTB ;GO=1                  staa KWIFJ ;clear flag
     clr  PORTB ;GO=0                  ldaa #1
;wait for rise of DONE                 staa PORTJ ;GO=1
Loop brclr 1,PORTB,Loop                clr  PORTJ ;GO=0
     lda  PORTA                   ;wait for rise of DONE
     rts                          Loop brclr KWIFJ,$02,Loop
                                       ldaa PORTH
                                       rts
```

The same algorithms in C are presented in Program 3.13.

Program 3.13
C language routines to initialize and read from an ADC.

```
// MC68HC705J1A                    // MC68HC11A8
void Init(void){ // PB1=DONE in    void Init(void){
   DDRB=0x01;   // PB0=GO out       // PortC=DATA STRA=DONE STRB=GO
   PORTB=0;     // GO=0             unsigned char dummy;
   DDRA=0x00;}  // PA=DATA in          PIOC=0x03;   // EGA=1 INVB=1
unsigned char In(void){               DDRC=0x00;   // PC inputs
   PORTB=1;     // GO pulse            dummy=PIOC;
   PORTB=0;                           dummy=PORTCL;} // clear STAF
   while((PORTB&0x02)==0);         unsigned char In(void){
// wait for PB1=1                      PORTB=0;   // GO pulse
   return(PORTA);}                     while ((PIOC & STAF) == 0);
                                       return(PORTCL); }
```

continued on p. 158

Program 3.13
C language routines to
initialize and read from
an ADC.

continued from p. 157

```
// MC68HC708XL36/MC68HC912B32        // MC68HC812A4
void Init(void){ // PB1=DONE in      void Init(void){ // PJ1=DONE in
    DDRB=0x01;     // PB0=GO out          DDRJ=0x01;     // PJ0=GO out
    PORTB=0;       // GO=0                KPOLJ=0x02;    // rise on PJ1
    DDRA=0x00;}    // PA=DATA in          DDRH=0x00;     // PH DATA in
unsigned char In(void){                  PORTJ=0;}      // GO=0
    PORTB=1;       // GO pulse        unsigned char In(void){
    PORTB=0;                              KWIFJ=0x02;  // clear flag
    while((PORTB&0x02)==0);               PORTJ=1;     // GO pulse
// wait for PB1=1                         PORTJ=0;
    return(PORTA);}                       while((KWIFJ&0x02)==0);
                                          return(PORTH);}
```

3.4.6
Gadfly External
Sensor Interface
Using Input
Handshake

To input an 8-bit sensor reading, the software first waits for the next sensor reading to be ready. The new data available condition is signified by the rise of READY. On the 6805 and 6808, the software waits for the rising edge of READY. After that, it sets the ACK low, signifying that the computer is processing the new input. Next, the computer reads the DATA. Then, the software sets the ACK high, signifying that the computer is done processing the current input and is ready to accept another. The 6812 implementation is similar to the 6805 and 6808 in approach except the 6812 uses the key wakeup feature on Port J to wait for the rising edge of READY (Figure 3.19).

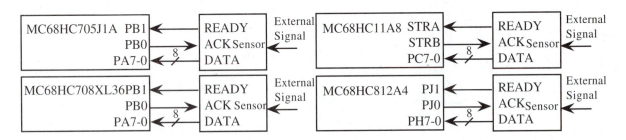

Figure 3.19
Handshaked hardware interface between a sensor and the microcomputer.

On the 6811, with the input handshake mode of Port C, the input data from the sensor is latched into Port CL on the active edge of STRA, the rising edge in this case. This rising edge of READY (STRA) also sets the flag STAF. The 6811 subroutine uses gadfly synchronization to wait for STAF to be set. The two-step sequence, read PIOC with STAF set followed by read Port CL, will clear the STAF flag. With PLS=0 and INVB=1, the rising edge of STRA automatically (without explicit software action) sets ACK (STRB) to 0 (opposite of INVB). The ACK (STRB) output goes to 1 (INVB) when the software reads Port CL. In this mode, the output ACK (STRB) is the complement of STAF flag. The falling edge of ACK is a signal from the 6811 to the sensor signifying that the 6811 has begun to process the current input (e.g., the data have been latched by the hardware but not yet read by the software). The rising edge of ACK signifies that the 6811 has finished processing the current input, and is ready to accept another.

Figure 3.20
Handshaked timing
diagram for a sensor.

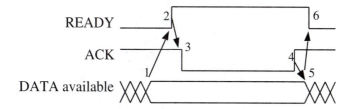

This interface is called *handshaked* or *interlocked* because each event (1, 2, 3, 4, 5, 6) follows in sequence, one event after the other. The arrows in Figure 3.20 represent causal events. In a handshaked interface, one event causes the next to occur, and the arrows create a "head to tail" sequence. There is no specific minimum or maximum time delays for these causal events, except they must occur in sequence.

1. New DATA available from sensor
2. Rising edge of READY signifies new DATA available (software waits for this event)
3. Falling edge of ACK signifies computer is starting to process the data; computer reads the DATA
4. Computer makes ACK=1, signifying computer is done
5. Sensor no longer needs to maintain DATA on its outputs
6. Sensor makes READY fall, meaning DATA is not valid

One of the issues involved in handshaked interfaces is whether or not to wait for the falling edge (step 6.) In general, it is good engineering practice to make the system robust and perform a lot of testing. Because the 6811 and 6812 solutions utilize edge-sensitive triggers (STRA and key wakeup), they automatically wait for both the READY=0 then READY=1 (Program 3.14). In the 6805/6808 implementations, we will

```
; MC68HC705J1A                        ; MC68HC11A8
; PortA=DATA PB1=READY PB0=ACK        ; PortC=DATA STRA=READY STRB=ACK
Init lda  #$01  ;PB1 input       Init    ldaa #$13   ;Input hndshake
     sta  DDRB  ;PB0 output              staa PIOC
     clr  DDRA  ;PA inputs               ldaa #$00   ;PC=input
     lda  #$01                           staa DDRC
     sta  PORTB ;ACK=1                    ldaa PIOC   ;STAF=0
     rts                                 ldaa PORTCL
In   brclr 1,PORTB,In                    rts
     clr  PORTB ;ACK=0              In    ldaa PIOC   ;wait for STAF
     lda  PORTA ;read DATA                bita #$80
     inc  PORTB ;ACK=1                    beq  loop
Loop brset 1,PORTB,Loop                  ldaa PORTCL ;read DATA
     rts                                 rts
```

continued on p. 160

Program 3.14 Handshaking assembly language routines to initialize and read from a sensor.

continued from p. 159

```
; MC68HC708XL36/MC68HC912B32        ; MC68HC812A4
; PortA=DATA PB1=READY PB0=ACK      ; PortH=DATA PJ1=READY PJ0=ACK
Init lda  #$01  ;PB1 input          Init ldaa #$01   ;PJ1 input
     sta  DDRB  ;PB0 output              staa DDRJ   ;PJ0 output
     clr  DDRA  ;PA inputs               ldaa #$02   ;rise on PJ1
     lda  #$01                           staa KPOLJ  ;key wakeup
     sta  PORTB ;ACK=1                   ldaa #$02
     rts                                 staa KWIFJ  ;clear flag1
                                         clr  DDRH   ;PH inputs
In   brclr 1,PORTB,In                    ldaa #$01
     clr  PORTB ;ACK=0                    staa PORTJ  ;ACK=1
     lda  PORTA ;read DATA                rts
     inc  PORTB ;ACK=1             In    brclr KWIFJ,$02,In
Loop brset 1,PORTB,Loop                  clr  PORTJ ;ACK=0
     rts                                 ldaa PORTH ;read DATA
                                         ldab #$02
                                         stab KWIFJ ;clear flag
                                         inc  PORTJ ;ACK=1
                                         rts
```

Program 3.14 Handshaking assembly language routines to initialize and read from a sensor.

explicitly wait for READY=0 and READY=1. If we did not wait for READY=0, then a subsequent call to In might find READY still high from the last call and return the same DATA a second time.

When initializing the bits in a control register like the PIOC, it is good programming practice to define the value for each bit and reason why that value was chosen. For example, after the 6811 program in Program 3.14 sets the PIOC to $13, a good comment would be as shown in Program 3.15.

Program 3.15 Example comment detailing what and why a value is used to initialize a control register.

```
; PortC ritual, Set PC7-PC0 inputs = sensor DATA
; PIOC ($1002)
; 7 STAF     Read Only Set on rise of STRA
; 6 STAI 0   Gadfly, no interrupts
; 5 CWOM 0   Normal outputs
; 4 HNDS 1   Input handshake
; 3 OIN  0
; 2 PLS  0   ACK=STRB goes to 0 on rise of READY
; 1 EGA  1   STAF set on rise of READY=STRA
; 0 INVB 1   ACK=STRB goes to 1 on a ReadCL
; STRB=ACK signifies 6811 status 0 means busy,1 means done
; STRA=READY rising edge when new sensor data is ready
```

These same algorithms can be implemented in C as shown in Program 3.16.

Observation: Programs written for embedded computers are tightly coupled (depend highly) on the hardware; therefore, it is good programming practice to document the hardware configuration in the software comments.

Program 3.16
Handshaking C language routines to initialize and read from a sensor.

```c
// MC68HC705J1A
void Init(void){ // PB1=READY in
    DDRB=0x01;   // PB0=ACK out
    PORTB=0x01;  // ACK=1
    DDRA=0x00;}  // PA=DATA in
unsigned char In(void){
unsigned char data;
    while((PORTB&0x02)==0);
    PORTB=0;     // ACK=0
    data=PORTA;  // read data
    PORTB=1;     // ACK=1
    while(PORTB&0x02);
    return(data);}
```

```c
// MC68HC11A8
void Init(void){
// PortC=DATA STRA=READY STRB=ACK
unsigned char dummy;
    PIOC=0x13;   // EGA=1 INVB=1
    DDRC=0x00;   // PC inputs
    dummy=PIOC;
    dummy=PORTCL;} // clear STAF

unsigned char In(void){
    while ((PIOC & STAF) == 0);
    return(PORTCL); }
```

```c
// MC68HC708XL36/MC68HC912B32
void Init(void){ // PB1=READY in
    DDRB=0x01;   // PB0=ACK out
    PORTB=0x01;  // ACK=1
    DDRA=0x00;}  // PA=DATA in

unsigned char In(void){
unsigned char data;
    while((PORTB&0x02)==0);
    PORTB=0;     // ACK=0
    data=PORTA;  // read data
    PORTB=1;     // ACK=1
    while(PORTB&0x02);
    return(data);}
```

```c
// MC68HC812A4
void Init(void){ // PJ1=READY in
    DDRJ=0x01;   // PJ0=ACK out
    KPOLJ=0x02;  // rise on PJ1
    DDRH=0x00;   // PH DATA in
    KWIFJ=0x02;  // clear flag1
    PORTJ=0X01;} // ACK=1
unsigned char In(void){
unsigned char data;
    while((KWIFJ&0x02)==0);
    PORTJ=0;     // ACK=0
    data=PORTH;  // read data
    KWIFJ=0x02;  // clear flag
    PORTJ=0x01;  // ACK=1
    return(data);}
```

Handshaking is a very reliable synchronization method when connecting devices from different manufacturers and at different speeds. It also allows you to upgrade one device (e.g., get a newer and faster sensor) without redesigning both sides of the interface. Handshaking is used for the SCSI and the IEEE488 instrumentation bus.

3.4.7
Gadfly Printer Interface Using Output Handshake

To output a character on this printer, the user first outputs the 7-bit ASCII code to the DATA, followed by a pulse on the signal START. The completion of the output operation is signified by the rise of READY. As with the earlier interfaces the 6805/6808 performs explicit operations on the control signals, the 6812 uses the key wakeup feature to wait for the rise of READY, and the 6811 uses the handshake mode to perform many of the functions directly in hardware (Figure 3.21).

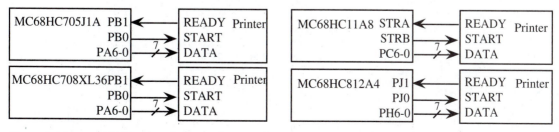

Figure 3.21 Handshaking hardware interface between a printer and the microcomputer.

The printer will latch the new DATA on the rising edge of START . The setup time is the time before the edge (100 ns before ↑START , in this case) at which time the DATA must be valid. The hold time is the time after that same edge (20 ns after ↑START , in this case) at which time the DATA must continue to be valid. The printer is finished and ready to accept another character on the rise of READY (Figure 3.22).

Figure 3.22
Handshaked timing diagram for a printer.

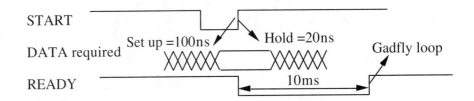

In the 6805/6808/6812 implementations, it is important to follow the sequence (1) set START=0, (2) output new DATA, and (3) set START=1. This sequence guarantees the setup and hold times for the printer. The 6811 solution uses output handshake that automatically satisfies the setup and hold (i.e., the Port C data first becomes valid, then the rising edge of STRB occurs). Again the 6805/6808 implementations explicitly wait for the READY to be low, then wait for READY to be high to detect the rising edge of READY (Program 3.17). These same algorithms can be implemented in C as shown in Program 3.18).

```
; MC68HC705J1A                        ; MC68HC11A8
; PA=DATA PB1=READY PB0=START         ; PC=DATA STRA=READY STRB=START
Init lda  #$01  ;PB1 input            ; 7 STAF     Set on rise of STRA
     sta  DDRB  ;PB0 output           ; 6 STAI 0  Gadfly, no interrpt
     lda  #$FF                        ; 5 CWOM 0  Normal outputs
     sta  DDRA  ;PA outputs           ; 4 HNDS 1  Output handshake
     lda  #$01                        ; 3 OIN  1
     sta  PORTB ;START=1              ; 2 PLS  1  Pulse on Write CL
     rts                              ; 1 EGA  1  STAF set on ↑ READY
                                      ; 0 INVB 0  Negative logic pulse
; Reg A is data to print             Init  ldaa #$1E   ;Output hndshke
Out  clr  PORTB ;START=0                    staa PIOC
     sta  PORTA ;write DATA                 ldaa #$FF   ;PC=output
     inc  PORTB ;START=1                    staa DDRC
; wait for the rising edge                 rts
Wait brset 1,PORTB,Wait              Out   staa PORTCL ;out, pulse
Loop brclr 1,PORTB,Loop              wait  ldab PIOC   ;wait for STAF
     rts                                   bitb #$80
                                           beq wait
                                           rts
```

```
; MC68HC708XL36/MC68HC912B32          ; MC68HC812A4
; PA=DATA PB1=READY PB0=START         ; PortH=DATA PJ1=READY PJ0=ACK
Init lda  #$01  ;PB1 input            Init ldaa #$01  ;PJ1 input
     sta  DDRB  ;PB0 output                staa DDRJ  ;PJ0 output
     lda  #$FF                             ldaa #$02  ;rise on PJ1
     sta  DDRA  ;PA outputs                staa KPOLJ ;key wakeup
     lda  #$01                             ldaa #$02
     sta  PORTB ;START=1                   staa KWIFJ ;clear flag
     rts                                   ldaa #$FF
                                           staa DDRH  ;PH outputs
; Reg A is data to print                   ldaa #$01
Out  clr  PORTB ;START=0                    staa PORTJ ;START=1
     sta  PORTA ;write DATA                 rts
     inc  PORTB ;START=1             Out   ldab #$02
; wait for the rising edge                stab KWIFJ ;clear flag
Wait brset 1,PORTB,Wait                   clr  PORTJ ;START=0
Loop brclr 1,PORTB,Loop                   staa PORTH ;write DATA
     rts                                   inc  PORTJ ;ACK=1
                                    Wait brclr KWIFJ,$02,Wait
                                          rts
```

Program 3.17 Handshaking assembly language routines to initialize and write to a printer.

```
// MC68HC705J1A                       // MC68HC11A8
void Init(void){ // PB1=READY in      void Init(void){
   DDRB=0x01;   // PB0=START out       // PortC=DATA STRA=READY STRB=START
   PORTB=0x01;  // ACK=1                   PIOC=0x1E;   // output handshake
   DDRA=0xFF;}  // PA=DATA out             DDRC=0xFF;}  // PC outputs
void Out(unsigned char data){
   PORTB=0;     // START=0            void Out(unsigned char data){
   PORTA=data;  // write data             PORTCL=data;
   PORTB=1;     // START=1                 while ((PIOC & STAF) == 0);}
   while(PORTB&0x02);
   while((PORTB&0x02)==0);}
```
```
// MC68HC708XL36/MC68HC912B32         // MC68HC812A4
void Init(void){ // PB1=READY in      void Init(void){ // PJ1=READY in
   DDRB=0x01;   // PB0=START out          DDRJ=0x01;   // PJ0=START out
   PORTB=0x01;  // ACK=1                  KPOLJ=0x02;  // rise on PJ1
   DDRA=0xFF;}  // PA=DATA out            DDRH=0xFF;   // PH DATA out
                                         PORTJ=0X01;} // START=1
void Out(unsigned char data){         void Out(unsigned char data){
   PORTB=0;     // START=0                KWIFJ=0x02;  // clear flag
   PORTA=data;  // write data             PORTJ=0;     // START=0
   PORTB=1;     // START=1                PORTH=data;  // write data
   while(PORTB&0x02);                     PORTJ=0x01;  // START=1
   while((PORTB&0x02)==0);}               while((KWIFJ&0x02)==0);}
```

Program 3.18 Handshaking C language routines to initialize and write to a printer.

**3.4.8
Gadfly
Synchronous
Serial Interface to
a Temperature
Sensor**

Many external modules use a synchronous serial interface because of its flexible design and low cost. Only four wires are required to connect an external sensor to the microcomputer. We introduce the Dallas Semiconductor DS1620 digital thermometer and thermostat as an example of a bit-banging serial interface. The purpose of presenting a serial interface is to compare and contrast it with the parallel interfacing in the rest of the chapter. The data sheet for this device is located on the accompanying CD (search for DS1620.pdf.) This device can be used as either a temperature sensor or a thermostat. We will develop another interface later in Chapter 7 that uses the SPI module. There are five types of communication between the computer and the DS1620. In all cases the RST and CLK lines are outputs from the computer and inputs to the DS1620 (Figure 3.23).

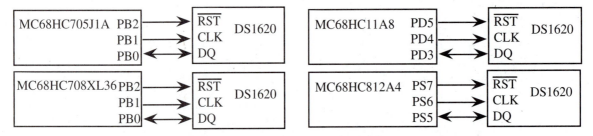

Figure 3.23 Hardware interface between a temperature sensor/controller and the microcomputer.

The ritual to initialize the DS1620 interface is presented first in Program 3.19. These same algorithms can be implemented in C as shown in Program 3.20.

Program 3.19
Assembly language
initialization of the
DS1620.

```
; MC68HC705J1A                        ; MC68HC11A8
; PB2=RST PB1=CLK PB0=DQ              ; PD5=RST PD4=CLK PD3=DQ
Init lda  #$07   ;PB2-0 output        Init ldaa #$38   ;PD5-3 output
     sta  DDRB                             staa DDRD
     lda  #$03   ;RST=0,CLK=1              ldaa #$18   ;RST=0,CLK=1
     sta  PORTB ;DQ=1                      staa PORTD ;DQ=1
     rts                                   rts
```

```
; MC68HC708XL36                       ; MC68HC812A4/MC68HC912B32
; PB2=RST PB1=CLK PB0=DQ              ; PS7=RST PS6=CLK PS5=DQ
Init lda  #$07   ;PB2-0 output        Init ldaa #$E0   ;PD5-3 output
     sta  DDRB                             staa DDRS
     lda  #$03   ;RST=0,CLK=1              ldaa #$60   ;RST=0,CLK=1
     sta  PORTB ;DQ=1                      staa PORTS ;DQ=1
     rts                                   rts
```

Program 3.20
C language initialization
of the DS1620.

```
// MC68HC705J1A                       // MC68HC11A8
void Init(void){ // PB2=RST=0         void Init(void){ // PD5=RST=0
  DDRB=0x07;    // PB1=CLK=1            DDRD=0x38;    // PD4=CLK=1
  PORTB=0x03;} // PB0=DQ=1             PORTD=0x18;} // PD3=DQ=1
```

```
// MC68HC708XL36                      // MC68HC812A4/MC68HC912B32
void Init(void){ // PB2=RST=0         void Init(void){ // PS7=RST=0
  DDRB=0x07;    // PB1=CLK=1            DDRS=0xE0;    // PS6=CLK=1
  PORTB=0x03;} // PB0=DQ=1             PORTS=0x60;} // PS5=DQ=1
```

The DS1620 is capable of sensing the current temperature, with a resolution of 0.5°C, a range of −55 to +125°C, and a conversion time of 1 s. The temperature data is encoded as 9-bit 2s complement signed binary fraction, with a ΔA of 0.5°C. The nine basis elements are −128, 64, 32, 16, 8, 4, 2, 1, and 0.5°C. In particular, some examples of temperature data are presented below in Table 3.7.

Table 3.7
Binary fixed point allows microcomputer to manipulate fractional values without a floating point.

Temperature, °C	Digital Value (Binary)	Digital Value (Hex)
+125.0°	011111010	$0FA
+64.0°	010000000	$080
+1.0°	000000010	$002
+0.5°	000000001	$001
0°	000000000	$000
−0.5°	111111111	$1FF
−16.0°	111100000	$1E0
−55.0°	110010010	$192

The chip also has two internal threshold EEPROM registers, TH and TL. When the temperature is above TH, the T_{HIGH} output goes to +5. When the temperature is below TL, the T_{LOW} output goes to +5. The third output, T_{COM}, implements a thermostat with hysteresis (Figure 3.24).

Figure 3.24
Thermostat control response with hysteresis.

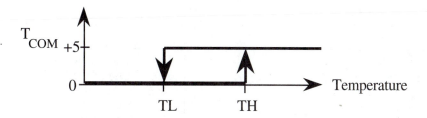

To send an 8-bit command to the DS1620, the microcomputer activates RST and clocks the 8-bit command code so that the data line DQ is stable during the 0 to 1 edge of the CLK. In this transmission, all three signals are outputs of the computer and inputs to the DS1620. The bits $c0, c1, c2, c3, c4, c5, c6,$ and $c7$ form the 8-bit command. Examples include `Start Convert Temperature` (\$EE) and `Stop Convert Temperature` (\$22). It takes 1 s to convert temperature (Figure 3.25).

Figure 3.25
Synchronous serial timing diagram for the DS1620 send command.

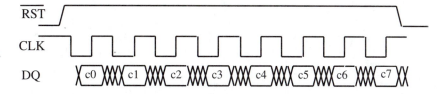

During the transmit phase the software will (1) set the CLK=0, (2) set the DQ output to the desired value, and (3) set the CLK to 1. In this way the DQ data are stable on the rising edge of the CLK. The functions to issue these simple commands are presented next. The command information is passed in RegA (Program 3.21). These same algorithms can be implemented in C as shown in Program 3.22.

```
; MC68HC705J1A
;output 8 bits, Reg A=data
out8  ldx   #8       ;8 bits
clop  bclr  1,PORTB  ;CLK=0
      lsra           ;lsb first
      bcc   set0
      bset  0,PORTB  ;DQ=1
      bra   next
set0  bclr  0,PORTB  ;DQ=0
next  bset  1,PORTB  ;CLK=1
      decx
      bne   clop
      rts

start bset  3,PORTB  ;RST=1
      lda   #$EE
      bsr   out8
      bclr  3,PORTB  ;RST=0
      rts
stop  bset  3,PORTB  ;RST=1
      lda   #$22
      bsr   out8
      bclr  3,PORTB  ;RST=0
      rts
```

```
; MC68HC11A8
;output 8 bits, Reg A=data
out8  ldab  #8        ;8 bits
clop  bclr  0,x,#$10  ;CLK=0
      lsra            ;lsb first
      bcc   set0
      bset  0,x,#$08  ;DQ=1
      bra   next
set0  bclr  0,x,#$08  ;DQ=0
next  bset  0,x,#$10  ;CLK=1
      decb
      bne   clop
      rts

start ldx   #PORTD
      bset  0,x,#$20  ;RST=1
      ldaa  #$EE
      bsr   out8
      bclr  0,x,#$20  ;RST=0
      rts
stop  ldx   #PORTD
      bset  0,x,#$20  ;RST=1
      ldaa  #$22
      bsr   out8
      bclr  0,x,#$20  ;RST=0
      rts
```

```
; MC68HC708XL36
out8  ldx   #8       ;8 bits
clop  bclr  1,PORTB  ;CLK=0
      lsra           ;lsb first
      bcc   set0
      bset  0,PORTB  ;DQ=1
      bra   next
set0  bclr  0,PORTB  ;DQ=0
next  bset  1,PORTB  ;CLK=1
      decx
      bne   clop
      rts
start bset  3,PORTB  ;RST=1
      lda   #$EE
      bsr   out8
      bclr  3,PORTB  ;RST=0
      rts
stop  bset  3,PORTB  ;RST=1
      ldaa  #$22
      bsr   out8
      bclr  3,PORTB  ;RST=0
      rts
```

```
; MC68HC812A4/MC68HC912B32
out8  ldab  #8           ;8 bits
clop  bclr  PORTS,#$40   ;CLK=0
      lsra               ;lsb first
      bcc   set0
      bset  PORTS,#$20   ;DQ=1
      bra   next
set0  bclr  PORTS,#$20   ;DQ=0
next  bset  PORTS,#$40   ;CLK=1
      decb
      bne   clop
      rts
start bset  PORTS,#$80   ;RST=1
      ldaa  #$EE
      bsr   out8
      bclr  PORTS,#$80   ;RST=0
      rts
stop  bset  PORTS,#$80   ;RST=1
      ldaa  #$22
      bsr   out8
      bclr  PORTS,#$80   ;RST=0
      rts
```

Program 3.21 Assembly language helper functions for the DS1620.

```
// MC68HC705J1A                          // MC68HC11A8
void out8(char code){ int n;            void out8(char code){ int n;
  for(n=0;n<8;n++){                       for(n=0;n<8;n++){
     PORTB &= 0xFD;    // PB1=CLK=0          PORTD &= 0xEF;    // PD4=CLK=0
     if(code&0x01)                           if(code&0x01)
       PORTB |= 0x01; // PB0=DQ=1              PORTD |= 0x08; // PD3=DQ=1
     else                                    else
       PORTB &= 0xFE; // PB0=DQ=0              PORTD &= 0xF7; // PD3=DQ=0
     PORTB |= 0x02;    // PB1=CLK=1          PORTD |= 0x10;    // PD4=CLK=1
     code = code>>1;}}                       code = code>>1;}}
void start(void){                       void start(void){
   PORTB |= 0x04;    // PB2=RST=1           PORTD |= 0x20;    // PD5=RST=1
   out8(0xEE);                              out8(0xEE);
   PORTD &= 0xFB;}   // PB2=RST=0           PORTD &= 0xDF;}   // PD5=RST=0
void stop(void){                        void stop(void){
   PORTB |= 0x04;    // PB2=RST=1           PORTD |= 0x20;    // PD5=RST=1
   out8(0x22);                              out8(0x22);
   PORTB &= 0xFB;}   // PB2=RST=0           PORTD &= 0xDF;}   // PD5=RST=0
```

```
// MC68HC708XL36                         // MC68HC812A4/MC68HC912B32
void out8(char code){ int n;            void out8(char code){ int n;
  for(n=0;n<8;n++){                       for(n=0;n<8;n++){
     PORTB &= 0xFD;    // PB1=CLK=0          PORTS &= 0xBF;    // PS6=CLK=0
     if(code&0x01)                           if(code&0x01)
       PORTB |= 0x01; // PB0=DQ=1              PORTS |= 0x20; // PS5=DQ=1
     else                                    else
       PORTB &= 0xFE; // PB0=DQ=0              PORTS &= 0xDF; // PS5=DQ=0
     PORTB |= 0x02;    // PB1=CLK=1          PORTS |= 0x40;    // PS6=CLK=1
     code = code>>1;}}                       code = code>>1;}}
void start(void){                       void start(void){
   PORTB |= 0x04;    // PB2=RST=1           PORTS |= 0x80;    // PS7=RST=1
   out8(0xEE);                              out8(0xEE);
   PORTD &= 0xFB;}   // PB2=RST=0           PORTS &= 0x7F;}   // PS7=RST=0
void stop(void){                        void stop(void){
   PORTB |= 0x04;    // PB2=RST=1           PORTS |= 0x80;    // PS7=RST=1
   out8(0x22);                              out8(0x22);
   PORTB &= 0xFB;}   // PB2=RST=0           PORTS &= 0x7F;}   // PS7=RST=0
```

Program 3.22 C language helper functions for the DS1620.

The second type of communication involves writing the 8-bit configuration register. In this case, bits c0, c1, c2, c3, c4, c5, c6, and c7 form the 8-bit `Write Config` command ($0C), and the bits d0, d1, d2, d3, d4, d5, d6, and d7 are the new data written to the configuration register. For both the command and data transmission, the data line DQ is stable during the 0 to 1 edge of the CLK. Again, all three signals are outputs of the computer and inputs to the DS1620. It takes 5 ms to save a new configuration value in its EEPROM (Figure 3.26).

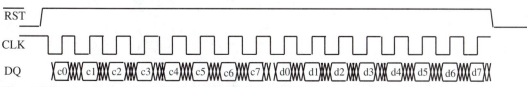

Figure 3.26 Synchronous serial timing diagram for the DS1620 send configuration.

Bits 7, 6, 5 of the configuration/status register are status bits (Table 3.8). The THF flag bit is set if the temperature ever goes above TH and cleared by writing a zero to the bit. Similarly, the TLF flag bit is set if the temperature ever goes below TL and is also cleared by writing a zero to the bit. The status bits are volatile, so the information is lost if the power is removed.

Table 3.8
The DS1620 has three status flags.

Bit	Mode	Configuration/status register meaning
7	DONE	1=Conversion done, 0=conversion in progress
6	THF	1=temperature above TH, 0=temperature below TH
5	TLF	1=temperature below TL, 0=temperature above TL

Bits 1, 0 of the configuration/status register are mode control bits (implemented in nonvolatile EEPROM) (Table 3.9).

Table 3.9
The DS1620 has two control bits.

Bit	Status	Configuration/status register meaning
1	CPU	1=CPU control, 0=stand alone operation
0	1SHOT	1=one conversion and stop, 0=continuous conversions

The functions to set the configuration register are presented next in Program 3.23. The information is passed in RegA. These same algorithms can be implemented in C as shown in Program 3.24.

Program 3.23
Assembly language functions to set the configuration register on the DS1620.

```
; MC68HC705J1A
config sta    temp       ;temporary
       bset  3,PORTB ;RST=1
       lda   #$0C
       bsr   out8
       lda   temp
       bsr   out8
       bclr  3,PORTB ;RST=0
       rts
```

```
; MC68HC11A8
config ldx   #PORTD
       psha
       bset  0,x,#$20 ;RST=1
       ldaa  #$0C
       bsr   out8
       pula
       bsr   out8
       bclr  0,x,#$20 ;RST=0
       rts
```

```
; MC68HC708XL36
config psha
       bset  3,PORTB ;RST=1
       lda   #$0C
       bsr   out8
       pula
       bsr   out8
       bclr  3,PORTB ;RST=0
       rts
```

```
; MC68HC812A4/MC68HC912B32
config psha
       bset  PORTS,#$80 ;RST=1
       ldaa  #$0C
       bsr   out8
       pula
       bsr   out8
       bclr  PORTS,#$80 ;RST=0
       rts
```

Program 3.24
C language functions to
set the configuration
register on the DS1620.

```
// MC68HC705J1A
void config(char data){
    PORTB |= 0x04;    // PB2=RST=1
    out8(0x0C);
    out8(data);
    PORTD &= 0xFB;}   // PB2=RST=0
```

```
// MC68HC11A8
void config(char data){
    PORTD |= 0x20;    // PD5=RST=1
    out8(0x0C);
    out8(data);
    PORTD &= 0xDF;}   // PD5=RST=0
```

```
// MC68HC708XL36
void config(char data){
    PORTB |= 0x04;    // PB2=RST=1
    out8(0x0C);
    out8(data);
    PORTD &= 0xFB;}   // PB2=RST=0
```

```
// MC68HC812A4/MC68HC912B32
void config(char data){
    PORTS |= 0x80;    // PS7=RST=1
    out8(0x0C);
    out8(data);
    PORTS &= 0x7F;}   // PS7=RST=0
```

The third type of communication involves writing a 9-bit temperature threshold register. In this case, bits c0, c1, c2, c3, c4, c5, c6, and c7 form either the Write TH command ($01) or the Write TL command ($02), and the bits d0, d1, d2, d3, d4, d5, d6, d7, and d8 are the new 9-bit data written to the corresponding threshold register. For this mode too, all three signals are outputs of the computer and inputs to the DS1620. It takes 5 ms to program a new threshold value in its EEPROM (Figure 3.27).

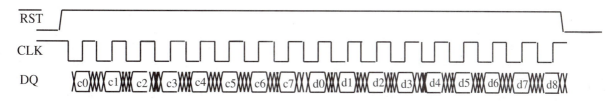

Figure 3.27
Synchronous serial timing diagram for the DS1620 write threshold.

The functions to issue these commands are presented next. It is easier to manipulate 9-bit data on the 6811/6812 versus the 6805/6808. The helper function is presented in Program 3.25.

```
; MC68HC705J1A
;output 9 bits, Reg X:A=data
out9  stx   temp
      ldx   #8        ;8 bits
olop  bclr  1,PORTB   ;CLK=0
      lsra            ;lsb first
      bcc   oset0
      bset  0,PORTB   ;DQ=1
      bra   onext
oset0 bclr  0,PORTB   ;DQ=0
onext bset  1,PORTB   ;CLK=1
      decx
      bne   olop
      lda   temp      ;msbit
```

```
; MC68HC11A8
;output 9 bits, Reg D=data
out9  ldx   #9        ;9 bits
olop  bclr  0,x,#$10  ;CLK=0
      lsrd            ;lsb first
      bcc   oset0
      bset  0,x,#$08  ;DQ=1
      bra   onext
oset0 bclr  0,x,#$08  ;DQ=0
onext bset  0,x,#$10  ;CLK=1
      dex
      bne   olop
      rts
```

continued on p. 170

Program 3.25 Assembly language 9-bit output helper function for the DS1620.

continued from p. 169

```
        bclr 1,PORTB ;CLK=0
        lsra          ;msb last
        bcc  mset0
        bset 0,PORTB ;DQ=1
        bra  mnext
mset0 bclr 0,PORTB ;DQ=0
mnext bset 1,PORTB ;CLK=1
        rts
```

```
; MC68HC708XL36
;output 9 bits, Reg X:A=data
out9  stx  temp
        ldx  #8        ;8 bits
olop  bclr 1,PORTB ;CLK=0
        lsra          ;lsb first
        bcc  oset0
        bset 0,PORTB ;DQ=1
        bra  onext
oset0 bclr 0,PORTB ;DQ=0
onext bset 1,PORTB ;CLK=1
        decx
        bne  olop
        lda  temp      ;msbit
        bclr 1,PORTB ;CLK=0
        lsra          ;msb last
        bcc  mset0
        bset 0,PORTB ;DQ=1
        bra  mnext
mset0 bclr 0,PORTB ;DQ=0
mnext bset 1,PORTB ;CLK=1
        rts
```

```
; MC68HC812A4/MC68HC912B32
out9    ldx  #9         ;9 bits
olop    bclr PORTS,#$40 ;CLK=0
        lsrd            ;lsb
        bcc  oset0
        bset PORTS,#$20 ;DQ=1
        bra  next
oset0   bclr PORTS,#$20 ;DQ=0
onext   bset PORTS,#$40 ;CLK=1
        dex
        bne  olop
        rts
```

Program 3.25 Assembly language 9-bit output helper function for the DS1620.

The routines shown in Program 3.26 access the threshold registers.

Program 3.26 Assembly language functions to set the threshold registers on the DS1620.

```
; MC68HC705J1A
WriteTH sta  wtemp
        stx  wtemp2
        bset 3,PORTB ;RST=1
        lda  #$01
        bsr  out8
        lda  wtemp
        ldx  wtemp2
        bsr  out9
        bclr 3,PORTB ;RST=0
        rts
WriteTL sta  wtemp
        stx  wtemp2
        bset 3,PORTB ;RST=1
        lda  #$02
        bsr  out8
        lda  wtemp
```

```
; MC68HC11A8
;Reg D is temperature value
WriteTH psha
        pshb
        ldx  #PORTD
        bset 0,x,#$20 ;RST=1
        ldaa #$01
        bsr  out8
        pulb
        pula
        bsr  out9
        bclr 0,x,#$20 ;RST=0
        rts
;Reg D is temperature value
WriteTL psha
        pshb
        ldx  #PORTD
```

```
        ldx   wtemp2                          bset  0,x,#$20  ;RST=1
        bsr   out9                            ldaa  #$02
        bclr  3,PORTB  ;RST=0                 bsr   out8
        rts                                   pulb
                                              pula
                                              bsr   out9
                                              bclr  0,x,#$20  ;RST=0
                                              rts
; MC68HC708XL36                         ; MC68HC812A4/MC68HC912B32
WriteTH psha                            ;Reg D is temperature value
        pshx                            WriteTH pshd
        bset  3,PORTB  ;RST=1                   bset  PORTS,#$80  ;RST=1
        lda   #$01                              ldaa  #$01
        bsr   out8                              bsr   out8
        pulx                                    puld
        pula                                    bsr   out9
        bsr   out9                              bclr  PORTS,#$80  ;RST=0
        bclr  3,PORTB  ;RST=0                   rts
        rts                             ;Reg D is temperature value
WriteTL psha                            WriteTL pshd
        pshx                                    bset  PORTS,#$80  ;RST=1
        bset  3,PORTB  ;RST=1                   ldaa  #$02
        lda   #$02                              bsr   out8
        bsr   out8                              puld
        pulx                                    bsr   out9
        pula                                    bclr  PORTS,#$80  ;RST=0
        bsr   out9                              rts
        bclr  3,PORTB  ;RST=0
        rts
```

These same algorithms can be implemented in C as shown in Program 3.27.

```
// MC68HC705J1A                              // MC68HC11A8
void out9(int code){ int n;                 void out9(int code){ int n;
  for(n=0;n<9;n++){                           for(n=0;n<9;n++){
    PORTB &= 0xFD;    // PB1=CLK=0               PORTD &= 0xEF;    // PD4=CLK=0
    if(code&0x01)                               if(code&0x01)
      PORTB |= 0x01; // PB0=DQ=1                   PORTD |= 0x08; // PD3=DQ=1
    else                                        else
      PORTB &= 0xFE; // PB0=DQ=0                   PORTD &= 0xF7; // PD3=DQ=0
    PORTB |= 0x02;   // PB1=CLK=1               PORTD |= 0x10;   // PD4=CLK=1
    code = code>>1;}}                           code = code>>1;}}
void WriteTH(int data){                     void WriteTH(int data){
  PORTB |= 0x04;    // PB2=RST=1               PORTD |= 0x20;    // PD5=RST=1
  out8(0x01);                                 out8(0x01);
  out9(data);                                 out9(data);
  PORTD &= 0xFB;}   // PB2=RST=0               PORTD &= 0xDF;}   // PD5=RST=0
void WriteTL(int data){                     void WriteTL(int data){
  PORTB |= 0x04;    // PB2=RST=1               PORTD |= 0x20;    // PD5=RST=1
  out8(0x02);                                 out8(0x02);
  out9(data);                                 out9(data);
  PORTB &= 0xFB;}   // PB2=RST=0.              PORTD &= 0xDF;}   // PD5=RST=0
```

Program 3.27 C language functions to set the threshold registers on the DS1620.

continued on p. 172

```
continued from p. 171
// MC68HC708XL36
void out9(int code){ int n;
  for(n=0;n<9;n++){
     PORTB &= 0xFD;    // PB1=CLK=0
     if(code&0x01)
        PORTB |= 0x01; // PB0=DQ=1
     else
        PORTB &= 0xFE; // PB0=DQ=0
     PORTB |= 0x02;    // PB1=CLK=1
     code = code>>1;}}
void WriteTH(int data){
  PORTB |= 0x04;    // PB2=RST=1
  out8(0x01);
  out9(data);
  PORTD &= 0xFB;}  // PB2=RST=0
void WriteTL(int data){
  PORTB |= 0x04;    // PB2=RST=1
  out8(0x02);
  out9(data);
  PORTB &= 0xFB;}  // PB2=RST=0
```

```
// MC68HC812A4/MC68HC912B32
void out9(int code){ int n;
  for(n=0;n<9;n++){
     PORTS &= 0xBF:    // PS6=CLK=0
     if(code&0x01)
        PORTS |= 0x20; // PS5=DQ=1
     else
        PORTS &= 0xDF; // PS5=DQ=0
     PORTS |= 0x40;    // PS6=CLK=1
     code = code>>1;}}
void WriteTH(int data){
  PORTS |= 0x80;    // PS7=RST=1
  out8(0x01);
  out9(data);
  PORTS &= 0x7F;}  // PS7=RST=0
void WriteTL(int data){
  PORTS |= 0x80;    // PS7=RST=1
  out8(0x02);
  out9(data);
  PORTS &= 0x7F;}  // PS7=RST=0
```

Program 3.27 C language functions to set the threshold registers on the DS1620.

The last two types of communication involve first sending an 8-bit command to the DS1620, followed by receiving information back from the sensor. In particular, this fourth type of communication receives an 8-bit status byte. In this case, bits c0, c1, c2, c3, c4, c5, c6, and c7 form the Read Config command ($AC), and the bits d0, d1, d2, d3, d4, d5, d6, and d7 are the current value of the configuration/status register. During the first 8 bits, the DQ line is output from the computer, while during the last 8 bits it is an input to the computer. Once again the DQ line changes on the 1 to 0 edge of the CLK and is stable during the 0 to 1 edge. We will have to change the direction register halfway through to implement this protocol (Figure 3.28).

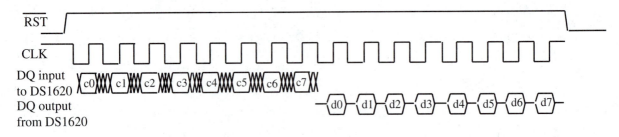

Figure 3.28
Synchronous serial timing diagram for the DS1620 read status.

During the receive phase of the communication, the software will (1) set the CLK=0, (2) read the DQ input, and (3) set the CLK to 1. The functions to issue these commands are presented next. The result is returned in RegA (Program 3.28). These same algorithms can be implemented in C as shown in Program 3.29.

```
; MC68HC705J1A
;input 8 bits, Reg A=data
in8    ldx  #8       ;8 bits
       lda  DDRB
       and  #$FE      ;DQ input
       sta  DDRB
ilop   bclr 1,PORTB ;CLK=0
       lsra           ;lsb first
       brclr 0,PORTB,inext
       ora  #$80      ;DQ=1
inext  bset 1,PORTB ;CLK=1
       decx
       bne  ilop
       tax
       lda  DDRB
       ora  #$01    ;DQ output
       sta  DDRB
       txa
       rts

;Reg A is config value
ReadConfig bset 3, PORTB ;RST=1
       lda  #$AC
       bsr  out8
       bsr  in8
       bclr 3, PORTB ;RST=0
       rts
```

```
; MC68HC11A8
;input 8 bits, Reg A=data
in8    ldy  #8         ;8 bits
       ldaa DDRD
       anda #$F7      ;DQ input
       staa DDRD
ilop   bclr 0,x,#$10 ;CLK=0
       lsra            ;lsb first
       brclr 0,x,#$08,inext
       oraa #$80 ;DQ=1
inext  bset 0,x,#$10 ;CLK=1
       dey
       bne  ilop
       ldab DDRD
       orab #$08       ;DQ output
       stab DDRD
       rts
;Reg A is config value
ReadConfig ldx #PORTD
       bset 0,x,#$20 ;RST=1
       ldaa #$AC
       bsr  out8
       bsr  in8
       bclr 0,x,#$20 ;RST=0
       rts
```

```
; MC68HC708XL36
;input 8 bits, Reg A=data
in8    ldx  #8       ;8 bits
       lda  DDRB
       and  #$FE      ;DQ input
       sta  DDRB
ilop   bclr 1,PORTB ;CLK=0
       lsra           ;lsb first
       brclr 0,PORTB,inext
       ora  #$80      ;DQ=1
inext  bset 1,PORTB ;CLK=1
       decx
       bne  ilop
       tax
       lda  DDRB
       ora  #$01    ;DQ output
       sta  DDRB
       txa
       rts
;Reg A is config value
ReadConfig bset 3,PORTB ;RST=1
       lda  #$AC
       bsr  out8
       bsr  in8
       bclr 3,PORTB ;RST=0
       rts
```

```
; MC68HC812A4/MC68HC912B32
;input 8 bits, Reg A=data
in8     ldx  #8         ;8 bits
        ldaa DDRS
        anda #$DF      ;DQ input
        staa DDRS
ilop    bclr PORTS,#$40 ;CLK=0
        lsra             ;lsb
        brclr PORTS,#$20,inext
        oraa #$80         ;DQ=1
inext   bset PORTS,#$40 ;CLK=1
        dex
        bne  ilop
        ldab DDRS
        orab #$20         ;DQ output
        stab DDRS
        rts
;Reg A is config value
ReadConfig
        bset PORTS,#$80 ;RST=1
        ldaa #$AC
        bsr  out8
        bsr  in8
        bclr PORTS,#$80 ;RST=0
        rts
```

Program 3.28 Assembly language functions to read the configuration register on the DS1620.

```
// MC68HC705J1A
unsigned char in8(void){ int n;
unsigned char result;
  DDRB &= 0xFE; // PB0=DQ input
  for(n=0;n<8;n++){
     PORTB &= 0xFD;   // PB1=CLK=0
     result = result>>1;
     if(PORTB&0x01)
        result |= 0x80; // PB0=DQ=1
     PORTB |= 0x02;}   // PB1=CLK=1
  DDRB |= 0x01;   // PB0=DQ=output
  return result;}
unsigned char ReadConfig(void){
unsigned char value;
  PORTB |= 0x04;   // PB2=RST=1
  out8(0xAC);
  value=in8();
  PORTD &= 0xFB;   // PB2=RST=0
  return value;}
```

```
// MC68HC11A8
unsigned char in8(void){ int n;
unsigned char result;
  DDRD &= 0xF7; // PD3=DQ input
  for(n=0;n<8;n++){
     PORTD &= 0xEF;   // PD4=CLK=0
     result = result>>1;
     if(PORTD&0x08)
        result |= 0x80; // PD3=DQ=1
     PORTD |= 0x10;}   // PD4=CLK=1
  DDRD |= 0x08; // PD3=DQ output
  return result;}
unsigned char ReadConfig(void){
unsigned char value;
  PORTD |= 0x20;   // PD5=RST=1
  out8(0xAC);
  value=in8();
  PORTD &= 0xDF;   // PD5=RST=0
  return value;}
```

```
// MC68HC708XL36
unsigned char in8(void){ int n;
unsigned char result;
  DDRB &= 0xFE; // PB0=DQ input
  for(n=0;n<8;n++){
     PORTB &= 0xFD;   // PB1=CLK=0
     result = result>>1;
     if(PORTB&0x01)
        result |= 0x80; // PB0=DQ=1
     PORTB |= 0x02;}   // PB1=CLK=1
  DDRB |= 0x01;   // PB0=DQ=output
  return result;}
unsigned char ReadConfig(void){
unsigned char value;
  PORTB |= 0x04;   // PB2=RST=1
  out8(0xAC);
  value=in8();
  PORTD &= 0xFB;   // PB2=RST=0
  return value;}
```

```
// MC68HC812A4/MC68HC912B32
unsigned char in8(void){ int n;
unsigned char result;
  DDRS &= 0xDF; // PS5=DQ input
  for(n=0;n<8;n++){
     PORTS &= 0xBF;   // PS6=CLK=0
     result = result>>1;
     if(PORTS&0x20)
        result |= 0x80; // PS5=DQ=1
     PORTS |= 0x40;}   // PS6=CLK=1
  DDRS |= 0x20; // PS5=DQ output
  return result;}
unsigned char ReadConfig(void){
unsigned char value;
  PORTS |= 0x80;   // PS7=RST=1
  out8(0xAC);
  value=in8();
  PORTS &= 0x7F;   // PS7=RST=0
  return value;}
```

Program 3.29 C language functions to read the configuration register on the DS1620.

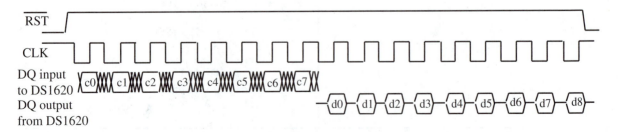

Figure 3.29 Synchronous serial timing diagram for the DS1620 read data.

The last type of communication involves reading temperature values from the DS1620. In this case, bits c0, c1, c2, c3, c4, c5, c6, and c7 form the `Read TH` command ($A1), the `Read TL` command ($A2), or the `Read Temperature` command ($AA). The bits d0, d1, d2, d3, d4, d5, d6, d7, and d8 are the current value of the corresponding temperature register (Figure 3.29).

The helper subroutine, `in9`, receives a 9-bit data frame from the DS1620 (Program 3.30).

Program 3.30

Assembly language 9-bit read helper function for the DS1620.

```
; MC68HC705J1A
;input 9 bits, Reg A:X=data
in9    ldx   #9        ;9 bits
       lda   DDRB
       and   #$FE      ;DQ input
       sta   DDRB
       clra            ;ms byte
jlop   bclr  1,PORTB   ;CLK=0
       lsra            ;lsb first
       ror   temp9
       brclr 0,PORTB,jnext
       ora   #$80      ;DQ=1
jnext  bset  1,PORTB   ;CLK=1
       decx
       bne   jlop
       tax
       lda   DDRB
       ora   #$01      ;DQ output
       sta   DDRB
       txa
       ldx   temp9
       rts
```

```
; MC68HC11A8
;input 9 bits, Reg D=data
in9    ldy   #9         ;9 bits
       ldaa  DDRD
       anda  #$F7       ;DQ input
       staa  DDRD
       clra
jlop   bclr  0,x,#$10  ;CLK=0
       lsrd             ;lsb first
       brclr 0,x,#$08,jnext
       oraa  #$01 ;DQ=1
jnext  bset  0,x,#$10  ;CLK=1
       dey
       bne   jlop
       psha
       ldaa  DDRD
       oraa  #$08       ;DQ output
       staa  DDRD
       pula
       rts
```

```
; MC68HC708XL36
;input 9 bits, Reg A:X=data
in9    ldx   #9        ;9 bits
       lda   DDRB
       and   #$FE      ;DQ input
       sta   DDRB
       clra            ;ms byte
jlop   bclr  1,PORTB   ;CLK=0
       lsra            ;lsb first
       ror   temp9
       brclr 0,PORTB,jnext
       ora   #$80      ;DQ=1
jnext  bset  1,PORTB   ;CLK=1
       decx
       bne   jlop
       tax
       lda   DDRB
       ora   #$01      ;DQ output
       sta   DDRB
       txa
       ldx   temp9
       rts
```

```
; MC68HC812A4/MC68HC912B32
;input 9 bits, Reg D=data
in9    ldx   #9         ;9 bits
       ldaa  DDRS
       anda  #$DF       ;DQ input
       staa  DDRS
       clra
jlop   bclr  PORTS,#$40 ;CLK=0
       lsrd             ;lsb
       brclr PORTS,#$20,jnext
       oraa  #$01       ;DQ=1
jnext  bset  PORTS,#$40 ;CLK=1
       dex
       bne   ilop
       ldab  DDRS
       orab  #$20       ;DQ output
       stab  DDRS
       rts
```

The last three assembly subroutines allow you to read 9-bit temperature values from the DS1620 (Program 3.31).

Program 3.31
Assembly language functions to read the temperatures from the DS1620.

```
; MC68HC705J1A
;Reg A:X returned as TH value
ReadTH bset 3,PORTB ;RST=1
       lda  #$A1
       bsr  out8
       bsr  in9
       bclr 3,PORTB ;RST=0
       rts
;Reg A:X returned as TL value
ReadTL bset 3,PORTB ;RST=1
       lda  #$A2
       bsr  out8
       bsr  in9
       bclr 3,PORTB ;RST=0
       rts
;Reg A:X returned temperature
ReadT  bset 3,PORTB ;RST=1
       lda  #$AA
       bsr  out8
       bsr  in9
       bclr 3,PORTB ;RST=0
       rts
```

```
; MC68HC11A8
;Reg D returned as TH value
ReadTH ldx  #PORTD
       bset 0,x,#$20 ;RST=1
       ldaa #$A1
       bsr  out8
       bsr  in9
       bclr 0,x,#$20 ;RST=0
       rts
;Reg D returned as TL value
ReadTL ldx  #PORTD
       bset 0,x,#$20 ;RST=1
       ldaa #$A2
       bsr  out8
       bsr  in9
       bclr 0,x,#$20 ;RST=0
       rts
;Reg D returned as Temperature
ReadT  ldx  #PORTD
       bset 0,x,#$20 ;RST=1
       ldaa #$AA
       bsr  out8
       bsr  in9
       bclr 0,x,#$20 ;RST=0
       rts
```

```
; MC68HC708XL36
;Reg A:X returned as TH value
ReadTH bset 3,PORTB ;RST=1
       lda  #$A1
       bsr  out8
       bsr  in9
       bclr 3,PORTB ;RST=0
       rts
;Reg A:X returned as TL value
ReadTL bset 3,PORTB ;RST=1
       lda  #$A2
       bsr  out8
       bsr  in9
       bclr 3,PORTB ;RST=0
       rts
;Reg A:X returned temperature
ReadT  bset 3,PORTB ;RST=1
       lda  #$AA
       bsr  out8
       bsr  in9
       bclr 3,PORTB ;RST=0
       rts
```

```
; MC68HC812A4/MC68HC912B32
;Reg D returned as TH value
ReadTH  bset PORTS,#$80 ;RST=1
        ldaa #$A1
        bsr  out8
        bsr  in9
        bclr PORTS,#$80 ;RST=0
        rts
;Reg D returned as TL value
ReadTL  bset PORTS,#$80 ;RST=1
        ldaa #$A2
        bsr  out8
        bsr  in9
        bclr PORTS,#$80 ;RST=0
        rts
;Reg D returned as temperature
ReadT   bset PORTS,#$80 ;RST=1
        lda  #$AA
        bsr  out8
        bsr  in9
        bclr PORTS, ;RST=0
        rts
```

The helper C function, `in9`, receives a 9-bit data frame from the DS1620 (Program 3.32).

```
// MC68HC705J1A
unsigned int in9(void){ int n;
unsigned int result=0;
  DDRB &= 0xFE; // PB0=DQ input
  for(n=0;n<9;n++){
    PORTB &= 0xFD;   // PB1=CLK=0
    result = result>>1;
    if(PORTB&0x01)
      result |= 0x0100; // PB0=DQ=1
    PORTB |= 0x02;}  // PB1=CLK=1
  DDRB |= 0x01;    // PB0=DQ=output
  return result;}
```

```
// MC68HC11A8
unsigned int in9(void){ int n;
unsigned int result=0;
  DDRD &= 0xF7; // PD3=DQ input
  for(n=0;n<9;n++){
    PORTD &= 0xEF;   // PD4=CLK=0
    result = result>>1;
    if(PORTD&0x08)
      result |= 0x0100; // PD3=DQ=1
    PORTD |= 0x10;}  // PD4=CLK=1
  DDRD |= 0x08; // PD3=DQ output
  return result;}
```

```
// MC68HC708XL36
unsigned int in9(void){ int n;
unsigned int result=0;
  DDRB &= 0xFE; // PB0=DQ input
  for(n=0;n<9;n++){
    PORTB &= 0xFD;   // PB1=CLK=0
    result = result>>1;
    if(PORTB&0x01)
      result |= 0x0100; // PB0=DQ=1
    PORTB |= 0x02;}  // PB1=CLK=1
  DDRB |= 0x01;    // PB0=DQ=output
  return result;}
```

```
// MC68HC812A4/MC68HC912B32
unsigned int in9(void){ int n;
unsigned int result=0;
  DDRS &= 0xDF; // PS5=DQ input
  for(n=0;n<9;n++){
    PORTS &= 0xBF;   // PS6=CLK=0
    result = result>>1;
    if(PORTS&0x20)
      result |= 0x0100; // PS5=DQ=1
    PORTS |= 0x40;}  // PS6=CLK=1
  DDRS |= 0x20; // PS5=DQ output
  return result;}
```

Program 3.32 C language 9-bit read helper function for the DS1620.

This last interface can be implemented in C as shown in Program 3.33.

Program 3.33
C language functions to read the temperatures from the DS1620.

```
// MC68HC705J1A
unsigned int ReadTH(void){
unsigned int value;
    PORTB |= 0x04;   // PB2=RST=1
    out8(0xA1);
    value=in9();
    PORTD &= 0xFB;   // PB2=RST=0
    return value;}
unsigned int ReadTL(void){
unsigned int value;
    PORTB |= 0x04;   // PB2=RST=1
    out8(0xA2);
    value=in9();
    PORTD &= 0xFB;   // PB2=RST=0
    return value;}
unsigned int ReadT(void){
unsigned int value;
    PORTB |= 0x04;   // PB2=RST=1
    out8(0xAA);
    value=in9();
    PORTD &= 0xFB;   // PB2=RST=0
    return value;}
```

```
// MC68HC11A8
unsigned int ReadTH(void){
unsigned int value;
    PORTD |= 0x20;   // PD5=RST=1
    out8(0xA1);
    value=in9();
    PORTD &= 0xDF;   // PD5=RST=0
    return value;}
unsigned int ReadTL(void){
unsigned int value;
    PORTD |= 0x20;   // PD5=RST=1
    out8(0xA2);
    value=in9();
    PORTD &= 0xDF;   // PD5=RST=0
    return value;}
unsigned int ReadT(void){
unsigned int value;
    PORTD |= 0x20;   // PD5=RST=1
    out8(0xAA);
    value=in9();
    PORTD &= 0xDF;   // PD5=RST=0
    return value;}
```

continued on p. 178

continued from p. 177

Program 3.33
C language functions to read the temperatures from the DS1620.

```
// MC68HC708XL36
unsigned int ReadTH(void){
unsigned int value;
   PORTB |= 0x04;    // PB2=RST=1
   out8(0xA1);
   value=in9();
   PORTD &= 0xFB;    // PB2=RST=0
   return value;}
unsigned int ReadTL(void){
unsigned int value;
   PORTB |= 0x04;    // PB2=RST=1
   out8(0xA2);
   value=in9();
   PORTD &= 0xFB;    // PB2=RST=0
   return value;}
unsigned int ReadT(void){
unsigned int value;
   PORTB |= 0x04;    // PB2=RST=1
   out8(0xAA);
   value=in9();
   PORTD &= 0xFB;    // PB2=RST=0
   return value;}
```

```
// MC68HC812A4/MC68HC912B32
unsigned int ReadTH(void){
unsigned int value;
   PORTS |= 0x80;    // PS7=RST=1
   out8(0xA1);
   value=in9();
   PORTS &= 0x7F;    // PS7=RST=0
   return value;}
unsigned int ReadTL(void){
unsigned int value;
   PORTS |= 0x80;    // PS7=RST=1
   out8(0xA2);
   value=in9();
   PORTS &= 0x7F;    // PS7=RST=0
   return value;}
unsigned int ReadT(void){
unsigned int value;
   PORTS |= 0x80;    // PS7=RST=1
   out8(0xAA);
   value=in9();
   PORTS &= 0x7F;    // PS7=RST=0
   return value;}
```

3.5 Glossary

bandwidth The information transfer rate, the amount of data transferred per second.

blind cycle A software/hardware synchronization method where the software waits a specified amount of time for the hardware operation to complete. The software has no direct information (blind) about the status of the hardware.

buffered I/O A FIFO queue is placed in between the hardware and software in an attempt to increase bandwidth by allowing both hardware and software to run in parallel.

busy waiting Same as gadfly.

CPU bound A situation where the input or output device is faster than the software. In other words, it takes less time for the I/O device to process data than for the software to process data. The hardware waits for the software.

device driver A collection of software routines that perform I/O functions.

direction register A bidirectional port configuration register that determines if the port will be an input, an output, or a combination of inputs and outputs.

DMA Direct memory access is a software/hardware synchronization method where the hardware itself causes a data transfer between the I/O device and memory at the appropriate time when data needs to be transferred. The software usually can perform other work while waiting for the hardware. No software action is required for each individual byte.

gadfly A software/hardware synchronization method where the software continuously reads the hardware status while waiting for the hardware operation to complete. The software usually performs no work while waiting for the hardware. Same as busy waiting.

handshake A software/hardware synchronization method where control and status signals go both directions between the transmitter and receiver. The communication is interlocked, meaning each device will wait for the other.

IEEE488 A medium-speed handshaking parallel I/O standard used for desktop instruments.

interrupt A software/hardware synchronization method where the hardware causes a special software program (interrupt handler) to execute when its operation is complete. The software usually can perform other work while waiting for the hardware.

I/O bound A situation where the input or output device is slower than the software. In other words, it takes longer for the I/O device to process data than for the software to process data. The software waits for the hardware.

I/O port A hardware device that connects the computer with external components.

latched input port An input port where the signals are latched with hardware (saved) on an edge of an associated strobe signal.

latency The response time of the computer to external events. For example, the time between new input becoming available and the time the input is read by the computer. For example, the time between an output device becoming idle and the time the input in the computer writes new data to it.

open collector A digital logic output that has two states, low and off.

parallel port A port where all signals are available simultaneously. In this book the parallel ports are 8 bits wide.

periodic polling A software/hardware synchronization method that is a combination of interrupts and gadfly. An interrupt occurs at a regular rate (periodic) independent of the hardware status. The interrupt handler checks the hardware device (polls) to determine if its operation is complete. The software usually can perform other work while waiting for the hardware.

port Same as I/O port.

priority When two requests for service are made simultaneously, priority determines in which order to process them.

real time A system that can guarantee an upper bound (worst case) on latency.

ritual Software, usually executed once at the beginning of the program, that defines the operational modes of the I/O ports.

SCSI Small Computer Systems Interface, a high-speed handshaking parallel I/O standard.

serial port A port where all signals are available 1 bit at a time.

throughput The information transfer rate, the amount of data transferred per second.

tristate The state of a tristate logic output when off or not driven.

tristate logic A digital logic device that has three output states: low, high, and off.

unbuffered I/O The hardware and software are tightly coupled so that both wait for each other during the transmission of data.

3.6 Exercises

3.1 a) In this problem you will write a function that outputs data to the printer in Figure 3.30 using a gadfly handshake protocol. You may write the software in assembly or C. The following sequence will print one ASCII character:

Figure 3.30
Simple printer interface.

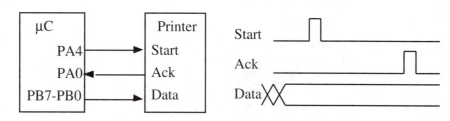

1. The microcomputer puts the 8-bit ASCII on the `Data` lines
2. The microcomputer issues a `Start` pulse (does not matter how wide)
3. The microcomputer waits for the `Ack` pulse (printer is done)

You do not have to save/restore registers. You may assume the `Ack` pulse is larger than 10 μs. The 8-bit ASCII data to print is *passed by value on the stack*. An example calling sequence is

```
ldab.#.V
pshb          The.ASCII  V  is pushed on the stack
jsr  Output
ins           The  V  is.discarded
```

Full credit will be given to the solution that includes the appropriate use of the `bset bclr brset` and `brclr` instructions. If you write in C, pass by value with the following prototype:

```
void Output(unsigned char);
```

b) How long is your `Start` pulse? Explain your calculation.

3.2 a) What is a trap?
b) Why do we use traps? Give an example.

3.3 What are five different methods for I/O synchronization?

3.4 What does DMA stand for?

3.5 What is the biggest disadvantage of blind cycle counting?

3.6 What is "gadfly"?

3.7 What is the difference between regular interrupts and periodic polling?

3.8 What are the interrupt-related bits in CCR?

3.9 Mention three instructions that can be used to disable or enable interrupts.

3.10 Can you explain what is meant by "saving program context"? Show the stack contents after context is saved, if an interrupt is recognized while executing a LSLA instruction located at address \$e010. The contents of the processor registers are: X=\$ab12, Y=\$cd34, A=\$55, B=\$66, CCR=\$04.

3.11 In all your programs, you included some assembler directives depositing the program start address to locations \$FFFE & \$FFFF. For example:

```
org $FFFE
fdb Start
```

where Start is the starting address of your program. Explain the purpose of doing this.

3.12 The objective of this problem is to interface an input device to a 6811 (Figure 3.31) and implement a binary tree command interpreter. To solve this problem with a different microcomputer, you could substitute any available I/O port for STRA STRB PC7-PC0. You may write the software in assembly or C. The sequence to input an ASCII character from the input device is as follows. When a new character is available, the input device puts its ASCII code on the 8-bit

Figure 3.31
An input device interfaced to a 6811.

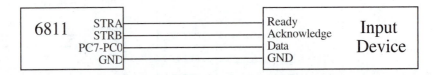

Figure 3.32
Timing diagram of the interface between an input device and a 6811.

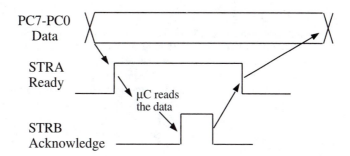

Data, then the input device makes   Ready=1. Next your software should read the 8-bit ASCII code, then acknowledge receipt of the data by pulsing Acknowledge. After the pulse, the input device will make Ready=0 again (Figure 3.32).

a) Show bit by bit your choice (what and why) for the parallel I/O control register.

b) Show the ritual that initializes the microcomputer.

c) Show the subroutine that inputs one data byte. The sequence of steps is described above. The input data are returned by value using RegB. Use gadfly synchronization. If you write in C, implement call by reference with the following prototype.

```
unsigned char Input(void);
```

To improve search time, your command interpreter will use a binary tree (Figure 3.33) instead of a table or single-linked list. The six subroutines NEXT, FORWARD, BACK, LEFT, RIGHT, STOP are given (you do not write these six subroutines.) Each node of the tree has one ASCII character (e.g., 'B', 'L', . . .), a 16-bit subroutine address (e.g., BACK, LEFT, . . .), and two pointers to its children. Notice that the tree is defined in alphabetical order. The command interpreter will input one character from the input device, search the tree, and if a match is found, the interpreter will execute the appropriate subroutine.

d) Show the pseudo-operations (e.g., FCC, FCB, FDB) that define the above binary tree. You may define a 'nil' pointer as any address that can not be a valid node pointer (e.g., 0, $1000, $FFFF)

e) Show the command interpreter that initializes Port C (calls the subroutine you wrote in part b, then repeats over and over the following sequence:

1. It inputs one ASCII character (calls the subroutine you wrote in part c)
2. It looks up that letter in the binary tree
3. If found then it executes the appropriate subroutine.

Figure 3.33
A linked binary tree data structure.

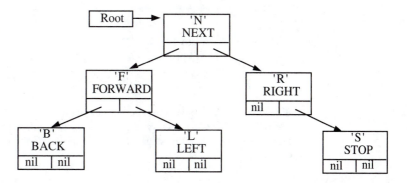

In this first example assume the input letter is 'L'. Your program initializes a pointer with the ROOT (it points to node N.) Since 'L'<'N', your program updates the pointer in the direction of nodes that are alphabetically before 'N' (now it points to node F). Since 'L'>'F', your program updates the pointer in the direction of nodes that are alphabetically after 'F' (now it points to node L). Since 'L'='L', your program executes the subroutine LEFT.

In this second example assume the input letter is 'P'. Your program initializes a pointer with the ROOT (it points to node N). Since 'P'>'N', your program updates the pointer in the direction of nodes that are alphabetically after 'N' (now it points to node R). Since 'P'<'R', your program updates the pointer in the direction of nodes that are alphabetically before 'R' (now it points to nil). Since the pointer is nil, there is no match.

3.13 The objective of this problem is to interface an input device to a 6811 single-chip computer using Port C (Figure 3.34). To solve this problem with a different microcomputer, you could substitute any available I/O port for STRA STRB PC7-PC0. You may write the software in assembly or C.

Figure 3.34
An input device
interfaced to a 6811.

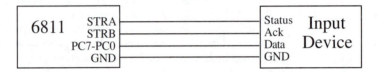

The timing shown in Figure 3.35 occurs when 1 byte is transferred.
a) Show bit by bit your choice (what and why) for the parallel I/O control register.
b) Show the ritual that initializes the microcomputer.
c) Show the subroutine that inputs one data byte. Use gadfly synchronization. If you write in assembly return the new data by reference using RegY. RegY points to the place to store the next data byte. If you write in C, implement call by reference with the following prototype:

```
void Input(unsigned char *);
```

Figure 3.35
Timing diagram of the
interface between an
input device and a 6811.

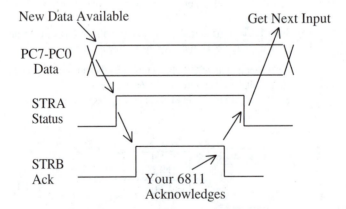

3.14 The objective of this problem is to interface an output device to a 6811 single-chip computer using Port C (Figure 3.36). To solve this problem with a different microcomputer, you could substitute any available I/O port for STRA STRB PC7-PC0. You may write the software in assembly or C.

Figure 3.36
An output device interfaced to a 6811.

First, the microcomputer outputs data on PC7-PC0. Then, the microcomputer makes STRB=1; 1 ms later the output device will make STRA=1 (the microcomputer waits for STRA=1). Then, the microcomputer makes STRB=0; 1 μs later the output device will make STRA=0 (the microcomputer need not check for STRA=0) (Figure 3.37).

  a) Show bit by bit your choice (what and why) for the parallel I/O control register.
  b) Show the ritual that initializes the microcomputer. Use assembly or C.
  c) Show the subroutine that outputs one data byte. The sequence of steps is described above. If you write in assembly, the output data are passed by value using Reg B. Use gadfly synchronization. If you write in C, implement call by value with the following prototype:

```
void OUTPUT(unsigned char);
```

Figure 3.37
Timing diagram of the interface between an output device and a 6811.

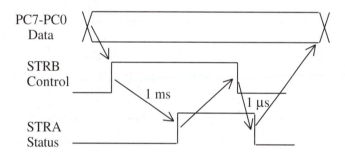

3.15 The objective of this problem is to interface a position transducer array to a microcomputer (Figure 3.38). You may use any available I/O port. Use gadfly synchronization. You may write the software in assembly or C. The sequence to read a 5-bit sensor position from the input device is as follows:

The microcomputer specifies which of eight sensors is to be read (sets N2,N1,N0)
The microcomputer tells the sensor array to read the position by a negative logic pulse on START

Figure 3.38
Timing diagram for a position sensor.

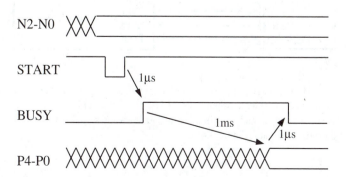

The sensor signals it is working by setting BUSY=1
After about 1 ms the 5-bit position is available on the lines P4, P3, P2, P1, P0
The sensor signals it is done by setting BUSY=0
The microcomputer reads the 5-bit position

a) Show the connections between the position transducer array and the microcomputer. Label chip numbers but not pin numbers of any required digital logic (Figure 3.39).

b) Show bit by bit your choice (what and why) for the parallel I/O control register. Specify 0 for bits that must be 0. Specify 1 for bits that must be 1. Specify X for bits that can be set in the ritual to be either 0 or 1. The following program illustrates the calling sequence of the RITUAL and SENSOR functions:

Figure 3.39
An eight-channel sensor array.

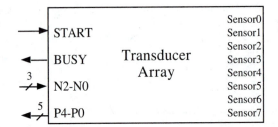

```
unsigned char data[8];              /* Room for all 8 results       */
void main(void) { unsigned char N;  /* Sensor position 0,1,2..., 7  */
     RITUAL();                       /* Initialize uC Part c)        */
     for (N=0;N<8;N++)
          data[N]=SENSOR(N);         /* Read sensor N Part d)        */
}
```

c) Show the RITUAL() procedure that initializes the microcomputer.

d) Show the SENSOR(N) function that reads one data byte. The sequence of steps is described above. Use gadfly synchronization.

3.16 The objective of this problem is to interface an I/O device to a 6811 (Figure 3.40). The same interface must be capable of both input and output. To solve this problem with a different microcomputer, you could substitute any available I/O port for STRA STRB PC7-PC0. You may write the software in assembly or C.

Figure 3.40
An I/O device interfaced to a 6811.

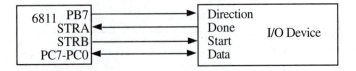

The sequence to receive an 8-bit data value from the input device to the microcomputer is as follows. First the microcomputer sets Direction=1 and issues a Start pulse. When the input device has the data ready, it will drive its data lines and signal with Done=1. The microcomputer should then read the data (Figure 3.41).

Figure 3.41
Input timing diagram of
the interface between
an I/O device and a
6811.

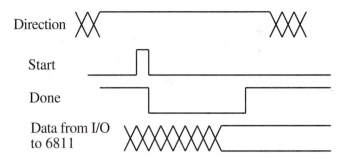

a) Show the C (or assembly language) procedure that initializes the microcomputer.

b) Show the C (or assembly language) procedure that inputs one data byte. Assume the ritual in part a has been called. The sequence of steps is described above. The input data is returned by value. Use gadfly synchronization. If you write in C, implement call by reference with the following prototype:

```
void Input(unsigned char *);
```

The sequence to transmit an 8-bit data value from the microcomputer to the output device is as follows. First the microcomputer sets Direction=0, places the desired 8-bit information on the data lines, and issues a Start pulse. When the output device is complete, it signals with Done=1. The microcomputer should then make its data output hiZ (tristate) (Figure 3.42).

Figure 3.42
Output timing diagram
of the interface between
an I/O device and a
6811.

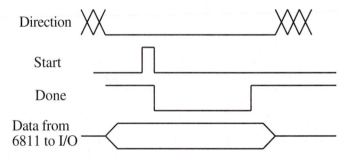

c) Show the C (or assembly language) procedure that outputs one data byte. Assume the ritual in part a has been called. The sequence of steps is described above. The output data is passed by value. Use gadfly synchronization. If you write in C pass by value with the following prototype:

```
void Output(unsigned char);
```

3.17 The objective of this problem is to interface an input device to a 6811 (Figure 3.43). To solve this problem with a different microcomputer, you could substitute any available I/O port for STRA STRB PC7-PC0. You may write the software in assembly or C.

Figure 3.43
An input device
interfaced to a 6811.

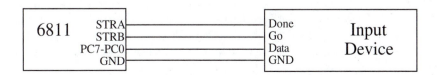

The sequence to input an ASCII character from the input device is as follows. When a new character is available, the input device puts its ASCII code on the 8-bit Data, then the input device makes Done=1. The hardware of the 6811 Port C should make Go=0 almost immediately. Recognizing that Done=1, your software should read the 8-bit ASCII code, then acknowledge receipt of the data by making Go=1 again. After the Go=1, the input device will make Done=0 again (Figure 3.44).

Figure 3.44
Timing diagram of the interface between an input device and a 6811.

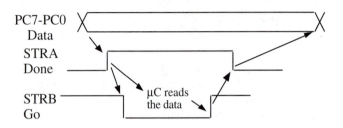

a) Show bit by bit your choice (what and why) for the parallel I/O control register. Specify that 1 for this bit must be 1, 0 for this bit must be 0, and X for this bit can be written with either 0 or 1.
b) Show the C program (ritual) that initializes the microcomputer.
c) Show the C procedure that inputs one data byte. The sequence of steps is described above. The input data is returned by value. Use gadfly synchronization. If you write in C, implement a function return value with the following prototype:

```
unsigned char Input(void);
```

3.18 The objective of this problem is to interface the following input device to a microcomputer single-chip computer using any available I/O port (Figure 3.45). The figure shows the connections to Port C. You may write the software in assembly or C.

Figure 3.45
An input device interfaced to a microcomputer.

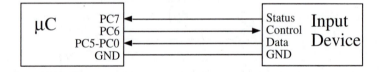

First, your microcomputer sets Control=1, meaning input is requested. Then, the input device will create a new 6-bit input and place it on Data. The input device will then make Status=1 signifying new data available. Next, your microcomputer program should read the 6-bit input value. Your program should then make Control=0, meaning it has read the data. The input device will then make Status=0, signifying the transfer is complete. Your software should wait until Status=0 before returning (Figure 3.46).

a) Show your choice in hexadecimal for data direction register, e.g., DDRC.
b) Show the assembly language subroutine that inputs one 6-bit data. The sequence of steps is described above. The input data is returned by value in RegA. Use gadfly synchronization. Full credit will be given to the subroutine that appropriately uses the bset bclr brset and/or brclr assembly language instructions. If you write in C, implement a function return value with the following prototype:

```
unsigned char Input(void);
```

Figure 3.46
Timing diagram of the interface between an input device and a microcomputer.

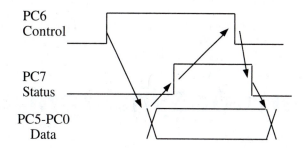

PC6
Control

PC7
Status

PC5-PC0
Data

3.19 These questions refer specifically to the 6811 PIOC.

   a) What are the different bits in the 6811 PIOC? Name each bit and describe the purpose of each bit.

   b) What is the address of PIOC?

   c) Which bit in PIOC enables handshaking mode?

   d) Which bit in PIOC is not writable? What operation sets this bit? What operation clears this bit?

   e) Which two pins on the 6811 are used specifically for handshaking?

   f) How can you generate a strobe pulse on the STRB pin on the 6811?

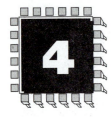

4 Interrupt Synchronization

Chapter 4 objectives are to:

❑ Introduce the concept of interrupt synchronization
❑ Discuss the issues involved in reentrant programming
❑ Implement and apply the FIFO circular queue
❑ Discuss the specific details of using interrupts on the 6805, 6808, 6811, and 6812
❑ Interface a keyboard and printer using external IRQ interrupts
❑ Show a high-priority/low-latency XIRQ power failure interface
❑ Define mechanisms to establish priority including round robin
❑ Design and implement background I/O for simple devices using periodic polling

There are many reasons to consider interrupt synchronization. The first consideration is that the software in a real-time system must respond to hardware events within a prescribed time. To illustrate the need for interrupts, consider a keyboard interface where the time between new keyboard inputs might be as small as 10 ms. In this situation, the software latency is the time from when the new keyboard input is ready until the time the software reads the new data. To prevent loss of data in this case, the software latency must be less than 10 ms. We can implement real-time software using gadfly polling only when the size and complexity of the system are very small. Interrupts are important for these real-time systems because they provide a mechanism to guarantee an upper bound on the software response time. Interrupts also give us a way to respond to infrequent but important events. Alarm conditions like low battery power and error conditions can be handled with interrupts. Periodic interrupts, generated by the timer at a regular rate, will be necessary to implement data acquisition and control systems. In the unbuffered interfaces of Chapter 3, the hardware and software took turns waiting for each other. Interrupts provide a way to buffer the data so that the hardware and software spend less time waiting. In particular, the buffer we will use is a FIFO queue placed between the interrupt routine and the main program to increase the overall bandwidth. We will begin our discussion with general issues, then present the specific details about the 6805, 6808, 6811, and 6812 microcomputers. After that, a number of simple interrupt examples will be presented, and at the end of the chapter we will discuss some advanced concepts like priority, round-robin polling, and periodic polling.

4.1   What Are Interrupts?

4.1.1
Interrupt
Definition

An *interrupt* is the automatic transfer of software execution in response to hardware that is asynchronous with the current software execution. The hardware can be either an external I/O device (like a keyboard or printer) or an internal event (like an opcode fault or a periodic timer). When the hardware needs service (busy-to-done state transition), it will request an interrupt. Recall from Chapter 2 that a *thread* is defined as the path of action of software as it executes. The execution of the interrupt service routine is called a *background* thread. This thread is created by the hardware interrupt request and is killed when the interrupt service routine executes the `rti` instruction. A new thread is created for each interrupt request. It is important to consider each individual request as a separate thread because local variables and registers used in the interrupt service routine are unique and separate from one interrupt event to the next. In a multithreaded system we consider the threads as cooperating to perform an overall task. Consequently we will develop ways for the threads to communicate (see Section 4.3 on FIFO queues) and synchronize (see the discussion of semaphores in Section 5.2) with each other. Most embedded systems have a single common overall goal. On the other hand, general-purpose computers can have multiple unrelated functions to perform. A *process* is also defined as the action of software as it executes. The difference is processes do not necessarily cooperate toward a common shared goal. Threads share access to global variables, while processes have separate globals.

The software has dynamic control over aspects of the interrupt request sequence. First, each potential interrupt source has a separate arm bit that the software can activate or deactivate. The software will set the arm bits for those devices it wishes to accept interrupts from, and it will deactivate the arm bits within those devices from which interrupts are not to be allowed. In other words, it uses the arm bits to individually select which devices will and which devices will not request interrupts. The second aspect that the software controls is the interrupt enable bit, I, which is in the condition code register. The software can enable all armed interrupts by setting I=0, or it can disable all interrupts by setting I=1. The disabled interrupt state (I=1) does not dismiss the interrupt requests; rather it postpones them until a later time, when the software deems it convenient to handle the requests. We will pay special attention to these enable/disable software actions. In particular we will need to disable interrupts when executing nonreentrant code, but disabling interrupts will have the effect of increasing the response time of software.

There are two general methods with which we configure external hardware so that it can request an interrupt. The first method is a shared negative-logic-level-active request like $\overline{IRQ}$. All the devices that need to request interrupts have an open-collector negative-logic interrupt request line. The hardware requests service by pulling the interrupt request $\overline{IRQ}$ line low. The line over the IRQ signifies negative logic. In other words, an interrupt is requested when $\overline{IRQ}$ is zero. Because the request lines are open-collector, a pull-up resistor is needed to make $\overline{IRQ}$ high when no devices need service (Figure 4.1).

Normally these interrupt requests share the same interrupt vector. This means that whichever device requests an interrupt, the same interrupt service routine is executed. Therefore the interrupt service routine must first determine which device requested the interrupt. The 6811 and 6812 have two of these shared negative-logic-level-active interrupts, $\overline{IRQ}$ and $\overline{XIRQ}$. XIRQ interrupts on the 6811 and 6812 are nonmaskable. In other words, once they are enabled (the X bit in the CCR is 0), they cannot be disabled.

Figure 4.1
Wire- or negative-logic
interrupt request line.

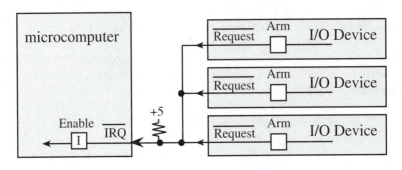

Figure 4.2
Dedicated edge-
triggered interrupt
request lines.

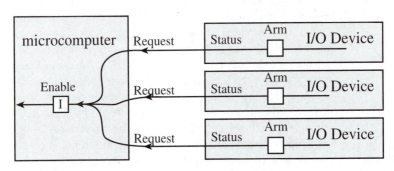

The second method uses multiple dedicated *edge-triggered* requests (Figure 4.2). In this method, each device has its own interrupt request line that is usually connected to a status signal in the I/O device. In this way, an interrupt is requested on the *busy-to-done* state transition. Edge-triggered means the interrupt is requested on the rise, the fall, or both the rise and the fall of the *Status* signal. With *vectored* interrupts these individual requests have a unique interrupt vector. This means there can be a separate interrupt service routine for each device. In this way the microcomputer will automatically execute the appropriate software interrupt handler when an interrupt is requested. Therefore the interrupt service routine does not need to determine which device requested the interrupt.

The original IBM PC had only eight dedicated edge-triggered interrupt lines, and the current IBM PC I/O bus has only 15. This small number can be a serious limitation in a computer system with many I/O devices.

Observation: Microcomputer systems running in expanded mode often use shared negative-logic-level-active interrupts for their external I/O devices.

Observation: Microcomputer systems running in single-chip mode often use dedicated edge-triggered interrupts for their I/O devices.

Observation: The number of interrupting devices on a system using dedicated edge-triggered interrupts is limited when compared to a system using shared negative-logic-level-active interrupts.

Observation: Most Motorola microcomputers support both shared negative-logic and dedicated edge-triggered interrupts.

The advantages of a wire- or negative-logic interrupt request are (1) additional I/O devices can be added without redesigning the hardware, (2) there is no fundamental limit to the number of interrupting I/O devices you can have, and (3) the microcomputer hardware is simple. The advantages of the dedicated edge-triggered interrupt request are (1) the software is simpler, therefore it will be easier to debug and it will run faster, (2) there will be less coupling between software modules, making it easier to debug and to reuse code, and (3) it will be easier to implement priority such that higher-priority requests are handled quickly, while lower-priority interrupt requests can be postponed. Because the Motorola microcomputers support both types of interrupt, the designer can and must address these considerations.

4.1.2
Interrupt Service
Routines

The interrupt service routine (ISR) is the software module that is executed when the hardware requests an interrupt. From Section 4.1.1 we see that there may be one large ISR that handles all requests (polled interrupts) or many small ISRs specific for each potential source of interrupt (vectored interrupts). The design of the ISR requires careful consideration of many factors that will be discussed in this chapter. When an interrupt is requested (and the device is armed and the I bit is zero), the microcomputer will service an interrupt:

1. The execution of the main program is suspended (the current instruction is finished[1])
2. The ISR, or background thread, is executed
3. The main program is resumed when the ISR executes `rti`

When the microcomputer accepts an interrupt request, it will automatically save the execution state of the main thread by pushing all its registers on the stack. After the ISR provides the necessary service, it will execute a `rti` instruction. This instruction pulls the registers from the stack, which returns control to the main program. Execution of the main program will then continue with the exact stack and register values that existed before the interrupt. Although interrupt handlers can allocate, access, then deallocate local variables, parameters passing between threads must be implemented using global memory variables. Global variables are also required if an interrupt thread wishes to pass information to itself, (e.g., from one interrupt instance to another). The execution of the main program is called the *foreground thread,* and the executions of ISRs are called *background threads* (Figure 4.3).

Figure 4.3
An interrupt causes the main thread to be suspended, and the interrupt thread is run.

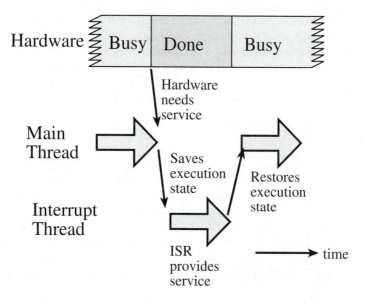

---

[1]The 6812 instructions `rev revw` and `wav` can be interrupted in the middle of their execution.

Table 4.1
Each synchronization
method has motivations
for its use.

Gadfly	Interrupts	DMA
Predicable	Variable arrival times	Low latency
Simple I/O	Complex I/O, different speeds	High bandwidth
Fixed load	Variable load	
Dedicated, single thread	Other functions to do	
Single process	Multithread or multiprocess	
Nothing else to do	Infrequent but important alarms	
	Program errors	
	Overflow, invalid opcode	
	Illegal stack or memory access	
	Machine errors	
	Power failure, memory fault	
	Breakpoints for debugging	
	Real-time clocks	
	Data acquisition and control	

4.1.3
When to Use
Interrupts

The factors listed in Table 4.1 should be considered when deciding the most appropriate mechanism to synchronize hardware and software. One should not always use gadfly just because one is too lazy to implement the complexities of interrupts. On the other hand, one should not always use interrupts just because they are fun and exciting. Direct Memory Access (DMA) will be covered in Chapter 10.

4.1.4
Interthread
Communication

For regular function calls we use the registers and stack to pass parameters, but interrupt threads have logically separate resisters and stack. In particular, all registers are automatically saved by the microcomputer[2] as it switches from the main program (foreground thread) to the ISR (background thread). The `rti` instruction will restore the registers (including the interrupt enable bits and the PC) back to their previous values. Thus, all parameter passing must occur through global memory. One cannot pass data from the main program to the ISR using registers or the stack. The classic producer/consumer problem has two threads. One thread produces data, and the other consumes data. For an input device, the background thread is the producer because it generates new data, and the foreground thread is the consumer because it uses the data up. For an output device, the data flows in the other direction so that the producer/consumer roles are reversed. It is appropriate to pass data from the producer thread to the consumer thread using a FIFO queue (Figure 4.4).

> **Observation:** For systems with interrupt-driven I/O on multiple devices, there will be a separate FIFO for each device.

An input device needs service (busy-to-done state transition) when new data are available. The ISR (background) will accept the data and process it. If more data is desired, the ISR will restart the input hardware (done-to-busy state transition).

An output device needs service (busy-to-done state transition) when the device is idle, ready to output more data. The ISR (background) will get more data and output it. The output function will restart the hardware (done-to-busy state transition). For this output device example, it would be appropriate to GET more data from a FIFO queue and send it to the device. Two particular problems with output device interrupts are:

---

[2]The 6808 does not save its Register H so that it can be consistent with the 6805, which has no Register H.

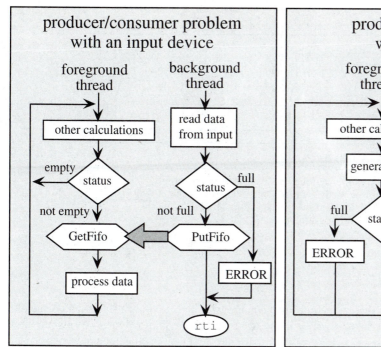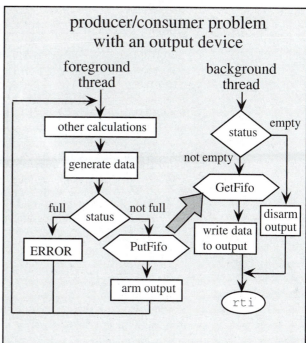

Figure 4.4   FIFO queues can be used to pass data between threads.

1. How does one generate the first interrupt? In other words, how does one start the output thread?

2. What does one do if an output interrupt occurs (device is idle) but there are no more data currently available (e.g., FIFO is empty)?

These problems can be solved by arming and disarming when needed.

The foreground thread (main program) executes a loop and accesses the FIFO when it needs to input or output data. The background threads (interrupts) are executed when the hardware needs service. For an input device, an interrupt is requested (causing the ISR to be executed) when new input data are available. The busy-to-done state transition causes an interrupt. These hardware state transitions are illustrated in Figures 4.5 and 4.8.

Figure 4.5
The input device
interrupts the computer
when it has new data.

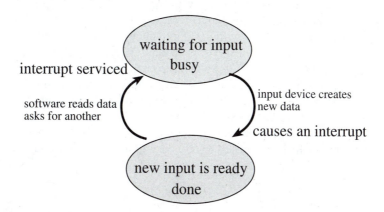

Figure 4.6
Hardware/software timing of an I/O-bound input interface.

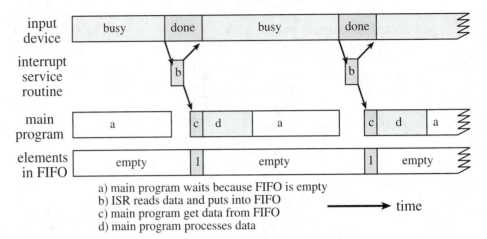

a) main program waits because FIFO is empty
b) ISR reads data and puts into FIFO
c) main program get data from FIFO
d) main program processes data

One way to visualize the interrupt synchronization is to draw a state versus time plot of the activities of the hardware and the two software modules (Figure 4.6). For an input device, the main thread begins by waiting for new input. When the input device is busy, it is in the process of creating new input. When the input device is done, new data are available and an interrupt is requested. The ISR will read the data and put them into the FIFO. Once data are in the FIFO, the main program is released to go on. The arrows from one graph to the other represent the synchronizing events. In this first example, the time for the software to read and process the data is less than the time for the input device to create new input. This situation is called *I/O-bound*. In this situation, the FIFO has either 0 or 1 entry, and the use of interrupts does not enhance the bandwidth over the gadfly implementations presented in Chapter 3. Even with an I/O-bound device it may be more efficient to utilize interrupts because it provides a straightforward approach to servicing multiple devices.

In this second example, the input device starts with a burst of high-bandwidth activity (Figure 4.7). As long as the ISR is fast enough to keep up with the input device and as long as the FIFO does not become full, no data are lost. In this situation, the overall bandwidth is higher than it would be with a gadfly implementation, because the input device does not have to wait for each data byte to be processed. This is the classic example of a *buffered* input, because data enter the system (via the interrupts), are temporarily stored in a buffer (put into the FIFO), and are processed later (by the main program, get from the FIFO). When the I/O device is faster than the software, the system is called *CPU-bound*. As we will see later, this system can work if the producer rate temporarily exceeds the consumer rate (a short burst of high-bandwidth input). If the external device sustained the high-bandwidth input rate, then the FIFO would become full and data would be lost.

For an output device, the interrupt is requested when the output is idle and ready to accept more data (Figure 4.8). The busy-to-done state transition causes an interrupt. The ISR gives the output device another piece of data to output.

Figure 4.7
Hardware/software
timing of an input
interface during a high-
bandwidth burst.

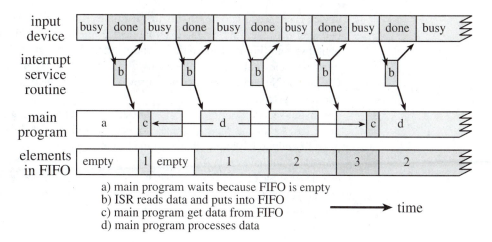

a) main program waits because FIFO is empty
b) ISR reads data and puts into FIFO
c) main program get data from FIFO
d) main program processes data

Figure 4.8
The output device
interrupts the computer
when it is idle and needs
new data.

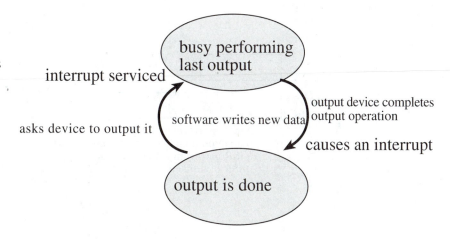

Again, we can visualize the interrupt synchronization by drawing a state versus time plot of the activities of the hardware and the two software modules. For an output device interface, the output device is initially disarmed and the FIFO is empty. The main thread begins by generating new data. After the main program puts the data into the FIFO, it arms the output interrupts. This first interrupt occurs immediately, and the ISR gets the data from the FIFO and outputs it to the external device. The output device becomes busy because it is in the process of outputting data. It is important to realize that it only takes the software on the order of 1 μs to write data to one of its output ports, but usually it takes the output device much longer to fully process the data. When the output device is done, it is ready to accept more data and an interrupt is requested. If the FIFO is empty at this point, the ISR will disarm the output device. If the FIFO is not empty, the ISR will get data from the FIFO and write them out to the output port. Once data are written to the output port, the output device is released to go on. In this first example, the time for the software to generate data is longer than the time for the external device to output it. This is an example of a CPU-bound system (Figure 4.9). In this situation, the FIFO has either 0 or 1 entry, and the use of interrupts does not enhance the bandwidth over the gadfly implementations presented in Chapter 3. Nevertheless, interrupts provide a well-defined mechanism for dealing with complex systems.

In this second example, the software starts with a burst of high-bandwidth activity (Figure 4.10). As long as the FIFO does not become full, no data are lost. In this situation, the overall bandwidth is higher than it would be with a gadfly implementation, because the software does not have to wait for each data byte to be processed by the hardware. This is the classic example of a *buffered* output, because data enter the system (via the main program), are temporarily stored in a buffer (put into the FIFO), and are processed later (by the ISR, get from the FIFO, write to external device). When the I/O device is slower than the soft-

Figure 4.9
Hardware/software timing of a CPU-bound output interface.

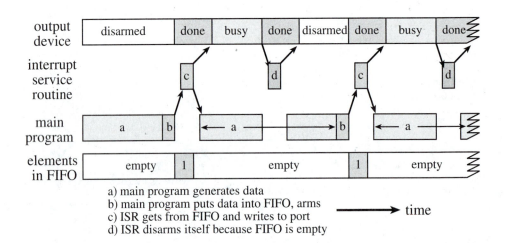

Figure 4.10
Hardware/software timing of an I/O-bound output interface.

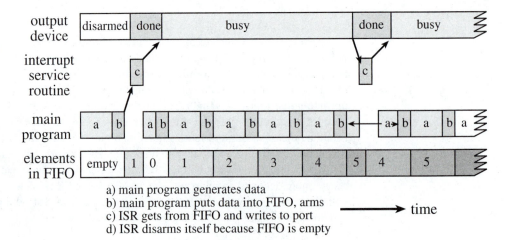

ware, the system is called I/O bound. Just like the input scenario, the FIFO might become full if the producer rate is too high for too long.

There are other types of interrupts that are not an input or an output. For example, we will configure the computer to request an interrupt on a periodic basis. This means an interrupt handler will be executed at fixed time intervals. This periodic interrupt will be essential for the implementation of real-time data acquisition and real-time control systems. For example, if we are implementing a digital controller that executes a control algorithm 100 times a second, then we will set up the timer hardware to request an interrupt every 10 ms. The ISR will execute the digital control algorithm and return to the main thread. We will learn how to create the period interrupt in Chapter 6 and use it in Chapters 12 and 13.

An axiom with interrupt synchronization is that the interrupt program should execute as fast as possible. The interrupt should occur when it is time to perform a needed function, and the ISR should come in clean, perform that function, and return right away. Placing backward branches (gadfly loops, iterations) in the interrupt software should be avoided if possible. The percentage of time spent executing interrupt software should be minimized. For an input device, the *interface latency* is the time between when new input are available and the time when the software reads the input data. We can also define *device latency* as the response time of the external I/O device. For example, if we request that a certain sector be read from a disk, then the device latency is the time it takes to find the correct track and spin the disk (seek) so that the proper sector is positioned under the read head. For an output device, the interface latency is the time between when the output device is idle and the time when the software writes new data. A *real-time* system is one that can guarantee an upper bound interface latency.

4.2   Reentrant Programming

A program segment is reentrant if it can be concurrently executed by two (or more) threads. This issue is very important when using interrupt programming. To implement reentrant software, we place local variables on the stack and avoid storing into global memory variables. When writing in assembly, we use *registers* or the *stack* for parameter passing to create reentrant subroutines. Typically each thread will have its own set of registers and stack. A nonreentrant subroutine will have a section of code called a *vulnerable window* or *critical section.* An error occurs if:

1. One thread calls the nonreentrant subroutine
2. That thread is executing in the "vulnerable" window when interrupted by a second thread
3. The second thread calls the same subroutine

A number of scenarios can happen next. In the first scenario, the second thread is allowed to complete the execution of the subroutine, control is then returned to the first thread, and the first thread finishes the subroutine. This first scenario is the usual case with interrupt programming. In the second scenario, the second thread executes part of it, is interrupted and then reentered by a third thread, the third thread finishes, the control is returned to the second thread and it finishes, and last the control is returned to the first thread and it finishes. This second scenario can happen in interrupt programming if interrupts are reenabled during the execution of the ISR. In the third scenario, the second thread executes part of it, is interrupted and the first thread continues, the first thread finishes, the control is returned to the second thread and it finishes. This third scenario will occur only when you use a thread scheduler like the ones described in the next chapter. Since most embedded systems do not have a thread scheduler, we will focus on the first two scenarios. A vulnerable window may exist when two different subroutines access and modify the same memory-resident data structure. In Figure 4.4, we saw two concurrent threads communicating via a FIFO queue. When processing input, it is possible for the main program to start to execute GetFifo, be interrupted, and the input interrupt routine calls PutFifo. To verify correctness of our system, we must consider what would happen if the PutFifo subroutine is executed in between any two assembly instructions of the GetFifo routine. A similar problem arises when processing outputs. In this situation the main program may start to execute PutFifo, be interrupted, and the output interrupt routine calls GetFifo. To verify correctness of this system, we must consider what would happen if the GetFifo subroutine is executed in between any two assembly instructions of the PutFifo routine. If we are processing both input and output, then two FIFOs would be used, so none of the individual functions can be reentered. Nevertheless, the PutFifo and GetFifo routines share access to global variables, so we must treat this as a potential problem. The three functions InitFifo PutFifo and GetFifo all manipulate the same global variables; therefore, to operate properly we need a mechanism to implement mutual exclusion, which simply means only one thread at a time is allowed access to the FIFO module. For most of this book, we will implement mutual exclusion by disabling interrupts. In the next chapter, we will describe how to create a preemptive thread scheduler and use semaphores to implement mutual exclusion.

This first example (Program 4.1) is a nonreentrant assembly subroutine that uses a memory variable Second. This subroutine could have been made reentrant by implementing Second as a local variable, but the purpose of the example is to illustrate what can go wrong when a nonreentrant subroutine is reentered.

Program 4.1 This subroutine is nonreentrant because of the read-modify-write access to a global.

```
Second  rmb  2     Temporary global variable
* Input parameters: Reg X,Y contain 2 16 bit numbers
* Output parameter: Reg X is returned with the average
AVE     sty  Second  Save the second number in memory
        xgdx           Reg D contains first number
        addd Second  Reg D=First+Second
        lsrd           (First+Second)/2
        adcb #0      round up?
        adca #0
        xgdx
        rts
```

A *vulnerable window* exists between the sty and the addd instructions. Assume there are two concurrent threads (Main, InterruptService) that both call this subroutine. Concurrent means that both threads are ready to run. Because there is only one computer, exactly one thread will be active (running) at a time. Typically, the operating system (OS) switches execution control back and forth using interrupts. For example, the Main thread might be executing when an interrupt causes the computer to switch over and execute the `InterruptService`. When the `InterruptService` is done, it executes a `rti`, and the control returns back to the `Main` thread. An error occurs if (Figure 4.11):

1. The `Main` thread calls `AVE`
2. The `Main` thread executes the `sty` instruction, saving its second number in `Second`
3. The OS halts the `Main` thread (using an interrupt) and starts the `Interrupt-Service` thread
4. The `InterruptService` thread calls `AVE`

   The `InterruptService` thread executes the `sty`, saving its second number in `Second`

   The `InterruptService` thread finishes `AVE`

5. The OS returns control back to the `Main` thread
6. The `Main` thread executes the `addd` instruction but gets the wrong `Second` number

Figure 4.11 One sequence of operations that results in the reentering of the subroutine `Ave`.

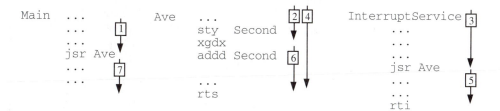

A *vulnerable window* exists in the following program between the `Result=y` and the `Result+x` instructions.

```
int Result;   /* Temporary global variable */
int AVE(int x,y);{
      Result=y;   /* Save the second number in global memory */
      Result=(Result+x)>>1;   /* (First+Second)/2 */
      return(Result);}
```

An error occurs if (Figure 4.12):

1. The `Main` thread calls `AVE`
2. The `Main` thread executes the `Result=y` instruction, saving its second number in `Result`
3. The OS halts the `Main` thread (using an interrupt) and starts the `Interrupt-Service` thread
4. The `InterruptService` thread calls `AVE`

   The `InterruptService` thread executes the `Result=y`, saving its number over other copy

   The `InterruptService` thread finishes `AVE`

5. The OS returns control back to the `Main` thread
6. The `Main` thread calculates `(Resulttx)>>1` but gets the wrong `Result` number

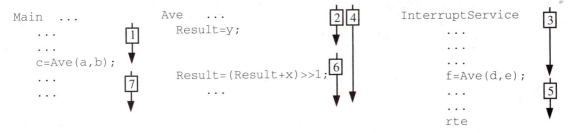

Figure 4.12
One sequence of operations that results in the reentering of the C function `Ave`.

The solution to this simple problem is to implement `Result` as a local variable either in a register or on the stack. But we saw earlier in the chapter that global variables are needed for interthread communication, so not all nonreentrant activity can be solved by using local variables. In this chapter we will disable interrupts during the vulnerable window. In the next chapter semaphores are used to provide mutual exclusive access to certain memory-resident data structures.

An *atomic operation* is one that once started is guaranteed to finish. In most computers, once an instruction has begun, the instruction must be finished before the computer can process an interrupt. Therefore, the following read-modify-write sequence is atomic because it cannot be halted in the middle of its operation:

```
inc counter   where counter is a global variable
```

On the other hand, this read-modify-write sequence is *nonatomic* because it can start, then be interrupted:

```
ldaa counter      where counter is a global variable
inca
staa counter
```

In general, nonreentrant code can be grouped into three categories, all involving nonatomic writes to global variables. We will classify I/O ports as global variables for the consideration of reentrancy. We will group registers into the same category as local variables because each thread will have its own registers and stack.

The first group is the *read-modify-write* sequence:

1. The software reads the global variable, producing a copy of the data.
2. The software modifies the copy (at this point the original variable is still unmodified).
3. The software writes the modification back into the global variable.

An assembly example is shown in Program 4.2. In C, this read-modify-write could be implemented as shown in Program 4.3.

Program 4.2 This subroutine is also nonreentrant because of the read-modify-write access to a global.

```
Money rmb   2        bank balance implemented as a global
* add $100 to the account
more  ldd   Money  where Money is a global variable
      addd  #100
      std   Money   Money=Money+100
      rts
```

Program 4.3 This C function is nonreentrant because of the read-modify-write access to a global.

```
unsigned int Money;  /* bank balance implemented as a global */
/* add 100 dollars */
void more(void){
        Money += 100;}
```

In the second group is the *write followed by read,* where the global variable is used for temporary storage:

1. The software writes to the global variable (this becomes the only copy of important information).
2. The software reads from the global variable, expecting the original data to still be there.

An assembly example is shown in Program 4.4. In C, Program 4.5 shows an illustrative example of this read-modify-write.

Program 4.4 This assembly subroutine is nonreentrant because of the write-read access to a global.

```
temp  rmb  2        temporary result implemented as a global
* calculate RegX=RegX+2*RegD
mac   stx  temp    Save X so that it can be added
      lsld         RegD=2*RegD
      addd temp    RegD=RegX+2*RegD
      xgdx         RegX=RegX+2*RegD
      rts
```

Program 4.5 This C function is nonreentrant because of the write-read access to a global.

```
int temp;  /* global temporary */
/* calculate x+2*d */
int mac(int x, int d){
        temp = x+2*d;    /* write to a global variable */
        return (temp);}  /* read from global */
```

In the third group, we have a *nonatomic multistep write* to a global variable:

1. The software writes part of the new value to a global variable.
2. The software writes the rest of the new value to a global variable.

Program 4.6 shows an assembly example, and in C, this multistep write sequence could be implemented as shown in Program 4.7.

Program 4.6 This assembly subroutine is nonreentrant because of the multistep write access to a global.

```
Info  rmb  4        32-bit data implemented as a global
* set the variable using RegX and RegY
set   stx  Info    Info is a 32 bit global variable
      sty  Info+2
      rts
```

Program 4.7 This C function is nonreentrant because of the multi-step write access to a global.

```
int info[2];  /* 32-bit global */
void set(int x, int y){
      info[0]=x;
      info[1]=y;}
```

Observation: When considering reentrant software and vulnerable windows, we classify accesses to I/O ports the same as accesses to global variables.

We can make a subroutine reentrant by using a stack variable

If we can eliminate the global variables, then the subroutine becomes reentrant. The example of Program 4.1 is redesigned by placing the temporary on the stack, as shown in Program 4.8. There are no "vulnerable" windows because each thread has its own registers and stack.

Program 4.8 This assembly subroutine is reentrant because it does not write to any globals.

```
* Input parameters: Reg X,Y contain 2 16 bit numbers
* Output parameter: Reg X is returned with the average
AVE      pshy        Save the second number on the stack
         tsy         Reg Y points the Second number
         xgdx        Reg D contains first number
         addd 0,Y    Reg D=First+Second
         lsrd        (First+Second)/2
         adcb #0     round up?
         adca #0
         xgdx
         puly
         rts
```

We can make a subroutine reentrant by disabling interrupts

Sometimes one must access global memory to implement the desired function. Consider the example of a message mailbox (Program 4.9). This mailbox has a status flag (0 means empty) and an 8-bit value. The subroutine SEND will check the status, and if empty it will store the message (passed in RegB) into the mailbox. In this example there are multiple concurrent threads that may call SEND. Never mind for now how the mailbox is emptied.

Program 4.9 This assembly subroutine is nonreentrant because of the read-modify-write access to a global.

```
Status   rmb  1    0 means empty, -1 means it contains something
Message  rmb  1    data to be communicated
* Input parameter: Reg B contains an 8 bit message
* Output parameter: Reg CC (C bit) is 1 for OK, 0 for busy error
SEND     tst  Status   check if mailbox is empty
         bmi  Busy     full, can't store, so return with C=0
         stab Message  store
         dec  Status   signify it now contains a message
         sec           stored OK, so return with C=1
Busy     rts
```

Clearly there is a *vulnerable window* or *critical section* between the tst and dec instructions that makes this subroutine nonreentrant. One can make this subroutine reentrant by disabling interrupts during the vulnerable window. It is important not to disable interrupts too long so as not to affect the dynamic performance of the other threads. Notice also that the interrupts are not simply disabled then enabled, but rather the interrupt status is saved, the interrupts disabled, then the interrupt status is restored. Consider what would happen (Programs 4.10, 4.11) if you simply added a sei at the beginning and a cli at the end of the above subroutine, then called it with the interrupts disabled.

Program 4.10 This
assembly subroutine is
reentrant because it
disables interrupts
during the critical
section.

```
Status    rmb   1      0 means empty, -1 means it contains something
Message   rmb   1      data to be communicated
* Input parameter: Reg B contains an 8 bit message
* Output parameter: Reg CC (C bit) is 1 for OK, 0 for busy error
SEND      clc          Initialize carry=0
          tpa          save current interrupt state
          psha
          sei          disable interrupts during vulnerable window
          tst   Status check if mailbox is empty
          bmi   Busy   full, so return with C=0
          staa  Message store
          dec   Status signify it now contains a message
          pula
          oraa  #1     OK, so return with C=1
          psha
Busy      pula         restore interrupt status
          tap
          rts
```

Program 4.11 This C
function is reentrant
because it disables
interrupts during the
critical section.

```
int Empty;   /* -1 means empty, 0 means it contains something */
int Message; /* data to be communicated */
int SEND(int data){ int OK;
char SaveSP;
    asm(" tpa\n staa %SaveSP\n sei"); /* make atomic, entering critical */
    OK=0;            /* Assume it is not OK */
    if(Empty){
       Message=data;
       Empty=0;  /* signify it now contains a message*/
       OK=-1;}    /* Successful */
    asm(" ldaa %SaveSP\n tap");  /* end critical section */
    return(OK);}
```

Program 4.12 This
assembly subroutine can
be used as part of a
binary semaphore.

```
* Global parameter: Semi4 is the memory location to test and set
* If the location is zero, it will set it (make it -1)
*       and return Reg CC (Z bit) is 1 for OK
* If the location is nonzero, it will return Reg CC (Z bit) = 0
Semi4 fcb  0           Semaphore is initially free
TAS   tst  Semi4       check if already set
      bne  Out         busy, operation failed, so return with Z=0
      dec  Semi4       signify it is now busy
      bita #0          operation successful, so return with Z=1
Out   rts
```

Some machines provide a *test and set* function to solve this reentrant problem. This single (nondivisible) operation is equivalent to the subroutine in Program 4.12. The test and set operation, once started, must be allowed to complete. More details about semaphores can be found in Chapter 5.

Reentrant programming is very important when writing high-level language software, too. Obviously, we minimize the use of global variables. But when global variables are necessary, we must be able to recognize potential sources of bugs due to nonreentrant code. We must study the assembly language output produced by the compiler. For example, we cannot determine whether the following read-modify-write operation is reentrant without knowing if it is atomic:

```
time++;
```

If the compiler generates the following object code, then `time++;` is atomic (therefore not critical):

```
inc time
```

If the compiler generates the following object code, then `time++;` is not atomic (therefore critical):

```
ldd  time
addd #1
std  time
```

4.3   First-In–First-Out Queue

4.3.1
Introduction to
FIFOs

As we saw earlier, the FIFO circular queue is quite useful for implementing a buffered I/O interface. It can be used for both buffered input and buffered output. The order-preserving data structure temporarily saves data created by the source (producer) before they are processed by the sink (consumer). The class of FIFOs studied in this section will be statically allocated global structures. Because they are global variables, it means they will exist permanently and can be carefully shared by more than one program. The advantage of using a FIFO structure for a data flow problem is that we can decouple the producer and consumer threads. Without the FIFO we would have to produce one piece of data, then process it, produce another piece of data, then process it. With the FIFO, the producer thread can continue to produce data without having to wait for the consumer to finish processing the previous data. This decoupling can significantly improve system performance (Figure 4.13).

Figure 4.13
The FIFO is used to buffer data between the producer and consumer.

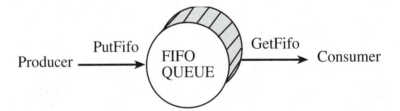

You have probably already experienced the convenience of FIFOs. For example, you can continue to type another command into the DOS command interpreter while it is still processing a previous command. The ASCII codes are put (calls *PutFifo*) in a FIFO whenever you hit the key. When the DOS command interpreter is free, it calls *GetFifo* for more keyboard data to process. A FIFO is also used when you ask the computer to print a file. Rather than waiting for the actual printing to occur character by character, the print command will *PUT* the data in a FIFO. Whenever the printer is free, it will *GET* data from the FIFO. The advantage of the FIFO is it allows you to continue to use your computer while the printing occurs in the background. To implement this magic of background printing we will need interrupts.

There are many producer/consumer applications. In Table 4.2 the tasks on the left are producers that create or input data, while the tasks on the right are consumers that process or output data.

Table 4.2
Producer-consumer examples.

Source/producer		Sink/consumer
Keyboard input	→	Program that interprets
Program with data	→	Printer output
Program sends message	→	Program receives message
Microphone and ADC	→	Program that saves sound data
Program that has sound data	→	DAC and speaker

The producer puts data into the FIFO. The PutFifo operation does not discard information already in the FIFO. If the FIFO is full and the user calls PutFifo, the PutFifo routine will return a full error, signifying the last (newest) data were not properly saved. The sink process removes data from the FIFO. The GetFifo routine will modify the FIFO. After a get, the particular information returned from the GetFifo routine is no longer saved on the FIFO. If the FIFO is empty and the user tries to get, the GetFifo routine will return an empty error, signifying no data could be retrieved. The FIFO is order-preserving, such that the information is returned by repeated calls of GetFifo in the same order as the data were saved by repeated calls of PutFifo.

There are many ways to implement a statically allocated FIFO. We can use either a pointer or an index to access the data in the FIFO. We can use either two pointers (or two indices) or two pointers (or two indices) and a counter. The counter specifies how many entries are currently stored in the FIFO. There are even hardware implementations of FIFO queues. We begin with the two-pointer implementation. It is a little harder to implement but does have some advantages over the other implementations.

4.3.2
Two-Pointer FIFO
Implementation

We will begin with the two-pointer implementation. If we were to have infinite memory, a FIFO implementation is easy (Figure 4.14). GetPt points to the data that will be removed by the next call to GetFifo, and PutPt points to the empty space where the data will be stored by the next call to PutFifo (Program 4.13).

Figure 4.14
The FIFO implementation with infinite memory.

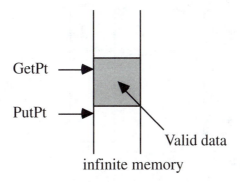

Program 4.13 Code
fragments showing the
basic idea of a FIFO.

```
GetPt   rmb 2    Pointer to oldest data, next to be removed by GET
PutPt   rmb 2    Pointer to free memory, place to PUT next data
* Reg A contains byte to store into the FIFO
PutFifo ldx   PutPt   Reg X points to free place
        staa  ,X       Store data into FIFO
        inx             Update pointer
        stx   PutPt
        rts
* Reg A returned with byte from FIFO
GetFifo ldx   GetPt   Reg X points to oldest data
        ldaa  ,X       Read data from FIFO
        inx             Update pointer
        stx   GetPt
        rts
```

Three modifications are required to the above subroutines. If the FIFO is full when `PutFifo` is called, then the subroutine should return a full error (e.g., V=1). Similarly, if the FIFO is empty when `GetFifo` is called, then the subroutine should return an empty error (e.g., V=1). The `PutPt` and `GetPt` must be wrapped back up to the top when they reach the bottom (Figures 4.15, 4.16).

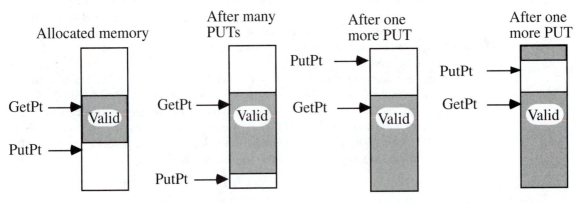

Figure 4.15   The FIFO PutFifo operation showing the pointer wrap.

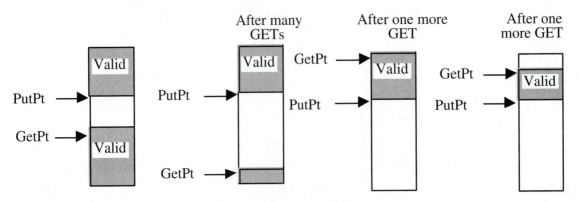

Figure 4.16   The FIFO GetFifo operation showing the pointer wrap.

Two mechanisms determine whether the FIFO is empty or full. A simple method is to implement a counter containing the number of bytes currently stored in the FIFO. Get-Fifo would decrement the counter, and PutFifo would increment the counter. The second method is to prevent the FIFO from being completely full. For example, if the FIFO had 10 bytes allocated, then the PutFifo subroutine would allow a maximum of 9 bytes to be stored. If there were 9 bytes already in the FIFO and another PutFifo were called, then the FIFO would not be modified and a full error would be returned. In this way if PutPt equals GetPt at the beginning of GetFifo, then the FIFO is empty. Similarly, if PutPt+1 equals GetPt at the beginning of PutFifo, then the FIFO is full. Be careful to wrap the PutPt+1 before comparing it to GetFifo. This second method does not require the length to be stored or calculated. The C implementation of this FIFO module is shown in Program 4.14.

```c
#define FifoSize 10     /* Number of 8 bit data in the Fifo less one */
char *PutPt;    /* Pointer of where to put next */
char *GetPt;    /* Pointer of where to get next */
                /* FIFO is empty if PutPt=GetPt */
                /* FIFO is full  if PutPt+1=GetPt (with wrap) */
char Fifo[FifoSize];     /* The statically allocated fifo data */
void InitFifo(void) { char SaveSP;
 asm(" tpa\n staa %SaveSP\n sei"); /* make atomic, entering critical */
 PutPt=GetPt=&Fifo[0];  /* Empty when PutPt=GetPt */
 asm(" ldaa %SaveSP\n tap");        /* end critical section */
}
int PutFifo (char data) { char *Ppt; /* Temporary put pointer */
char SaveSP;
   asm(" tpa\n staa %SaveSP\n sei"); /* make atomic, entering critical */
   Ppt=PutPt;              /* Copy of put pointer */
   *(Ppt++)=data;          /* Try to put data into fifo */
   if (Ppt == &Fifo[FifoSize]) Ppt = &Fifo[0];  /* Wrap */
   if (Ppt == GetPt ) {
      asm(" ldaa %SaveSP\n tap"); /* end critical section */
      return(0);}     /* Failed, fifo was full */
    else{
      PutPt=Ppt;
      asm(" ldaa %SaveSP\n tap");  /* end critical section */
      return(-1);    /* Successful */
     }
 }
int GetFifo (char *datapt) { char SaveSP;
   if (PutPt == GetPt ) {
       return(0);}    /* Empty if PutPt=GetPt */
   else{
      asm(" tpa\n staa %SaveSP\n sei"); /* make atomic, entering critical */
      *datapt=*(GetPt++);
      if (GetPt == &Fifo[FifoSize]) GetPt = &Fifo[0];
      asm(" ldaa %SaveSP\n tap");  /* end critical section */
      return(-1); }
}
```

Program 4.14 Two-pointer implementation of a FIFO.

The assembly language implementation of the initialization is shown in Program 4.15.

Program 4.15 Initialization of a two-pointer FIFO.

```
* Two pointer implementation of the FIFO
FifoSize  equ  10    Number of 8 bit data in the Fifo, less one
PutPt     rmb  2     Pointer of where to put next */
GetPt     rmb  2     Pointer of where to get next
* FIFO is  empty if PutPt=GetPt
* FIFO is full  if PutPt+1=GetPt (with wrap)
Fifo      rmb  FifoSize   The statically allocated fifo data
**********Initialize FIFO*****************

* No parameters
InitFifo  ldx #Fifo
          tpa
          sei         make atomic, entering critical section
          stx PutPt
          stx GetPt   Empty when PutPt=GetPt
          tap         end critical section, restore CCR
          rts
```

To check for FIFO full, the `PutFifo` routine in Program 4.16 attempts to put using a temporary `PutPt`. If putting makes the FIFO look empty, then the temporary `PutPt` is discarded and the routine is exited without saving the data. This is why a FIFO with 10 allocated bytes can hold only 9 data points. If putting doesn't make the FIFO look empty, then the temporary `PutPt` is stored into the actual `PutPt`, saving the data as desired.

Program 4.16 Assembly routine to put into a two-pointer FIFO.

```
**********Put a byte into the FIFO*****************
* Input   RegA contains 8 bit data to put
* Output  RegA is -1 if successful, 0 if data not stored
PutFifo   psha
          tpa
          tab
          pula
          pshb              save old CCR
          sei               make atomic, entering critical section
          ldx  PutPt        RegX is Temporary put pointer
          staa 0,x          Try to put data into fifo
          inx
          cpx  #Fifo+FifoSize
          bne  PutNoWrap    skip if no wrapping needed
          ldx  #Fifo        Wrap
PutNoWrap clra              assume it will fail
          cpx  GetPt        Full if now the same
          beq  PutDone
          coma              RegA=-1 means OK
          stx  PutPt
PutDone   tab               end critical section
          pula
          tpa               restore CCR to previous value
          tba
          rts
```

To check for FIFO empty, the `GetFifo` routine in Program 4.17 simply checks to see if `GetPt` equals `PutPt`. If they match at the start of the routine, then `GetFifo` returns with the "empty" condition signified.

Program 4.17 Assembly routine to get from a two-pointer FIFO.

```
***********Get a byte from the FIFO******************
* Input   RegX points to place for 8 bit data from Get
* Output  RegA is -1 if successful, 0 if Fifo was empty when called
GetFifo   tpa
          psha              save old CCR
          sei               make atomic, entering critical section
          clra              assume it will fail
          ldy  GetPt
          cpx  PutPt        Empty if initially the same
          beq  GetDone
          coma              RegA=-1 means OK
          ldab 0,y          Data from FIFO
          stab 0,x          Return by reference
          iny
          cpy  #Fifo+FifoSize
          bne  GetNoWrap    skip if no wrapping needed
          ldy  #Fifo        Wrap
GetNoWrap sty  GetPt
GetDone   tab               end critical section
          pula
          tpa               restore CCR to previous value
          tba
          rts
```

Since `PutFifo` and `GetFifo` have read-modify-write accesses to global variables, they themselves are not reentrant. Similarly, `InitFifo` has a multiple-step write-access to global variables. Therefore `InitFifo` is not reentrant. Consequently, interrupts are temporarily disabled to prevent one thread from reentering these FIFO functions. Notice that at the end of the critical section, interrupts are not enabled, but rather the interrupt status is restored to its previous state (enabled or disabled). This method of disabling interrupts allows you to nest a function call from one nonreentrant function to another.

One advantage of this pointer implementation is that if you have a single thread that calls the `GetFifo` (e.g., the main program) and a single thread that calls the `PutFifo` (e.g., the serial port receive interrupt handler) as shown in Figure 4.4, then this `PutFifo` function can interrupt this `GetFifo` function without loss of data. So in this particular situation, interrupts would not have to be disabled. It would also operate properly if there were a single interrupt thread calling `GetFifo` (e.g., the serial port transmit interrupt handler) and a single thread calling `PutFifo` (e.g., the main program). On the other hand, if the situation is more general, and multiple threads could call `PutFifo` or multiple threads could call `GetFifo`, then the interrupts would have to be temporarily disabled as shown.

4.3.3 Two-Pointer/ Counter FIFO Implementation

The other method to determine if a FIFO is empty or full is to implement a counter. In the code of Program 4.18, `Size` contains the number of bytes currently stored in the FIFO. The advantage of implementing the counter is that FIFO quarter-full and three-quarter-full conditions are easier to implement. If you were studying the behavior of a system, it might be informative to measure the current `Size` as a function of time.

Program 4.18 C
language routines to
implement a two-pointer
with counter FIFO.

```
/* Pointer,counter implementation of the FIFO */
#define FifoSize 10    /* Number of 8 bit data in the Fifo */
char *PutPt;     /* Pointer of where to put next */
char *GetPt;     /* Pointer of where to get next */
unsigned char Size;  /* Number of elements currently in the FIFO */
                  /* FIFO is empty if Size=0 */
                  /* FIFO is full  if Size=FifoSize */
char Fifo[FifoSize];    /* The statically allocated fifo data */
void InitFifo(void) { char SaveSP;
 asm(" tpa\n staa %SaveSP\n sei");   /* make atomic, entering critical*/
 PutPt=GetPt=&Fifo[0];    /* Empty when Size==0 */
 Size=0;
 asm(" ldaa %SaveSP\n tap");  /* end critical section */
}
int PutFifo (char data) { char SaveSP;
   if (Size == FifoSize ) {
      return(0);}    /* Failed, fifo was full */
    else{
      asm(" tpa\n staa %SaveSP\n sei");   /* make atomic, critical*/
      Size++;
      *(PutPt++)=data;    /* put data into fifo */
      if (PutPt == &Fifo[FifoSize]) PutPt = &Fifo[0];  /* Wrap */
      asm(" ldaa %SaveSP\n tap");  /* end critical section */
      return(-1);    /* Successful */
      }
}
int GetFifo (char *datapt) { char SaveSP;
   if (Size == 0 ){
      return(0);}        /* Empty if Size=0 */
    else{
      asm(" tpa\n staa %SaveSP\n sei");   /* make atomic, critical*/
      *datapt=*(GetPt++); Size--;
      if (GetPt == &Fifo[FifoSize]) GetPt = &Fifo[0];
      asm(" ldaa %SaveSP\n tap");  /* end critical section */
      return(-1); }
}
```

The assembly language implementation of the initialization is shown in Program 4.19.

Program 4.19 Initialization of a two-pointer
with counter FIFO.

```
* Pointer,counter implementation of the FIFO
FifoSize    equ  10     Number of 8 bit data in the Fifo
PutPt       rmb  2      Pointer of where to put next
GetPt       rmb  2      Pointer of where to get next
Size        rmb  1
* FIFO is empty if Size=0
* FIFO is full  if Size=FifoSize
Fifo        rmb  FifoSize    The statically allocated fifo data
```

```
***********Initialize FIFO*****************
* No parameters
InitFifo  tpa
          sei            make atomic, entering critical section
          ldx   #Fifo
          stx   PutPt
          stx   GetPt   Empty when Size == 0
          clr   Size
          tap            end critical section
          rts
```

To check for FIFO full, the PutFifo routine in Program 4.20 simply compares Size to the maximum allowed value. If the FIFO is already full, then the routine is exited without saving the data. With this implementation a FIFO with 10 allocated bytes actually can hold 10 data points.

Program 4.20 Assembly routine to put into a two-pointer with counter FIFO.

```
***********Put a byte into the FIFO******************
* Input   RegA contains 8 bit data to put
* Output  RegA is -1 if successful, 0 if data not stored
PutFifo      psha
             tpa
             tab
             pula
             pshb            save old CCR
             sei             make atomic, entering critical section
             ldab  Size
             cmpb  #FifoSize  Full if Size==FifoSize
             bne   PutNotFull
             clra
             bra   PutDone
PutNotFull   incb
             stab  Size        Size++
             ldx   PutPt
             staa  0,x         Put data into fifo
             inx
             cpx   #Fifo+FifoSize
             bne   PutNoWrap  skip if no wrapping needed
             ldx   #Fifo       Wrap
PutNoWrap    ldaa  #-1         success means OK
             stx   PutPt
PutDone      tab             end critical section
             pula
             tpa             restore CCR to previous value
             tba
             rts
```

To check for FIFO empty, the GetFifo routine of Program 4.21 simply checks to see if Size equals 0. If Size is zero at the start of the routine, then GetFifo returns with the "empty" condition signified.

Program 4.21 Assembly
routine to get from a
two-pointer with counter
FIFO.

```
***********Get a byte from the FIFO*****************
* Input  RegX points to place for 8 bit data from Get
* Output RegA is -1 if successful, 0 if Fifo was empty when called
GetFifo tpa
        psha     save old CCR
        sei      make atomic, entering critical section
        clra        assume it will fail
        tst  Size
        beq  GetDone
        dec  Size
        ldy  GetPt
        coma            RegA=-1 means OK
        ldab 0,y        Data from FIFO
        stab 0,x        Return by reference
        iny
        cpy  #Fifo+FifoSize
        bne  GetNoWrap  skip if no wrapping needed
        ldy  #Fifo      Wrap
GetNoWrap sty GetPt
GetDone tab         end critical section
        pula
        tpa             restore CCR to previous value
        tba
        rts
```

4.3.4
Index FIFO
Implementation

Another method to implement a statically allocated FIFO is to use indices instead of point-ers. Except for the situation where the compiler does not support pointers, there is no par-ticular advantage of indices over pointers. The purpose of this example is to illustrate the use of arrays and indices. Just like the previous FIFOs, this FIFO is used for order-preserving temporary storage. The function PutFifo will enter one 8-bit byte into the queue, and GetFifo will remove 1 byte. If you call PutFifo while the FIFO is full (Size is equal to FifoSize), the routine will return a 0. Otherwise, PutFifo will save the data in the queue and return a 1. The index PutI specifies where to put the next 8-bit data. The routine GetFifo actually returns two parameters. The queue status is the regular function return parameter, while the data removed from the queue is return by reference. That is, the calling routine passes in a pointer, and GetFifo stores the removed data at that address. If you call GetFifo while the FIFO is empty (Size is equal to zero), the routine will return a 0. Otherwise, GetFifo will return the oldest data from the queue and return a 1. The index GetI specifies where to get the next 8-bit data. Program 4.22 implements a FIFO in C language using two indices and a counter.

```
/* Index,counter implementation of the FIFO */
#define FifoSize 10    /* Number of 8 bit data in the Fifo */
unsigned char PutI;    /* Index of where to put next */
unsigned char GetI;    /* Index of where to get next */
unsigned char Size;    /* Number of elements currently in the FIFO */
            /* FIFO is empty if Size=0 */
            /* FIFO is full  if Size=FifoSize */
```

```
char Fifo[FifoSize];     /* The statically allocated fifo data */
void InitFifo(void) {char SaveSP;
  asm(" tpa\n staa %SaveSP\n sei");   /* make atomic, entering critical*/
  PutI=GetI=Size=0;       /* Empty when Size==0 */
  asm(" ldaa %SaveSP\n tap");  /* end critical section */
}
int PutFifo (char data) { char SaveSP;
  if (Size == FifoSize )
     return(0);      /* Failed, fifo was full */
   else{
     asm(" tpa\n staa %SaveSP\n sei");   /* make atomic, entering critical*/
     Size++;
     Fifo[PutI++]=data;       /*  put data into fifo */
     if (PutI == FifoSize) PutI = 0;  /* Wrap */
     asm(" ldaa %SaveSP\n tap");  /* end critical section */
     return(-1);   /* Successful */
    }
}
int GetFifo (char *datapt) { char SaveSP;
   if (Size == 0 )
       return(0);     /* Empty if Size=0 */
   else{
       asm(" tpa\n staa %SaveSP\n sei"); /* make atomic, entering critical*/
       *datapt=Fifo[GetI++]; Size--;
       if (GetI == FifoSize) GetI = 0;
       asm(" ldaa %SaveSP\n tap");  /* end critical section */
       return(-1); }
}
```

Program 4.22 C implementation of two indices with counter FIFO.

The assembly language implementation of the initialization is shown in Program 4.23.

Program 4.23 Initialization of two indices with counter FIFO.

```
* Index,counter implementation of the FIFO
FifoSize equ  10     Number of 8 bit data in the Fifo
PutI     rmb  1      Index of where to put next
GetI     rmb  1      Index of where to get next
Size     rmb  1
* FIFO is empty if Size=0
* FIFO is full  if Size=FifoSize
Fifo        rmb  FifoSize    The statically allocated fifo data
**********Initialize FIFO******************

* No parameters
InitFifo    tpa
            sei        make atomic, entering critical section
            clr  PutI
            clr  GetI
            clr  Size  Empty when Size == 0
            tap        end critical section
            rts
```

The assembly language implementations of the `PutFifo` and `GetFifo` are shown in Programs 4.24 and 4.25. To check for FIFO full, the `PutFifo` routine of Program 4.24 simply compares `Size` to the maximum allowed value. If the FIFO is already full, then the routine is exited without saving the data. With this implementation a FIFO with 10 allocated bytes can actually hold 10 data points.

Program 4.24 Assembly routine to put into two indices with counter FIFO.

```
***********Put a byte into the FIFO******************
* Input   RegA contains 8 bit data to put
* Output  RegA is -1 if successful, 0 if data not stored
PutFifo     psha
            tpa
            tab
            pula
            pshb        save old CCR
            sei         make atomic, entering critical section
            ldab Size
            cmpb #FifoSize Full if Size==FifoSize
            beq  PutNotFull
            clra
            bra  PutDone
PutNotFull  incb
            stab Size       Size++
            ldx  #Fifo
            ldab PutI
            abx
            staa 0,x        Put data into fifo
            incb
            cmpb #FifoSize
            bne  PutNoWrap  skip if no wrapping needed
            clrb            Wrap
PutNoWrap   clra            success
            coma            RegA=-1 means OK
            stab PutI
PutDone     tab         end critical section
            pula
            tpa         restore CCR to previous value
            tba
            rts
```

Program 4.25 Assembly routine to get from two indices with counter FIFO.

```
***********Get a byte from the FIFO******************
* Input   RegX points to place for 8 bit data from Get
* Output  RegA is -1 if successful, 0 if Fifo was empty when called
GetFifo     tpa
            psha        save old CCR
            sei         make atomic, entering critical section
            clra        assume it will fail
            tst  Size
            beq  GetDone
            ldy  #Fifo
            ldab GetI
            aby
```

```
                 coma                RegA=-1 means OK
                 ldab 0,y            Data from FIFO
                 stab 0,x            Return by reference
                 ldab GetI
                 incb
                 cmpb #FifoSize
                 bne  GetNoWrap skip if no wrapping needed
                 clrb                Wrap
GetNoWrap        stab GetI
GetDone          tab             end critical section
                 pula
                 tpa             restore CCR to previous value
                 tba
                 rts
```

To check for FIFO empty, the GetFifo routine of Program 4.25 simply checks to see if Size equals 0. If Size is zero at the start of the routine, then GetFifo returns with the "empty" condition signified.

Just like the other FIFO examples, interrupts are temporarily disabled to prevent one thread from reentering these nonreentrant functions.

4.3.5
FIFO Dynamics

As you recall, the FIFO passes the data from the producer to the consumer. In general, the rates at which data are produced and consumed can vary dynamically. Human beings do not enter data into a keyboard at a constant rate. Even printers require more time to print color graph1cs versus black and white text. Let t_p be the time (in seconds) between calls to PutFifo and r_p be the arrival rate (producer rate in bytes per second) into the system. Similarly, let t_g be the time (in seconds) between calls to GetFifo and r_g be the service rate (consumer rate in bytes per second) out of the system.

$$r_g = \frac{1}{t_g} \qquad r_p = \frac{1}{t_p}$$

If the minimum time between calls to PutFifo is greater than the maximum time between calls to GetFifo,

$$\min t_p \geq \max t_g$$

then a FIFO is not necessary and the data flow program could be solved with a simple global variable. On the other hand, if the time between calls to PutFifo becomes less than the time between calls to GetFifo because either:

- The arrival rate temporarily increases
- The service rate temporarily decreases

then information will be collected in the FIFO. For example, a person might type very fast for a while, followed by a long pause. The FIFO could be used to capture without loss all the data as they come in very fast. Clearly on average the system must be able to process the data (the consumer thread) at least as fast as the average rate at which the data arrives (producer thread). If the average producer rate is larger than the average consumer rate,

$$\bar{r}_p > \bar{r}_g$$

then the FIFO will eventually overflow no matter how large the FIFO. If the producer rate is temporarily high and that causes the FIFO to become full, then this problem can be solved by increasing the FIFO size.

There is a fundamental difference between an empty error and a full error. Consider the application of using a FIFO between your computer and its printer. This is a good idea because the computer can temporarily generate data to be printed at a very high rate, followed by long pauses. The printer is like a turtle. It can print at a slow but steady rate (e.g., 10 characters/s). The computer will put a byte into the FIFO that it wants printed. The printer will get a byte out of the FIFO when it is ready to print another character. A full error occurs when the computer calls `PutFifo` at too fast a rate. A full error is serious, because if ignored data will be lost. On the other hand, an empty error occurs when the printer is ready to print but the computer has nothing in mind. An empty error is not serious, because in this case the printer just sits there doing nothing.

4.4   General Features of Interrupts on the 6805/6808/6811/6812

In this section we will present the specific details for each of our four microcomputers. As you develop experience using interrupts, you will come to notice a few common aspects that most computers share. The following paragraphs outline three essential mechanisms that are needed to utilize interrupts. Although every computer that uses interrupts includes all three mechanisms, there is a wide spectrum of implementation methods.

All interrupting systems have the *ability for the hardware to request action from the computer.* The interrupt requests can be generated by using a separate connection to the microprocessor for each device or by using a negative-logic wire-or requests employing open-collector logic. The Motorola microcomputers support both types.

All interrupting systems must have the *ability for the computer to determine the source.* A vectored interrupt system employs separate connections for each device so that the computer can give automatic resolution. You can recognize a vectored system because each device has a separate interrupt vector address. With a polled interrupt system, the software must poll each device, looking for the device that requested the interrupt.

The third necessary component of the interface is the *ability for the computer to acknowledge the interrupt.* Normally there is a flag in the interface that is set on the busy-to-done state transition. Acknowledging the interrupt involves clearing the flag that caused the interrupt. It is important to shut off the request so that the computer will not mistakenly request a second (and inappropriate) interrupt for the same condition. Some Intel systems use a hardware acknowledgment that automatically clears the request. Most Motorola microcomputers use a software acknowledge. So when we identify the flag, it will be important to know exactly what conditions will set the flag (and request an interrupt) and how the software will clear it (acknowledge) in the ISR.

Even though there are no standard definitions for the terms *mask, enable,* and *arm* in the computer science or computer engineering communities, in this book we will assign the following specific meanings. To *arm (disarm)* a device means to enable (shut off) the source of interrupts. One arms (disarms) a device if one is (is not) interested in interrupts at all. For example, the STAI bit in the 6811 PIOC register is the arm bit for the 6811 STRA pin. Similarly the 6811/6812 TMSK1 has eight arm bits for the output compare and input capture interrupts. The Motorola literature calls the arm bit an "interrupt enable mask." To *enable (disable)* means to allow interrupts at this time (postpone interrupts until a later time). We disable interrupts if it is currently not convenient to process. In particular, to disable interrupts we set the I bit in 6805/6808/6811/6812 condition code register using the `sei` instruction. There are some interrupts that cannot be disabled, such as XIRQ, SWI, illegal opcode, reset. For most of this book (except in Chapter 5) we will disable interrupts while executing in a vulnerable window.

> **Common Error:** The system will crash if the ISR does not either acknowledge or disarm the device requesting the interrupt.

> **Common Error:** The ISR does not have to explicitly disable interrupts at the beginning (`sei`) or explicitly reenable interrupts at the end (`cli`).

The sequence of events that occurs during an interrupt service is quite similar on the four Motorola microcomputers 6805, 6808, 6811, and 6812. The sequence begins with the *Hardware needs service (busy-to-done) transition.* This signal is connected to an input of the microcomputer that can generate an interrupt. For example, the STRA (6811 only), key wakeup, input capture, SCI, and SPI systems support interrupt requests. Some interrupts are internally generated like output compare, real-time interrupt (RTI), and timer overflow.

The second event is the *setting of a flag* in one of the I/O status registers of the microcomputer. This is the same flag that a gadfly interface would be polling on. Examples include the STRA (*STAF*), key wakeup (*KWIFJ2*), input capture (*IC3F*), SCI (*RDRF*), SPI (*SPIF*), output compare (*OC5F*), RTI (*RTIF*), and timer overflow (*TOF*). For an interrupt to be requested the appropriate *flag* bit must be *armed.* Examples include the STRA (*STAI*), key wakeup (*KWIEJ2*), input capture (*IC3I*), SCI (*RIE*), SPI (*SPIE*), output compare (*OC5I*), RTI (*RTII*), and timer overflow (*TOI*). In summary, three conditions must be met simultaneously for an interrupt service to occur:

The interrupting event occurs that sets the flag	STAF=1
The device is armed	STAI=1
The microcomputer interrupts are enabled	I=0.

The third event in the interrupt processing sequence is the *thread switch.* The thread switch is performed by the microcomputer hardware automatically. The specific steps include:

1. The microcomputer will finish the current instruction[3].
2. All the registers are pushed on the stack,[4] the **CCR** is on top, with the **I** bit still equal to 0.
3. The microcomputer will get vector address from memory and put it into the **PC.**
4. The microcomputer will set **I = 1.**

The fourth event is the software *execution of the ISR.* For a polled interrupt configuration the ISR must poll each possible device and branch to a specific handler for that device. The polling order establishes device priority. For a vectored interrupt configuration, you could poll anyway to check for run-time hardware/software errors. The ISR must either acknowledge or disarm the interrupt. We acknowledge an interrupt by clearing the flag that was set in the second event shown above. We will see later in the chapter that there is an optional step at this point for low-priority interrupt handlers. After we acknowledge a low-priority interrupt, we may reenable interrupts (`cli`) to allow higher-priority devices to go first. All ISRs must perform the necessary operations (read data, write data, etc.) and pass parameters through global memory (e.g., FIFO queue).

---

[3]Again, the 6812 allows the instructions `rev` `revw` and `wav` to be interrupted.
[4]Again, the 6808 does not save its Register H.

The last event is another thread switch to *return control back to the thread that was running* when the interrupt was processed. In particular, the software executes a rti at the end of the ISR, which will pull all the registers off the stack. Since the CCR was pushed on the stack in step 2 above with I=0, the execution of rti automatically reenables interrupts. After the ISR executes rti, the stack is restored to the state it was before the interrupt (there may be one more or less entry in the FIFO however).

4.4.1
6805 Interrupts

Like most Motorola microcomputers, the 6805 automatically saves all registers, and the rti instruction restores registers. When the 6805 processes an interrupt, it temporarily sets I during the ISR to postpone other interrupt requests, including itself. Figure 4.17 shows the 6805 stack before and after an interrupt is processed.

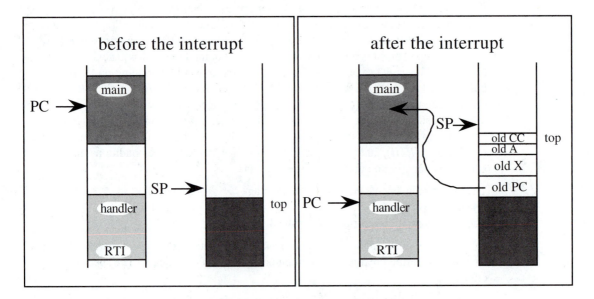

Figure 4.17
6805/6808 stack before and after an interrupt.

4.4.2
6808 Interrupts

Unlike most Motorola microcomputers, the 6808 does not automatically save all its registers. To make the 6808 object code compatible with the 6805, the 6808 does not save and restore Register H automatically. If you use Register H in the ISR, then you must explicitly save and restore it using the pshh and pulh instructions. The stack picture for the 6808 is identical to the 6805.

4.4.3
6811 Interrupts

The 6811 interrupt hardware will automatically save all registers on the stack during the thread switch. The thread switch is the process of stopping the foreground (main) thread and starting the background (interrupt handler). The "oldPC" value on the stack points to the place in the foreground thread to resume once the interrupt is complete. At the end of the interrupt handler, another thread switch occurs as the rti instruction restores registers from the stack (including the PC) (Figure 4.18).

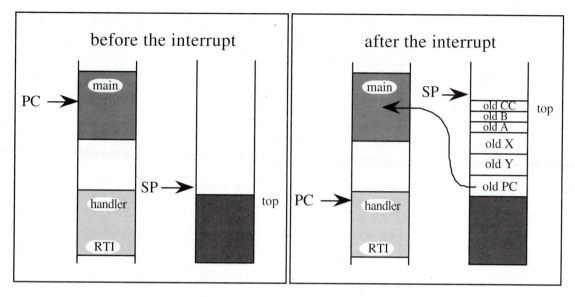

Figure 4.18
6811 stack before and after an interrupt.

The 6811 has two external requests $\overline{\text{IRQ}}$ and $\overline{\text{XIRQ}}$ that are level-zero active. Many of the internal I/O devices can generate interrupt requests based on external events (e.g., STRA, input capture, SCI, SPI). Most of the interrupt requests will temporarily set the I bit in the CC during the interrupt program to prevent other interrupts (including itself). On the other hand, the XIRQ request temporarily sets both the I and X bits in the CC during the interrupt program to postpone all other interrupts sources. The MC68HC11A8 can support:

A STRA interrupt
Three input capture interrupts
Five output compare interrupts
Three timer interrupts (timer overflow, RTI, pulse accumulator)
Two serial port interrupts (SCI and SPI)

The interrupts have a fixed priority, but you can elevate one request to highest priority using the hardware priority interrupt (HPRIO) register ($103C).

We typically use XIRQ to interface a single highest-priority device. XIRQ has a separate interrupt vector ($FFF4) and a separate enable bit (X). Once the X bit is cleared (enabled), the software cannot disable it. A XIRQ interrupt is requested when the external XIRQ pin is low and the X bit in the CCR is 0. XIRQ processing will automatically set $X = I = 1$ (an IRQ cannot interrupt an XIRQ service) at the start of the XIRQ handler. Just like regular interrupts, the X and I bits will be restored to their original values by the `rti` instruction.

4.4.4
6812 Interrupts

CPU12 exceptions include resets and interrupts. Each exception has an associated 16-bit vector that points to the memory location where the routine that handles the exception is located. Vectors are stored in the upper 128 bytes of the standard 64-kbyte address map.

4.4.4.1
6812 Exceptions

A hardware priority hierarchy determines which reset or interrupt is serviced first when simultaneous requests are made. Six sources are not maskable. The remaining sources have a mask bit that can be enabled (armed) or turned off (disarmed). The priorities of the nonmaskable sources are:

1. Power-on-reset (POR) or regular hardware RESET pin
2. Clock monitor reset
3. COP watchdog reset
4. Unimplemented instruction trap
5. Software interrupt instruction (`swi`)
6. XIRQ signal (if X bit in CCR = 0)

Maskable interrupt sources include on-chip peripheral systems and external interrupt service requests. Interrupts from these sources are recognized when the global interrupt mask bit (I) in the CCR is cleared. The default state of the I bit out of reset is 1, but it can be written at any time.

The 6812 interrupt hardware also automatically saves all registers on the stack during the thread-switch. The "oldPC" value on the stack points to the place in the foreground thread to resume once the interrupt is complete. At the end of the interrupt handler, another thread switch occurs as the `rti` instruction restores registers from the stack (including the PC) (Figure 4.19).

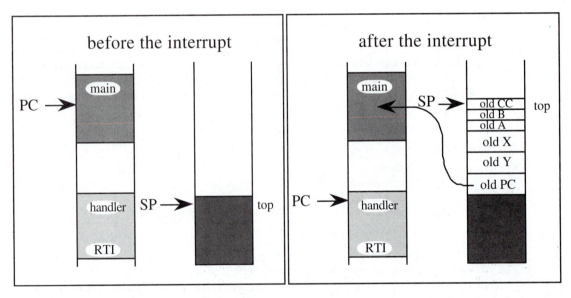

Figure 4.19
6812 stack before and after an interrupt.

Like the 6811, the 6812 has two external requests $\overline{\text{IRQ}}$ and $\overline{\text{XIRQ}}$ that are level-zero active. Many of the internal I/O devices can generate interrupt requests based on external events (e.g., key wakeup, input capture, SCI, SPI). These interrupt requests will temporarily set the I bit in the CC during the interrupt program to prevent other interrupts (including itself). On the other hand, the XIRQ request temporarily sets both the I and X bits in the CC during the interrupt program to postpone all other interrupts sources. The MC68HC812A4 can support:

24 key wakeup interrupts (Ports D, H, and J)
Eight input capture/output compare interrupts
An ADC interrupt
Three timer interrupts (timer overflow, RTI, pulse accumulator),
Three serial port interrupts (two SCIs and SPI).

The interrupts have a fixed priority, but you can elevate one request to highest priority using the HPRIO register ($001F). The relative priorities of the other interrupt sources remain the same.

The 6812 XIRQ interrupt system is similar to the one on the 6811. We typically use XIRQ to interface a single highest-priority device. XIRQ has a separate interrupt vector ($FFF4) and a separate enable bit (X). Once the X bit is cleared (enabled), the software cannot disable it. A XIRQ interrupt is requested when the external XIRQ pin is low and the X bit in the CCR is 0. XIRQ processing will automatically set $X = I = 1$ (an IRQ cannot interrupt an XIRQ service) at the start of the XIRQ handler. Just like regular interrupts, the X and I bits will be restored to their original values by the `rti` instruction.

4.4.4.2
MC68HC812A4 Key
Wakeup Interrupts

One of the simplest interrupt mechanisms on the 6812 is the key wakeup interrupt. On the MC68HC812A4, key wakeups are available on Ports D, H, J. Any or all of these 24 pins can be configured as a key wakeup interrupt. Each of the 24 wakeup lines has a separate I/O pin (PORTD PORTH PORTJ),

	7	6	5	4	3	2	1	0	
PORTD	PD7	PD6	PD5	PD4	PD3	PD2	PD1	PD0	$0005
PORTH	PH7	PH6	PH5	PH4	PH3	PH2	PH1	PH0	$0024
PORTJ	PJ7	PJ6	PJ5	PJ4	PJ3	PJ2	PJ1	PJ0	$0028

a separate direction bit (DDRD DDRH DDRJ that we configure as an input to use key wakeup),

	7	6	5	4	3	2	1	0	
DDRD			0 means input 1 means output						$0007
DDRH			0 means input 1 means output						$0025
DDRJ			0 means input 1 means output						$0029

a separate flag bit (KWIFD, KWIFH, KWIFJ),

	7	6	5	4	3	2	1	0		
KWIFD		set on fall of corresponding bit clear on a write 1 to this register							$0021
KWIFH		set on fall of corresponding bit clear on a write 1 to this register							$0027
KWIFJ		set on active edge of corresponding bit clear on a write 1 to this register							$002B

and a separate arm bit (KWIED, KWIEH, KWIEJ).

	7	6	5	4	3	2	1	0		
KWIED			0 means disarm 1 means arm						$0020
KWIEH			0 means disarm 1 means arm						$0026
KWIEJ			0 means disarm 1 means arm						$002A

For Ports D and H, a falling edge on the input pin will set the corresponding flag bit. Wakeup interrupts on Port J can be configured on either the rising or falling edge. If the corresponding bit in the KPOLJ is 0, then a falling edge on Port J is active. Conversely, if the bit in the KPOLJ register is 1, then a rising edge on Port J will set the flag. Another convenience of Port J is the available pull-up (Table 4.3) or pull-down resistors via the PUPSJ and PULEJ configuration registers. Each of the eight pins of port J can be configured separately.

	7	6	5	4	3	2	1	0	
PUPSJ									$002D
PULEJ									$002E

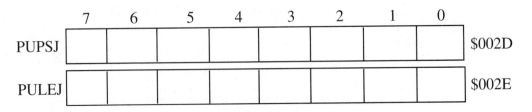

Table 4.3
Pull-up modes of Port J.

DDRJ	PUPSJ bit	PULEJ bit	Port J mode
1	0/1	0/1	Regular output
0	0/1	0	Regular input
0	0	1	Input with passive pull-down
0	1	1	Input with passive pull-up

Figure 4.20
6812 Port J supports internal passive pull-up and pull-down.

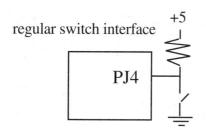

regular switch interface

+5

PJ4

with internal pull up
PUPSJ.4=1 and PULEJ.4=1

PJ4

A typical application of pull-up is the interface of simple switches. Using pull-up mode eliminates the need for an external resistor (Figure 4.20). A detailed discussion of interfacing switches and keyboards will be presented in Chapter 8 on parallel port interfaces.

Three conditions must be simultaneously true for a key wakeup interrupt to be requested:

- The flag bit is set
- The arm bit is set
- The I bit in the 6812 CCR is 0

Even though there are 24 key wakeup lines, there are only three interrupt vectors. So, if two or more wakeup interrupts are used on the same port, it will be necessary to poll. Interrupt polling is the software function to look and see which of the potential sources requested the interrupt.

4.5 Interrupt Vectors and Priority

4.5.1 MC6805 Interrupt Vectors and Priority

This book covers two MC6805 microcomputers, the MC68HC705J1A and the MC68HC05C8. The priority is fixed in the order shown in Table 4.4, with the top of the list having the lowest priority and *Reset* having the highest. Some of the potential interrupt requests share the same interrupt vector (e.g., both RTI and TO use the vector at $07F8). Therefore, when this request is processed, the software must determine which of the two possible conditions caused the interrupt. The entire list of potential MC68HC705J1A interrupts is included in the table. Table 4.5 shows the list of potential MC68HC05C8 interrupts.

Table 4.4
Interrupt vectors for the MC68HC705J1A.

Vector address	Interrupt source	CCR mask	Local mask
$07F8	Timer functions: RTI: real-time interrupt TOF: Timer overflow	 I bit I bit	 TSCR.RTIE TSCR.TOIE
$07FA	IRQ external interrupt	I bit	ISCR.IRQE
$07FC	Software interrupt	None	None
$07FE	RESET	None	None

Table 4.5
Interrupt vectors for the
MC68HC05C8

Vector address	Interrupt source	CCR mask	Local mask
$1FF6	SCI:		
	TC: transmit complete	I bit	SCCR2.TCIE
	TDRE: transmit data register empty	I bit	SCCR2.TIE
	RDRF: receive data register full	I bit	SCCR2.RIE
	IDLE: idle line detect	I bit	SCCR2.ILIE
	OR: overrun	I bit	SCCR2.RIE
$1FF8	Timer functions:		
	ICF: input capture flag	I bit	TCR.ICIE
	OCF: output compare flag	I bit	TCR.OCIE
	TOF: timer overflow flag	I bit	TCR.TOIE
$1FFA	IRQ external interrupt	I bit	ISCR.IRQE
$1FFC	Software interrupt	None	None
$1FFE	RESET	None	None

4.5.2 MC68HC708XL36 Interrupt Vectors and Priority

The priority is fixed in the order shown in Table 4.6, with *IRQ2/Keypad* having the lowest priority and Reset having the highest. Many of the potential interrupt requests share the same interrupt vector (e.g., there are nine possible interrupt sources [eight keypad plus IRQ2] that all use the vector at $FFE0). Therefore, when this request is processed, the software must determine which of the nine possible conditions caused the interrupt. The entire list of potential MC68HC708XL36 interrupts is included in the table.

Table 4.6
Interrupt vectors for the
MC68HC708XL36.

Vector address	Interrupt source	CCR mask	Local mask
$FF00–$FFDE	Reserved	—	—
$FFE0	IRQ2/Keypad	I Bit	ISCR.IMASK2
$FFE2	SCI transmit buffer empty	I Bit	SCC2.SCTIE
$FFE2	SCI transmit buffer complete	I Bit	SCC2.CTIE
$FFE4	SCI receive buffer full	I Bit	SCC2.SCRIE
$FFE4	SCI receive buffer idle	I Bit	SCC2.ILIE
$FFE6	SCI receive buffer overrun	I Bit	SCC3.ORIE
$FFE6	SCI receive buffer noise error	I Bit	SCC3.NEIE
$FFE6	SCI receive buffer frame error	I Bit	SCC3.FEIE
$FFE6	SCI receive buffer parity error	I Bit	SCC3.PEIE
$FFE8	SPI transmit buffer empty	I Bit	SPCR.SPTIE
$FFEA	SPI receive buffer full	I Bit	SPCR.SPRIE
$FFEC	Timer overflow	I Bit	TSC.TOIE
$FFEE	Timer channel 3	I Bit	TSC3.CH3IE
$FFF0	Timer channel 2	I Bit	TSC2.CH2IE
$FFF2	Timer channel 1	I Bit	TSC1.CH1IE
$FFF4	Timer channel 0	I Bit	TSC0.CH0IE
$FFF6	DMA channel 2 complete	I Bit	DC1.IEC2
$FFF6	DMA channel 1 complete	I Bit	DC1.IEC1
$FFF6	DMA channel 0 complete	I Bit	DC1.IEC0
$FFF8	PLL	I Bit	PCTL.PLLIE
$FFFA	IRQ1 external interrupt	I Bit	ISCR.IMASK1
$FFFC	Software interrupt	None	None
$FFFE	RESET	None	None

4.5.3
MC68HC11A8
Interrupt Vectors
and Priority

The priority is fixed in the order shown in Table 4.7, with *SCI* having the lowest priority and Reset having the highest. For some interrupt sources, such as the SCI interrupts, flags are automatically cleared during the response to the interrupt requests. For example, the RDRF flag in the SCI system is cleared by the automatic clearing mechanism, consisting of a read of the SCI status register while RDRF is set, followed by a read of the SCI data register. The normal response to an RDRF interrupt request is to read the SCI status register to check for receive errors, then to read the received data from the SCI data register. These two steps satisfy the automatic clearing mechanism without requiring any special instructions. Many of the potential interrupt requests share the same interrupt vector (e.g., there are multiple possible interrupt sources [STAF and all external connections to IRQ] that all use the vector at $FFF2). Therefore, when this request is processed, the software

Table 4.7

Interrupt vectors for the 6811.

Vector	Interrupt source	Enable	Arm
Start:			
$FFFE	Power on reset	Always	Always highest priority
$FFFE	Hardware reset	Always	Always
COP:			
$FFFC	COP clock monitor fail	Always	OPTION.CME=1
$FFFA	COP failure	Always	CONFIG.NOCOP=0
External interrupts:			
$FFF4	Nonmaskable XIRQ	X=0	External hardware
$FFF2	External IRQ	I=0	External hardware (e.g., 6821, 6840, 6850)
Parallel I/O:			
$FFF2	Parallel I/O handshake, STAF	I=0	PIOC.STAI=1
Programmable timer:			
$FFF0	Real time interrupt, RTIF	I=0	TMSK2.RTII=1
$FFEE	Timer input capture 1, IC1F	I=0	TMSK1.IC1I=1
$FFEC	Timer input capture 2, IC2F	I=0	TMSK1.IC2I=1
$FFEA	Timer input capture 3, IC3F	I=0	TMSK1.IC3I=1
$FFE8	Timer output compare 1, OC1F	I=0	TMSK1.OC1I=1
$FFE6	Timer output compare 2, OC2F	I=0	TMSK1.OC2I=1
$FFE4	Timer output compare 3, OC3F	I=0	TMSK1.OC3I=1
$FFE2	Timer output compare 4, OC4F	I=0	TMSK1.OC4I=1
$FFE0	Timer output compare 5, OC5F	I=0	TMSK1.OC5I=1
$FFDE	Timer overflow, TOF	I=0	TMSK2.TOI=1
Pulse accumulator:			
$FFDC	Pulse accumulator overflow	I=0	TMSK2.PAOVI=1
$FFDA	Pulse accumulator input edge	I=0	TMSK2.PAII=1
SPI:			
$FFD8	SPI serial transfer complete, SPIF	I=0	SPCR.SPIE=1
SCI:			
$FFD6	Receive data register full, RDRF	I=0	SCCR2.RIE=1
$FFD6	Receive overrun, OVRN	I=0	SCCR2.RIE=1
$FFD6	Transmit data register empty, TDRE	I=0	SCCR2.TIE=1
$FFD6	Transmit complete, TC	I=0	SCCR2.TCIE=1
$FFD6	Idle line detect, IDLE	I=0	SCCR2.ILIE=1
Software interrupts:			
$FFF8	Illegal opcode trap	Always	Always
$FFF6	Software interrupt SWI	Always	Always lowest priority

must determine which of the possible signals caused the interrupt. The entire list of potential 6811 interrupts is included in the table.

How we establish the vector depends on the memory configuration. For a stand-alone
single-chip 6811 system or multichip systems with your own PROM at $C000 to $FFFF,
we can use assembly language to set the vectors.

```
org    $FFF0
fdb    RTIHAN      Pointer to real time interrupt handler
org    $FFF2
fdb    IRQHAN      Pointer to external IRQ and STRA handler
org    $FFF4
fdb    XIRQHAN     Pointer to external XIRQ handler
org    $FFFE
fdb    RESETHAN    Pointer to hardware reset and power on reset handler
```

With the Motorola 6811 EVB, we usually use a $C000 to $FFFF 8K PROM containing the
debugger called BUFFALO. Since the interrupt vector addresses are in this PROM (something you do not wish to change while developing 6811 EVB software), BUFFALO creates
jumps into RAM. You must place a 3-byte `jmp` instruction at these reserved RAM memory
locations. Table 4.8 lists all BUFFALO RAM interrupt jump locations.

Table 4.8
Interrupt vectors for the 6811 EVB using BUFFALO.

Interrupt source	6811 vector	BUFFALO jump location
SCI	$FFD6	$00C4–$00C6
SPI	$FFD8	$00C7–$00C9
Pulse accumulator input edge	$FFDA	$00CA–$00CC
Pulse accumulator overflow	$FFDC	$00CD–$00CF
Timer overflow	$FFDE	$00D0–$00D2
Timer output compare 5	$FFE0	$00D3–$00D5
Timer output compare 4	$FFE2	$00D6–$00D8
Timer output compare 3	$FFE4	$00D9–$00DB
Timer output compare 2	$FFE6	$00DC–$00DE
Timer output compare 1	$FFE8	$00DF–$00E1
Timer input capture 3	$FFEA	$00E2–$00E4
Timer input capture 2	$FFEC	$00E5–$00E7
Timer input capture 1	$FFEE	$00E8–$00EA
RTI	$FFF0	$00EB–$00ED
IRQ (6821,6840,6850, and STAF)	$FFF2	$00EE–$00F0
XIRQ (external hardware)	$FFF4	$00F1–$00F3
Software interrupt (SWI)	$FFF6	$00F4–$00F6
Illegal opcode	$FFF8	$00F7–$00F9
COP	$FFFA	$00FA–$00FC
Clock monitor	$FFFC	$00FD–$00FF
Reset	$FFFE	

In each case, a 3-byte value is stored in the jump location similar to the following examples. The jump opcode ($7E) is stored in the first byte, followed by the 16-bit address of
the device handler. For systems with the BUFFALO monitor add this code to the ritual

```
ldaa  #$7E      Op code for JMP
staa  $00EB
ldx   #RTIHAN
```

```
stx  $00EC    JMP RTIHAN
ldaa #$7E     Op code for JMP
staa $00EE
ldx  #IRQHAN
stx  $00EF    JMP IRQHAN
ldaa #$7E     Op code for JMP
staa $00F1
ldx  #XIRQHAN
stx  $00F2    JMP XIRQHAN
```

4.5.4
MC68HC812A4
Interrupt Vectors
and Priority

The priority is fixed in the order shown in Table 4.9, with *Key Wakeup H* having the lowest priority and *Reset* having the highest. Any one particular system usually uses just a few interrupts, but the entire list of potential MC68HC812A4 interrupts is included in the table.

Table 4.9
Interrupt vectors for the
MC68HC812A4.

Vector Address	Interrupt source	CCR mask	Local enable
$FFFE	Reset	None	none
$FFFC	COP clock monitor fail reset	None	COPCTL.CME
		None	COPCTL.FCME
$FFFA	COP failure reset	None	COP rate selected
$FFF8	Unimplemented instruction trap	None	none
$FFF6	SWI	None	none
$FFF4	XIRQ	X bit	none
$FFF2	IRQ or	I bit	INTCR.IRQEN
	Key wakeup D	I bit	KWIED,[7:0]
$FFF0	RTI, RTIF	I bit	RTICTL.RTIE
$FFEE	Timer channel 0, C0F	I bit	TMSK1.C0I
$FFEC	Timer channel 1, C1F	I bit	TMSK1.C1I
$FFEA	Timer channel 2, C2F	I bit	TMSK1.C2I
$FFE8	Timer channel 3, C3F	I bit	TMSK1.C3I
$FFE6	Timer channel 4, C4F	I bit	TMSK1.C4I
$FFE4	Timer channel 5, C5F	I bit	TMSK1.C5I
$FFE2	Timer channel 6, C6F	I bit	TMSK1.C6I
$FFE0	Timer channel 7, C7F	I bit	TMSK1.C7I
$FFDE	Timer overflow, TOF	I bit	TMSK2.TOI
$FFDC	Pulse accumulator overflow	I bit	PACTL.PAOVI
$FFDA	Pulse accumulator input edge	I bit	PACTL.PAI
$FFD8	SPI serial transfer complete, SPIF	I bit	SP0CR1.SPI0E
$FFD6	SCI 0 transmit buffer empty, TDE	I bit	SC0CR2.TIE
	SCI 0 transmit complete, TC	I bit	SC0CR2.TCIE
	SCI 0 receiver buffer full, RDF	I bit	SC0CR2.RIE
	SCI 0 receiver idle	I bit	SC0CR2.ILIE
$FFD4	SCI 1 transmit buffer empty, TDE	I bit	SC1CR2.TIE
	SCI 1 transmit complete, TC	I bit	SC1CR2.TCIE
	SCI 1 receiver buffer full, RDF	I bit	SC1CR2.RIE
	SCI 1 receiver idle	I bit	SC1CR2.ILIE
$FFD2	ATD sequence complete	I bit	ATDCTL2.ASCIE
$FFD0	Key wakeup J (stop wakeup)	I bit	KWIEJ.[7:0]
$FFCE	Key wakeup H (stop wakeup)	I bit	KWIEH.[7:0]
$FF80–$FFCD	Reserved	I bit	

For some interrupt sources, such as the SCI interrupts, flags are automatically cleared during the response to the interrupt requests. For example, the RDRF flag in the SCI system is cleared by the automatic clearing mechanism, consisting of a read of the SCI status register while RDRF is set, followed by a read of the SCI data register. The normal response to an RDRF interrupt request is to read the SCI status register to check for receive errors, then to read the received data from the SCI data register. These two steps satisfy the automatic clearing mechanism without requiring any special instructions. Many of the potential interrupt requests share the same interrupt vector. (e.g., there are eight possible interrupt sources [PJ7-PJ0] that all use the vector at $FFD0). Therefore, when this request is processed, the software must determine which of the eight possible signals caused the interrupt.

4.5.5
MC68HC912B32
Interrupt Vectors
and Priority

The priority is fixed in the order shown in Table 4.10, with *BDLC* having the lowest priority and Reset having the highest. Many of the potential interrupt requests share the same interrupt vector. For example, there are four possible SCI interrupt sources (TDRE, TC, RDRF, IDLE) that all use the vector at $FFD6. Therefore, when this request is processed, the software must determine which of the four possible signals caused the interrupt. Any one particular system usually uses just a few interrupts, but the entire list of potential MC68HC912B32 interrupts is included in the table.

Table 4.10

Interrupt vectors for the MC68HC912B32.

Vector address	Interrupt source	CCR mask	Local enable
$FFFE, $FFFF	Reset	None	None
$FFFC, $FFFD	COP clock monitor fail reset	None	COPCTL.CME
		None	COPCTL.FCME
$FFFA, $FFFB	COP failure reset	None	COP rate selected
$FFF8, $FFF9	Unimplemented instruction trap	None	None
$FFF6, $FFF7	SWI	None	None
$FFF4, $FFF5	XIRQ	X bit	None
$FFF2, $FFF3	IRQ	I bit	INTCR.IRQEN
$FFF0, $FFF1	RTI, RTIF	I bit	RTICTL.RTIE
$FFEE, $FFEF	Timer channel 0, C0F	I bit	TMSK1.C0I
$FFEC, $FFED	Timer channel 1, C1F	I bit	TMSK1.C1I
$FFEA, $FFEB	Timer channel 2, C2F	I bit	TMSK1.C2I
$FFE8, $FFE9	Timer channel 3, C3F	I bit	TMSK1.C3I
$FFE6, $FFE7	Timer channel 4, C4F	I bit	TMSK1.C4I
$FFE4, $FFE5	Timer channel 5, C5F	I bit	TMSK1.C5I
$FFE2, $FFE3	Timer channel 6, C6F	I bit	TMSK1.C6I
$FFE0, $FFE1	Timer channel 7, C7F	I bit	TMSK1.C7I
$FFDE, $FFDF	Timer overflow, TOF	I bit	TMSK2.TOI
$FFDC, $FFDD	Pulse accumulator overflow	I bit	PACTL.PAOVI
$FFDA, $FFDB	Pulse accumulator input edge	I bit	PACTL.PAI
$FFD8, $FFD9	SPI serial transfer complete	I bit	SP0CR1.SPIE
$FFD6, $FFD7	SCI 0 transmit buffer empty, TDE	I bit	SC0CR2.TIE
	SCI 0 transmit complete, TC	I bit	SC0CR2.TCIE
	SCI 0 receiver buffer full, RDF	I bit	SC0CR2.RIE
	SCI 0 receiver idle	I bit	SC0CR2.ILIE
$FFD4, $FFD5	Reserved	I bit	—
$FFD2, $FFD3	ATD	I bit	ATDCTL2.ASCIE
$FFD0, $FFD1	BDLC	I bit	BCR1.IE
$FF80–$FFCF	Reserved	I bit	—

4.6 External Interrupt Design Approach

This short section deals with the design steps that occur when interfacing an I/O device with interrupt synchronization. Because the details of specific I/O devices can vary considerably, this section only serves as a framework, listing the issues we should consider when using interrupts. Computer engineering design, like all disciplines, is an iterative process. We will begin with the external hardware design.

First, we identify the busy-to-done state transition to cause interrupt. We ask the question, what status signal from the I/O device signifies when the input device has new data and needs the software to read them. For an output device, we look for a signal that specifies when it is finished and needs the software to give it more data. As shown in Figure 4.21 this signal will be an output of the I/O device and an input to the microcomputer.

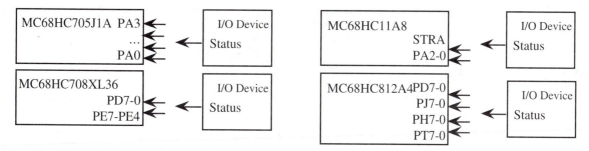

Figure 4.21
We connect external devices to microcomputer lines that can generate interrupts.

Next, we connect I/O status signal to a microcomputer input that can generate interrupts. On the MC68HC705J1A, we could use the keyboard scan interrupts on Port A. On the MC68HC708XL36, we could use the keyboard inputs on Port D or the input capture on Port E. On the MC68HC11, we could use STRA or input capture on Port A. On the MC68HC812A4, we could use the Key wakeup on Ports D, J, H, or input capture on Port T. On the MC68HC912B32, we could use input capture on Port T.

There are four major components to the interrupting software system. In no particular order, they are the ritual, the main program, the ISR(s), and the interrupt vectors.

The *ritual* is executed once on start-up. It is probably a good idea to disable interrupts at the beginning of the ritual so that interrupts are not requested before the entire initialization sequence is allowed to finish. The ritual usually initializes the global data structures (e.g., InitFifo), sets the I/O port direction register(s) as needed, and sets the I/O port interrupt hardware control register, specifying the proper conditions to request the interrupt. An optional step is to clear the flag (the one that when set will request an interrupt). We like to add this optional step so that the first interrupt to occur will be the result of activity that occurred after the ritual, and not the result of activities prior to executing the ritual. For example, it is possible for the power-on sequence to different hardware modules to be different, causing edges to occur on digital lines that falsely mimic actual I/O activity. The last steps of the ritual are to arm the device and enable interrupts.

The *main program* executes in the foreground. At the start the SP is initialized, and the ritual is executed. The main program generally performs tasks that are not time-critical. The main program interacts with the ISRs via global memory data structures (e.g., FIFO queue).

The *interrupt handler* executes in the background. For a polled interrupt, it must determine the source by polling potential devices. The polling order will establish a priority, although later in the chapter we discuss more sophisticated methods for implementing priority. The ISR must acknowledge (clear the flag that requested the interrupt) or disarm. We acknowledge if we are interested in more interrupts and disarm if we are no longer interested in interrupts. Information is exchanged with the main program and other interrupt handlers (including itself) via global memory. The ISR executes rti to return control back to the program previously executing. Because of the real-time nature of interrupting devices, debugging tools must be minimally intrusive. Examples of good debugging tools to use for interrupts are (1) instrumentation that dumps into an array, (2) instrumentation that dumps into an array with filtering, and (3) instrumentation using an output port. The following are examples of bad debugging tools because they significantly affect the dynamic response of the interface (i.e., they require too much time to execute): (1) single-stepping, (2) breakpoints, and (3) print statements. These techniques were discussed in Section 2.11.

The last component is the *interrupt vectors*. On general-purpose computers, interrupt vectors are in RAM, and the software dynamically attaches and unattaches interrupt han-

Program 4.26 ICC12 C code to set interrupt vectors for the 6812.

```
extern void _start(); /* entry point in crt12.s */
extern void SCIhandler();
extern void TC4handler();
extern void TOFhandler();
#define DUMMY_ENTRY    (void (*)())0xF000
#pragma abs_address:0xffce
void (*interrupt_vectors[])() = {
        DUMMY_ENTRY,    /* ffce 812 KeyWakeUpH */
        DUMMY_ENTRY,    /* ffd0 912 BDLC, 812 KeyWakeUpJ */
        DUMMY_ENTRY,    /* ffd2 ATD */
        DUMMY_ENTRY,    /* ffd4 812 SCI1 */
        SCIhandler,     /* ffd6 SCI, 812 SCI0 */
        DUMMY_ENTRY,    /* ffd8 SPI */
        DUMMY_ENTRY,    /* ffda PAIE */
        DUMMY_ENTRY,    /* ffdc PAO */
        TOFhandler,     /* ffde TOF */
        DUMMY_ENTRY,    /* ffe0 TC7 */
        DUMMY_ENTRY,    /* ffe2 TC6 */
        DUMMY_ENTRY,    /* ffe4 TC5 */
        TC4handler,     /* ffe6 TC4 */
        DUMMY_ENTRY,    /* ffe8 TC3 */
        DUMMY_ENTRY,    /* ffea TC2 */
        DUMMY_ENTRY,    /* ffec TC1 */
        DUMMY_ENTRY,    /* ffee TC0 */
        DUMMY_ENTRY,    /* fff0 RTI  */
        DUMMY_ENTRY,    /* fff2 IRQ, 812 KeyWakeUpD*/
        DUMMY_ENTRY,    /* fff4 XIRQ  */
        DUMMY_ENTRY,    /* fff6 SWI   */
        DUMMY_ENTRY,    /* fff8 ILLOP */
        DUMMY_ENTRY,    /* fffa COP   */
        DUMMY_ENTRY,    /* fffc CLM   */
        _start          /* fffe RESET, entry point into ICC12 */
        };
#pragma end_abs_address
```

dlers to these vectors. On most embedded systems, the interrupt vectors reside in PROM or ROM and their values are determined at compile time and initialized by the ROM programmer. Nevertheless, we must establish interrupt vectors to point to the appropriate ISR. The syntax for setting vectors is compiler-dependent. Program 4.26 establishes some interrupt vectors (including the reset vector) for the MC68HC812A4 using the ImageCraft ICC12.

4.7   Polled Versus Vectored Interrupts

As we defined earlier, when more than one source of interrupt exists, the computer must have a reliable method to determine which interrupt request has been made. There are two common approaches, and the computers studied in this book (6805/6808/6811/6812) apply a combination of both methods. The first approach is called vectored interrupts. With a vectored interrupt system each potential interrupt source has a unique interrupt vector address (Program 4.27). You simply place the correct handler address in each vector, and the hardware automatically calls the correct software when an interrupt is requested. One external interrupt request is shown in Figure 4.22 and another interrupt is generated by a timer overflow.

```
; MC68HC705J1A
TimeHan lda #$08  ;TOFR
        sta TSCR  ;clear TOF
;*Timer interrupt calculations*
        rti
ExtHan  lda ISCR
        ora #$02  ;IRQR
        sta ISCR  ;clear IRQ
;*External interrupt calculations*
        rti
        org $07F8  ;timer overflow
        fdb TimeHan
        org $07FA  ;IRQ external
        fdb ExtHan
```

```
; MC68HC11A8
TimeHan ldaa #$80  ;TOF is bit 7
        staa TFLG2 ;clear TOF
;*Timer interrupt calculations*
        rti
ExtHan  ldaa PIOC
        ldaa PORTCL ;clear STAF
;*External interrupt calculations*
        rti
        org  $FFDE  ;timer overflow
        fdb  TimeHan
        org  $FFF2  ;IRQ external
        fdb  ExtHan
```

```
; MC68HC708XL36
TimeHan lda TSC  ;read status
        sta TSC  ;clear TOF
;*Timer interrupt calculations*
        rti
ExtHan  lda ISCR
        ora #$40  ;ACK2
        sta ISCR  ;clear IRQ2
;*External interrupt calculations*
        rti
        org $FFEC  ;timer overflow
        fdb TimeHan
        org $FFE0  ;IRQ2/Keypad
        fdb ExtHan
```

```
; MC68HC812A4
TimeHan ldaa #$80  ;TOF is bit 7
        staa TFLG2 ;clear TOF
;*Timer interrupt calculations*
        rti
ExtHan  ldaa #$01  ;Ack
        staa KWIFJ  ;clear flag
;*External interrupt calculations*
        rti
        org  $FFDE  ;timer overflow
        fdb  TimeHan
        org  $FFD0  ;Key wakeup J
        fdb  ExtHan
```

Program 4.27 Example of a vectored interrupt.

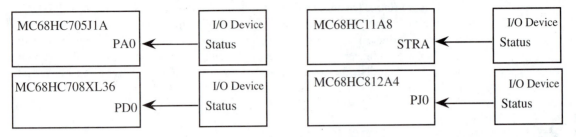

Figure 4.22 An external I/O device is connected to the microcomputer.

Since the two sources have separate vectors, the timer interrupt hardware (timer over-flow TOF) will automatically activate `TimeHan` and the external interrupt hardware will automatically activate `ExtHan`.

The second approach is called polled interrupts. With a polled interrupt system, multi-ple interrupt sources share the same interrupt vector address (Program 4.28). Once the in-terrupt has occurred, the ISR software must poll the potential devices to determine which device needs service. Two external interrupt requests are shown in Figure 4.23.

```
; MC68HC705J1A
ExtHan brset 0 PORTA PA0Han
       brset 1 PORTA PA1Han
       bset  1,ISCR ;IRQR
       rti          ;no more

PA0Han
;*PA0 interrupt calculations*
       bra ExtHan ;other
PA1Han
;*PA1 interrupt calculations*
       bra ExtHan ;other

       org $07FA  ;IRQ external
       fdb ExtHan
```

```
; MC68HC11A8
ExtHan   ldaa PIOC    ;which one
         bita #$80    ;STAF?
         bne  STAFHan
         ldaa OtherStatus
         bita #$80    ;External?
         bne  OtherHan
         swi          ;error
STAFHan ldaa PORTCL ;clear STAF
;*STAF interrupt calculations*
         rti
OtherHan ldaa OtherData
;*Other interrupt calculations*
         rti
         org  $FFF2  ;IRQ external
         fdb  ExtHan
```

```
; MC68HC708XL36
ExtHan brset 0,PORTD,PD0Han
       brset 1,PORTD,PD1Han
       bset  6,ISCR ;ACK2
       rti          ;no more
PD0Han
;*PD0 interrupt calculations*
       bra ExtHan ;other
PD1Han
;*PD1 interrupt calculations*
       bra ExtHan ;other

       org $FFE0  ;IRQ2/Keypad
       fdb ExtHan
```

```
; MC68HC812A4
ExtHan brset KWIFJ,$01,KJ0Han
       brset KWIFJ,$02,KJ1Han
       swi          ;error

KJ0Han ldaa #$01    ;Ack flag0
       staa KWIFJ ;clear flag0
;*KJ0 interrupt calculations*
       rti

KJ1Han ldaa #$02    ;Ack flag1
       staa KWIFJ ;clear flag1
;*KJ1 interrupt calculations*
       rti
       org  $FFD0  ;Key wakeup J
       fdb  ExtHan
```

Program 4.28 Example of a polled interrupt.

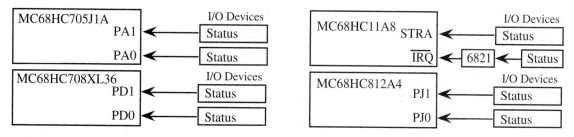

Figure 4.23
When two or more devices share an interrupt line, then the ISR must poll.

Since the two sources have the same vector, the ISR software must first determine which one caused the interrupt. The 6811/6812 systems have a separate acknowledgment so that if both interrupts are pending, acknowledging one will not satisfy the other, so the second device[5] will request a second interrupt and get serviced. On the other hand, the 6805/6808 systems share a common acknowledgment so that a return from interrupt is executed only after both are polled and neither needs service.

It is possible to loose a 6805/6808 interrupt if (1) PA0/PD0 interrupt comes, (2) the PA0/PD0 is serviced, (3) the PA1/PD1 is polled but it is not ready (4) a PA1/PD1 interrupt occurs between the `brset` 1 and `bset` instructions, and (5) the PA1/PD1 interrupt is acknowledged without it getting service. On the other hand, the sequence of requests to the 6811/6812 does not matter because each request has a separate acknowledgement.

> **Common Error:** If two interrupts were requested, it would be a mistake to service just one and acknowledge them both.

> **Observation:** External events are often asynchronous to program execution, so careful thought is required to consider the effect if an external interrupt request were to come in between each pair of instructions.

> **Observation:** The computer automatically sets the I bit during processing so that an interrupt handler will not interrupt itself.

Two polling techniques are presented in this book. The first is *minimal* polling. This method performs a minimally sufficient check to determine the source. Usually this entails simply checking the particular flag bit that caused the interrupt. The above interrupt polling are examples of minimal polling. A more robust polling method is called *polling for 0s and 1s*. This verifies as much information in the control/status register as possible and usually entails checking for the presence of both 1s and 0s in the control/status register. For example, assume the interrupt flag bit is in bit 7, bit 6 is unknown, and bits 5–0 should be 000111. We write this expected condition as 1x000111. Simple polling only checks the flag bit, while polling for 0s and 1s compares the actual 7 bits with their expected values. One implementation is

```
ldaa CSR        read control/status register
anda #%1011111  clear bits that are indeterminate
cmpa #%1000111  expected value if this device active
beq  Handler    execute if this device is requesting
```

It is good software engineering practice to install consistency checks during the prototype and debugging stages. Once debugged, the system can be optimizing by removing them.

---

[5]This other device could be a 6821 parallel port connected to the 6811 in expanded mode.

4.8 Keyboard Interface Using Interrupts

The interface in Figure 4.24 uses interrupts to input characters from the keyboard. When the user types a key on this keyboard, the 7-bit ASCII code becomes available on the **DATA,** followed by a rise in the signal **STROBE** that causes an interrupt. The data remain available until the next key is typed (Figure 4.25). The 6811 Port C hardware will latch the data in on the rising edge of **STROBE.** The software will poll to verify the interrupt has occurred, then call the handler that inputs the data, acknowledges the interrupt, and puts the character into a FIFO (Figure 4.26).

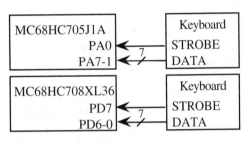
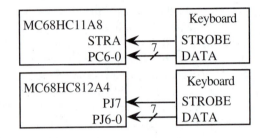

Figure 4.24
A keyboard is interfaced to the microcomputer.

Figure 4.25
The keyboard timing diagram.

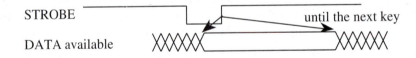

Figure 4.26
A flowchart of the interrupting keyboard interface.

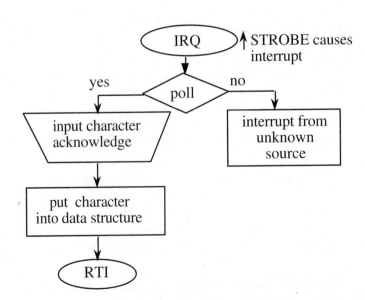

The ritual and interrupt programs in assembly language are shown in Program 4.29.

```
; MC68HC705J1A
; PA7-PA1 inputs = keyboard DATA
; PA0=STROBE interrupt on rise
Init sei          ; make atomic
     clr  DDRA    ;all inputs
     mov  #$82,ISCR  ;arm,clr IRQ
     jsr  InitFifo
     cli          ;Enable IRQ
     rts

ExtHan brset 0,PORTA,KeyHan
       swi           ;error

KeyHan lda  PORTA ;data
       lsra
       jsr  PutFifo
       bset 1,ISCR ;Ack
       rti

       org $07FA    ;IRQ external
       fdb ExtHan
```

```
; MC68HC11A8
; PC6-PC0 inputs = keyboard DATA
; STRA=STROBE interrupt on rise
; 6 STAI 1  Interrupts armed
; 5 CWOM 0  Normal outputs
; 4 HNDS 0  No handshake
; 3 OIN  0
; 2 PLS  0  STRB not used
; 1 EGA  1  STAF set on rise of READY
; 0 INVB 0  STRB not used
Init    sei          ;Make this atomic
        ldaa #$80    ;PC7 is an output
        staa DDRC    ;PC6-0 inputs
        ldaa #$42
        staa PIOC
        ldaa PIOC    ;clears STAF
        ldaa PORTCL
        clr  PORTC   ;Make PC7=0
        jsr  InitFifo
        cli          ;Enable IRQ
        rts
ExtHan ldaa PIOC     ;poll STAF
        bmi KeyHan
        swi          ;error
KeyHan ldaa PORTCL   ;clear STAF
        jsr PutFifo
        rti
        org $FFF2    ;IRQ external
        fdb ExtHan
```

```
; MC68HC708XL36
; PD6-PD0 inputs = keyboard DATA
; PD7=STROBE interrupt on rise
Init sei          ; make atomic
     clr DDRD     ;all inputs
     mov #$80,KBICR ;rising PD7
     mov #$40,ISCR  ;arm,clr IRQ2
     jsr InitFifo
     cli          ;Enable IRQ
     rts

ExtHan brset 7,PORTD,KeyHan
       swi           ;error

KeyHan lda PORTD
       and #$7F
       jsr PutFifo
       bset 6,ISCR ;ACK2
       rti
       org $FFE0    ;IRQ2/Keypad
       fdb Exthan
```

```
; MC68HC812A4
; PJ6-PJ0 inputs = keyboard DATA
; PJ7=STROBE interrupt on rise
Init    sei          ; make atomic
        clr  DDRJ    ;all inputs
        ldaa #$80
        staa KPOLJ   ;rising edge PJ7
        staa KWIEJ   ;arm PJ7
        staa KWIFJ   ;clear flag7
        jsr  InitFifo
        cli          ;Enable IRQ
        rts
ExtHan brset KWIFJ,$80,KeyHan
        swi          ;error
KeyHan ldaa #$80     ;Ack flag7
        staa KWIFJ   ;clear flag7
        ldaa PORTJ
        anda #$7F
        jsr  PutFifo
        rti
        org  $FFD0   ;Key wakeup J
        fdb  ExtHan
```

Program 4.29 Interrupting keyboard software.

Performance Tip: In the early stages of a development project, inserting consistency checks (like polling in the above example when it was not necessary) can identify hardware and software bugs. Once the system is thoroughly tested, you could remove the checks for enhanced speed.

Observation: When power is applied to a system, each device turns on separately; therefore, it is possible to get initial rising and/or falling edges that do not represent actual I/O events. Thus, it is good software practice to clear interrupt flags in the initialization so that the first interrupt represents the first I/O event.

The latency for this system is the time between the rise of **STROBE,** and the time when the software reads the input data. The logical analyzer output during the 6811 interrupt request would be as follows. *The 14 italicized cycles represent the thread switch performed automatically in hardware by the 6811.* Latency is 28 cycles (+1 instruction) for the 6811. When running on a 6811 EVB with BUFFALO, you have to add the `jmp` instruction.

```
STRA R/W Address Data Comment
 0    1   F100   B7   STAA $1234 instruction in MAIN
 1    1   F101   12   Key typed rise of STROBE
 1    1   F102   34   Finish instruction (IRQ requested)
 1    0   1234   55   Write to memory
 1    1   F103        Start to process IRQ, ??dummy PC??
 1    1   F104          ??dummy PC read??
 1    0   SP     03   PCL push all registers
 1    0   SP-1   F1   PCH
 1    0   SP-2        YL
 1    0   SP-3        YH
 1    0   SP-4        XL
 1    0   SP-5        XH
 1    0   SP-6        A
 1    0   SP-7        B
 1    0   SP-8        CC (with I=0)
 1    1   SP-8        ??dummy stack read??, then I=1
 1    1   FFF2   00   High byte of IRQ vector fetch
 1    1   FFF3   EE   Low byte of IRQ vector fetch
 1    1   00EE   7E   op code jmp INTHAN            (if using BUFFALO)
 1    1   00EF   81      operand    high byte of INTHAN (if using BUFFALO)
 1    1   00F0   23      operand    low byte of INTHAN  (if using BUFFALO)
 1    1   8123   B6   op code for LDAA
 1    1   8124   10      operand high byte of 1002 extended addr
 1    1   8125   02      operand low byte of 1002
 1    1   1002   C2      Read PIOC with STAF set
 1    1   8126   28   Op code for BMI
 1    1   8127   03   PC relative offset
 1    1   FFFF        Null
 1    1   812B   B6   Op code for LDAA
 1    1   812C   10      operand high byte of 1005 extended addr
 1    1   812D   05      operand low byte of 1005
 1    1   1005   55      Read CL data, STAF=0, Acknowledge
```

Observation: The CycleView mode of the TExaS simulator allows you to observe the bus activity during an interrupt service, both the thread switch from main to ISR and from ISR to main.

The interrupt programs, written in ICC11/ICC12 syntax, for the 6811/6812 are shown in Program 4.30. For a single-chip system, the interrupt vectors are specified in the **VECTOR.C** file shown earlier as Program 4.26.

```
// MC68HC11A8
// PC6-PC0 inputs = keyboard DATA
// STRA=STROBE interrupt on rise
void Init(void){
unsigned char dummy;
asm(" sei");
    PIOC=0x42;    // EGA=1, STAI
    DDRC=0x80;    // STRA=STROBE
    PORTC=0x00;   // PC7=0
    dummy=PIOC;   dummy=PORTCL;
    InitFifo();
asm(" cli");}
#pragma interrupt_handler ExtHan()
void ExtHan(void){
    if((PIOC & STAF)==0)asm(" swi");
    PutFifo(PORTCL);} // ack
```
```
// MC68HC812A4
// PJ6-PJ0 inputs = keyboard DATA
// PJ7=STROBE interrupt on rise
void Init(void){
asm(" sei");
    DDRJ=0x00;    // PJ6-0 DATA
    KPOLJ=0x80;   // rise on PJ7
    KWIEJ=0x80;   // arm PJ7
    KWIFJ=0x80;   // clear flag7
    InitFifo();
asm(" cli");}
#pragma interrupt_handler ExtHan()
void ExtHan(void){
    if((KWIFJ&0x80)==0)asm(" swi");
    KWIFJ=0x80;  // clear flag
    PutFifo(PORTJ&0x7F);}
```

Program 4.30 Interrupting keyboard software.

Observation: Data are lost when the FIFO gets full.

4.9 Printer Interface Using IRQ Interrupts

To output a character on this printer (Figure 4.27), the user first outputs the 7-bit ASCII code to the **DATA,** followed by a pulse on the signal **START.** The completion of the output operation is signified by the rise of **READY,** which causes an interrupt. This printer interrupt example will output one line of characters, then stop. A FIFO could have been used to create a continuous data flow.

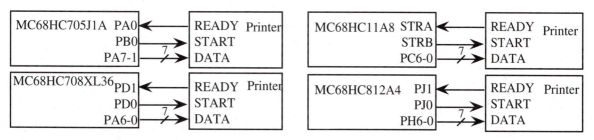

Figure 4.27   The hardware interface of a printer to the microcomputer.

For the first character, which is started in the ritual, the timing (Figure 4.28) looks similar to the gadfly solution presented in Chapter 3. But for subsequent characters, the timing (Figure 4.29) begins with the busy-to-done state transition, which in this case is the rise of **READY.** The rise of **READY** causes an interrupt. The interrupt software outputs new data to the printer and issues another **START** pulse. For the 6811, the acknowledge function is the same as the service function (STAA PORTCL). The other implementations will have an explicit acknowledge operation.

Figure 4.28
Timing diagram of a printer showing the steps to print a character.

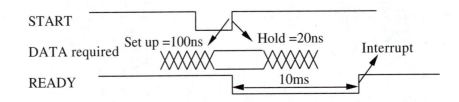

Figure 4.29
Timing diagram of a printer showing the steps during an interrupt service.

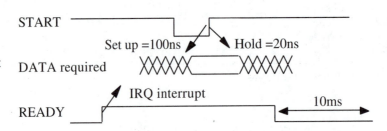

We begin this interface with a single-buffer data structure, which contains ASCII characters terminated by the null (0) character. Fill is used to copy new data into the Line buffer during initialization, and Get will be called within the interrupt handler to get more data. This data structure could be enhanced by checking for pointer overflow within both the Fill and Get routines (Program 4.31).

```
; MC68HC705J1A                          ; MC68HC11A8
;****goes in RAM*************            ;****goes in RAM*************
OK     rmb 1    ;0=busy, 1=done         OK     rmb 1    ;0=busy, 1=done
Line   rmb 20   ;ASCII, end with 0      Line   rmb 20   ;ASCII, end with 0
Pt     rmb 2    ;pointer to Line        Pt     rmb 2    ;pointer to Line
Pt2    rmb 2    ;temporary              ;****goes in ROM*************
;****goes in ROM*************           ;Input RegX=>string
;Input RegX=>string                     Fill   ldy  #Line ;RegX=>string
Fill   stx Pt2 ;Pt2=>string                    sty  Pt    ;initialize pointer
       ldx #Line                        Floop  ldaa 0,X   ;copy data
       stx Pt     ;init pointer                staa 0,Y
```

```
Floop ldx Pt2
      lda 0,X    ;read data
      incx
      stx Pt2
      ldx Pt
      sta 0,X    ;write data
      incx
      stx Pt
      tst        ;end?
      bne Floop
      ldx #Line
      stx Pt     ;init pointer
      clr OK
      rts
;Return RegA=data
Get   ldx Pt
      lda 0,X    ;return RegA=data
      incx
      stx Pt
      rts
```

```
      inx
      iny
      tsta       ;end?
      bne Floop
      clr OK
      rts

;Return RegA=data
Get   ldx Pt
      ldaa 0,X   ;read data
      inx
      stx Pt
      rst
```

```
; MC68HC708XL36
;*****goes in RAM*************
OK    rmb  1    ;0=busy, 1=done
Line  rmb  20   ;ASCII, end with 0
Pt    rmb  2    ;pointer to Line
Pt2   rmb  2    ;temporary
;*****goes in ROM*************
;Input RegH:X=>string
Fill  sthx Pt2 ;Pt2=>string
      ldhx #Line
      sthx Pt    ;init pointer
Floop ldhx Pt2
      lda  0,X   ;read data
      aix  #1
      sthx Pt2
      ldhx Pt
      sta  0,X   ;write data
      aix  #1
      sthx Pt
      tst        ;end?
      bne Floop
      ldhx #Line
      sthx Pt    ;init pointer
      clr OK
      rts
;Return RegA=data
Get   ldhx Pt
      lda  0,X   ;return RegA=data
      aix  #1
      sthx Pt
      rts
```

```
; MC68HC812A4
;*****goes in RAM*************
OK    rmb  1    ;0=busy, 1=done
Line  rmb  20   ;ASCII, end with 0
Pt    rmb  2    ;pointer to Line
;*****goes in ROM*************
;Input RegX=>string
Fill  ldy  #Line ;RegX=>string
      sty  Pt     ;initialize pointer
Floop ldaa 1,X+  ;copy data
      staa 1,Y+
      tsta        ;end?
      bne  Floop
      clr  OK
      rts
;Return RegA=data
Get   ldx  Pt
      ldaa 1,X+  ;read data
      stx  Pt
      rts
```

Program 4.31 Helper routines for the printer interface.

Next, we have the ritual `Init` that initializes the data structure, arming the device so that the rise of READY causes an interrupt. The system is started by writing the first data from the ritual (Program 4.32).

Last, we have the interrupt handler that gets called on the rise of READY (Figure 4.30). The handler will first poll, executing an error routine if the interrupt is from some other unknown source. It will then acknowledge the interrupt, get a character from the data structure, and output it to the printer. If the data are the null (0) character, then the global flag OK is set and the system is disarmed (Program 4.33).

```
; MC68HC705J1A
; PA7-PA1 outputs = printer DATA
; PB0=START
; PA0=READY interrupt on rise
;Input RegX=>string
Init sei        ; make atomic
     bsr Fill   ;Init global
     mov #$01,DDRB  ;PB0 output
     mov #$FE,DDRA  ;PA outputs
     mov #$01,PORTB ;START=1
     mov #$82,ISCR  ;arm,clr IRQ
     bsr Get
     bsr Out    ;start first
     cli        ;Enable IRQ
     rts
Out  clr PORTB  ;START=0
     lsla       ;bits7:1
     sta PORTA  ;write DATA
     inc PORTB  ;START=1
     rts
```

```
; MC68HC11A8
; PC6-PC0 outputs = printer DATA
; STRA=READY interrupt on rise
; 6 STAI 1  Interrupts armed
; 5 CWOM 0  Normal outputs
; 4 HNDS 1  Output handshake
; 3 OIN  1
; 2 PLS  1  START=STRB pulse
; 1 EGA  1  STAF set on rise of READY
; 0 INVB 0  STRB not used
;Input RegX=>string
Init  sei            ;Make this atomic
      bsr Fill       ;Init global
      ldaa #$FF      ;PC7 is an output
      staa DDRC      ;PC6-0 outputs
      ldaa #$5E
      staa PIOC
      bsr Get        ;start first
      staa PORTCL
      cli            ;Enable IRQ
      rts
```

```
; MC68HC708XL36
; PA6-PA0 outputs = printer DATA
; PD0=START
; PD1=READY interrupt on rise
;Input RegX=>string
;Input RegH:X=>string
Init sei        ; make atomic
     bsr Fill   ;Init global
     mov #$01,DDRD  ;PD0 output
     mov #$FF,DDRA  ;PA outputs
     mov #$01,PORTB ;START=1
     mov #$02,KBICR ;rising PD1
     mov #$42,ISCR  ;arm,clr IRQ2
     bsr Get
     bsr Out    ;start first
     cli        ;Enable IRQ
     rts
Out  clr PORTD  ;START=0
     sta PORTA  ;write DATA
     inc PORTD  ;START=1
     rts
```

```
; MC68HC812A4
; PH6-PH0 outputs = printer DATA
; PJ1=READY interrupt on rise
;Input RegX=>string
Init sei            ; make atomic
     bsr Fill       ;Init global
     ldaa #$FF
     staa DDRH      ;PH6-0 outputs
     ldaa #$01      ;PJ0 output START
     staa DDRJ      ;PJ1 input
     ldaa #$02
     staa KPOLJ     ;rising edge PJ1
     staa KWIEJ     ;arm PJ1
     staa KWIFJ     ;clear flag1
     bsr  Get
     bsr  Out       ;start first
     cli            ;Enable IRQ
     rts
Out  clr  PORTJ ;START=0
     staa PORTH ;write DATA
     inc  PORTJ ;START=1
     rts
```

Program 4.32 Initialization routines for the printer interface.

Figure 4.30
Flowchart of the printer
ISR.

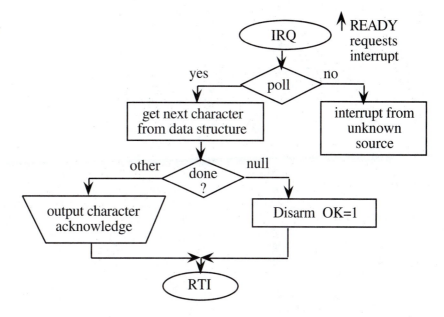

Program 4.33 ISR routines for the printer interface.

```
; MC68HC705J1A
ExtHan brset 0,PORTA,PrtHan
       swi          ;error
PrtHan bset 1,ISCR ;IRQR
       bsr   Get
       tsta
       beq   Disarm
       bsr   Out    ;start next
       bra   Done
Disarm clr   ISCR   ;disarm IRQ2
       inc   OK     ;line complete
Done   rti

       org   $07FA  ;IRQ external
       fdb   ExtHan
```

```
; MC68HC11A8
ExtHan ldaa PIOC     ;poll
       bmi   PrtHan
       swi          ;error
PrtHan bsr   Get
       tsta
       beq   Disarm
       staa PORTCL  ;start next
       bra   Done
Disarm ldaa #$1E    ;STAI=0
       staa PIOC
       inc   OK     ;line complete
Done   rti
       org   $FFF2  ;IRQ external
       fdb   ExtHan
```

```
; MC68HC708XL36
ExtHan brset 1,PORTD,PrtHan
       swi          ;error

PrtHan bset 6,ISCR ;ACK2
       bsr   Get
       tsta
       beq   Disarm
       bsr   Out    ;start next
       bra   Done
Disarm mov #$22,ISCR ;disarm IRQ2
       inc   OK      ;line complete
Done   rti
       org   $FFE0  ;IRQ2/Keypad
       fdb   ExtHan
```

```
; MC68HC812A4
ExtHan brset KWIFJ,$02,PrtHan
       swi          ;error
PrtHan ldaa #$02    ;Ack flag1
       staa KWIFJ   ;clear flag1
       bsr   Get
       tsta
       beq   Disarm
       bsr   Out    ;start next
       bra   Done
Disarm clr   KWIEJ  ;disarm PJ1
       inc   OK     ;line complete
Done   rti
       org   $FFD0  ;Key wakeup J
       fdb   ExtHan
```

In C, the 6811/6812 solutions are shown in Program 4.34.

```
// MC68HC11A8                          // MC68HC812A4
// PC6-PC0 outputs = printer DATA      // PH6-PH0 outputs = printer DATA
// STRA=READY interrupt on rise        // PJ1=READY interrupt on rise
// STRB=START pulse out                // PJ0=START pulse out
unsigned char OK;   // 0=busy, 1=done  unsigned char OK;   // 0=busy, 1=done
unsigned char Line[20]; //ASCII data   unsigned char Line[20]; //ASCII data
unsigned char *Pt;  // pointer to line  unsigned char *Pt;  // pointer to line
void Fill(unsigned char *p){            void Fill(unsigned char *p){
   Pt=&Line[0];                            Pt=&Line[0];
   while((*Pt++)=(*p++)); // copy          while((*Pt++)=(*p++)); // copy
   Pt=&Line[0];   // initialize pointer    Pt=&Line[0];   // initialize pointer
   OK=0;}                                  OK=0;}
unsigned char Get(void){                unsigned char Get(void){
   return(*Pt++);}                         return(*Pt++);}
void Init(unsigned char *thePt){        void Out(unsigned char data){
asm(" sei");   // make atomic              PORTJ=0;    // START=0
   Fill(thePt); // copy data into global   PORTH=data; // write DATA
   DDRC=0xFF;  // Port C outputs           PORTJ=1;}   // START=1
   PIOC=0x5E;  // arm out handshake     void Init(unsigned char *thePt){
   PORTCL=Get(); // start first         asm(" sei");    // make atomic
asm(" cli");}                              Fill(thePt); // copy data into global
#pragma interrupt_handler ExtHan()         DDRH=0xFF;  // PH6-0 output DATA
void ExtHan(void){ unsigned char data;     DDRJ=0x01;  // PJ0=START output
   if((PIOC &STAF)==0)asm(" swi");         KPOLJ=0x02; // rise on PJ1
   if(data=Get())                          KWIEJ=0x02; // arm PJ1
     PORTCL=data;  // start next           KWIFJ=0x02; // clear flag1
   else{                                   Out(Get()); // start first
     PIOC=0x1E;    // disarm            asm(" cli");}
     OK=1;}}       // line complete     #pragma interrupt_handler ExtHan()
                                        void ExtHan(void){
                                           if((KWIFJ&0x02)==0)asm(" swi");
                                           KWIFJ=0x02; // clear flag1
                                           if(data=Get())
                                             Out(data);    // start next
                                           else{
                                             KWIEJ=0x00;   // disarm
                                             OK=1;}}        // line complete
```

Program 4.34 C language software interrupt for the printer.

Two problems with output interrupts were introduced earlier: (1) How does one generate the first interrupt? (2) What does one do if an output interrupt occurs (device is idle) but there are no more data currently available (e.g., the data structure is empty)? In this example, the ritual could start the output because the first character to print was available at that time. If the first data were not available at the time of the ritual, the output initiation would have to be postponed until the first data are available. Also, in this case, the system disarms when there are no more data to print. The system is rearmed (by calling the ritual again) when more output is desired.

Another solution involves the use of nonprinting dummy characters. In this technique, a special nonprinting character like NULL ($00) or SYN ($16) is transmitted when the device is ready but there is nothing to print. This method is a little simpler because one does not have to arm, disarm, and rearm. The disadvantage is that software overhead is required to process interrupts when no real function is being performed. In other words, the main thread will execute slower. In addition, the printer must discard these dummy characters. The arm/disarm technique was presented at the beginning of the chapter during the discussion of the producer/consumer problem. It will be presented again in Chapter 7 on the serial communications chapter.

4.10 Power System Interface Using $\overline{\text{XIRQ}}$ Synchronization

The objective of this section is to use $\dot{\overline{\text{XIRQ}}}$ interrupts to create a very low latency interface (Figure 4.31). This power system will monitor the voltage level from the regular power supply. If the level drops below a safe threshold, it will trigger an XIRQ interrupt by signaling the problem with a rising edge on **Too Low.** This rising edge will clear the 74HC74 flip-flop, making its output, XIRQ, low. The XIRQ handler will service the crisis by enabling the backup power system by making PB1 = 1. The handler can acknowledge the XIRQ interrupt by toggling PB0=0, then PB0=1 again. Assuming the program is running with XIRQ interrupts enabled, the latency of this interface is quite low (Program 4.35). The C language implementation is shown in Program 4.36.

Figure 4.31
Hardware interface of an XIRQ interrupting device.

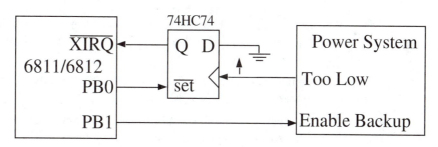

```
* Called to initialize the power system
RITUAL ldaa #$FF
       staa DDRB     Port B outputs (6812 only)
       ldaa #0       Backup power initially off
       staa PORTB    Set the flip flop, make XIRQ=1
       ldaa #1
       staa PORTB    Flip flop ready to receive rising edge of TooLow
       ldaa #$10     Enable XIRQ, Disable IRQ
       tap
       rts           Back to main, foreground thread

*Note that the software can only enable XIRQ and cannot disable XIRQ.
* In this way, XIRQ is nonmaskable.
```

continued on p. 244

Program 4.35 Assembly software for the XIRQ interrupt.

continued from p. 243

```
XIRQHAN ldaa #2
        staa PORTB    Enable BackUp power, acknowledge XIRQ
        ldaa #3
        staa PORTB    Will thread another rising edge of TooLow
        rti

        org  $FFF4
        fdb  XIRQHAN  XIRQ interrupt vector
```

Program 4.35 Assembly software for the XIRQ interrupt.

Program 4.36 C language software for the XIRQ interrupt.

```
/* Power System interface
   XIRQ requested on a rise of TooLow
   PB0, negative logic pulse, will acknowledge XIRQ
   PB1=1 will activate backup power */
#pragma interrupt_handler PowerLow()
void PowerLow(void){ PORTB=2; PORTB=3; }  /* Ack, turn on backup power */
void Ritual(void){
     DDRB=0xFF;          // Port B outputs (6812 only)
     PORTB=0; PORTB=1;  // Make XIRQ=1
asm(" ldaa #0x10\n"
    "  tap");
}
```

Observation: Some older computers have a nonmaskable interrupt (NMI) that is always enabled. These computers have problems when a NMI is requested before the system has been initialized, because the system is not yet ready to handle the NMI request.

Common Error: It is a mistake to enable XIRQ interrupts before the system has been initialized to handle the XIRQ request.

The latency of this interface is defined as the time from the rising edge of **TooLow** to when the software enables the backup power. On the 6811 it includes the 2–41 cycles required to finish the current instruction, the 14 cycles to process the XIRQ, and the six cycles needed to execute the first two instructions of the XIRQHAN. The logic analyzer output during the interrupt request would be as follows. *The italicized portion is the thread switch performed automatically in hardware by the 6811.*

TooLow	$\overline{\text{XIRQ}}$	R/W	Address	Data	Comment
0	1	1	F200	B7	STAA $5678 instruction in MAIN
1	1	1	F201	56	Power Failure occurs
1	0	1	F202	78	Finish instruction (XIRQ requested)
1	0	0	5678	12	Write to memory
1	*0*	*1*	*F203*		*Start to process XIRQ, dummy PC*
1	*0*	*1*	*F204*		*??dummy PC read??*
1	*0*	*0*	*SP*	*03*	*PCL push all registers*
1	*0*	*0*	*SP-1*	*F2*	*PCH*
1	*0*	*0*	*SP-2*		*YL*
1	*0*	*0*	*SP-3*		*YH*

1	*0*	*0*	*SP-4*		*XL*
1	*0*	*0*	*SP-5*		*XH*
1	*0*	*0*	*SP-6*		*A*
1	*0*	*0*	*SP-7*		*B*
1	*0*	*0*	*SP-8*		*CC (with X=0 and the old I)*
1	*0*	*1*	*SP-8*		*??dummy stack read??, then , X=1, I=1*
1	*0*	*1*	*FFF4*	*E6*	*High byte of XIRQHAN (vector fetch)*
1	*0*	*1*	*FFF5*	*00*	*Low byte of XIRQHAN*
1	0	1	E600	86	Op code for LDAA
1	0	1	E601	02	immediate operand
1	0	1	E602	B7	Op code for STAA
1	0	1	E603	10	extended mode operand
1	0	1	E604	04	PORTB address is $1004
1	0	0	1004	02	**set**=0 acknowledge, enable backup
1	1	1	E605	86	Op code for LDAA
1	1	1	E606	03	immediate operand
1	1	1	E607	B7	Op code for STAA
1	1	1	E608	10	extended mode operand
1	1	1	E604	04	PORTB address is $1004
1	1	0	1004	03	**set**=1
1	1	1	E609	3B	Op code for RTI
1	1	1	E60A		dummy PC read
1	1	1	SP-9		dummy stack read
1	1	1	SP-8		CC (with X=0 and the old I)
1	1	1	SP-7		B
1	1	1	SP-6		A
1	1	1	SP-5		XH
1	1	1	SP-4		XL
1	1	1	SP-3		YH
1	1	1	SP-2		YL
1	1	1	SP-1	F2	PCH
1	1	1	SP	03	PCL pull all registers
1	1	1	F203		Continue execution of previous function

4.11 Interrupt Polling Using Linked Lists

The process of polling involves software at the beginning of the ISR that checks one by one the list of possible devices that might have caused the interrupt. When the interrupt polling software detects a device that needs service, it executes the appropriate device driver to service the interrupt. There are two considerations when we decide whether or not to poll at the beginning of the interrupt handler. First, if there are two or more potential sources of interrupt that operate through the same interrupt vector, then we must poll to determine which one is interrupting. Examples of this situation are (1) external IRQ and STAF and (2) the asynchronous serial port, SCI. Second, even if we do not have to poll because our interrupting device has its own dedicated vector, sometimes we will poll anyway just to add a layer of robustness to our software. In this way, if the software arrives at the first location of the interrupt handler due to a software or hardware error, our software can determine an error has occurred because the device that should have been ready is not.

Figure 4.32
Linked list data structure used to implement polling.

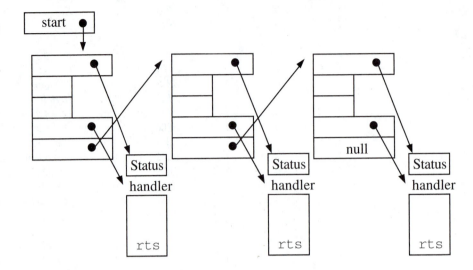

In this example, we implement polling using linked lists (Figure 4.32). The purpose of using linked lists is to make it easier to debug, change the polling order, add devices, or subtract devices. Although this example uses statically allocated linked list, if the linked list was created at run time, then changing polling order or adding/subtracting devices could be performed dynamically.

4.11.1
6811 Interrupt
Polling Using
Linked Lists

In this 6811 implementation, we have three potential interrupt sources using the same interrupt vector. The first is STAF, the second is Port A of an external 6821, and the third is Port B of the 6821. The details of the 6821 are not important for this example. Rather the purpose of this example is to illustrate the use of linked lists when implementing interrupt polling. There will be one node for each potential device. Each node of the statically allocated linked list has enough information to poll the device. If successful, the node also has a handler address that will be called. Program 4.37 shows the 6811 assembly language.

Program 4.37 6811
assembly structure for
interrupt polling using
linked lists.

```
start   fdb   llSTAF    place to start polling
Sreg    equ   0         Index to Status Register
Amask   equ   2            and mask
Cmask   equ   3            compare mask
DevHan  equ   4            device handler
NextPt  equ   6            next pointer
num     fcb   3         number of devices
llSTAF  fdb   $1002     address of PIOC
        fcb   $ff       look at all the bits in PIOC
        fcb   $C0       expect exact match with $C0
        fdb   STAFhan   device handler
        fdb   llCA1     pointer to next device to poll
```

```
        llCA1   fdb  $2011      address of 6821 Port A Control/Status
                fcb  $87        look at bits 7,2,1,0
                fcb  $85        expect bits 7,2,1,0 to be 1,1,0,1
                fdb  CA1han     device handler for CA1
                fdb  11CB2      pointer to next device to poll
        llCB2   fdb  $2013      address of 6821 Port B Control/Status
                fcb  $7C        look at bits 6,5,4,3,2
                fcb  $4C        expect bits 6,5,4,3,2 to be 1,0,0,1,1
                fdb  CB2han     device handler for CB2
                fdb  0          no more
```

The interrupt handler performs the following four steps: (1) read status register (ldaa ,y), (2) eliminate irrelevant bits (anda), (3) determine if this device is requesting (cmpa), and (4) execute the handler (jsr) if it is requesting. This particular implementation does not check for an interrupt from an unknown source. We will see in Section 4.14 how to modify this solution to implement round-robin polling. We assume each device handler terminates with a RTS instruction and saves registers B and X (Program 4.38).

Program 4.38 6811 assembly implementation of interrupt polling using linked lists.

```
IrqHan  ldx   start      Reg X points to linked list place to start
        ldab  num        number of possible devices
next    ldy   Sreg,x     Reg Y points to status reg
        ldaa  ,y         read status
        anda  Amask,x    clear bits that are indeterminate
        cmpa  Cmask,x    expected value if this device active
        bne   Notyet     skip if this device not requesting
        ldy   DevHan,x   Reg Y points to device handler
        jsr   ,y         call device handler, will return here
Notyet  ldx   NextPt,x   Reg X points to next entry
        decb             device counter
        bne   next       check next device
        rti
```

In C, this linked list polling system is shown in Program 4.39.

```
const struct    Node{
        unsigned char *StatusPt;     /* Pointer to status register */
        unsigned char Amask;         /* And Mask */
        unsigned char Cmask;         /* Compare Mask */
        void (*Handler)(void);       /* Handler for this task */
        const struct Node *NextPt;       /* Link to Next Node */
};
unsigned char CLdata,PIAAdata,PIABdata;
void STRAHan(void){     // regular functions that return (rts) when done
    CLdata=PORTCL;}
void PIAHanA(void){
    PIAAdata=ADATA;}
void PIAHanB(void){
    PIABdata=BDATA;}
typedef const struct Node NodeType;
```

continued on p. 248

Program 4.39 C language implementation of interrupt polling on the 6811 using linked lists.

continued from p. 247

```
typedef NodeType * NodePtr;
NodeType sys[3]={
    {&PIOC, 0xFF, 0xC0, STRAHan, &sys[1]},
    {&ACNT, 0x87, 0x85, PIAHanA, &sys[2]},
    {&BCNT, 0x7C, 0x4C, PIAHanB, 0} };
#pragma interrupt_handler IRQHan()
void IRQHan(void){ NodePtr Pt; unsigned char Status;
 Pt=&sys[0];
 while(Pt){               // executes device handlers for all requests
    Status=*(Pt->StatusPt);
    if((Status&(Pt->Amask))==(Pt->Cmask)){
        (*Pt->Handler)();}      /* Execute handler */
    Pt=Pt->NextPt; } }  // returns after all devices have been polled
void main(void){
    while(1);}
```

Program 4.39 C language implementation of interrupt polling on the 6811 using linked lists.

4.11.2
6812 Interrupt Polling Using Linked Lists

In this 6812 implementation, we have three potential interrupt sources using the same interrupt vector. In this system we will have three key wakeup interrupts on PJ2, PJ1, and PJ0. The purpose of this example is to illustrate the use of linked lists when implementing interrupt polling. There will be one node for each potential device. Each node of the statically allocated linked list has enough information to poll the device. If successful, the node also has a handler address that will be called. This solution is simpler than in the 6811 because all key wakeups on Port J have the same status register, and we only need to test for 1s and no 0s. Program 4.40 shows the 6812 assembly language.

Program 4.40 6812 assembly structure for interrupt polling using linked lists.

```
start    fdb    llPJ2    place to start polling
Mask     equ    0              and mask
DevHan   equ    1              device handler
NextPt   equ    3              next pointer
num      fcb    3        number of devices
llPJ2    fcb    $04      look at bit 2
         fdb    PJ2han   device handler
         fdb    llPJ1    pointer to next device to poll
llPJ1    fcb    $02      look at bit 1
         fdb    PJ1han   device handler
         fdb    llPJ0    pointer to next device to poll
llPJ0    fcb    $01      look at bit 0
         fdb    PJ0han   device handler
         fdb    0        end of list
```

The interrupt handler performs the following four steps: (1) read status register (ldaa,y), (2) determine if this device is requesting (anda), and (3) execute the handler (jsr) if it is requesting (Program 4.41). This particular implementation does not check for an interrupt from an unknown source. We will see in Section 4.14 how to modify this solution to implement round-robin polling. We assume each device handler terminates with a RTS instruction and saves registers B and X. In C, this linked list polling system is shown in Program 4.42.

Program 4.41 6812 assembly implementation of interrupt polling using linked lists.

```
IrqHan   ldx   start        Reg X points to linked list place to start
         ldab  num          number of possible devices
next     ldaa  KWIFJ        read status
         anda  Mask,x       check if proper bit is set
         beq   Notyet       skip if this device not requesting
         jsr   [DevHan,x]   call device handler, will return here
Notyet   ldx   NextPt,x     Reg X points to next entry
         decb               device counter
         bne   next         check next device
         rti
```

```c
const struct Node{
     unsigned char Mask;              /* And Mask */
     void (*Handler)(void);           /* Handler for this task */
     const struct Node *NextPt;          /* Link to Next Node */
};
unsigned char Counter2,Counter1,Counter0;
void PJ2Han(void){    // regular functions that return (rts) when done
   KWIFJ=0x04;          // acknowledge
   Counter2++;}
void PJ1Han(void){    // regular functions that return (rts) when done
   KWIFJ=0x02;          // acknowledge
   Counter1++;}
void PJ0Han(void){    // regular functions that return (rts) when done
   KWIFJ=0x01;          // acknowledge
   Counter0++;}
typedef const struct Node NodeType;
typedef NodeType * NodePtr;
NodeType sys[3]={
   {0x04, PJ2Han, &sys[1]},
   {0x02, PJ1Han, &sys[2]},
   {0x01, PJ0Han,   0   } };
#pragma interrupt_handler IRQHan()
void IRQHan(void){ NodePtr Pt; unsigned char Status;
 Pt=&sys[0];
 while(Pt){          // executes device handlers for all requests
    if(KWIFJ&(Pt->Mask)){
        (*Pt->Handler)();}     /* Execute handler */
    Pt=Pt->NextPt; } }  // returns after all devices have been polled
void main(void){
     while(1);}
```

Program 4.42 C language implementation of interrupt polling on the 6812 using linked lists.

4.12 Fixed Priority Implemented Using One Interrupt Line

In this example (Figure 4.33), Device 1 has higher priority than Device 2. Since they share a common interrupt vector, this is an example of a polled interrupt. The hardware is shown with external parallel ports, but the discussion would be identical if two key wakeups were used.

Figure 4.33
A polled interrupt where Device 1 has higher priority than Device 2.

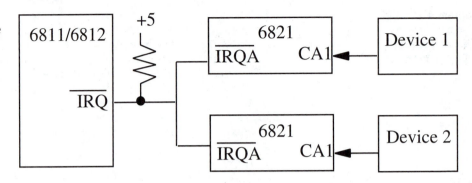

There are two steps that we need to perform to let Device 1 go ahead of Device 2. First, if both simultaneously request interrupts, then obviously Device 1 goes first. The more difficult situation is when Device 2 requests an interrupt, is allowed to start its function, then Device 1 requests service. For this scenario, we need a mechanism to postpone the service of Device 2, serve Device 1 right now, then finish the service for Device 2 after we are done with Device 1. We can accomplish this effect by reenabling interrupts in the Device 2 handler after the Device 2 interrupt has been acknowledged. The software will crash if we reenable interrupts in the Device 2 handler before we acknowledge it (the three conditions flag set, flag armed, I=0 would cause Device 2 to interrupt itself over and over). Obviously we would not want to reenable interrupts while servicing the higher-priority device. This *selective reenabling* approach works for two devices but does not work quite the same way for three or more devices. If we were to have three devices, clearly we would reenable interrupts for the lowest-priority device and not reenable interrupts for the highest-priority device. The difficult question is whether to reenable interrupts while servicing the middle-priority device. If we do reenable interrupts while servicing the middle device, then we implement a two-level priority system, where 1 is higher than 2 and 3 but 2 and 3 essentially have the same priority. If we do not reenable interrupts while servicing the middle device, then we implement a different two-level priority system, where 1 and 2 are higher than 3 but 1 and 2 have the same priority. Luckily, when we are interested in priority on an embedded system, it usually fits the two-level situation, where one or two devices need to have priority over all the rest. So, the flowchart shown in Figure 4.34 can be applied in general to solve a two-level priority system. We enable interrupts for the lower-priority devices and do not reenable interrupts for the higher-priority devices.

4.13 Fixed Priority Implemented Using XIRQ

One of the disadvantages of the priority implementation of Section 4.12 is that it takes some time to poll and acknowledge a low-priority device. If we wish to reduce the latency and increase the priority of one device over all the other interrupt sources, then XIRQ can be used (Figure 4.35). Once the X bit is 0, the software including ISRs cannot postpone an XIRQ request. It doesn't matter which request comes first, an XIRQ interrupt will be processed at the completion of the current instruction.

Figure 4.34
A flowchart of the ISR
where Device 1 has
higher priority than
Device 2.

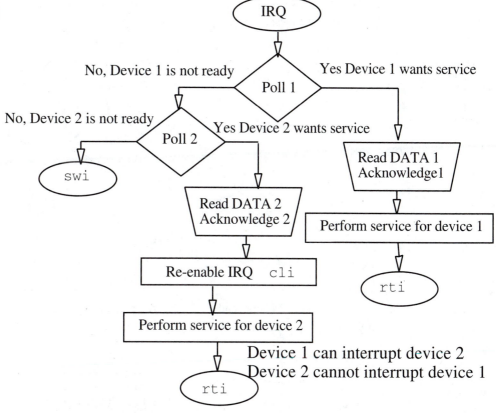

Device 1 can interrupt device 2
Device 2 cannot interrupt device 1

Figure 4.35
A vectored interrupt
where Device 1 has
higher priority than
Device 2.

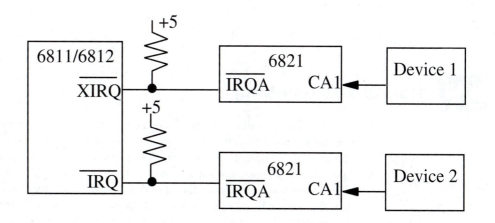

The flowchart in Figure 4.36 shows the operations of the XIRQ and IRQ interrupt service routines.

Figure 4.36 Two flowcharts of the ISRs where Device 1 has higher priority than Device 2.

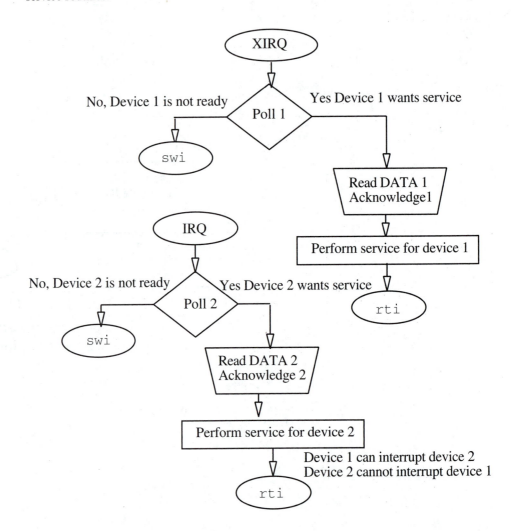

4.14 Round-Robin Polling

Rather than implement priority, sometimes we wish to implement *no priority*. This means we can guarantee service under heavy load for equally important devices. For interrupts that occur on an infrequent basis, priority doesn't really matter. But as the number of interrupts increase and the interrupt request rates increase, we may wish to use round-robin polling. This implementation works only for polled interrupts and does not apply to vectored interrupts. Two round-robin schemes are shown for a situation with three devices called A, B, C. The particular example sequence of events is shown for the situation that B and C always want service and A never does.

Scheme 1 Start polling after device that last got service. Example sequence of events:

Interrupt, poll A, B, C (B, C need service)

Service B

Interrupt, poll C, A,B (B, C need service)

Service C

Interrupt, poll A, B, C (B, C need service)

Service B

Interrupt, poll C, A, B

Scheme 2 Cycle through list of devices independent of which device last got service.
Example sequence of events:

Interrupt, poll A, B, C

Interrupt, poll B, C, A

Interrupt, poll C, A, B

Interrupt, poll A, B, C, etc.

Our previous polling examples in Section 4.11 can be modified to implement either round-robin scheme. Programs 4.43 and 4.44 implement the second simple method that rotates the polling order independent of which devices request service.

```
NodeType sys[3]={
    {&PIOC, 0xFF, 0xC0, STRAHan, &sys[1]},
    {&ACNT, 0x87, 0x85, PIAHanA, &sys[2]},
    {&BCNT, 0x7C, 0x4C, PIAHanB, &sys[0]} };
NodePtr Pt=&sys[0];  // points to the one that got polled first at last interrupt
#pragma interrupt_handler IRQHan()
void IRQHan(void){  unsigned char Counter,Status;
 Counter=3;        // quit after three devices checked
 Pt=Pt->NextPt;    // rotates ABC BCA CAB polling orders
 while(Counter--){
    Status=*(Pt->StatusPt);
    if((Status&(Pt->Amask))==(Pt->Cmask)){
        (*Pt->Handler)();}       /* Execute handler */
    Pt=Pt->NextPt; } }
```

Program 4.43 C language implementation of round-robin polling on the 6811.

```
NodeType sys[3]={
    {0x04, PJ2Han, &sys[1]},
    {0x02, PJ1Han, &sys[2]},
    {0x01, PJ0Han, &sys[0]} };
#pragma interrupt_handler IRQHan()
NodePtr Pt=&sys[0];  // points to the one that got polled first at last interrupt
void IRQHan(void){ unsigned char Counter,Status;
 Counter=3;        // quit after three devices checked
 Pt=Pt->NextPt;    // rotates ABC BCA CAB polling orders
 while(Counter--){
    if(KWIFJ&(Pt->Mask)){
        (*Pt->Handler)();}       /* Execute handler */
    Pt=Pt->NextPt; } }  // returns after all devices have been polled
```

Program 4.44 C language implementation of round-robin polling on the 6812.

4.15 Periodic Polling

The purpose of this section is to present alternative methods to create a real time interrupt (RTI). A RTI is one that is requested on a fixed time basis. This interfacing technique is required for data acquisition and control systems, because software servicing must be performed at accurate time intervals. For a data acquisition system, it is important to establish an accurate sampling rate. The time in between ADC samples must be equal (and known) for the digital signal processing to function properly. Similarly for microcomputer-based control systems, it is important to maintain both the ADC and DAC timing.

Another application of RTIs is called *intermittent* or *periodic polling*. In regular gadfly, the main program polls the I/O devices continuously (refer back to Section 3.3). With intermittent polling, the I/O devices are polled on a regular basis (established by the RTI). If no device needs service, then the interrupt simply returns. This method frees the main

Figure 4.37
An ISR flowchart that implements periodic polling.

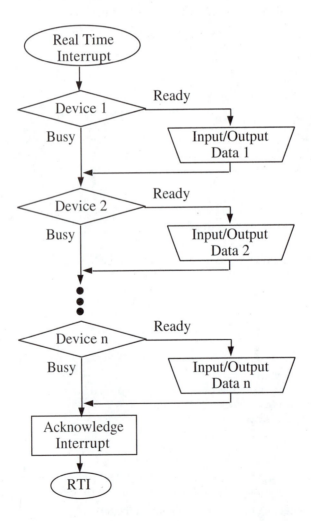

program from the I/O tasks. Older IBM PC computers use an 18-Hz RTI to interface its keyboard. We use periodic polling (Figure 4.37) if the following two conditions apply:

1. The I/O hardware cannot generate interrupts directly.
2. We wish to perform the I/O functions in the background.

4.15.1 Real Time Interrupt Using a 6811 STRA

Although most microcomputers have internal clocks that can be used to create periodic interrupts, there are applications where it is desirable to use an external clock (Figure 4.38). For example, if you had multiple microcomputers and wished to create simultaneous interrupt requests, you could use this approach. The clock frequency of a 555 timer is determined by the resistor and capacitor values. The data sheet for various 555 timers can be found on the accompanying CD. An astable multivibrator creates the 1-kHz squarewave. The period of a 555 timer is $0.693 \cdot C_T \cdot (R_A + 2R_B)$. In our circuit, R_A is $4.4\,R\Omega$, R_B is $5\,R\Omega$, and C_T is $0.1\,\mu F$ (Program 4.45).

Figure 4.38
A periodic interrupt implemented with an external clock.

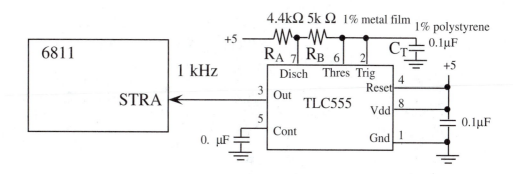

Program 4.45 6811 assembly implementation of a periodic interrupt using an external clock.

```
TIME    rmb  2        incremented every 1ms
RITUAL  sei           disable interrupts during RITUAL
        ldaa #$42     PIOC STAI=1, HNDS=0, EGA=1
        staa $1002    Arm interrupt on rise of STRA
        ldd  #0
        std  TIME     initialize variable
        ldaa $1005    initially clear STRA
        cli           enable
        rts
* POLL for zeros and ones
IRQHAN  ldaa $1002    11000010 if interrupting
        cmpa #$C2
        beq  CLKHAN
        swi
CLKHAN  ldaa $1005    Acknowledge
        ldx  TIME
        inx
        stx  TIME
        rti
```

4.15.2
6811 RTI

The RTI feature can be used to generate interrupts at a fixed rate. Two bits (RTR1 and RTR0) in the PACTR register ($1026) determine the interrupt rate (Figure 4.39). For an E clock of 2 MHz, Table 4.11 lists RTI rates.

   The assembly language implementation of a 30.517 Hz perodic interrupt is shown in Program 4.46. Program 4.47 is the C implementation.

Figure 4.39
6811 registers used to configure the RTI.

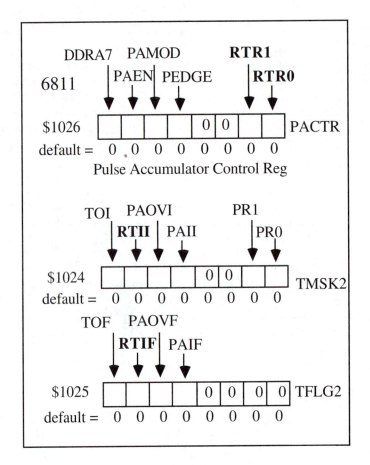

Table 4.11 RTI rates.	RTR1	RTR0	Divide E by	Period (μs)	Frequency (Hz)
	0	0	2^{13}	4,096	244.14
	0	1	2^{14}	8,192	122.07
	1	0	2^{15}	16,384	61.035
	1	1	2^{16}	32,768	30.517

Program 4.46 6811
assembly implemen-
tation of a periodic
interrupt using RTI.

```
RITUAL sei            disable interrupts during RITUAL
       ldaa  #3       Set RTR1,RTR0 = 11 Interrupt period = 32.768ms
       staa  $1026
       ldaa  #$40     Set RTII=1
       staa  $1024
       cli            Enable IRQ interrupts
       rts
```

```
RTIHAN  ldaa  $1025       Polling for zeros and ones expect=X1XX0000
        anda  #$4F        ignore TOF, PAOVF, and PAIF
        cmpa  #$40        RTIF should equal 1
        beq   OK
        swi                Error
OK      ldaa  #$40        RTIF is cleared by writing to TFLG2
        staa  $1025             with bit 6 set
* service occurs every 32.768ms or about 30.517Hz
        rti
```

Program 4.47 6811 C language implementation of a periodic interrupt using RTI.

```
unsigned int Time;
#pragma interrupt_handler RTIHan()
void RTIHan(void){
    if((TFLG2&0x4F)!=0x40)asm(" swi"); /*  Illegal interrupt */
    TFLG2=RTIF;      /* Acknowledge by clearing RTIF */
    Time++;
  }
void Ritual(void){
    asm(" sei");     /* Make ritual atomic */
    PACTL=(0xFC&TMSK2)|2;    /* Set RTR to 2, 61.035Hz  */
    TMSK2|=RTII;             /* Arm RTI  */
    Time=0;       /* Initialize global data structures */
    asm(" cli");
  }
```

4.15.3 6812 RTI

The RTI feature can be used to generate interrupts at a fixed rate. Three bits (RTR2, RTR1 and RTR0) in the RTICTL register ($0014) determine the interrupt rate. For an E clock of 8 MHz, Table 4.12 lists the RTI rates.

Table 4.12 RTI rates.

RTR2	RTR1	RTR0	Divide E by	Period (μs)	Frequency (Hz)
0	0	0	off	off	off
0	0	1	2^{13}	1,024	976.56
0	1	0	2^{14}	2,048	488.28
0	1	1	2^{15}	4,096	244.14
1	0	0	2^{16}	8,192	122.07
1	0	1	2^{17}	16,384	61.035
1	1	0	2^{18}	32,768	30.517
1	1	1	2^{19}	65,536	15.259

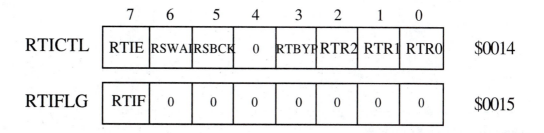

	7	6	5	4	3	2	1	0	
RTICTL	RTIE	RSWAI	RSBCK	0	RTBYP	RTR2	RTR1	RTR0	$0014
RTIFLG	RTIF	0	0	0	0	0	0	0	$0015

An assembly language implementation of a 30.517 Hz periodic interrupt is shown in Program 4.48. Program 4.49 is a C language implementation.

Program 4.48 6812 assembly implementation of a periodic interrupt using RTI.

```
RITUAL sei            disable interrupts during RITUAL
       ldaa #$86      Set RTR2-0 = 110 Interrupt period = 32.768ms
       staa RTICTL    Set RTIE=1 arm RTI interrupts
       cli            Enable IRQ interrupts
       rts
RTIHAN ldaa RTIFLG    Polling for zeros and ones expect=10000000
       cmpa #$80      RTIF should equal 1
       beq  OK
       swi            Error
OK     ldaa #$80      RTIF is cleared by writing to RTIFLG
       staa RTIFLG          with bit 7 set
* service occurs every 32.768ms or about 30.517Hz
       rti
```

Program 4.49 6812 C language implementation of a periodic interrupt using RTI.

```
// Real Time Interrupt example 4.14.2
unsigned int Time;
#pragma interrupt_handler RTIHan()
void RTIHan(void){
    if(RTIFLG!=0x80) asm(" swi"); //  Illegal interrupt
    RTIFLG=0x80;         // Acknowledge by clearing RTIF
    Time++;}
void Ritual(void){
    asm(" sei");     // Make ritual atomic
    RTICTL=0x86;     // Arm, Set RTR to 6, 30.517Hz
    Time=0;          // Initialize global data structures
    asm(" cli");}
```

4.15.4 6812 Timer Overflow Interrupt

The timer overflow interrupt feature can also be used to generate interrupts at a fixed rate. The 16-bit TCNT register is incremented at a fixed rate. The TOF flag is set when the counter overflows and wraps back around (automatically) to 0. If armed, TOF can generate an interrupt. Three bits (PR2, PR1, and PR0) in the TMSK2 register ($008D) determine the rate at which the counter will increment, hence will determine the TOF interrupt rate. For an E clock of 8 MHz, Table 4.13 lists overflow rates.

PR2	PR1	PR0	Divide by	TCNT clock	TOF interrupt period	TOF interrupt rate
0	0	0	1	125 ns	8.192 ms	122.07 Hz
0	0	1	2	250 ns	16.384 ms	61.035 Hz
0	1	0	4	500 ns	32.768 ms	30.517 Hz
0	1	1	8	1 μs	65.536 ms	15.259 Hz
1	0	0	16	2 μs	131.072 ms	7.63 Hz
1	0	1	32	4 μs	262.144 ms	3.81 Hz
1	1	0	Reserved	-	-	-
1	1	1	Reserved	-	-	-

Table 4.13 Timer overflow rates.

Observation: With a PR2-0 value of 010, the TCNT will clock at the same rate as a 2-Mhz 6811.

	7	6	5	4	3	2	1	0	
TSCR	TEN	TSWAI	TSBCK	TFFCA	0	0	0	0	$0086
TMSK2	TOI	0	TPU	TDRB	TCRE	PR2	PR1	PR0	$008D
TFLG2	TOF	0	0	0	0	0	0	0	$008F

To create a TOF periodic interrupt, we enable the timer (TEN=1), arm the timer overflow (TOI), and set the rate (PR2-0). The other bits in these registers will be presented in the chapter on the timer. Example periodic interrupt programs are shown in Programs 4.50 and 4.51.

Program 4.50 6812 assembly implementation of a periodic interrupt using timer overflow.

```
RITUAL  sei             disable interrupts during RITUAL
        ldaa #$B2       Set PR2-0 = 010 Interrupt period = 32.768ms
        staa TMSK2      Set TOI=1 arm TOF interrupts
        ldaa #$80       TEN=1
        staa TSCR       enable TCNT
        cli             Enable IRQ interrupts
        rts
TOFHAN  ldaa TFLG2      Polling for zeros and ones expect=10000000
        cmpa #$80       RTIF should equal 1
        beq  OK
        swi             Error
OK      ldaa #$80       TOF is cleared by writing to TFLG2
        staa TFLG2          with bit 7 set
* service occurs every 32.768ms or about 30.517Hz
        rti
```

Program 4.51 6812 C language implementation of a periodic interrupt using timer overflow.

```
// TOF Interrupt example
unsigned int Time;
#pragma interrupt_handler TOFHan()
void TOFHan(void){
   if(TFLG2!=0x80) asm(" swi"); //  Illegal interrupt
   TFLG2=0x80;        // Acknowledge by clearing TOF
   Time++;
 }
void Ritual(void){
    asm(" sei");      // Make ritual atomic
    TMSK2=0xB2 ;      // Arm, Set PR to 010, 30.517Hz
    TSCR=0x80;        // enable counter
    Time=0;           // Initialize global data structures
    asm(" cli");
 }
```

4.16 Glossary

atomic Software execution that cannot be divided or interrupted. Once started, an atomic operation will run to its completion without interruption. On most computers the assembly language instructions are atomic.

bandwidth The information transfer rate, the amount of data transferred per second. Same as throughput.

critical section Locations within a software module, which if an interrupt were to occur at one of these locations, then an error could occur (e.g., data lost, corrupted data, program crash). Same as vulnerable window.

handshake A software/hardware synchronization method where control and status signals go both directions between the transmitter and receiver. The communication is interlocked, meaning each device will wait for the other.

interrupt A software/hardware synchronization method where the hardware causes a special software program (interrupt handler) to execute when its operation is complete. The software usually can perform other work while waiting for the hardware.

latency In this book latency usually refers to the response time of the computer to external events (e.g., the time between new input becoming available and the time the input is read by the computer, or the time between an output device becoming idle and the time the input in the computer writes new data to it). There can also be a latency for an I/O device, which is the response time of the external I/O device hardware to a software command.

multithreaded A system with multiple threads (e.g., main program and interrupt service routines) that cooperate toward a common overall goal.

nonatomic Software execution that can be divided or interrupted. Most lines of C code require multiple assembly language instructions to execute; therefore an interrupt may occur in the middle of a line of C code.

nonreentrant A software module that once started by one thread cannot be interrupted and executed by a second thread. A nonreentrant module usually involves nonatomic accesses to global variables or I/O ports: read-modify-write, write followed by read, or a multistep write.

periodic polling A software/hardware synchronization method that is a combination of interrupts and gadfly. An interrupt occurs at a regular rate (periodic) independent of the hardware status. The interrupt handler checks the hardware device (polls) to determine if its operation is complete. The software usually can perform other work while waiting for the hardware.

polling A software function to look and see which of the potential sources requested the interrupt.

priority When two requests for service are made simultaneously, priority determines which order to process them.

process The execution of software that does not necessarily cooperate with other processes.

real time A system that can guarantee an upper bound (worst case) on latency.

reentrant A software module that can be started by one thread and interrupted and executed by a second thread. A reentrant module allow both threads to properly execute the desired function.

thread The execution of software that cooperates with other threads.

throughput The information transfer rate, the amount of data transferred per second. Same as bandwidth.

vulnerable window Locations within a software module, which if an interrupt were to occur at one of these locations, then an error could occur (e.g., data lost, corrupted data, program crash.) Same as critical section.

4.17 Exercises

4.1 The objective of this problem is to use interrupts to interface the input device in Figure 4.40 to a 6811. To solve this problem with a different microcomputer, you could substitute any available I/O port for STRA STRB PC7-PC0. You may write the software in assembly or C.

Figure 4.40
Input device interface.

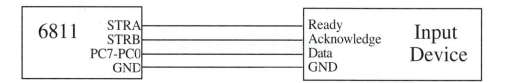

The sequence to input an ASCII character from the input device is as follows. When a new character is available, the input device puts its ASCII code on the 8-bit **Data,** then the input device makes **Ready**=1. Next your interrupt software should read the 8-bit ASCII code, then acknowledge receipt of the data by pulsing **Acknowledge.** After the pulse, the input device will make **Ready**-0 again (Figure 4.41).

Figure 4.41
Input device timing.

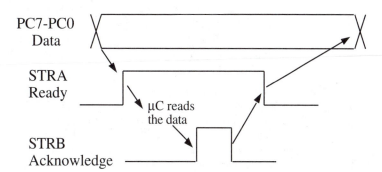

a) Show bit by bit your choice (what and why) for the PIOC register.

b) Show the ritual that initializes the 6811. You should call the function `InitFifo()`, which initializes the FIFO. You may assume the FIFO routines already exist. Do not write the FIFO routines. You may write your answers in assembly or C.

c) Show the interrupt handler that inputs one data byte. The sequence of steps is described above. Store the input data in the FIFO, call `Put(unsigned char)`. Ignore FIFO full errors. No polling is required. You do not have to write the main program that gets and processes the data.

4.2 The objective of this problem is to interface an output device (Figure 4.42) to a 6811 single-chip computer using Port C using interrupt synchronization. To solve this problem with a different microcomputer, you could substitute any available I/O port for STRA STRB PC7-PC0. You may write the software in assembly or C.

Figure 4.42
Output device interface.

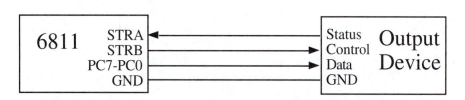

Figure 4.43
Output device timing.

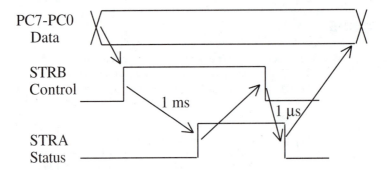

First, your 6811 outputs data on PC7–PC0. Then, your 6811 makes STRB=1; 1 ms later the output device will make STRA=1 (causes an interrupt). Then, your 6811 makes STRB=0; 1 μs later the output device will make STRA=0 (your 6811 need not check for STRA=0) (Figure 4.43).

a) Show bit by bit your choice (what and why) for the PIOC register.

b) Write a C program with a prototype `void Print(unsigned char *pt);` that begins the operations required to print the null (0) terminated ASCII string pointed to by `pt` on the output device. Clear a global variable DONE (which will be set in the interrupt routine when the system is done printing). This routine includes the ritual that initializes the 6811.

c) Show the ISR that outputs one data byte. The sequence of steps is described above. Disarm and set the DONE flag when the string is finished. Do not output the null. Execute `swi` if you get an illegal interrupt. You do not have to write the main program that generates the data and calls `Print()`.

4.3 *Read the entire question before starting.* The objective is to interface a standard IBM PC keyboard to a microcomputer using STRA, Port C, and output compare (Figure 4.44). The TTL level **Clock** I/O is connected to both **STRA** and **PC0**. The **Clock** is sometimes an input and sometimes an open-collector output of the 6811. The TTL level **Data** is connected to **PC7**. The **Data** also is sometimes an input and sometimes an open-collector output of the 6811. To solve this problem with a different microcomputer, you could substitute any available I/O port for STRA PC7–PC0. You may write the software in assembly or C.

Figure 4.44
Keyboard interface.

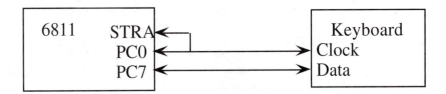

You will *not* use FIFOs in this problem, although your system could be significantly enhanced with FIFOs that decouple the main and background threads. Rather, there will be two global variables (which in actuality behave like a 1-byte FIFO):

```
unsigned char info;    // 8 bit data from the keyboard
unsigned char flag;    // true means info contains valid information
```

You will implement a simple half-duplex solution. In other words, your microcomputer software will be operating in either receive or transmit mode, but never both at the same time. You will use info and flag (no other globals are allowed). You can assume the main program leaves the system mostly in receive mode, and the operator does not type a key during the short intervals when the main program activates transmit mode. A real IBM PC keyboard interface must handle the situation where communication is attempted in both directions simultaneously (but you don't have to worry about it here).

Receive mode is where data is being received from the keyboard into the microcomputer. When the operator types a key on the keyboard, one or more (up to six) *scan codes* (8 bits each) are sent from the keyboard to the microcomputer. You will write a background interrupt activated by STRA that performs communication with the keyboard. You will also write a C function with the following prototype:

```
void ReceiveMode(void);   // place the keyboard system in receive mode
```

The main program, which you do not write, is free to access the info and flag to collect and process the scan codes. In the Figure 4.45 timing diagram, the thin lines represent keyboard outputs that are open-collector with resistor pull-up, and the thick lines represent computer outputs that are also open-collector.

Figure 4.45
Keyboard receive timing.

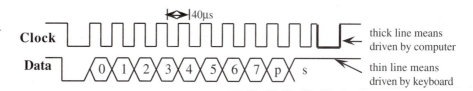

To receive a byte, your microcomputer first places both the **Clock** and **Data** in input mode. In the idle state, the keyboard makes both **Clock** and **Data** high (actually the keyboard signals are also open-collector, and the high levels are generated by resistor pull-ups on the keyboard side). When the keyboard wishes to send a scan code (remember one to six scan codes are sent after the operator types/releases a key), the keyboard will first set the **Data** low, then create a high-to-low transition on the **Clock.** This is like the start bit we saw in the asynchronous serial interface. You will accept and ignore the start bit. Then, every 40 μs the keyboard will put the next information bit on the **Data** line and create another high-to-low transition on **Clock.** Notice the **Data** changes on the low-to-high transition, and you should input the binary bits on the high-to-low edges. It will send 8 bits of information (bit 0 first), which you will read and then create an 8-bit byte from the 8 individual bits). It also sends an odd parity bit (p) and stop bit (s) that you will read and ignore. It would be a good idea to do some error checking (returning with an error on a timeout, making sure the start bit is low, the odd parity is correct, and the stop bit is high), but these features are *not* required in this problem. Since the main program is performing other unrelated tasks, you will implement these receive functions in a background thread. If the flag is 0 when the stop bit is received (meaning the info is currently empty), put the new byte into info and set the flag (like a FIFO put). If the flag is still set when the stop bit is received (because the main program hasn't read the previous transmission), simply discard the new data (like an overrun on the SCI serial port or like a FIFO full error). As an acknowledgment back to the keyboard, your computer should make the **Clock** an open-collector output and drive **Clock** low for 40 μs. Use OC2 (gadfly) to create this 40-μs timing. At the end of the STRA interrupt handler, your microcomputer system should leave **Clock** and **Data** in input mode, ready to accept the next scan code. There will be exactly one interrupt request for each scan code received.

Figure 4.46 Keyboard
transmit timing.

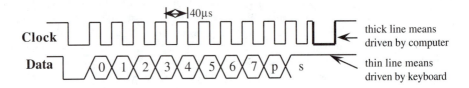

Clock —— thick line means
driven by computer

Data —— thin line means
driven by keyboard

Transmit mode is where commands are sent from the microcomputer to the keyboard. When the
main program wishes to modify the keyboard functions (turn on LEDs, enable/disable certain functions) it sends a *command code* (8 bits each) to the keyboard (Figure 4.46). You will write a gadfly C
function (with *no* background interrupt threads and *no* globals) with the following prototype:

```
void SendCommand(unsigned char);  // transmit one command to the keyboard
```

The main program, which you do not write, can call this function to transmit command codes to the
keyboard. *Notice that this main program does not disable interrupts as it accesses the shared globals.* For example, the main program could first initialize the keyboard using transmit mode, then
switch to receive mode to input and process scan codes:

```
void main(void){ unsigned char ScanCode;
    OtherInitialization();   // unrelated to your keyboard (don't write this)
    SendCommand(0x25);       // specify keyboard operation mode
    SendCommand(0x42);       // specify keyboard operation mode
    ReceiveMode();           // your function that initializes the keyboard
    while(1){
      if(flag) {
          ScanCode=info;     // read next byte (like a FIFO get)
          flag=0;            // means can accept another receive transmission
          Process(ScanCode);} // don't write this Process
      OtherProcess();}}        // unrelated to your keyboard (don't write this)
```

To transmit a byte, the microcomputer first disables the receive mode functions, places **Clock** and
Data in open-collector output mode with the output high (floating), and waits for **Clock** to be high.
Next the computer sets **Clock** low for 400 μs. Again use OC2 (gadfly) to create this timing delay.
Then you pull the **Data** low (while the **Clock** is still low). Next, you release the **Clock** by making it
an input (it should float high because of the pull-up in the keyboard). The keyboard now should capture control of the **Clock** and drive it low. On the next nine low-to-high transitions you will set a new
binary bit on the **Data** line. On these nine high-to-low transitions, the binary bit is latched into the
keyboard. The first 8 bits are the command code (bit 0 first), and the ninth bit is odd parity. After the
odd parity bit is received by the keyboard, both the **Clock** and **Data** lines are held low by the keyboard until it is ready to accept another command. At the end, your microcomputer system should
leave both **Clock** and **Data** in input mode.

4.4 You will design and implement a FSM using the 6812 *key wakeup interrupts* (Figure 4.47). One
of the limitations of the previous FSM implementations is that they require 100% of the processor time and run in the foreground. In this system, there are two inputs and two outputs. We will
implement a FSM where state transitions only occur on the rising edges of one of the two inputs.
These rising edges should cause a key wakeup interrupt, and the FSM controller will be run in
the interrupt handler. Your system will use a statically allocated linked data structure, with a
one-to-one correspondence to the FSM. machine. You should be able to change the data structure to accommodate other two-input, two-output FSMs of similar structure without changing
the program controller. Even though you could shut off the controller once it gets to the "End"
state, continue to accept key wakeup interrupts. You can assume that the inputs occur independently, and many occur at the same time.

Figure 4.47
Finite-state machine.

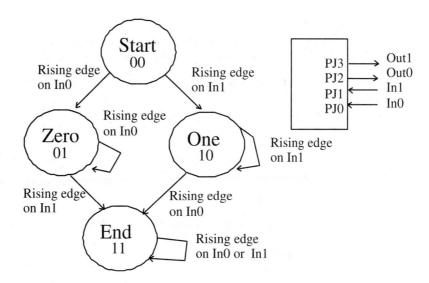

The FSM controller sequence is:

1. Wait for a rising edge on **In1** or **In0** (causing key wakeup interrupt)
2. Go to the next state depending on the current state and which rising edge occurred
3. Output the pattern of the new state on **Out1** and **Out0**
    a) Show the linked data structure. You may write this in C or assembly.
    b) Show the ritual that is executed once at the beginning. There will be a main program (which you will not write) and possibly other interrupts, but this is the only device using Port J. The initial state is **Start** with its initial output of 00.
    c) Show the key wakeup interrupt handler. Don't worry about how the interrupt vector is set.

4.5 Design the hardware/software interface that receives data from the following fully interlocked sensor device. The sensor has **status** and **data** outputs and an **ack** input.

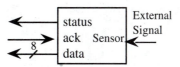

The following timing diagram illustrates the sequence of events required to input a sensor reading into the computer. The procedure begins with the software setting its **ack** output high. Second, the sensor will perform a conversion, place the 8-bit result on its **data** lines, and set its **status** low. Third, the software will read the **data,** then set its **ack** low. Fourth, the sensor will bring its **status** high again.

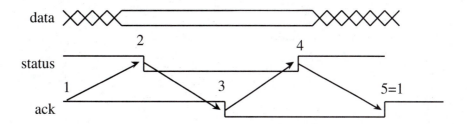

The fifth step (really the same as the first step for the next data transfer) is for the software to set its **ack** high, meaning it wishes to collect another sensor reading.

The interface timing has the following additional constraints/observations:

- Sensor **data** outputs are valid when **status** is low (software reads **data** after 2 and before 3).
- Timing events 1, 2, 3, 4 occur exactly in this order.
- Delays from 1 to 2 and from 3 to 4 are determined by the sensor, varying from 10 µs to 1 s.
- Delays from 2 to 3 and from 4 to 1 are system response times that should be minimized.

Your software must follow these conditions:

- Both the fall and rise of **status** will cause interrupts.
- These two interrupts will have separate vectors (vectored interrupt).
- If your computer doesn't support vectored interrupt, then you may use polled interrupts.
- A FIFO queue will link your producer (background interrupt) with a consumer (main program).
- You will not write the main program that gets from the FIFO and processes the data.
- You may use any one of the FIFOs in Chapter 4 without showing its implementation.
- You may ignore COP and TOF interrupts.
- Don't worry about how the interrupt vector is set.
- You may ignore FIFO full errors (although it would have been possible to prevent them).

   a) Show the hardware interface between the sensor and your computer. You may use any available port, but make sure two interrupts can be requested that have separate vectors.
   b) Show the ritual that will be called at the beginning of the main program.
   c) Show the interrupt handler that is executed on the fall of status (timing event 2).
   d) Show the interrupt handler that is executed on the rise of status (timing event 4).

5 Threads

Chapter 5 objectives are to:

❑ Define a thread control block
❑ Design and implement a preemptive thread scheduler
❑ Design and implement spin-lock semaphores
❑ Design and implement block semaphores
❑ Present applications that employ semaphores

In Chapter 4 we used interrupts to create a multithreaded environment. In that configuration we had a single foreground thread (the main program) and multiple background threads (the ISRs). These threads are run using a simple algorithm. The ISR of an input device is invoked when new input is available. The ISR of an output device is invoked when the output device is idle and needs more data. Last, the ISR of a periodic task is run at a regular rate. The main program runs in the remaining intervals. Because most embedded applications are small in size, and static in nature, this configuration is usually adequate. The limitation of a single foreground thread comes as the size and complexity of the system grows. As we saw in Chapter 2, it is appropriate to partition a large project into modules, then piece together those modules in either a layered or hierarchical manner. It is during this "piece together" phase of a large project that the simple single foreground thread environment breaks down. It is easy to combine each module's initialization routine into one common initialization sequence. The only difficulty during initialization is to prevent the initialization of one module from undoing the initialization of a previously initialized module. For example, you must be careful when two modules initialize the same I/O control register. Similarly, it is straightforward to combine the ISRs of the modules. If the modules use separate vectors, then no additional effort is required. If two modules use the same interrupt vector, then one of the polling schemes described in Chapter 4 can be used. The difficulty arises when combining the foreground functions of the modules. With a single foreground thread, we are forced into a sequential program structure similar to Figure 3.10. For the projects where the modules are tightly coupled (interdependent), this sequential model is both natural and appropriate. On the other hand, the projects where the modules are more loosely coupled (independent) may more naturally fit a multiple foreground thread configuration. The goal of this chapter is to develop the software techniques to implement multiple foreground threads (the scheduler) and provide synchronization tools (semaphores) that allow the threads to interact with each other. Systems that implement a thread scheduler still may employ regular I/O driven interrupts. In this way, the system supports multiple foreground threads and multiple background threads.

5.1   Multithreaded Preemptive Scheduler

We define a thread as the execution of a software task that has its own stack and registers (Figure 5.1). Another name for thread is lightweight process. Since each thread has a separate stack, its local variables are private, which means it alone has access. Multiple threads cooperate to perform an overall function. Since threads interact for a common goal, they do share resources, such as global memory, and I/O devices (Figure 5.2).

Figure 5.1
Each thread has its own registers and stack.

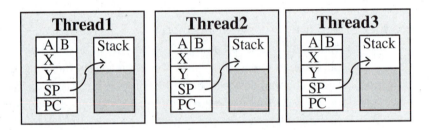

Figure 5.2
Threads share global memory and I/O ports.

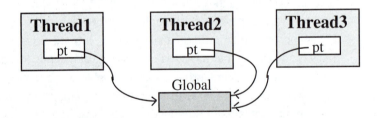

Some simple examples of multiple threads are the interrupt-driven I/O examples of Chapter 4. In each of these examples, the background thread (ISR) executes when the I/O device is done performing the required I/O operation. The foreground thread (main program) executes during the times when no interrupts are needed. A global data structure is used to communicate between threads. Notice that the information stored on the stack or in the microcomputer registers by one thread is not accessible by another thread.

A thread can be in one of three states (Figure 5.3). A thread is in the *blocked state* when it is waiting for some external event like I/O (keyboard input available, printer ready, I/O device available.) If a thread communicates with other threads, then it can be blocked wait-

Figure 5.3
A thread can be in one
of three states.

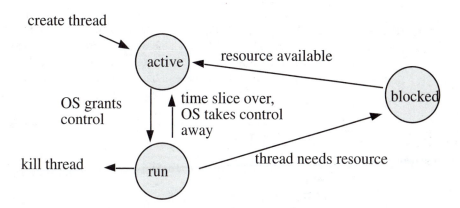

ing for an input message or waiting for another thread to be ready to accept its output message. If a thread wishes to output to the display, but another thread is currently outputting, it will block. We will use a semaphore mechanism to implement the sharing of the display output among multiple threads. If a thread needs information from a FIFO (calls `GetFifo`), then it will be blocked if the FIFO is empty (because it cannot retrieve any information). On the other hand, if a thread outputs information to a FIFO (calls `PutFifo`), then it will be blocked if the FIFO is full (because it cannot save its information).

A thread is in the *active state* if it is ready to run but waiting for its turn. A thread is in the *run state* if it is currently executing. With a single instruction stream computer like the 6811 and 6812, at most one thread can be in the run state at a time. A good implementation is to use linked list data structures to hold the ready and blocked threads. We can create a separate blocked linked list for each reason why the thread cannot execute. For example, one blocked list for waiting for the output display to be free, one for full during a call to `PutFifo`, and one for empty during a call to `GetFifo`. In general, we will have one blocked list with each blocking semaphore.

In Figure 5.4, thread 5 is running, threads 1 and 2 are ready to run, thread 6 is blocked waiting for the printer, and threads 3 and 4 are blocked because the FIFO is empty.

Figure 5.4
Thread 5 is running,
threads 1 and 2 are read
to run, and the rest are
blocked.

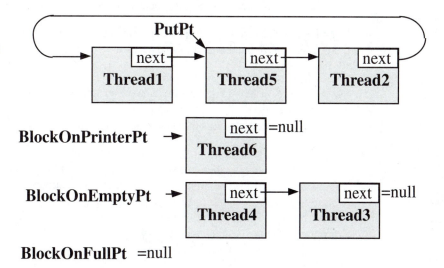

In this section we will develop a simple multithreaded round-robin scheduler. In particular, there will be three statically allocated threads that are each allowed to execute 10 ms in a round-robin fashion. Even though there are two programs, `ProgA` and `ProgB`, there will be three threads (Figure 5.5). Recall that a thread is not simply the software but the execution of the software. In this way, we will have two threads executing the same program, `ProgA`.

Figure 5.5
The circular linked list allows the scheduler to run all three threads equally.

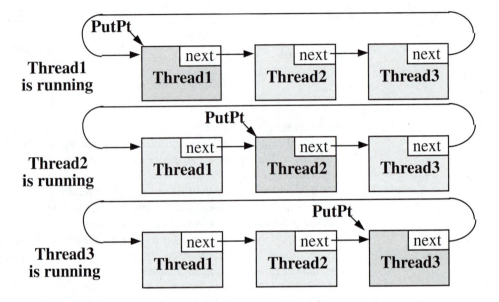

The thread control block (TCB) will store the information private to each thread. There will be a 64-byte TCB structure for each thread. The TCB must contain:

  1. A pointer so that it can be chained into a linked list
  2. The value of its stack pointer
  3. A stack area that includes local variables and other registers

While a thread is running, it uses the actual 6811 hardware registers CCR, B, A, X, Y, PC, SP (Figure 5.6). In addition to these necessary components, the TCB might also contain:

  4. Thread number, type, or name
  5. Age, or how long this thread has been active
  6. Priority
  7. Resources that this thread has been granted

To illustrate the concept of a preemptive scheduler, we will implement a simple system first in 6811 assembly then in 6811 C. The three statically allocated threads are arranged in a circular linked list (Program 5.1). This example illustrates the difference between a program (e.g., `ProgA` and `ProgB`) and a thread (e.g., Thread1, Thread2, and Thread3). Notice that Threads 1 and 2 both execute `ProgA`. There are many applications where the same program is being executed multiple times.

Figure 5.6
The running thread uses the actual registers, while the other threads have their register values saved on the stack.

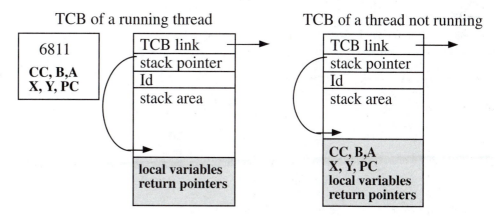

Program 5.1 Assembly code that statically allocates three threads.

```
TCB1  fdb  TCB2         TCB2  fdb  TCB3         TCB3  fdb  TCB1  link
      fdb  IS1                fdb  IS2                fdb  IS3   SP
      fcb  1                  fcb  2                  fcb  4     Id
      rmb  49                 rmb  49                 rmb  49
IS1   rmb  1           IS2   rmb  1                  IS3   rmb  1
      fcb  $40                fcb  $40                fcb  $40   CCR
      fdb  0,0,0             fdb  0,0,0              fdb  0,0,0  DXY
      fdb  ProgA             fdb  ProgA              fdb  ProgB  PC
```

Even though the thread has not yet been allowed to run, it is created with an initial stack area that "looks like" it had been previously suspended by an OC5 interrupt. Notice that the initial value loaded into the CCR ($40) when the thread runs for the first time has XIRQ disabled (X=1) and IRQ enabled (I=0). When the thread is launched for the first time, it will execute the program specified by the value in the "initial PC" location. The Id is used to visualize the active process. On the 6812, change rmb 49, rmb1 to rmb50, rmb0.

The round-robin preemptive scheduler simply switches to a new thread every 10 ms (Programs 5.2, 5.3).

```
Next  equ  0          pointer to next TCB
SP    equ  2          Stack pointer for this thread
Id    equ  4          Used to visualize which thread is current running
RunPt rmb  2          pointer to thread that is currently running
Main  ldaa #$FF
      staa DDRC       PortC displays which program is executing
      ldx  #TCB1      First thread to run
      jmp  Start
```

continued on p. 272

Program 5.2 Assembly code that implements the preemptive thread switcher.

continued from p. 271

```
* Suspend thread which is currently running
OC5Han ldx   RunPt
       sts   SP,x      save Stack Pointer in TCB
* launch next thread
       ldx   Next,x
Start  stx   RunPt
       ldaa  Id,x
       staa  PORTB     visualizes running thread
       lds   SP,x      set SP for this new thread
       ldd   TOC5
       addd  #20000    interrupts every 10 ms
       std   TOC5
       ldaa  #$08      ($20 on the 6812)
       staa  TFLG1     acknowledge OC5
       rti
```

Program 5.3 Assembly code for the two main programs and shared subroutine.

```
ProgA pshx              ProgB pshx              Sub pshx
      tsx                     tsx                   tsx
      ldd  #5                 ldd  #5               std  0,x
      std  0,x               std  0,x               ldaa #1
LoopA ldaa #2          LoopB ldaa #4               staa PORTC
      staa PORTC              staa PORTC            ldd  0,x
      ldd  0,x               ldd  0,x               addd #1
      jsr  sub               jsr  sub               pulx
      std  0,x               std  0,x               rts
      bra  LoopA             bra  LoopB
```

Figure 5.7 Profile of the three threads running two main programs.

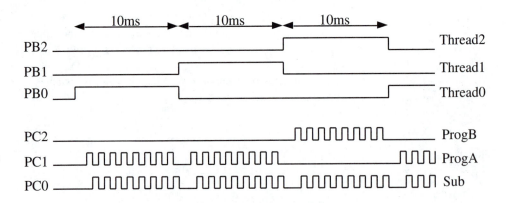

Next we will implement a simple multithreaded system in 6811 C, also with three statically allocated threads running two programs (Figure 5.7). The syntax of these programs, like all 6811 C programs in this book, conforms to the ImageCraft ICC11 version 4.5 or later. The thread.c example that is installed with the TExaS simulator (look for it in the ICC11 subfolder) conforms with the freeware version ICC11 and has slightly different syntax (Program 5.4).

Program 5.4 C code for
the two main programs
and shared subroutine.

```
int Sub(int j){ int i;
    PORTC=1;  /* Port C=program is being executed */
    i=j+1;
    return(i);}
void ProgA(){ int i;
    i=5;
    while(1) { PORTC=2; i=Sub(i);}}
void ProgB(){ int i;
    i=6;
    while(1) { PORTC=4; i=Sub(i);}}
```

One of the tricky parts about defining the TCB is to choose the size of `MoreStack` so
that RegA and RegX are contiguous. Most compilers will try to allocate 16-bit variables
(e.g., RegX) on an even address. Notice that TCB starts at an even address CCR and RegA
will be at odd addresses and the rest will be even. If you change the 49 to 50, then the
compiler will leave a 1-byte gap between RegA and RegX, and the program will not work
(Programs 5.5, 5.6).

Program 5.5 6811 C
code for the thread
control block.

```
struct TCB
{   struct TCB *Next;       /* Link to Next TCB */
    unsigned char *SP;      /* Stack Pointer when not running  */
    unsigned int  Id;       /* output to PortB visualizing active thread */
    unsigned char MoreStack[49];  /* more stack */
    unsigned char CCR;      /* Initial CCR */
    unsigned char RegB;     /* Initial RegB */
    unsigned char RegA;     /* Initial RegA */
    unsigned int RegX;      /* Initial RegX */
    unsigned int RegY;      /* Initial RegY */
    void (*PC)(void);       /* Initial PC */
};
typedef struct TCB TCBType;
typedef TCBType * TCBPtr;
TCBType sys[3]={
    { &sys[1],               /* Pointer to Next */
      &sys[0].MoreStack[49],    /* Initial SP */
      1,                    /* Id */
      { 0},
      0x40,0,0,0,0,         /* CCR,B,A,X,Y */
      ProgA, },             /* Initial PC */
    { &sys[2],               /* Pointer to Next */
      &sys[1].MoreStack[49],    /* Initial SP */
      2,                    /* Id */
      { 0},
      0x40,0,0,0,0,         /* CCR,B,A,X,Y */
      ProgA, },             /* Initial PC */
    { &sys[0],               /* Pointer to Next */
      &sys[2].MoreStack[49],    /* Initial SP */
      4,                    /* Id */
      { 0},
      0x40,0,0,0,0,         /* CCR,B,A,X,Y */
      ProgB, } };           /* Initial PC */
```

```
TCBPtr RunPt;   /* Pointer to current thread */
#pragma interrupt_handler ThreadSwitch()
void ThreadSwitch(){
asm(" ldx _RunPt\n"
    " sts 2,x");
     RunPt=RunPt->Next;
     PORTB=RunPt->Id;  /* PortB=active thread */
asm(" ldx _RunPt\n"
    " lds 2,x");
     TOC3=TCNT+20000;  /* Thread runs for 10 ms */
     TFLG1=0x20; }     /* ack by clearing TOC3F */
void main(void){ DDRC=0xFF;    /* PortC outputs specify that program is running */
     RunPt=&sys[0];  /* Specify first thread */
asm(" sei");
     TOC3vector=&ThreadSwitch;
     TFLG1 = 0x20;   /* Clear OC3F */
     TMSK1 = 0x20;   /* Arm TOC3 */
     TOC3=TCNT+20000;
     PORTB=RunPt->Id;
asm(" ldx _RunPt\n"
    " lds 2,x\n"
    " cli\n"
    " rti");}     /* Launch First Thread */
```

Program 5.6 6811 C code for the thread switcher.

The 6812 implementation is virtually identical, except the 6812 SP points to the top stack entry, whereas the 6811 SP+1 points to the top stack entry. Assembly and C implementations of this example for both the 6811 and 6812 can be found as part of the TExaS simulator. Look for the Thread.* files after the simulator has been installed. The TExaS examples include a logic analyzer connection so that you can create profiles like the one presented in Figure 5.7.

5.1.2
Other Scheduling
Algorithms

A non-preemptive (cooperative) scheduler trusts each thread to voluntarily release control on a periodic basis. Although easy to implement, because it doesn't require interrupts, it is not appropriate for real-time systems.

A priority scheduler assigns each thread a priority number (e.g., 1 is the highest). Normally, we add priority to a system that implements blocking semaphores and not to one that uses spin-lock semaphores. Two or more threads can have the same priority. A priority 2 thread is run only if no priority 1 threads are ready to run. Similarly, we run a priority 3 thread only if no priority 1 or priority 2 threads are ready. If all threads have the same priority, then the scheduler reverts to a round-robin system. The advantage of priority is that we can reduce the latency (response time) for important tasks by giving those tasks a high priority. The disadvantage is that on a busy system, low-priority threads may never be run. This situation is called *starvation*.

5.1.3
Dynamic
Allocation of
Threads

In the above examples, the number of threads was specified at assembly/compile time. We can use the compiler's memory manager to dynamically allocate the deallocate threads. The Program 5.7 code fragment dynamically allocates space for a new TCB and initializes its contents.

Program 5.7 6811 C function to create a new thread.

```
void create(void (*program)(void), int TheId){
  TCBPtr NewPt;       // pointer to new thread control block
  NewPt=(TCBPtr)malloc(sizeof(TCBType)); // space for new TCB
  if(NewPt==0)return;
  NewPt->SP=&(NewPt->CCR-1);  /* 6811 Stack Pointer when not running  */
  NewPt->Id=TheId;            /* used to visualize active thread */
  NewPt->CCR=0x40;            /* Initial CCR, I=0 */
  NewPt->RegB=0;              /* Initial RegB */
  NewPt->RegA=0;              /* Initial RegA */
  NewPt->RegX=0;              /* Initial RegX */
  NewPt->RegY=0;              /* Initial RegY */
  NewPt->PC=program;          /* Initial PC */
  if(RunPt){
    NewPt->Next=RunPt->Next;
    RunPt->Next=NewPt;}       /* will run Next */
  else
    RunPt=NewPt;              /* the first and only thread */
}
```

5.2 Semaphores

We will use semaphores to implement synchronization, sharing, and communication between threads. A semaphore is a counter and has two atomic functions (methods) apart from initialization that operate on the counter. The operations are called P or wait (derived from the Dutch *proberen,* "to test"), and V or signal (derived from the Dutch *verhogen,* "to increment"). When we use a semaphore, we usually can assign a meaning or significance to the counter value. A binary semaphore has a global variable that can be 1 for free and 0 for busy.

5.2.1 Spin-Lock Semaphore Implementation

There are many implementations of semaphores, but the simplest is called *spin-lock* (Figure 5.8). If the thread calls `wait` with the counter less than or equal to 0 it will "spin" (do nothing) until the counter goes above zero (Program 5.8). In the context of the previous round-robin scheduler, a thread that is "spinning" will perform no useful work but eventually will be suspended by the OC5 handler, and other threads will execute. It is

Figure 5.8
Flowcharts of a spin-lock counting semaphore.

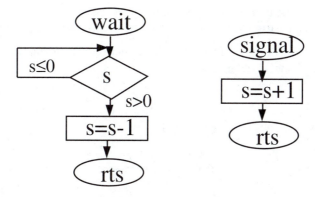

Program 5.8 6811/6812
assembly code for a spin-
lock counting semaphore.

```
S        fcb  1       semaphore counter initialized to 1
wait     sei          make read-modify-write atomic
         ldaa S       current value of semaphore
         bhi  OK      available if >0
         cli
         bra  wait    **interrupts can occur here**
OK       deca
         staa S       S=S-1
         cli
         rts
signal   inc  S       S=S+1, this is atomic
         rts
```

important to allow interrupts to occur while the thread is spinning so that the computer does not hang up. The read-modify-write operation on the global semaphore must be made atomic, because the scheduler might switch threads in between any two instructions that execute with the interrupts enabled.

The 6812 `minm` instruction can be used to implement a binary semaphore without disabling interrupts (Program 5.9). In C, these functions are shown in Program 5.10.

```
S        fcb  1      semaphore flag initialized to 1
bWait    clra        new value
loop     minm S      in either case S is now 0, carry set if S used to 1
         bcc  loop   loop until S=1
         rts
bSignal  ldaa #1
         staa S      S=1
         rts
```

Program 5.9 6812 assembly code for a spin-lock binary semaphore.

Program 5.10 6812 C
code for a spin-lock
binary semaphore.

```
void bWait(char *semaphore){
asm(" clra\n"            // new value for semaphore
"loop: minm [2,x]\n"     // test and set (ICC12 version 5)
"      bcc loop\n"); }
void bSignal(char *semaphore){
      (*semaphore)=1;} // compiler makes this atomic
```

Observation: If the semaphores can be implemented without disabling interrupts, then the latency in response to external events will be improved.

We can use three binary semaphores and a counter to implement a counting semaphore again without disabling interrupts (Program 5.11).

Program 5.11 C code
for a counting
semaphore.

```
struct sema4    // counting semaphore based on 3 binary semaphores
{   int value;  // semaphore value
    char s1;    // binary semaphore
    char s2;    // binary semaphore
    char s3;    // binary semaphore
};
```

```
typedef struct sema4 sema4Type;
typedef sema4Type * sema4Ptr;
void Wait(sema4Ptr semaphore){
     bWait(&semaphore->s3);  // wait if other caller to Wait gets here first
     bWait(&semaphore->s1);  // mutual exclusive access to value
     (semaphore->value)--;   // basic function of Wait
     if((semaphore->value)<0){
         bSignal(&semaphore->s1); // end of mutual exclusive access to value
         bWait(&semaphore->s2);   // wait for value to go above 0
         }
     else
         bSignal(&semaphore->s1); // end of mutual exclusive access to value
     bWait(&semaphore->s3);        // let other callers to Wait get in
}
void Signal(sema4Ptr semaphore){
     bWait(&semaphore->s1);  // mutual exclusive access to value
     (semaphore->value)++;   // basic function of Signal
     if((semaphore->value)<=0)
         bSignal(&semaphore->s2);   // allow S2 spinner to continue
     bSignal(&semaphore->s1); // end of mutual exclusive access to value
}
void Initialize(sema4Ptr semaphore, int initial){
     semaphore->s1=1;   // first one to bWait(s1) continues
     semaphore->s2=0;   // first one to bWait(s2) spins
     semaphore->s3=1;   // first one to bWait(s3) continues
     semaphore->value=initial;}
```

**5.2.2
Blocking
Semaphore
Implementation**

For this implementation, each semaphore has an integer global variable (an up/down counter) and a blocked **tcb** linked list. This linked list contains the threads that are blocked (not ready to run) because they called the Wait function when the semaphore counter was less than or equal to 0. Our semaphore "object" also has three "methods": Initialize, Wait, and Signal (Figure 5.9). The proper way to make a multistep sequence atomic is to first save the interrupt status (I bit) on the stack, then disable interrupts. At the end of the atomic sequence, we restore the I bit back to its original value rather than simply assuming it was enabled to begin with. This procedure to disable/enable interrupts handles the situation of nested critical sections. We must be careful to block and wake up the correct number of threads.

Figure 5.9
Flowcharts of a blocking
counting semaphore.

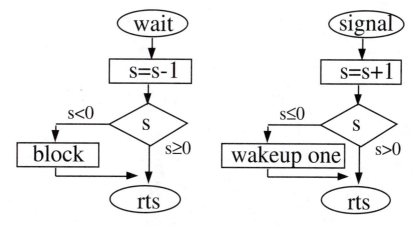

Initialize (Program 5.12):

1. Set the counter to its initial value
2. Clear the associated blocked **tcb** linked list

Wait:

1. Disable interrupts to make atomic:

```
tpa
psha    save old interrupt status
sei
```

2. Decrement the semaphore counter, $S=S-1$
3. If the semaphore counter is less than 0, then
>   Block this thread onto the associated **tcb** blocked linked list by executing SWI
>   The SWI operation performs stack operations similar to an OC5 interrupt
>   The SWI handler suspends the current thread
>>   Moves the **tcb** of this thread from the active list to the blocked list
>   Launches another thread from the active list

4. Restore interrupt status:

```
pula
tap
```

Signal:

1. Disable interrupts to make atomic:

```
tpa
psha    save old interrupt status
sei
```

2. Increment the semaphore counter, $S=S+1$
3. If the semaphore counter is less than or equal to 0, then
>   Wake up one thread from the **tcb** linked list
>   Do not suspend execution of this thread
>   Simply move the **tcb** of the "wakeup" thread from the blocked list to the active list

4. Restore interrupt status:

```
pula
tap
```

Program 5.12 Assembly code to initialize a blocking semaphore.

```
S       rmb  1    semaphore counter
BlockPt rmb  2    Pointer to threads blocked on S
Init    tpa
        psha           Save old value of I
        sei            Make atomic
        ldaa #1
        staa S         Init semaphore value
        ldx  #Null
        stx  BlockPt   empty list
        pula
        tap            Restore old value of I
        rts
```

The implementation in Program 5.13 assumes the system only has one semaphore. In a more general implementation, there would be multiple semaphores, each with its own counter and blocked list. In this case, a pointer (call by reference) to the semaphore structure (counter and blocked list) would be passed to the SWI handler.

Program 5.13 Assembly code helper function to block a thread, used to implement a blocking semaphore.

```
        sty   Next,x   link "to be blocked"
        stx   BlockPt
* Launch next thread
        ldx   RunPt
        lds   SP,x     set SP for this new thread
        ldd   TCNT     Next thread gets a full 10ms time slice
        addd  #20000   interrupt after 10 ms
        std   TOC5
        ldaa  #$08     ($20 on the 6812)
        staa  TFLG1    clear OC5F
        rti


* To block a thread on semaphore S, execute SWI
SWIhan  ldx   RunPt   running process "to be blocked"
        sts   SP,x    save Stack Pointer in its TCB
* Unlink "to be blocked" thread from RunPt list
        ldy   Next,x  find previous thread (Figure 5.10)
```

Figure 5.10
Linked list of threads ready to run before the thread is blocked.

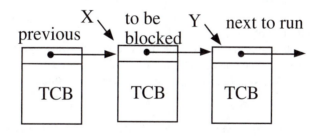

```
        sty RunPt   next one to run
look    cpx Next,y  search to find previous
        beq found
        ldy Next,y
        bra look
found   ldd RunPt   one after blocked (Figure 5.11)
```

Figure 5.11
Linked list at this point in the program.

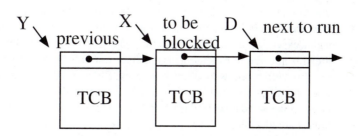

```
                      std Next,y link previous to next to run
              *  Put "to be blocked" thread on block list
                      ldy BlockPt  (Figure 5.12)
```

Figure 5.12
TCBs after the thread is
blocked (ready-to-run
TCBs are not shown).

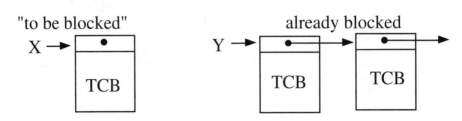

5.3   Applications of Semaphores

This section can be used in two ways. First, it provides a short introduction to the kinds of problems that can be solved using semaphores. In other words, if you have a problem similar to one of these examples, then you should consider a thread scheduler with blocking semaphores as one possible implementation. Second, this section provides the basic approach to solving these particular problems. As stated in the introduction to this chapter, we will generally find that large software systems with more-or-less independent modules can use semaphores when synchronization is needed. On the other hand, for smaller systems with more tightly coupled modules, then the overhead of a thread scheduler will probably not be justified.

5.3.1
Thread
Synchronization
or Rendezvous

The objective of this example is to synchronize threads 1 and 2. In other words, whichever thread gets to this part of the code first will wait for the other. Initially semaphores S1 and S2 are both 0. The two threads are said to *rendezvous* at the code following the signal and wait calls. The significance of the two semaphores is illustrated in the following table:

S1	S2	Meaning
0	0	Neither thread has arrived at the rendezvous location or both have passed
−1	+1	Thread 2 arrived first and is waiting for thread 1
+1	−1	Thread 1 arrived first and is waiting for thread 2

Thread 1	Thread 2
signal(&S1);	signal(&S2);
wait(&S2);	wait(&S1);

5.3.2
Resource Sharing,
Nonreentrant
Code or Mutual
Exclusion

The objective of this example is to share a common resource on a one-at-a-time basis. The critical section (or vulnerable window) of nonreentrant software is that region that should be executed only by one thread at a time. Mutual exclusion means that once a thread has begun executing in its critical section (the printf(); in this example), then no other thread is allowed to execute its critical section. In other words, whichever thread starts to print first will be allowed to finish printing. Either a binary or a counting semaphore can be used. Initially, the semaphore is 1. If S=1, it means the printer is free. If S=0, it means the printer is busy and no thread is waiting. If S<0, it means the printer is busy and one or more threads are blocked.

Thread 1	**Thread 2**	**Thread 3**
`bwait(&S);`	`bwait(&S);`	`bwait(&S);`
`printf("bye");`	`printf("tchau");`	`printf("ciao");`
`bsignal(&S);`	`bsignal(&S);`	`bsignal(&S);`

5.3.3 Thread Communication Between Two Threads Using a Mailbox

The objective of this example is to send mail from thread 1 to thread 2. The `send` semaphore allows the producer to tell the consumer that new mail is available. The `ack` semaphore is a mechanism for the consumer to tell the producer that the mail was received. Initially, semaphores `send` and `ack` are both 0. The significance of the two semaphores is illustrated in the following table. In this example, the two threads will rendezvous at this point.

Send	Ack	Meaning
0	0	No mail available, consumer is not waiting
−1	0	No mail available, consumer is waiting for mail
+1	−1	Mail available and producer is waiting for acknowledgment

Producer thread	**Consumer thread**
`Mail=4;`	`wait(&send);`
`signal(&send)`	`read(Mail);`
`wait(&ack)`	`signal(&ack)`

5.3.4 Thread Communication Between Many Threads Using a FIFO Queue

In the *bounded buffer* problem, we can have multiple threads putting data into a finite-size FIFO and multiple threads getting data out of this FIFO. Bounded buffer is simply another name for the standard FIFO queue we used in Chapter 4 to solve the producer/consumer application. We need two counting semaphores called `CurrentSize` and `RoomLeft` that contain the number of entries currently stored in the FIFO and the number of empty spaces left in the FIFO, respectively. `CurrentSize` is initialized to zero, and `RoomLeft` is initialized to the maximum allowable number of elements in the FIFO. The `PutFifo` routine executes the following steps:

```
Wait(&RoomLeft);        /* This will block the thread when FULL */
asm(" sei");            /* mutually exclusive access to FIFO */
Enter information into the FIFO structure
asm(" cli");
Signal(&CurrentSize);   /* Wake up a thread if blocked on empty */
```

The `GetFifo` routine executes the following steps:

```
Wait(&CurrentSize);        /* This will block the thread when EMPTY */
asm(" sei");               /* mutually exclusive access to FIFO */
Remove information from the FIFO structure
asm(" cli");
Signal(&RoomLeft);         /* Wake up a thread if blocked on full */
```

Another implementation of this bounded buffer problem adds the semaphore called `Mutex` to implement the mutually exclusive access to the FIFO. `Mutex` is initialized to 1. The significance of this semaphore is illustrated in the following table:

Mutex	Meaning
1	No thread is currently entering or removing data
0	One thread is entering or removing data, none are blocked
−1	One thread is entering or removing data, one is blocked

The `PutFifo` routine executes the following steps:

```
Wait(&RoomLeft);          /* This will block the thread when FULL */
Wait(&Mutex);             /* Access to FIFO is critical code */
Enter information into the FIFO structure
Signal(&Mutex);
Signal(&CurrentSize);     /* Wake up a thread blocked on empty? */
```

The `GetFifo` routine executes the following steps:

```
Wait(&CurrentSize);       /* This will block the thread when EMPTY */
Wait(&Mutex);             /* Access to FIFO is critical code */
Remove information from the FIFO structure
Signal(&Mutex);
Signal(&RoomLeft);        /* Wake up a thread blocked on full? */
```

5.4 Glossary

active A thread that is in the ready-to-run circular linked list. It is either running or ready to run.

binary semaphore A semaphore that can have two values. The value=1 means OK and the value=0 means busy. Compare to counting semaphore.

blocked A thread that is not scheduled for running because it is waiting for an external event.

blocking semaphore A semaphore where the threads will block (so other threads can perform useful functions) when they execute wait on a busy semaphore. Contrast to spin-lock semaphore.

cooperative multitasking A scheduler that cannot suspend execution of a thread without the thread's permission. The thread must cooperate and suspend itself. Same as nonpreemptive scheduler.

counting semaphore A semaphore that has a signed integer value. The value>0 means OK and the value≤0 means busy. Compare to binary semaphore.

critical section Locations within a software module, which if an interrupt were to occur at one of these locations, then an error could occur (e.g., data lost, corrupted data, program crash). Same as vulnerable window.

deadlock A scenario that occurs when two or more threads are all blocked, each waiting for the other with no hope of recovery.

dynamic allocation Data structures like the TCB that are created at run time by calling `malloc()` and exist until the software releases the memory block back to the heap by calling `free()`. Contrast to static allocation.

lightweight process Same as a thread.

multithreaded A system with multiple threads that cooperates toward a common overall goal.

mutual exclusion or **mutex** Thread synchronization where at most one thread at a time is allowed to enter.

nonpreemptive scheduler A scheduler that cannot suspend execution of a thread without the thread's permission. The thread must cooperate and suspend itself. Same as cooperative multitasking.

preemptive scheduler A scheduler that has the power to suspend execution of a thread without the thread's permission.

priority When two or more threads are ready to run, priority determines which order to run them.

response time Similar to latency, it is the delay between the time an event occurs and the time the software responds to the event.

round-robin scheduler A scheduler that runs each active thread equally.

scheduler System software that suspends and launches threads.

semaphore A system function with two operations (wait and signal) that provide for thread synchronization and resource sharing.

spin-lock semaphore A semaphore where the threads will spin (run but do no useful function) when they execute wait on a busy semaphore. Contrast to blocking semaphore.

starvation A condition that occurs with a priority scheduler where low-priority threads are never run.

static allocation Data structures like the TCB that are defined at assembly or compile time and exist throughout the life of the software. Contrast to dynamic allocation.

thread The execution of software that cooperates with other threads.

thread control block (TCB) Information about each thread.

vulnerable window Locations within a software module, which if an interrupt were to occur at one of these locations, then an error could occur (e.g., data lost, corrupted data, program crash). Same as critical section.

5.5 Exercises

5.1. In this exercise you will extend the preemptive scheduler to support priority. This system should support three levels of priority: 1 will be the highest. You can solve this problem on the 6811 or 6812 using either assembly or C.

   a) Redesign the TCB to include a 16-bit integer for the priority (although the values will be restricted to 1, 2, 3). Show the static allocation for the three threads from the example in this chapter, assuming the first two are priority 2 and the last is priority 3. There are no priority 1 threads in this example, but there might be in the future.

   b) Redesign the scheduler to support this priority scheme.

   c) In the chapter it said "*Normally, we add priority to a system that implements blocking semaphores and not to one that uses spin-lock semaphores.*" What specifically will happen here if the system is run with spin-lock semaphores?

   d) Even when the system supports blocking semaphores, starvation might happen to the low-priority threads. Describe the sequence of events that causes starvation.

   e) Suggest a solution to the starvation problem.

5.2. We can use semaphores to limit access to resources. In the following example, both threads need access to a printer and a SPI port. The binary semaphore sPrint provides mutual exclusive access to the printer, and the binary semaphore sSPI provides mutual exclusive access to the SPI port. The following scenario has a serious flaw:

Thread 1	**Thread 2**
bwait(&sPrint);	bwait(&sSPI);
bwait(&sSPI);	bwait(&sPrint);
printf("bye");	OutSPI(5);
OutSPI(6);	printf("tchau");
bsignal(&sPrint);	bsignal(&sSPI);
bsignal(&sSPI);	bsignal(&sPrint);

a) Describe the sequence of events that would lead to the condition where both threads are stuck forever.

b) How could you change the above program to prevent the deadlock?

5.3 You are given four identical I/O ports to manage on the MC68HC12A4: Port A, Port B, Port C, and Port D. You may assume there is a preemptive thread scheduler and blocking semaphores.

a) Look up the address of each port and its direction register.

b) Create a data structure to hold an address of the port and the address of the data direction register. Assume the type of this structure is called `PortType`.

c) Design and implement a manager that supports two functions. The first function is called `NewPort`. Its prototype is

```
PortType *NewPort(void);
```

If a port is available when a thread calls `NewPort`, then a pointer to the structure, defined in part b, is returned. If no port is available, then the thread will block. When a port becomes available, this thread will be awakened and the pointer to the structure will be returned. You may define and use blocking semaphores without showing the implementation of the semaphore or scheduler. The second function is called `FreePort`, and its prototype is

```
void FreePort(PortType *pt);
```

This function returns a port so that it can be used by the other threads. Include a function that initializes the system, where all five ports are free. (*Hint:* The solution is very similar to the FIFO queue example shown in Section 5.3.4.)

5.4. Some computers have a `swap` instruction. Modify the subroutines shown in Program 5.9, written in 6812 assembly language, to implement a binary spin-lock semaphore using this fictional instruction. The system must provide for mutual exclusion. You may not disable interrupts. Do not use the `minm` instruction. The syntax of this fictional atomic operation is

```
swap      Operand,r
```

where `Operand` is any standard 6812 addressing mode and `r` is any of the 6812 registers. This instruction exchanges the 8- or 16-bit contents of the `Operand` and data register. It does not set any condition code bits.

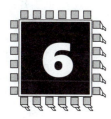

6 Timing Generation and Measurements

Chapter 6 objectives are to use:

❑ Input capture to generate interrupts and measure period or pulse width
❑ Output compare to create periodic interrupts, generate square waves, and measure frequency
❑ Both input capture and output compare to make flexible and robust measurement systems

The timer systems on the Motorola microcomputers are very versatile. Over the last 20 years, the evolution of these timer functions has paralleled the growth of new applications possible with these embedded computers. In other words, inexpensive yet powerful embedded systems have been made possible by the capabilities of the timer system. In this chapter we will introduce these functions, then use them throughout the remainder of the book. If we review the applications introduced in the first chapter (see Section 1.1.1), we will find that virtually all of them make extensive use of the timer functions developed in this chapter.

6.1 Input Capture

6.1.1 Basic Principles of Input Capture

We can use input capture to measure the period or pulse width of TTL-level signals (Figure 6.1). The input capture system can also be used to trigger interrupts on rising or falling transitions of external signals. TCNT is a 16-bit counter incremented at a fixed rate. Each input capture module has:

> An external input pin, ICn
> A flag bit
> Two edge control bits, EDGnB and EDGnA (only one control bit on the MC68HC05C8)
> An interrupt mask bit (arm)
> A 16-bit input capture register

The various members of the Motorola HC05, HC08, HC11, and HC12 families have from zero to eight input capture modules. For example, the MC68HC705JA1 has none, the

Figure 6.1
Basic components of
input capture.

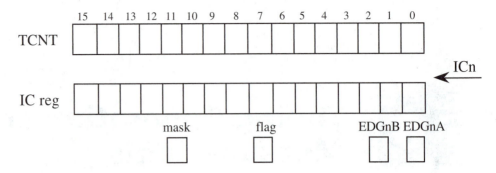

MC68HC05C8 has one, the MC68HC708XL36 has four, the MC68HC11A8 has three, the MC68HC812A4 has eight, and the MC68HC912B32 has eight. An external input signal is connected to the input capture pin. The EDGnB, EDGnA bits specify whether the rising, falling, or both rising and falling edges of the external signal will trigger a capture event. Two or three actions result from a capture event: (1) the current TCNT value to be copied into the input capture register, (2) the input capture flag is set and (3) an interrupt is requested (occurs only when the mask is 1). You can set up the MC68HC708XL36 so that a DMA cycle occurs on an input capture event.

The module is armed when the mask bit is 1. This means an interrupt will be requested on a capture event. The input capture mechanism has many uses. Three common applications are:

1. Arm the flag bit so that an interrupt is requested on the active edge of the external signal
2. Perform two rising edge input captures and subtract the two measurements to get period
3. Perform a rising edge capture, then a falling edge capture, and subtract the two measurements to get pulse width

6.1.2
Input Capture
Details

Next we will overview the specific input capture functions on particular Motorola microcomputers. This section is intended to supplement rather than replace the Motorola manuals. When designing systems with input capture, please refer also to the reference manual of your specific Motorola microcomputer.

6.1.2.1
MC68HC05C8 Input
Capture Details

On this 6805, TCNT is a 16-bit unsigned counter that is incremented every four E clocks. If the basic bus clock is 2 MHz, then TCNT is incremented every 2 μs. This counter cannot be stopped or reset. The fundamental approach to input capture involves connecting an external TTL-level signal(s) to the TCAP input. Since there are no direction register bits associated with TCAP, this pin is always an input. The input capture event occurs on either the rising or falling edge of this external signal. The 6805 will set the input capture flag (ICF) and latch the current 16 bit TCNT value into the input capture latch (ICR). If the input capture flag is armed (ICIE=1), then an interrupt will be requested when the flag is set (Figure 6.2).

Figure 6.2
Input capture interface
on the MC68HC05C8.

6805 input capture   TCAP ◄───

	15	14	13	12	11	10	9	8	7	6	5	4	3	2	1	0	
TCNT																	$0018
ALTCNT																	$001A
ICR																	$0014
OCR																	$0016

We specify the active edge (i.e., the edge that latches TCNT and sets the flag) by initializing the TCR register. We can arm or disarm the input capture interrupts by specifying the ICIE bit.

	7	6	5	4	3	2	1	0	
TCR	**ICIE**	OCIE	TOIE	0	0	0	**IEDG**	OLVL	$0012

ICIE: Input capture interrupt enable
    0 = ICF interrupts disabled
    1 = input capture interrupt requested when ICF status flag is set
OCIE: Output compare interrupt enable
    0 = OCF interrupts disabled
    1 = output compare interrupt requested when OCF status flag is set
TOIE: Timer overflow interrupt enable
    0 = TOF interrupts disabled
    1 = timer overflow interrupt requested when TOF status flag is set
IEDG: Input capture edge
    0 = rising edge of TCAP sets ICF
    1 = falling edge of TCAP sets ICF
OLVL: Output level
    0 = set TCMP=0 on an output compare event (when TCNT equals OCR)
    1 = set TCMP=1 on an output compare event (when TCNT equals OCR)

Our software can determine if an input capture event has occurred by checking the ICF bit in the TSR register. The ICF flag is cleared by the software when it first reads TSR with ICF set, followed by a read of the low byte of ICR.

	7	6	5	4	3	2	1	0	
TSR	**ICF**	OCF	TOF	0	0	0	0	0	$0013

Every time the TCNT register overflows from $FFFF to 0, the TOF flag in the TSR register is set. The TOF condition will cause an interrupt if the mask TOIE=1. The TOF flag is cleared by the software when it first reads TSR with TOF set, followed by a read of the low byte of TCNT. To prevent the inadvertent clearing of TOF, regular reading of TCNT to measure elapsed time should use the alternative TCNT. Reads of ALTCNT return the same value as TCNT but do not affect the TOF flag.

6.1.2.2
MC68HC708XL36 Input
Capture Details

On the MC68HC708XL36, TCNT is a 16-bit unsigned counter that is incremented at a rate determined by three bits (PS2, PS1, and PS0) in the TSC register ($0020). This counter can be stopped and/or reset using the TSTOP and TRST bits in the TSC register. The fundamental approach to input capture involves connecting an external TTL-level signal(s) to PTE7, PTE6, PTE5, and/or PTE4. There is a direction register associated with PTE. The corresponding bits in DDRE should be zero to signify inputs. The input capture event occurs on either the rising, falling, or both rising and falling edges of this external signal. The 6808 will set the input capture flag (CH0F, CH1F, CH2F, or CH3F) and latch the current 16-bit TCNT value into the input capture latch (TCH0, TCH1, TCH2, or TCH3). If the input capture flag is armed (CH0IE, CH1IE, CH2IE, or CH3IE), then an interrupt will be requested when the flag is set (Figure 6.3).

Figure 6.3
Input capture interface
on the
MC68HC708XL36.

We make the pin an input capture by clearing MSnB and MSnA. We can arm the input capture interrupts by setting the corresponding CHnIE bit.

We specify the active edge (i.e., the edge that latches TCNT and sets the flag) by initializing the corresponding ELSnB, ELSnA bits in the TSCn register (Table 6.1).

	7	6	5	4	3	2	1	0	
TSC0	CH0F	CH0IE	MS0B	MS0A	ELS0B	ELS0A	TOV0	CH0MAX	$0026
TSC1	CH1F	CH1IE	0	MS1A	ELS1B	ELS1A	TOV1	CH1MAX	$0029
TSC2	CH2F	CH2IE	MS2B	MS2A	ELS2B	ELS2A	TOV2	CH2MAX	$002C
TSC3	CH3F	CH3IE	0	MS3A	ELS3B	ELS3A	TOV3	CH3MAX	$002F

Table 6.1
Four control bits define the mode and the active edge used for input capture.

MSnB	MSnA	ELSnB	ELSnA	Active edge
X	X	0	0	None, regular I/O on Port E
0	0	0	1	Capture on rising
0	0	1	0	Capture on falling
0	0	1	1	Capture on both rising and falling

Our software can determine if an input capture event has occurred by checking the CHnF flag bit. To clear the CHnF flag bit, the software first reads the TSCn register with the bit set, followed by writing a logic 0 to the CHnF bit. Every time the TCNT register overflows from $FFFF to 0, the TOF flag in the TSC register is set. The TOF condition will cause an interrupt if the mask TOIE=1. To clear the TOF flag bit, the software first reads the TSC register with the bit set, followed by writing a logic 0 to the TOF bit (Table 6.2).

	7	6	5	4	3	2	1	0	
TSC	TOF	TOIE	TSTOP	TRST	0	PS2	PS1	PS0	$0020

Table 6.2
Three control bits define TCNT clock rate.

PS2	PS1	PS0	Divide by	TCNT clock
0	0	0	1	125 ns
0	0	1	2	250 ns
0	1	0	4	500 ns
0	1	1	8	1 μs
1	0	0	16	2 μs
1	0	1	32	4 μs
1	1	0	64	8 μs
1	1	1	PTE3/TCLK	External

The TDMA register is used to activate DMA transfers on input capture or output compare events. If both the DMAnS bit and the CHnIE bits are both set, then a DMA request is generated whenever the CHnF flag bit is set. See Chapter 10 for more details about DMA.

	7	6	5	4	3	2	1	0	
TDMA	0	0	0	0	DMA3S	DMA2S	DMA1S	DMA0S	$0021

6.1.2.3
MC68HC11A8 Input Capture Details

On the 6811, TCNT is a 16-bit unsigned counter that is incremented at a rate determined by two bits (PR1 and PR0) in the TMSK2 register ($1024). This counter cannot be stopped or reset. The fundamental approach to input capture involves connecting an external TTL-level signal(s) to PA2, PA1, and/or PA0. Since there are no direction register bits associated with PA2-0, these bits are always inputs (Figure 6.4). The input capture event occurs on either the rising, falling, or both rising and falling edges of this external signal. The 6811 will set the input capture flag (IC1F, IC2F, or IC3F) and latch the current 16-bit TCNT value into the input capture latch (TIC1, TIC2, or TIC3). If the input capture flag is armed (IC1I, IC2I, or IC3I), then an interrupt will be requested when the flag is set.

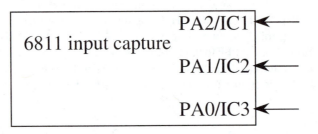

Figure 6.4
Input capture interface
on the MC68HC11.

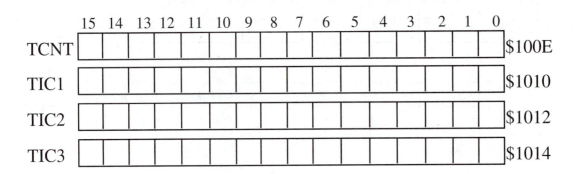

We specify the active edge (i.e., the edge that latches TCNT and sets the flag) by initializing the TCTL2 register (Table 6.3).

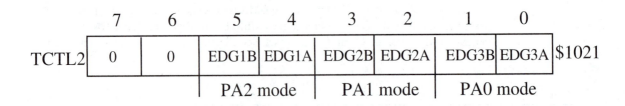

Table 6.3
Two control bits define
the active edge used for
input capture.

EDGnB	EDGnA	Active edge
0	0	None
0	1	Capture on rising
1	0	Capture on falling
1	1	Capture on both rising and falling

We can arm or disarm the input capture interrupts by initializing the TMSK1 register.

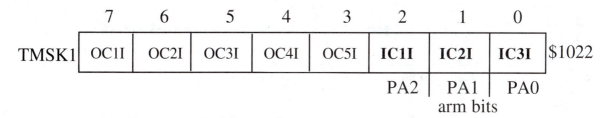

Our software can determine if an input capture event has occurred by reading the TFLG1 register.

Every time the TCNT register overflows from $FFFF to 0, the TOF flag in the TFLG2 register is set. The TOF condition will cause an interrupt if the mask TOI=1.

The flags in the TFLG1 register are cleared by writing a 1 into the specific flag bit we wish to clear. For example, writing a $FF into TFLG1 will clear all eight flags. The following are valid examples for clearing IC3F. In other words, these acknowledge sequences clear the IC3F flag without affecting the other seven flags in the TFLG1 register.

```ldy #$1000```
```ldaa #$01```
```staa $23,Y```
	```ldaa #$01```
	```staa $1023```
		```TFLG1=0×01;```

It is inappropriate to use the `bset` instruction (e.g., `bset $23,X,$01`) to clear flags, because this read-modify-write instruction will also clear all the other bits in the flag register. For example, assume the TFLG1 initially equals $19. The execution of the `bset $23,X,$01` instruction produces the following two accesses:

```
read  $1023 $19     the $19 is ored with $01 to get $19
write $1023 $19     clears OC4F, OC5F and IC3F
```

Similarly, the following C code will mistakenly clear all flag bits in TFLG1,

```
TFLG1 |= 0x01;
```

For an E clock of 2 MHz, Table 6.4 lists the control bits that define the TCNT clock rate.

Table 6.4
Two control bits define
the TCNT clock rate.

PR1	PR0	Divide by	TCNT clock
0	0	1	500 ns
0	1	2	1 μs
1	0	4	2 μs
1	1	8	4 μs

6.1.2.4
6812 Input Capture
Details

On the 6812, TCNT is a 16-bit unsigned counter that is incremented at a rate determined by three bits (PR2, PR1, and PR0) in the TMSK2 register ($008D).

	7	6	5	4	3	2	1	0	
TSCR	TEN	TSWAI	TSBCK	TFFCA	0	0	0	0	$0086
TMSK2	TOI	0	TPU	TDRB	TCRE	PR2	PR1	PR0	$008D

For an E clock of 8 MHz, Table 6.5 lists the control bits that define the TCNT clock rate.

Table 6.5
Three control bits define
the TCNT clock rate.

PR2	PR1	PR0	Divide by	TCNT clock
0	0	0	1	125 ns
0	0	1	2	250 ns
0	1	0	4	500 ns
0	1	1	8	1 μs
1	0	0	16	2 μs
1	0	1	32	4 μs
1	1	0	Reserved	—
1	1	1	Reserved	—

Observation: With a PR2-0 value of 010, the TCNT will clock at the same rate as a 2 MHz 6811.

TEN: Timer enable
  0 = disables timer TCNT
  1 = allows TCNT to increment
TSWAI: Timer stops while in wait
  0 = TCNT counts in wait mode
  1 = TCNT stops in wait mode

TSBCK: Timer stops while in background
  0 = TCNT counts in background mode
  1 = TCNT stops in background mode
TFFCA: Timer fast flag clear all
  0 = flags cleared normally
  1 = any access to TCNT ($0084–$0085) clears timer overflow flag, TOF
    = read from input capture ($0090–$009F) clears input capture flag, CnF
    = write to output compare ($0090–$009F) clears output compare flag, CnF
    = read from PACNT ($00A2–$00A3) clears pulse accumulator flags, PAOVF, PAIF
TOI: Timer overflow interrupt enable
  0 = no interrupts
  1 = TOF will cause an interrupt
TPU: Timer pull-up resistor enable
  0 = no pull-up on Port T inputs
  1 = pull-up resistor enabled on Port T inputs
TDRB: Timer drive reduction
  0 = normal current on Port T outputs
  1 = reduced current on Port T outputs
TCRE: Timer counter reset enable
  0 = free-running counter
  1 = reset TCNT=0 on a successful output compare 7

TCNT can be reset with a special mode of output compare 7 (TCRE=1). But in this book, we will consider TCNT as a simple 16-bit counter clocked at a continuous rate (we will always set TCRE=0). The fundamental approach to input capture involves connecting an external TTL-level signal to one of the eight input capture pins on Port T (Figure 6.5). The pin is selected as input capture by placing a 0 in the corresponding bit of the TIOS register. There is a direction register, DDRT, for Port T, which means we should clear the corresponding bits for the input capture inputs. The input capture event occurs on either the rising, falling, or both the rising and falling edges of this external signal. The 6812 will set the input capture flag (CnF in the TFLG1 register) and latch the current 16-bit TCNT value into the input capture latch (TCn). If the input capture flag is armed (CnI in the TMSK1 register), then an interrupt will be requested when the flag is set.

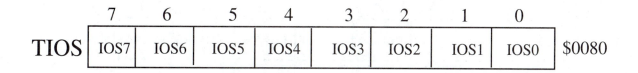

TIOS

7	6	5	4	3	2	1	0
IOS7	IOS6	IOS5	IOS4	IOS3	IOS2	IOS1	IOS0

$0080

Figure 6.5
Input capture interface on the MC68HC812A4 or MC68HC912B32.

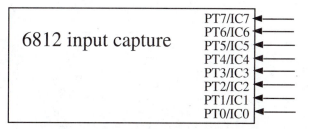

There is a separate 16-bit input capture register for each of the eight input capture modules.

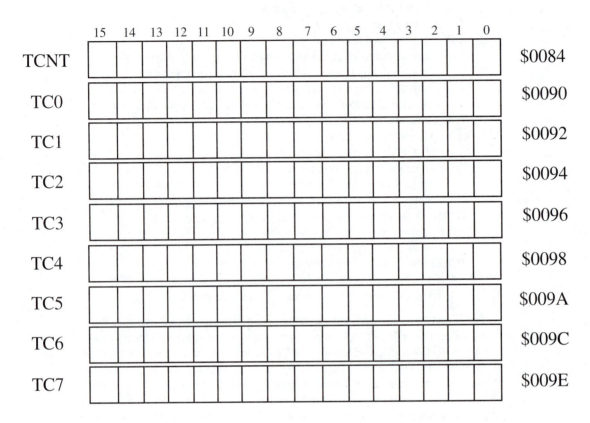

We specify the active edge (i.e., the edge that latches TCNT and sets the flag) by initializing the TCTL3 and TCTL4 registers (Table 6.6).

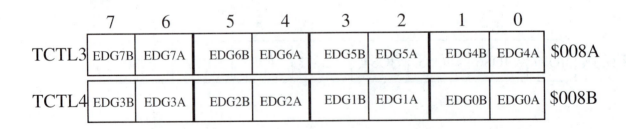

Table 6.6
Two control bits define the active edge used for input capture.

EDGnB	EDGnA	Active edge
0	0	None
0	1	Capture on rising
1	0	Capture on falling
1	1	Capture on both rising and falling

We can arm or disarm the input capture interrupts by initializing the TMSK1 register. Our software can determine if an input capture event has occurred by reading the TFLG1 register.

	7	6	5	4	3	2	1	0	
TMSK1	C7I	C6I	C5I	C4I	C3I	C2I	C1I	C0I	$008C
TFLG1	C7F	C6F	C5F	C4F	C3F	C2F	C1F	C0F	$008E
TFLG2	TOF	0	0	0	0	0	0	0	$008F

The flags in the TFLG1 and TFLG2 registers are cleared by writing a 1 into the specific flag bit we wish to clear. For example, writing a $FF into TFLG1 will clear all eight flags. The following are valid examples for clearing C3F. That is, these acknowledge sequences clear the C3F flag without affecting the other seven flags in the TFLG1 register.

```
movb #$08,TFLG1          TFLG1 = 0x08;
```

Just like the 6811, the following assembly and C codes will mistakenly clear all flag bits in TFLG1:

```
bset TFLG1,#$08          TFLG1 |= 0x08;
```

6.1.3 Real Time Interrupt Using an Input Capture

In this example, we create a periodic interrupt using an external clock and input capture (Figure 6.6). Although most microcomputers have internal clocks that can be used to create periodic interrupts, there are applications where it is desirable to use an external clock. For example, if you had multiple microcomputers and wished to create simultaneous interrupt requests, you could use this approach. The clock frequency of a TLC555 timer is determined by the resistor and capacitor values. The data sheet for various 555 timers can be found on the accompanying CD. This astable multivibrator creates a 1-kHz square wave. The period of a TLC555 timer is $0.693 \times C_T \times (R_A + 2R_B)$. In our circuit, R_A is 4.4kΩ, R_B is 5kΩ, and C_T is 0.1μF.

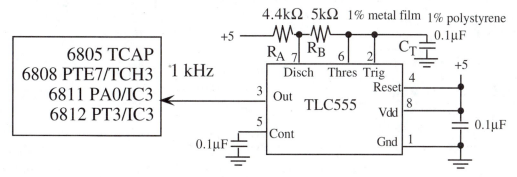

Figure 6.6 An external signal is connected to the input capture.

An input capture interrupt occurs on each rise of the square wave generated by the TLC555. The latency of the system is defined as the time delay between the rise of the input capture signal to the increment of TIME. Assuming there are no other interrupts, and assuming the Main program does not disable interrupts, the delay includes the time to (1) finish the current instruction, (2) process the interrupt, and (3) execute the interrupt handler up to and including changing TIME (Table 6.7).

Table 6.7
Components of latency and their values for the assembly language implementation.

Component	6805	6808	6811	6812
Longest instruction (cycles/μs)	11=5.5μs	9=1.125μs	41=20.5μs	13=1.625μs
Process the interrupt (cycles/μs)	9=4.5μs	9=1.125μs	14=7μs	9=1.125μs
Execute handler (cycles/μs)	33=16.5μs	24=3μs	28=14μs	15=1.875μs
Max latency (μs)	26.5	5.25	41.5	4.625

The latency may be larger if there are other sections of code that execute with the interrupts disabled, like other interrupt handlers. The ritual Init sets input capture to interrupt on the rise and initializes the global TIME (Program 6.1).

The interrupt software performs a poll, acknowledges the interrupt, and increments the global variable. The 6805/6808 interrupt handlers "poll for 0s and 1s." The 6811/6812 inter-

```
; MC68HC05C8
;external signal to TCAP
TIME rmb   2        ;every 1 ms
Init sei            ;make atomic
     lda #$80
     sta TCR        ;arm, rising
     lda TSR        ; clear ICF
     lda ISR+1      ; read low byte
     clr TIME
     clr TIME+1
     cli            ;enable
     rts
```

```
; MC68HC708XL36
;external signal to PTE7/TCH3
TIME rmb   2        ;every 1 ms
Init sei            ;make atomic
     lda TSC3
     mov #$44,TSC3   ;arm, rising
; write 0 to bit 7, clears CH3F
     clr TIME
     clr TIME+1
     cli            ;enable
     rts
```

```
; MC68HC11A8
;external signal to PA0/IC3
TIME rmb   2        ;every 1 ms
Init sei            ;make atomic
     ldaa TCTL2     ;Old value
     anda #$FC      ;Clear EDG3B=0
     oraa #$01      ;EDG3BA =01
     staa TCTL2     ;on rise of PA0
     ldaa TMSK1     ;Old value
     oraa #$01      ;IC3I=1
     staa TMSK1     ;Arm IC3F
     ldd  #0
     std  TIME      ;init global
     ldaa #$01      ;clear IC3F
     staa TFLG1
     cli            ;enable
     rts
```

```
; MC68HC812A4
;external signal to PT3/IC3
TIME rmb   2        ;every 1 ms
Init sei            ;make atomic
     bclr TIOS,$08 ;PT3=input capture
     clr  DDRT      ;PT3 is input
     movb #$80,TSCR ;enable TCNT
     movb #$32,TMSK2 ; 500 ns clk
     ldaa TCTL4    ;old value
     anda #$7F     ;Clear EDG3B=0
     oraa #$40     ;EDG3BA =01
     staa TCTL4    ;on rise of IC3
     bset TMSK1,$08  ;Arm IC3F
     movw #0,TIME    ;init global
     movb #$08,TFLG1 ;clear IC3F
     cli            ;enable
     rts
```

Program 6.1 Initialization of a periodic interrupt using input capture and an external clock.

rupt handlers cannot "poll for 0s and 1s" because none of the bits in the status register is guaranteed to be zero (Program 6.2). The same algorithm written in C is presented in Program 6.3.

```
; MC68HC05C8
ICHan  lda TSR ;expect 1XXX0000  [3]
       and #$8F                  [2]
       cmp #$80                  [2]
       beq  ClkHan               [3]
       swi
ClkHan lda ISR+1 ; Acknowledge   [4]
       lda TIME+1                [3]
       inca                      [3]
       sta TIME+1                [4]
       lda TIME                  [3]
       adc #0                    [2]
       sta TIME                  [4]
       rti
       org $1FF8
       fdb ICHan
```

```
; MC68HC708XL36
CH3Han lda TSC3 ;expect 11000100 [3]
       and #$C4                  [2]
       beq  ClkHan               [3]
       swi
ClkHan pshh          ;H not saved [2]
       bclr 7,TSC3  ;Acknowledge [4]
       ldhx TIME                 [4]
       aix  #1                   [2]
       sthx TIME                 [4]
       pulh
       rti
       org $FFEE   ;timer channel 3
       fdb  CH3Han
```

```
; MC68HC11A8
IC3Han ldaa TFLG1 ;is XXXXXXX1   [4]
       anda #$01                 [2]
       bne  ClkHan               [3]
       swi
ClkHan ldaa #$01  ;clear IC3F    [2]
       staa TFLG1 ;Acknowledge   [4]
       ldx  TIME                 [5]
       inx                       [3]
       stx  TIME                 [5]
       rti
       org  $FFEA
       fdb  IC3Han
```

```
; MC68HC812A4
IC3Han brset #$08,TFLG1,ClkHan   [4]
       swi
ClkHan movb #$08,TFLG1 ;clr IC3F [4]
       ldx  TIME                 [3]
       inx                       [1]
       stx  TIME                 [3]
       rti
       org  $FFE8  ;timer channel 3
       fdb  IC3Han
```

Program 6.2 Periodic interrupt using input capture and an external clock.

```
// MC68HC11A8
// PAO input = external signal
unsigned int TIME;   // incremented

void Init(void){
asm(" sei");   // make atomic
   TCTL2 = (TCTL2&0xFC)|0x01;
   TMSK1 |= 0x01;  // Arm IC3
   TFLG1=0x01; // initially clear
   TIME=0;
asm(" cli");}

#pragma interrupt_handler IC3Han
void IC3Han(void){
   if((TFLG1&0x01)==0) asm(" swi");
   TFLG1=0x01;   // acknowledge
   TIME++;}
```

```
// MC68HC812A4
// PT3 input = external signal
unsigned int TIME;   // incremented
void Init(void){
asm(" sei");   // make atomic
   TIOS &= 0xF7; // PT3 input capture
   DDRT &= 0xF7; // PT3 is input
   TSCR = 0x80;   // enable TCNT
   TMSK2= 0x32;   // 500ns clock
   TCTL4 = (TCTL4&0x3F)|0x40;
   TMSK1 |= 0x08;  // Arm IC3
   TFLG1=0x08; // initially clear
   TIME=0;
asm(" cli");}
#pragma interrupt_handler IC3Han
void IC3Han(void){
   if((TFLG1&0x08)==0) asm(" swi");
   TFLG1=0x08;   //acknowledge
   TIME++;}
```

Program 6.3 Periodic interrupt using input capture and an external clock.

Common Error: When two software modules both need to set the same configuration register, a poorly written initialization by one software module undoes the initialization performed by the other.

6.1.4
Period
Measurement

Before one implements a system that measures period, it is appropriate to consider the issues of resolution, precision, and range. The *resolution* of a period measurement is defined as the smallest change in period that can reliably be detected. In the first example, if the period increases by 500 ns, then there will be one more TCNT clock between the first rising edge and the second rising edge. In this situation, the period calculated by `TIC1-First` will increase by 1; therefore the period measurement resolution is 500 ns. The resolution is also the significance or the units of the measurement. In this first example, if the calculation of `Period` results in 1000, then it represents a period of 1000 · 500 ns, or 500 μs. In the example presented later in Section 6.4, the period must increase at least 1 ms for the measurement to be able to reliably detect the change. In the example of Section 6.4, the period measurement resolution is 1 ms. The *precision* of the period measurement is defined as the number of separate and distinguishable measurements. In the first example, a 16-bit counter is used, so there are about 65,536 different periods that can be measured. We can specify the precision in alternatives (e.g., 65536) or in bits (e.g., 16 bits). The precision of the example in Section 6.1.4.2 is 32 bits. The last issue to consider is the *range* of the period measurement, which is defined as the minimum and maximum values that can reliably be measured. We are concerned what happens if the period is too small or too large. A good measurement system should be able to detect overflows and underflows. In addition, we would not like the system to crash, or hang up if the input period is out of range. Similarly, it is desirable if the system can detect when there is no period.

The default TCNT rate for the 6805 and 6811 is 2 MHz. For the 6808 implementations, we will set the three bits (PS2, PS1, and PS0) in the TSC register to 010, making TCNT clock every 500 ns. Similarly for the 6812 implementations, we will set the three bits (PR2, PR1, and PR0) in the TMSK2 register to 010, making TCNT clock every 500 ns. In this way, all four implementations will be equivalent.

6.1.4.1
16-bit Period
Measurement with a
500-ns Resolution

In this example, the TTL-level input signal is connected to an input capture pin. Each rising edge will generate an input capture interrupt (Figure 6.7). The period is calculated as the difference in TIC1 latch values from one rising edge to the other (Figure 6.8). For example, if the period is 8192 μs, the IC1 interrupts will be requested every 16,384 cycles, and the difference between TIC1 latch values will be 16384=$4000. This subtraction re-

Figure 6.7
To measure period we connect the external signal to an input capture.

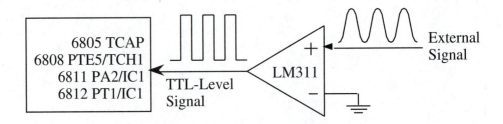

Figure 6.8
Timing example showing counter rollover during 16-bit period measurement.

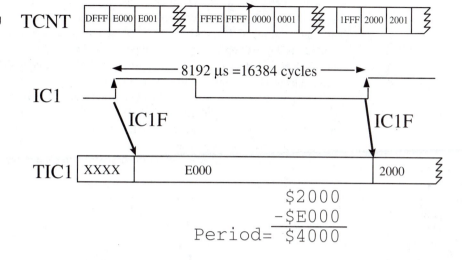

mains valid even if the TCNT overflows and wraps around in between IC1 interrupts. On the other hand, this method will not operate properly if the period is larger than 65,535 cycles, or 32,767 μs.

The resolution is 500 ns because the period must increase by at least this amount before the difference between TIC1 measurements will reliably change. Even though a 16-bit counter is used, the precision is a little less than 16 bits, because the shortest period that can be handled with this interrupt-driven approach is shown in Table 6.8. This factor is determined by counting all the cycles of the assembly version of IC1han and adding the time needed for the microcomputer to process the interrupt. In other words, if the interrupts are requested at a rate faster than the minimum period specified in Table 6.8, then some interrupts will be lost. Also notice that as the period approaches this minimum, a higher and higher percentage of the computer execution is utilized just in the IC1han itself. For example, on the 6811, see Table 6.9.

Table 6.8
Calculation of the smallest period that can be handled by the assembly language input capture interrupt.

Component	6805	6808	6811	6812
Process the interrupt (cycles/μs)	9=4.5μs	9=1.125μs	14=7μs	9=1.125μs
Execute entire handler (cycles/μs)	49=24.5μs	45=5.625μs	50=25μs	31=3.875μs
Minimum period (cycles/μs)	58=29μs	54=6.75μs	64=32μs	50=5μs

Table 6.9
Calculation of the percentage time in the 6811 interrupt handler as a function of input period.

Period (μs)	Cycles/interrupt	Percentage time in handler (%)
32	64	100
64	64	50
320	64	10
P	64	3200/P

The C code executes slower than the assembly, so the percentage overhead is more, but the general trend (percentage overhead approaches 100% as the period approaches zero) still holds. As mentioned earlier, this implementation cannot detect if the period is larger than 32 ms. For example, a signal with period of 40,960 μs gives the same result as a signal with period of 8192 μs. This limitation will be corrected in the other two period measurement examples presented later in this chapter.

The period measurement system written in assembly is presented in Program 6.4 and Program 6.5. The 16-bit subtraction calculates the number of TCNT clocks between rising edges. Since the ritual does not wait for the first edge, the first period measurement will be incorrect and should be neglected. Because the input capture interrupt has a separate vector, the software does not poll. An interrupt is requested on each rising edge of the input signal.

```
; MC68HC05C8
;external signal to TCAP
Period rmb  2  ;units 500 ns
First  rmb  2  ;TCNT at first edge
Done   rmb  1  ;set each rising
Init sei           ;make atomic
     lda #$80
     sta TCR       ;arm,rising
     lda TSR       ;clear ICF
     lda ISR+1     ;read low byte
     lda TCNT
     sta First
     lda TCNT+1
     sta First+1
     clr Done
     cli           ;enable
     rts
```

```
; MC68HC708XL36
;external signal to PTE5/TCH1
Period rmb  2  ;units 500 ns
First  rmb  2  ;TCNT at first edge
Done   rmb  1  ;set each rising
Init sei           ;make atomic
     mov #$02,TSC   ;500ns clk
     lda TSC1
     mov #$44,TSC1  ;arm,rising
; write 0 to bit 7, clears CH1F
     ldhx TCNT
     sthx First
     clr Done
     cli           ;enable
     rts
```

```
; MC68HC11A8
;external signal to PA2/IC1
Period rmb  2  ;units 500 ns
First  rmb  2  ;TCNT at first edge
Done   rmb  1  ;set each rising
Init sei           ;make atomic
     ldaa TCTL2    ;Old value
     anda #$CF     ;Clear EDG1B=0
     oraa #$10     ;EDG1BA =01
     staa TCTL2    ;on rise of PA2
     ldd  TCNT
     std  First    ;init global
     clr  Done
     ldaa #$04     ;clear IC1F
     staa TFLG1
     ldaa TMSK1    ;Old value
     oraa #$04     ;IC1I=1
     staa TMSK1    ;Arm IC1F
     cli           ;enable
     rts
```

```
; MC68HC812A4
;external signal to PT1/IC1
Period rmb  2  ;units 500 ns
First  rmb  2  ;TCNT at first edge
Done   rmb  1  ;set each rising
Init sei           ;make atomic
     bclr TIOS,$02  ;PT1=input capture
     bclr DDRT,$02  ;PT1 is input
     movb #$80,TSCR ;enable TCNT
     movb #$32,TMSK2 ;500ns clk
     ldaa TCTL4    ;old value
     anda #$F3     ;Clear EDG1B=0
     oraa #$04     ;EDG1BA =01
     staa TCTL4    ;on rise of IC1
     movw TCNT,First ;init global
     clr  Done
     movb #$02,TFLG1 ;clear ICIF
     bset TMSK1,$02 ;Arm IC1F
     cli           ;enable
     rts
```

Program 6.4 Initialization for period measurement.

```
; MC68HC05C8
ICHan  lda TSR ; Acknowledge   [3]
       lda ICR+1               [3]
       sub First+1             [3]
       sta Period+1            [4]
       lda ICR                 [3]
       sbc First               [3]
       sta Period              [4]
       lda ICR+1               [3]
       sta First+1             [4]
       lda ICR                 [3]
       sta First               [4]
       lda #$FF                [2]
       sta Done                [4]
       rti                     [6]
       org $1FF8
       fdb ICHan
```

```
; MC68HC708XL36
CH1Han bclr 7,TSC1  ;Acknowledge [4]
       lda TCH1+1               [3]
       sub First+1              [3]
       sta Period+1             [4]
       lda TCH1                 [3]
       sbc First                [3]
       sta Period               [4]
       mov #$FF,Done            [4]
       mov TCH1,First           [5]
       mov TCH1+1,First+1       [5]
       rti                      [7]
       org $FFF2  ;timer channel 1
       fdb CH1Han
```

```
; MC68HC11A8
IC1Han ldaa #$01  ;clear IC3F   [2]
       staa TFLG1 ;Acknowledge  [4]
       ldd  TIC1                [5]
       subd First               [6]
       std  Period              [5]
       ldd  TIC1                [5]
       std  First               [5]
       ldaa #$FF  ;set flag     [2]
       staa Done                [4]
       rti                      [12]
       org  $FFEE
       fdb  IC1Han
```

```
; MC68HC812A4
IC1Han movb #$08,TFLG1 ;clear C3F [4]
       ldd  TC1                  [3]
       subd First                [3]
       std  Period               [3]
       movw TC1,First            [6]
       movb #$FF,Done            [4]
       rti                       [8]
       org  $FFEC  ;timer channel 1
       fdb  IC1Han
```

Program 6.5 ISR for period measurement.

The period measurement system written in C is presented in Program 6.6. The code Period=TIC1-First; calculates the number of E clocks between rising edges. The first period measurement will be incorrect and should be neglected.

```
// MC68HC11A8
// PA2/IC1 input = external signal
// rising edge to rising edge
// resolution = 500ns
// Range = 36 µs to 32 ms,
// no overflow checking
// IC1 interrupt each period,
unsigned int Period;  // units of 500 ns
unsigned int First;   // TCNT first edge
unsigned char Done;   // Set each rising
```

```
// MC68HC812A4
// PT1/IC1 input = external signal
// rising edge to rising edge
// resolution = 500ns
// Range = 36 µs to 32 ms,
// no overflow checking
// IC1 interrupt each period,
unsigned int Period;  // units of 500 ns
unsigned int First;   // TCNT first edge
unsigned char Done;   // Set each rising
```

continued on p. 302

Program 6.6 C language period measurement.

```
continued from p. 301                     void Ritual(void){
void Ritual(void){                            asm(" sei");    // make atomic
    asm(" sei");    // make atomic           TIOS &= 0xFD;   // PT1 input capture
    TCTL2 = (TCTL2&0xCF)|0x10; // rising     DDRT &= 0xFD;   // PT1 is input
    First = TCNT;   // first will be wrong   TSCR = 0x80;    // enable TCNT
    Done=0;         // set on subsequent     TMSK2= 0x32;    // 500ns clock
    TFLG1 = 0x04;   // Clear IC1F            TCTL4 = (TCTL4&0xF3)|0X04; // rising
    TMSK1 |= 0x04;  // Arm IC1               First = TCNT;   // first will be wrong
    asm(" cli");}                            Done=0;         // set on subsequent
                                             TFLG1 = 0x02;   // Clear C1F
#pragma interrupt_handler TIC1handler()      TMSK1 |= 0x02;  // Arm IC1
void TIC1handler(void){                       asm(" cli");}
    Period=TIC1-First;                   #pragma interrupt_handler TIC1handler()
    First=TIC1;  // Setup for next       void TIC1handler(void){
    TFLG1=0x04;  // ack by clearing IC1F      Period=TC1-First;
    Done=0xFF;}                               First=TC1;    // Setup for next
                                             TFLG1=0x02;   // ack by clearing C1F
                                             Done=0xFF;}
```

Program 6.6 C language period measurement.

For the 6808 implementation, we can adjust the three bits (PS2, PS1, and PS0) in the TSC register to specify one of seven possible period measurement resolutions varying from 125 ns to 8 μs. We can also connect an external clock to PTE3/TCLK on the 6808 and use it for the period measurement. For the 6811 implementation, we can adjust the two bits (PR1 and PR0) in the TMSK2 register to specify one of four possible period measurement resolutions varying from 500 ns to 4 μs. For the 6812 implementation, we can adjust the three bits (PR2, PR1, and PR0) in the TMSK2 register to specify one of six possible period measurement resolutions varying from 125 ns to 4 μs.

6.1.4.2
32-bit Period
Measurement with a
500-ns Resolution

In this example, we measure the period of an external signal with a precision of 32 bits and a resolution of 500-ns. The TTL-level signal connected to an input capture pin (Figure 6.7). Every time the TCNT register overflows from $FFFF to 0, the TOF flag is set. We can increase the precision of the period measurement as well as implementing period-too-long error checking by counting the number of TOF flag setting events during one period. To implement the period measurement task in the background, we will arm both the input capture and the timer overflow interrupts.

If we use a 16-bit counter, Count, for the number of times the TOF is set during one period, the precision of the period measurement is 32 bits. Let T_1 be the 32-bit time of the first rising edge of the input signal. The high 16 bits of T_1 are 0, and the low 16 bits are the input capture latch value at the time of the first rising edge. Let T_2 be the 32-bit time of the second rising edge of the input signal. The high 16 bits of T_2 are derived from the value of Count, and the low 16 bits are the input capture latch value at the time of the second rising edge. The period is calculated from the 32-bit subtraction:

$$\text{Period} = T_2 - T_1$$

Figure 6.9 illustrates a typical 32-bit measurement of period that is 73728=$00012000 cycles long. On the first rising edge of IC1, the TIC1 value of $4000 is copied into the global variable First, and Count is cleared. In this situation, T_1 is $00004000 cycles. Next, the TCNT overflows causing a TOF interrupt. The TOhandler() increments Count to 1. On the second rising edge of IC1, the time T_2 is measured as Count concatenated with TIC1. In this case, $T_2 = \$00016000$. The period is the difference between T_2 and T_1, that is, 73728 $=$ $\$12000$. Notice, how the C program below handles the borrow from the least significant word into the most significant word during the 32-bit subtraction, using the statement `if (TIC1<First) MsPeriod--; .`

The tricky part about implementing this 32-bit precision measurement is when the TOF and IC1F flags are both set approximately at the same time. At the time of the first rising edge of IC1, if TOF is not set, then the time T_1 is simply TIC1. If TOF is set, then it could have occurred just before IC1, in which case the next TOF interrupt should not increment `Count` or it could have occurred just after IC1, in which case the next TOF interrupt should increment `Count`.

Figure 6.10 illustrates the situation when the TOF is set just before the IC1F is set. Even though the TOF flag is set before the IC1F flag, they both could occur during the same

Figure 6.9
Simple illustration of the 32-bit period measurement.

Figure 6.10
Situation when TOF is set just before the first rising edge of IC1.

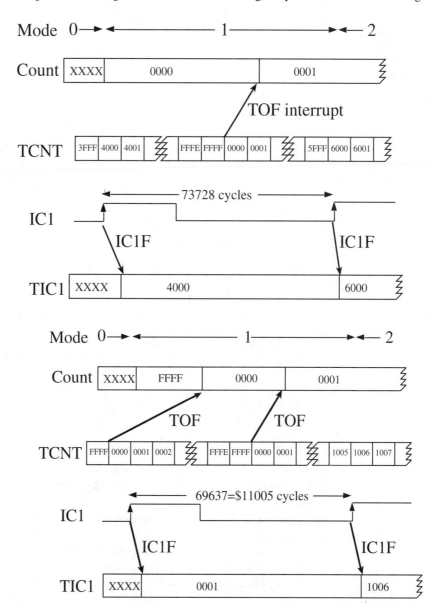

instruction. Since IC1F is a higher-priority interrupt than TOF, the `TIC1handler()` will be executed before the `T0handler()`. If the TOF flag setting occurred just before the IC1, then the most significant bit of TIC1 will be 0. In this situation, the statement `if (((TIC1&0x8000)==0)&&(TFLG2&0x80))Count--;}` will initialize `Count` to $FFFF so that the first TOF interrupt will effectively not be counted. In this case, T_1 is $00000001 cycles and T_2 is $00011006 cycles, and the period is $11005, or 69,637 cycles.

Figure 6.11 illustrates the situation when a TOF occurs just after the IC1F. We assume the TOF flag is set before the line `if(((TIC1&0x8000)==0)&&(TFLG2&0x80))Count--;}` is executed. If the TOF flag is set after this line, then we have a case like Figure 6.8, and there is no problem. Since the TOF flag setting occurred just after the IC1, then the most significant bit of TIC1 will be 1. In this situation, the statement `if (((TIC1&0x8000)==0)&&(TFLG2&0x80))Count--;}` will leave `Count` at 0 so that the first TOF interrupt will be properly counted. In this case, T_1 is $0000FFFE cycles and T_2 is $00021006 cycles, and the period is $11008, or 69,640 cycles.

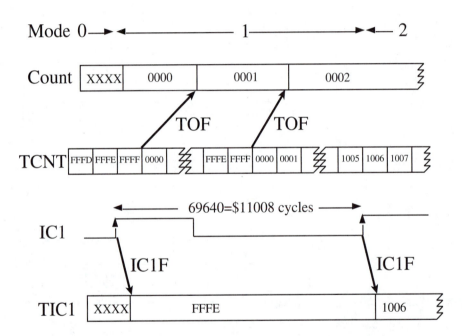

There is a similar problem that exists if the TOF is set just before or just after the second setting of the IC1F. In this case, we will count the TOF, by incrementing `Count` if the TOF occurs just before IC1F. The statement `if(((TIC1&0x8000)==0)&& (TFLG2&0x80)) Count++;` will correct for this situation.

Notice also that if the `Count` reaches 65535 either in `Mode=0` (searching for first rise) or in `Mode=1` (search for second rise), the measurement is terminated. This mechanism allows the system to detect a period-too-long. This measurement system does not measure every period, but rather it could measure every other one. This fairly complicated solution was derived from a 6811 assembly language program found in the Motorola 68HC11 Reference Manual, Example 10-3.

The following programs measure period from rising edge to rising edge with a resolution of 500 ns (E clock). The precision is 32 bits. The range varies from 0 to 35 min (E clock*4billion). An input capture interrupt is generated on each rising edge. The signal is connected to IC1 (Figure 6.7). The timer overflow interrupt counts the global variable `Count`. The globals `MsPeriod` and `LsPeriod` compose the 32-bit period measurement. `First` is the TCNT at the first rising edge. The global `Mode` means:

```
0 means looking for first edge of IC1
1 means looking for second edge
2 means measurement is done
```

The 6805/6808 assembly in Program 6.7 initializes the 32-bit period measurement. The next part (Program 6.8) is the input capture ISR, which is called for every new measurement. The 6805 IC and TOF share the same vector, so polling is required for the 6805.

```
; MC68HC05C8                              ; MC68HC708XL36
;external signal to TCAP                  ;external signal to PTE5/TCH1
Period rmb   4   ;units 500 ns            Period rmb   4   ;units 500 ns
First  rmb   2   ;TCNT at first edge      First  rmb   2   ;TCNT at first edge
Count  rmb   2   ;number of TOF's         Count  rmb   2   ;number of TOF's
Mode   rmb   1                            Mode   rmb   1
Init sei         ;make atomic             Init sei            ;make atomic
     lda #$A0    ;arm TOF, ICF                 lda TSC        ;arm clear TOF
     sta TCR     ;ICF rising                   mov #$42,TSC   ;500ns clk
     lda TSR     ;part of clear                lda TSC1
     lda ISR+1   ;clear ICF                    mov #$44,TSC1  ;arm,rising
     lda TCNT+1  ;clear TOF             ; write 0 to bit 7, clears CH1F
     clr Mode                                  clr  Mode
     cli         ;enable                       cli            ;enable
     rts                                       rts
```

Program 6.7 Assembly initialization for 32-bit period measurement.

```
; MC68HC05C8                              ; MC68HC708XL36
TimHan lda TSR   ;poll for ICF            CH1Han pshh
       bpl ChkTOF                                bclr 7,TSC1   ;Ack, CH1F=0
ICHan  lda Mode                                  lda Mode
       bne Mode1  ;skip if mode=1                bne Mode1     ;skip if mode=1
Mode0  lda ICR1                          Mode0  mov TCH1,First
       sta First                                mov TCH1+1,First+1
       lda ICR+1   ;ack ICF                      clr Count
       sta First+1                              clr Count+1
       clr Count                                inc Mode      ;change 0 to 1
       clr Count+1                              lda TCH1
       inc Mode      ;change 0 to 1             bmi done      ;skip if TCH1 bit 15
       lda ICR                                  lda TSC
       bmi ChkTOF ;skip if ICR:15               bpl done      ;skip if not TOF
       lda TSR                                  ldhx Count    ;both TOF and CH1F
       and #$40      ;check TOF                  aix #-1       ;but TOF was first
       beq ChkTOF ;skip if not TOF              sthx Count    ;so TOF not counted
       ida TCNT+1 ;TOF and ICF                   bra done
       bra TOFok  ;TOF was first          Mode1  lda TCH1
```
continued on p. 306

Program 6.8 Assembly input capture ISR for 32-bit period measurement.

```
continued from p. 305                              bmi  skip     ;skip if TCH1 bit 15
Model  lda  ICR                                     lda  TSC
       bmi  skip     ;skip if ICR:15                bpl  skip     ;skip if not TOF
       lda  TSR                                     ldhx Count    ;both TOF and CH1F
       and  #$40     ;check TOF                      aix  #1       ;but TOF was first
       beq  skip     ;skip if not TOF               sthx Count    ;so TOF is counted
       lda  Count+1  ;TOF and ICF           skip    inc  Mode     ;change 1 to 2
       add  #1       ;TOF was first                  lda  TCH1+1
       sta  Count+1                                  sub  First+1
       lda  Count                                    sta  Period+3  ;bottom 8 bits
       adc  #0                                       lda  TCH1
       sta  Count    ;TOF is counted                 sbc  First
skip   inc  Mode     ;change 1 to 2                  sta  Period+2
       lda  ICR+1    ;ack ICF                        lda  Count+1
       sub  First+1                                  sbc  #0
       sta  Period+3 ;bottom 8 bits                  sta  Period+1
       lda  ICR                                      lda  Count
       sbc  First                                    sbc  #0
       sta  Period+2                                 sta  Period   ;top 16 bits
       lda  Count+1                                  mov  #$02,TSC    ;disarm TOF
       sbc  #0                                       mov  #$04,TSC1   ;disarm CH1F
       sta  Period+1                        done     pulh
       lda  Count                                    rti
       sbc  #0                                       org  $FFF2    ;timer channel 1
       sta  Period   ;top 16 bits                    fdb  CH1Han
       clr  TCR      ;disarm TOF,ICF
       rti
       org  $1FF8
       fdb  TimHan   ;TOF and ICF
```

Program 6.8 Assembly input capture ISR for 32-bit period measurement.

The TOF interrupt performs the upper 16 bits of the measurement (Program 6.9). Program 6.10 implements 32-bit period measurement in C for the 6811 and 6812.

```
; MC68HC05C8                                  ; MC68HC708XL36
ChkTOF lda  TSR
       and  #$40     ;check TOF               TOFHan bclr 7,TSC    ;Ack TOF=0
       beq  TOFok    ;skip if not TOF                lda  Count+1
       lda  Count+1                                  add  #1
       add  #1                                       sta  Count+1
       sta  Count+1                                  lda  Count
       lda  Count                                    adc  #0
       adc  #0                                       sta  Count
       sta  Count                                    bcc  TOFok
       bcc  TOFok                                    mov  #2,Mode ;error no first
       lda  #2                                       lda  #$FF
       sta  Mode ;error no first                     sta  Period
       lda  #$FF                                     sta  Period+1
```

```
        sta  Period                           sta  Period+2
        sta  Period+1                         sta  Period+3
        sta  Period+2                         mov  #$02,TSC    ;disarm TOF
        sta  Period+3                         mov  #$04,TSC1   ;disarm CH1F
        clr  TCR    ;disarm TOF,ICF    TOFok  rti
TOFok   rti                                   org  $FFEC  ;timer overflow
                                              fdb  TOFHan
```

Program 6.9 Assembly timer overflow ISR for 32-bit period measurement.

```
// MC68HC11A8                           // MC68HC812A4
unsigned int MsPeriod,LsPeriod;        unsigned int MsPeriod,LsPeriod;
unsigned int First;                    unsigned int First;
unsigned int Count;                    unsigned int Count;
unsigned char Mode;                    unsigned char Mode;
#pragma interrupt_handler TOhandler()  #pragma interrupt_handler TOhandler()
void TOhandler(void){                   void TOhandler(void){
    TFLG2=0x80;                             TFLG2=0x80;
    Count++;                                Count++;
    if(Count==65535){ // 35 minutes        if(Count==65535){ // 35 minutes
        MsPeriod=LsPeriod=65535;            MsPeriod=LsPeriod=65535;
        TMSK1=0x00;                         TMSK1=0x00;
        TMSK2=0x00;  // Disarm               TMSK2=0x00;  // Disarm
        Mode=2;}}                            Mode=2;}}
#pragma interrupt_handler TIC1handler()  #pragma interrupt_handler TIC1handler()
void TIC1handler(void){                 void TIC1handler(void){
 if(Mode==0){                            if(Mode==0){
   First = TIC1; Count=0; Mode=1;          First = TC1; Count=0; Mode=1;
   if(((TIC1&0x8000)==0)                    if(((TC1&0x8000)==0)
     &&(TFLG2&0x80)) Count--;)               &&(TFLG2&0x80)) Count--;}
   else {                                  else {
     if(((TIC1&0x8000)==0)                    if(((TC1&0x8000)==0)
       &&(TFLG2&0x80))Count++;                 &&(TFLG2&0x80))Count++;
     MsPeriod=Count; Mode=2;                 MsPeriod=Count; Mode=2;
     LsPeriod=TIC1-First;                    LsPeriod=TC1-First;
     if (TIC1<First) MsPeriod--;             if (TC1<First) MsPeriod--;
     TMSK1=0x00; TMSK2=0x00;} // Disarm      TMSK1=0x00; TMSK2=0x00;} // Disarm
 TFLG1=0x04;} // ack, clear IC1F         TFLG1=0x02;} // ack, clear C1F
void Ritual(void){                      void Ritual(void){
 asm(" sei");    // make atomic           asm(" sei");    // make atomic
 TFLG1 = 0x04;   // Clear IC1F            TIOS &= 0xFD;   // PT1 input capture
 TMSK1 |= 0x04;  // Arm  IC1              DDRT &= 0xFD;   // PT1 is input
 TCTL2 = (TCTL2&0xCF)|0x10;  //rising     TMSK2=0xB2 ;    // Arm, TOF 30.517Hz
 TFLG2 = 0x80;   // Clear TOF             TSCR=0x80;      // enable counter
 TMSK2 |= 0x80;  // Arm TOF               TFLG1 = 0x02;   // Clear C1F
 Mode1=0                                  TMSK1 |= 0x02;  // Arm IC1, C1I=1
 asm(" cli"); }                           TCTL4 = (TCTL4&0xFC)|0x04;  // rising
                                          TFLG2 = 0x80;   // Clear TOF
                                          Mode=0;
                                          asm(" cli"); }
```

Program 6.10 C language 32-bit period measurement.

6.1.5
Pulse-Width
Measurement

The basic idea of pulse-width measurement is to cause an input capture event on both the rising and falling edges of an input signal. The difference between these two times will be the pulse width. Just like period measurement, the resolution is determined by the rate at which TCNT is incremented.

6.1.5.1
Gadfly Pulse-Width
Measurement

The objective is to use *input capture pulse-width measurement* to measure resistance (Figure 6.12). This basic approach is employed by most joystick interfaces. The resistance measurement range is $0 \le R \le 1\ M\Omega$. The desired resolution is $1\ k\Omega$. We will use gadfly synchronization.

Figure 6.12
To measure resistance
using pulse width we
connect the external
signal to an input
capture.

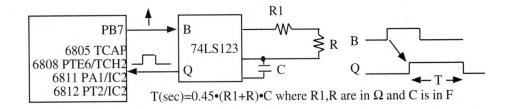

$T(sec)=0.45 \cdot (R1+R) \cdot C$ where R1,R are in Ω and C is in F

Below is the hardware interface between the unknown resistance R and the input capture pin of the microcomputer. A rising edge on PB7 causes a monostable positive logic pulse on the "Q" pin of the 74LS123. We choose R_1 and C so that the resistance resolution maps into a pulse-width measurement resolution of 500 ns, and the resistance range $0 \le R \le 1\ M\Omega$ maps into $500 \le T \le 1000\ \mu s$. The following equation describes the pulse width generated by the 74LS123 monostable as a function of the resistances and capacitance.

$$T=0.45 \times (R+R_1) \times C$$

For a linear system, with x as input and y as output, we can use calculus to relate the measurement resolution of the input and output.

$$\Delta y = \frac{\partial y}{\partial x}\Delta x$$

Therefore, the relationship between the pulse-width measurement resolution ΔT and the resulting resistance measurement resolution is determined by the value of the capacitor.

$$\Delta T=0.45\Delta R \cdot C$$

To make a ΔT of 500 ns correspond to a ΔR of $1 k\Omega$, we choose

$$C=\Delta T/(0.45 \cdot \Delta R)=500\ ns/0.45\ k\Omega=1111 pF$$

To study the range of pulse widths, we look at R=0,

$$T=0.45 \cdot R1 \cdot C$$

so

$$R_1=T/(0.45 \cdot C)=500\mu s/(0.45 \cdot 1111pF)=1M\Omega$$

As a check, notice at $R=1M\Omega$, $T= 0.45 \cdot (2M\Omega) \cdot 1111pF=1ms$.

The measurement subroutine including ritual returns in RegD the current resistance R in kΩ. For example, if the resistance R is 123 kΩ, then RegD will be 123. We will not worry about resistances R greater than 1 MΩ or if R is disconnected. With the 6812 fast clear (TFFCA=1), reading the input capture register TC2 clears the flag C2F.

First we present the ritual that will configure the system in Program 6.11. Next, we present the subroutine used to perform the input capture measurement in Program 6.12. The

```
; MC68HC05C8
; B=PB7,  Q=TCAP, set global R
Init bset 7,DDRB  ;PB7 output
     rts
```

```
; MC68HC708XL36
; B=PB7,  Q=PTE6/TCH6, set global R
Init bset 7,DDRB   ;PB7 output
     mov #$02,TSC  ;500ns clk
     rts
```

```
; MC68HC11A8
; B=PB7,  Q=PA1/IC2
Init  ldx  #$1000 ;I/O registers
      ldaa #$00  ;gadfly
      staa $22,X ;TMSK1 IC2I=0
      rts
```

```
; MC68HC812A4
; B=PB7,  Q=PT2/IC2
Init  bclr TIOS,$04  ;PT2=input capt.
      movb #$90,TSCR ;enable, fast clr
      movb #$32,TMSK2 ;500 ns clk
      clr  TMSK1      ;gadfly, C2I=0
      bclr DDRB,$80
      rts
```

Program 6.11 Ritual written in assembly language for the pulse-width measurement.

```
; MC68HC05C8
Rising rmb 2
R      rmb 2     ;resistance in Kohm
Meas lda #$00  ;disarm TOF, ICF
     sta TCR   ;ICF rising
     lda TSR
     lda ICR+1 ;clear ICF
     bclr 7,PORTB  ;B=0
     bset 7,PORTB  ;trigger LS123
First brclr 7,TSR,First ;wait
     lda #$02
     sta TCR   ;ICF falling
     lda TSR
     lda ICR+1 ;clear ICF
     lda TCH2
     sta Rising
     lda TCH2+1
     sta Rising+1
Secnd brclr 7,TSR,Secnd ;wait
     lda ICR+1
     sub Rising+1
     sta R+1   ;R+ICR-Rising
     lda ICR
     sbc Rising
     sta R
     lda R+1
     sub #232  ;1000=3*256+232
```

```
; MC68HC708XL36
Rising rmb 2
R      rmb 2     ;resistance in Kohm
Meas mov #$02,TSC  ;500ns clk
     lda TSC1
     mov #$04,TSC2 ;rising,clear CH2F
     bclr 7,PORTB  ;B=0
     bset 7,PORTB  ;B=1, trigger LS123
First brclr 7,TSC2,First ;wait rising
     mov #$08,TSC2 ;falling,clear CH2F
     ldhx TCH2
     sthx Rising
Secnd brclr 7,TSC2,Secnd ;wait falling
     lda TCH2+1
     sub Rising+1
     sta R+1     ;R=TCH2-Rising
     lda TCH2
     sbc Rising
     sta R
     lda R+1
     sub #232   ;1000=3*256+232
     sta R+1     ;R=R-1000
     lda R
     sbc #3
     sta R
     rts
```

continued on p. 310

Program 6.12 Assembly language pulse-width measurement.

<table>
<tr><td>

continued from p. 309

```
        sta R+1    ;R=R-1000
        lda R
        sbc #3
        sta R
        rts
```
</td><td></td></tr>
<tr><td>

```
MC68HC11A8
; return Reg D as R in Kohm
Rising equ   0       ;First TCNT
Meas    ldx  #$1000 ;I/O registers
        ldaa #$04    ;Rising edge
        staa $21,X   ;Set TCTL2
        bclr $23,X,$FD  ;IC2F=0
        bclr $04,X,$80  ;PB7=0
        bset $04,X,$80  ;PB7=1
First   brclr $23,X,$02,First
;Wait for first rising edge
        ldy  $12,X   ;TCNT at rising
        ldaa #$08    ;Falling edge
        staa $21,X   ;Set TCTL2
        bclr $23,X,$FD  ;IC2F=0
        pshy          ;Save on stack
Second brclr $23,X,$02,Second;
;Wait for next falling edge
        ldd  $12,X  ;TCNT at falling
        tsy
        subd Rising,Y
;RegD=pulse width 1000 to 2000 cyc
        subd #1000   ;0<=R<=1000Kohm
        puly
        rts
```
</td><td>

```
; MC68HC812A4
; return Reg D as R in Kohm
Meas    movb #$10,TCTL4 ;Rising edge
        movb #$04,TFLG1 ;C2F=0
        bclr PORTB,$80  ;PB7=0
        bset PORTB,$80  ;PB7=0
First   brclr TFLG1,$04,First
;Wait for first rising edge
        ldy  TC2  ;TCNT at rising, C2F=0
        movb #$20,TCTL4 ;Falling edge
        pshy          ;Save on stack
Second brclr TFLG1,$04,Second
;Wait for next falling edge
        ldd  TC2 ;TCNT at falling
        subd 2,SP+
;RegD=pulse width 1000 to 2000 cyc
        subd #1000   ;0,=R,=1000Kohm
        rts
```
</td></tr>
</table>

Program 6.12 Assembly language pulse-width measurement.

```
// MC68HC11A8
unsigned int Measure(void) {
unsigned int Rising;
  TCTL2=(TCTL2&0xF3)|0x04; // Rising edge
  TFLG1=0x02;        // clear IC2F
  PORTB&=0x7F;
  PORTB|=0x80;       // rising edge on PB7
  while(TFLG1&0x02==0){}; // wait rise
  Rising=TIC2;       // TCNT at rising edge
  TCTL2=(TCTL2&0xF3)|0x08; // Falling edge
  TFLG1=0x02;        // clear IC2F
  while(TFLG1&0x02==0){}; // wait fall
  return(TIC2-Rising-1000); }
void Init(void){
  TMSK1=0x00;}       // no interrupts
```

```
// MC68HC812A4
unsigned int Measure(void) {
unsigned int Rising;
  TCTL4=(TCTL4&0xCF)|0x10; // Rising edge
  TFLG1=0x04;        // clear C2F
  PORTB&=0x7F;
  PORTB|=0x80;       // rising edge on PB7
  while(TFLG1&0x04==0){}; // Wait for rise
  Rising=TC2;        // TCNT at rising edge
  TCTL4=(TCTL4&0xCF)|0x20; // Falling edge
  while(TFLG1&0x04==0){}; // Wait for fall
  return(TC2-Rising-1000); }
void Init(void){
  DDRB|=0x80;   // PB7 is output
  TIOS&=0xFB;   // clear bit 2
  DDRT&=0xFB;   // PT2 is input capture
  TSCR=0x90;    // enable, fast clear
  TMSK2=0x32;   // 500 ns clock
  TMSK1=0x00;}  // no interrupts
```

Program 6.13 C language pulse-width measurement.

6805/6808 versions use global variables for temporary storage and return parameter, while the 6811/6812 implementations use registers and the stack. This algorithm written in C is presented in Program 6.13. Again the function returns resistance in units of $k\Omega$.

6.1.5.2
Interrupt-Driven Pulse-
Width Measurement

In this example, the TTL-level input signal is connected to an input capture (Figure 6.7). Both the rising and falling edges will generate an input capture interrupt.[1] The pulse width is calculated as the difference in TIC1 latch values from a rising edge to the next falling edge (Figure 6.13). In this example the background thread sets the global variable Rising on the rising edge and calculates the pulse width PW on the falling edge. The interrupt handler simply reads the current value on the input capture pin to determine if this interrupt is a rising or falling edge. If the first interrupt is a falling edge, then the first pulse width measured will be inaccurate.

Figure 6.13
Pulse width is the time between the rising and falling edges.

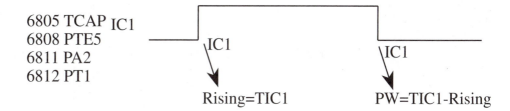

The pulse-width measurement is performed from rising edge to falling edge. The resolution is 500 ns (determined by E clock). The range is about 50 μs to 32 ms, with no overflow checking. The low end of the range is determined by the software overhead to process the interrupt (see the discussion about minimum period in Section 6.1.4.1). IC1 interrupts occur on both the rising and falling edges. The global PW contains the most recent measurement. Rising is a private[2] global variable containing the TCNT value at the rising edge. Done is set at the falling edge of IC1, signifying a new measurement is available. The main program reads Done and PW to process the measurements. The main program can clear Done after it has processed the measurement. The interrupt handler could (but doesn't in this simple example) check Done and trigger an error if a new measurement is complete, but the previous measurement has not been processed yet. Because the Ritual() sets Rising equal to the current time (and not the time of a rising edge), if the first interrupt is falling edge, then the first measurement will be incorrect. If the first interrupt after the Ritual() is executed is a rising edge, then the first measurement will be correct (Programs 6.14 and 6.15).

```
; MC68HC05C8
;external signal to TCAP
PW       rmb   2   ;units 500 ns
Rising   rmb   2   ;TCNT at rising
Done     rmb   1   ;set each fall
```

```
; MC68HC708XL36
;external signal to PTE5/TCH1
PW       rmb   2   ;units 500 ns
Rising   rmb   2   ;TCNT at rising edge
Done     rmb   1   ;set each fall
```

continued on p. 312

Program 6.14 Assembly language pulse width measurement.

[1]The 6805 does not have an interrupt on both rising and falling modes.

[2]A global variable only accessed by one module.

continued from p. 311

```
Init sei            ;make atomic
     lda #$80
     sta TCR        ;arm,rising
     lda TSR        ;clear ICF
     lda ISR+1      ;read low byte
     clr Done
     cli            ;enable
     rts
ICHan lda TSR ; Acknowledge
      lda TCR ;which one?
      and #$02 ;IEDG
      bne fall
rise  lda ICR+1
      sta Rising+1
      lda ICR
      sta Rising
      bra ICrti
fall  lda ICR+1
      sub Rising+1
      sta PW+1
      lda ICR
      sbc Rising
      sta PW
      dec Done  ;set to $FF
ICrti rti
      org  $1FF8
      fdb  ICHan
```

```
Init sei            ;make atomic
     mov #$02,TSC   ;500ns clk
     lda TSC1
     mov #$4C,TSC1 ;arm,both,clr CH1F
     clr Done
     cli            ;enable
     rts

CH1Han bclr 7,TSC1   ;Acknowledge
       lda PORTE
       and #$20    ;check PTE5
       beq fall    ;falling if PTE5=0
rise   mov TCH1,Rising
       mov TCH1+1,Rising+1
       bra CH1rti
fall   lda TCH1+1
       sub Rising+1
       sta PW+1
       lda TCH1
       sbc Rising
       sta PW
       mov #$FF,Done
CH1rti rti
       org  $FFF2   ;time channel 1
       fdb  CH1Han
```

Program 6.14 Assembly language pulse-width measurement.

```
// MC68HC11A8
unsigned int PW;       // units of 500 ns
unsigned int Rising;   // TCNT at rising
unsigned char Done;    // Set each falling
#define PA2 0x04       // the input signal
#pragma interrupt_handler TIC1handler()
void TIC1handler(void){
  if(PORTA&PA2){       // PA2=1 if rising
    Rising=TIC1;}      // Setup for next
  else{
    PW=TIC1-Rising;    // the measurement
    Done=0xFF;}
  TFLG1=0x04;}         // ack, IC1F=0
void Ritual(void){
    asm(" sei");       // make atomic
    TCTL2 |= 0x30;
// IC1F set on both rising and falling
    Rising = TCNT;     // current TCNT
    Done=0;            // set on falling
    TFLG1 = 0x04;      // Clear IC1F
    TMSK1|= 0x04;      // Arm IC1
    asm(" cli");)
```

```
// MC68HC812A4
unsigned int PW;       // units of 500 ns
unsigned int Rising;   // TCNT at rising
unsigned char Done;    // Set each falling
#define PT1 0x02       // the input signal
#pragma interrupt_handler TIC1handler()
void TIC1handler(void){
  if(PORTT&PT1){       // PT1=1 if rising
    Rising=TC1;}       // Setup for next
  else{
    PW=TC1-Rising;     // the measurement
    Done=0xFF;}        // ack, fast clr C1F=0
void Ritual(void) {
asm(" sei");   // make atomic
TIOS&=~PT1;    // clear bit 1
DDRT&=~PT1;    // PT1 is input capture
TSCR=0x90;     // enable, fast clear
TMSK2=0x32;    // 500 ns clock
TCTL4|=0x0C;   // Both edges IC1
TMSK1|=0x02;   // arm IC1
TFLG1=0x02;    // clear C1F
Done=0;
asm(" cli");}
```

Program 6.15 C language pulse-width measurement.

One of the limitations of the previous measurement technique is that the lower bound on the range is determined by the software overhead to process the input capture interrupt. If the period is below that minimum, the error goes undetected. To solve this problem, the TTL-level input signal is connected to both IC1 and IC2 (Figure 6.14). The rising-edge time will be measured by IC2 without the need of an interrupt, and the falling-edge interrupts will be handled by IC1.

Figure 6.14
To measure pulse we could connect the external signal to two input captures.

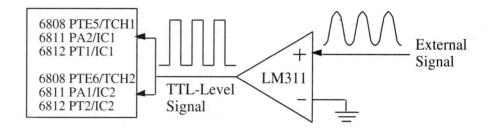

The pulse width is calculated as the difference in TIC1–TIC2 latch values. In this example the IC1 interrupt handler simply sets the global variable PW at the time of the falling edge. Because no software is required to process the IC2 measurement, there is no minimum pulse width. On the other hand, software processing is required to handle the IC1 signal, so there is a minimum period (e.g., there must be more than 50 μs from one falling edge to the next falling edge). Again, the first measurement may or may not be accurate (Figure 6.15).

Figure 6.15
The rising edge is measured with IC2, and the falling edge is measured with IC1.

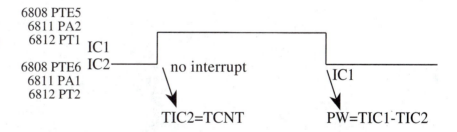

The pulse-width measurement is performed from rising edge to falling edge. The resolution is 500 ns (determined by E clock). The range is about 500 ns to 32 ms, with no overflow checking. IC1 interrupts only occur on the falling edges. The global PW contains the most recent measurement. Done is set at the falling edge of IC1, signifying a new measurement is available. If the first edge after the Ritual() is executed is a falling edge, then the first measurement will be incorrect (because TIC2 is incorrect). If the first edge after the Ritual() is executed is a rising edge, then the first measurement will be correct. Compared to the last example, notice how little software overhead is required to perform these measurements (Program 6.16).

Notice how the C code in Program 6.17 sets two bits of TCTL2 (6811) and TCTL4 (6812) without modifying the other six bits. We call this a "friendly" ritual because it does not undo other initializations. On the other hand, the rituals all have to get together and agree on a common value for the 6812 TMSK2.

```
; MC68HC05C8                    ; MC68HC708XL36
; not available                ;signal to PTE5/TCH1 and PTE6/TCH2
; because only one input capture  PW     rmb   2  ;units 500 ns
                               Done   rmb   1  ;set each fall
                               Init sei          ;make atomic
                                     mov #$02,TSC  ;500ns clk
                                     lda TSC1
                                     mov #$48,TSC1 ;arm, fall
                                     mov #$08,TSC2 ;disarm, rise
                                     clr Done
                                     cli          ;enable
                                     rts
                               CH1Han bclr 7,TSC1   ;Acknowledge
                                     lda  TCH1+1
                                     sub  TCH2+1
                                     sta  PW+1
                                     lda  TCH1
                                     sbc  TCH2
                                     sta  PW
                                     mov  #$FF,Done
                                     rti
                                     org  $FFF2   ;timer channel 1
                                     fdb  CH1Han
```

Program 6.16 Assembly language pulse-width measurement using two input captures.

```
// MC68HC11A8                          // MC68HC812A4
unsigned int PW;     // units of 500 ns   unsigned int PW;     // units of 500 ns
unsigned char Done; // Set each falling   unsigned char Done; // Set each falling
#pragma interrupt_handler TIC1handler()   #pragma interrupt_handler TIC1handler()
void TIC1handler(void){                   void TIC1handler(void){
  PW=TIC1-TIC2; // time from rise to fall    TFLG1=0x02;  // ack C1F
  Done=0xFF;                                 PW=TC1-TC2;  // the measurement
  TFLG1=0x04;} // ack by clearing IC1F       Done=0xFF;}
                                          void Ritual(void) {
void Ritual(void){                          asm(" sei");  // make atomic
  asm(" sei");     // make atomic           TIOS&=0xF9;   // clear bits 2,1
  TCTL2=(TCTL2&0xCF)|0x20;                   DDRT&=0xF9;   // PT2,PT1 input captures
// falling edges of IC1, TCNT->TIC1         TSCR=0x80;    // enable
  TCTL2=(TCTL2&0xF3)|0x04;                   TMSK2=0x32;   // 500 ns clock
// rising edges of IC2, TCNT->TIC2          TCTL4=(TCTL4&0xCF)|0x10; // IC2 Rise
  Done=0;   // set on the falling edge      TCTL4=(TCTL4&0xF3)|0x08; // IC1 Fall
  TFLG1 = 0x04;     // Clear IC1F           Done=0;   // set on the falling edge
  TMSK1|= 0x04;     // Arm IC1, not IC2     TMSK1|=0x02;  // arm IC1, not IC2
  asm(" cli");}                             TFLG1=0x02;   // clear C1F
                                            asm(" cli");}
```

Program 6.17 C language pulse-width measurement using two input captures.

6.2 Output Compare

6.2.1
General Concepts

Similar to our introduction to input capture, we begin our discussion of output compare with some general comments. Output compare will be used to create square waves, generate pulses, implement time delays, and execute periodic interrupts. A common output technique used in computer-based control systems is a variable duty cycle actuator. Another name for this is pulse-width modulation (PWM). In Figure 6.16, the shaded areas represent times when power is applied. The computer generates a square wave with a variable duty cycle using the output compare interface. To apply a large amount of power to the control system it issues a square wave that is mostly high. To apply a small amount of power, it outputs a square wave that is mostly low.

Figure 6.16
The output power is controlled by varying the duty cycle of a square wave.

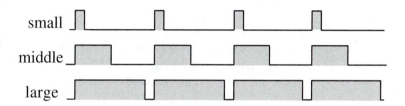

We will also use output compare together with input capture to measure frequency. Output compare and input capture can also be combined to measure period and frequency over a wide range of ranges and resolutions. The same 16-bit TCNT, incremented at a fixed rate, is used for input capture and output compare. Each output compare module has:

An external output pin OCn
A flag bit
A force compare control bit FOCn (6811 and 6812 only)
Two control bits OMn, Oln (only one control bit on the MC68HC05C8)
An interrupt mask bit
A 16-bit output compare register

The various microcomputers from Motorola have from zero to eight output compare modules (Figure 6.17). For example, the MC68HC705JA1 has none, the MC68HC05C8 has one, the MC68HC708XL36 has four, the MC68HC11 has five, the MC68HC812A4 has eight, and the MC68HC912B32 has eight. The maximum number of input capture modules plus output compare modules on the 6812 is eight.

Figure 6.17
The basic components of output compare.

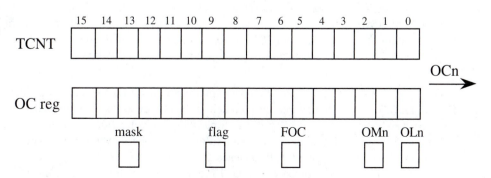

The output compare pin is an output of the computer, hence can be used to control an external device. An output compare event occurs, setting the flag bit, when either:

1. The 16-bit TCNT matches the 16-bit OC register
2. The software writes a 1 to the FOC bit (6811 and 6812 only)

The OMn, OLn bits specify what effect the output compare event will have on the output pin. Two or three actions result from an output compare event: (1) the OCn output bit changes, (2) the output compare flag is set, and (3) an interrupt is requested (occurs only when mask=1). Just like the input capture, the output compare flag on the 6811/6812 is cleared by writing a 1 to it.

The module is armed when the mask bit is 1. One simple application of output compare is to create a fixed time delay. Let `fixed` be the number of cycles you wish to wait. The steps to create the delay are:

1. Read the current 16-bit TCNT
2. Calculate TCNT + `fixed`;
3. Set the 16-bit output compare register to TCNT + `fixed`
4. Clear the output compare flag
5. Wait for the output compare flag to be set (gadfly or interrupt)

If you perform the addition TCNT + `fixed` with 16-bit integer math (e.g., `addd` instruction), then this approach will function properly even if the counter rolls over from $FFFF to 0. For example, assume the current TCNT is $F000, and we wish to delay 8192 ($2000) cycles. The 16-bit addition `$F000+$2000` will result in a sum of `$1000`. If we put the $1000 into the output compare register, then it will take the correct number (i.e., 8192 cycles) for the TCNT to go from $F000 to $1000. The time for the software to execute steps 1 to 4, will determine the minimum delay that can be created with this approach. For example, if the delay is so short that the TCNT hits TCNT + `fixed` before the flag is cleared in step 4, then the delay will be a 65536+ `fixed` delay. For obvious reasons, the maximum delay with this simple method is 65,536 cycles.

One of the output compare modules, PA7/OC1 on the 6811 and PT7/OC7 on the 6812, can be configured such that an output compare event on it will cause changes on some or all of the other output compare pins. This coupled behavior can be used to create synchronized signals. For example, we can create pulses that start together or end together. In a similar way, two output compares on the 6808 can be coupled so that short pulses can be created. In this section, we will focus on basic issues and the features that are common to all output compare modules.

6.2.2
Output Compare
Details

In this subsection we will overview the specific output compare functions on particular Motorola microcomputers. This section is intended to supplement rather than replace the Motorola manuals. When designing systems with output compare, please refer also to the reference manual of your specific Motorola microcomputer.

6.2.2.1
MC68HC05C8 Output
Compare Details

Since the 6805 input capture and output compare modules share many registers, we have already introduced much of the output compare components. The TCNT register is incremented every 500 ns. Since there is no direction register bit associated with TCMP pin, this bit is always output. An output compare event occurs when the 16-bit output compare register (OCR) matches the 16-bit TCNT register (Figure 6.18). This event will set the output compare flag (OCF) in the TSR register and make the TCMP output equal to OLVL.

Figure 6.18
The basic components of output compare on the MC68HC05C8.

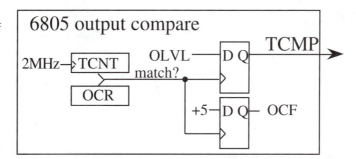

The OCF flag is cleared by a read of TSR with OCF set, followed by an access to the low byte of OCR. If TOIE is set, then an interrupt will be requested when the OCF flag is set. The output compare modules on the 6808, 6811, and 6812 provide a toggle output compare mode, but this 6805 module has only two functions:

Set TCMP to 0 (if OLVL=0)
Set TCMP to 1 (if OLVL=1)

6.2.2.2
MC68HC708XL36
Output Compare Details

Since the MC68HC708XL36 input capture and output compare modules share I/O pins and many registers, we have already introduced much of the output compare components. TCNT is a 16-bit unsigned counter that is incremented at a rate determined by three bits (PS2, PS1, and PS0) in the TSC register ($0020). There is a direction register associated with PTE. The corresponding bits in DDRE should be 1 to signify outputs. An output compare event occurs when one of the 16-bit OCRs (TCH0, TCH1, TCH2, or TCH3) matches the 16-bit TCNT register (Figure 6.19). This event will set the corresponding OCF (CH0F, CH1F, CH2F, or CH3F).

Figure 6.19
Output compare on the MC68HC708XL36.

MC68HC708XL36
output compare

PTE7/TCH3 →
PTE6/TCH2 →
PTE5/TCH1 →
PTE4/TCH0 →

We make the pin an output compare by setting MSnA. We can arm the output compare interrupts by setting the corresponding CHnIE bit.

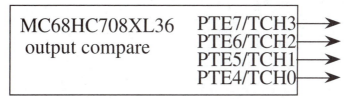

	7	6	5	4	3	4	1	0	
TSC0	CH0F	CH0IE	MS0B	MS0A	ELS0B	ELS0A	TOV0	CH0MAX	$0026
TSC1	CH1F	CH1IE	0	MS1A	ELS1B	ELS1A	TOV1	CH1MAX	$0029
TSC2	CH2F	CH2IE	MS2B	MS2A	ELS2B	ELS2A	TOV2	CH2MAX	$002C
TSC3	CH3F	CH3IE	0	MS3A	ELS3B	ELS3A	TOV3	CH3MAX	$002F

Table 6.10

Four bits determine the action caused by a 6808 output compare event.

MSnB	MSnA	ELSnB	ELSnA	Active edge
X	X	0	0	None, regular I/O on Port E
0	0	0	1	Capture on rising
0	0	1	0	Capture on falling
0	0	1	1	Capture on both rising and falling
0	1	0	1	Toggle output on output compare
0	1	1	0	Clear output on output compare
0	1	1	1	Set output on output compare
1	X	0	1	Buffered toggle output
1	X	1	0	Buffered clear output
1	X	1	1	Buffered set output

We specify the output compare action (i.e., when TCNT matches TCHn and sets the flag CHnF) by initializing the corresponding MSnB, MSnA, ELSnB, ELSnA bits in the TSCn register. Recall that some of these settings were previously specified in the input capture section. They are repeated for clarity in Table 6.10.

TOVn: Toggle on overflow bit
    0 = Timer overflow does not affect the output TCHn
    1 = When TCNT overflows, the output TCHn is toggled
CHnMAX: Channel n maximum duty cycle bit (used with the TOVn=1 mode)
    0 = regular operation
    1 = If TOVn=1, then the output TCHn is forced to 100% duty cycle on overflow

Since the 6808 output compare has a special feature for simplifying PWM, a specific 6808 PWM program is shown in Program 6.18.

```
;external signal to PTE7/TCH3
Init ldhx #9999   generate a 100 Hz square wave, 10000 cycles 1µs each
     sthx TMOD    when TCNT=TMOD, then TOF is set and TCNT=0 again
     mov  #$03,TSC no TOF interrupts, 1µs counting
     mov  #$1A,TSC3
;          6=0     no interrupts
; 5,4,3,2=0110 clear on output compare
;          1=1     TOV3 toggle when TOF set (will set to one)
;          0=0     CH3MAX off, regular PWM
     ldhx #5000 50% duty cycle
     sthx TCH3
     rts
```

Program 6.18
6808 assembly language PWM output.

To change the duty cycle the software sets TCH3 to any integer less than 9999. When TCH3=TCNT, the output is cleared; when TCNT=TMOD, the output is toggled (set) and TCNT is cleared. To make the output 100%, the software can set the CH3MAX bit (Figure 6.20).

Figure 6.20
Pulse-width modulation
using output compare
on the
MC68HC708XL36.

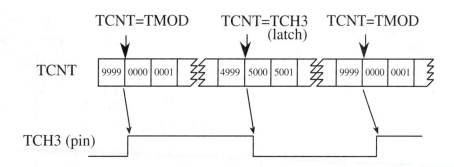

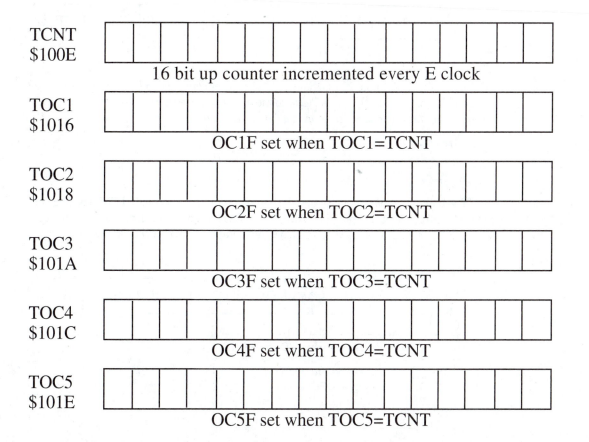

6.2.2.3
6811 Output Compare
Details

TCNT is a 16-bit unsigned counter that is incremented at a rate determined by two bits (PR1 and PR0) in the TMSK2 register. Since there are no direction register bits associated with PA6-3, these bits are always outputs. There is a direction register bit for PA7/OC1 (DDR7 bit in the PACTL register) that should be set to 1 if OC1 is to be used as an output. An output compare event occurs when one of the 16-bit OCRs (TOC1, TOC2, TOC3, TOC4, or TOC5) matches the 16-bit TCNT register. This event will set the corresponding OCF (OC1F, OC2F, OC3F, OC4F, or OC5F) in the TFLG1 register. If the output pin is activated, as specified in the TCTL1 register, then this output compare event can also modify the output value (set to 1, clear to 0, or toggle) (Figure 6.21).

Figure 6.21
Available signals for
output compare on the
MC68HC11A8.

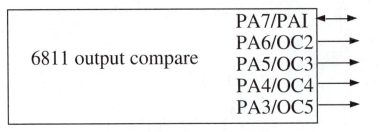

We can arm or disarm the individual output compare interrupts by initializing the TMSK1 register.

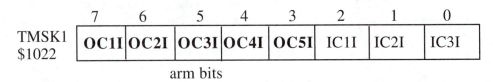

Our software can determine if an output compare event has occurred by reading the TFLG1 register.

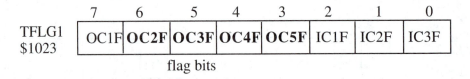

Recall that the flags in the TFLG1 and TFLG2 registers are cleared by writing a 1 into the specific flag bit we wish to clear. These acknowledge sequences clear the OC3F flag without affecting the other seven flags in the TFLG1 register.

```
ldaa #$20
staa $1023        Clears OC3F in TFLG1

TFLG1=0x20;       /* Clears OC3F in TFLG1 */
```

The output compare event (when the output compare latch equals TCNT) can be configured to affect an output pin (Table 6.11). The TCTL1 register determines what effect the OC2, OC3, OC4, and OC5 event will have (none, toggle, clear, or set) on the output pin.

	7	6	5	4	3	2	1	0
TCTL1 $1020	OM2	OL2	OM3	OL3	OM4	OL4	OM5	OL5

Table 6.11
Two bits determine the
action caused by a 6811
output compare event.

OMn	OLn	Effect of when TOCn=TCNT
0	0	Does not affect OCn
0	1	Toggle OCn
1	0	Clear OCn=0
1	1	Set OCn=1

The output compare 1, OC1, operates differently. PA7 can also be used as the pulse accumulator mechanism. On a successful OC1 event (TOC1=TCNT), the 6811 can be programmed to set or clear any of the output compare pins. The OC1M register selects which pin(s) will be affected by the OC1 event. Clear the corresponding bit(s) of this register to 0 to disconnect OC1 from the output pin(s). Conversely, set the corresponding bit(s) of this register to 1 to attach the OC1 event to the other output pin(s).

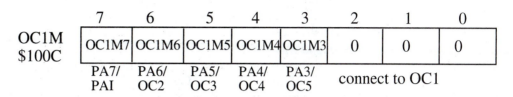

For every bit in the OC1M register that is 1, the corresponding bit in the OC1D register will specify the resulting value of the output pin(s) after an OC1 event.

	7	6	5	4	3	2	1	0
OC1D $100D	OC1D7	OC1D6	OC1D5	OC1D4	OC1D3	0	0	0
	PA7/ PAI	PA6/ OC2	PA5/ OC3	PA4/ OC4	PA3/ OC5	value after OC1		

This mode can be used to synchronized outputs or short pulses.

6.2.2.4
6812 Output Compare Details

Since the 6812 input capture and output compare modules share I/O pins and many registers, we have already introduced much of the output compare components. The TEN bit in the TSCR register must be enabled. The TFFCA bit in the TSCR register can be set to activate fast clear (access to the 16-bit register TCn clears flag bit CnF). TCNT is a 16-bit unsigned counter that is incremented at a rate determined by three bits (PR2, PR1, and PR0) in the TMSK2 register ($008D). The fundamental approach to output compare involves connecting one of the eight output compare pins on Port T to an external TTL-level device. The pin is selected as output compare by placing a 1 in the corresponding bit of the TIOS register. With input capture, we also set the direction bit in DDRT to zero, but selecting output compare in the TIOS register automatically makes the bit an output. An output compare event occurs when one of the OCRs (TCn) matches the TCNT register. This event will set the corresponding OCF (CnF) in the TFLG1 register. If the output pin is activated, as specified in the TCTL1 and TCTL2 registers, then this output compare event can also modify the output value (set to 1, clear to 0, or toggle) (Figure 6.22).

Figure 6.22
Available signals for output compare on the MC68HC812A4 and MC68HC912B32.

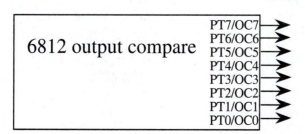

We can arm or disarm the individual output compare interrupts by initializing the TMSK1 register. Our software can determine if an output compare event has occurred by reading the TFLG1 register. Recall that the flags in the TFLG1 and TFLG2 registers are cleared by writing a 1 into the specific flag bit we wish to clear. For example, writing a $FF into TFLG1 will clear all eight flags. The following are valid examples for clearing C3F. That is, these acknowledge sequences clear the C3F flag without affecting the other seven flags in the TFLG1 register.

`movb #$08,TFLG1`	`TFLG1=0x08;`

The output compare event (when the 16-bit OCR equals the 16-bit TCNT) can be configured to affect an output pin (Table 6.12). The TCTL1 and TCTL2 registers determine what effect each of the eight possible output compare events will have (none, toggle, clear, or set) on the output pin.

	7	6	5	4	3	2	1	0	
TCTL1	OM7	OL7	OM6	OL6	OM5	OL5	OM4	OL4	$0088
TCTL2	OM3	OL3	OM2	OL2	OM1	OL1	OM0	OL0	$0089

Table 6.12

Two bits determine the action caused by 6812 output compare event.

OMn	OLn	Effect of when TOCn=TCNT
0	0	Does not affect OCn
0	1	Toggle OCn
1	0	Clear OCn=0
1	1	Set OCn=1

Like the 6811, there is one output compare, OC7, which operates differently. On a successful OC7 event (TC7=TCNT), the 6812 can be programmed to set or clear any of the other output compare pins. The OC7M register selects which pin(s) will be affected by the OC7 event. Clear the corresponding bit(s) of the OC7M register to 0 to disconnect OC7 from the output pin(s). Conversely, set the corresponding bit(s) of the OC7M register to 1 to attach the OC7 event to the output pin(s).

	7	6	5	4	3	2	1	0	
OC7M	OC7M7	OC7M6	OC7M5	OC7M4	OC7M3	OC7M2	OC7M1	OC7M0	$0082
OC7D	OC7D7	OC7D6	OC7D5	OC7D4	OC7D3	OC7D2	OC7D1	OC7D0	$0083

For every bit in the OC7M register that is 1, the corresponding bit in the OC7D register will specify the resulting value of the output pin(s) after an OC7 event.

6.2.3
Periodic Interrupt
Using Output
Compare

The output compare feature is a convenient mechanism to create an interrupt-driven real-time clock. This first output compare example simply increments a global variable, TIME, every 1 ms. This is an important example, because we will use this type of background processing in our real-time data acquisition and control systems. A hard real-time system is one in which we can guarantee an absolute worst-case latency. The latency of this particular system is defined as the time delay between the setting of the output compare flag to the increment of TIME. In this example, output compare interrupts are requested exactly every 2000 cycles. Because of uncertainties about which instruction is currently being executed at the time of the output compare event, and the possibility of the other software executing with interrupts disabled, we cannot know exactly when TIME will be incremented. On the other hand, if no other software executes with interrupts disabled for longer than 1 ms, we can guarantee the output compare interrupt will be requested exactly every 1 ms and serviced on average exactly 1000 times/s. In this situation, the accuracy of the software TIME will be as good as the stability of the crystal used to create the E clock. Assuming all other software does not execute with the interrupts disabled for longer than T, where $T < 1$ ms, then this system qualifies as hard real time because we can calculate the absolute worst-case latency (Table 6.13).

Component	6805	6808	6811	6812
Time with interrupts disabled	T	T	T	T
Longest instruction (cycles/μs)	11=5.5μs	9=1.125μs	41=20.5μs	13=1.625μs
Process the interrupt (cycles/μs)	9=4.5μs	9=1.125μs	14=7μs	9=1.125μs
Execute handler (cycles/μs)	19=9.5μs	12=1.5μs	13=6.5μs	6=0.75μs
Max latency (μs)	T+19.5	T+3.75	T+34	T+3.5

Table 6.13 Components of latency calculated for different assembly language implementations.

The 6808 and 6812 implementations have a shorter latency because they are faster than the 6805 and 6811. The assembly implementation of this real-time clock is as follows. TIME is a shared global, so read and write accesses by the main program should be atomic (e.g., ldd TIME and not ldaa TIME, ldab TIME+1). Notice that the read-modify-write access to TIME (ldx inx stx) is atomic because interrupts are disabled automatically before the interrupt handler is started.

Program 6.19 contains the assembly language ritual to initialize the system; Program 6.20 contains the interrupt handlers.

```
; MC68HC05C8                        ; MC68HC708XL36
TIME rmb  2      ;every 1 ms        TIME rmb  2       ;every 1 ms
Init sei         ;make atomic       Init sei          ;make atomic
     mov #$40,TCR ;arm OCF               mov #$02,TSC   ;500ns clk
     lda TSR     ;clear OCF              lda TSC3
     ldx TCNT    ;latches LSB too        mov #$54,TSC3 ;arm,toggle OC
     lda TCNT+1  ;read low byte     ; write 0 to bit 7, clears CH3F
     add #$D0    ;$07D0=2000             ldx TCNT
     sta OCR+1                           lda TCNT+1 ;low byte
     txa                                 add #$D0   ;2000=$07D0
     adc #$07                            sta TCH3+1 ;TCH3=TCNT+2000
     sta OCR     ; first in 1ms          txa
     clr TIME                            adc #$07
```

continued on p. 324

Program 6.19 Assembly language periodic interrupt using output compare.

```
continued from p. 323
      clr TIME+1                                  sta TCH3   ;setup next
      cli          ;enable                        clr  TIME
      rts                                         clr  TIME+1
                                                  cli           ;enable
                                                  rts
```

`; MC68HC11A8`
`TIME    rmb  2      ;inc every 1ms`
`Init    sei         ;make atomic`
`        ldaa TMSK1 ;Old value`
`        oraa #$08  ;TMSK1 OC5I=1`
`        staa TMSK1 ;Arm OC5F`
`        ldd  #0`
`        std  TIME  ;initialize`
`        ldaa #$08  ;clear OC5F`
`        staa TFLG1`
`        ldd  TCNT  ;current time`
`        addd #2000 ;first in 1 ms`
`        std  TOC5`
`        cli        ;enable`
`        rts`
	`; MC68HC812A4`
	`TIME    rmb  2       ;inc every 1ms`
	`Init    sei         ;make atomic`
	`        bset TIOS,$20   ;OC5`
	`        movb #$80,TSCR  ;enable`
	`        movb #$32,TMSK2 ;500 ns clk`
	`        bset TMSK1,$20  ;Arm OC5`
	`        movw #0,TIME    ;initialize`
	`        movb #$20,TFLG1 ;clear C5F`
	`        ldd  TCNT   ;current time`
	`        addd #2000 ;first in 1 ms`
	`        std  TC5`
	`        cli        ;enable`
	`        rts`

Program 6.19 Assembly language periodic interrupt using output compare.

`; MC68HC05C8`
`OCHan lda TIME+1        [3]`
`      inca             [3]`
`      sta TIME+1       [4]`
`      lda TIME         [3]`
`      adc #0           [2]`
`      sta TIME         [4]`
`      lda TSR    ;part of Ack`
`      lda OCR+1  ;low byte`
`      add #$D0   ;$07D0=2000`
`      sta OCR+1`
`      lda OCR`
`      adc #$07`
`      sta OCR    ;next in 1ms`
`      rti`
`      org $1FF8`
`      fdb OCHan`
	`; MC68HC708XL36`
	`CH3Han pshh                 [2]`
	`       ldhx TIME            [4]`
	`       aix  #1              [2]`
	`       sthx TIME            [4]`
	`       bclr 7,TSC3   ;Acknowledge`
	`       lda TCH3+1 ;low byte`
	`       add #$D0    ;2000=$07D0`
	`       sta TCH3+1 ;TCH3=TCH3+2000`
	`       lda TCH3`
	`       adc #$07`
	`       sta TCH3   ;setup next`
	`       pulh`
	`       rti`
	`       org  $FFEE  ;timer channel 3`
	`       fdb  CH3Han`

`; MC68HC11A8`
`OC5HAN ldx  TIME             [5]`
`       inx                  [3]`
`       stx  TIME            [5]`
`       ldaa #$08  ;clear OC5F`
`       staa TFLG1 ;Acknowledge`
`       ldd  TOC5`
`       addd #2000 ;next`
`       std  TOC5`
`       rti`
`       org  $FFE0`
`       fdb  OC5HAN`
	`; MC68HC812A4`
	`OC5HAN ldx  TIME             [3]`
	`       inx                  [1]`
	`       stx  TIME            [2]`
	`       movb #$20,TFLG1 ;Acknowledge`
	`       ldd  TC5`
	`       addd #2000 ;next`
	`       std  TC5`
	`       rti`
	`       org  $FFE4`
	`       fdb  OC5HAN`

Program 6.20 Assembly language ISR for periodic interrupt using output compare.

In C, this periodic RTI is shown in Program 6.21. In this example the 6812 does not use fast clear, but if you were to use fast clear, then the access to TC5 would automatically clear the C5F flag.

```
// MC68HC11A8                              // MC68HC812A4
#define Rate 2000                          #define Rate 2000
#define OC5  0x08                          #define OC5  0x20
unsigned int Time;  // Inc every 1ms       unsigned int Time;  // Inc every 1ms
#pragma interrupt_handler TOC5handler()    #pragma interrupt_handler TOC5handler()
void TOC5handler(void){                    void TOC5handler(void){
   TFLG1=OC5;       // Ack interrupt          TFLG1=OC5;       // ack OC5F
   TOC5=TOC5+Rate; // Executed every 1 ms     TC5=TC5+Rate;    // Executed every 1 ms
   Time++; }                                  Time++; }

void ritual(void) {                        void ritual(void) {
asm(" sei");     // make atomic             asm(" sei");       // make atomic
   TMSK1|=OC5;  // Arm output compare 5        TIOS|=OC5;      // enable OC5
   Time = 0;                                   TSCR|=0x80;     // enable
   TFLG1=OC5;   // Initially clear OC5F        TMSK2=0x32;     // 500 ns clock
   TOC5=TCNT+Rate; // First one in 1 ms        TMSK1|=OC5;     // Arm output compare 5
asm(" cli"); }                                 Time = 0;
                                               TFLG1=OC5;      // Initially clear C5F
                                               TC5=TCNT+Rate;  // First one in 1 ms
                                            asm(" cli"); }
```

Program 6.21 C language periodic interrupt using output compare.

Common Error: If the interrupt handler were to execute `TOC5=TCNT+Rate;` instead of `TOC5=TOC5+Rate;`, then interrupts would be requested at a fluctuating rate a little bit slower than every `Rate` cycles.

Observation: With 6812 fast clear active, the access to TCNT also clears the TOF flag, so if both TOF and input capture/output compare interrupts are used, then fast clear should be deactivated.

6.2.4
Square-Wave
Generation

This example generates a 50% duty cycle square wave using output compare. The output is high for `Period` cycles, then low for `Period` cycles. Output compare interrupts will be requested at a rate twice as fast as the resulting square-wave frequency. One interrupt is required for the rising edge on the output compare pin and another for the falling edge. Toggle mode is used to create the 50% duty cycle square wave (Figure 6.23). In this mode, the output compare pin is toggled whenever the output compare latch matches TCNT.

Figure 6.23
Square-wave generation using output compare.

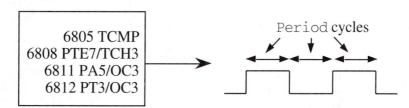

The output compare interrupt handler simply acknowledges the interrupt and calculates the time for the next signal transition. This implementation will create an exact square wave independent of software execution delays as long as the interrupt is serviced within `Period` cycles. If the software delay is more than `Period` cycles between the setting of the output compare flag (hardware event latch=TCNT) and the acknowledgment (software clears the flag), then the next edge will not occur for another 65,536 cycles.

In assembly, Program 6.22 contains the ritual and Program 6.23 contains the interrupt handlers.

```
; MC68HC05C8
Period rmb  2       ;unitsµsec
Init sei            ;make atomic
     lda #$40
     sta TCR        ;arm OCF
     lda TSR        ;clear OCF
     ldx TCNT       ;latches LSB too
     lda TCNT+1     ;read low byte
     add #50        ;first in 25µs
     sta OCR+1
     txa
     adc #0
     sta OCR
     cli            ;enable
     rts
```

```
; MC68HC708XL36
Period rmb  2       ;units µsec
Init sei            ;make atomic
     bset 7,DDRE ;PTE7 output
     mov #$02,TSC    ;500ns clk
     lda TSC3
     mov #$54,TSC3  ;arm,toggle OC
; write 0 to bit 7, clears CH3F
     ldhx TCNT
     aix  #50       ;first in 25µs
     sthx TCH3
     cli            ;enable
     rts
```

```
; MC68HC11A8
Period rmb  2       ;units µsec
Init    sei            ;make atomic
        ldaa TMSK1 ;Old value
        oraa #$20  ;TMSK1 OC3I=1
        staa TMSK1 ;Arm OC3F
        ldaa TCTL1
        anda #$CF  ;OM3=0
        oraa #$10  ;OL3=1
        staa TCTL1
        ldaa #$20  ;clear OC3F
        staa TFLG1
        ldd  TCNT  ;current time
        addd #2000 ;first in 1 ms
        std  TOC3
        cli            ;enable
        rts
```

```
; MC68HC812A4
Period rmb  2       ;units µsec
Init    sei            ;make atomic
        bset TIOS,$08   ;OC3
        bset DDRT,$08   ;PT3 output
        movb #$80,TSCR  ;enable
        movb #$32,TMSK2 ;500 ns clk
        bset TMSK1,$08  ;Arm OC3
        bset TCTL2,$40  ;OL3=1
        bclr TCTL2,$80  ;OM3=0
        movb #$08,TFLG1 ;clear C3F
        ldd  TCNT  ;current time
        addd #50   ;first in 25µs
        std  TC3
        cli            ;enable
        rts
```

Program 6.22 Ritual for the assembly language implementation of the square wave using output compare.

To determine the fastest square wave that can be created we count the time it takes to process an interrupt. The fastest square wave will occupy 100% of the computer execution and request interrupts at twice the frequency (Table 6.14). In C, this square-wave software is shown in Program 6.24.

```
; MC68HC05C8                                    ; MC68HC708XL36
OCHan lda TSR ;part of ACK      [3]             CH3Han bclr 7,TSC3  ;Ack           [4]
      lda OCR+1  ;low byte       [3]                   lda TCH3+1  ;low byte      [3]
      add Period+1              [3]                    add Period+1              [3]
      sta OCR+1 ;OCR=OCR+Period [4]                    sta TCH3+1 ;TCH3=TCH3+Period[3]
      lda OCR                   [3]                    lda TCH3                  [3]
      adc Period                [3]                    adc Period                [3]
      sta OCR    ;setup next    [4]                    sta TCH3   ;setup next    [3]
      lda TCR                   [3]                    rti                       [7]
      eor #1     ;toggle OLVL   [2]
      sta TCR                   [4]
      rti                       [6]                    org  $FFEE  ;timer channel 3
      org $1FF8                                        fdb  CH3Han
      fdb OCHan
```
```
; MC68HC11A8                                    ; MC68HC812A4
OC3HAN ldaa #$20  ;clear OC3F [2]               OC3HAN movb #$08,TFLG1 ;Ack         [4]
       staa TFLG1 ;Ack          [4]                    ldd  TC3                    [3]
       ldd  TOC3                [5]                    addd Period ;next           [3]
       addd Period ;next        [6]                    std  TC3                    [2]
       std  TOC3                [5]                    rti                         [8]
       rti                     [12]                    org  $FFE8
       org  $FFE4                                      fdb  OC3HAN
       fdb  OC3HAN
```

Program 6.23 Assembly language squarewave using output compare.

Table 6.14
Total time in the handler calculated for different assembly language implementations.

Component	6805	6808	6811	6812
Process the interrupt (cycles/μs)	9=4.5μs	9=1.125μs	14=7μs	9=1.125μs
Execute entire handler (cycles/μs)	38=19μs	29=9.625μs	34=17μs	20=2.5μs
Total time (μs)	23.5	4.75	24	3.625

```
// MC68HC11A8                                    // MC68HC812A4
unsigned int Period;  // Period in usec          unsigned int Period;  // Period in usec
// Number of Cycles Low and number High          // Number of Cycles Low and number High
#pragma interrupt_handler TOC3handler()          #pragma interrupt_handler TOC3handler()
void TOC3handler(void){                          void TOC3handler(void){
    TOC3=TOC3+Period;  // calculate Next              TFLG1=0x08;      // ack OC3F
    TFLG1=0x20;}       // ack, OC3F=0                 TC3=TC3+Period;} // calculate Next
                                                 void ritual(void) {
void ritual(void){                               asm(" sei");      // make atomic
    asm(" sei");     // make atomic                  TIOS|=0x08;   // enable OC3
    TFLG1 = 0x20;    // clear OC3F                    DDRT|=0x08;   // PT3 is output
    TMSK1|= 0x20;    // arm OC3                       TSCR=0x80;    // enable
    TCTL1 = (TCTL1&0xCF)|0x10;                        TMSK2=0x32;   // 500 ns clock
// PA5 toggle on each interrupt                       TCTL2 = (TCTL2&0x3F)|0x40;
    TOC3 = TCNT+50; // first right away          // PT3 toggle on each interrupt
    asm(" cli"); }                                   TMSK1|=0x08;  // Arm output compare 3
                                                     TFLG1|=0x08;  // Initially clear C3F
                                                     TC3=TCNT+50;  // First one in 25 us
                                                 asm(" cli"); }
```

Program 6.24 C language squarewave using output compare.

The following is the assembly listing output of the `TOC3handler()` generated by the ICC11 compiler (Version 4.5). We later added the cycles shown on the right.

```
             13 ; void TOC3handler(void){
             14 ;      TOC3=TOC3+Period;
0000 FC 10 1A  15        ldd  4122          [5]
0003 F3 00 00  16        addd _Period        [6]
0006 FD 10 1A  17        std  4122          [5]
0009          18 SW.12:: ;     TFLG1=0x20;}
0009 C6 20    19        ldab #32            [2]
000B F7 10 23 20        stab  4131          [4]
000E 3B       21        rti                [12]
```

The total time to process this C language OC3 interrupt includes (1) 14 cycles for the 6811 to process the interrupt and (2) 34 cycles to complete the `TOC3handler()`. Therefore, each OC3 interrupt requires 48 cycles to process. In this case, we did not include the time to finish the current instruction because we will calculate the overhead required to generate this square wave (Table 6.15). Since an OC3 interrupt is requested every `Period` cycle and each one takes exactly 48 cycles to complete, the overhead required to produce the square wave can be calculated as a percentage of the total available 6811 execution cycles.

Table 6.15
Percentage overhead calculated for a 6811 C language implementation versus frequency.

Frequency	Period	Interrupt every (cycles)	Time to process (cycles)	Overhead (%)
10 Hz	100 ms	50,000	48	0.1
100 Hz	10 ms	5,000	48	1
1 kHz	1 ms	500	48	10
5 kHz	200 μs	200	48	24
1/P	P (μs)	P	48	4800/P

Just like the previous example of the real-time clock, as long as no other software executes with the interrupts disabled for longer than `Period` cycles, this solution will create the perfect square wave. Remember the toggle event (i.e., when PA5/OC3 changes) occurs automatically in hardware at the time when TOC3=TCNT, and not during the software execution of the `TOC3handler()`.

The slowest square wave that can be generated by this implementation has a period of 65,535 μs—that is, about 15.3 Hz. On the 6808/6811/6812 you can set up the TCNT to clock every 4 μs. Thus, the slowest square wave you can generate is about 2 Hz. Less accurate square waves can be made using an interrupt real-time clock and a software counter. Using the real-time clock, we could interrupt at a fixed interval **T** and decrement a software counter. When that counter hits 0, we could toggle an output bit and reset the counter to **N**. The period of the resulting square wave would be 2·**N**·**T**. These functions could be added to the implementations presented earlier in Program 6.19 to Program 6.21.

**6.2.5
Pulse-Width
Modulation**

This example generates a variable duty cycle square wave using output compare (Figure 6.24). Pulse-width modulation is an effective and thus popular mechanism for the embedded microcomputers to control external devices. The output is 1 for `High` cycles, then 0 for `Low` cycles. Output compare interrupts will again be requested at a rate twice as fast as the resulting square-wave frequency. One interrupt is required for the rising edge and another

Figure 6.24
Pulse-width modulation
using output compare.

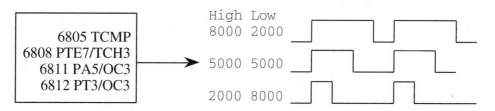

for the falling edge. Clear to 0 and set to 1 modes are used to create the variable duty cycle square wave. In the examples below, High plus Low will always equal 10,000, so in each case the square-wave period will be 10,000 cycles, or 5 ms. By adjusting the ratio of

$$\text{Duty cycle} = \frac{\text{high}}{\text{high} + \text{low}}$$

the software can control the duty cycle. This implementation cannot generate waves close to 0 or 100% duty cycle. The upper and lower limits can be calculated by counting the cycles required to process the output compare interrupt like in the previous example. If **T** is the maximum number of cycles to process the output compare interrupt, then both High and Low must be greater than **T.**

In assembly language, the ritual is in Program 6.25 and the interrupt handler in Program 6.26 (there will be an interrupt for both the rise and fall).

```
; MC68HC05C8
High rmb  2 ;number of cycles high
Low  rmb  2 ;number of cycles low
Init sei         ;make atomic
        lda #$41
        sta TCR      ;arm OCF
        lda TSR      ;clear OCF
        ldx TCNT     ;latches LSB too
        lda TCNT+1   ;read low byte
        add #50      ;first in 25µs
        sta OCR+1
        txa
        adc #0
        sta OCR
        cli          ;enable
        rts
```

```
; MC68HC708XL36
High rmb  2  ;number of cycles high
Low  rmb  2  ;number of cycles low
Init sei           ;make atomic
        bset 7,DDRE ;PTE7 output
        mov #$02,TSC   ;500ns clk
        lda TSC3
        mov #$5C,TSC3  ;arm,set OC
; write 0 to bit 7, clears CH3F
        ldhx TCNT
        aix  #50     ;first in 25µs
        sthx TCH3
        cli          ;enable
        rts
```

```
; MC68HC11A8
High rmb  2 ;number of cycles high
Low  rmb  2 ;number of cycles low
RITUAL sei           ;make atomic
        ldaa TMSK1 ;Old value
        oraa #$20  ;TMSK1 OC3I=1
        staa TMSK1 ;Arm OC3F
        ldaa TCTL1
        oraa #$30  ;OM3=1, OL3=1
```

```
; MC68HC812A4
High rmb  2  ;number of cycles high
Low  rmb  2  ;number of cycles low
RITUAL sei           ;make atomic
        bset TIOS,$08   ;OC3
        bset DDRT,$08   ;PT3 output
        movb #$80,TSCR  ;enable
        movb #$32,TMSK2 ;500 ns clk
        bset TMSK1,$08  ;Arm OC3
```

continued on p. 330

Program 6.25 Assembly language initialization PWM squarewave using output compare.

continued from p. 329

```
        staa  TCTL1                          bset  TCTL2,$C0  ;OL3=1,OM3=1
        ldaa  #$20   ;clear OC3F             movb  #$08,TFLG1 ;clear C3F
        staa  TFLG1                          ldd   TCNT   ;current time
        ldd   TCNT   ;current time           addd  #50    ;first in 25µs
        addd  #50    ;first in 25µs          std   TC3
        std   TOC3                           cli          ;enable
        cli          ;enable                 rts
        rts
```

Program 6.25 Assembly language initialization PWM squarewave using output compare.

```
; MC68HC05C8                              ; MC68HC708XL36
OCHan lda TSR ;part of ACK     [3]        CH3Han bclr 7,TSC3   ;Ack          [4]
      brclr 0,TCR,zero         [5]               brclr 2,TSC3,zero          [3]
one   lda OCR+1  ;low byte     [3]        one    lda TCH3+1  ;low byte       [3]
      add High+1               [3]               add High+1                 [3]
      sta OCR+1 ;OCR=OCR+High  [4]               sta TCH3+1 ;TCH3=TCH3+High  [3]
      lda OCR                  [3]               lda TCH3                    [3]
      adc High                 [3]               adc High                   [3]
      sta OCR    ;setup next   [4]               sta TCH3   ;setup next      [3]
      lda #$40   ;next is 0    [2]               bclr 2,TSC2                 [4]
      sta TCR                  [4]               bra done                    [3]
      bra done                 [3]        zero   lda TCH3+1  ;low byte       [3]
zero lda OCR+1  ;low byte      [3]               add Low+1                   [3]
      add Low+1                [3]               sta TCH3+1  ;TCH3=TCH3+Low   [3]
      sta OCR+1 ;OCR=OCR+Low   [4]               lda TCH3                    [3]
      lda OCR                  [3]               adc Low                     [3]
      adc Low                  [3]               sta TCH3    ;setup next     [3]
      sta OCR    ;setup next   [4]               bset 2,TSC2                 [4]
      lda #$41   ;next is 1    [2]        done   rti                        [7]
      sta TCR                  [4]
done rti                       [6]               org  $FFEE  ;timer channel 3
      org $1FF8                                  fdb  CH3Han
      fdb OCHan
```

```
; MC68HC11A8                              ; MC68HC812A4
OC3HAN ldaa #$20  ;clear OC3F [2]         OC3HAN movb #$08,TFLG1 ;Ack         [4]
       staa TFLG1 ;Ack        [4]                ldaa TCTL2   ;rise/fall?     [3]
       ldaa TCTL2 ;rise/fall? [4]                bita #$40                    [1]
       bita #$10              [2]                beq  zero                   [1/3]
       beq  zero              [3]         one    ldd  TC3                     [3]
one    ldd TOC3               [5]                addd High ;now PT3 is 1      [3]
       addd High ;OC3 is 1    [6]                std  TC3                     [2]
       std  TOC3              [5]                bclr TCTL2,$40               [4]
       ldaa TCTL2             [4]                bra  done                    [3]
       anda #$BF              [2]         zero   ldd  TC3                     [3]
       staa TCLT2             [4]                addd Low  ;now PT3 is 0      [3]
       bra  done              [3]                std  TC3                     [2]
zero   ldd TOC3               [5]                bset TCTL2, $40              [4]
       addd Low  ;OC is 0     [6]         done   rti                         [8]
       std  TOC3              [5]
```

```
          ldaa  TCTL2            [4]                        org   $FFE8
          oraa  #$40             [2]                        fdb   OC3HAN
          staa  TCLT2            [4]
done      rti                   [12]
          org   $FFE4
          fdb   OC3HAN
```

Program 6.26 Assembly language ISR for PWM square wave using output compare.

To determine the fastest square wave that can be created we count the time it takes to process an interrupt. The cycle times shown in brackets, e.g., [2], in Program 6.26 are executed only when the output goes high. The two numbers in Table 6.16 represent the two possible branch patterns of the interrupt execution. This table presents the **T** parameter discussed previously.

In C, this PWM software is shown in Program 6.27. To determine the **T** parameter for the C implementation, you could observe the assembly listing from the compiler, count cycles, and perform a calculation like Table 6.16.

Table 6.16
Total time in the handler calculated for different assembly language implementations.

Component	6805	6808	6811	6812
Process the interrupt (cycles)	9	9	14	9
Execute entire handler (cycles)	40–43	36–39	53–56	31–32
Total time **T** (μs)	49–52	45–48	67–70	40–41

```
// MC68HC11A8                                    // MC68HC812A4
unsigned int High; // Num of Cycles High         unsigned int High; // Num of Cycles High
unsigned int Low;  // Num of Cycles Low          unsigned int Low;  // Num of Cycles Low
// Period is High+Low Cycles                     // Period is High+Low Cycles
#pragma interrupt_handler TOC3handler()          #pragma interrupt_handler TOC3handler()
void TOC3handler(void){                          void TOC3handler(void){
    if(TCTL1&0x10){    // PA5 is now high             TFLG1=0x08;        // ack OC3F
       TOC3=TOC3+High; // 1 for High cyc             if(TCTL2&0x40){     // PT3 is now high
       TCTL1&=0xEF;}   // clear on next                 TC3=TC3+High;   // 1 for High cyc
    else {             // PA5 is now low              TCTL2&=0xBF;}   // clear on next
       TOC3=TOC3+Low;  // 0 for Low cycles          else {             // PT3 is now low
       TCTL1|=0x10;}   // set on next int              TC3=TC3+Low;    // 0 for Low cycles
    TFLG1=0x20;}       // ack, clear OC3F             TCTL2|=0x40;}}  // set on next int
                                                 void ritual(void){
void ritual(void){                                   asm(" sei");   // make atomic
    asm(" sei");     // make atomic                  TIOS|=0x08;    // enable OC3
    TFLG1 = 0x20;    // initially OC3F=0             DDRT|=0x08;    // PT3 is output
    TMSK1|= 0x20;    // arm OC3                      TSCR=0x80;     // enable
    TCTL1|= 0x30;    // PA5 set on next int          TMSK2=0x32;    // 500 ns clock
    TOC3 = TCNT+50; // first right away             TMSK1|=0x08;   // Arm output compare 3
    asm(" cli"); }                                   TFLG1=0x08;    // Initially clear C3F
                                                 TCTL2|= 0xC0;   // PT3 set on next int
void main(void){                                     TC3 = TCNT+50; // first right away
    High=8000; Low+2000;                            asm(" cli"); }
    ritual();                                    void main(void){
    while(1);}                                        High=8000; Low=2000;
                                                 ritual();
                                                     while(1);}
```

Program 6.27 C language PWM square wave using output compare.

6.2.6 Delayed Pulse Generation

One application of the coupled output compare mechanism is a delayed output pulse (Figure 6.25). One pulse will be generated each time the function `Pulse()` is called. We will supply two parameters to this pulse generation function, `Delay` and `Width`. The first parameter is the delay (in cycles), and the second is the width of the pulse. Delayed pulse generation could have been created with a single output compare module, but the use of two output compare modules allows the `Width` to be as short as 1 cycle. We can set 6811 TOC1 (or 6812 TC7) to the time we want the pulse to go high and set 6811 TOC3 (or 6812 TC3) to the time we want the pulse to go low. The OC3 interrupt is disarmed in the interrupt handler so that the output pulse occurs only once and is not repeated every 65,536 counts of TCNT. Although the `Width` can be as short as 1 cycle, the `Delay` must be big enough so that TCNT does not reach the TOC3 value before the OC3F flag is cleared (i.e., the execution time of this function must be less than the `Delay` time).

Figure 6.25
Delayed pulse generation using output compare.

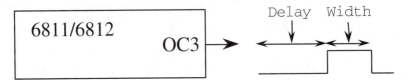

On the 6811, we set both OC1M5 and OC1D5 to 1. When TCNT equals TOC1, PA5/OC3 will go high. We set OM3 to 1 and OL3 to 0. When TCNT=TOC3, PA5/OC3 will go low. Similarly on the 6812, we set both OC7M3 and OC7D3 to 1. When TCNT equals TC7, PT3/OC3 will go high. We set OM3 to 1 and OL3 to 0. When TCNT=TC3, PT3/OC3 will go low (Program 6.28).

```
// MC68HC11A8
void Pulse(unsigned int Delay,
    unsigned int Width){
  asm(" sei");  // make atomic
  TOC1=TCNT+Delay;
  TOC3=TOC1+Width;
  OC1M=0x20;     // connect OC1 to PA5/OC3
  OC1D=0x20;     // PA5=1 when TOC1=TCNT
  TCTL1=(TCTL1&0xCF)|0x20;
// PA5=0 when TOC3=TCNT
  TFLG1 = 0x20;     // Clear OC3F
  TMSK1|= 0x20;     // Arm OC3F
  asm(" cli");}


#pragma interrupt_handler TOC3handler()
void TOC3handler(void){
  OC1M=0;  // disconnect OC1 from PA5
  OC1D=0;
  TCTL1&=0xCF;  // disable OC3
  TMSK1&=0xDF;} // disarm OC3F
```

```
// MC68HC812A4
void Pulse(unsigned int Delay,
    unsigned int Width){
  asm(" sei");  // make atomic
  TIOS|=0x08;   // enable OC3
  DDRT|=0x08;   // PT3 is output
  TSCR=0x90;    // enable, fast clear
  TMSK2=0x32;   // 500 ns clock
  TC7=TCNT+Delay;
  TC3=TC7+Width;
  OC7M=0x08;    // connect OC7 to PT3
  OC7D=0x08;    // PT3=1 when TC7=TCNT
  TCTL2=(TCTL2&0x3F)|0x80;
// PT3=0 when TC3=TCNT
  TFLG1 = 0x08;     // Clear C3F
  TMSK1|= 0x08;     // Arm C3F
  asm(" cli");}
#pragma interrupt_handler TOC3handler()
void TOC3handler(void){
  OC7M=0;  // disconnect OC7 from PT3/OC3
  OC7D=0;
  TCTL2&=0x3F;  // disable OC3
  TMSK1&=0xF7;} // disarm C3F
```

Program 6.28 C language delayed pulse output using output compare.

Observation: We acknowledge an interrupt (clear its flag) within the interrupt handler when we are interested in subsequent interrupts, and we disarm an interrupt (clear its mask) when we are not interested in any more interrupts from this source.

6.3   Frequency Measurement

6.3.1
Frequency
Measurement
Concepts

The direct measurement of frequency involves counting input pulses for a fixed amount of time. The basic idea is to use Input Capture to count pulses and use Output Compare to create the fixed time interval. For example, we could initialize Input Capture to interrupt on every rising edge of our input signal. During the Input Capture handler, we could increment a Counter. At the beginning of our fixed time interval, the Counter is initialized to zero, and at the end of the interval, we can calculate frequency:

$$f = \frac{counter}{fixed\ time}$$

The frequency resolution, Δf, is defined to be the smallest change in frequency that can be reliably measured by the system. For the system to detect a change, the frequency must increase (or decrease) enough so that there is one more (or one less) pulse during the fixed time interval. Therefore, the frequency resolution is

$$\Delta f = \frac{1}{fixed\ time}$$

This frequency resolution also specifies the units of the measurement.

6.3.2
Frequency
Measurement
with $\Delta f = 100Hz$

If we count pulses in a 10-ms time interval, then the number of pulses represents the signal frequency with units 1/10 ms, or 100 Hz. For example, if there are 5 pulses during the 10-ms interval, then the frequency is 500 Hz. For this system, the measurement resolution is 100 Hz, so the frequency would have to increase to 600 Hz (or decrease to 400 Hz) for the change to be detected (Figure 6.26).

Figure 6.26
Basic timing involved in frequency measurement using both input capture and output compare.

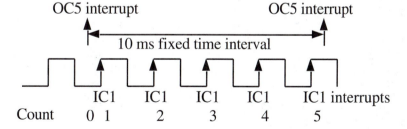

The highest frequency that can be measured will be determined by how fast the Input Capture interrupt handler can count pulses. We should select a counter precision (in this case only 8 bits is needed) to hold this maximum number. In this example, the TTL-level input signal is connected to IC1 (PA2 on the 6811 or PT1 on the 6812). The rising edge will generate an input capture interrupt (Figure 6.27).

Figure 6.27
Frequency measurement
using both input capture
and output compare.

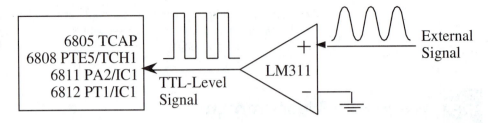

In C, the frequency measurement software is as follows. The frequency measurement counts the number of rising edges in a 10-ms interval. The measurement resolution is 100 Hz (determined by the 10-ms interval). An IC1 interrupt occurs on each rising edge, and an OC5 interrupt occurs every 10 ms. The TTL-level input signal is connected to IC1 (6811 PA2 or 6812 PT1). The foreground/background threads communicate via two shared globals. The background thread will update Freq with a new measurement and set Done. When Done is set, the foreground thread will read the global Freq and clear Done.

```
unsigned int Freq;    /* Frequency with units of 100 Hz */
unsigned char Done;   /* Set each measurement, every 10 ms */
```

There is a private global (only accessed by the background threads and not the foreground).

```
unsigned int Count;   /* Number of rising edges */
```

The 6805 and 6808 implementation are shown in assembly in Program 6.29. If we assume the fastest period is 50 μs, then the largest frequency will be 20 kHz. Therefore, the maximum values of Count and Freq will be 200, so 8-bit variables can be used. A frequency of 0 will result in no input capture interrupts, and the system will properly report the frequency of 0. The 6811 and 6812 implementation are shown in C in Program 6.30.

```
; MC68HC05C8                           ; MC68HC708XL36
Count rmb 1 ;in progress              Count rmb 1   ;measurement in progress
Freq rmb 1  ;in 100Hz units          Freq rmb 1    ;in 100Hz units
Done rmb 1  ;$FF when ready           Done rmb 1    ;set to $FF when ready
Init sei        ;make atomic          Init sei           ;make atomic
     lda #$C0   ;arm ICF rising            mov #$02,TSC    ;500ns clk
     sta TCR    ;arm OCF                   lda TSC1
     lda TSR    ;clear ICF                 mov #$44,TSC1   ;arm,rising
     lda ISR+1  ;read low byte             clr Done
     clr Done                             clr Count
     clr Count                            lda TSC3
     ldx TCNT   ;latches LSB too           mov #$54,TSC3   ;arm,toggle OC
     lda TCNT+1 ;read low byte             ldhx TCNT
     add #50    ;first in 25μs             aix #50     ;first in 25μs
     sta OCR+1                            sthx TCH3
     txa                                  cli             ;enable
     adc #0                               rts
     sta OCR                         OC3Han bclr 7,TSC3 ;called every 10ms
     cli        ;enable                      lda TCH3+1 ;low byte
     rts                                     add #$20    ;20000=$4E20
IHan lda TSR                                 sta TCH3+1 ;TCH3=TCH3+20000
     bit #$40   ;poll OCF                    lda TCH3
     bne ChIC                                adc #$4E
```

```
OChan lda  OCR+1                              sta  TCH3    ;setup next
      add  #$20                               lda  Count
      sta  OCR+1 ;OCR=OCR+20000               sta  Freq
      lda  OCR                                mov  #$FF,Done
      adc  #$4E                               clr  Count   ;setup for next
      sta  OCR                                rti
      lda  Count                       IC1Han bclr 7,TSC1  ;Ack
      sta  Freq                               inc  Count
      lda  #$FF                               rti
      sta  Done                               org  $FFF2   ;timer channel 1
      clr  Count   ;setup for next            fdb  IC1Han
ChIC  lda  TSR ;poll ICF                      org  $FFEE   ;timer channel 3
      bit  #$80                               fdb  OC3Han
      bne  Iret
ICHan lda  ICR+1 ;ack
      inc  Count
Iret  rti
      org  $1FF8
      fdb  Ihan   ;both ICF and OCF
```

Program 6.29 Assembly language frequency measurement.

```
// MC68HC11A8                                 // MC68HC812A4
#define IC1F 0x04   // connected here         #pragma interrupt_handler TIC1handler()
#pragma interrupt_handler TIC1handler()        void TIC1handler(void){
void TIC1handler(void){                          Count++;      // number of rising edges
  Count++;       // number of rising edges       TFLG1=0x02;}  // ack, clear C1F
  TFLG1=IC1F;}  // ack, clear IC1F             #define Rate 20000  //  10 ms
                                               #pragma interrupt_handler TOC5handler()
#define Rate 20000  //  10 ms                  void TOC5handler(void){
#define OC5F 0x08                                  TFLG1= 0x20;      // Acknowledge
#pragma interrupt_handler TOC5handler()            TC5 = TC5+Rate;  // every 10 ms
void TOC5handler(void){                            Freq = Count;    // 100 Hz units
    TFLG1= OC5F;       // Acknowledge              Done = 0xff;
    TOC5 = TOC5+Rate;  // every 10 ms              Count = 0; }     // Setup for next
    Freq = Count;      // 100 Hz units         void ritual(void) {
    Done = 0xff;                               asm(" sei");         // make atomic
    Count = 0; }       // Setup for next           TIOS|=0x20;   // enable OC5
void ritual(void) {                                TSCR=0x80;    // enable
asm(" sei");           // make atomic              TMSK2=0x32;   // 500 ns clock
    TMSK1|=OC5F+IC1F; // Arm OC5 and IC1           TMSK1|=0x22;  // Arm OC5 and IC1
    TOC5=TCNT+Rate;    // First in 10 ms           TC5=TCNT+Rate;  // First in 10 ms
    TCTL2 = (TCTL2&0xCF)|0x10;                     TCTL4 = (TCTL4&0xF3)|0x04;
/* IC1F set on rising edges */                 /* CIF set on rising edges */
    Count = 0;         // Set up for first         Count = 0;        // Set up for first
    Done=0;                                        Done=0;
/* Set on the subsequent measurements */       /* Set on the subsequent measurements */
    TFLG1=OC5F+IC1F;  // clear OC5F, IC1F          TFLG1=0x22;  // clear OC5F
asm(" cli"): }                                 asm(" cli"); }
```

Program 6.30 C language frequency measurement.

6.4   Conversion Between Frequency and Period

6.4.1 Using Period Measurement to Calculate Frequency

Period and frequency are obviously related, so when faced with a problem that requires frequency information we could measure period and calculate frequency from the period measurement.

$$f = \frac{1}{p}$$

Assume we use the 16-bit period measurement interface described earlier in Section 6.1.4.1. In this system a global variable Period contains the measured period, with a range from 36 μs^1 to 32 ms and a resolution of 500 ns. This corresponds to a frequency range of 31 to 27,778 Hz. If we add another global variable, Freq, that will hold the calculated frequency in hertz, then a 32-bit by 16-bit divide (a 6805/6808/6811 subroutine or 6812 opcode) can be used to calculate

$$Freq = \frac{2000000}{Period}$$

It is easy to see how the 36-μs to 32-ms period range maps into the 31- to 27,778-Hz frequency range, but mapping the 500-ns period resolution into an equivalent frequency resolution is a little more tricky. If the frequency is f, then the frequency must change to f+Δf such that the period changes by at least Δp=500 ns. 1/f is the initial period, and 1/(f+Δf) is the new period. These two periods differ by 500 ns. In other words,

$$\Delta p = \frac{1}{f} - \frac{1}{f + \Delta f}$$

We can rearrange this equation to relate Δf as a function of Δp and f.

$$\Delta f = \frac{1}{(1/f) - \Delta p} - f$$

This very nonlinear relationship, shown in Table 6.17, illustrates that although the period resolution is fixed at 500 ns, the equivalent frequency resolution varies from 500 Hz to

Table 6.17
Relationship between frequency resolution and frequency when calculated using period measurement.

Frequency (Hz)	Period (μs)	Δf (Hz)
31,250	32	500
20,000	50	200
10,000	100	50
5,000	200	13
2,000	500	2
1,000	1,000	0.5
500	2,000	0.13
200	5,000	0.02
100	10,000	0.005
50	20,000	0.001
31.25	32,000	0.0005

---

[1]Each of the four microcomputers had a separate lower bound on period; 36 μs was the fastest period that could be measured by the 2-MHz 6811.

0.0005 Hz. If the signal frequency is restricted to values below 1413 Hz, then we can say the frequency resolution will be better than 1 Hz.

6.4.2
Using Frequency
Measurement to
Calculate Period

Similarly, when faced with a problem that requires a period measurement, we could measure frequency and calculate period from the frequency measurement.

$$p = \frac{1}{f}$$

A similar nonlinear relationship exists between the frequency resolution and period resolution. In general, the period measurement approach will be faster, but the frequency measurement approach will be more robust in the face of missed edges or extra pulses. See Exercise 6.3 for a comparison between frequency and period measurement.

6.5 Measurements Using Both Input Capture and Output Compare

6.5.1
Period
Measurement
with $\Delta p = 1ms$

The objective is to measure period with a resolution of 1 ms. The TTL-level input signal is connected to an input capture. Each rising edge will generate an input capture interrupt. In addition, output compare is used to increment a software counter, Time, every 1 ms. The period is calculated as the number of 1-ms output compare interrupts between one rising edge of the input capture pin to the other rising edge of the input capture pin. For example, if the period is 8192 μs, there will be eight output compare interrupts between successive input capture interrupts (Figure 6.27).

Assembly language versions for the 6805/6808 and C language implementations for the 6811/6812 are presented. The period measurement counts the number of 1-ms intervals between successive rising edges. The period measurement resolution is 1 ms, because the period must increase by at least 1 ms for there to be a different number of counts. The range is 0 to 65 s, and the precision of 16 bits is determined by the size of the Time counter. If the gadfly loop in the ritual is removed, the first measurement will be inaccurate, but the subsequent measurements will be okay. There is an IC1 interrupt each period, and the external signal is connected to IC1. A real-time clock with a 1-ms rate is created with OC3. The foreground/background threads communicate via three shared globals. The background thread will update Period with a new measurement and set Done. When Done is set, the foreground thread will read the global Period and clear Done. OverFlow is set by the background and read by the foreground if the period is larger than 65 s. The ritual will wait for the first rising edge. In this way the first measurement will be correct. The only problem is that if no signal exists (the input capture pin is a constant level), this ritual software will hang at this wait.

```
unsigned int Period;    /* Period in msec */
unsigned char OverFlow; /* Set if Period is too big*/
unsigned char Done;     /* Set each rising edge of IC1 */
```

There is a private global (only accessed by the background threads and not the foreground).

```
unsigned int Cnt;       /* number of msec in one period */
```

The first part of the assembly implementation initializes the globals then configures the timer modes (Program 6.31).

```
; MC68HC05C8                          ; MC68HC708XL36
Init sei          ;make atomic        Init sei          ;make atomic
     lda #$C0     ;arm ICF rising          mov #$02,TSC   ;500ns clk
     sta TCR      ;arm OCF                 lda TSC1
     lda TSR      ;clear ICF               mov #$44,TSC1  ;arm,rising
     lda ISR+1    ;read low byte           clr Done
     clr Done     ;set when ready          ldhx #0
     clr Cnt                               sthx Cnt       ;number of msec
     clr Cnt+1                             clr OverFlow ;set if too big
     clr OverFlow ;set if too big     wait brclr 7,TSC1,wait ;wait for first
wait brclr 7,TSR,wait ;wait                bclr 7,TSC1  ;clear flag
     lda ISR+1    ;read low byte           lda TSC3
     ldx TCNT     ;latches LSB too         mov #$54,TSC3  ;arm,toggle OC
     lda TCNT+1   ;read low byte           ldhx TCNT
     add #50      ;first in 25µs           aix  #50     ;first in 25µs
     sta OCR+1                             sthx TCH3
     txa                                   cli          ;enable
     adc #0                                rts
     sta OCR
     cli          ;enable
     rts
```

Program 6.31 Assembly language initialization for period measurement.

The second part includes two ISRs. The output compare handler simply counts 1-ms time intervals. The input capture interrupt occurs on the external input. This ISR occurs when a new measurement is ready (Programs 6.32 and 6.33).

```
; MC68HC05C8                          ; MC68HC708XL36
IHan lda TSR                          OC3Han pshh          ;called every 10ms
     bit #$40     ;poll OCF                  bclr 7,TSC3  ;Ack
     bne ChIC                                lda TCH3+1 ;low byte
OChan lda OCR+1                              add #$D0     ;2000=$07D0
     add #$D0                                sta TCH3+1 ;TCH3=TCH3+2000
     sta OCR+1 ;OCR=OCR+2000                 lda TCH3
     lda OCR                                 adc #$07
     adc #$07                                sta TCH3   ;setup next
     sta OCR                                 ldhx Cnt
     lda Cnt+1                               aix  #1
     add #1                                  sthx Cnt
     sta Cnt+1                               cphx #0
     lda Cnt                                 bne  ok
     adc #0                                  mov #$FF,OverFlow ;too big
     sta Cnt                          ok     pulh
     bne ChIC                                rti
     lda Cnt+1                        IC1Han pshh          ;called every rising
     bne ChIC ;Is Cnt equal to 0             bclr 7,TSC1  ;Ack
     lda #$FF                                lda OverFlow
     sta OverFlow ;too big                   beq good
ChIC lda TSR ;poll ICF               bad    sta Period
```

```
ICHan lda ICR+1 ;ack                         bra set
      lda OverFlow            good    ldhx Cnt
      beq good                        sthx Period ;units in msec
bad   sta Period              set     mov #$FF,Done
      sta Period+1 ;Period=65535      clr Cnt
      clr OverFlow                    clr Cnt+1
      bra set                        pulh
good  lda Cnt+1                      rti
      sta Freq+1                     org $FFF2  ;timer channel 1
      lda Cnt                        fdb IC1Han
      sta Freq                       org $FFEE  ;timer channel 3
set   lda #$FF                       fdb OC3Han
      sta Done
      clr Cnt
      clr Cnt+1
Iret  rti
      org $1FF8
      fdb Ihan  ;both ICF and OCF
```

Program 6.32 Assembly language ISRs for period measurement.

```c
// MC68HC11A8                         // MC68HC812A4
#define resolution 2000              #define resolution 2000
#pragma interrupt_handler TOC3handler()   #pragma interrupt_handler TOC3handler()
void TOC3handler(void){              void TOC3handler(void){
    TOC3=TOC3+resolution;  // every 1 ms     TFLG1= 0x08;        // Acknowledge
    TFLG1=0x20;       // ack, clear OC3F      TC3=TC3+resolution; // every 1 ms
    Cnt++;                                    Cnt++;
    if(Cnt==0) OverFlow=0xFF;}                if(Cnt==0) OverFlow=0xFF;}
#pragma interrupt_handler TIC1handler()   #pragma interrupt_handler TIC1handler()
void TIC1handler(void){              void TIC1handler(void){
    TFLG1=0x04;   // ack, clear IC1F         TFLG1=0x02;   // ack, clear C1F
    if(OverFlow){                            if(OverFlow){
        Period=65535;                            Period=65535;
        OverFlow=0;}                             OverFlow=0;}
    else                                     else
        Period=Cnt;                              Period=Cnt;
    Cnt=0;                                   Cnt=0;
    Done=0xFF;}                              Done=0xFF;}
void Ritual(void){                   void Ritual(void){
    asm(" sei");  // make atomic             asm(" sei");  // make atomic
    TFLG1 = 0x24;    // Clear OC3F,IC1F      TIOS| = 0x08;   // enable OC3
    TMSK1 = 0x24;    // Arm OC3 and IC1      TSCR  = 0x80;   // enable
    TCTL2 = 0x10;    // rising edges         TMSK2 = 0x32;   // 500 ns clock
    while((TFLG1&0x04)==0);                  TFLG1 = 0x0A;   // Clear C3F,C1F
// wait for first rising                     TMSK1 = 0x0A;   // Arm OC3 and IC1
    TFLG1 = 0x04;    // clear IC1F           TCTL4 = (TCTL4&0xF3)|0x04;
    TOC3=TCNT+resolution;            /* C1F set on rising edges */
    Cnt=0; OverFlow=0; Done=0;               while((TFLG1&0x02)==0); // wait rising
    asm(" cli"); }                           TFLG1 = 0x02;   // Clear C1F
                                             TC3=TCNT+resolution;
                                             Cnt=0; OverFlow=0; Done=0;
                                             asm(" cli"); }
```

Program 6.33 C language implementation of period measurement.

6.5.2
Frequency
Measurement
with Δf=0.1Hz

If we count pulses in a 10-s time interval, then the number of pulses represents the signal frequency with units 1/10 s, or 0.1 Hz (e.g., if there are 12,345 pulses during the 10-s interval, then the frequency is 1234.5 Hz). For this system, the measurement resolution is 0.1 Hz. For example, the frequency would have to increase to 1234.6 Hz (or decrease to 1234.4 Hz) for the change (a different number of IC1 interrupts) to be detected. OC5 interrupts every 25 ms, and it takes 400 OC5 interrupts to create the 10-s time delay. The number of IC1 interrupts in the 10-s interval represents the input frequency with units of 0.1 Hz (Figure 6.28).

The highest frequency that can be measured will be determined by how fast the Input Capture interrupt handler can count pulses. We should select a counter precision (e.g., 8, 16, or 24 bits) so that the highest frequency does not overflow the counter. The TTL-level input signal is connected to IC1 (PA2 on the 6811 or PT1 on the 6812). Each rising edge will generate an input capture interrupt.

Figure 6.28
Basic timing involved in
frequency measurement.

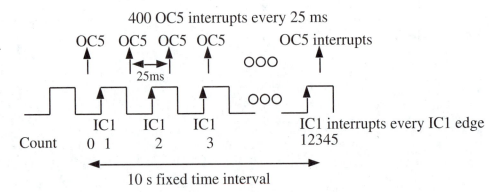

Assembly language versions for the 6805/6808 and C language implementations for the 6811/6812 are presented. To simplify the implementation, we will assume the maximum frequency is 6.5 KHz so that a 16-bit counter can be used. The frequency measurement is the number of rising edges in a 10-s interval. A periodic interrupt is generated with OC5. The foreground/background threads communicate via two shared globals. The background thread will update `Freq` with a new measurement and set `Done`. When `Done` is set, the foreground thread will read the global `Freq` and clear `Done`. Just like the other frequency measurement system, a frequency of 0 will result in no input capture interrupts, and the system will properly report the frequency of 0.

```
unsigned int Freq;   /* Frequency with units of 0.1 Hz */
unsigned char Done;  /* Set each measurement, every 10 s */
```

There are two private globals (only accessed by the background threads and not the foreground.)

```
unsigned int FourHundred; // Used to create 10sec interval
unsigned int Count;       // Number of rising edges
```

We begin with the 6805/6808 assembly code for initialization (Program 6.34).

```
; MC68HC05C8
Init sei          ;make atomic
     lda #$C0      ;arm ICF rising
     sta TCR       ;arm OCF
     lda TSR       ;clear ICF
     lda ISR+1     ;read low byte
     clr Done
     clr Count
     clr Count+1
     clr FourHundred
     clr FourHundred+1
     ldx TCNT      ;latches LSB too
     lda TCNT+1    ;read low byte
     add #50       ;first in 25µs
     sta OCR+1
     txa
     adc #0
     sta OCR
     cli           ;enable
     rts
```

```
; MC68HC708XL36
Init sei            ;make atomic
     mov #$02,TSC   ;500ns clk
     lda TSC1
     mov #$44,TSC1   ;arm,rising
     clr Done
     ldhx #0
     sthx Count
     sthx FourHundred
     lda TSC3
     mov #$54,TSC3   ;arm,toggle OC
     ldhx TCNT
     aix  #50       ;first in 25µs
     sthx TCH3
     cli            ;enable
     rts
```

Program 6.34 Assembly language initialization for frequency measurement.

The input capture interrupt counts the number of cycles, and the output compare interrupt is used to mark the end of the fixed time measurement (Program 6.35).

```
; MC68HC05C8
IHan lda TSR
     bit #$40    ;poll OCF
     bne ChIC
OChan lda OCR+1
     add #$50
     sta OCR+1 ;OCR=OCR+50000
     lda OCR
     adc #$C3
     sta OCR
     lda FourHundred+1
     add #1
     sta FourHundred+1
     lda FourHundred
     adc #0
     sta FourHundred
     lda FourHundred+1 ;400=$190
     cmp #$90
     bne ChIC
     lda FourHundred
     cmp #1
```

```
; MC68HC708XL36
OC3Han pshh        ;called every 25ms
       bclr 7,TSC3   ;Ack
       lda TCH3+1 ;low byte
       add #$50    ;50000=$C350
       sta TCH3+1 ;TCH3=TCH3+50000
       lda TCH3
       adc #$C3
       sta TCH3   ;setup next
       lda FourHundred+1
       add #1
       sta FourHundred+1
       lda FourHundred
       adc #0
       sta FourHundred
       lda FourHundred+1 ;400=$190
       cmp #$90
       bne skip
       lda FourHundred
       cmp #1
       bne skip
```

continued on p. 342

Program 6.35 Assembly language ISRs for frequency measurement.

```
continued from p. 341
            bne  ChIC
            lda  Count+1   ;every 10sec
            sta  Freq+1
            lda  Count
            sta  Freq
            lda  #$FF
            sta  Done
            clr  Count  ;setup for next
            clr  Count+1
            clr  FourHundred
            clr  FourHundred+1
ChIC  lda  TSR ;poll ICF
            bit  #$80
            bne  Iret
ICHan lda  ICR+1 ;ack
            lda  Count+1
            add  #1
            sta  Count+1
            lda  Count
            adc  #0
            sta  Count
Iret  rti
            org  $1FF8
            fdb  Ihan  ;both ICF and OCF
```

```
            ldhx Count  ;every 10sec
            sthx Freq
            mov  #$FF,Done
            ldhx #0
            sthx Count    ;setup for next
            sthx FourHundred
skip   pulh
            rti
IC1Han pshh        ;called every rising
            bclr 7,TSC1  ;Ack
            ldhx Count
            aix  #1
            sthx Count
            pulh
            rti
            org  $FFF2  ;timer channel 1
            fdb  IC1Han
            org  $FFEE  ;timer channel 3
            fdb  OC3Han
```

Program 6.35 Assembly language ISRs for frequency measurement.

Similar to the assembly language versions, the input capture counts cycles and the output compare creates the 10-s fixed time interval (Program 6.36).

```
// MC68HC11A8
#define IC1F 0x04    // connected here
#pragma interrupt_handler TIC1handler()
void TIC1handler(void){
   Count++;       // number of rising edges
   TFLG1=IC1F;}  // ack, clear IC1F

#define Rate 50000  //  25 ms
#define OC5F 0x08
#pragma interrupt_handler TOC5handler()
void TOC5handler(void){
   TFLG1= OC5F;       // Acknowledge
   TOC5 = TOC5+Rate; // every 25 ms
   if (++FourHundred==400){
      Freq = Count;   //  0.1 Hz units
      FourHundred=0;
      Done = 0xff;
      Count = 0; }}  // Setup for next
void ritual(void) {
asm(" sei");     // make atomic
   TMSK1|=OC5F+IC1F; // Arm OC5 and IC1
   TOC5=TCNT+Rate;  // First in 25 ms
   TCTL2 = (TCTL2&0xCF)|0x10;
```

```
// MC68HC812A4
#pragma interrupt_handler TIC1handler()
void TIC1handler(void){
   Count++;        // number of rising edges
   TFLG1=0x02;}  // ack, clear C1F
#define Rate 50000  //  25 ms
#pragma interrupt_handler TOC5handler()
void TOC5handler(void){
   TFLG1= 0x20;      // Acknowledge
   TC5 = TC5+Rate;   // every 25 ms
   if (++FourHundred==400){
      Freq = Count;   //  0.1 Hz units
      FourHundred=0;
      Done = 0xff;
      Count = 0; }}  // Setup for next
void ritual(void) {
asm(" sei");     // make atomic
   TIOS| = 0x20;   // enable OC5
   TSCR  = 0x80;   // enable
   TMSK2 = 0x32;   // 500 ns clock
   TMSK1 = 0x22;   // Arm OC5 and IC1
   TCTL4 = (TCTL4&0xF3)|0x04;
/* C1F set on rising edges */
```

```
/* IC1F set on rising edges */              TC5=TCNT+Rate;  // First in 25 ms
    Count = 0;         // Set up for first      Count = 0;          // Set up for first
    Done+0;  // Set on subsequent meas         Done=0;  // Set on subsequent meas
    FourHundred=0;                             FourHundred=0;
    TFLG1 = OC5F+IC1F; // Clear OC5F IC1F      TFLG1 = 0x22;    // Clear C5F,C1F
asm(" cli"); }                              asm(" cli"); }
```

Program 6.36 C language implementation of frequency measurement.

6.6   Glossary

Acknowledge Clearing the interrupt flag bit that requested the interrupt.

Arm Activate so that interrupts are requested.

Atomic A software action that once started will complete without interruption.

Background mode A 6812 mode with the BDM active.

Disarm Deactivate so that interrupts are not requested.

Duty cycle For a periodic digital wave, it is the percentage of time the signal is high.

Fast clear A 6812 timer mode where the associated flag is automatically cleared when the timer register is accessed.

Input capture A mechanism to set a flag and capture the current time (TCNT value) on the rising, falling, or rising and falling edges of an external signal. The input capture event can also request an interrupt.

Interrupt flag A status bit that is set by the timer hardware to signify an external event has occurred.

Interrupt mask A control bit that, if programmed to 1, will cause an interrupt request when the associated flag is set. Same as **arm.**

Latch As a noun, it means a register. As a verb, it means to store data into the register.

Latency Time delay from hardware event requiring service and the execution of the software service.

Output compare A mechanism to cause a flag to be set and an output pin to change when the TCNT matches a preset value. The output compare event can also request an interrupt.

Overflow When TCNT increments from $FFFF back to $0000, setting the TOF flag. This overflow event can also request an interrupt.

Poll for 0s and 1s An interrupt handler that checks both for the interrupt flag and the presence of other 1s and 0s in the status register.

Precision For an input signal, it is the number of distinguishable input signals that can be reliably detected by the measurement. For an output signal, it is the number of different output parameters that can be produced by the system.

Private variable A global variable that is used by a single thread and not shared with other threads.

Pulse-width modulation (PWM) A technique to deliver a variable signal (voltage, power, energy) using an on/off signal with a variable percentage of time the signal is on (duty cycle). Same as **variable duty cycle.**

Range Includes both the smallest possible and the largest possible signal (input or output).

Resolution For an input signal, it is the smallest change in the input parameter that can be reliably detected by the measurement. For an output signal, it is the smallest change in the output parameter that can be produced by the system.

Simple poll An interrupt handler that simply checks the interrupt flag.

6.7 Exercises

6.1 The objective of this problem is to measure the frequency of a square wave connected to IC1. You may only use IC1 and OC5 (i.e., no other 6811/6812 I/O feature or external device is allowed). The frequency range is 0 to 2000 Hz, and the *resolution is 0.1 Hz*. For example, if the frequency is 567.83 Hz, then your software will set the global Freq to 5678. The C program in Section 6.3.2 measures frequency with units of 100 Hz. Make modifications to this program so that the resolution is improved to 0.1 Hz. Don't worry about frequencies above 2000 Hz.

6.2 When a debugger wishes to single-step a program that exists in RAM, it can replace opcodes one at a time with SWI (it will have to replace two opcodes when single-stepping a conditional branch). This approach is not feasible when testing software stored in ROM or PROM. In this exercise you will use output compare interrupts to implement single-step debugging. Your approach will work for programs in RAM or ROM.

a) Write a debugging function that initializes the OC5 interrupt then calls the UserRoutine. The first OC5 interrupt should occur after exactly one instruction of the UserRoutine has been executed. You may assume the UserRoutine has no I/O parameters. You may also assume that UserRoutine has no interrupts of its own and it does not disable interrupts. When the UserRoutine returns back to your function, you should shut off OC5. You can start with the following syntax and add the OC5 code. You may write your answer in assembly or C.

```
// Single step in C                      * Single Step in assembly
void debug(void (*UserRoutine)(void)){   debug    * X points to UserRoutine
// add stuff here to initialize OC5       * stuff here to initialize OC5
   (*UserRoutine)();                              jsr 0,X    call UserRoutine
// add stuff here to stop OC5             * add stuff here stop OC5
}                                          rts
```

You are given two functions that you will call but do not need to write.

```
char GetChar(void); // waits for keyboard input and returns the ASCII code
void Display(void); // displays debugging information like registers
```

b) Write the OC5 interrupt handler that calls Display then GetChar. In a real debugger we would process the keyboard input and interact with the user, but in this simple solution the keyboard input is used only to pause. Before returning from interrupt you should set up OC5 so that the 6811/6812 will execute exactly one more instruction of UserRoutine before another OC5 interrupt is generated.

6.3 A microcomputer-based PID controller requires a frequency measurement to be performed every 20 ms. This problem will investigate three techniques:

A. Direct frequency measurement
B. Period measurement
C. A combined frequency/period measurement

The frequency range is 100 to 1000 Hz, and we wish to obtain the best frequency resolution possible under the constraint that a new frequency measurement must be available every 20 ms for the PID controller. Each method has two interrupt handlers:

1. A 20-ms real time clock using OC5
2. Falling edge of the square wave using 6811 STRA, or 6812 PJ0 key wake up.

There are five 16-bit unsigned global variables used by the three techniques:

<table>
<tr><td>FREQ is the frequency in hertz calculated during the OC5 handler</td><td>used by A, B, C</td></tr>
<tr><td>CNT is the number of falling edges in the current 20-ms interval</td><td>used by A, C</td></tr>
<tr><td>FIRST is the TCNT value at the time of the first falling edge</td><td>used by C</td></tr>
<tr><td>PREVIOUS is the TCNT value at the time of the previous falling edge</td><td>used by B</td></tr>
<tr><td>LAST is the TCNT value at the time of the last falling edge</td><td>used by B, C</td></tr>
</table>

A. Direct frequency measurement   The ritual performs:

```
CNT=0;
```

The 6811 STRA or 6812 key wake-up interrupt handler performs:

```
// 6811 code                              // 6812 code
   Read PIOC,Read Port CL /* Ack */          KWIFJ=0x01; // ack
   CNT=CNT+1;                                 CNT=CNT+1;
```

The OC5 interrupt handler performs:

```
Write OC5F=1;                    /* Acknowledge OC5 */
TOC5=TOC5+40000;
Calculates FREQ=50·CNT;
CNT=0;
```

The frequency resolution is 50 Hz regardless of the square-wave frequency.

B. Period measurement   The ritual performs:

```
PREVIOUS =LAST =0;
```

The 6811 STRA or 6812 key wake-up interrupt handler performs:

```
// 6811 code                              // 6812 code
   Read PIOC,Read Port CL /* Ack */          KWIFJ=0x01; // ack
   PREVIOUS=LAST                             PREVIOUS=LAST
   LAST=TCNT;                                LAST=TCNT;
```

The OC5 interrupt handler performs:

```
Write OC5F=1;                    /* Acknowledge OC5 */
TOC5=TOC5+40000;
Calculates FREQ=???? (answered in part a)
PREVIOUS =LAST =0;
```

C. Combined frequency/period measurement   The ritual performs:

```
FIRST=LAST=0;
CNT=-1;
```

The 6811 STRA or 6812 key wake-up interrupt handler performs:

```
// 6811 code                              // 6812 code
   Read PIOC,Read Port CL /* Ack */          KWIFJ=0x01; // ack
   If (CNT==-1)   FIRST=TCNT;                If (CNT==-1)   FIRST=TCNT;
   else           LAST=TCNT;                 else           LAST=TCNT;
   CNT=CNT+1;                                CNT=CNT+1;
```

The OC5 interrupt handler performs:

```
Write OC5F=1;   /* Acknowledge OC5 */
TOC5=TOC5+40000
Calculates FREQ=????(answered in part d)
FIRST=LAST=0;
CNT=-1;
```

The hardware for each method is shown in Figure 6.29.

Figure 6.29
Interface for Exercise 6.3.

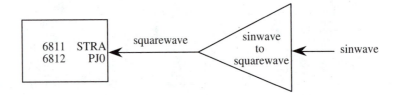

a) Specify the equation used to calculate `FREQ` from `PREVIOUS` and `LAST` in the **Period measurement** OC5 handler (no software is required).
b) What is the frequency resolution with units when the square-wave frequency is 100 Hz?
c) What is the frequency resolution with units when the square-wave frequency is 1000 Hz?
d) Specify the equation used to calculate `FREQ` from `FIRST LAST` and `CNT` in the **Combined frequency/period measurement** OC5 handler (no software is required).
e) What is the frequency resolution with units when the square-wave frequency is 100 Hz?
f) What is the frequency resolution with units when the square-wave frequency is 1000 Hz?

6.4 The objective of this problem is to measure body temperature using input capture (pulse-width measurement) (Figure 6.30). A shunt resistor is placed in parallel with a thermistor. The thermistor-shunt combination R has the following linear relationship for temperatures from 90 to 110°F.

$$R = 100k\Omega - (T - 90°F)\cdot 1k\Omega/°F \quad \text{where R is the resistance of the thermistor-shunt}$$

Figure 6.30
Interface for Exercise 6.4.

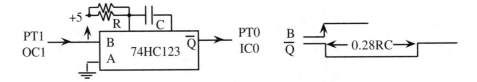

In other words, the resistance varies from 100kΩ to 80kΩ as the temperature varies from 90 to 110°F. The range of your system is 90 to 110°F and the resolution should be better than 0.01°F. You will use a 74HC123. On the rise of the B trigger input, the 123 creates a negative logic pulse on its output. The width of the pulse is about 0.28·R·C. Temperature will be measured 100 times a second. For the 6811 connect to PA3/OC5 and PA2/IC1.

a) The pulse-width measurement resolution will be 500 ns for the 6811 and 125 ns for the 6812. Choose the capacitor value so that this pulse width measurement resolution matches the desired temperature resolution of 0.01°F.
b) Given this value of C, what is the pulse width at 90°F? Give the answer in both microseconds and E clock cycles.

c) Given this value of C, what is the pulse width at 110°F? Give the answer in both microseconds and E clock cycles.

d) Write the ritual that configures an output compare on PT1 as a periodic interrupt. A rising edge should occur every 10 ms. Configure an input capture on PT0 to measure the pulse width.

e) Show the interrupt handler(s) that perform the temperature measurement tasks in the background and sets a global, `Temperature`, 100 times a second. The units of this unsigned decimal fixed point number are 0.01°F. For example, `Temperature` will vary from 9000 to 11000 as temperature varies from 90 to 110°F. The output compare interrupt will start the pulse (You may assume the interrupt vectors are properly established, but please use simple interrupt handler names like `TOC1handler()` and `TIC0handler()`.

6.5. Design a wind direction measurement instrument using the input capture technique. Again, you are given a transducer that has a resistance that is linearly related to the wind direction. As the wind direction varies from 0 to 360 degrees, the transducer resistance varies from 0 to 1000Ω. The frequencies of interest are 0 to 0.5 Hz, and the sampling rate will be 1 Hz. One way to interface the transducer to the computer is to use an astable multivibrator like the 555. The period of a 555 timer is $0.693 \cdot C_T \cdot (R_A + 2R_B)$. The 555 output could be connected to the Input Capture Port channel 7. (See Exercise 12.10.)

a) Show the hardware interface.

b) Write the ritual and gadfly function/subroutine that measures the wind direction and returns a 16-bit unsigned result with units of degrees (i.e., the value varies from 0 to 359). (You do not have to write software that samples at 1 Hz, simply a function that measures wind direction once.)

6.6 The objective of this problem is to design an underwater ultrasonic ranging system. The distance to the object, **d,** can vary from 1 to 100 m. The ultrasonic transducer will send a short 5-μs sound pulse into the water in the direction of interest. The sound wave will travel at 1500 m/s and reflect off the first object it runs into. The reflected wave will also travel at 1500 m/s back to the transducer. The reflected pulse is sensed by the same transducer. Your system will trigger the electronics (give a 5-μs digital pulse), measure the time of flight, then calculate the distance to the object. Using periodic interrupts, the software will issue a 5-μs pulse out PT1 once a second. Using interrupting input capture, the software will measure the time of flight, Δt. The input capture interrupt handler will calculate distance **d** as a decimal fixed-point value with units of 0.01 m and enter it into a FIFO queue. The main program will call the ritual, then get data out of the FIFO queue. The main program will call `Alarm()` if the distance is less than 15 m. You do not have to give the implementation of `Alarm()`. You may use any of the FIFOs in Chapter 4 without showing its implementation. To solve this problem on other microcomputers, you may modify the hardware in Figure 6.31 to conform to the available ports. As shown in the figure the time of flight, Δt, is measured from the rise of PT1 (6812 output) to the rise of PT0 (6812 input).

Figure 6.31
Interface for Exercise 6.6.

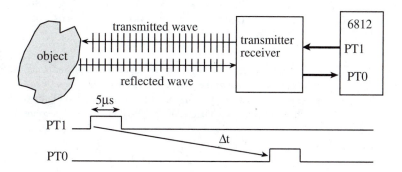

a) Derive an equation that relates the distance **d** to the time of flight **Δt**.

b) Use this equation to calculate the minimum and maximum possible time of flight **Δt**.

c) Choose the TCNT rate that will satisfy the range and resolution requirements of this problem.

d) Give the ritual that initializes the interface, including PORTT timer channels 0 and 1. Don't worry about the other six timer channels.

e) Give the main program that first calls the ritual. The main program will empty the FIFO and call Alarm() if the distance drops below 15 m. Decide whether it is better to convert time (TCNT counts) to distance (decimal fixed point, 0.01m) here in the main program or in the interrupt handler.

f) Give the TC1handler() periodic interrupt handler that issues pulses of about 5-μs duration on PT1 every 1 sec. Good interrupt software has no backward jumps.

g) Give the TC0handler() interrupt handler that measures the time of flight and puts the result (either the count or the calculated distance) into the FIFO. Good interrupt software has no backward jumps. There may be additional sonic echoes, so only calculate the range of the first one and ignore the others.

6.7 The objective of this exercise is to interface a silicon-controlled rectifier (SCR) using input capture and/or output compare. A 120-Hz digital logic waveform (**Sync**) is available that is synchronized to the zero crossings of the 60-Hz alternating current wave. A 100-μs pulse on the digital logic **Control** signal will turn on the SCR. The SCR will automatically shut off on the next zero crossing of the 60-Hz wave. **Sync** will be an input to the microcomputer, and **Control** will an output (Figure 6.32).

Figure 6.32
Timing for Exercise 6.7.

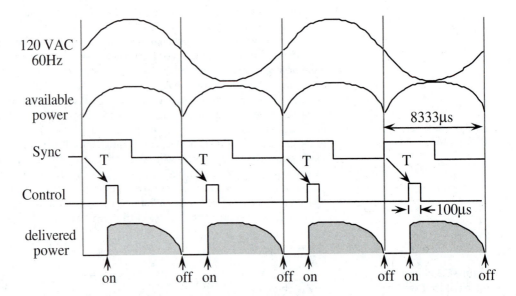

The software controls the amount of delivered power by adjusting the time **T**, which is the delay from the rising edge of the **Sync** input to the rising edge of **Control** output. To solve this problem on other microcomputers, you may modify the hardware in Figure 6.33 to conform to the available ports. On the 6811 Sync is connected to IC2/PA1 and Control is connected to OC2/PA6.

Figure 6.33
Interface for Exercise
6.7.

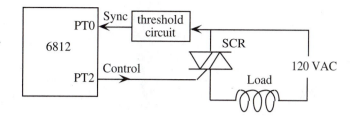

The delay **T** will vary from 300 to 8000 μs. When **T** is 300 μs, full power is being delivered. When **T** is 8000 μs, almost no power is delivered to the load. A 16-bit unsigned global variable

```
unsigned int T; // delay 2400 to 64000 in 125 ns clock cycles
```

will be set by the main program (which you will not write) and read by the interrupt software (which you will write).

   a) Give the ritual which initializes the interface. You may use any available feature of your microcomputer.
   b) Give the `TC2handler()` and `TC0handler()` interrupt handlers that implement this interface. Good interrupt software has no backward jumps.
   c) This is indeed an example of a real-time system. Give the upper bound on the latency of the `TC0handler()`. Don't calculate what it actually is but rather determine theoretically how fast it must be to work properly.

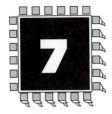

7 Serial I/O Devices

Chapter 7 objectives are to:

❏ Discuss fundamental concepts associated with serial communication systems: asynchronous, synchronous, bandwidth, full-duplex, half-duplex and simplex
❏ Present the RS232 and RS422 communication protocols, signals, and interface chips
❏ Write low-level device drivers that perform basic I/O with serial ports
❏ Discuss interfacing issues associated with performing the I/O as a background interrupt thread

In many applications, a single dedicated microcomputer is insufficient to perform all the tasks of the embedded system. One solution would be to use a larger and more powerful microcomputer, and another approach would be to distribute the tasks among multiple microcomputers. This second approach has the advantages of modularity and expandability. To implement a distributed system, we need a mechanism to send and receive information between the microcomputers. A second scenario that requires communication is a central general-purpose computer linked to multiple remote embedded systems for the purpose of distributed data collection or distributed control. For situations where the required bandwidth is less than about 1000 bytes/s, the built-in serial ports of the microcomputer can be used.

Chapter 7 deals with external devices that we connect to the serial I/O ports of our computer. In particular, we will interface terminals, keyboards, displays, printers, and other computers. In this chapter we will focus on serial channels that employ a direct physical connection between the microcomputers; later (in Chapter 14) we expand the communication system to include networks and modems.

7.1 Introduction and Definitions

Serial communication involves the transmission of one bit of information at a time. One bit is sent, a time delay occurs, then the next bit is sent. This section will introduce the use of serial communication as an interfacing technique for various microcomputer peripherals. Since many peripheral devices such as printers, keyboards, scanners, and mice have their own computers, the communication problem can be generalized to one of transmitting information between two computers. The *universal asynchronous receiver/transmitter (UART)* is the interface chip that implements the serial data transmission. Motorola calls it an *asynchronous communications interface adapter (ACIA)* or *SCI*. The *serial channel* is the collection of signals (or wires) that implement the communication. To improve bandwidth, remove noise, and increase range, we place interface logic between the digital logic UART device and the serial channel. We define the *data terminal equipment (DTE)* as the computer or a terminal and the *data communication equipment (DCE)* as the modem or printer (Figure 7.1).

Figure 7.1
A serial channel connects a DTE to a DCE.

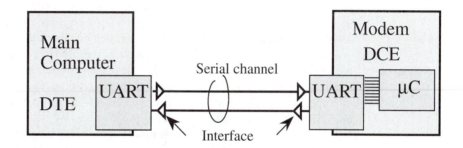

The interface logic (e.g., MC1488/MC1489, MAX232, or MC145407) converts between TTL/MOS/CMOS logic levels and RS232 logic levels. In this protocol, a typical bidirectional channel requires three wires (RxD, TxD, ground). We use RS422 voltage levels when we want long cable lengths and high bandwidths. The binary signal is encoded on the RS422 line as a voltage difference (MC3486/3487, MAX485, 3691, 3695, 8921/8922/8923, 75176, or 78120). In this protocol, a typical bidirectional channel requires five wires (RxD^+, RxD^-, TxD^+, TxD^-, ground). Typical voltage levels are shown in Table 7.1.

Table 7.1
Voltage levels for the CMOS RS232 and RS422 protocols.

		Typical CMOS Level	Typical RS232 Level	Typical RS422 Level
True	Mark	+5 V	TxD = -12 V	$(TxD^+ - TxD^-) = -3$ V
False	Space	+0.1 V	TxD = $+12$ V	$(TxD^+ - TxD^-) = +3$ V

A *frame* is a complete and nondivisible packet of bits. A frame includes both information (e.g., data, characters) and overhead (start bit, error checking, and stop bits). A frame is the smallest packet that can be transmitted. The RS232 and RS422 protocols have one

Figure 7.2
A RS232 frame showing one start, seven data, one parity, and two stop bits.

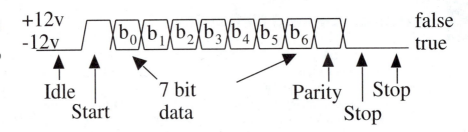

start bit, seven/eighth data bits, no/even/odd *parity,* and one/1.5/two *stop bits* (Figure 7.2). The RS232 idle level is true (-12 V). The start bit is false ($+12$ V). A true data bit is -12 V, and a false data bit is $+12$ V.

> **Observation:** RS232 and RS422 data channels are in negative logic because the true voltage is less than the false voltage.

Parity is generated by the transmitter and checked by the receiver. Parity is used to detect errors. For *even parity,* the number of 1s in the data plus parity is an even number. For *odd parity,* the number of 1s in the data plus parity is an odd number. If errors are unlikely, then operating without parity is faster and simpler. The *bit time* is the basic unit of time used in serial communication. It is the time between each bit. The transmitter outputs a bit, waits one bit time, then outputs the next bit. The start bit is used to synchronize the receiver with the transmitter. The receiver waits on the idle line until a start bit is first detected. After the true-to-false transition, the receiver waits a half a bit time. The half a bit time wait places the input sampling time in the middle of each data bit, giving the best tolerance to variations between the transmitter and receiver clock rates. We will discuss the detailed timing later in the chapter. Next, the receiver reads one bit every bit time. The *baud rate* is the total number of bits (information, overhead, and idle) per time that is transmitted in the serial communication. Later in Chapter 14, we will define the baud rate of a modem as the number of sounds per second. But for now, we define,

$$\text{Baud rate} = \frac{1}{\text{bit time}}$$

We will define *information* as the data that the "user" intends to be transmitted by the communication system. Examples of information include:

- Characters to be printed on your printer
- A picture file to be transmitted to another computer
- A digitally encoded voice message communicated to your friend
- The object code file to be downloaded from the PC to the 6805/6808/6811/6812

We will define *overhead* as signals added by the "operating system" to the communication to affect reliable transmission. Examples of overhead include:

- Start bit(s), start byte(s), or start code(s)
- Stop bit(s), stop byte(s), or stop code(s)
- Error checking bits such as parity, cyclic redundancy check (CRC), and checksum
- Synchronization messages like ACK, NAK, XON, XOFF

Although, in a general sense overhead signals contain "information," overhead signals are not included when calculating bandwidth or considering full-duplex, half-duplex, or simplex.

An important parameter in all communication systems is *bandwidth.* We will use the three terms bandwidth, *bit rate,* and *throughput* interchangeably to specify the number of information bits per time that is transmitted. These terms apply to all forms of communication:

- Parallel
- Serial
- Mixed parallel/serial

For serial communication systems, we can calculate:

$$\text{Bandwidth} = \frac{\text{number of information bits/frame}}{\text{total number of bits/frame}} \times \text{baud rate}$$

A *full-duplex communication system* allows information (data, characters) to transfer simultaneously in both directions. A *full-duplex channel* allows bits (information, error checking, synchronization, or overhead) to transfer simultaneously in both directions (Figure 7.3).

Figure 7.3
A full-duplex serial channel connects two DTEs (computers).

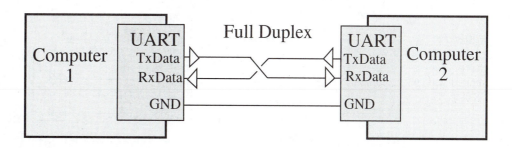

A *half-duplex communication* system allows information to transfer in both directions, but in only one direction at a time. Half-duplex is a term usually defined for modem communications, but in this book we will expand its meaning to include any serial protocol that allows communication in both directions, but in only one direction at a time. A fundamental problem with half-duplex is the detection and recovery from a collision. A collision occurs when both computers simultaneously transmit data. Fortunately, every transmission frame is echoed back into its own receiver. The transmitter program can output a frame, wait for the frame to be transmitted (which will be echoed into its own receiver), then check the incoming parity and compare the data to detect a collision. If a collision occurs, then it probably will be detected by both computers. After a collision, the transmitter can wait awhile and retransmit the frame. The two computers need to decide which one will transmit first after a collision so that a second collision can be avoided. The first hardware mechanism to implement half-duplex utilizes tristate logic. In this system, the transmitter will enable the driver ($\overline{\text{RTS}}$ =0) before transmission, then disable the driver after complete transmission ($\overline{\text{RTS}}$ =1).

Observation: People communicate in half-duplex.

When using the SCI (built into most of the 6805, 6808, 6811, and 6812 computers), complete transmission occurs when the SCI **Receive Data Register Full** flag is set (RDRF=1) and not when the SCI **Transmit Data Register Empty** flag is set (TDRE=1). The SCI Transmit Complete Flag (TC) should be set approximately at the same time as the

Figure 7.4
A half-duplex serial channel can be implemented with tristate logic.

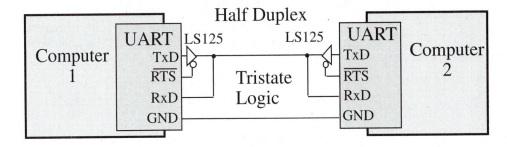

RDRF, because TC=1 means a frame has been shifted out, and RDRF=1 means a frame has been shifted in (Figure 7.4).

Another hardware mechanism for half-duplex utilizes open-collector logic (Figure 7.5). The 7407 driver has two output states: zero and hiZ. The logic high is created with the passive pull-up. With open collector, the half-duplex channel is the logical AND of the two TxD outputs. In this system, the transmitter simply transmits its frame without needing to enable or disable the driver.

Figure 7.5
A half-duplex serial channel can be implemented with open-collector logic.

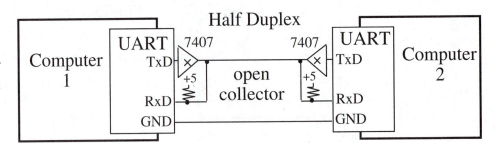

One application of the half-duplex protocol is the desktop serial bus, or multidrop network (Figure 7.6). In the following example, each microcomputer can send a message to another microcomputer. The 6811 Port D outputs can be made open-collector by setting DWOM=1. The 6812 serial port output can be made open-collector by setting WOMS=1. All the grounds are tied together. Assume there are less than 128 microcomputers in this network so that each microcomputer can have a unique 7-bit address. A frame with bit 7 equal to 1 is defined as an address; otherwise, the frame is data. The transmitting microcomputer first sends the address of the destination microcomputer, followed by multiple data frames. All the receiver microcomputers listen for their particular address. Once a microcomputer receiver recognizes its address, it will accept the data frames that follow. There are various options for determining the end of the message:

1. Define each message to be a fixed length
2. Send the message length as the second character
3. Define a special character that specifies end of message

A *simplex communication* system allows information to transfer in only one direction. The XON/XOFF protocol that we will cover later is an example of a communication system that has a full-duplex channel but implements simplex communication. This is because with XON/XOFF information (characters) are transmitted from the computer to the printer,

Figure 7.6
A desktop network is created using a half-duplex serial channel implemented with open-collector logic.

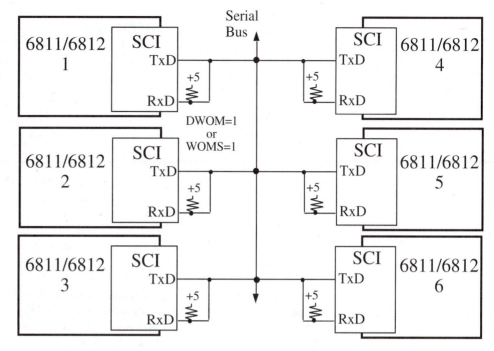

but only XON/XOFF (error checking) flags are sent from the printer back to the computer. In this case, no information (data, characters) is sent from the printer to the computer. Figure 7.7 shows a simplex serial channel.

One application of simplex channels is the ring network. Rather than using a common bus like Figure 7.6, the six microcomputers could be connected in a ring topology using six simplex channels like Figure 7.7. The system could use the same address/data format described for the multidrop half-duplex connection. If a microcomputer receives a message addressed to a different node, it simply retransmits it along the ring. This system is slower than the multidrop system, but it does not have to contend with collision errors.

Figure 7.7
A simplex serial channel between two computers.

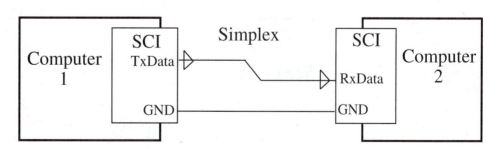

To transfer information correctly, both sides of the channel must operate at the same baud rate. In an *asynchronous* communication system, the two devices have separate and distinct clocks (Figure 7.8). Because these two clocks are generated separately (one on each side), they will not have exactly the same frequency or be in phase. If the two baud rate clocks have different frequencies, the phase between the clocks will also drift over time.

Figure 7.8
Nodes on an
asynchronous channel
run at the same
frequency but have
separate clocks.

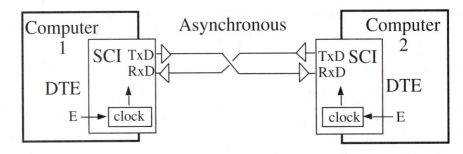

Transmission will occur as long as the periods of the two baud rate clocks are close enough. The −12 V to +12 V edge at the beginning of the start bit is used to synchronize the receiver with the transmitter. If the two periods in a RS232 system differ by less than 5%, then after 10 bits the receiver will be off by less than half a bit time (and no error will occur). Any larger difference between the two periods will cause an error.

In a *synchronous* communication system, the two devices share the same clock (Figure 7.9). Typically a separate wire in the serial cable carries the clock. In this way, very high baud rates can be obtained. Another advantage of synchronous communication is that very long frames can be transmitted. Larger frames reduce the OS overhead for long transmissions because fewer frames need be processed per message. Even though in this chapter we will design various low-bandwidth synchronous systems using the SPI, synchronous communication is best applied to systems that require bandwidths above 1 Mb/s. The cost of this increased performance is the additional wire in the cable. The clock must be interfaced with channel drivers (e.g., RS232, RS422, optocouplers) similar to the transmit and receiver data signals.

Figure 7.9
Nodes on a synchronous
channel operate off a
common clock.

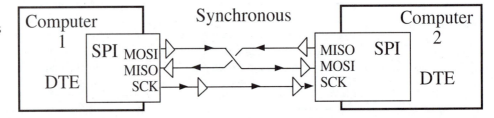

7.2 RS232 Specifications

The baud rate in a RS232 system can be as high as 20,000 bits/s. Because a typical cable has 50 pF/ft, the maximum distance for RS232 transmission is limited to 50 ft. There are 21 signals defined for full modem (*MO*dulate/*DEM*odulate) communication (Figure 7.10).

Figure 7.10
DB25 (RS232), DB9 (EIA-574), and RJ45 (EIA-561) connectors used in many serial applications.

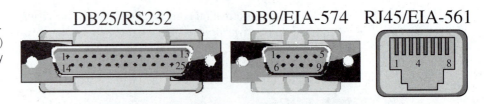

DB25 Pin	RS232 Name	DB9 Pin	EIA-574 Name	RJ45 Pin	EIA-561 Name	Signal	Description	True (V)	DTE	DCE
1						FG	Frame Ground/Shield			
2	BA	3	103	6	103	TxD	Transmit Data	−12	Out	In
3	BB	2	104	5	104	RxD	Receive Data	−12	In	Out
4	CA	7	105/133	8	105/133	RTS	Request to Send	+12	Out	In
5	CB	8	106	7	106	CTS	Clear to Send	+12	In	Out
6	CC	6	107			DSR	Data Set Ready	+12	In	Out
7	AB	5	102	4	102	SG	Signal Ground			
8	CF	1	109	2	109	DCD	Data Carrier Detect	+12	In	Out
9							Positive Test Voltage			
10							Negative Test Voltage			
11							Not Assigned			
12						sDCD	Secondary DCD	+12	In	Out
13						sCTS	Secondary CTS	+12	In	Out
14						sTxD	Secondary TxD	−12	Out	In
15	DB					TxC	Transmit Clk (DCE)		In	Out
16						sRxD	Secondary RxD	−12	In	Out
17	DD					RxC	Receive Clock		In	Out
18	LL						Local Loopback			
19						sRTS	Secondary RTS	+12	Out	In
20	CD	4	108	3	108	DTR	Data Terminal Rdy	+12	Out	In
21	RL					SQ	Signal Quality	+12	In	Out
22	CE	9	125	1	125	RI	Ring Indicator	+12	In	Out
23						SEL	Speed Selector DTE		In	Out
24	DA					TCK	Speed Selector DCE		Out	In
25	TM					TM	Test mode	+12	In	Out

Table 7.2 Pin assignments for the RS232 EIA-574 and EIA-561 protocols.

Table 7.2 shows the entire set of RS232 signals. The RS232 standard uses a DB25 connector that has 25 pins. The EIA-574 standard uses RS232 voltage levels and a DB9 connector that has only nine pins. The EIA-561 standard also uses RS232 voltage levels but with a RJ45 connector that has only eight pins. The most commonly used signals of the full RS232 standard are available with the EIA-561/EIA-574 protocols.

The frame ground is connected on one side to the *ground shield* of the cable. The shield will provide protection from electric field interference. The *twisted cable* has a small area between the wires. The smaller the area, the less the magnetic field pickup. There is one disadvantage to reducing the area between the connectors. The capacitance to ground is inversely related to the separation distance between the wires. Thus as the area decreases, the capacitance will increase. This increased capacitive load will limit both the distance and the baud rate (Figure 7.11).

Figure 7.11
The simplest RS232 cable uses just T×D, R×D, and ground (with an optional ground shield).

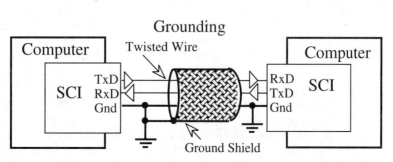

Grounding

The signal ground is connected on both sides to the supply return. A separate wire should be used for signal ground (do not use the ground shield to connect the two signal grounds). The noise immunity will be degraded if the ground shield is connected on both sides. There are many available RS232 driver chips, but the Maxim MAX232 (Figure 7.12) and Sipex SP232A are popular devices because of their low cost and simple implementation. These chips employ a charge pump (using the 100-nF capacitors) to create the standard +12 and −12 output voltages with only a +5-V supply.

Figure 7.12
RS232 interface to Motorola microcomputers.

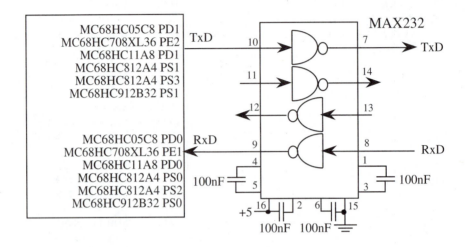

Figure 7.13 overviews the RS232 specifications for an output signal. Similarly, Figure 7.14 overviews the RS232 specifications for an input signal.

Figure 7.13
RS232 output specifications.

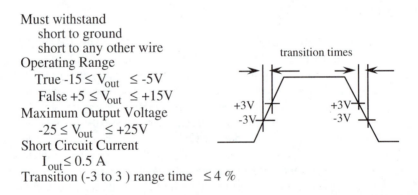

Must withstand
    short to ground
    short to any other wire
Operating Range
    True $-15 \leq V_{out} \leq -5V$
    False $+5 \leq V_{out} \leq +15V$
Maximum Output Voltage
    $-25 \leq V_{out} \leq +25V$
Short Circuit Current
    $I_{out} \leq 0.5$ A
Transition (-3 to 3) range time $\leq 4\%$

On the data lines, a true or mark voltage is negative. Conversely, for the control lines, a true signal has a positive voltage. *Request to Send (RTS)* is a signal from the computer to the modem requesting transmission be allowed. *Clear to Send (CTS)* is the acknowledge signal back from the modem signifying transmission can proceed. *Data Set Ready (DSR)* is a modem signal specifying that I/O can occur. *Data Carrier Detect (DCD)* is a modem signal specifying that the carrier frequencies have been established on its telephone line. *Ring Indicator (RI)* is a modem signal that is true when the phone rings. The *Receive Clock* and *Transmit Clock* are used to establish synchronous serial communication. *Data Terminal Ready (DTR)* is a printer signal specifying the status of the printer. This signal is sometimes called $\overline{Busy}$. When DTR is +12 V, the printer is ready and can accept more charac-

Figure 7.14
RS232 input
specifications.

Maximum Slew Rate
$dV_{in}/dt \leq 30V/\mu s$
Operating Range
True $-15 \leq V_{in} \leq -3V$
Transition $-3 < V_{in} < +3V$
False $+3 \leq V_{in} \leq +15V$
Input Resistance
$3000\Omega \leq R_{in} \leq 7000\ \Omega$
Input Capacitance including cable
$C_{in} \leq 2500\ pF$
Input open circuit voltage
$E_{in} \leq 2\ V$

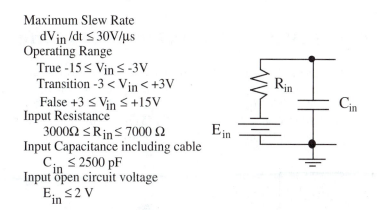

ters. Conversely when DTR is -12 V, the printer is busy and cannot accept more characters. None of the built-in serial ports of the Motorola microcomputer discussed in this book supports these control lines explicitly. On the other hand, it is straightforward to implement these hardware handshaking signals using simple I/O lines. If we wish to generate interrupts on edges of these lines, then we would use input capture or key wake-up features.

A typical sequence for initiating communication between a computer (DTE) and a modem (DCE) begins with the turning on of the power in both devices. The computer activates the DTR line, and the modem responds by activating the DSR signal. Next, the modem establishes a link across the telephone line with the other modem. A functional link requires a separate carrier frequency from both modems. See Section 14.9 for additional details about the modem/modem link. The modem signals to the computer that a proper link has been established by activating DCD. When the computer wishes to transmit it, it activates RTS. If okay, the modem responds by activating CTS. After receiving the CTS, the computer can transmit data. The computer can continuously activate RTS if multiple frames are to be sent, but it must postpone transmission if CTS becomes false.

Faced with the design of a RS232 interface, "real engineers" would use one of the many RS232 chip available. The MC1488/MC1489 require ± 12-V supplies. Some devices (e.g., MAX232, MC145407) use a charge-pump mechanism so that they can operate on a single $+5$-V power supply; other devices (e.g., MC145406) have multiple bidirectional interfaces (Figure 7.15).

Figure 7.15
More RS232 interface
chips.

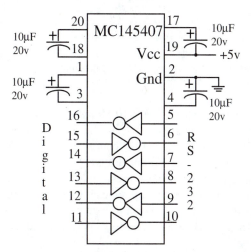
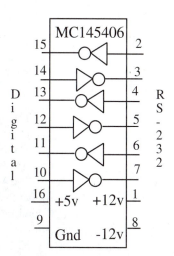

7.3   RS422/AppleTalk/RS423/RS485 Balanced Differential Lines

To increase the baud rate and maximum distance, the balanced differential line protocols were introduced. The RS422 signal is encoded in a differential signal, A−B. There are many RS422 interface chips, such as SP301, SP304, MAX486, MAX488, MC3486/3487, MC3691, MC3695, 8921/8922/8923, 75176, or 78120. A full-duplex RS422 channel is implemented in Figure 7.16 with Sipex SP301 drivers.

Figure 7.16
RS422 serial channel.

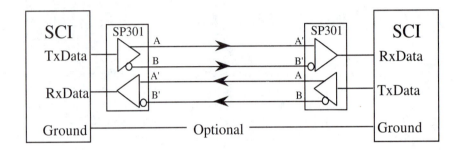

Because each signal requires two wires, five wires (ground included) are needed to implement a full-duplex channel. With RS232 one typically connects one receiver to one transmitter. But with RS422, up to ten receivers can be connected to one transmitter. Table 7.3 summarizes four common EIA standards.

Table 7.3
Specifications for the RS232, RS423A, RS422, and RS485 protocols.

Specification	RS232D	RS423A	RS422	RS485
Mode of operation	Single-ended	Single-ended	Differential	Differential
Drivers on one line	1	1	1	32
Receivers on one line	1	10	10	32
Maximum distance (ft)	50	4,000	4,000	4,000
Maximum data rate	20 Kbits/sec	100 Kbits/sec	10 Mbits/sec	10 Mbits/sec
Maximum driver output	±25 V	±6 V	-0.25 to $+6$ V	-7 to $+12$ V
Driver output (loaded)	±5 V	±3.6 V	±2 V	±1.5 V
Driver output (unloaded)	$\pm$ 15V	±6 V	±5 V	±5 V
Driver load impedance	$3k\Omega$ to $7k\Omega$	450Ω min	100Ω	54Ω
Receiver input voltage	±15 V	±12 V	±7 V	-7 to $+12$ V
Receiver input sensitivity	±3 V	±200 mV	±200 mV	±200 mV
Receiver input resistance	$3k\Omega$ to $7k\Omega$	$4k\Omega$ min	$4k\Omega$ min	$12k\Omega$ min

The maximum baud rate at 40 ft is 10 Mbits/sec. At 4000 ft, the baud rate can only be as high as 100 kbits/sec. Table 7.4 shows two implementations of the RS422 protocol.

The EIA-530 standard uses a DB25 connector that has 25 pins. The RS449 standard uses a DB37 connector that has 37 pins (Figure 7.17).

The Macintosh (except the iMac) uses AppleTalk for its serial interface. AppleTalk is similar to the RS422 standard in that the signal is encoded as a voltage difference. The DIN-8 connector for the Macintosh is shown in Figure 7.18. The AppleTalk TxD−, TxD+ are similar to the RS422 A,B signals, and the AppleTalk RxD−, RxD+ are similar to the RS422 A',B' signals.

Table 7.4
Pin assignments for the
EIA-530 and RS449
protocols.

DB25 Pin	EIA-530 Name	DB37 Pin	RS449 Name	Signal	Description	DTE	DCE
1		1		FG	Frame Ground/Shield		
2	BA (A)	4	SD (A)	T×D	Transmit Data	Out	In
14	BA (B)	22	SD (B)	T×D	Transmit Data	Out	In
3	BB (A)	6	RD (A)	R×D	Receive Data	In	Out
16	BB (B)	24	RD (B)	R×D	Receive Data	In	Out
4	CA (A)	7	RS (A)	RTS	Request to Send	Out	In
19	CA (B)	25	RS (B)	RTS	Request to Send	Out	In
5	CB (A)	9	CS (A)	CTS	Clear to Send	In	Out
13	CB (B)	27	CS (B)	CTS	Clear to Send	In	Out
6	CC (A)	11	DM (A)	DSR	Data Set Ready	In	Out
22	CC (B)	29	DM (B)	DSR	Data Set Ready	In	Out
20	CD (A)	12	TR (A)	DTR	Data Terminal Rdy	Out	In
23	CD (B)	30	TR (B)	DTR	Data Terminal Rdy	Out	In
7	AB	19	SG	SG	Signal Ground		
8	CF (A)	13	RR (A)	DCD	Data Carrier Detect	In	Out
10	CF (B)	31	RR (B)	DCD	Data Carrier Detect	In	Out
15	DB (A)	5	ST (A)	TxC	Transmit Clk (DCE)	In	Out
12	DB (B)	23	ST (B)	TxC	Transmit Clk (DCE)	In	Out
17	DD (A)	8	RT (A)	RxC	Receive Clock	In	Out
9	DD (B)	26	RT (B)	RxC	Receive Clock	In	Out
18	LL	10	LL		Local Loopback	Out	In
21	RL	14	RL	RL	Remote Loopback	Out	In
24	DA (A)	17	TT (A)	TCK	Speed Selector DCE	Out	In
11	DA (B)	35	TT (B)	TCK	Speed Selector DCE	Out	In
25	TM	18	TM	TM	Test mode	In	Out

Figure 7.17
DB25 (EIA-530), and
DB37 (RS-449)
connectors used in
RS422 applications.

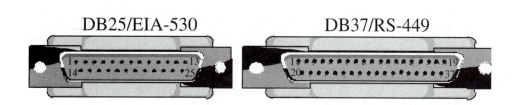

DB25/EIA-530   DB37/RS-449

Figure 7.18
Macintosh AppleTalk
DIN8 connector.

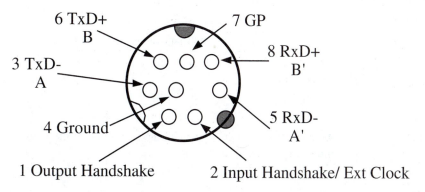

Looking into the male DIN-8 Connector

Table 7.5

Pin connections to interface an AppleTalk channel to a standard RS232 DB9 channel.

AppleTalk Pins	RS232 (DB-9) Pins	Comment
8−RxD+ 4-Gnd	5-Gnd	Two grounds connected and RxD+ is grounded
3−TxD−	2−RxD	Serial data from the Macintosh to the external device
5−RxD−	3−TxD	Serial data from the external device to the Macintosh

Table 7.5 shows the three-wire serial interface between a Macintosh AppleTalk DIN8 and a standard RS232 DB9 connector. This simple interface does not support hardware handshaking. Notice that the three pins DIN8-Pin8, DIN8-Pin4, and DB9-Pin5 are all connected together, but only one wire exists in the cable. This example illustrates how we interface a differential protocol with a single-ended protocol.

7.3.1
RS422 Output Specifications

The output voltage levels are shown in Table 7.6. A key RS422 specification is that the output impedances should be balanced. If the I/O impedances are balanced, then added noise in the cable creates a common-mode voltage, and the common-mode rejection of

Table 7.6

Output voltage levels for the RS422 differential line protocol.

	Output Voltage
True or Mark	$-6 \leq A-B \leq -2\,V$
Transition	$-2 \leq A-B \leq +2\,V$
False or Space	$+2 \leq A-B \leq +6\,V$

the input will eliminate it. More details about common mode are presented later in Chapters 11 and 12.

$$R_{Aout} = R_{Bout} \cdot 100\Omega$$

The time in the transition region must be less than

10% for baud rates above 5 Mb/s
20 ns for baud rates below 5 Mb/s

7.3.2
RS422 Input Specifications

The input voltage levels are as shown in Table 7.7.

Table 7.7

Input voltage thresholds for the RS422 differential line protocol.

	Input Voltage
True or Mark	$A-B \leq -0.2\,V$
Transition	$-0.2\,V \leq A-B \leq +0.2\,V$
False or Space	$+0.2\,V \leq A-B$

As mentioned earlier, to provide noise immunity the common-mode input impedance's must also be balanced

$$4k\Omega \leq R_{A'in} = R_{B'in}$$

The balanced nature of the interface produces good noise immunity. The differential input impedance is specified by the plot in Figure 7.19. Any point within the shaded region

Figure 7.19
RS422 input current versus input voltage relationship.

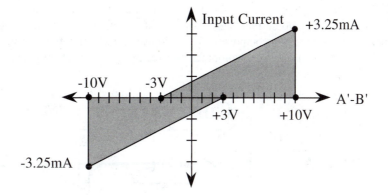

is allowed. Even though the ground connection in the RS422 cable is optional, it is assumed the grounds are connected somewhere. In particular, the interface will operate with a common-mode voltage up to 7 V.

$$\frac{|A' + B'|}{2} \leq + 7 \text{ V}$$

7.3.3
RS485 Half-Duplex Channel

RS485 can be either half-duplex or full-duplex. The RS485 protocol, illustrated in Figure 7.20, implements a half-duplex channel using differential voltage signals. The Sipex SP483 or Maxim MAX483 implements the half-duplex RS485 channel. One of the advantages of RS485 is that up to 32 devices can be connected onto a single serial bus. When more than one transmitter can drive the serial bus, the protocol is also called *multidrop*. To transmit the computer enables the driver by making DE active, then sends the serial frame from the TxD output of the SCI port. If RE is also active during transmission, the transmitted frame is echoed into the serial receiver of the SCI RxD line. To receive a frame the computer simply enables its receiver (by making RE active) and accepts a serial frame on the RxD line in the usual manner. Be careful when selecting the resistances on a half-duplex network so that the total driver impedance is about 54Ω.

Figure 7.20
A half-duplex serial channel is implemented with RS485 logic.

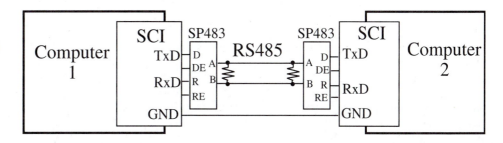

7.4 Other Communication Protocols

There are a variety of communication protocols for implementing communications channels. They vary in cost, distance, bandwidth, and noise immunity. Some protocols require the two computers to have a common ground, while others are isolated from each other. Isolation is important to prevent noise on one system from creating errors on another.

7.4.1
Current Loop
Channel

Current loop is an old standard where true is encoded as a 20-mA current and false is sig-nified by no current. The advantage of current loop is its inherent electrical isolation be-tween the two computers. We could use current loop in applications where we wished to prevent noise in one computer from coupling into the electronics of the other (Figure 7.21).

Figure 7.21
Current loop serial interface.

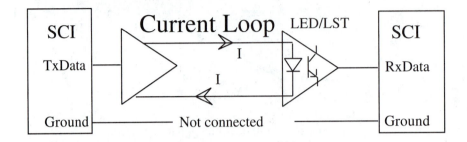

7.4.2
Introduction to
Modems

Detailed information about modems is presented later in Chapter 14. *Frequency-shift key-ing* (FSK) modems encode the binary data as frequencies that are transmitted as sounds on standard phone lines (Figure 7.22). The modulator converts the TxD data from the trans-mission computer into sounds. The phone line transmits the sounds to the receiver modem. The demodulator on the receiver converts the sound frequencies back into a regular true/false digital line. The SCI on the receiver accepts the serial frame in the usual way (start bit, data bits, stop bit).

Figure 7.22
Modem serial interface.

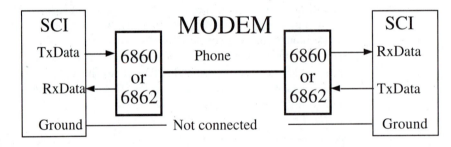

Two pairs of frequencies implement full-duplex at 300 bits/s (Table 7.8).

Table 7.8
Frequency parameters
for a 300 bits FSK
modem protocol.

Logic	Originate	Answer
True	1270 Hz	2225 Hz
False	1070 Hz	2025 Hz

The *phase-encoded modem* also uses standard phone lines. Phase-encoded modems provide for increased bandwidth by encoding multiple bits into periodic phase shifts of the sound signal. More details can be found in Section 14.9. Even higher bandwidth is imple-mented by combining phase and amplitude information. A significant amount of digital signal processing is required to produce reliable transmission at high speeds.

7.4.3
Optical Channel

A *fiber-optic light* channel uses a LED transmitter, a fiber-optic cable, and a light sensor. Binary information is encoded as the presence or absence of light in the cable (Figure 7.23). Fiber-optic cable is used in applications requiring long distances and/or high bandwidth. Similar to the current loop channel, fiber-optic cables provide electrical isolation between the two computers. Techniques for interfacing LEDs can be found in Section 8.2. Another advantage of the optical channel is its noise immunity. All the other protocols are to some degree susceptible to electric field noise. Communication using fiber optics is not affected by electric fields along the cable.

Figure 7.23
Fiber-optic serial interface.

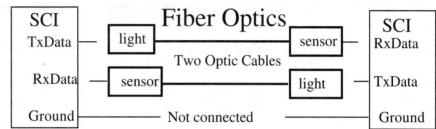

7.4.4
Digital Logic Channel

The simplest channel to implement uses standard *digital logic*. When the two computers are located in the same box, it is appropriate to send information directly from one to the other without special hardware circuits (Figure 7.24). Although standard digital logic will not be appropriate for long distances and/or in the presence of strong electric fields, it certainly is cheap and simple and should at least be considered for communication across short distances.

Figure 7.24
Simple digital logic serial interface.

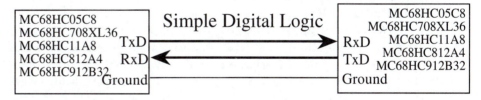

If isolation is required, then optoisolators like the 6N139 can be used (Figure 7.25).

Figure 7.25
Isolated digital logic serial interface.

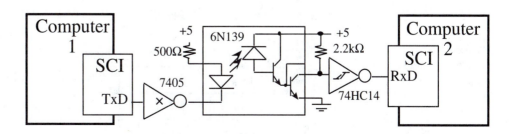

<table>
<tr><td>**7.5**</td><td></td></tr>
</table>

7.5 Serial Communications Interface

Most of the Motorola embedded microcomputers support at least one SCI. Before discussing the detailed operation of particular devices, we will begin with general features common to all devices. Common features that affect the entire SCI module include:

A baud rate control register used to select the transmission rate
A mode bit M used to select 8-bit (M=0) or 9-bit (M=1) data frames

Each device is capable of creating its own serial port clock with a period that is an integer multiple of the E clock period. The programmer will select the baud rate by specifying the integer divide-by used to convert the E clock into the serial port clock.

Common Error: If you change the E clock frequency without changing the baud rate register, the SCI will operate at an incorrect baud rate.

Table 7.9 lists the I/O port locations of the serial ports for the various microcomputers discussed in this book. The MC68HC812A4 has two serial ports.

Table 7.9
Serial port pins available on various Motorola microcomputers.

Microcomputer	Pin for TxD	Pin for RxD
MC68HC05C8	PD1	PD0
MC68HC708XL36	PTE2	PTE1
MC68HC11	PD1	PD0
MC68HC812A4	PS3/PS1	PS2/PS0
MC68HC912B32	PS1	PS0

7.5.1
Transmitting in Asynchronous Mode

We will begin with transmission, because it is straightforward. Common features that affect the transmitter portion of the SCI include (Figure 7.26):

TxD data output pin, with TTL voltage levels
10- or 11-bit shift register, which cannot be directly accessed by the programmer; this shift register is separate from the receive shift register
Serial Communications Data Register (SCDR), which is write only, even though at the same address this data register is separate from the receive data register
T8 data bit that you set before writing to SCDR when 9-bit data mode (M=1) is used

The control bits that affect the transmitter are:

Transmit Enable control bit (TE), which you initialize to 1 to enable the transmitter
Send Break control bit (SBK), which you set to 1 to send blocks of 10 or 11 zeros
Transmit Interrupt Enable control bit (TIE), which you set to 1 to arm the TDRE flag
Transmit Complete Enable control bit (TCIE), which you set to 1 to arm the TC flag

Figure 7.26
Data register and shift register used to implement the transmit serial interface.

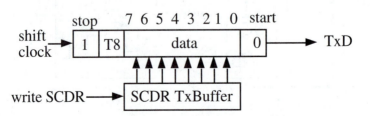

The status bits generated by transmitter activity are:

Transmit Data Register Empty flag (TDRE), which is set when the transmit SCDR is empty; TDRE is cleared by reading the TDRE flag (with it set), then writing to the SCDR

Transmit Complete flag (TC), which is set when the transmit shift register is done shifting; TC is cleared by reading the TC flag (with it set), then writing to the SCDR

When new data (8 bits) are loaded in the SCDR, they are copied into the 10- or 11-bit transmit shift register. Next, the start bit, T8 (if M=1), and stop bits are added. Then, the frame is shifted out one bit at a time at a rate specified by the baud rate register. If there are already data in the shift register when the SCDR is written, it will wait until the previous frame is transmitted before it, too, is transferred. In the timing diagrams of Figures 7.27 and 7.28, the "T" arrows refer to times when the transmit shift register is shifted, causing a change in the TxD output pin.

Figure 7.27
Transmit data frames for M=0 and M=1.

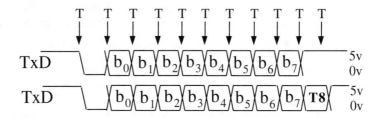

Figure 7.28
Start bit timing during a transmit data frame.

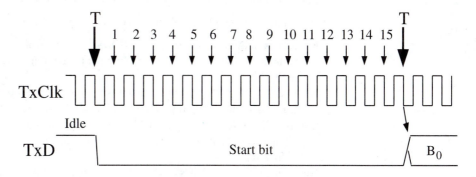

The serial port hardware is actually controlled by a clock that is 16 times (32 times on the MC68HC05C8) faster than the baud rate. The digital hardware in the SCI counts 16 times in between changes to the TxD output line. Pseudocode for the transmission process is shown below (these operations occur automatically in hardware).

```
TRANSMIT  Set TxD=0              Output start bit
          Wait 16 clock times    Wait 1 bit time
          Set n=0                Bit counter
TLOOP     Set TxD=bn             Output data bit
          Wait 16 clock times    Wait 1 bit time
          Set n=n+1
          Goto TLOOP if n≤7
          Set TxD=T8             Output T8 bit (optional if M=1)
          Wait 16 clock times    Wait 1 bit time (optional if M=1)
          Set TxD=1              Output a stop bit
          Wait 16 clock times    Wait 1 bit time
```

In essence, the SCDR and transmit shift register behave together like a two-element FIFO, with writing into the SCDR analogous to the PutFifo operation and the shifting data out analogous to the GetFifo operation. In fact, the serial port interface chip used in most PCs has a 16-byte hardware FIFO between the data register and the shift register. A PC that has a 16C550-compatible UART supports this hardware FIFO function. This FIFO reduces the latency requirements of the OS to service the serial port hardware.

7.5.2 Receiving in Asynchronous Mode

Receiving data frames is a little trickier than transmission because we have to synchronize the receive shift register with the incoming data. Common features that affect the receiver portion of the SCI include:

RxD data input pin, with TTL voltage levels
10- or 11-bit shift register, which cannot be directly accessed by the programmer; this shift register is separate from the transmit shift register
Serial Communications Data Register (SCDR), which is read only, even though at the same address this data register is separate from the transmit data register
R8 bit that you can read after receiving a frame in 9-bit data mode ($M=1$)

The control bits that affect the receiver are:

Receiver Enable control bit (RE), which you initialized to 1 to enable the receiver
Receiver Wakeup control bit (RWU), which you set to 1 to allow a receiver input to wake up the computer
Receiver Interrupt Enable control bit (RIE), which you set to 1 to arm the RDRF flag
Idle Line Interrupt Enable control bit (ILIE), which you set to 1 to arm the IDLE flag

The status bits generated by receiver activity are (Figure 7.29):

Receive Data Register Full flag (RDRF), which is set when new input data is available; RDRF is cleared by reading the RDRF flag (with it set), then reading the SCDR
Receiver Idle flag (IDLE), which is set when the receiver line becomes idle; IDLE is cleared by reading the IDLE flag (with it set), then reading the SCDR
Overrun flag (OR), which is set when input data is lost because previous data frames had not been read; OR is cleared by reading the OR flag (with it set), then reading the SCDR
Noise flag (NF), which is set when the input is noisy; NF is cleared by reading the NF flag (with it set), then reading the SCDR
Framing Error (FE), which is set when the stop bit is incorrect; FE is cleared by reading the FE flag (with it set), then reading the SCDR.

Figure 7.29
Data register and shift register used to implement the receive serial interface.

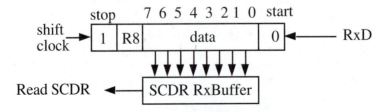

The receiver waits for the 1 to 0 edge signifying a start bit, then shifts in 10 or 11 bits of data one at a time from the RxD line. The start and stop bits are removed (checked for noise and framing errors), the 8 bits of data are loaded into the SCDR, the ninth data bit is put in

R8 (if M=1), and the RDRF flag is set. If there are already data in the SCDR when the shift register is finished, it will wait until the previous frame is read by the software before it is transferred. An overrun occurs when there is one receive frame in the SCDR, one receive frame in the receive shift register, and a third frame comes into RxD. In the timing diagrams of Figures 7.30 and 7.31, the "S" arrow refers to the time when the receiver detects the 1 to 0 edge of the start bit, and the "R" arrows refer to the times the receiver shift register is shifted, causing the current value of the RxD input pin to be recorded.

Figure 7.30
Receive data frames for M=0 and M=1.

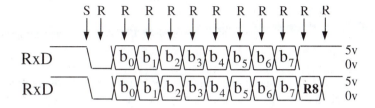

Figure 7.31
Start bit timing during a receive data frame.

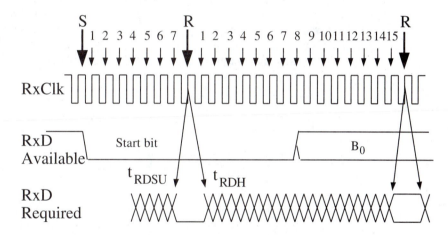

If "R" refers to the time the RxD pin is recorded, the setup time is the time before "R" the input must be valid. The hold time is the time after the "R" the data must continue to be valid. Pseudocode for the receive process is shown below (these operations occur automatically in hardware). Because the receiver must wait a half a bit time after the start 1 to 0 edge, it requires a clock faster than the baud rate. In particular, with a serial clock 16 times faster than the baud rate, it waits a half a bit time by waiting 8 serial clock periods. All of the operations (e.g., waiting for RxD=0) are synchronized to the serial clock.

```
RECEIVE   Goto RECEIVE if RxD=1    Wait for start bit
          Wait 8 clock times       Wait half a bit time
          Goto RECEIVE if RxD=1    False start? (NF)
          Set n=0                   Bit counter
RLOOP     Wait 16 clock times      Wait 1 bit time
          Set bn=RXD                Input data bit
          Set n=n+1
          Goto RLOOP if n≤7
          Wait 16 clock times      Wait 1 bit time (optional if M=1)
          Set R8= RxD               Read R8 bit (optional if M=1)
          Wait 16 clock times      Wait 1 bit time
          Set FE=1 if RxD =0        Framing error if no stop bit
```

An overrun occurs when there is one receive frame in the SCDR, one receive frame in the receive shift register, and a third frame comes into RxD. To avoid overrun, we can design a real-time system (i.e., one with a maximum latency). The latency of a SCI receiver is the delay between the time when new data arrives in the receiver SCDR and the time the software reads the SCDR. If the latency is always less than 10 (11 if M=1) bit times, then overrun will never occur.

> **Observation:** With a serial port that has a shift register and one data register (no additional FIFO buffering), the latency requirement of the input interface is the time it takes to transmit one data frame.

In the following example, assume the SCI receive shift register and receive data register are initially empty. Three incoming serial frames occur one right after another, but the software does not respond. At the end of the first frame, the $31 goes into the receive SCDR and the RDRF flag is set. In this scenario, the software is busy doing other things and does not respond to the setting of RDRF. Next, the second frame is entered into the receive shift register. At the end of the second frame, there is the $31 in the SCDR and the $32 in the shift register. If the software were to respond at this point, then both characters would be properly received. If the third frame begins before the first is read by the software, then an overrun error occurs and a frame is lost (Figure 7.32). We can see from this worst-case scenario that the software must read the data from SCDR within 10 bit times of the setting of RDRF.

Figure 7.32
Three receive data frames result in an overrun (OR) error.

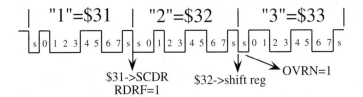

Next we will overview the specific SCI functions on particular Motorola microcomputers. This section is intended to supplement rather than replace the Motorola manuals. When designing systems with a SCI, please also refer to the reference manual of your specific Motorola microcomputer.

7.5.3
MC68HC05C8 SCI
Details

The MC68HC05C8 has one asynchronous serial port using Port D bits 1, 0. We use the BAUD register to select the baud rate. There are two integer factors that divide the E clock to get the baud rate. The 2 bits SCP1 and SCP0 determine the prescale factor. The 3 bits SCR2, SCR1, and SCR0 specify the other factor (Table 7.10).

Table 7.10
Baud rate selection bits for the MC68HC05C8.

SCP1	SCP0	Prescaler, P
0	0	1
0	1	3
1	0	4
1	1	13

SCR2	SCR1	SCR0	Divider, BR
0	0	0	1
0	0	1	2
0	1	0	4
0	1	1	8
1	0	0	16
1	0	1	32
1	1	0	64
1	1	1	128

	7	6	5	4	3	2	1	0	
BAUD	0	0	SCP1	SCP0	0	SCR2	SCR1	SCR0	$000D

The baud rate can be calculated using the values in Table 7.10 and the equation

$$\text{SCI baud rate} = \frac{\text{ECLK}}{16 \times \text{P} \times \text{BR}}$$

Table 7.11 shows the BAUD register values for common baud rates when the E clock is 2 MHz.

Table 7.11
Baud rate selection bits to create common protocols.

BAUD	Baud Rate (bits/sec)
$30	9600
$31	4800
$32	2400
$33	1200

	7	6	5	4	3	2	1	0	
SCCR1	R8	T8	0	M	WAKE	0	0	0	$000E

R8: Receive data bit 8
    If M bit is set, R8 stores ninth bit in receive data character.
T8: Transmit data bit 8
    If M bit is set, T8 stores ninth bit in transmit data character.
M: Mode (select character format)
    0 creates a 10-bit frame with 1 start bit, 8 data bits, 1 stop bit
    1 creates an 11-bit frame with 1 start bit, 9 data bits, 1 stop bit (the ninth data bit is
       in T8/R8)
WAKE: Wake up by address mark/idle
    0 = Wake up by IDLE line recognition
    1 = Wake up by address mark (most significant data bit set)

	7	6	5	4	3	2	1	0	
SCCR2	TIE	TCIE	RIE	ILIE	TE	RE	RWU	SBK	$000F

TIE: Transmit interrupt enable
    0 = TDRE interrupts disabled
    1 = SCI interrupt requested when TDRE status flag is set
TCIE: Transmit complete interrupt enable
    0 = TC interrupts disabled
    1 = SCI interrupt requested if TC is set to 1

RIE: Receiver interrupt enable
  0 = RDRF and OR interrupts disabled
  1 = SCI interrupt requested when RDRF flag or the OR status flag is set
ILIE: Idle line interrupt enable
  0 = IDLE interrupts disabled
  1 = SCI interrupt requested when IDLE status flag is set
TE: Transmitter enable
  0 = Transmitter disabled
  1 = Transmitter enabled
RE: Receiver enable
  0 = Receiver disabled
  1 = Receiver enabled
RWU: Receiver wake-up control
  0 = Normal SCI receiver
  1 = Wake-up enabled and receiver interrupts inhibited
SBK: Send break
  0 = Break generator off
  1 = Break codes generated as long as SBK is set to 1

	7	6	5	4	3	2	1	0	
SCSR	TDRE	TC	RDRF	IDLE	OR	NF	FE	0	$0010

TDRE: Transmit data register empty flag
  Set if transmit data can be written to SCDR; if TDRE is 0, transmit data register
    contains previous data that have not yet been moved to the transmit shift
    register. Writing into the SCDR when TDRE is set will result in a loss of data.
    Cleared by SCSR read with TDRE set, followed by SCDR write
TC: Transmit complete flag
  Set if transmitter is idle (no data, preamble, or break transmission in progress).
    Cleared by SCSR read with TC set, followed by SCDR write
RDRF: Receive data register full flag
  Set if a received character is ready to be read from SCDR. Cleared by SCSR read
    with RDRF set, followed by SCDR read
IDLE: Idle line detected flag
  Set if the RxD line is idle (10 or 11 consecutive logic 1s). IDLE flag is inhibited
    when RWU is set to 1. Cleared by SCSR read with IDLE set, followed by
    SCDR read. Once cleared, IDLE is not set again until the R×D line has been
    active and becomes idle again
OR: Overrun error flag
  Set if a new character is received before a previously received character is read
    from SCDR. Cleared by SCSR read with OR set, followed by SCDR read
NF: Noise error flag
  Set if majority sample logic detects anything other than a unanimous decision.
    Cleared by SCSR read with NF set, followed by SCDR read
FE: Framing error
  Set if a 0 is detected where a stop bit was expected. Cleared by SCSR read with
    FE set, followed by SCDR read

	7	6	5	4	3	2	1	0	
SCDR	R7/T7	R6/T6	R5/T5	R4/T4	R3/T3	R2/T2	R1/T1	R0/T0	$0011

7.5.4
MC68HC708XL36
SCI Details

The MC68HC708XL36 has one asynchronous serial port using Port E bits 2, 1. We use the SCBR register to select the baud rate. There are two integer factors that divide the E clock to get the baud rate. The 2 bits SCP1 and SCP0 determine the prescale factor. The 3 bits SCR2, SCR1, and SCR0 specify the other factor (Table 7.12).

Table 7.12
Baud rate selection bits for the MC68HC708XL36.

SCP1	SCP0	Prescaler, P
0	0	1
0	1	3
1	0	4
1	1	13

SCR2	SCR1	SCR0	Divider, BR
0	0	0	1
0	0	1	2
0	1	0	4
0	1	1	8
1	0	0	16
1	0	1	32
1	1	0	64
1	1	1	128

	7	6	5	4	3	2	1	0	
SCBR	0	0	SCP1	SCP0	0	SCR2	SCR1	SCR0	$0019

CGMXCLK is the crystal oscillator frequency (e.g., 16 MHz for an 8 MHz E clock). The baud rate can be calculated using the values in Table 7.12 and the equation

$$\text{SCI baud rate} = \frac{\text{CGMXCLK}}{64 \times P \times BR}$$

Table 7.13 shows the BAUD register values for common baud rates when the E clock is 8 MHz.

Table 7.13
Baud rate selection bits to create common protocols.

BAUD	Baud rate (bits/s)
$30	19,200
$31	9,600
$32	4,800
$33	2,400

	7	6	5	4	3	2	1	0	
SCC1	LOOPS	ENSCI	TXINV	M	WAKE	ILTY	PEN	PTY	$0013

LOOPS: Loop mode select bit
  In loop mode both the receiver and transmit shift register are connected to the transmitter pin, PTE2. In loop mode output transmissions are echoed into the receiver shift register

ENSCI: Enable SCI bit

    Set this bit to 1 to enable the SCI and activate the baud rate register

TXINV: Transmit inversion bit

    Set this bit to 1 to invert the polarity of transmissions including start, stop, and break

M: Mode (select character format)

    0 creates a 10-bit frame with 1 start bit, 8 data bits, 1 stop bit

    1 creates an 11-bit frame with 1 start bit, 9 data bits, 1 stop bit (the ninth data bit is in T8/R8); if parity is active, the ninth bit is automatically created on transmission and tested on reception

WAKE: Wake-up by address mark/idle

    0 = Wake up by IDLE line recognition

    1 = Wake up by address mark (most significant data bit set)

ILTY: Idle line type

    0 = Idle line counting begins after the start bit

    1 = Idle line counting begins after the stop bit

PE: Parity enable

    0 = Parity is disabled

    1 = Parity is enabled

PT: Parity type

    0 = Even parity is selected. An even number of 1s in the data character causes the parity bit to be 0, and an odd number of 1s causes the parity bit to be 1

    1 = Odd parity is selected. An odd number of 1s in the data character causes the parity bit to be 0, and an even number of 1s causes the parity bit to be 1

	7	6	5	4	3	2	1	0	
SCC2	SCTIE	TCIE	SCRIE	ILIE	TE	RE	RWU	SBK	$0014

SCTIE: SCI transmit interrupt enable

    0 = SCTE interrupts/DMA disabled

    1 = SCI interrupt or DMA requested when SCTE status flag is set

TCIE: Transmit complete interrupt enable

    0 = TC interrupts disabled

    1 = SCI interrupt requested if TC is set to 1

SCRIE: SCI Receiver interrupt enable

    0 = SCRF interrupts/DMA disabled

    1 = SCI interrupt or DMA requested when SCRF flag status flag is set

ILIE: Idle line interrupt enable

    0 = IDLE interrupts disabled

    1 = SCI interrupt requested when IDLE status flag is set

TE: Transmitter enable

    0 = Transmitter disabled

    1 = Transmitter enabled

RE: Receiver enable

    0 = Receiver disabled

    1 = Receiver enabled

RWU: Receiver wake-up control

    0 = Normal SCI receiver

    1 = Wake-up enabled and receiver interrupts inhibited

SBK: Send break

    0 = Break generator off

    1 = Break codes generated as long as SBK is set to 1

	7	6	5	4	3	2	1	0	
SCC3	R8	T8	DMARE	DMATE	ORIE	NEIE	FEIE	PEIE	$0015

R8: Receive data bit 8

    If M bit is set, R8 stores ninth bit in receive data character

T8: Transmit data bit 8

    If M bit is set, T8 stores ninth bit in transmit data character

DMARE: DMA receiver enable

    0 = SCRF DMA disabled. If SCRIE set, SCRF flag causes an interrupt

    1 = If SCRIE also set, DMA requested when SCRF flag status flag is set

DMATE: DMA transmitter enable

    0 = SCTE DMA disabled. If SCTIE set, SCTE flag causes an interrupt

    1 = If SCTIE also set, DMA requested when SCTE flag status flag is set

ORIE: Receiver overrun interrupt enable

    0 = OR interrupts disabled

    1 = SCI interrupt requested if OR is set to 1

NEIE: Receiver noise error interrupt enable

    0 = NF interrupts disabled

    1 = SCI interrupt requested when NF status flag is set

FEIE: Receiver framing error interrupt enable

    0 = FE interrupts disabled

    1 = SCI interrupt requested when FE status flag is set

PEIE: Receiver parity error interrupt enable

    0 = PE interrupts disabled

    1 = SCI interrupt requested when PE status flag is set

	7	6	5	4	3	2	1	0	
SCS1	SCTE	TC	SCRF	IDLE	OR	NF	FE	PE	$0016

SCTE: SCI transmit data register empty flag

    Set if transmit data can be written to SCDR; if SCTE is 0, transmit data register
        contains previous data that have not yet been moved to the transmit shift
        register. Writing into the SCDR when SCTE is set will result in a loss of data.
        Cleared by SCSR read with SCTE set, followed by SCDR write

TC: Transmit complete flag

    Set if transmitter is idle (no data, preamble, or break transmission in progress).
        Cleared by SCSR read with TC set, followed by SCDR write

SCRF: SCI receive data register full flag

Set if a received character is ready to be read from SCDR. Cleared by SCSR read with SCRF set, followed by SCDR read

IDLE: Idle line detected flag

Set if the RxD line is idle (10 or 11 consecutive logic 1s). IDLE flag is inhibited when RWU is set to 1. Cleared by SCSR read with IDLE set, followed by SCDR read. Once cleared, IDLE is not set again until the RxD line has been active and becomes idle again

OR: Receiver overrun error flag

New byte is ready to be transferred from the receive shift register to the receive data register and the receive data register is already full (RDRF bit is set). Data transfer is inhibited until this bit is cleared

0 = No overrun

1 = Overrun detected

NF: Receiver noise error flag

Set during the same cycle as the RDRF bit but not set in the case of an overrun (OR)

0 = Unanimous decision

1 = Noise on a valid start bit, any of the data bits, or on the stop bit

FE: Receiver framing error flag

Set when a 0 is detected where a stop bit was expected. Clear the FE flag by reading SCS1 with FE set and then reading SCDR

0 = Stop bit detected

1 = Zero detected rather than a stop bit

PF: Receiver parity error flag

Indicates if received data's parity matches parity bit. This feature is active only when parity is enabled. The type of parity tested for is determined by the PT (parity type) bit in SCC1

0 = Parity correct

1 = Incorrect parity detected

	7	6	5	4	3	2	1	0	
SCS2	0	0	0	0	0	0	BKF	RPF	$0017

BKF: Receiver break flag

Set when a break (10 or 11 zeros) is detected on the receiver input. Clear the BKF flag by reading SCS2 with BKF set and then reading SCDR. A second break will be detected only after the receiver goes high, then low for another 10 or 11 zeros.

0 = no break detected

1 = break detected

RPF: Reception in progress flag

This bit is controlled by the receiver front end. It is set during the RT1 time period of the start bit search. It is cleared when an idle state is detected or when the receiver circuitry detects a false start bit (generally due to noise or baud rate mismatch)

0 = A character is not being received

1 = A character is being received

	7	6	5	4	3	2	1	0	
SCDR	R7/T7	R6/T6	R5/T5	R4/T4	R3/T3	R2/T2	R1/T1	R0/T0	$0018

7.5.5
MC68HC11A8 SCI
Details

Most versions of the 6811 have one asynchronous serial port using Port D bits 1, 0. We use the BAUD register to select the baud rate. There are two integer factors that divide the E clock to get the baud rate. The 2 bits SCP1 and SCP0 determine the prescale factor. The 3 bits SCR2, SCR1, and SCR0 specify the other factor (Table 7.14).

Table 7.14
Baud rate selection bits for the MC68HC11A8.

SCP1	SCP0	Prescaler, P
0	0	1
0	1	3
1	0	4
1	1	13

SCR2	SCR1	SCR0	Divider, BR
0	0	0	1
0	0	1	2
0	1	0	4
0	1	1	8
1	0	0	16
1	0	1	32
1	1	0	64
1	1	1	128

	7	6	5	4	3	2	1	0	
BAUD	TCLR	0	SCP1	SCP0	RCKB	SCR2	SCR1	SCR0	$102B

The baud rate can be calculated using the values in Table 7.14 and the equation

$$\text{SCI baud rate} = \frac{\text{ECLK}}{16 \times \text{P} \times \text{BR}}$$

Table 7.15 shows the BAUD register values for common baud rates when the E clock is 2 MHz.

Table 7.15
Baud rate selection bits to create common protocols.

BAUD	Baud Rate (bits/s)
$30	9600
$31	4800
$32	2400
$33	1200

	7	6	5	4	3	2	1	0	
SCCR1	R8	T8	0	M	WAKE	0	0	0	$102C

R8: Receive data bit 8
If M bit is set, R8 stores ninth bit in receive data character

T8: Transmit data bit 8

    If M bit is set, T8 stores ninth bit in transmit data character

M: Mode (select character format)

    0 creates a 10-bit frame with 1 start bit, 8 data bits, 1 stop bit

    1 creates an 11-bit frame with 1 start bit, 9 data bits, 1 stop bit (the ninth data bit is in T8/R8)

WAKE: Wake up by address mark/idle

    0 = Wake up by IDLE line recognition

    1 = Wake up by address mark (most significant data bit set)

	7	6	5	4	3	2	1	0	
SCCR2	TIE	TCIE	RIE	ILIE	TE	RE	RWU	SBK	$102D

TIE: Transmit interrupt enable

    0 = TDRE interrupts disabled

    1 = SCI interrupt requested when TDRE status flag is set

TCIE: Transmit complete interrupt enable

    0 = TC interrupts disabled

    1 = SCI interrupt requested if TC is set to 1

RIE: Receiver interrupt enable

    0 = RDRF and OR interrupts disabled

    1 = SCI interrupt requested when RDRF flag or the OR status flag is set

ILIE: Idle line interrupt enable

    0 = IDLE interrupts disabled

    1 = SCI interrupt requested when IDLE status flag is set

TE: Transmitter enable

    0 = Transmitter disabled

    1 = Transmitter enabled

RE: Receiver enable

    0 = Receiver disabled

    1 = Receiver enabled

RWU: Receiver wake-up control

    0 = Normal SCI receiver

    1 = Wake-up enabled and receiver interrupts inhibited

SBK: Send break

    0 = Break generator off

    1 = Break codes generated as long as SBK is set to 1

	7	6	5	4	3	2	1	0	
SCSR	TDRE	TC	RDRF	IDLE	OR	NF	FE	0	$102E

TDRE: Transmit data register empty flag

    Set if transmit data can be written to SCDR; if TDRE is 0, transmit data register contains previous data that have not yet been moved to the transmit shift register. Writing into the SCDR when TDRE is set will result in a loss of data. Cleared by SCSR read with TDRE set, followed by SCDR write

TC: Transmit complete flag

Set if transmitter is idle (no data, preamble, or break transmission in progress). Cleared by SCSR read with TC set, followed by SCDR write

RDRF: Receive data register full flag

Set if a received character is ready to be read from SCDR. Cleared by SCSR read with RDRF set, followed by SCDR read

IDLE: Idle line detected flag

Set if the RxD line is idle (10 or 11 consecutive logic 1s). IDLE flag is inhibited when RWU is set to 1. Cleared by SCSR read with IDLE set, followed by SCDR read. Once cleared, IDLE is not set again until the RxD line has been active and becomes idle again

OR: Overrun error flag

Set if a new character is received before a previously received character is read from SCDR. Cleared by SCSR read with OR set, followed by SCDR read

NF: Noise error flag

Set if majority sample logic detects anything other than a unanimous decision. Cleared by SCSR read with NF set, followed by SCDR read

FE: Framing error

Set if a zero is detected where a stop bit was expected. Cleared by SCSR read with FE set, followed by SCDR read

	7	6	5	4	3	2	1	0	
SCDR	R7/T7	R6/T6	R5/T5	R4/T4	R3/T3	R2/T2	R1/T1	R0/T0	$102F

7.5.6 MC68HC812A4 SCI Details

The MC68HC812A4 has two asynchronous serial ports using Port S bits 3, 2 and bits 1, 0. The MC68HC912B32 has only one serial port using Port S bits 1, 0. Each serial port has a 16-bit baud rate register.

	15	14	13	12	11	10	9	8	7	6	5	4	3	2	1	0	
SC0BD																	$00C0
SC1BD																	$00C8

The least significant 13 bits determine the baud rate for the receive and transmit operations. If BR is the value written to bits 12:0 and MCLK is the module clock, then the baud rate is

$$\text{SCI baud rate} = \frac{\text{MCLK}}{(16 \times \text{BR})}$$

The MCLK frequency is determined by the 2 bits MCSB and MCSA in the clock control (CLKCTL) register.

	7	6	5	4	3	2	1	0	
CLKCTL	LCKF	PLLON	PLLS	BCSC	BCSB	BCSA	**MCSB**	**MCSA**	$0047

The PCLK is the internal bus rate clock, typically 8 MHz (Table 7.16).

Table 7.16
MCLK rate selection bits
for the MC68HC812A4.

MCSB	MCSA	MCLK Rate
0	0	MCLK = PCLK
0	1	Divide by 2
1	0	Divide by 4
1	1	Divide by 8

	7	6	5	4	3	2	1	0	
SC0CR1 SC1CR1	LOOPS	WOMS	RSRC	M	WAKE	PE	ILT	PT	$00C2 $00CA

LOOPS: SCI LOOP mode/single wire mode enable

   0 = SCI transmit and receive sections operate normally

   1 = SCI receive section is disconnected from the RxD pin, and the RxD pin is available as general-purpose I/O. The receiver input is determined by the RSRC bit. The transmitter output is controlled by the associated DDRS bit. Both the transmitter and the receiver must be enabled to use the LOOP or the single-wire mode. If the DDRS bit associated with the TxD pin is set during the LOOPS = 1, the TxD pin outputs the SCI waveform. If the DDRS bit associated with the TxD pin is clear during the LOOPS = 1, the TxD pin becomes high (IDLE line state) for RSRC = 0 and high impedance for RSRC = 1

WOMS: Wired-OR mode for serial pins for TxD and RxD

   0 = Pins operate in a normal mode with both high and low drive capability. To affect the RxD bit, that bit would have to be configured as an output (via DDRS0/2) that is the single-wire case when using the SCI. WOMS bit still affects general-purpose output on TxD and RxD pins when SCIx is not using these pins

   1 = Each pin operates in an open-drain fashion if that pin is declared as an output.

RSRC: Receiver source

   When LOOPS = 1, the RSRC bit determines the internal feedback path for the receiver

   0 = Receiver input is connected to the transmitter internally (not TxD pin)

   1 = Receiver input is connected to the TxD pin

Table 7.17 shows the function of Port S bits 1 and 3.

LOOPS	RSRC	DDRS1(3)	WOMS	Function of Port S bit 1(3)
0	x	x	x	Normal operations
1	0	0	0	LOOP mode without TxD output
1	0	0	1	LOOP mode without TxD output (TxD = HiZ)
1	0	1	0	LOOP mode with TxD output (CMOS)
1	0	1	1	LOOP mode with TxD output (open-drain)
1	1	0	x	Single-wire mode without TxD output
1	1	1	0	Single-wire mode with TxD output
1	1	1	1	Single-wire mode for receive and transmit (open-drain)

Table 7.17 Serial port modes for the MC68HC812A4.

M: Mode (select character format)

    0 = One start, eight data, one stop bit

    1 = One start, eight data, ninth data, one stop bit

WAKE: Wake up by address mark/idle

    0 = Wake up by IDLE line recognition

    1 = Wake up by address mark (last data bit set)

ILT: Idle line type

    Determines which of two types of idle line detection will be used by the SCI receiver

    0 = Short idle line mode is enabled

    1 = Long idle line mode is detected

PE: Parity enable

    0 = Parity is disabled

    1 = Parity is enabled

PT: Parity type

    0 = Even parity is selected. An even number of 1s in the data character causes the parity bit to be 0, and an odd number of 1s causes the parity bit to be 1

    1 = Odd parity is selected. An odd number of 1s in the data character causes the parity bit to be 0, and an even number of 1s causes the parity bit to be 1

	7	6	5	4	3	2	1	0	
SC0CR2 SC1CR2	TIE	TCIE	RIE	ILIE	TE	RE	RWU	SBK	$00C3 $00CB

TIE: Transmit interrupt enable

    0 = TDRE interrupts disabled

    1 = SCI interrupt will be requested whenever the TDRE status flag is set

TCIE: Transmit complete interrupt enable

    0 = TC interrupts disabled

    1 = SCI interrupt will be requested whenever the TC status flag is set

RIE: Receiver interrupt enable

    0 = RDRF and OR interrupts disabled

    1 = SCI interrupt will be requested whenever the RDRF status flag or the OR status flag is set

ILIE: Idle line interrupt enable

    0 = IDLE interrupts disabled

    1 = SCI interrupt will be requested whenever the IDLE status flag is set

TE: Transmitter enable

    0 = Transmitter disabled

    1 = SCI transmit logic is enabled

RE: Receiver enable

    0 = Receiver disabled

    1 = Enables the SCI receive circuitry

RWU: Receiver wake-up control

    0 = Normal SCI Receiver

    1 = Enables the wake-up function and inhibits further receiver interrupts

SBK: Send break

    0 = Break generator off

    1 = Generate a break code (at least 10 or 11 contiguous zeros)

	7	6	5	4	3	2	1	0	
SC0SR1 SC1SR1	TDRE	TC	RDRF	IDLE	OR	NF	FE	PF	$00C4 $00CC

TDRE: Transmit data register empty flag

    Set if transmit data can be written to SCDR; if TDRE is 0, transmit data register contains previous data that have not yet been moved to the transmit shift register. Writing into the SCDR when TDRE is set will result in a loss of data. Cleared by SCSR read with TDRE set, followed by SCDR write

    0 = SCxDR busy

    1 = Any byte in the transmit data register is transferred to the serial shift register so that new data may now be written to the transmit data register

TC: Transmit complete flag

    Flag is set when the transmitter is idle (no data, preamble, or break transmission in progress). Clear by reading SCxSR1 with TC set and then writing to SCxDR

    0 = Transmitter busy

    1 = Transmitter is idle

RDRF: Receive data register full

    Flag once cleared, IDLE is not set again until the RxD line has been active and becomes idle again. RDRF is set if a received character is ready to be read from SCxDR. Clear the RDRF flag by reading SCxSR1 with RDRF set and then reading SCxDR

    0 = SCxDR empty

    1 = SCxDR full

IDLE: Idle line detected flag

    Receiver idle line is detected (the receipt of a minimum of 10 or 11 consecutive 1s). This bit will not be set by the idle line condition when the RWU bit is set. Once cleared, IDLE will not be set again until after RDRF has been set (after the line has been active and becomes idle again)

    0 = RxD line is idle

    1 = RxD line is active

OR: Receiver overrun error flag

    New byte is ready to be transferred from the receive shift register to the receive data register and the receive data register is already full (RDRF bit is set). Data transfer is inhibited until this bit is cleared

    0 = No overrun

    1 = Overrun detected

NF: Receiver noise error flag

    Set during the same cycle as the RDRF bit but not set in the case of an overrun (OR)

    0 = Unanimous decision

    1 = Noise on a valid start bit, any of the data bits, or on the stop bit

FE: Receiver framing error flag

    Set when a 0 is detected where a stop bit was expected. Clear the FE flag by reading SCxSR1 with FE set and then reading SCxDR

    0 = Stop bit detected

    1 = Zero detected rather than a stop bit

PF: Receiver parity error flag

Indicates if received data's parity matches parity bit. This feature is active only when parity is enabled. The type of parity tested for is determined by the PT (parity type) bit in SCxCR1

0 = Parity correct

1 = Incorrect parity detected

	7	6	5	4	3	2	1	0	
SC0SR2	0	0	0	0	0	0	0	RAF	$00C5
SC1SR2									$00CD

RAF: Receiver active flag

This bit is controlled by the receiver front end. It is set during the RT1 time period of the start bit search. It is cleared when an idle state is detected or when the receiver circuitry detects a false start bit (generally due to noise or baud rate mismatch)

0 = A character is not being received

1 = A character is being received

	7	6	5	4	3	2	1	0	
SC0DRH	R8	T8	0	0	0	0	0	0	$00C6
SC1DRH									$00CE

R8: Receive bit 8

Read anytime. Write has no meaning or effect. This bit is the ninth serial data bit received when the SCI system is configured for nine-data-bit operation

T8: Transmit bit 8

Read or write anytime. This bit is the ninth serial data bit transmitted when the SCI system is configured for nine-data-bit operation. When using 9-bit data format, this bit does not have to be written for each data word. The same value will be transmitted as the ninth bit until this bit is rewritten

	7	6	5	4	3	2	1	0	
SC0DRL	R7/T7	R6/T6	R5/T5	R4/T4	R3/T3	R2/T2	R1/T1	R0/T0	$00C7
SC1DRL									$00CF

R7T7–R0T0: Receive/transmit data bits 7 to 0

Reads access the 8 bits of the read-only SCI receive data register (RDR). Writes access the 8 bits of the write-only SCI transmit data register (TDR). SCxDRL:SCxDRH form the 9-bit data word for the SCI. If the SCI is being used with a 7- or 8-bit data word, only SCxDRL needs to be accessed. If a 9-bit format is used, the upper register should be written first to ensure that it is transferred to the transmitter shift register with the lower register

7.6 SCI Applications

**7.6.1
SCI Hardware
Interface**

A Maxim converter chip is used to generate the RS232 voltage levels (Figure 7.33). The RS232 timing is generated automatically by the SCI. A device driver for this chip was shown in Section 2.4.2.

Figure 7.33
Hardware interface
implementing an
asynchronous RS232
channel.

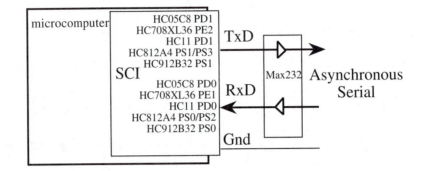

The ritual and input routine were presented earlier in Section 2.7.2, but there are three possibilities for implementing the output routine. The first implementation waits in a gadfly loop until TDRE is set, then writes data to the SCDR. The function accepts the new ASCII character as a call by value in Register A. As we saw previously in Section 3.3, this is the most efficient way to implement an output gadfly loop (Program 7.1).

```
; MC68HC05C8
; RegA is data to output
OutChar brclr 7,SCSR,OutChar
        sta SCDR ;output
        rts
```

```
; MC68HC11A8
OutChar ldab SCSR     ;status
        bitb #$80     ;tdre?
        beq  OutChar
        staa SCDR     ;output
        rts
```

```
; MC68HC708XL36
OutChar brclr 7,SCS1,OutChar
        sta SCDR  ;output
        rts
```

```
; MC68HC812A4
OutChar brclr SC0SR1,#$80,OutChar
        staa  SC0DRL ;sci data
        rts
```

Program 7.1 Assembly subroutine to output a character using the SCI port.

```
; MC68HC05C8
; RegA is data to output
OutChar sta SCDR ;output
wait2   brclr 7,SCSR,wait2
        rts
```

```
; MC68HC11A8
OutChar staa SCDR     ;output
wait2   ldab SCSR     ;status
        bitb #$80     ;tdre?
        beq  wait2
        rts
```

```
; MC68HC708XL36
OutChar sta SCDR   ;output
wait2   brclr 7,SCS1,wait2
        rts
```

```
; MC68HC812A4
OutChar staa SC0DRL ;sci data
wait2   brclr SC0SR1,#$80,wait2
        rts
```

Program 7.2 A second assembly subroutine to output a character using the SCI port.

The second implementation writes the data to the SCDR, then waits in a gadfly loop until TDRE is set. Notice that TDRE is not set after the transmission of the data but when the SCDR register becomes empty (Program 7.2).

The third implementation writes the data to the SCDR, then waits in a gadfly loop until TC is set. Notice that in this case the function will not return until the entire frame has been shifted out (transmit complete). This would be an appropriate synchronization method if the software needed to synchronize the output transmission with other I/O functions. This is the only one of the three methods where the software actions are synchronized with the hardware operations. In particular, the serial transmission begins immediately after `OutChar` is called, and immediately after the completion of the serial transmission the software returns from `OutChar` (Program 7.3). These routines for the 6811 and 6812 in C are shown in Program 7.4.

```
; MC68HC05C8                          ; MC68HC11A8
; RegA is data to output     OutChar  staa  SCDR       ;output
OutChar sta SCDR ;output     wait3    ldab  SCSR       ;status
wait3   brclr 6,SCSR,wait3             bitb  #$40       ;TC?
        rts                           beq   wait3
                                      rts

; MC68HC708XL36                       ; MC68HC812A4
OutChar sta SCDR   ;output   OutChar  staa  SC0DRL ;sci data
wait3   brclr 6,SCS1,wait3   wait3    brclr SC0SR1,#$40,wait3
        rts                           rts
```

Program 7.3 A third assembly subroutine to output a character using the SCI port.

```
// 68HC11A8 SCI routines              // 68HC812A4 SCI routines
void init(void) {                     void init(void) {
    BAUD=0x33;  // 1200 baud              SC0BD=417;    // 1200 baud
    SCCR1=0x00; // 8data, 1stop           SC0CR1=0x00;  // 8data, 1stop
    SCCR2=0x0C;}// enable gadfly          SC0CR2=0x0C;} // enable gadfly
#define RDRF 0x20                     #define RDRF 0x20
#define TDRE 0x80                     #define TDRE 0x80
#define TC 0x40                       #define TC 0x40
unsigned char insci(void){            unsigned char insci(void){
    while ((SCSR & RDRF) == 0);           while ((SC0SR1 & RDRF) == 0);
    return(SCDR); }                       return(SC0DRL); }
void OutChar(unsigned char letter){   void OutChar(unsigned char letter){
/* Wait for TDRE then output */       /* Wait for TDRE then output */
    while ((SCSR & TDRE) == 0);           while ((SC0SR1 & TDRE) == 0);
    SCDR=letter; }                        SC0DRL=letter; }
void outsci2(unsigned char letter){   void outsci2(unsigned char letter){
/* Output then wait for TDRE y */     /* Output then wait for TDRE y */
    SCDR=letter;                          SC0DRL=letter;
    while ((SCSR & TDRE) == 0);           while ((SC0SR1 & TDRE) == 0); }
void outsci3(unsigned char letter){   void outsci3(unsigned char letter){
/* Output then wait for TC*/          /* Output then wait for TC */
    SCDR=letter;                          SC0DRL=letter;
    while ((SCSR & TC) == 0);   /}        while ((SC0SR1 & TC) == 0);  }
```

Program 7.4 Three C functions to output a character using the SCI port.

The TExaS simulator has example serial port functions for the 6805, 6808, 6811 and 6812 written in assembly and C. Once TExaS is installed, the gadfly (busy-waiting) SCI functions can be found in the TUT2.* files. The TUT4.* files implement similar serial port operations using interrupt synchronization.

7.6.2
SCI Receive Only
Interrupt Interface

In this example, the objective is to input characters only, so just the receiver is armed and not the transmitter. Assume we have a serial port connected to the SCI as in the previous section. An IRQ interrupt will occur whenever the receiver data register is full (i.e., when RDRF=1). The interrupt is acknowledged by reading the data. The ISR will place the incoming data into the FIFO. Refer back to the left-hand side of Figure 4.4 for an overview of this system. The main program or some other interrupt routine will get the data out of the FIFO. The SCCR2 initialization is shown in Table 7.18.

Table 7.18
SCCR2 initialize value.

Bit	SCCR2
7 0	TIE=0
6 0	TCIE=0
5 1	RIE=1, arm
4 0	ILIE=0
3 1	TE enable TxD
2 1	RE enable RxD
1 0	RWU no wake up
0 0	SBK no break

The baud rate is 4800 bits/s. One start, eight data, no parity, one stop bit mode is used. The assembly language implementations of the rituals are shown in Program 7.5.

```
; MC68HC05C8
RITSCI sei         ;make atomic
       lda #$31 ;4800 baud
       sta BAUD
       lda #00  ;M=0, 8 bit
       sta SCCR1 ;1 stop
       lda #$2C
       sta SCCR2
       bsr CLRQ  ;Init FIFO
       cli        ;Enable
       rts
```

```
; MC68HC11A8
RITSCI sei         ;make atomic
       ldaa #$31 ;4800 baud
       staa BAUD
       ldaa #00  ;M=0, 8 bit data
       staa SCCR1 ;1 stop
       ldaa #$2C
       staa SCCR2
       bsr  CLRQ ;Initialize FIFO
       cli        ;Enable
       rts
```

```
; MC68HC708XL36
RITSCI sei         ;make atomic
       lda #$32 ;4800 baud
       sta SCBR
       lda #00  ;M=0, 8 bit
       sta SCCR1 ;1 stop
       lda #$2C
       sta SCCR2
       bsr CLRQ  ;Init FIFO
       cli        ;Enable
       rts
```

```
; MC68HC812A4 and MC68HC912B32
RITSCI sei         ;make atomic
       ldd  #104 ;4800 baud
       std  SC0BD
       ldaa #00  ;M=0, 8 bit data
       staa SC0CR1 ;1 stop
       ldaa #$2C
       staa SC0CR2
       bsr  CLRQ ;Initialize FIFO
       cli        ;Enable
       rts
```

Program 7.5 Assembly ritual to initialize the SCI port to accept receiver interrupts.

There are three possible methods to determine the source of interrupt:

- No polling required because this is the only source of SCI interrupts.[1] Simply store the device handler address (e.g., `INSCI`) in the proper place
- Simple polling (Program 7.6)
- Polling for "1s and 0s" (there are no guaranteed 0s in the 6808 and 6812) (Program 7.7)

Program 7.6 Simple polling for receiver interrupts.

```
; MC68HC05C8
SCIHAN  brset 5,SCSR,INSCI
```
```
; MC68HC11A8
SCIHAN  ldaa SCSR
             anda #$20   RDRF set?
             bne  INSCI
```
```
; MC68HC708XL36
SCIHAN  brset 5,SCS1,INSCI
```
```
; MC68HC812A4 and MC68HC912B32
SCIHAN  brset SC0SR1,#$20,INSCI
```

Program 7.7 Polling for 1s and 0s for receiver interrupts.

```
; MC68HC05C8
SCIHAN  lda SCSR   XX1XXXX0
             and #$21   RDRF set?
             cmp #$20   Bit0=0?
             beq INSCI
```
```
; MC68HC11A8
SCIHAN  ldaa SCSR   XX1XXXX0
             anda #$21   RDRF set?
             cmpa #$20   Bit0=0?
             beq  INSCI
```

Recall from Chapter 4 that "polling for 0s and 1s" verifies as much information in the control/status register as possible. In particular it means checking for the presence of both 1s and 0s in the control/status register. The device handler, `INSCI`, performs the specific tasks (1) acknowledge, (2) read data, and (3) pass data to other threads via a shared global data structure (FIFO.) ERROR will be called on either a communication error or FIFO full. Assume the `PutFifo` subroutine takes Register A as a call by value input parameter and returns C=1 if the FIFO is full (Program 7.8). In C, these functions are shown in Program 7.9.

```
; MC68HC05C8
INSCI lda SCSR ;status
      and #$0E ;OR NF FE
      bne ERROR
      lda SCDR ;data ack
      bsr PutFifo ;Communicate
      bcs ERROR ;FIFO full?
      rti
```
```
; MC68HC11A8
INSCI ldaa SCSR ;status
      anda #$0E ;OR, NF, FE
      bne  ERROR
      ldaa SCDR ;data, ack
      bsr  PutFifo ;Communicate
      bcs  ERROR ;FIFO full?
      rti
```
```
; MC68HC708XL36
INSCI lda SCS1 ;status
      and #$0E ;OR NF FE
      bne ERROR
      lda SCDR ;data ack
      bsr PutFifo ;Communicate
      bcs ERROR ;FIFO full?
      rti
```
```
; MC68HC812A4 and MC68HC912B32
INSCI ldaa SC0SR1 ;status
      anda #$0E ;OR, NF, FE
      bne  ERROR
      ldaa SCDR ;data, ack
      bsr  PutFifo ;Communicate
      bcs  ERROR ;FIFO full?
      rti
```

Program 7.8 Assembly ISR for receiver interrupts.

[1]The SCI requires that you first read the status register with the RDRF bit set, then read the data register to clear the RDRF.

```
// MC68HC11A8                            // MC68HC812A4 and MC68HC912B32
#pragma interrupt_handler SCIHAN()       #pragma interrupt_handler SCIHAN()
#define RDRF 0x20                         #define RDRF 0x20
// Executed on RDRF (not TDRE)           // Executed on RDRF (not TDRE)
void SCIhandler(void){                   void SCIhandler(void){
    if((SCSR&0x21)!= RDRF) error();          if((SCOSR1&0x20)!= RDRF) error();
    if(SCSR&0x0E) error();                   if(SCOSR1&0x0F) error();
    if(PutFifo(SCDR)) error(); }             if(PutFifo(SCODRL)) error(); }
void RitualSCI(void){                    void RitualSCI(void){
asm(" sei");                             asm(" sei");
    BAUD=0x31;     // 4800 bits/sec          SCOBD=104;    // 4800 baud
    SCCR1=0;       // 8 bit 1 stop          SCOCR1=0x00;  // 8data, 1stop
    SCCR2=0x2C;    // Receiver intrpt       SCOCR2=0x2C;  // Receiver intrpt
    CLRQ();        // Clear FIFO            CLRQ();       // Clear FIFO
asm(" cli");}                            asm(" cli");}
```

Program 7.9 C language implementation of receiver interrupts.

7.6.3 SCI Transmit Only Interrupt Interface

There are two problems when using interrupts to synchronize output devices. The first problem is how and when to generate the first interrupt. The second problem is what to do if the device is ready but there are no data to output. The first technique utilizes a nonprinting "null" character. Examples include NULL ($00) and SYN ($16). The OutChar function does not actually perform output but rather stores the data in a FIFO circular queue. Whenever the SCI transmitter is idle (TDRE set), a SCI interrupt will occur and the interrupt handler will get a character from the FIFO and output it to the SCDR. The "null" character is transmitted whenever the device is ready but no real data are available. The system on the other end should ignore the "null" transmissions. It is a simple and reliable technique but it does waste computer and network time transmitting these idle characters (Figure 7.34).

Figure 7.34
A simple technique for implementing serial output with interrupt synchronization.

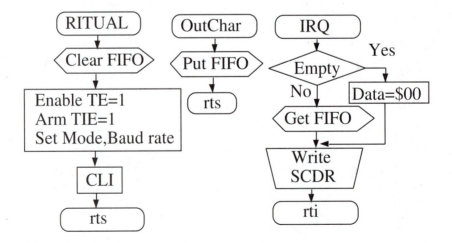

The second technique requires modification of the OutChar subroutine. In this approach, the SCI transmit device is armed after every `PutFifo`. (If the SCI were already armed, then rearming would have no effect). Now, if the SCI transmit process needs service (TDRE interrupt) and the FIFO is empty, the device handler will disarm itself. This

Figure 7.35
An improved technique
for implementing serial
output with interrupt
synchronization.

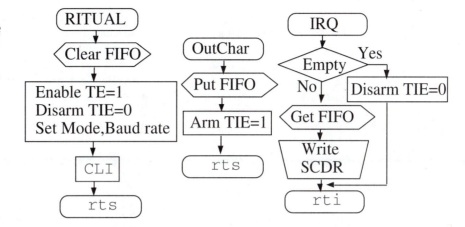

technique is more complicated but reduces software and network overhead (Figure 7.35).
Refer back to the right-hand side of Figure 4.4 for an overview of this approach.

7.6.4
Input and Output
Interrupts on the
SCI

To implement simultaneous input and output, two FIFOs are required. The RxFifo will pass
data from the InSCI interrupt handler to the main thread like the system shown in Section 7.6.2.
A second FIFO, TxFifo, will pass data from the main thread to the OutSCI interrupt handler
like the system shown in Section 7.6.3. Because the TxFifo is empty at the time of the ritual,
the transmit interrupts are initially disarmed. Each time the main thread calls OutChar, the
transmit interrupts are armed. The background interrupt handler will disarm the transmit interrupts when the TxFifo becomes empty. Both sides of Figure 4.4 describe this system.

The initialization routine enables the SCI and establishes the 9600 bits/s baud rate. The
mode is one start bit, eight data bits (M=0), no parity, and one stop bit. The two FIFOs are
initialized, and RDRF is armed (Program 7.10).

Program 7.10
Assembly ritual to
initialize the SCI port to
accept receiver and
transmitter interrupts.

```
; MC68HC05C8
Init lda #$30   ;9600 baud
     sta BAUD
     lda #$00   ;mode
     sta SCCR1
     jsr RxInitFifo ;empty
     jsr TxInitFifo ;empty
     lda #$2c   ;just RDRF
     sta SCCR2 ;enable SCI
     cli
     rts
```

```
; MC68HC11A8
Init ldaa #$30   ;9600 baud
     staa BAUD
     ldaa #$00   ;mode
     staa SCCR1
     jsr  RxInitFifo ;empty
     jsr  TxInitFifo ;empty
     ldaa #$2c   ;just RDRF
     staa SCCR2 ;enable SCI
     cli
     rts
```

```
; MC68HC708XL36
Init mov #$31,SCBR ;9600
     mov #$00,SCC1
     jsr RxInitFifo ;empty
     jsr TxInitFifo ;empty
     mov #$2c,SCC2 ;enable
     mov #$00,SCC3 ;no DMA
     cli
     rts
```

```
; MC68HC812A4/MC68HC912B32
Init  movw #104,SC0BD   ;9600
      movb #$00,SC0CR1 ;mode
      movb #$2c,SC0CR2 ;RDRF
      jsr  RxInitFifo  ;empty
      jsr  TxInitFifo  ;empty
      cli
      rts
```

Next we show the two functions called by the main program when it wishes to perform I/O. The input routine waits in a gadfly loop until there are data in the FIFO, then returns the data. The output routine simply puts the data into the FIFO. OutChar will wait only when the FIFO is full. Both InChar and OutChar pass the ASCII character by value in Register A. Assume the get functions return the Z bit set if the FIFO is empty. Similarly assume the put functions return the Z bit set if the FIFO is full (Program 7.11).

Program 7.11
Assembly subroutines called by the main program to perform serial I/O.

```
; MC68HC05C8                          ; MC68HC11A8
InChar jsr  RxGetFifo                 InChar  jsr   RxGetFifo
       beq  InChar                            beq   InChar
       rts                                    rts
OutChar jsr  TxPutFifo ;save          OutChar jsr   TxPutFifo ;save
        beq  OutChar  ;full?                  beq   OutChar  ;full?
        lda #$AC    ;arm both                 ldab #$AC      ;arm both
        sta SCCR2   ;TDRE, RDRF               stab SCCR2     ;TDRE, RDRF
        rts                                   rts
```

```
; MC68HC708XL36                       ; MC68HC812A4/MC68HC912B32
InChar   jsr RxGetFifo                InChar  jsr   RxGetFifo
         beq InChar                           beq   InChar
         rts                                  rts
OutChar jsr TxPutFifo ;save           OutChar jsr   TxPutFifo ;save
        beq OutChar  ;full?                   beq   OutChar  ;full?
        mov #$AC,SCC2 ;arm                    movb #$AC,SC0CR2 ;arm
        rts                                   rts
```

Since both the receiver and transmitter generate interrupts via the same interrupt vector, we must poll at the start of the interrupt handler. If it is a receiver interrupt, we read the data and put it into the RxFifo. If it is a transmitter interrupt, then we try to get from the TxFifo. If there are data, then they are output to the SCI. If the TxFifo is empty (no more to output), then we disarm the transmitter. The receiver function is checked first because it has an upper bound on its latency. There is no real-time urgency for processing the transmitter function, so it is performed second (Program 7.12).

Program 7.12
Assembly ISR for receiver and transmitter interrupts.

```
; MC68HC05C8                          ; MC68HC11A8
SCIhdlr lda                           SCIhdlr ldaa
        and #$20                              anda #$20      ;check
        beq ChkTDRE                           beq  ChkTDRE ;Not
InSCI   lda SCDR     ;ASCII           InSCI   ldaa SCDR      ;ASCII
        bsr                                   bsr
ChkTDRE lda                           ChkTDRE ldaa SCSR
        and #$80                              anda #$80      ;check TDRE
        beq SCIdone                           beq  SCIdone ;Not TDRE
OutSCI  bsr                           OutSCI  bsr  TxGetFifo
        beq nomore                            beq  nomore
        sta SCDR    ;start                    staa SCDR      ;start next
        bra SCIdone                           bra  SCIdone
nomore  lda #$2C                      nomore  ldaa #$2C
        sta SCCR2   ;disarm                   staa SCCR2     ;disarm TDRE
SCIdone rti                           SCIdone rti
```

```
; MC68HC708XL36                      ; MC68HC812A4/MC68HC912B32
SCIhdlr lda                          SCIhdlr ldaa
        and #$20                             anda #$20      ;check
        beq ChkSCTE ;Not                     beq  ChkTDRE ;Not RDRF
InSCI   lda SCODRL  ;ASCII           InSCI   ldaa SCODRL   ;ASCII
        bsr                                  bsr
ChkSCTE lda                          ChkTDRE ldaa SC0SR1
        and #$80                             anda #$80      ;check TDRE
        beq SCIdone                          beq  SCIdone ;Not TDRE
OutSCI  bsr                          OutSCI  bsr  TxGetFifo
        beq                                  beq  nomore
        sta SCDR   ;start                    staa SCODRL    ;start next
        bra SCIdone                          bra  SCIdone
nomore  mov #$2C,SCC2 disarm         nomore  movb #$2C,SC0CR2 ;disarm
SCIdone rti                          SCIdone rti
```

Common Error: Notice that the above transmit device driver either acknowledges the interrupt by sending another character or disarms itself because the T×Fifo is empty. The software will crash (infinite loop) if it returns from interrupt without acknowledging or disarming.

7.6.5 Serial Port Printer Interfaces

One problem with printers is that the printer bandwidth (the actual number of characters per second that can be printed) may be less than the maximum bandwidth supported by the serial channel. Five conditions support this situation: (1) Special characters may require more time to print (e.g., carriage return, line feed, tab, form feed, and graphics); (2) most printers have internal FIFOs that could get full; (3) the printer cable may be disconnected; (4) the printer may be deselected; (5) the printer power may be off. The output interfaces shown previously provide no feedback from the printer that could be used to detect/correct these five problems. There are two mechanisms, called flow control, to synchronize the computer with a variable rate output device. These two flow control protocols are called DTR and XON/XOFF.

7.6.5.1 Use of Data Terminal Ready (DTR) to Interface a Printer

The first method uses a hardware signal, DTR (pin 4 on the DB9 connector or pin 20 on the DB25 connector), as feedback from the printer to the microcomputer. DTR is -12 V if the printer is busy and is not currently able to accept transmission. DTR is $+12$ V if the printer is ready and able to accept transmission. This mechanism can handle all five of the above situations. The computer input (or input capture if interrupts are needed[2]) mechanism will handle the DTR protocol using additional software checking. With a standard RS232 interface when DTR is -12 V, the input capture line will be $+5$ V (which will stop the gadfly loop even if the TDR register is empty). Thus, when DTR is -12 V, transmission is temporarily suspended. When DTR is $+12$ V, the input capture line will be 0 V, and transmission can proceed normally (Figures 7.36 and 7.37).

---

[2]The *key wake up* could also be used to generate interrupts on the high-to-low and low-to-high transitions.

Figure 7.36
Hardware interface
implementing a RS232
simplex channel with
DTR handshaking.

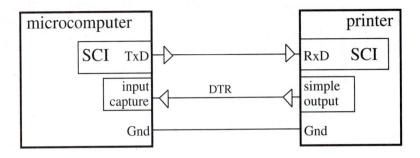

Figure 7.37
Software flowchart
implementing a simplex
channel with DTR
handshaking.

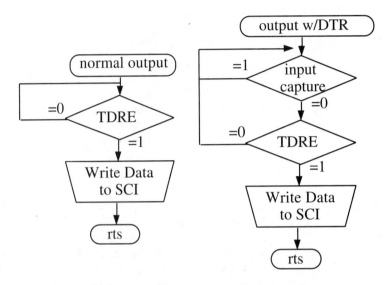

The assembly language implementation for this gadfly approach is shown in Program 7.13. The same ritual used in the previous section can be used in this gadfly implementation. In each case the data to print are in Register A. In C, these functions are shown in Program 7.14.

Program 7.13 Assembly
functions for serial
output using DTR
synchronization.

```
; MC68HC05C8
; PA0 is DTR
; RegA is data to output
OutChar brset 0,PORTA,OutChar
        brclr 7,SCSR,OutChar
        sta SCDR ;output
        rts
```

```
; MC68HC11A8
;PA0/IC3 is DTR
OutChar ldab PORTA    ;wait for
        bitb #$01     ; DTR=0
        bne  OutChar ;
        ldab SCSR     ;status
        bitb #$80     ;tdre?
        beq  OutChar
        staa SCDR     ;output
        rts
```

```
; MC68HC708XL36
; PA0 is DTR
OutChar brset 0,PORTA,OutChar
        brclr 7,SCS1,OutChar
        sta SCDR   ;output
        rts
```

```
; MC68HC812A4 or MC68HC912B32
;PT3/IC3 is DTR
OutChar brclr PORTT,#$08,OutChar
        brclr SC0SR1,#$80,OutChar
        staa SC0DRL ;sci data
        rts
```

```
// MC68HC11A8                            // MC68HC812A4 or MC68HC912B32
// PA0/IC3 is DTR                        // PT3/IC3 is DTR
void OutChar(unsigned char data) {       void OutChar(unsigned char data) {
   while((PORTA&0x01)||((SCSR&0x80)==0));    while((PORTT&0x08)||((SCOSR1&0x80)==0));
   SCDR= data; }                            SCODRL= data; }
```

Program 7.14 C functions for serial output using DTR synchronization.

Next, we will rework this problem using interrupts to handle the DTR hardware synchronization. The baud rate still is 1200 bits/s. The RS232 protocol remains one start bit, eight bit, no parity, and one stop bit. The DTR signal from the printer provides feedback information about the printer status. When DTR is −12 V (6811 PA0 or 6812 PT3 is high), the printer is not ready to accept more data. In this case, our computer will postpone transmitting more frames. When DTR is +12 V (6811 PA0 or 6812 PT3 is low), the printer is ready to accept more data. At this point, the computer will resume transmission. The `OutChar(data)`, shown below, is called by the main program when it wishes to print. We assume there are three FIFO functions. The implementation was presented in Chapter 4.

`InitFifo()`	initializes the FIFO
`flag=PutFifo(data)`	enters a byte `data` into the FIFO
	returns a `flag=true` if the FIFO is full and the data were not saved
`flag=GetFifo(&data)`	returns with a byte and `flag` is false if successful
	returns `flag` true if the FIFO was empty at the time of the call

This helper function will turn off transmit interrupts if DTR is −12 V (6811 PA0 or 6812 PT3 is high) (Program 7.15). The C function `OutChar(data)` initiates an output. The actual output occurs in the background. `OutChar` returns a true flag if the operation cannot be performed because the FIFO was full (Program 7.16). The ISRs perform the serial output operations in the background (Program 7.17). The ritual software is executed once. The interrupt vector locations should also be initialized (Program 7.18).

```
// MC68HC11A8                            // MC68HC812A4 or MC68HC912B32
// PA0/IC3 is DTR                        // PT3/IC3 is DTR
void checkIC3(void) {                    void checkIC3(void) {
   if(PORTA&0x01)   // PA0=1 if DTR=-12     if(PORTT&0x08)   // PT3=1 if DTR=-12
      SCCR2=0x0C;   // SCI TxD disarmed       SCOCR2=0x0C;  // SCI TxD disarmed
   else                                    else
      SCCR2=0x8C;} // SCI TxD armed          SCOCR2=0x8C;} // SCI TxD armed
```

Program 7.15 C language helper function for serial output using DTR synchronization.

Program 7.16
C language output
function using DTR
synchronization.

```
int OutChar(unsigned char data) { unsigned char flag;
    flag=PutFifo(data^0x80);  /* Bit7=1 is the first stop bit */
    checkIC3();               /* Arm SCI if DTR=+12 */
    return(flag);}            /* error if FIFO is full */
```

```
// MC68HC11A8                              // MC68HC812A4 or MC68HC912B32
// PA0/IC3 is DTR                          // PT3/IC3 is DTR
 #pragma interrupt_handler IC3Han()         #pragma interrupt_handler IC3Han()
void IC3Han(void) {                        void IC3Han(void) {
    TFLG1=0x01;   // Ack clear IC3F            TFLG1=0x08;   // Ack clear C3F
    checkIC3();}  // Arm SCI if DTR=+12        checkIC3();}  // Arm SCI if DTR=+12
#pragma interrupt_handler SCIHAN()         #pragma interrupt_handler SCIHAN()
void SCIHan(void) { unsigned char data;    void SCIHan(void) { unsigned char data;
  if(GetFifo(&data);)                        if(GetFifo(&data);)
    SCCR2=0x0C; // disarmed, empty             SCOCR2=0x0C; // disarmed, empty
  else                                       else
    SCDR=data;} // output, ack interrupt       SCODRL=data;} // output, ack
```

Program 7.17 C language ISR using DTR synchronization.

```
// MC68HC11A8                              // MC68HC812A4 or MC68HC912B32
// PA0/IC3 is DTR                          // PT3/IC3 is DTR
void Ritual(void){                         void Ritual(void){
    asm(" sei");                               asm(" sei");
    InitFifo();                                InitFifo();
    BAUD=0x33;     // 1200 bits/sec            SCOBD=417;      // 1200 baud
    SCCR1=0x00;    // 8 bit, 1 stop           SCOCR1=0x00;    // 8 bit, 1 stop
    SCCR2=0x0C;                                SCOCR2=0x0C;
/* SCI disarmed because FIFO is empty */      TIOS &= 0xF7;   // PT3 input capture
    TCTL2=0x03;    // both rise or fall       DDRT &= 0xF7;   // PT3 is input
    TMSK1=0x01;    // Arm IC3                  TSCR = 0x80;    // enable TCNT
    TFLG1=0x01;    // clear IC3F */           TMSK2= 0x32;    // 500ns clock
    asm(" cli");}                             TCTL4 |= 0xC0;  // both rise, fall
                                              TMSK1 |= 0x08;  // Arm IC3
                                              TFLG1=0x08;     // initially clear
                                              asm(" cli");}
```

Program 7.18 C language initialization using DTR synchronization.

7.6.5.2
Use of XON/XOFF to
Interface a Printer

Another flow control technique is the XON/XOFF protocol (Figures 7.38 and 7.39). This technique allows the printer to signal back to the computer that it cannot currently accept more data. The advantage of XON/XOFF is that it uses a standard full-duplex serial channel. This is an example of a simplex communication *system* constructed on top of a full-duplex *channel*.

In this system we must assume that the printer power is on and the cable is properly connected. Most printers maintain a FIFO between the incoming serial data from the computer and the actual printing hardware. When the printer FIFO is almost full (e.g., 80% full) or paper is out, the printer will send a single XOFF ($13) transmission back to the computer. Once the FIFO has more room (e.g., 20% full) or more paper, the printer will send a single XON ($11) to the computer. Typically the computer assumes the printer FIFO is empty, so it can initially begin transmission without first receiving an XON. But, once the computer receives an XOFF, it must wait for an XON before it continues transmission. One disadvantage of XON/XOFF is that it cannot handle cable disconnection or printer power-off problems.

Figure 7.38
Hardware interface
implementing a serial
channel with XON/XOFF
handshaking.

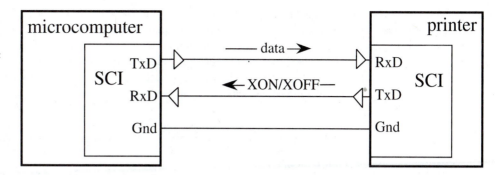

Figure 7.39
Software flowchart
implementing a serial
channel with XON/XOFF
handshaking.

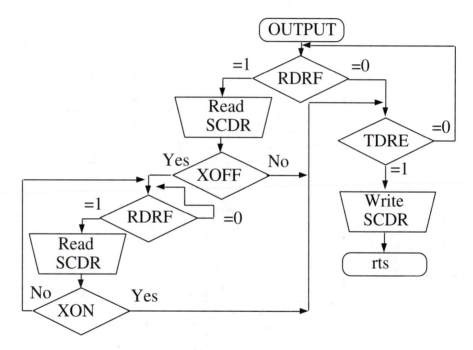

7.7 Synchronous Transmission and Receiving Using the SPI

7.7.1 SPI Fundamentals

Many of the Motorola embedded microcomputers include a SPI. The fundamental difference between a SCI, which implements an asynchronous protocol, and a SPI, which implements a synchronous protocol, is the manner in which the clock is implemented. Two devices communicating with asynchronous serial interfaces (SCI) operate at the same frequency (baud rate) but have two separate (not synchronized) clocks. Two devices communicating with synchronous serial interfaces (SPI) operate the same clock (synchronized). Typically, the master device creates the clock, and the slave device(s) uses the clock to latch the data in or out. Before discussing the detailed operation of particular devices, we will begin with general features common to all devices. The Motorola SPI includes four I/O lines. The slave select ($\overline{SS}$) is an optional negative logic control signal from master to slave signifying the channel is active. The second line, SCK, is a 50% duty cycle clock generated by the master. The master-out slave-in (MOSI) is a data line driven by the master and received by the slave. The master-in slave-out (MISO) is a data line driven by the slave and received by the master. To work properly, the transmitting device uses one edge of the clock to change its output, and the receiving device uses the other edge to accept the data. Table 7.19 lists the I/O port locations of the synchronous serial ports for the various microcomputers discussed in this book.

Table 7.19
Synchronous serial port pins on various Motorola microcomputers.

Microcomputer	Pin for $\overline{SS}$	Pin for SCK	Pin for MOSI	Pin for MISO
MC68HC05C8	PD5	PD4	PD3	PD2
MC68HC708XL36	PTF0	PTF1	PTF2	PTF3
MC68HC11	PD5	PD4	PD3	PD2
MC68HC812A4	PS7	PS6	PS5	PS4
MC68HC912B32	PS7	PS6	PS5	PS4

The SPI allows the computer to communicate synchronously with peripheral devices and other microprocessors. The SPI system in the microcomputer can operate as a master or as a slave. In the SPI system the 8-bit data register, SPDR, in the master and the 8-bit data register in the slave, also SPDR, are linked to form a distributed 16-bit register. Figure 7.40 illustrates communication between master and slave. The interface logic shown in the figure can implement any of the physical channels discussed earlier in the chapter (e.g., simple CMOS digital logic, RS232, RS422, optically isolated).

Figure 7.40
A synchronous serial interface between two Motorola microcomputers.

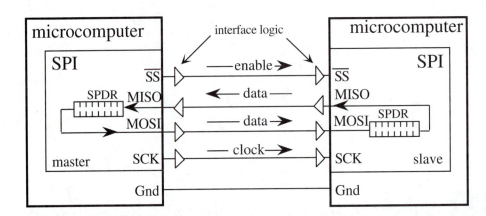

When a data transfer operation is performed, this 16-bit register is serially shifted eight bit positions by the SCK clock from the master so that the data are effectively exchanged between the master and the slave. Data written to the SPDR register of the master are transmitted to the slave. Data written to the SPDR register of the slave are transmitted to the master. The SPI is also capable of interprocessor communications in a multiple master system.
Common control features for the SPI module include:

A baud rate control register used to select the transmission rate
A mode bit in the control register to select master versus slave, clock polarity, clock phase
Interrupt arm bit
Ability to make the outputs open-drain (open-collector)

Common status bits for the SPI module include:

SPIF, transmission complete
WCOL, write collision
MODF, mode fault

Observation: Because the clocks are shared, if you change the E clock frequency, the transfer rate will change, but the SPI still should operate properly.

The key to proper transmission is to select one edge of the clock (shown as T in Figure 7.41) to be used by the transmitter to change the output, and use the other edge (shown as R) to latch the data in the receiver. In this way data are latched during the time when they are stable. Data available is the time when the output data is actually valid, and data required is the time when the input data must be valid. For the communication to occur without error, the data available from the device that is driving the data line must overlap (start before and end after) the data required by the other device that is receiving the data. It is this overlap that will determine the maximum frequency at which synchronous serial communication can occur. More discussion of the concepts of data available and data required can be found in Section 9.3.

Figure 7.41
Synchronous serial timing showing that the data available interval overlaps the data required interval.

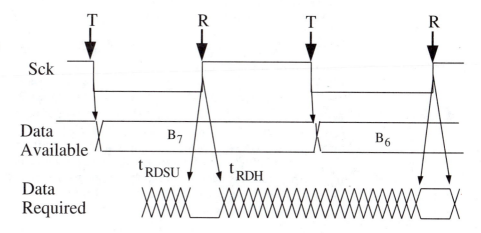

The general idea of SPI communication can be illustrated by the following pseudocodes. Although it is possible to implement synchronous serial transmission on any computer with simple I/O pins in software using the bit-banging approach (see the DS1620 example in Section 3.4.8), these operations are implemented in hardware by the SPI interface.

```
TRANSMIT   Set n=7                              Bit counter
TLOOP         On the fall of Sck, set Data=bn   Output data bit
              Set n=n-1
              Goto TLOOP if n≥0
              set Data=1                         Idle output

RECEIVE    Set n=7                              Bit counter
RLOOP         On the rise of Sck, read Data
              Set bn=Data                        Input data bit
              Set n=n-1
              Goto RLOOP if n≥0
```

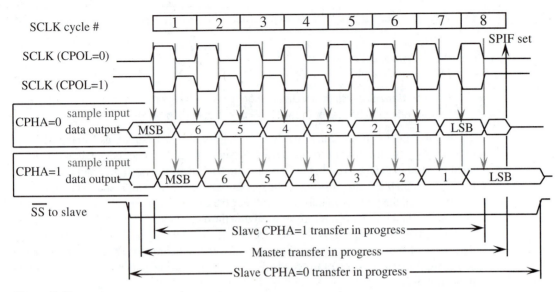

Figure 7.42
Synchronous serial modes of the Motorola SPI interface.

The SPI timing is shown in Figure 7.42. Three mode control bits affect the transmission protocol. If the device is a master, it generates the SCK, and data are output on the MOSI pin and input on the MISO pin. If the device is a slave, the SCK is an input, and data are input on the MOSI pin and output on the MISO pin. The CPOL control bit specifies the polarity of the SCK. The CPHA bit affects the timing of the first bit transferred and received. In Figure 7.42, the data are shown with MSB transferred first, but the 6812 has an option where the bits are transferred in the other order (LSB first).

Next we will overview the specific SPI functions on particular Motorola microcomputers. This section is intended to supplement rather than replace the Motorola manuals. When designing systems with a SPI, please also refer to the reference manual of your specific Motorola microcomputer.

**7.7.2
MC68HC05C8 SPI
Details**

The SPI port on the MC68HC05C8 uses the four pins PD5= $\overline{SS}$, PD4=SCLK, PD3=MOSI, and PD2=MISO.

	7	6	5	4	3	2	1	0	
SPCR	SPIE	SPE	0	MSTR	CPOL	CPHA	SPR1	SPR0	$000A

SPIE: Serial peripheral interrupt enable
    0 = SPI interrupts disabled
    1 = SPI interrupts enabled
SPE: Serial peripheral system enable
    0 = SPI off
    1 = SPI on

MSTR: Master mode select
  0 = Slave mode
  1 = Master mode
CPOL, CPHA: SPI clock polarity, clock phase
These two bits are used to specify the clock format to be used in SPI operations

The SPR1 and SPR0 control bits determines the transfer rate. Table 7.20 shows the bit rate selection when the E clock frequency is 2 MHz.

Table 7.20
Bit rate selection for the synchronous serial port on the MC68HC05C8.

SPR1	SPR0	Divisor	Transfer frequency	Bit time
0	0	2	1 MHz	1 µs
0	1	4	500 kHz	2 µs
1	0	16	125 kHz	8 µs
1	1	32	62.5 kHz	16 µs

	7	6	5	4	3	2	1	0	
SPSR	SPIF	WCOL	0	MODF	0	0	0	0	$000B

SPIF: SPI transfer complete flag
Set when an SPI transfer is complete. Cleared by reading SPSR with SPIF set, followed by SPDR access
WCOL: Write collision
Set when SPDR is written while transfer is in progress. Cleared by SPSR with WCOL set, followed by SPDR access
MODF: Mode fault (a Mode Fault terminates SPI operation)
Set when $\overline{SS}$ is pulled low while MSTR = 1. Cleared by SPSR read with MODF set, followed by SPCR write.

	7	6	5	4	3	2	1	0	
SPDR	bit 7	bit 6	bit 5	bit 4	bit 3	bit 2	bit 1	bit 0	$000C

7.7.3 MC68HC708XL36 SPI Details

The SPI port on the MC68HC708XL36 uses the four pins PTF0= $\overline{SS}$, PTF1=SCLK, PTF2=MOSI, and PTF3=MISO.

	7	6	5	4	3	2	1	0	
SPCR	SPRIE	DMAS	SPMSTR	CPOL	CPHA	SPWOM	SPE	SPTIE	$0010

SPRIE: SPI receiver interrupt enable
  0 = SPI receiver interrupts disabled
  1 = SPI receiver interrupts or DMA enabled

DMAS: DMA select

    0 = SPI DMA disabled

    1 = SPI DMA enabled (DMAS+SPRIE is receive DMA, DMAS+SPTIE is transmit DMA)

SPMSTR: SPI master mode select

    0 = Slave mode

    1 = Master mode

CPOL, CPHA: SPI clock polarity, clock phase

    These two bits are used to specify the clock format to be used in SPI operations

SPWOM: SPI Wired-OR mode

    SPWOM affects PTF1/SPCLK, PTF2/MOSI, PTF3/MISO

    0 = Normal CMOS outputs

    1 = Open-drain outputs

SPE: Serial peripheral system enable

    0 = SPI off

    1 = SPI on

SPTIE: SPI transmitter interrupt enable

    0 = SPI transmitter interrupts disabled

    1 = SPI transmitter interrupts or DMA enabled

	7	6	5	4	3	2	1	0	
SPSCR	SPRF	0	OVRF	MODF	SPTE	0	SPR1	SPR0	$0011

SPRF: SPI receive transfer complete flag

    Set when an SPI receive transfer is complete. Cleared by reading SPSCR with SPRF set, followed by a read SPDR

OVRF: Overflow bit

    Set when SPDR is not read and another receiver transfer occurs. Cleared by reading SPSCR with OVRF set, followed by a read SPDR

MODF: Mode fault (a Mode Fault terminates SPI operation)

    Set when SS is pulled low while MSTR = 1. Cleared by reading SPSCR with MODF set, followed by SPCR write.

SPTE: SPI transmit empty flag

    Set when an SPI transmit transfer is complete. Cleared by reading SPSCR with SPTE set, followed by a write SPDR.

The SPR1 and SPR0 control bits determine the transfer rate. Table 7.21 shows the bit rate selection when the CGMOUT clock frequency is 4 MHz.

Table 7.21
Bit rate selection for the synchronous serial port on the MC68HC708XL36.

SPR1	SPR0	Divisor	Transfer frequency	Bit time
0	0	2	2 MHz	0.5 μs
0	1	8	500 kHz	2 μs
1	0	32	125 kHz	8 μs
1	1	128	31.25 kHz	32 μs

	7	6	5	4	3	2	1	0	
SPDR	bit 7	bit 6	bit 5	bit 4	bit 3	bit 2	bit 1	bit 0	$0012

7.7.4
6811 SPI Details

The SPI port on the MC68HC11A8 uses the four pins PD5= $\overline{SS}$, PD4=SCLK, PD3=MOSI, and PD2=MISO.

	7	6	5	4	3	2	1	0	
SPCR	SPIE	SPE	DWOM	MSTR	CPOL	CPHA	SPR1	SPR0	$1028

SPIE: Serial peripheral interrupt enable
    0 = SPI interrupts disabled
    1 = SPI interrupts enabled
SPE: Serial peripheral system enable
    0 = SPI off
    1 = SPI on
DWOM: Port D wired-OR mode
    DWOM affects all six port D pins.
    0 = Normal CMOS outputs
    1 = Open-drain outputs
MSTR: Master mode select
    0 = Slave mode
    1 = Master mode
CPOL, CPHA: SPI clock polarity, clock phase
    These two bits are used to specify the clock format to be used in SPI operations.

The SPR1 and SPR0 control bits determine the transfer rate. Table 7.22 shows the bit rate selection when the E clock frequency is 2 MHz.

Table 7.22
Bit rate selection for the synchronous serial port on the MC68HC11A8.

SPR1	SPR0	Divisor	Transfer frequency	Bit time
0	0	2	1 MHz	1 μs
0	1	4	500 kHz	2 μs
1	0	16	125 kHz	8 μs
1	1	32	62.5 kHz	16 μs

	7	6	5	4	3	2	1	0	
SPSR	SPIF	WCOL	0	MODF	0	0	0	0	$1029

SPIF: SPI transfer complete flag
    Set when an SPI transfer is complete. Cleared by reading SPSR with SPIF set, followed by SPDR access

WCOL: Write collision

Set when SPDR is written while transfer is in progress. Cleared by reading SPSR read with WCOL set, followed by SPDR access

MODF: Mode fault (a Mode Fault terminates SPI operation)

Set when $\overline{SS}$ is pulled low while MSTR = 1. Cleared by reading SPSR with MODF set, followed by SPCR write.

	7	6	5	4	3	2	1	0	
SPDR	bit 7	bit 6	bit 5	bit 4	bit 3	bit 2	bit 1	bit 0	$102A

7.7.5
6812 SPI Details

The SPI port on the MC68HC12 uses the four pins PS7 = $\overline{SS}$, PS6=SCLK, PS5=MOSI, and PS4=MISO. When the SPI is enabled, all pins that are defined by the configuration as inputs will be inputs regardless of the state of the DDRS bits for those pins. All pins that are defined as SPI outputs will be outputs only if the DDRS bits for those pins are set. Any SPI output whose corresponding DDRS bit is cleared can be used as a general-purpose input. A bidirectional serial pin is possible using the DDRS as the direction control. The E clock is input to a divider series, and the resulting SPI clock rate may be selected to be E divided by 2, 4, 8, 16, 32, 64, 128, or 256. Three bits in the SP0BR register control the SPI clock rate.

	7	6	5	4	3	2	1	0	
SP0CR1	SPIE	SPE	SWOM	MSTR	CPOL	CPHA	SSOE	LSBF	$00D0

SPIE: SPI interrupt enable

1 = Hardware interrupt sequence is requested each time the SPIF or MODF status flag is set

0 = SPI interrupts are inhibited

SPE: SPI system enable

0 = SPI internal hardware is initialized and SPI system is in a low-power disabled state

1 = PS[4:7] are dedicated to the SPI function When MODF is set, SPE always reads 0. SP0CR1 must be written as part of a mode fault recovery sequence

SWOM: Port S wired-OR mode

Controls not only SPI output pins but also the general-purpose output pins (PS[4:7]) that are not used by SPI

0 = SPI and/or PS[4:7] output buffers operate normally

1 = SPI and/or PS[4:7] output buffers behave as open-drain outputs

MSTR: SPI master/slave mode select

0 = Slave mode

1 = Master mode

CPOL, CPHA: SPI clock polarity, clock phase

These two bits are used to specify the clock format to be used in SPI operations

SSOE: Slave select output enable

The $\overline{SS}$ output feature is enabled only in the master mode by asserting the SSOE and DDRS7

LSBF: SPI LSB first enable
0 = Data is transferred most significant bit first
1 = Data are transferred least significant bit first. Normally data are transferred most significant bit first. This bit does not affect the position of the MSB and LSB in the data register. Reads and writes of the data register will always have MSB in bit 7.

The control bit SSOE affects the operation of the SS pin as defined in Table 7.23.

Table 7.23
Mode selection for the synchronous serial port on the MC68HC812A4.

DDRS7	SSOE	Master mode	Slave mode
0	0	$\overline{SS}$ input with MODF feature	$\overline{SS}$ input
0	1	Reserved	$\overline{SS}$ input
1	0	General-purpose output	$\overline{SS}$ input
1	1	$\overline{SS}$ output	$\overline{SS}$ input

	7	6	5	4	3	2	1	0	
SP0CR2	0	0	0	0	PUPS	RDS	0	SPC0	$00D1

PUPS: Pull-up port S enable
0 = No internal pull-ups on port S
1 = All port S input pins have an active pull-up device. If a pin is programmed as output, the pull-up device becomes inactive
RDS: Reduce drive of Port S
0 = Port S output drivers operate normally
1 = All port S output pins have reduced drive capability for lower power and less noise
SPC0: Serial pin control 0
This bit decides serial pin configurations with MSTR control bit

The SPC0 and MSTR control bits determine the I/O configuration of the PS4 and PS5, as illustrated in Figure 7.43.

Figure 7.43
Synchronous serial modes of the Motorola 6812 SPI interface.

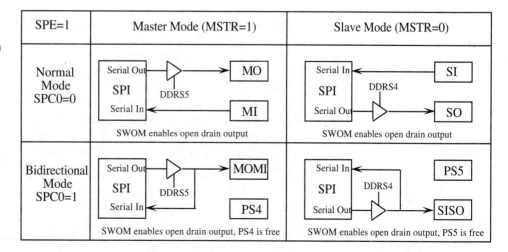

	7	6	5	4	3	2	1	0	
SP0BR	0	0	0	0	0	SPR2	SPR1	SPR0	$00D2

The SP0BR register determines the transfer rate. Table 7.24 shows the bit rate selection when the E clock frequency is 8 MHz.

Table 7.24
Bit rate selection for the synchronous serial port on the MC68HC812A4.

SPR2	SPR1	SPR0	Divisor	Transfer frequency	Bit time
0	0	0	2	4 MHz	250 ns
0	0	1	4	2 MHz	500 ns
0	1	0	8	1 MHz	1 μs
0	1	1	16	500 kHz	2 μs
1	0	0	32	250 kHz	4 μs
1	0	1	64	125 kHz	8 μs
1	1	0	128	62.5 kHz	16 μs
1	1	1	256	31.25 kHz	32 μs

	7	6	5	4	3	2	1	0	
SP0SR	SPIF	WCOL	0	MODF	0	0	0	0	$00D3

SPIF: SPI Interrupt request
> SPIF is set after the eighth SCK cycle in a data transfer, and it is cleared by reading the SP0SR register (with SPIF set), followed by an access (read or write) to the SPI data register

WCOL: Write collision status flag
> The MCU write is disabled to avoid writing over the data being transferred. No interrupt is generated because the error status flag can be read upon completion of the transfer that was in progress at the time of the error. Automatically cleared by a read of the SP0SR (with WCOL set), followed by an access (read or write) to the SP0DR register
> 0 = No write collision
> 1 = Indicates that a serial transfer was in progress when the MCU tried to write new data into the SP0DR data register

MODF: SPI mode error interrupt status flag
> This bit is set automatically by SPI hardware if the MSTR control bit is set and the slave select input pin becomes 0. This condition is not permitted in normal operation. In the case where DDRS bit 7 is set, the PS7 pin is a general-purpose output pin or SS output pin rather than being dedicated as the SS input for the SPI system. In this special case the mode fault function is inhibited and MODF remains cleared. This flag is automatically cleared by a read of the SP0SR (with MODF set), followed by a write to the SP0CR1 register

	7	6	5	4	3	2	1	0	
SP0DR	bit 7	bit 6	bit 5	bit 4	bit 3	bit 2	bit 1	bit 0	$00D5

The SP0DR 8-bit register is both the input and output register. Reads of this register are double-buffered but writes cause data to be written directly into the serial shifter. Note that some slave devices are very simple and either accept data from the master without returning data to the master or pass data to the master without requiring data from the master.

7.7.6
SPI Applications

7.7.6.1
Digital-to-Analog Converter

This first example shows the synchronous serial interface between the computer and an Analog Devices DAC8043 DAC. A DAC accepts a digital input (in our case a number between 0 and 4095) and creates an analog output (in our case a voltage between 0 and −5). Detailed discussion of DACs will be presented later in Chapter 11. Here in this section we will focus on the digital hardware and software aspects of the serial interface (Figure 7.44).

Figure 7.44
A 12-bit DAC interfaced to the SPI port.

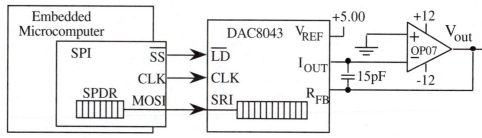

As with any SPI interface, there are basic interfacing issues to consider.

1. *Word size.* In this case we need to transmit 12 bits to the DAC. The DAC8043 data sheet suggests that we can embed the 12-bit data right-justified (it will ignore the first 4 bits that are sent) into a 16-bit transmission.
2. *Bit order.* The DAC8043 requires the most significant bits first.
3. *Clock phase, clock polarity.* There are two issues to resolve. Since the DAC8043 samples its serial input data on the rising edge of the clock, the SPI must change the data on the falling edge. CPOL=CPHA=0 and CPOL=CPHA=1 both satisfy this requirement. The second issue is which edge comes first, the rise or the fall. In this interface it probably doesn't matter.
4. *Bandwidth.* We look at the timing specifications of the DAC8043. The minimum clock low width of 120 ns means the shortest SPI period we can use is 250 ns.

The first 4 bits sent will be 0, followed by the 12 data bits that specify the analog output (Figure 7.45).

Figure 7.45
DAC8043 DAC serial timing.

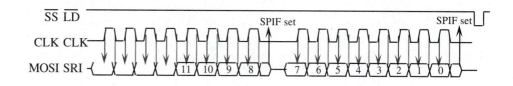

Because we want the $\overline{LD}$ signal to remain high for the entire 16-bit transfer then pulse low, we will implement it using the regular I/O pin functions. The ritual initializes the direction register, SPI mode, and bandwidth (Program 7.19). To change the digital-to-analog (D/A) output, two 8-bit transmissions are sent (Program 7.20).

```
// MC68HC05C8
void DACInit(void){ // PD5=LD=1
   DDRD|=0x38;  // PD4=CLK=SPI clock out
   PORTD|=0x20; // PD3=SRI=SPI master out
/* bit SPCR
 7 SPIE = 0    no interrupts
 6 SPE  = 1    enable SPI
 4 MSTR = 1    master
 3 CPOL = 0    output changes on fall,
 2 CPHA = 0    clock normally low
 1 SPR1 = 0    1 MHz operation
 0 SPR0 = 0     */
   SPCR=0x50;}
```

```
// MC68HC708XL36
void DACInit(void){ // PTF0=LD=1
   DDRF|=0x07;  // PTF1=CLK=SPI clock out
   PORTF|=0x01; // PTF2=SRI=SPI master out
/* bit SPCR
 7 SPRIE  = 0    no receive interrupts
 6 DMAS   = 0    no DMA
 5 SPMSTR = 1    master
 4 CPOL   = 0    output changes on fall,
 3 CPHA   = 0    clock normally low
 2 SPWOM  = 0    regular outputs
 1 SPE    = 1    enable SPI
 0 SPTIE  = 0    no transmit interrupts */
   SPCR=0x42;
   SPSCR=0x00;} // 2 MHz clock
```

```
// MC68HC11A8
void DACInit(void){ // PD5=LD=1
   DDRD|=0x38;  // PD4=CLK=SPI clock out
   PORTD|=0x20; // PD3=SRI=SPI master out
/* bit SPCR
 7 SPIE = 0    no interrupts
 6 SPE  = 1    enable SPI
 5 DWOM = 0    regular outputs
 4 MSTR = 1    master
 3 CPOL = 0    output changes on fall,
 2 CPHA = 0    clock normally low
 1 SPR1 = 0    1 MHz operation
 0 SPR0 = 0     */
   SPCR=0x50;}
```

```
// MC68HC812A4 or MC68HC912B32
void DACInit(void){ // PS7=LD=1
   DDRS=0xE0;   // PS6=CLK=SPI clock out
   PORTS=0x80;  // PS5=SRI=SPI master out
/* bit SPOCR1
 7 SPIE = 0    no interrupts
 6 SPE  = 1    enable SPI
 5 SWOM = 0    regular outputs
 4 MSTR = 1    master
 3 CPOL = 0    output changes on fall,
 2 CPHA = 0    clock normally low
 1 SSOE = 0    PS7 regular output, LD
 0 LSBF = 0    most sign bit first */
   SPOCR1=0x50;
/* bit SPOCR2
 3 PUPS = 0    no internal
 2 RDS  = 0    regular drive
 0 SPCO = 0    normal mode */
   SPOCR2=0x00;
   SPOBR=0x00;} // 4Mhz
```

Program 7.19 C language initialization for a D/A interface using the SPI.

```
// MC68HC05C8
#define SPRF 0x80
void DACout(unsigned int code){
unsigned char dummy;
   SPDR=0x00FF &(code>>8); // msbyte
   while((SPSR&SPIF)==0);  // gadfly wait
   dummy=SPDR;             // clear SPIF
   SPDR=0x00FF& code;      // lsbyte
   while((SPSR&SPIF)==0);  // gadfly wait
```

```
// MC68HC708XL36
#define SPRF 0x80
void DACout(unsigned int code){
unsigned char dummy;
   SPDR=0x00FF &(code>>8); // msbyte
   while((SPSCR&SPRF)==0);  // gadfly wait
   dummy=SPDR;             // clear SPIF
   SPDR=0x00FF& code;      // lsbyte
   while((SPSCR&SPRF)==0);  // gadfly wait
```

dummy=SPDR;	// clear SPIF	dummy=SPDR;	// clear SPIF
PORTD &= ~0x20;	// PD5=LD=0	PORTF &= ~0x01;	// PTF0=LD=0
PORTD |= 0x20; }	// PD5=LD=1	PORTF |= 0x01; }	// PTF0=LD=1

```
// MC68HC11A8
#define SPIF 0x80
void DACout(unsigned int code){
unsigned char dummy;
  SPDR=0x00FF &(code>>8); // msbyte
  while((SPSR&SPIF)==0);  // gadfly wait
  dummy=SPDR;             // clear SPIF
  SPDR=0x00FF& code;      // lsbyte
  while((SPSR&SPIF)==0);  // gadfly wait
  dummy=SPDR;             // clear SPIF
  PORTD &= ~0x20;         // PD5=LD=0
  PORTD |= 0x20; }        // PD5=LD=1
```

```
// MC68HC812A4 or MC68HC912B32
#define SPIF 0x80
void DACout(unsigned int code){
unsigned char dummy;
  SPODR=0x00FF &(code>>8); // msbyte
  while((SPOSR&SPIF)==0);  // gadfly wait
  dummy=SPODR;             // clear SPIF
  SPODR=0x00FF& code;      // lsbyte
  while((SPOSR&SPIF)==0);  // gadfly wait
  dummy=SPODR;             // clear SPIF
  PORTS &= ~0x80;          // PS7=LD=0
  PORTS |= 0x80; }         // PS7=LD=1
```

Program 7.20 C language function for a D/A interface using the SPI.

<table>
<tr><td>7.7.6.2
Analog-to-Digital
Converter</td><td>This second SPI example shows the synchronous serial interface between the computer and a Maxim MAX1247 ADC. An ADC accepts an analog input (in our case a voltage between 0 and +2.5 V on one of the four analog inputs CH3–CH0) and creates a digital output (in our case a number between 0 and 4095). Detailed discussion of ADCs will also be presented later in Chapter 11. Here in this section we will focus on the hardware and software aspects of the serial interface (Figure 7.46).</td></tr>
</table>

Figure 7.46
A four-channel 12-bit ADC interfaced to the SPI port.

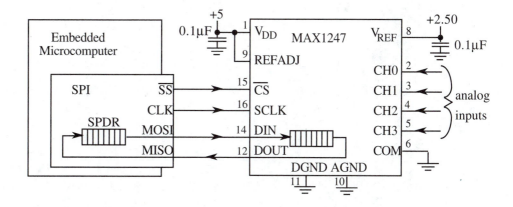

Again, there are basic interfacing issues to consider for this interface:

1. *Word size.* In this case we need to first transmit 8 bits to the ADC, then receive 12 bits back from the ADC. The MAX1247 data sheet suggests that it will embed the 12-bit data into two 8-bit transmissions.
2. *Bit order.* The MAX1247 requires the most significant bits first.
3. *Clock phase, clock polarity.* Since the MAX1247 samples its serial input data on the rising edge of the clock, the SPI must change the data on the falling edge. CPOL=CPHA=0 and CPOL=CPHA=1 both satisfy this requirement. We will use the CPOL=CPHA=0 mode as suggested in the Maxim data sheet.

4. *Bandwidth.* We look at the timing specifications of the MAX1247. The maximum SCLK frequency is 2 MHz, and the minimum clock low/high widths is 200 ns, so the shortest SPI period we can use is 500 ns.

The first 8 bits sent will specify the channel and A/D mode. After conversion, the 12 data bits result is returned. Notice that bit 7 of the mode select is always high, and the 12-bit A/D result is embedded into the middle of the two 8-bit transmissions. The software will shift the 16-bit data 3 bits to the right to produce the 0 to 4095 result (Figure 7.47).

Figure 7.47
DAC8043 D/A serial
timing.

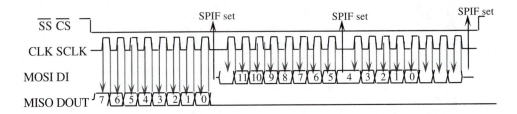

Because we want the $\overline{CS}$ signal to remain low for the entire 24-bit transfer, we will implement it using the regular I/O pin functions. The ritual initializes the direction register, SPI mode, and bandwidth (Program 7.21).

```
// MC68HC05C8
void ADCInit(void){ // PD5=CS=1
   DDRD|=0x38;  // PD4=SCLK=SPI clock out
   PORTD|=0x20; // PD3=DIN=SPI master out
/* bit SPCR         PD2=DOUT=SPI master in
 7 SPIE = 0   no interrupts
 6 SPE  = 1   enable SPI
 4 MSTR = 1   master
 3 CPOL = 0   output changes on fall,
 2 CPHA = 0   clock normally low
 1 SPR1 = 0   1 MHz operation
 0 SPR0 = 0    */
   SPCR=0x50;}
```

```
// MC68HC708XL36
void ADCInit(void){ // PTF0=CS=1
   DDRF|=0x07;  // PTF1=SCLK=SPI clock out
   PORTF|=0x01; // PTF2=DIN=SPI master out
/* bit SPCR         PTF3=DOUT=SPI master in
 7 SPRIE  = 0   no receive interrupts
 6 DMAS   = 0   no DMA
 5 SPMSTR = 1   master
 4 CPOL   = 0   output changes on fall,
 3 CPHA   = 0   clock normally low
 2 SPWOM  = 0   regular outputs
 1 SPE    = 1   enable SPI
 0 SPTIE  = 0   no transmit interrupts */
   SPCR=0x42;
   SPSCR=0x00;} // 2 MHz clock
```

```
// MC68HC11A8
void ADCInit(void){ // PD5=CS=1
   DDRD|=0x38;  // PD4=SCLK=SPI clock out
   PORTD|=0x20; // PD3=DIN=SPI master out
/* bit SPCR         PD2=DOUT=SPI master in
 7 SPIE = 0   no interrupts
 6 SPE  = 1   enable SPI
 5 DWOM = 0   regular outputs
 4 MSTR = 1   master
 3 CPOL = 0   output changes on fall,
 2 CPHA = 0   clock normally low
 1 SPR1 = 0   1 MHz operation
 0 SPR0 = 0    */
   SPCR=0x50;}
```

```
// MC68HC812A4 or MC68HC912B32
void ADCInit(void){ // PS7=CS=1
   DDRS=0xE0;   // PS6=SCLK=SPI clock out
   PORTS=0x80;  // PS5=DIN=SPI master out
/* bit SPOCR1       PS4=DOUT=SPI master in
 7 SPIE = 0   no interrupts
 6 SPE  = 1   enable SPI
 5 SWOM = 0   regular outputs
 4 MSTR = 1   master
 3 CPOL = 0   output changes on fall,
 2 CPHA = 0   clock normally low
 1 SSOE = 0   PS7 regular output, LD
 0 LSBF = 0   most sign bit first */
   SPOCR1=0x50;
```

```
/* bit SP0CR2
 3 PUPS = 0    no internal pullups
 2 RDS  = 0    regular drive
 0 SPC0 = 0    normal mode */
    SP0CR2=0x00;
    SP0BR=0x01;} // 2Mhz
```

Program 7.21 C language initialization for an A/D interface using the SPI.

Recall that when the software outputs to the SPI data register, the 8-bit register in the SPI is exchanged with the 8-bit register in the ADC. To read the ADC, three 8-bit transmissions are exchanged (Program 7.22). On the first exchange, the software specifies the channel and A/D mode, then the 12-bit A/D data are returned during the second and third transmission. All the A/D modes in the following `#define` statements implement unipolar voltage range, single-ended, and external clock:

```
#define CH0 0×9F
#define CH1 0×DF
#define CH2 0×AF
#define CH3 0×EF
```

```
// MC68HC05C8
unsigned int ADCin(unsigned char code){
unsigned int data;
   PORTD &= ~0x20; // PD5=CS=0
   SPDR=code;     // set channel,mode
   while((SPSR&0x80)==0); // gadfly wait
   data=SPDR;     // clear SPIF
   SP0DR=0;       // start SPI
   while((SPSR&0x80)==0); // gadfly wait
   data=SPDR<8; // msbyte of A/D
   SPDR=0;        // start SPI
   while((SPSR&0x80)==0); // gadfly wait
   data+SPDR;     // lsbyte of A/D
   PORTD |= 0x20; // PD5=CS=1
   return data>>3;} // right justify
```

```
// MC68HC708XL36
unsigned int ADCin(unsigned char code){
unsigned int data;
   PORTF &= ~0x01; // PTF0=CS=0
   SPDR=code;     // set channel,mode
   while((SPSCR&0x80)==0); // gadfly wait
   data=SPDR;     // clear SPIF
   SPDR=0;        // start SPI
   while((SPSCR&0x80)==0); // gadfly wait
   data=SPDR<<8; // msbyte of A/D
   SPDR=0;        // start SPI
   while((SP0SR&0x80)==0); // gadfly wait
   data+SPDR;     // lsbyte of A/D
   PORTF |= 0x01; // PTF0=CS=1
   return data>>3;} // right justify
```

```
// MC68HC11A8
unsigned int ADCin(unsigned char code){
unsigned int data;
   PORTD &= ~0x20; // PD5=CS=0
   SPDR=code;     // set channel,mode
   while((SPSR&SPIF)==0); // gadfly wait
   data=SPDR;     // clear SPIF
   SP0DR=0;       // start SPI
   while((SPSR&SPIF)==0); // gadfly wait
   data=SPDR<<8; // msbyte of A/D
   SPDR=0;        // start SPI
   while((SPSR&SPIF)==0); // gadfly wait
   data+=SODR;    // lsbyte of A/D
   PORTD |= 0x20; // PD5=CS=1
   return data>>3;} // right justify
```

```
// MC68HC812A4 or MC68HC912B32
unsigned int ADCin(unsigned char code){
unsigned int data;
   PORTS&= ~0x80; // PS7=CS=0
   SP0DR=code;    // set channel,mode
   while((SP0SR&0x80)==0); // gadfly wait
   data=SP0DR;    // clear SPIF
   SP0DR=0;       // start SPI
   while((SP0SR&0x80)==0); // gadfly wait
   data=SP0DR<<8; // msbyte of A/D
   SP0DR=0;       // start SPI
   while((SP0SR&0x80)==0); // gadfly wait
   data+=SPD0R;   // lsbyte of A/D
   PORTS |=0.80; // PS7=CS=1
   return data>>3;} // right justify
```

Program 7.22 C language function for an A/D interface using the SPI.

One of the basic building components of a microprocessor-based control system is the sensor. In this bang-bang control example, we will interface a DS1620 to the computer and use it as part of the temperature controller. The DS1620 chip automatically implements a bang-bang digital control system that applies heat to the room to maintain the temperature as close to a desired temperature setpoint T* as possible. The microcomputer will communicate with the DS1620 to set the two threshold temperatures, T_{HIGH} and T_{LOW}.

Once programmed with the two setpoint temperatures T_{HIGH} and T_{LOW}, the DS1620 will perform the above bang-bang algorithm automatically. Figure 7.48 shows the DS1620-based controller. The DS1620 can be programmed at the factory before installing the chip into a system. Here it is shown with a microcomputer, which allows the operator to adjust the setpoints.

Figure 7.48
DS1620-based
temperature controller.

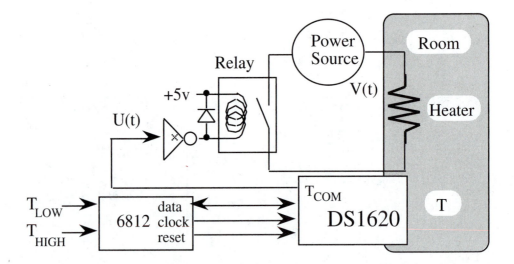

To run the controller, the microcomputer simply sends the two setpoints to the DS1620 and puts it in continuous mode. The DS1620 was previously interfaced to the computer in Chapter 3 using simple output commands; in this section we will interface it using the 6812 SPI (Program 7.23). The 6812 SPI module has a few unique features that allow it to be used with the DS1620. First, it allows data to be transferred least significant bit first. Second, and more important, it has a bidirectional half-duplex mode that allows data to be transmitted and received on the same pin. The basic idea is that to transmit 9 bits of temperature data to the DS1620 (Program 7.24), the 6812:

1. Sets the RST signal=1
2. Outputs the 8-bit command code for write TH (or write TL) to the DS1620 using one SPI transfer
3. Outputs the 9-bit data value to the DS1620 using two SPI transfers, least significant bit first and right-justified (i.e., most significant 7 bits are 0)
4. Sets the RST signal=0

Program 7.23
C language initialization of a temperature sensor interface using the SPI.

```
// Interface between MC68HC812A4/MC68HC912B32 and DS1620 using the SPI
// bitstatus  Configuration/Status Register meaning
// 7  DONE    1=Conversion done, 0=conversion in progress
// 6  THF     1=temperature above TH, 0=temperature below TH
// 5  TLF     1=temperature below TL, 0=temperature above TL
// 1  CPU     1=CPU control, 0=stand alone operation
// 0  1SHOT   1=one conversion and stop, 0=continuous conversions
// temperature         digital value (binary) digital value (hex)
// +125.0 C   011111010                   $0FA
//  +64.0 C   010000000                   $080
//   +1.0 C   000000010                   $002
//   +0.5 C   000000001                   $001
//      0 C   000000000                   $000
//   -0.5 C   111111111                   $1FF
//  -16.0 C   111100000                   $1E0
//  -55.0 C   110010010                   $192
void DS1620Init(void){ // PS7=RST=0
    DDRS=0xE0;   // PS6=CLK=SPI clock out
    PORTS=0x60;  // PS5=DQ=SPI bidirectional data
/* bit SPOCR1
7 SPIE = 0  no interrupts
6 SPE  = 1  enable SPI
5 SWOM = 0  regular outputs?
4 MSTR = 1  master
3 CPOL = 1  output changes on fall
2 CPHA = 1  and input clocked in on rise
1 SSOE = 0  PS7 regular output DS1620 RST
0 LSBF = 1  least significant bit first */
    SPOCR1=0x5D;
/* bit SPOCR2
3 PUPS = 0  no internal pullups
2 RDS  = 0  regular drive
0 SPC0 = 1  bidirectional mode */
    SPOCR2=0x01;
    SPOBR=0x02;} // 1MHz could be 2Mhz
```

Program 7.24
C language helper functions for a temperature sensor interface using the SPI.

```
#define SPIF 0x80
void out8(char code){ unsigned char dummy;
// assumes DDRS bit 5 is 1, output
    SPODR=code;
    while((SPOSR&SPIF)==0);  // gadfly wait for SPIF
    dummy=SPODR;}            // clear SPIF
void out9(int code){ unsigned char dummy;
    SPODR=0x00FF & code;     // lsbyte
    while((SPOSR&SPIF)==0);  // gadfly wait for SPIF
    dummy=SPODR;             // clear SPIF
    SPODR=0x00FF&(code>>8);  // msbyte
    while((SPOSR&SPIF)==0);  // gadfly wait for SPIF
    dummy=SPODR;}            // clear SPIF
```

continued on p. 412

Program 7.24
C language helper
functions for a
temperature sensor
interface using the SPI.

continued from p. 411

```
unsigned char in8(void){ int n; unsigned char result;
   DDRS &= 0xDF; // PS5=DQ input
   SPODR=0;        // start shift register
   while((SPOSR&SPIF)==0); // gadfly wait for SPIF
   result=SPODR;            // get data, clear SPIF
   DDRS |= 0x20; // PS5=DQ output
   return result;}
unsigned int in9(void){ unsigned int result;
   DDRS &= 0xDF; // PS5=DQ input
   SPODR=0;        // start shift register
   while((SPOSR&SPIF)==0); // gadfly wait for SPIF
   result=SPODR;            // get LS data, clear SPIF
   SPODR=0;        // start shift register
   while((SPOSR&SPIF)==0); // gadfly wait for SPIF
   if(SPODR&0x01)            // get MS data, clear SPIF
      result |= 0xFF00;     // negative
   else
      result &= 0x00FF;     // positive
   DDRS |= 0x20; // PS5=DQ output
   return result;}
```

The basic idea is that to receive 9 bits of temperature data from the DS1620, the 6812:

1. Sets the RST signal=1
2. Outputs the 8-bit command code for read T to the DS1620 using one SPI transfer
3. Switches the direction bit for the data line to input
4. Inputs the 9-bit data value to the DS1620 using two SPI transfers, least significant bit first and right-justified (i.e., most significant 7 bits are sign extended)
5. Switches the direction bit for the data line to output
6. Sets the RST signal=0

On top of these low-level DS1620 driver functions, we implement the higher-level routines of Program 7.25.

Program 7.25
C language functions for
a temperature sensor
interface using the SPI.

```
#define RST1    PORTS|=0x80;
#define RST0    PORTS= 0x7F;
void DS1620Start(void){
   RST1              // RST=1
   out8(0xEE);
   RST0}             // RST=0
void DS1620WriteConfig(char data){
   RST1              // RST=1
   out8(0x0C);
   out8(data);
   RST0}             // RST=0
void DS1620WriteTH(int data){
   RST1              // RST=1
   out8(0x01);
   out9(data);
   RST0}             // RST=0
```

```
void DS1620WriteTL(int data){
    RST1              // RST=1
    out8(0x02);
    out9(data);
    RST0}             // RST=0
unsigned char DS1620ReadConfig(void){ unsigned char value;
    RST1              // RST=1
    out8(0xAC);
    value=in8();
    RST0              // RST=0
    return value;}
unsigned int DS1620ReadT(void){ unsigned int value;
    RST1              // RST=1
    out8(0xAA);
    value=in9();
    RST0              // RST=0
    return value;}
```

7.8 Glossary

asynchronous communications interface adapter (ACIA) Device to transmit data with asynchronous serial communication protocol (same as UART and SCI).

asynchronous protocol A protocol where the two devices have separate and distinct clocks.

bandwidth The number of information bits per time that is transmitted (same as bit rate and throughput).

baud rate The total number of bits (information, overhead, and idle) per time that is transmitted in the serial communication.

bit rate The number of information bits per time that is transmitted (same as bandwidth and throughput).

bit time The basic unit of time used in serial communication.

data communication equipment (DCE) Modem or printer.

data terminal equipment (DTE) Computer or a terminal.

even parity The number of 1s in the data plus parity is an even number.

frame A complete and distinct packet of bits.

framing error An error when the receiver expects a stop bit (1) and the input is 0.

full-duplex channel Hardware that allows bits (information, error checking, synchronization, or overhead) to transfer simultaneously in both directions.

full-duplex communication A system that allows information (data, characters) to transfer simultaneously in both directions.

half-duplex channel Hardware that allows bits (information, error checking, synchronization, or overhead) to transfer in both directions but only in one direction at a time.

half-duplex communication A system that allows information to transfer in both directions but only in one direction at a time.

hold time When latching data into a register with a clock, it is the time after an edge that the data must continue to be valid (see setup time).

mark True, logic 1 (see space).

odd parity The number of 1s in the data plus parity is an odd number.

overrun error Occurs when the receiver gets a new frame but the data register and shift register already have information.

serial communication Involves the transmission of one bit of information at a time.

serial communications interface (SCI) Device to transmit data with asynchronous serial communication protocol (same as UART and ACIA).

serial peripheral interface (SPI) Device to transmit data with synchronous serial communication protocol.

space False, logic 0 (see mark).

simplex channel Hardware that allows bits (information, error checking, synchronization, or overhead) to transfer only in one direction, used to synchronize the receiver shift register with the transmitter clock.

simplex communication A system that allows information to transfer only in one direction.

start bit Overhead bit(s) specifying the beginning of the frame, used to synchronize the receiver shift register with the transmitter clock.

setup time When latching data into a register with a clock, it is the time before an edge that input must be valid (see hold time).

stop bit Overhead bit(s) specifying the end of the frame, used to separate one frame from the next.

synchronous protocol A system where the two devices share the same clock.

throughput The number of information bits per time that is transmitted (same as bit rate and bandwidth).

universal asynchronous receiver/transmitter (UART) Device to transmit data with asynchronous serial communication protocol (same as SCI and ACIA).

XON/XOFF A protocol used by printers to feedback the printer status to the computer. XOFF is sent from the printer to the computer to stop data transfer, and XON is sent from the printer to the computer to resume data transfer.

7.9   Exercises

7.1 The objective is to build the equivalent of the SCI receive function using input capture and output compare. Assume the baud rate is 1000 bits/s, resulting in a bit time of 1 ms. The TTL-level serial input is connected to an input capture pin like the 6811 PA1/IC2. You can also use an output compare as well. The hardware in Figure 7.49 is one possibility. To implement this problem on another computer, simply connect to an available input capture.

Figure 7.49
Interface used to create a serial input channel using input capture.

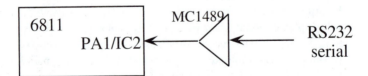

You are given (use without writing these three functions) a FIFO with prototypes:

```
int InitFifo(unsigned int);  // Create and initialize a FIFO,
   // pass in the size, returns true if successful
int PutFifo(char);  // Enter 8 bit data into FIFO,
   // returns true if successful, false if full and data not stored
int GetFifo(&char);  // Remove 8 bit data from FIFO,
   // returns true if successful, false if empty, no data removed
```

The main program, which you do not write, will call `GetFifo` to receive data from your serial interface. Your input capture/output compare interrupt handler(s) will accept the one start, eight data, one stop bit transmission. There is a global error flag, `FLAG`, which should be set on any of the following three conditions. Set the flag if the FIFO becomes full and data are lost (overrun). Also set the flag if the start bit is incorrect (false start). Also set the flag if the stop bit is incorrect (framing error). If there are no false start or framing errors, put the 8-bit data into the FIFO using `PutFifo`. The single-chip 6811 has 256 bytes of RAM, 2 kbytes of EEPROM, and an E clock of 2 MHz.

a) Carefully draw the input signal on PA1/IC2 when the ASCII character '5' is received. Draw the signal to scale. Assume the TCNT value is 1000 (decimal) when the first 1 to 0 transition occurs. Give the TCNT value for each transition of the input signal. Place arrows ($\uparrow$) on the drawing exactly when each IC and OC interrupt will occur, giving the TCNT value at each interrupt.

b) Define all global variables you need. Assume the compiler will place them in the 6811 single-chip RAM.

```
char FLAG;  // set when a serial input error occurs
```

c) Show the ritual that initializes global variables, input capture IC2, and output compare OC2. Assume that the 6811 is running in single-chip mode and that the two interrupt vectors are already initialized in ROM. Carefully choose an appropriate size when calling `InitFifo()` to create/initialize the FIFO.

d) Show the IC2 and OC2 interrupt handlers. *No backward jumps* are allowed.

7.2 The objective of this exercise is to design a printer interface that handles DTR hardware synchronization. To implement this problem on another computer, simply connect to an available input capture. For the 6811 connect: (1) SCI transmit to the printer receive, (2) the DTR pin 20 to input capture IC3/PA0, and (3) ground to ground. Your software must be *interrupt-driven and written in C*. The baud rate is 1200 bits/s. The RS232 protocol is one start bit, seven bit, no parity, and two stop bits. There is a DTR signal from the printer. When DTR is -12 V (PA0=high), your 6811 should not begin transmitting any more frames because the printer is not ready to accept it. When DTR is $+12$ V (PA0=low), the printer is ready to accept more data. There are three FIFO functions given (you do *not* write them). Your `OutChar(data)` is called by the main program (again, you do not write the main program). To implement this problem on another computer, simply connect DTR to an available input capture.

`InitFifo()`   initializes the FIFO
`flag=PutFifo(data)`   enters a byte `data` into the FIFO
                              returns a `flag`=true if the FIFO is full, and the data was not saved.
`flag=GetFifo(&data)` returns with a byte and `flag` is false if successful
                              returns `flag` true if the FIFO was empty at the time of the call

a) Show the ritual software that is executed once. Assume the interrupt vectors will be initialized automatically by the compiler.

b) Show the C function `OutChar(data)` that initiates an output. The actual output occurs in the background. `OutChar` should return a true flag if the operation cannot be performed.

c) Give the interrupt routine(s) that performs the serial output operations in the background. Give careful thought as to when and how to interrupt. Poll anyway you want. *Good software has* no *backward jumps.* Use labels `IC3handler()` and `SCIhandler()` for the two interrupt routines. Comments will be graded.

7.3 Design an automatic baud rate selector. The input capture system will be used to measure the first three positive logic pulse widths and the first three negative logic widths. To implement this problem on another computer, simply connect to an available input capture. Your software will then choose the smallest pulse width as the communication bit time. For example, if the smallest bit time is 400 E cycles (which falls between 625 and 313), then the baud rate should be 4800. Finally, your software will initialize the SCI to communicate at that correct baud rate (eight bit data, one start, one stop and no parity). Specify the E clock rate. You may assume the baud rate is 600, 1200, 2400, 4800, or 9600.

a) Show the hardware connections between a three-wire ±12 V RS232 full-duplex line (T×D, R×D, and Gnd) and the microcomputer. Any RS232 interface chip may be used. Give chip numbers but not pin numbers.

b) Show the appropriate data structure that holds the information of the above table. The organization of this information should be easy to understand and easy to change (e.g., add more baud rates, change of E clock frequency). This data structure will eliminate the need for a complex `if then` or `switch code` in part c.

c) Show the C software function that automatically selects the correct baud rate and initializes the SCI. Gadfly synchronization will be used. Input/output functions for the SCI are not required.

7.4 You will build a low-cost computer-based logic analyzer. It will be able to collect 2048 8-bit digital samples at 80 Mhz triggered by your software. You are given the following hardware components (you may use other 74HC digital devices, too).

> A 20-MHz clock generator (black box with 50% duty cycle digital output)
> A 2048 9-bit hardware FIFO (CY7C429, IDT7203, AM7203, or LH540203)

The FIFO contains a 2048 by 9 bit RAM and implements the classic FIFO functions. You will only use eight of the nine data pins. To reset the FIFO, you pull its ***RS** line low. This will clear the FIFO. When the reset ***RS** is high, you can perform a put operation by toggling the ***W** line low, then high. The 9 bits on the input data lines **D8–D0** are stored (put) in the FIFO on the rising edge of the ***W** line. When the reset ***RS** is high, you can perform a get operation by toggling the ***R** line low, then high. The 9 bits are removed (get) from the FIFO on the rising edge of the ***R** line. These 9 bits are available on the output data lines **O8–O0** when the *R line is low. There are three negative logic FIFO status outputs that are available:

full flag	***FF**	0 means full
half flag	***HF**	0 means more than half full
empty flag	***EF**	0 means empty

a) Show the hardware connections to your single-chip microcomputer. A RS232 channel will connect your logic analyzer system to a personal computer. The only external connections to your logic analyzer system are the eight FIFO data inputs. For some cool features that are possible with this approach, see the Circuit Cellar, *INK,* vol. 89, December 1997, pp. 46–49. Other features discussed in this article, but not to be implemented here, are computer-controlled sampling rate, optional external clock, and external reset to the device under test so that the FIFO and external circuit are started together.

b) Show the main program that initializes the SCI to 9600 baud, one stop, no parity, then loops:

- Resets the FIFO
- Waits for any character to be received on the SCI input (start command from the PC)
- Allows the FIFO to fill with 8-bit data at 20 MHz
- Waits for the FIFO to be full
- Reads all 2048 bytes from the FIFO and transmits them out the SCI output one at a time

This computer is dedicated to this task, so you must show all the software for the computer. The program never quits, repeats the loop over and over. Use gadfly or interrupt synchronization, whichever is most appropriate.

8 Parallel Port Interfaces

Chapter 8 objectives are to:

❑ Develop fundamental concepts associated with the physical behavior of the device

❑ Design the hardware interface between the device and the parallel port

❑ Write low-level device drivers that perform basic I/O with our device

❑ Discuss interfacing issues associated with performing the I/O as a background interrupt process

Chapter 8 deals with external devices that we connect to the parallel I/O port of our computer. In particular, we will interface switches, keyboards, single LEDs, LED displays, single LCDs, LCD displays, relays, and stepper motors. Starting with this chapter and running through Chapter 11, we make the shift away from the details of the microcomputer itself and discuss devices external to the computer. Later, in Chapters 12 through 14, the pieces (microcomputer and external devices) are combined to design embedded systems.

8.1 Input Switches and Keyboards

8.1.1 Interfacing a Switch to the Computer

We begin with the simple switch interface (Figure 8.1). To convert the mechanical signal into an electric signal, a resistor pull-up is used. When the switch is open, the output is pulled to +5 V. The amount of current this output can source (its equivalent I_{OH}) is determined by the resistor value. The smaller the resistor, the larger the I_{OH}. When the switch is closed, the output is forced to ground. The amount of current this output can sink (its equivalent I_{OL}) is huge, limited only by the capacity of the switch to conduct current. The resistor does not affect the equivalent I_{OL}. A smaller resistor does waste current when the switch contact is closed.

In digital logic we usually pull-up to +5 V rather than pulling down to zero because in most situations I_{OL} needs to be larger than I_{OH}. CMOS digital logic is an exception to this rule. Either pull-up or pull-down could be used for CMOS. When the switch in the pull-down circuit of Figure 8.2 is open, the output is pulled to ground. The amount of current this output can sink (its equivalent I_{OL}) is determined by the resistor value. The smaller the resistor, the

Figure 8.1
A simple switch interface.

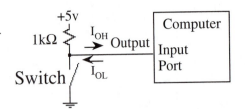

Switch	Output
Open	+5v
Closed	0v

Figure 8.2
Another simple switch interface.

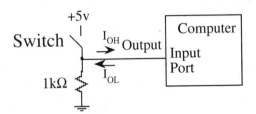

Switch	Output
Open	0v
Closed	+5v

Figure 8.3
The MC68HC812A4 Port J supports internal pull-up or pull-down.

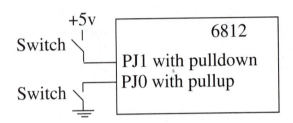

larger the I_{OL}. When the switch is closed, the output is forced to +5 V. The amount of current this output can source (its equivalent I_{OH}) is huge, limited only by the capacity of the switch to conduct current. The resistor does not affect the equivalent I_{OH}. Notice that the polarity of the switch is reversed in the pull-down interface as compared to the pull-up case.

Port J on the MC68HC812A4 supports both internal pull-ups and pull-downs. That is, either of the two circuits in Figures 8.1 and 8.2 could be implemented on the 6812 without the resistor, as, for example, in Figure 8.3.

The software initialization sets bits in the PUPEJ register to enable pull-up or pull-down. For each Port J pin that is enabled for pull-up or pull-down, the corresponding bit in the PUPSJ register determines if it is pull-up (1) or pull-down (0). It is good programming practice to first set the PUPSJ register, then set the PUPEJ register so that temporary glitches are avoided. Program 8.1 will initialize Port J with the appropriate pull-up and pull-down.

```
// MC68HC812A4
// PortJ bit 1 is connected to a switch to +5, using internal pull-down
// PortJ bit 0 is connected to a switch to 0, using internal pull-up
void Initialization(void){
    DDRJ  &= 0xFC;   // PJ1-0 inputs
    KPOLJ |= 0x03;   // flags set on the rise of PJ1, PJ0
    KWIEJ &= 0xFC;   // disarm PJ1, PJ0
    KWIFJ =  0x03;   // clear flags
    PUPSJ =  (PUPSJ&0xFC)|0x01; // pull-down on PJ1, pull-up on PJ0
    PUPEJ |= 0x03;}  // enable pull-up and pull-down
```

Program 8.1 The MC68HC812A4 Port J initialization.

8.1.2
Hardware Debouncing Using a Capacitor

The mechanical properties of a switch strongly affect the mechanical response to an applied force according to the following second-order differential equation. The three terms arise from Newton's Second Law, friction, and the spring constant, respectively.

$$ F = m\frac{d^2x}{dt^2} + K_d\frac{dx}{dt} + Kx $$

where m is the switch mass, K_d is the friction coefficient, and K is the spring constant. The damping ratio can be calculated as

$$ \xi = \frac{K_d}{2\sqrt{K{\cdot}m}} $$

Depending on the relative values for m, K_d, and K, the step response (i.e., what happens when you put your finger on the button) will be underdamped ($\xi > 1$ causes ringing) or overdamped ($\xi < 1$ causes a delayed but smooth rise.) Most inexpensive switches are underdamped. Hence, they will bounce both when touched and when released. Typical bounce times range from 1 to 25 ms. Ideally, the switch resistance is zero (actually about 0.1Ω) when closed and infinite when open (Figure 8.4).

Figure 8.4
Switch timing showing bounce on touch and release.

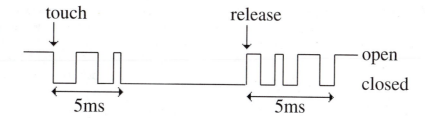

More expensive switches can be purchased with proper damping. These oil-filled switches do not bounce. Because it is easy to add software to solve the bounce problem, most keyboard devices use inexpensive switches that bounce. With the circuit having just a pull-up resistor, the electric output will bounce because the mechanical input bounces (Figure 8.5).

Figure 8.5
Switch bounce can be seen on the voltage signal.

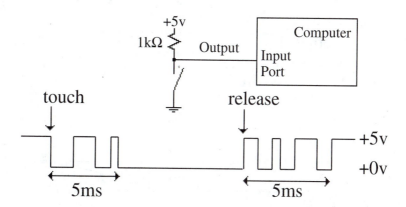

It may or may not be important to debounce the switch. For example, if we are counting the number of times the operator pushes the button, then we must debounce so that one push results in incrementing just once. We will debounce our standard computer keyboard so that when the operator types the letter A into her word processor, only one A is entered into the file. On the other hand, if the switch position specifies some static condition, and the operator sets the switch before she turns on the computer, then debouncing is not necessary.

We will study both hardware and software mechanisms to debounce the switch. The simplest hardware method is to use a capacitor across the switch (Figure 8.6). The capacitor value is chosen large enough so that the input voltage does not exceed the 0.7-V threshold of the NOT gate while it is bouncing. A detailed discussion about selecting the capacitor value will be presented later in this section.

Figure 8.6
Switch bounce removed with a capacitor (this is a bad circuit).

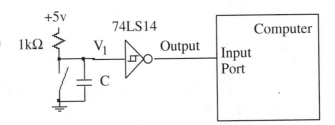

The touch timing with and without the capacitor is as shown in Figure 8.7. Notice that there is no delay between the touching of the switch and the transition of the signal output. The release timing with and without the capacitor is shown in Figure 8.8. There is a significant delay from the release of the switch until the fall of the output. To repeat, the capacitor is chosen such that the input voltage does not exceed threshold during the bouncing (Figure 8.9).

Figure 8.7
Switch touch bounce is removed by the capacitor.

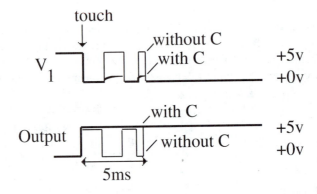

Figure 8.8
Switch release bounce is also removed by the capacitor.

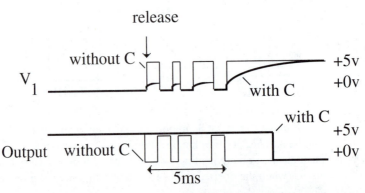

Figure 8.9
Timing used to calculate
capacitor value.

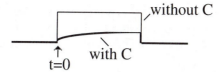

The voltage rise during a bounce interval when the switch is open is given by

$$V \geq 5 - 5e^{-t/RC}$$

In this example, R=1KΩ, and the bounce time is 5 ms. Thus, we will choose C such that

$$0.7 \geq 5 - 5e^{-5 \text{ ms}/(1k\Omega \cdot C)}$$
$$0.86 \leq e^{-5ms/(1k\Omega \cdot C)}$$
$$\ln(1.16) \geq 5ms/(1k\Omega \cdot C)$$
$$C \geq 5ms/(1k\Omega \cdot \ln(1.16)) = 33\mu F$$

One problem with the above interface is the instantaneous current that occurs when the switch bounces closed. At $t=0^-$, there is a charge on the capacitor. At $t=0^+$, the energy has been discharged and the voltage is zero. Theoretically, this can only occur if the current at $t=0$ is infinite (Figure 8.10).

Figure 8.10
A spark will occur
because a large current
occurs when the switch
is touched.

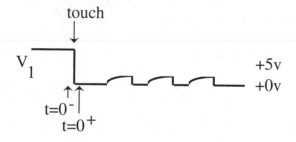

In reality, the current is not infinite, but it is large enough to cause a spark. These sparks will produce carbon deposits on the switch that will build up until the switch no longer works. To limit the current (thereby eliminating the sparks) a 22Ω resistor is placed in series with the switch. The value 22Ω was chosen to be much smaller than the 1kΩ, but much larger than the 0.1Ω contact resistance of the switch.

If an input switch is closed, its resistance will be about 0.1Ω, and the output of the 74LS14 will be high (a logic 1). If an input switch is open, its resistance will be infinite, and the output of the 74LS14 will be low (a logic 0). If you connect the output of this switch interface to a computer parallel port input, then the computer can read the state (on/off) of the switch (Figure 8.11).

Figure 8.11
A hardware interface
that removes the bounce
(good circuit compared
to Figure 8.6).

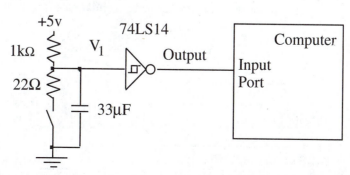

The Schmitt trigger 74LS14 is used to prevent multiple transitions when the switch is released. The difference between a regular NOT gate (74LS04) and the Schmitt trigger NOT gate (74LS14) is hysteresis (Figure 8.12). Hysteresis is required because the input of the NOT gate in the transition range is

$$0.7 \le V_1 \le 2.0 \text{ V}$$

Figure 8.12
Input/output relationship showing the difference between a 74LS04 and a 74LS14.

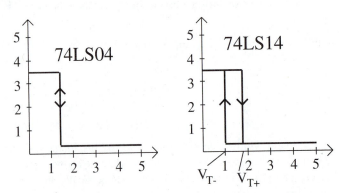

for a long period. Normally the time during which a digital input is in transition is on the order of a few nanoseconds. Recall that the rise and fall time for a 2-MHz E clock is 20 to 25 ns. For faster clocks, the transition times become even shorter. But in this application because of the 33 μF capacitor, the time in the transition range, Δt, is a huge 12 ms! In particular,

$$\Delta t = t_2 - t_1$$

where

$$0.7 = 5 - 5\, e^{-t_1/(1k\Omega \cdot 33\ \mu F)}$$

and

$$2.0 = 5 - 5e^{-t_2/(1k\Omega \cdot 33\ \mu F)}$$

Thus

$$t_1 = 1k\Omega \cdot 33\ \mu F \cdot \ln(1.16) = 5 \text{ ms}$$

and

$$t_2 = 1k\Omega \cdot 33\ \mu F \cdot \ln(1.67) = 17 \text{ ms}$$

Thus

$$\Delta t = 12 \text{ ms}$$

If the input stays in the transition range for a long time with just a little noise, then the output of a regular digital gate will toggle with the noise. The hysteresis removes the extra transitions that might occur with a regular gate (Figure 8.13).

Figure 8.13
Timing showing why a 74LS14 is used instead of a regular digital gate.

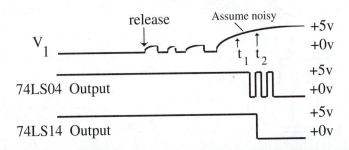

Observation: With a capacitor-based debounced switch, there is no delay between the closing of the switch and the rising edge at the computer input.

Observation: With a capacitor-based debounced switch, there is a large delay (over 10 ms) between the opening of the switch and the falling edge at the computer input.

8.1.3
Software
Debouncing

8.1.3.1
Software Debouncing
with Gadfly
Synchronization

It will be less expensive to remove the bounce using software methods. It is appropriate to use a software approach because the software is fast compared to the bounce time. Typically we use the pull-up resistor to convert the switch position into a TTL-level digital signal. The MC68HC812A4 also supports internal pull-up resistors, which can reduce the component count and simplify manufacturing. The 6812 key wake up could have been used in place of the input capture (Figure 8.14).

Figure 8.14
Switch interface.

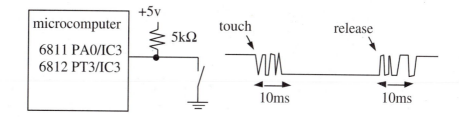

In these examples, we assume the switch bounce is less than 10 ms. The first two routines implement software debouncing using gadfly synchronization. The computer is dedicated to the interface and does not perform any other functions while the routines are running. The first routine (Program 8.2) waits for the switch to be pressed (PA0/PT3 low) and return 10 ms after the switch is pressed (Figure 8.15). The advantages of using output compare for implementing the wait loop are (1) it is easier to understand and

```
; MC68HC11A8
WaitPress: ldaa PORTA
;PA0=0 if switch pressed
           anda #$01
           bne  WaitPress
;loop until switch is pressed
           ldd  TCNT
           addd #20000   ;10ms delay
           std  TOC5
           ldaa #$08
           staa TFLG1    ;clear OC5F
loopP:     ldaa TFLG1
;10ms for switch to stop bouncing
           anda #$08     ;OC5F set?
```

```
; MC68HC812A4
WaitPress: ldaa PORTT
;PT3=0 if switch pressed
           anda #$08
           bne  WaitPress
;loop until switch is pressed
           ldd  TCNT
           addd #20000        ;10ms delay
           std  TC5           ; fast clear
loopP:     ldaa TFLG1
;wait 10ms for switch to stop bouncing
           anda #$20          ;C5F set?
           beq  loopP
           rts
```

continued on p. 424

Program 8.2 Switch debouncing using assembly software.

continued from p. 423

```
          beq   loopP
          rts
WaitRelease: ldaa PORTA
;PA0=1 if switch released
          anda #$01
          beq WaitRelease
;loop until switch is released
          ldd TCNT
          addd #20000  ;10ms delay
          std TOC5
          ldaa #$08
          staa TFLG1   ;clear OC5F
loopR:   ldaa TFLG1
;10ms for switch to stop bouncing
          anda #$08    ;OC5F set?
          beq loopR
          rts
```

```
WaitRelease: ldaa PORTT
;PT3=1 if switch released
          anda #$08
          beq  WaitRelease
;loop until switch is released
          ldd   TCNT
          addd #20000      ;10ms delay
          std  TC5         ; fast clear
loopR: ldda TFLG1
;wait 10ms for bouncing to stop
          anda #$20        ;C5F set?
          beq  loopR
          rts
Init: bset #$20,TIOS  ;enable OC5
      movb #$90,TSCR  ;enable fast clr
      movb #$32,TMSK2 ;500 ns clk
      bclr #$08,DDRT  ;PT3 is input
      rts
```

Program 8.2 Switch debouncing using assembly software.

Figure 8.15
Software flowcharts for
debouncing the switch.

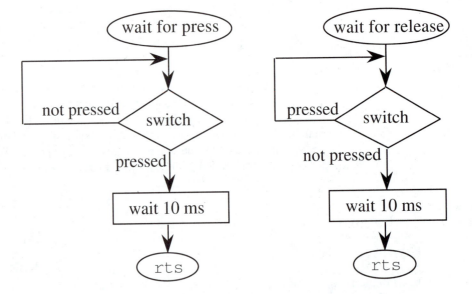

change if needed and (2) it is more accurate if background interrupts are occurring. No 6811 initialization is required because PA0 is always an input and OC5 output compare is active by default. In C, these routines could be written as shown in Program 8.3.

This second routine reads the current value of the switch (Program 8.4). If the switch is currently bouncing, it will wait for stability (Figure 8.16). A return value of 0 means pressed (PA0/PT3=0), and 1 means not pressed (PA0/PT3=1). Notice that the software always waits in a "do nothing" loop for 10 ms. This inefficiency can be eliminated by

```
// MC68HC11A8                          // MC68HC812A4
void WaitPress(void){                  void WaitPress(void){
    while (PORTA&0x01);                    while (PORTT&0x$08);
// Loop here until switch is pressed   // Loop here until switch is pressed
    TOC5=TCNT+20000;   // 10ms delay       TC5=TCNT+20000;    // 10ms delay
    TFLG1=0x08;        // Clear OC5F        while((TFLG1&0x20)==0);}
    while((TFLG1&0x08)==0);}            // wait for switch to stop bouncing
// wait for switch to stop bouncing

                                       void WaitRelease(void){
void WaitRelease(void){                    while ((PORTT&0x$08)==0);
    while ((PORTA&0x01)==0);           // Loop here until switch is released
// Loop here until switch is released      TC5=TCNT+20000;    // 10ms delay
    TOC5=TCNT+20000;   // 10ms delay        while((TFLG1&0x20)==0);}
    TFLG1=0x08;        // Clear OC5F    // wait for switch to stop bouncing
    while((TFLG1&0x08)==0); }
//wait for switch to stop bouncing     void ritual (void) {
                                         TIOS|=0x20        // enable OC5
// no ritual required                    TSCR=0x90;        // enable, fast clear
                                         TMSK2=0x32;       // 500 ns clock
                                         DDRT &= 0xF7;}    // PT3 is input
```

Program 8.3 Switch debouncing using C software.

```
; MC68HC11A8                           ; MC68HC812A4
* Reg B is the return value           * Reg B is the return value
ReadPA0: ldd  TCNT                     ReadPT3: ldd  TCNT
         addd #20000 ;10ms delay                addd #20000 ;10ms delay
         std  TOC5                              std  TC5
         ldaa #$08                              ldab PORTT ;0 if pressed
         staa TFLG1 ;clear OC5F                 andb #$08  ;B=old value
         ldab PORTA ;0 if pressed      Same:    ldaa TFLG1 ;10ms bouncing
         andb #$01  ;B=old value                anda #$20  ;C5F set?
Same:    ldaa TFLG1 ;10ms bouncing              bne  Done
         anda #$08  ;OC5F set?                  ldaa PORTT ;0 if pressed
         bne  Done                              anda #$08  ;A=new value
         ldaa PORTA ;0 if pressed               cba        ;same as before
         anda #$01  ;A=new value                beq  Same
         cba        ;same as before             bra  ReadPT3; different
         beq  Same                     Done:    rts
         bra  ReadPA0                  Init: bset #$20,TIOS ;enable OC5
;start over because different                movb #$90,TSCR ;enable fast clr
Done:    rts                                  movb #$32, TMSK2 ;500 ns clk
                                             bclr #$08,DDRT   ;PT3 is input
                                             rts
```

Program 8.4 Another example of switch debouncing using assembly software.

Figure 8.16
Another software
flowchart for
debouncing the switch.

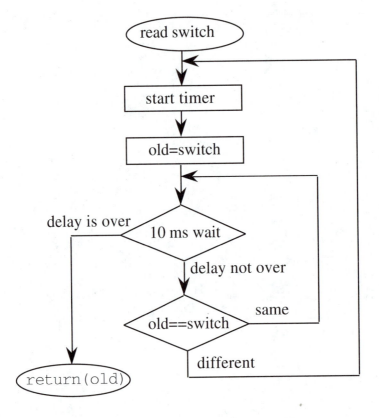

```
// MC68HC11A8                                   // MC68HC812A4
unsigned char ReadPA0(void){                    unsigned char ReadPT3(void){
unsigned char old;                              unsigned char old;
  old=PORTA&0x01;    // Current value             old=PORTT&0x08;    // Current value
  TOC5=TCNT+20000;   // 10ms delay                TC5=TCNT+20000;    // 10ms delay
  TFLG1=0x08;        // Clear OC5F                 while((TFLG1&0x20)==0){
  while((TFLG1&0x08)==0){                        // unchanged for 10ms?
// unchanged for 10ms?                               if(old<>(PORTT&0x08)){ // changed?
    if(old<>(PORTA&0x01)){ // changed?                 old=PORTT&0x08;    // New value
      old=PORTA&0x01;    // New value                  TC5=TCNT+20000;}   // restart delay
      TOC5=TCNT+20000;   // restart delay          }
      TFLG1=0x08; }      // Clear OC5F             return(old);}
  }                                             void ritual(void) {
  return(old);}                                   TIOS|=0x20;    // enable OC5
                                                  TSCR=0x90;     // enable, fast clear
// no ritual required                             TMSK2=0x32;    // 500 ns clock
                                                  DDRT &= 0xF7;} // PT3 is input
```

Program 8.5 Another example of switch debouncing using C software.

placing the switch I/O in a background interrupt-driven thread. In C, these routines could be written as shown in Program 8.5.

Observation: With a software-based debounced switch, the signal arrives at the computer input without delay, but software delays may occur at either touch or release.

Input capture, which is available on most Motorola microcomputers, is a convenient mechanism to detect changes on the digital signal. The key wake-up mechanism on the 6812 provides an alternative to input capture. First we will present the interface using input capture. The input capture can be configured to interrupt either on the rise, the fall, or both the rise and the fall. Because of the bounce, any of these modes will generate an interrupt request when the key is touched or released (Figure 8.17).

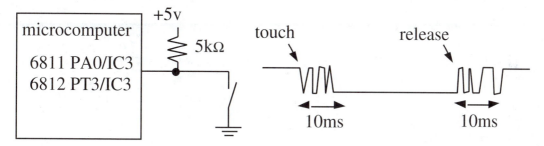

Figure 8.17 Switch interface.

A combination of input capture and output compare interrupts allows the switch interface to be performed in the background. This example simply counts the number of times the switch is pressed. The IC3 interrupt occurs immediately after the switch is pressed and released. Because the IC3 handler disarms itself, the bounce will not cause additional interrupts. The OC5 interrupt occurs 10 ms after the switch is pressed and 10 ms after the switch is released. At this time the switch position is stable (no bounce) (Figure 8.18).

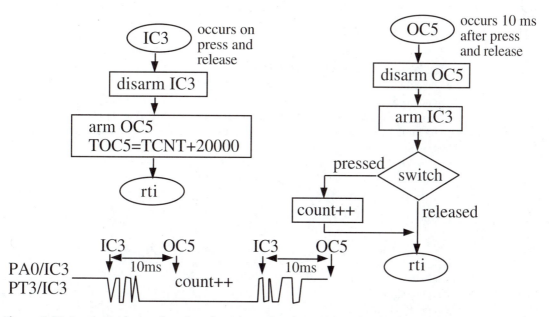

Figure 8.18 Another software flowchart for debouncing the switch with interrupts.

The first IC3 interrupt occurs when the switch is first touched. The first OC5 interrupt occurs 10 ms later. At this time the global variable, count, is incremented. The second IC3 interrupt occurs when the switch is released. The second OC5 interrupt does not increment the count but simply rearms the input capture system. The ritual initializes the system with IC3 armed and OC5 disarmed (Program 8.6).

```
// MC68HC11A8
// counts the number of button pushes
// signal connected to IC3=PA0
unsigned int count; // times pushed
#define wait 20000  // bounce wait (cyc)
#pragma interrupt_handler TOC5handler()
void TOC5handler(void){
    TMSK1=0x01;      // Arm IC3, disarm OC5
    TFLG1=0x01;      // clear IC3F
    if(PORTA&0x01==0) count++;}
// new count if PA0=0
#pragma interrupt_handler TIC3handler()
void TIC3handler(void){
    TMSK1=0x08;    // Disarm IC3, Arm OC5
    TOC5=TCNT+wait;
    TFLG=0x08;}      // clear OC5F
void Ritual (void){
    asm(" sei");   //make atomic
    TMSK1=0x01;      //Arm IC3, disarm OC5
    TFLG1=0x01;      //clear IC3F
    TCTL2 = 0x01; // IC3F set on rising
    count=0;
    asm(" cli"); }
```

```
// MC68HC812A4
// counts the number of button pushes
// signal connected to IC3=PT3
unsigned int count; // times pushed
#define wait 20000  // bounce wait (cyc)
#pragma interrupt_handler TOC5handler()
void TOC5handler(void){
    TMSK1=0x08;      // Arm IC3, disarm OC5
    TFLG1=0x08;      // clear C3F
    if(PORTT&0x08==0) count++;}
// new count if PT3=0
#pragma interrupt_handler TIC3handler()
void TIC3handler(void){
    TMSK1=0x20;      // Disarm IC3, Arm OC5
    TC5=TCNT+wait;} // clear CSF
void Ritual(void){
    asm(" sei");  // make  atomic
    TIOS=(TIOS&0xF7)|0x20; // enable OC5, IC3
    TSCR=0x90;        // enable, fast clear
    TMSK2=0x32;       // 500 ns clock
    DDRT=(DDRT&0xF7)|0x20; // PT3 is input
    TMSK1=0x08;      // Arm IC3, disarm OC5
    TFLG1=0x08;      // clear C3F
    TCTL4 = (TCTL4&0x3F)|0x40;  // rising
    count=0;
    asm(" cli"); }
```

Program 8.6 Switch debouncing using interrupts in C software.

The latency of this interface is defined as the time when the switch is touched until the time when the count is incremented. Because of the delay introduced by the OC5 interrupt, the latency is 10 ms. If we assume the switch is not bouncing (currently being touched or released) at the time of the ritual, we can reduce this latency to less than 50 μs by introducing a global variable. If LastState is true, then the switch is currently not pressed and the software is searching for a touch. If LastState is false, then the switch is currently pressed and the software is searching for a release (Program 8.7).

```
// MC68HC11A8
// counts the number of button pushes
// signal connected to IC3=PA0
unsigned int count; // times pushed
char LastState; // looking for touch?
// true means open, looking for a touch
// false means closed, looking for release
#define wait 20000  // bounce wait (cyc)
```

```
// MC68HC812A4
// counts the number of button pushes
// signal connected to IC3=PT3
unsigned int count; // times pushed
char LastState; // looking for touch?
// true means open, looking for a touch
// false means closed, looking for release
#define wait 20000  // bounce wait (cyc)
```

```
#pragma interrupt_handler TOC5handler()        #pragma interrupt_handler TOC5handler()
void TOC5handler(void){                         void TOC5handler(void){
    TMSK1=0x01;    // Arm IC3, disarm OC5           TMSK1=0x08;    // Arm IC3, disarm OC5
    TFLG1=0x01;}   // clear IC3F                     TFLG1=0x08;}   // clear C3F
#pragma interrupt_handler TIC3handler()         #pragma interrupt_handler TIC3handler()
void TIC3handler(void){                         void TIC3handler(void){
    if(LastState){                                  if(LastState){
       count++; // a touch has occurred              count++; // a touch has occurred
       LastState=0;}                                 LastState=0;}
    else                                            else
       LastState=1; // release occurred             LastState=1; // release occurred
    TMSK1=0x08;    // Disarm IC3, Arm OC5        TMSK1=0x20;    // Disarm IC3, Arm OC5
    TOC5=TCNT+wait;                              TC5=TCNT+wait;} // clear C5F
    TFLG1=0x08;}   // clear OC5F                 void Ritual (void){
void Ritual (void){                                 asm(" sei");   //make atomic
    asm(" sei");   //make atomic                TIOS=(TIOS&0xF7)|0x20;  // enable OC5, IC3
    TMSK1=0x01;    //Arm IC3, disarm OC5        TSCR=0x90;         // enable, fast clear
    TFLG1=0x01;    //clear IC3F                 TMSK2=0x32;    // 500 ns clock
    TCTL2 = 0x01;  // IC3F set on rising        DDRT=(DDRT&0xF7)|0x20; // PT3 is input
    count=0;                                    TMSK1=0x08;    // Arm IC3, disarm OC5
    LastState=PORTA&0x01;                       TFLG1=0x08;    // clear C3F
    asm(" cli"); }                              TCTL4 = (TCTL4&0x3F)|0x40;  // rising
                                                count=0;
                                                LastState=PORTT&0x08;
                                                asm(" cli"); }
```

Program 8.7 Another example of switch debouncing using interrupts in C software.

Now the latency is simply the time required for the microcomputer to recognize and process the input capture interrupt. Assuming there are no other interrupts, this time is less than 20 cycles.

8.1.4
Basic Approaches
to Interfacing
Multiple Keys

In this section we attempt to interface as many switches as possible to a single 8-bit parallel port, and we will consider three interfacing schemes (Figure 8.19). In a *direct interface* we connect each switch to a separate microcomputer input pin. For example, using just one 8-bit parallel port, we can connect eight switches using the direct scheme. An advantage of this interfacing approach is that the software can recognize all 256 (2^8) possible switch

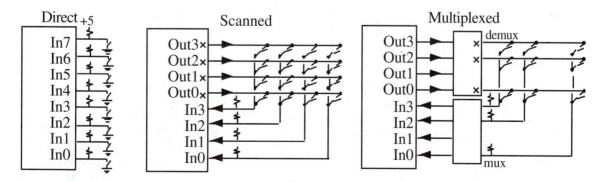

Figure 8.19 Three approaches to interfacing multiple keys.

patterns. If the switches were remote from the microcomputer, we would need a nine-wire cable to connect it to the microcomputer. In general, if there are n switches, we would need $n/8$ parallel ports and $n+1$ wires in our remote cable. This method will be used when there are a small number of switches or when we must recognize multiple and simultaneous key presses (Figure 8.20). Examples include music keyboards and the shift and control option keys. As illustrated in the Music Keyboard example in Section 8.15, implementing an interrupt-driven interface requires a lot of hardware. Therefore, if interrupt synchronization is required, it may be more appropriate to utilize periodic polling interrupt synchronization.

Figure 8.20
Multiple keys are implemented by placing the switches in a matrix. (Notice there are fewer wires in the cable than there are keys.)

In a *scanned interface* the switches are placed in a row/column matrix. The $\times$ at the four outputs signifies an open collector (an output with two states, HiZ and low). The computer drives one row at a time to zero while leaving the other rows at HiZ. By reading the column, the software can detect if a key is pressed in that row. The software "scans" the device by checking all rows one by one. For the 6811, the CWOM=1 feature of Port C can be used to implement this HiZ/0 binary output. For the other microcomputers, the open-collector functionality will be implemented by toggling the direction register. Table 8.1 illustrates the sequence to scan the four rows.

Table 8.1
Scanning patterns for a 4 by 4 matrix keyboard.

Row	Out3	Out2	Out1	Out0
3	0	HiZ	HiZ	HiZ
2	HiZ	0	HiZ	HiZ
1	HiZ	HiZ	0	HiZ
0	HiZ	HiZ	HiZ	0

For computers without an open-collector output mode, the direction register can be toggled to simulate the two output states, HiZ/0, of open-collector logic (see Exercise 1.10). This method can interface many switches with a small number of parallel I/O pins. In our example situation, the single 8-bit I/O port can handle 16 switches with only an eight-wire cable. The disadvantage of the scanned approach over the direct approach is that it can only handle situations where zero, one, or two switches are simultaneously pressed. This method is used for most of the switches in our standard computer keyboard. Recall that the shift, alt, and control keys are interfaced with the direct method. We can "arm" this interface for interrupts by driving all the rows to 0. The key wake-up mechanism can be used to generate interrupts on touch and release. Because of the switch bounce, a key wake-up interrupt will occur when any of the keys changes. For microcomputers like the 6811 without a key wake-up, we can add a simple NAND gate connected to all columns. For the 6811 interface, the output of the NAND gate will change when any of the keys are pressed.

In a *multiplexed interface,* the computer outputs the binary value defining the row number, and a hardware demultiplexer will output the 0 on the selected row and HiZ's on the other rows. The demultiplexer must have open-collector outputs (illustrated again by the $\times$ in the Figure 8.21 circuit). The computer outputs the sequence $00, $10, $20, $30 . . . , $F0 to scan the 16 rows, as shown in Table 8.2.

Table 8.2
Scanning patterns for a multiplexed 16 by 16 matrix keyboard.

Row	Computer output				Demultiplexer output			
	Out3	Out2	Out1	Out0	15	14	. . .	0
15	1	1	1	1	0	HiZ	. . .	HiZ
14	1	1	1	0	HiZ	0	. . .	HiZ
. . .	. . .	. . .	. . .	. . .	. . .	. . .	. . .	. . .
1	0	0	0	1	HiZ	HiZ	. . .	HiZ
0	0	0	0	0	HiZ	HiZ	. . .	0

In a similar way, the column information is passed to a hardware multiplexer that calculates the column position of any 0 found in the selected row. One additional signal is necessary to signify the condition that no keys are pressed in that row. Since this interface has 16 rows and 16 columns, we can interface up to 256 keys! We could sacrifice one of the columns to detect the no key pressed in this row situation. In this way, we can interface 240 (15 · 16) on the single 8-bit parallel port. If more than one key is pressed in the same row, this method will detect only one of them. Therefore, we classify this scheme as being able to handle only 0 or one key pressed. Applications that can utilize this approach include touch screens and touch pads because they have a lot of switches but are interested only in the 0 or 1 touch situation. Implementing an interrupt-driven interface would require too much additional hardware. In this case, periodic polling interrupt synchronization would be appropriate.

8.1.5
Sixteen-Key
Electronic Piano

In this direct interface, we connect each key up to a separate computer input pin so that we can distinguish all 2^{16} possible combinations. Again, we assume a 10-ms switch bounce time (Figure 8.21).

Figure 8.21 Multiple keys are interfaced directly to input ports.

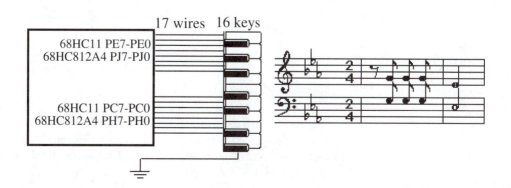

The computer inputs require an external pull-up resistor. Each bit in the global, KEY, is affected by a key (1 for not pressed, 0 for pressed) (Program 8.8).

```
; MC68HC11A8
KEY:    ds  2       ;current value
RITUAL: clr DDRC ; all inputs
        ldaa #$54
; Bit signal value comment
;  6  STAI    1    arm STAF
;  5  CWOM    0    not applicable
;  4  HNDS    1    Input handshake
;  3  OIN     0
;  2  PLS     1    pulse out
;  1  EGA     0    falling edge
;  0  INVB    0    negative pulse
        staa PIOC
        bsr  KeyBoard ;set 74LS374's
        cli           ;Enable IRQ
        rts
KeyBoard: ldaa PIOC ;part of clear
        ldaa PORTE   ;Read MSB
        ldab PORTCL  ;Read LSB, ack
        std  KEY
        rts
IRQHan: ldaa #$14   ;Disarm STAF
        staa PIOC
        ldd  TCNT
        addd #20000  ;OC5 10ms later
        std  TOC5
        ldaa #$08    ;Arm OC5
        staa TMSK1
        staa TFLG1   ;Clear OC5F
        rti
OC5Han: clr TMSK1    ;Disarm OC5
        ldaa #$54    ;Rearm STAF
        staa $PIOC
        bsr  KeyBoard ;read keyboard
        rti
```

```
; MC68HC812A4
KEY:    ds  2       ;current value
RITUAL: bset #$20,TIOS ;enable OC5
        movb #$90,TSCR  ;enable fast clr
        movb #$32,TMSK2 ;500 ns clk
        bclr #$20,TMSK1 ;disarm OC5
        clr  DDRH       ;key inputs
        clr  DDRJ       ;key inputs
        movb #$FF,KWIEH ;arm key wakeup
        movb #$FF,KWIEJ ;arm key wakeup
        clr  KPOLJ      ;falling edge
        clr  PUPEJ      ;regular inputs
        bsr  KeyBoard ;set 74LS374's
        cli             ;Enable IRQ
        rts
KeyBoard: movb #$FF,KWIFH ;clr flags
        movb #$FF,KWIFJ ;clear flags
        ldaa PORTJ   ;Read MSB
        ldab PORTH   ;Read LSB
        std  KEY
        rts
KeyHHan: clr KWIEH ;disarm key wakeup
        clr KWIEJ    ;disarm key wakeup
        bset #$20,TMSK1 ;arm OC5
        ldd  TCNT    ;TCNT
        addd #20000 ;OC5 10ms later
        std  TC5     ;fast clear
        rti
KeyJHan: clr KWIEH ;disarm key wakeup
        clr KWIEJ    ;disarm key wakeup
        bset #$20,TMSK1 ;arm OC5
        ldd  TCNT    ;TCNT
        addd #20000 ;OC5 10ms later
        std  TC5     ;fast clear
        rti
OC5Han: bclr #$20,TMSK1 ;disarm OC5
        movb #$FF, KWIEH ;arm key wakeup
        movb #$FF, KWIEJ ;arm key wakeup
        bsr  KeyBoard   ;read keyboard
        rti
```

Program 8.8 Assembly software interface of a direct connection keyboard.

The 6811 interface connects the 16 individual switches to Ports E and C. 6811 interrupts will be requested with STRA (Figure 8.22). To generate interrupts on the touch and release, the 6811 hardware requires a 16-bit latch and a 16-bit "equals" gate (Figure 8.23). In this way, a 6811 STRA interrupt is requested whenever the switch pattern changes.

Figure 8.22
Multiple keys are
interfaced directly to
input ports.

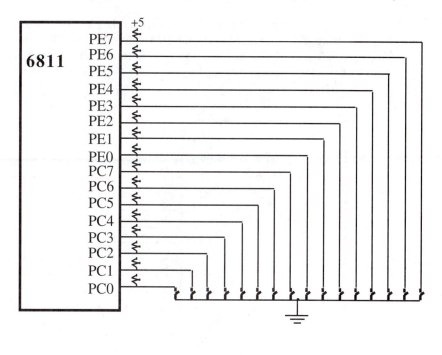

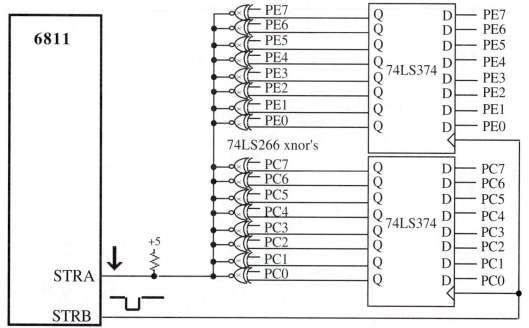

Figure 8.23 Hardware for generating interrupts on the 6811.

The read PIOC followed by the read CL data will cause a STRB pulse that will save
the current keyboard values into the 16 flip-flips built with two 74LS374s. The 74LS266
has open-collector outputs that are "wire-ANDed" together. Four 74LS266 chips have 16
exclusive-NOR gates that create a 16 by 16 "equals" function. If the current keyboard

values equal the values set by the interrupt handler, then STRA will equal 1. If any key changes since the last interrupt, STRA falls, causing an IRQ interrupt.

The 6812 interface does not require the 16-bit latch and 16-bit equals gate, because key wake-up interrupts can be generated on the touch or release of any of the 16 switches (Figure 8.24).

Figure 8.24
Hardware interface for the 6812.

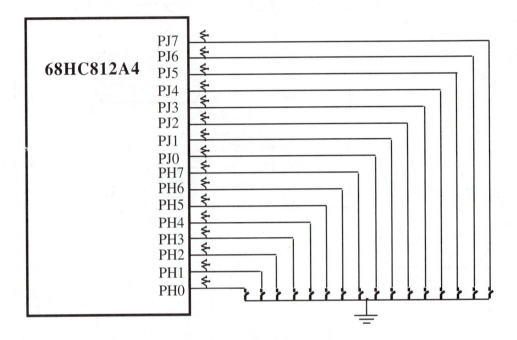

In C, the interface software would be as shown in Program 8.9. It would be possible to set both the 6812 Key WakeupH and Key WakeupJ interrupt vectors to point to the same ISR because they perform identical functions.

```
// MC68HC11A8
unsigned int KEY;      // current pattern
#define wait 20000     // bounce time (cyc)
void Ritual(void){
    asm(" sei");       // make atomic
    TMSK1=0x00;   // Disarm OC5
    DDRC=0;
// PortC are inputs from the keyboard
    PIOC=0x54;
// Input hndshk, fall, arm, STRB neg pulse
    KeyBoard();        // Initially read
    asm(" cli"); }

void KeyBoard(void){
    dummyRead=PIOC;   // read status
    KEY=(PORTE<<8)+PORTCL;}
// Set global, clear STAF

#pragma interrupt_handler IRQhandler()
```

```
// MC68HC812A4
unsigned int KEY;      // current pattern
#define wait 20000     // bounce time (cyc)
void Ritual(void){
    asm(" sei");       // make atomic
    TIOS|=0x20;        // enable OC5
    TSCR=0x90;         // enable, fast clear
    TMSK2=0x32;        // 500 ns clock
    TMSK1&=0xDF;       // Disarm OC5
    DDRH=DDRJ=0;       // H,J are key inputs
    KWIEH=KWIEJ=0xFF;  // all 16 are armed
    KPOLJ=0;           // falling edge
    PUPEJ=0;           // regular input
    KeyBoard();        // Initially read
    asm(" cli"); }
void KeyBoard(void){
    KWIFH=KWIFJ=0xFF;          // clear flags
    KEY=(PORTJ<<8)+PORTH;}  // Set global
#pragma interrupt_handler KeyHhandler()
```

```
void IRQhandler (void){                    void KeyHhandler (void){
    PIOC=0x14;      // Disarm STAF             KWIEH=KWIEJ=0x00; // all 16 are disarmed
    TMSK1=0x08;     // Arm OC5                 TMSK1|=0x20;        // Arm OC5
    TOC5=TCNT+wait;                           TC5=TCNT+wait;
    TFLG1=0x08;}    // clear OC5F              TFLG1=0x20;}        // clear C5F
                                           #pragma interrupt_handler KeyJhandler()
#pragma interrupt_handler TOC5handler()    void KeyJhandler (void){
void TOC5handler (void){                       KWIEH=KWIEJ=0x00; // all 16 are disarmed
  TMSK1=0x00;      // Disarm OC5               TMSK1|=0x20;        // Arm OC5
  PIOC=0x54;       // Rearm STAF               TC5=TCNT+wait;
  KeyBoard();}     // Read keys, set LS374's   TFLG1=0x20;}        // clear C5F
                                           #pragma interrupt_handler TOC5handler()
                                           void TOC5handler (void){
                                               TMSK1&=0xDF;  // Disarm OC5
                                               KWIEH=KWIEJ=0xFF; // all 16 are rearmed
                                               KeyBoard();}       // Read keys
```

Program 8.9 C software interface of a direct connection keyboard.

The four 74LS266 and two 74LS374 chips can be eliminated if periodic polling interrupt synchronization is used instead. With periodic polling, we establish a periodic interrupt (one that occurs at a regular fixed rate). In this example we will interrupt every 10 ms whether or not there has been a change in the keyboard. To handle the bounce, we set the periodic interrupt interval longer than the bounce time. If the switch happens to be bouncing at the time of the interrupt, global variable Key may change during this interrupt or maybe the next. The advantage of the previous hardware-based interrupt interface is that the global variable is changed exactly 10 ms after the key is pressed (or released). In this example there is a ±10-ms time jitter between the key press and when the global variable is changed. This interface, like all periodic polling systems, continuously generates interrupts that are wasted (inefficient) when no key change has occurred (Program 8.10).

```
// MC68HC11A8                              // MC68HC812A4
unsigned int Key;      // current pattern  unsigned int Key;       // current pattern
#define period 20000  // 10ms polling      #define period 20000   // 10ms polling
unsigned int KeyBoard(void){               unsigned int KeyBoard(void){
    return((PORTE<<8)+PORTC);} // pattern       return((PORTH<<8)+PORTJ);} // pattern
                                           void Ritual(void){
void Ritual(void){                             asm(" sei");      // make atomic
    asm(" sei");  // make atomic               TIOS|=0x20;       // enable OC5
    DDRC=0;       // inputs from keyboard      TSCR=0x90;        // enable, fast clear
    Key=KeyBoard();  // read 16 keys           TMSK2=0x32;       // 500 ns clock
    TMSK1=0x08;      // Arm OC5                 TMSK1|=0x20;      // Arm OC5
    TOC5=TCNT+period;                          DDRH=DDRJ=0;      // H,J are key inputs
    TFLG1=0x08;      // clear OC5F              PUPEJ=0;          // regular input
    asm(" cli"); }                             Key=KeyBoard();   // read 16 keys
                                               TC5=TCNT+period;
#pragma interrupt_handler TOC5handler()        asm(" cli"); }
void TOC5handler (void){                    #pragma interrupt_handler TOC5handler()
    Key=KeyBoard();    // Current pattern   void TOC5handler (void){
    TOC5=TOC5+period;                          Key=KeyBoard();   // Current pattern
    TFLG1=0.08;}       // ack OC5F              TC5=TC5+period;   // ack OC5F
```

Program 8.10 C software interface of a direct connection keyboard using periodic polling.

To save I/O lines, the scanned approach is used. In this approach the keys are divided into rows and columns. The computer can drive the rows with open-collector outputs and read the columns. The 6811 interface will use open-collector logic on PC7-PC4 by setting CWOM=1 and DDRC=$F0. Interrupts will be requested on both the touch and release using the STRA mechanism (Figure 8.25).

Figure 8.25
A matrix keyboard
interfaced to a 6811.

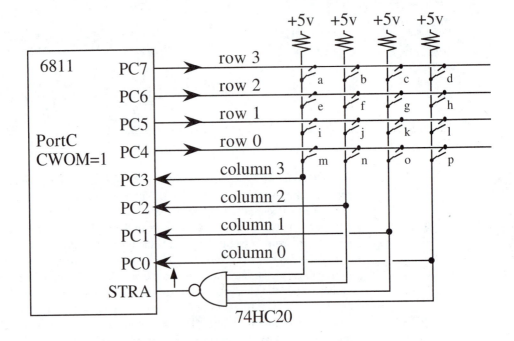

The 6812 interface will produce the open-collector function (outputs have two states, HiZ, 0) by toggling the direction register. The resistor pull-ups on the PJ3-PJ0 inputs will be configured internally. Interrupts will be requested on the touch and release using the key wake-up mechanism (Figure 8.26).

Figure 8.26
A matrix keyboard
interfaced to the
microcomputer.

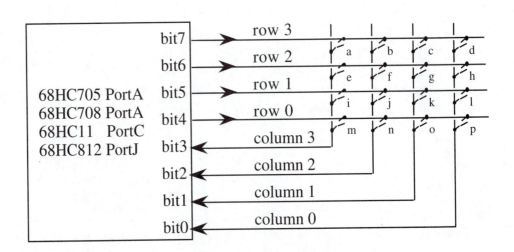

There are two steps to scan a particular row:

1. Select that row by driving it low, shown in Table 8.3 (make output 0), while the other rows are left not driven (make output HiZ)
On the 6811 we have CWOM=1, so the port C outputs will be HiZ when output is 1
On the 6812 we set the direction register, so only one row is output
2. Read the columns to see if any keys are pressed in that row (Table 8.3)
0 means the key is pressed
1 means the key is not pressed

Table 8.3
Patterns for a 4 by 4 matrix keyboard.

Row 3	Row 2	Row 1	Row 0	Column 3	Column 2	Column 1	Column 0
0	1	1	1	a	b	c	d
1	0	1	1	e	f	g	h
1	1	0	1	i	j	k	l
1	1	1	0	m	n	o	p

When all rows are 0, the 6811 STRA input will rise when any of the 16 keys is pressed. When all rows are 0, one of the 6812 key wake-up inputs will fall when any of the 16 keys is pressed. The scanned keyboard operates properly if:

1. No key is pressed
2. Exactly one key is pressed
3. Exactly two keys are pressed

This method can not be used to detect three or more keys simultaneously pressed. If three keys are pressed in an L shape, then the fourth key that completes the rectangle will appear to be pressed. Therefore special keys like shift, control, option, and alt are not placed in the scanned matrix but rather are interfaced directly, each to a separate input port, like the piano keyboard example. The 6812 implementation sets the direction register to inputs to create the open-collector output. The assembly language software to scan this keyboard is shown in Program 8.11. The C language software to scan this keyboard is shown in Program 8.12

```
; MC68HC11A8
Ritual: ldaa #$40
        staa PIOC   ;CWOM=1
        ldaa #$F0   ;PC7-PC4 outputs
        staa DDRC   ;PC3-PC0 inputs
        rts
ScanTab: dc.b %01110000    PC7 row
        dc.b "abcd"       ;characters
        dc.b %10110000    PC6 row
        dc.b "efgh"       ;characters
        dc.b %11010000    PC5 row
        dc.b "ijkl"       ;characters
```

```
; MC68HC812A4
Ritual: clr  DDRJ ;PJ3-PJ0 inputs
        rts          ;PJ7-PJ0 oc outputs
ScanTab: dc.b %01110000 ;PJ7 row data
        dc.b %10000000 ;PJ7 row DDRJ
        dc.b "abcd"    ;characters
        dc.b %10110000 ;PJ6 row
        dc.b %01000000 ;PJ6 row DDRJ
        dc.b "efgh"    ;characters
        dc.b %11010000 ;PJ5 row
        dc.b %00100000 ;PJ5 row DDRJ
        dc.b "ijkl"    ;characters
```

continued on p. 438

Program 8.11 Assembly software interface of a matrix scanned keyboard.

continued from p. 437

```
        dc.b %11100000  ;PC4 row
        dc.b "mnop"      ;characters
        dc.b 0
; Returns RegA ASCII key pressed,
;    RegY number of keys pressed
;    Y=0 if no key pressed
Scan:   ldy  #0   ;Number pressed
        ldx  #ScanTab
Loop:   ldab 0,x
        beq  Done
        stab PORTC ;select row
        ldab PORTC ;read columns
        lsrb         ;PC0 into carry
        bcs  NotPC0
        ldaa 4,x
        iny
NotPC0: lsrb         ;PC1 into carry
        bcs  NotPC1
        ldaa 3,x
        iny
NotPC1: lsrb         ;PC2 into carry
        bcs  NotPC2
        ldaa 2,x
        iny
NotPC2: lsrb         ;PC3 into carry
        bcs  NotPC3
        ldaa 1,x
        iny
NotPC3: ldab #5    ;Size of entry
        abx
        bra  Loop
Done:   rts
```

```
        dc.b %11100000  ;PJ4 row
        dc.b %00010000  ;PJ4 row DDRJ
        dc.b "mnop"      ;characters
        dc.b 0
; Returns RegA ASCII key pressed,
;    RegY number of keys pressed
;    Y=0 if no key pressed
Scan:   ldy  #0     ;Number pressed
        ldx  #ScanTab
Loop:   ldab 0,x     ;row select
        beq  Done
        stab PORTJ  ;select row
        ldab 1,x     ;DDRJ value
        stab DDRJ
        brset PORTJ,#$01,NotPJ0
        ldaa 5,x   ;code for column 0
        iny
NotPJ0: brset PORTJ,#$02,NotPJ1
        ldaa 4,x   ;code for column 1
        iny
NotPJ1: brset PORTJ,#$04,NotPJ2
        ldaa 3,x   ;code for column 2
        iny
NotPJ2: brset PORTJ,#$08,NotPJ3
        ldaa 2,x   ;code for column 3
        iny
NotPJ3: leax 6,x    ;Size of entry
        bra  Loop
Done:   rts
```

Program 8.11 Assembly software interface of a matrix scanned keyboard.

```
// MC68HC11A8
const struct Row
{   unsigned char out;
    unsigned char keycode[4];}
#typedef const struct Row RowType;
RowType ScanTab[5]={
{   0x70, "abcd" },
{   0xB0, "efgh" },
{   0xD0, "ijkl" },
{   0xE0, "mnop" },
{   0x00, "    " }};
void Ritual(void){ // PC3-PC0 are inputs
```

```
// MC68HC812A4
const struct Row
{   unsigned char out;
    unsigned char direction;
    unsigned char keycode[4];}
#typedef const struct Row RowType;
RowType ScanTab[5]={
{   0x70, 0x80, "abcd" }, // row 3
{   0xB0, 0x40, "efgh" }, // row 2
{   0xD0, 0x20, "ijkl" }, // row 1
{   0xE0, 0x10, "mnop" }, // row 0
{   0x00, 0x00, "    " }};
```

```
    PIOC=0x40;      // CWOM=1                    void Ritual(void){ // PJ3-PJ0 are inputs
    DDRC=0xF0;}     // PC7-PC4 are outputs         DDRJ=0x00;}    // PJ7-PJ4 are oc outputs
/* Returns ASCII code for key pressed,         /* Returns ASCII code for key pressed,
   Num is the number of keys pressed              Num is the number of keys pressed
   both equal zero if no key pressed */           both equal zero if no key pressed */
unsigned char Scan(unsigned int *Num) {        unsigned char Scan(unsigned int *Num){
RowType *pt; unsigned char column, key;        RowType *pt; unsigned char column,key; int j;
int j;                                           (*Num)=0; key=0;    // default values
  (*Num)=0; key=0;      // default values        pt=&ScanTab[0];
  pt=&ScanTab[0];                                while(pt->out){
  while(pt->out) {                                 PORTJ=pt->out;   // select row
    PORTC=pt->out;   // select row                 DDRJ=pt->direction;  // one output
    column=PORTC;    // read columns              column=PORTJ;    // read columns
    for(j=3; j>=0; j--) {                          for(j=3; j>=0; j--){
      if((column&0x01)==0){                          if((column&0x01)==0){
        key=pt->keycode[j];                            key=pt->keycode[j];
        (*Num)++;}                                     (*Num)++;}
      column>>=1;} // shift into position          column>>=1;} // shift into position
    pt++; }                                        pt++; }
  return key;}                                   return key;}
```

Program 8.12 C software interface of a matrix scanned keyboard.

A debounced scanned interface is created by combining the scanning software from this section with the debouncing software from the last section. In particular remove the simple `KeyBoard()` function in Program 8.9 and replace it with the `Scan()` function of Program 8.12.

8.1.7 Multiplexed/ Demultiplexed Scanned Keyboard

To interface many switches, the multiplexed interface approach is used. In this approach the keys are again divided into rows and columns. In this interface there are 16 rows and 9 columns, giving 144 keys. The computer specifies the row number (0 to 15) by outputting to the four most significant bits. The 4 to 16 demultiplexer, 74159, has open-collector outputs and will drive exactly one row to 0 and the other rows will be HiZ. The 74147 is a 10 to 4 line priority encoder. If all nine inputs of the 74147 are high (i.e., no key pressed in this row), then the least significant 4 bits signals will be 1111. If exactly one key is pressed in the row, then exactly one 0 will exist on the 74147's inputs, and the 74147 output will be the negative logic location of the column number. For example, if the key in column 4 is pressed, then the 74147 "4" input will be 0 and bit 3–0 will be 1011. If two or more keys are pressed in the same row, then bits 3–0 will be the column number (again in negative logic), the column with the highest number. For example, if the keys in columns 3 and 7 are both pressed, then the 74147 "3" and "7" inputs will both be 0 and bits 3–0 will only signify the "7" 1000. As mentioned earlier, a multiplexed interface cannot handle more than one key pressed simultaneously (Figure 8.27). The assembly language software to scan this keyboard is shown in Program 8.13.

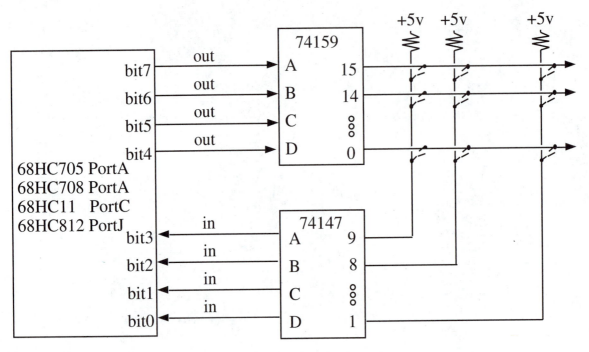

Figure 8.27 A multiplexed matrix keyboard interfaced to the microcomputer.

```
; MC68HC11A8
Key:      ds 1  ;current key
PrevKey:  ds 1  ;previous key
Ritual: sei
        ldaa #$F0   ;PC7-PC4 outputs
        staa DDRC   ;PC3-PC0 inputs
        bsr KeyScan
        staa PrevKey
        staa Key
        ldaa TMSK1
        oraa #$08   ; arm OC5
        staa TMSK1
        ldd  TCNT
        addd #20000
        std  TOC5   ; first in 10ms
        ldaa #$08
        staa TFLG1 ; clear OC5F
        cli
        rts
; returns RegA=code (0 for none)
KeyScan: clra       ;=0 means no
        clr  PORTC ;row=0
loop:   ldab PORTC ;read columns
        andb #$0F
        cmpb #$0F   ;$0F means no
        beq  none
```

```
; MC68HC812A4
Key:      ds 1  ;current key
PrevKey:  ds 1  ;previous key
Ritual: sei
        ldaa #$F0   ;PJ7-PJ4 outputs
        staa DDRJ   ;PJ3-PJ0 inputs
        bsr KeyScan
        staa PrevKey
        staa Key
        bset #$20,TIOS   ;activate OC5
        bset #$20,TMSK1 ;arm OC5
        ldaa #$32
        staa TMSK2  ;500ns clock
        ldd  TCNT
        addd #20000
        std  TC5    ; first in 10ms
        ldaa #$20
        staa TFLG1 ; clear C5F
        cli
        rts
; returns RegA=code (0 for none)
KeyScan: clra       ;=0 means no
        clr  PORTJ ;row=0
loop:   ldab PORTJ ;read columns
        andb #$0F
        cmpb #$0F   ;$0F means no
```

```
        eorb  #$0F  ;code 1-9                          beq   none
        tba           ;found one                       eorb  #$0F  ;code 1-9
none:   ldab  PORTC                                     tba           ;found one
        addb  #$10  ;next row                    none:  ldab  PORTJ
        stab  PORTC                                     addb  #$10   ;next row
        bne   loop                                      stab  PORTJ
        rts                                             bne   loop
TOC5handler:                                            rts
        bsr   KeyScan                           TOC5handler:
        cmpa  PrevKey ;same as last?                    bsr   KeyScan
        bne   skip                                      cmpa  PrevKey  ;same as last?
        staa  Key     ;new value                        bne   skip
skip:   staa  PrevKey                                   staa  Key        ;new value
        ldd   TOC5                              skip:   staa  PrevKey
        addd  #20000                                    ldd   TC5
        std   TOC5  ;every 10ms                         addd  #20000
        ldaa  #$08                                      std   TC5        ;every 10ms
        staa  TFLG1 ;ack OC5                            ldaa  #$08
        rti                                             staa  TFLG1      ;ack OC5
                                                        rti
```

Program 8.13 Assembly software interface of a multiplexed keyboard.

Periodic polling interrupt synchronization is appropriate for this interface. The background thread will maintain an 8-bit global variable, Key, storing in it the current key code from the matrix. To handle the bounce, we set the periodic interrupt interval longer than the bounce time. If the switch happens to be bouncing at the time of the interrupt, the 74147 outputs may contain incorrect patterns (e.g., representing patterns that are neither the old nor the new status). Therefore we will update Key only after the KeyScan() result is the same for two consecutive interrupts. In this example, there is a 10- to 20-ms time delay (latency) between the key press and when the global variable is changed. The C language software to scan this keyboard is shown in Program 8.14.

```
// MC68HC11A8                                    // MC68HC812A4
// PC7-PC4 row output                            // PJ7-PJ4 row output
// PC3-PC0 column inputs                         // PJ3-PJ0 column inputs
unsigned char Key;    // current pattern         unsigned char Key;    // current pattern
unsigned char PreviousKey;  // 10ms ago          unsigned char PreviousKey;  // 10ms ago
#define period 20000        // 10 ms             #define period 20000        // 10 ms
unsigned char KeyScan(void){                     unsigned char KeyScan(void){
    unsigned char key,row;                           unsigned char key,row;
    key=0;    // means no key pressed               key=0;    // means no key pressed
    for(row=0;row<16;row++){                        for(row=0;row<16;row++){
        PORTC=row<<4;  // Select row                    PORTJ=row<<4;  // Select row
        if((PORTC&0x0F)!=0x0F){                         if((PORTJ&0x0F)!=0x0F){
            key=PORTC^0x0F; }}                              key=PORTJ^0x0F; }}
    return(key);}                                    return(key);}
```
continued on p. 442

Program 8.14 C software interface of a multiplexed keyboard.

continued from p. 441

```
void Ritual(void){
    asm(" sei");   // make ritual atomic
    DDRC=$F0;
    PreviousKey=Key=KeyScan(); // read
    TMSK1|=0x08;        // Arm OC5
    TOC5=TCNT+wait;
    TFLG1=0x08;         // clear OC5F
    asm(" cli"); }
#pragma interrupt_handler TOC5handler()
void TOC5handler(void){
unsigned char NewKey;
    NewKey=KeyScan(); // Current pattern
    if(NewKey==PreviousKey) Key=NewKey;
    PreviousKey=NewKey;
    TOC5=TOC5+period;
    TFLG1=0x08;}        // ack OC5F
```

```
void Ritual(void){
    asm(" sei");   // make ritual atomic
    DDRJ=$F0;
    PreviousKey=Key=KeyScan(); // read
    TMSK1|=0x20;        // Arm OC5
    TIOS|=OC5;     // enable OC5
    TSCR|=0x80;    // enable
    TMSK2=0x32;    // 500 ns clock
    TC5=TCNT+wait;
    TFLG1=0x20;        // clear OC5F
    asm(" cli"); }
#pragma interrupt_handler TOC5handler()
void TOC5handler(void){
unsigned char NewKey;
    NewKey=KeyScan();  // Current pattern
    if(NewKey==PreviousKey) Key=NewKey;
    PreviousKey=NewKey;
    TOC5=TOC5+period;
    TFLG1=0x20;}            // ack OC5F
```

Program 8.14 C software interface of a multiplexed keyboard.

8.2 Output LEDs

Similar to the keyboard interfaces in the previous section, we will develop LED interfaces in the direct, scanned, and multiplexed categories (Figure 8.28). A direct interface has a unique computer output pin for each LED. In this way, a single output port can control eight LED segments. Once the output port is set, no software action is required to maintain the direct interface. All possible "on/off" patterns can be generated by the direct interface. The scanned interface organizes the LEDs in a matrix with rows and columns, and each row and each column has a unique output pin. In Figure 8.29 current sources are used to drive the rows and current sinks are connected to the columns. If the LEDs are configured in a 4 by 4 matrix, a single output port can control 16 LED segments. Software executed on a continuous and regular basis will be required to maintain the display. All possible on/off patterns can be generated by the scanned interface. The multiplexed display is also organized in a matrix with rows and columns but has multiplexing hardware so that there are more rows and columns than there are computer output pins. Theoretically, as shown in Figure 8.29, it is possible for a single output port to control $16 \cdot 16$, or 256, LED segments. Again, software executed on a continuous and regular basis will be required to maintain the display. Depending on the configuration of the multiplexer, not all possible on/off patterns can be generated by the multiplexed interface. Because of the maximum allowable LED current limitations (the details to be presented later in the section), it will not be practical to have a 16 by 16 LED matrix. Nevertheless, the multiplexed approach is common, especially for 7-segment and 15-segment LED displays. We will also develop an LED interface that uses a multiplexed approach but performs the scanning operations in hardware so that periodic software maintenance is not required.

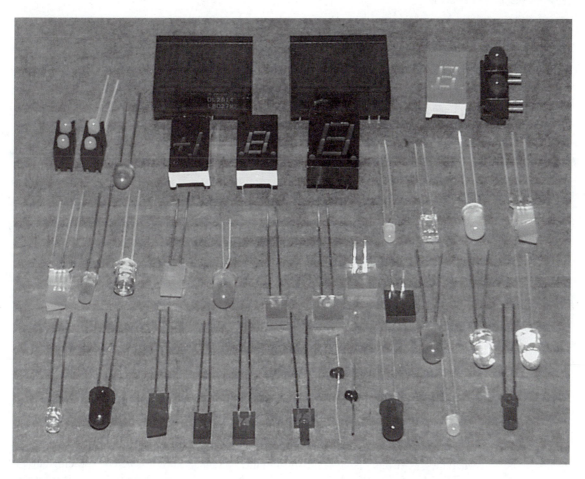

Figure 8.28
LEDs come in a wide variety of shapes, sizes, colors and configurations.

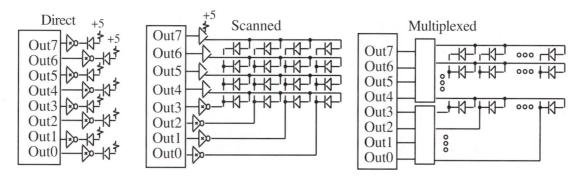

Figure 8.29
Three approaches to interfacing multiple LEDs.

8.2.1
Single LED
Interface

If you connect a microcomputer output pin to the LED interface, then the microcomputer can control the state (on/off) of the LED. If you make the microcomputer output high (a logic 1), the output of the 7405 will be low, current will flow through the LED, and it will be lit. The LED voltage will be about 2 V when it is lit. If you make the computer output low (a logic 0), the output of the 7405 will be off (HiZ, not driven, high impedance, disconnected), no current will flow through the LED, and it will be dark. The use of the 7405 provides the necessary current (about 10 mA) to activate the light. The 250Ω resistor is selected to control the brightness of the light (Figure 8.30).

Figure 8.30
A single LED interface.

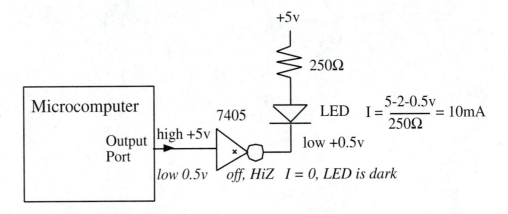

We can use either open-collector logic or open-emitter logic as current switches (Figure 8.31). When the open-collector output is 0, it will sink current to ground (turning on the LED, relay, stepper coil, solenoid, etc.). When the open-collector output is floating, it will not sink any current (turning off the LED, relay, stepper coil, solenoid, etc.). Similarly, when the 75491 input is high, the open-emitter output transistor is active, sourcing current into the LED. When the 75491 input is low, the open-emitter output transistor is off, making the LED current zero.

Figure 8.31
Two approaches to controlling the current to an LED.

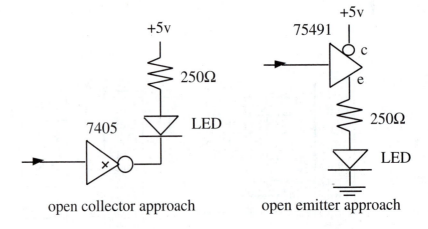

LEDs can be controlled either by using an open-collector gate (like the 7405, 7406, 75492, 75451, or NPN transistors) to sink current to ground or by using an open-emitter gate (like the 75491 or PNP transistors) to source current from the power supply. Table 8.4

Table 8.4
Output parameters for various open-collector gates.

Family	Example	V_{OL}*	I_{OL}
Standard TTL	7405	0.4 V	16 mA
Schottky TTL	74S05	0.5 V	20 mA
Low-power Schottky TTL	74LS05	0.5 V	8 mA
High-speed CMOS	74HC05	0.33 V	4 mA
High-voltage output TTL	7406	0.7 V	40 mA
High-voltage output TTL	7407	0.7 V	40 mA
Silicon monolithic IC	75492	0.9 V	250 mA
Silicon monolithic IC	75451-75454	0.5 V	300 mA
Darlington switch	ULN-2074	1.4 V	1.25 A
MOS field-effect transistor (MOSFET)	IRF-540	Varies	28 A

* Voltage at maximum I_{OL}.

Table 8.5
Output parameters for various open-emitter gates.

Family	Example	V_{CE}	I_{CE}
Silicon monolithic IC	75491	0.9 V	50 mA
Darlington switch	ULN-2074	1.4 V	1.25 A
MOSFET	IRF-540	Varies	28 A

provides the output low currents for some typical open-collector devices. Table 8.5 provides the output source currents for some typical open-emitter devices.

Darlington switches like the ULN-2061 through ULN-2077 and MOSFETs like the IRF-540 can be used either in open-collector mode to sink current or in open-emitter mode to source current. For all the devices the actual output voltage depends on the output current. What is shown in Tables 8.4 and 8.5 is the output voltage at maximum output current. For the transistor devices, the output voltages/currents also depend on the input currents.

For scanned and multiplexed LED interfaces, we will use both current sources and sinks. In the following interface circuits we assume the desired LED setpoint is 2 V and 10 mA. Since the LED is a diode, the voltage and current relationship is quite nonlinear. The voltage/current relationship for the LTP-1057A and LTP-1157A dot matrix LED displays is plotted in Figure 8.32.

Figure 8.32
Typical voltage/current response of a LED.

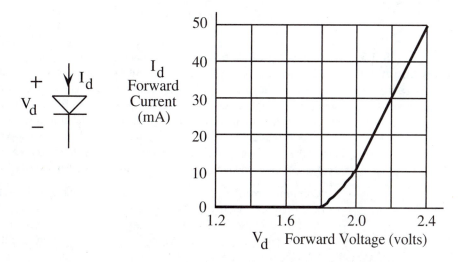

Table 8.6
Absolute maximum
rating for LTP-1057A
and LTP-1157A 5 by 7
dot matrix displays.

Parameter	red	green	yellow	orange	units
Maximum power	55	75	60	75	mW
Peak forward current	160	100	80	100	mA
Max continuous current	25	25	20	25	mA

To prevent the LED from overheating, we must limit the electric power by using a current-limiting series resistor as shown in Figure 8.32. Table 8.6 illustrates typical LED specifications.

The LED power can be calculated from its voltage and current.

$$P_d = V_d \cdot I_d$$

The resistor value is chosen to establish the desired voltage/current (V_d/I_d) operating point for the LED. For the 75492 open-collector circuit, the resistor is calculated as (Figure 8.33)

$$R = (5 - V_d - V_{OL})/I_d = (5 - 2 - 0.9)/10 \text{ mA} = 210\Omega$$

Similarly for the 75491 open-emitter circuit, the resistor is calculated as (Figure 8.35)

$$R = (5 - V_d - V_{ce})/I_d = (5 - 2 - 0.9)/10 \text{ mA} = 210\Omega$$

Figure 8.33
Calculating the resistor
used in the LED
interface.

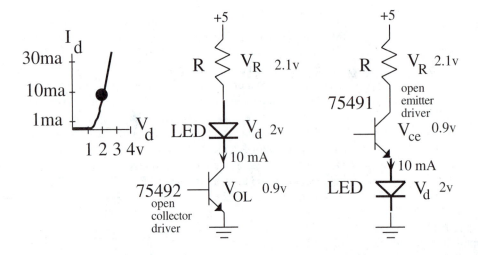

8.2.2 Seven-Segment LED Interfaces

When creating LEDs that display numbers, it is appropriate to use common-cathode or common-anode seven-segment LED modules. Some modules have an eighth or ninth segment to display a right-side and/or left-side decimal point. The circuit in Figure 8.34 shows a direct interface to a single seven-segment common-cathode LED display. It is called common cathode because the seven cathodes are connected. We label it direct because there is a separate computer output pin for each LED segment. The software simply writes to the output port to control the seven segments. A current sink is required to interface a common-anode display. It is called common anode because the seven anodes are connected (Figure 8.35).

> **Common Error:** If you try to replace the seven individual resistors in either of the two circuits in Figures 8.36 and 8.37 with a single resistor on the "common" side, then the brightness of each LED segment will be a function of how many LEDs are on. For example, if only one segment is on, it will be very bright; if all seven are on, then they will be dim.

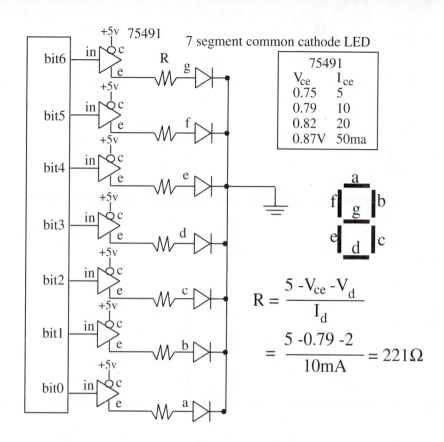

Figure 8.34
Seven-segment common-cathode LED interface.

7 segment common cathode LED

75491	
V_{ce}	I_{ce}
0.75	5
0.79	10
0.82	20
0.87V	50ma

$$R = \frac{5 - V_{ce} - V_d}{I_d}$$

$$= \frac{5 - 0.79 - 2}{10\text{mA}} = 221\Omega$$

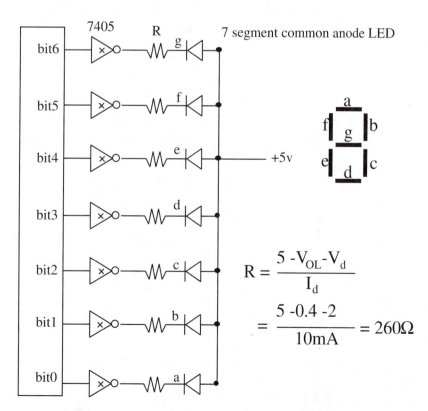

Figure 8.35
Seven-segment common-anode LED interface.

7 segment common anode LED

$$R = \frac{5 - V_{OL} - V_d}{I_d}$$

$$= \frac{5 - 0.4 - 2}{10\text{mA}} = 260\Omega$$

Observation: The current-limiting resistors should be placed on the individual LED connections and not on the "common" connection.

8.2.3
Scanned Seven-
Segment LED
Interface
We will design a three-digit LED display capable of displaying the numbers from 000 to 999. The 21 LED segments will be interfaced in a 3 column by 7 row rectangular matrix. There will be a column for each digit and a row for each of the segments a, b, c, d, e, f, g. The open-collector outputs of the 7406 will sink current from the seven rows. The ULN2074, in open-emitter mode, will source current for the three columns. The software will utilize a simple 3-byte variable to contain the current value to be displayed. For example, if the value "456" is to be displayed, the main program will set the 24-bit global to $666D7C, as explained in Table 8.7. Our interrupt software will read this global and output the appropriate signals to the output ports (Figure 8.36).

Figure 8.36
Seven Δ segment
common anode LED's.

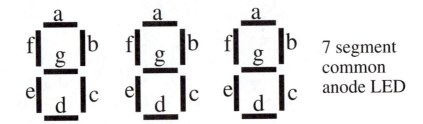

7 segment
common
anode LED

Table 8.7 gives the conversion between decimal digit and seven-segment binary code assuming segments g, f, e, d, c, b, a are mapped into bits 6, 5, 4, 3, 2, 1, 0. This interface can also be used to create hexadecimal displays.

Table 8.7
Patterns for a seven-segment display.

Digit	Segments	Binary code
0	f, e, d, c, b, a	%00111111=$3F
1	b, c	%00000110=$06
2	g, e, d, b, a	%01011011=$5B
3	g, d, c, b, a	%01001111=$4F
4	g, f, c, b	%01100110=$66
5	g, f, d, c, a	%01101101=$6D
6	g, f, e, d, c	%01111100=$7C
7	c, b, a	%00000111=$07
8	g, f, e, d, c, b, a	%01111111=$7F
9	g, f, c, b, a	%01100111=$67
A	g, f, e, c, b, a	%01101111=$6F
b	g, f, e, d, c	%01111100=$7C
C	f, e, d, a	%00111001=$39
d	g, e, d, c, b	%01011110=$5E
E	g, f, e, d, a	%01111001=$79
F	g, f, e, a	%01110001=$71

Figure 8.37
Timing used to scan a
LED interface.

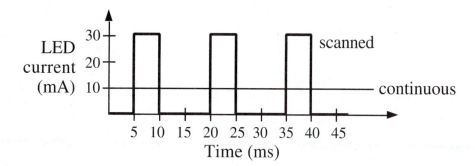

Assume the desired LED operating point is 10 mA, 2 V. Since the display will be scanned, we will operate each active LED at 30 mA with a 33% duty cycle. We will change the display every 5 ms. In this way, we will scan through the entire display every 15 ms, and each active LED will run at about 66 Hz, faster than the human eye can detect. The LED will "look like" it is continuously on at 10 mA (Figure 8.37).

The software will have to convert the decimal digits into the seven-segment codes using Table 8.7. The 7406 open-collector driver can sink the required 30 mA. If all seven segments are on, the source current will be 7 · 30 mA, or 210 mA. The ULN2074 can supply the necessary current (Figure 8.38).

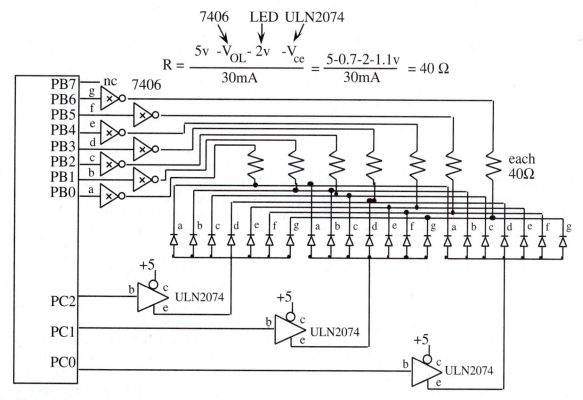

Figure 8.38
Circuit used to scan a LED interface.

It is important to place the resistors on the segment select side rather than on the digit select side. In this way the current through each diode is not a function of the number of diodes on. The software that initializes/maintains the display is shown in Program 8.15.

```
// MC68HC11A8
// PB7-PB0 output, 7 bit pattern
// PC2-PC0 output, selects LED digit
unsigned char code[3]; // binary codes
static unsigned char select[3]={4,2,1};
unsigned int index;      // 0,1,2
#define OC5F 0x08
#pragma interrupt_handler TOC5handler()
void TOC5handler(void){
    TFLG1=OC5F;        // Acknowledge
    TOC5=TOC5+10000; // every 5 ms
    PORTC=select[index]; // which LED?
    PORTB=code[index];   // enable
    if(++index==3) index=0;}
void ritual(void) {
asm(" sei");       // make atomic
    index=0;
    DDRC=0xFF;      // outputs
    TMSK1|=OC5F;  // Arm OC5
    TFLG1=OC5F;   // clear OC5F
    TOC5=TCNT+10000;
asm(" cli"); }
```

```
// MC68HC812A4
// PB7-PB0 output, 7 bit pattern
// PC2-PC0 output, selects LED digit
unsigned char code[3]; // binary codes
static unsigned char select[3]={4,2,1};
unsigned int index;      // 0,1,2
#define C5F 0x20
#pragma interrupt_handler TOC5handler()
void TOC5handler(void){
    TFLG1=C5F;         // Acknowledge
    TC5=TOC5+10000; // every 5 ms
    PORTC=select[index]; // which LED?
    PORTB=code[index];   // enable
    if(++index==3) index=0;}
void ritual(void) {
asm(" sei");       // make atomic
    index=0;
    DDRC=0xFF;        // outputs 7 segment code
    DDRB=0xFF;        // outputs select LED
    TMSK1|=C5F;     // Arm OC5
    TIOS|=C5F;      // enable OC5
    TSCR|=0x80;     // enable
    TMSK2=0x32;     // 500 ns clock
    TC5=TCNT+10000;
asm(" cli"); }
```

Program 8.15 C software interface of a scanned LED display.

8.2.4 Scanned LED Interface Using the 7447 Seven-Segment Decoder

The purpose of the previous example was to illustrate the details involved in scanning a LED display. For practical designs we will use special LED display decoder logic. The simplest of the LED drivers is the 7447 seven-segment decoder. Again we will design a three-digit LED display interfaced only to a single output port. The 21 LED segments will again be interfaced in a 3 column by 7 row rectangular matrix. There will be a column for each digit and a row for each of the segments a, b, c, d, e, f, g. The open-collector outputs of the 7447 will sink current from the seven rows. The ULN2074, in open-emitter mode, will source current for the three columns. The software will utilize a 12-bit packed BCD global variable to contain the current value to be displayed. For example, if the value "456" is to be displayed, the main program will set the 16-bit global to $0456. Our interrupt software will read this global and output the appropriate signals to the output port.

Again, we assume the desired LED operating point is 10 mA, 2 V. Since the display will be scanned, we will operate each active LED at 30 mA with a 33% duty cycle. The 7447A will simplify the generation of the seven-segment codes. This seven-segment LED driver can sink the required 30 mA. Just like the previous example, if all seven segments are on, the source current will be 7 · 30 mA, or 210 mA (Figure 8.39). The software that initializes/maintains the display is shown in Program 8.16.

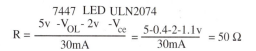

$$R = \frac{5v - V_{OL} - 2v - V_{ce}}{30mA} = \frac{5 - 0.4 - 2 - 1.1v}{30mA} = 50\ \Omega$$

7447 LED ULN2074

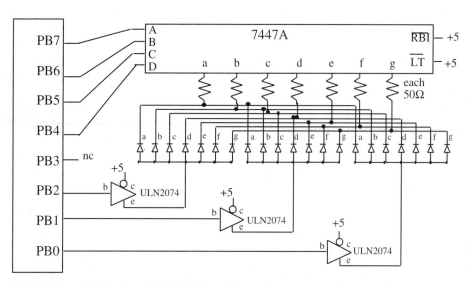

```
// MC68HC11A8
unsigned int global; // 12 bit packed BCD
const struct LED
{  unsigned char enable; // select
   unsigned char shift;  // bits to shift
   const struct LED *Next; };  // Link
#typedef const struct LED LEDType;
#typedef LEDType * LEDPtr;
LEDType LEDTab[3]={
{  0x04, 8, &LEDTab[1] },  // Most sig
{  0x02, 4, &LEDTab[2] },
{  0x01, 0, &LEDTab[0] }}; // least sig
LEDPtr Pt;   // Points to current digit
#define OC5F 0x08
#pragma interrupt_handler TOC5handler()
void TOC5handler(void){
    TFLG1=OC5F;        // Acknowledge
    TOC5=TOC5+10000;  // every 5 ms
    PORTB=(Pt->enable)
      +(global>>(pt->shift))<<4);
    Pt=Pt->Next; }
void ritual(void) {
asm(" sei");        // make atomic
    global=0;
    TMSK1=OC5F;      // Arm OC5
    Pt=&LEDTab[0];
    TFLG1=OC5F;      // clear OC5F */
    TOC5=TCNT+10000;
asm(" cli"); }
```

```
// MC68HC812A4
unsigned int global;  // 12 bit packed BCD
const struct LED
{  unsigned char enable; // select
   unsigned char shift;  // bits to shift
   const struct LED *Next; };  // Link
#typedef const struct LED LEDType;
#typedef LEDType * LEDPtr;
LEDType LEDTab[3]={
{  0x04, 8, &LEDTab[1] },  // Most sig
{  0x02, 4, &LEDTab[2] },
{  0x01, 0, &LEDTab[0] }}; // least sig
LEDPtr Pt;   // Points to current digit
#define C5F 0x20
#pragma interrupt_handler TOC5handler()
void TOC5handler(void){
    TFLG1=C5F;        // Acknowledge
    TC5=TC5+10000;  // every 5 ms
    PORTB=(Pt->enable)
      +(global>>(pt->shift))<<4);
    Pt=Pt->Next; }
void ritual(void) {
asm(" sei");      // make atomic
    DDRB=0xFF;  // outputs to LED's
    global=0;
    Pt=&LEDTab[0];
    TMSK1|=C5F;     // Arm OC5
    TIOS|=C5F;      // enable OC5
    TSCR|=0x80;     // enable
    TMSK2=0x32;     // 500 ns clock
    TC5=TCNT+10000;
asm(" cli"); }
```

Program 8.16 C software interface of a multiplexed LED display.

If more digits are needed, it would be easy to extend this approach by adding additional ULN2074 current drivers. There are two issues to consider as the number of digits is added. The first is the scan frequency. For the display to "look" continuous, each digit must be updated faster than 60 Hz. (If you look closely into the specifications of your computer monitor and television sets, you will see this same constraint.) So as you increase the number of digits the interrupt rate must increase so that each individual digit is scanned faster than 60 Hz. If there are five digits, then the periodic interrupt rate must be increased to at least 300 Hz for each digit to be updated at a rate of 60 Hz. The second issue to consider is the duty cycle for each digit. As the number of digits is increased, the duty cycle for each digit decreases. This makes it necessary to increase the instantaneous current. For example, if there were five digits, the duty cycle would be 20% and the current would have to increase to 50 mA. This design would require an open-collector driver capable of sinking 50 mA and an open emitter capable of sourcing 350 mA. Another limitation is the maximum instantaneous current allowed by the LED. There is an upper bound to the instantaneous LED current even if the duty cycle decreases, leaving the average power the same. Each LED is different, but this parameter is about 100 mA, as shown in Table 8.6.

Observation: The ratio of the maximum instantaneous current divided by the desired LED current determines the maximum number of columns in the LED matrix.

8.2.5
Integrated LED
Interface Using
the MC14489
Display Driver

For LED displays with many segments, one attractive solution is to use an integrated LED display driver. There are three advantages of a chip like the MC14489 over the other LED designs in this section. The first advantage is chip count. A single MC14489 20-pin chip and one resistor are all the components required to interface a five-digit LED display. The second advantage is that the scanning functions occur in hardware, eliminating the need for the software to execute periodic interrupts. Normally, when we think of the hardware/software trade-off, we often select the software solution because of its low cost and flexibility. This is a situation where a hardware solution could be cheaper. The third advantage is that it is easy to cascade multiple MC14489 drivers so that larger displays can be interfaced without requiring additional microcomputer output ports. Each of the five eight-segment LED devices has seven segments ("a–g") for creating the decimal digit plus one more segment ("h") for the decimal point. The packed BCD format is shifted serially into the MC14489. The SPI port is a convenient solution for the hardware/software interface between the computer and MC14489(s). Once the BCD pattern is loaded into the MC14489, the hardware will continuously generate the 8 by 5 matrix scanning signals that display the five digits. The **Rx** resistor controls the LED voltage/current operating point for the display. The software also has the ability to select the LED brightness: "regular" or "dim." The SPI SS pin (6811 PD5, 6812 PS7) will be configured as a simple output because the ENABLE pin will be low for 8 clock cycles when transmitting a command and for 24 clock cycles when transmitting data (Figure 8.40).

The MC14489 can handle multiple formats. This interface utilizes a packed BCD format, as shown in Figure 8.41. The clock signal (output of computer, input to MC14489) is used to shift 24 bits of data into the LED display interface. For data to be properly transferred, the computer will change the data on the falling edge of the CLOCK and the MC14489 will shift the data on the rising edge. This timing can be achieved with the SPI in master mode and CPHA=0, CPOL=0. For ENABLE to be low for 24 bits, we will configure it as a simple output. The 20 bits that control the five banks are encoded as packed

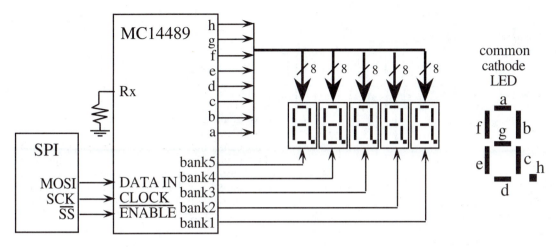

Figure 8.40 An integrated IC used to interface five seven-segment common-cathode LED digits.

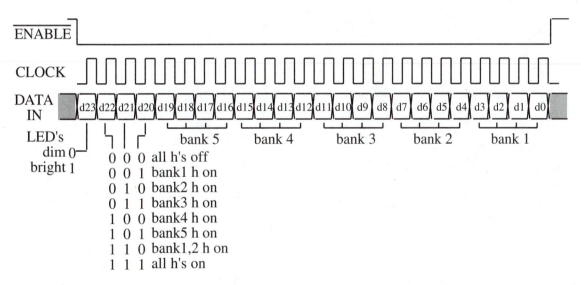

Figure 8.41 Data timing of an integrated LED controller.

BCD. The MC14469 will also display BCD digits AbCdEF, so the system can be used to show five decimal digits or five hexadecimal digits.

There is also an 8-bit command transmission that is used to configure the display. To implement the standard decimal (or hexadecimal) display, we will send a command of $01. This enables the device and puts all five banks in hexadecimal format (Figure 8.42). The software to control the interface is simplified when using the SPI. The transmission rate is configured for 1 MHz, but it actually could operate as fast as 2 MHz. Interrupt synchronization is not needed because each 8-bit data frame requires only 8 μs to complete (Program 8.17).

Figure 8.42
Configuration timing of
an integrated LED
controller.

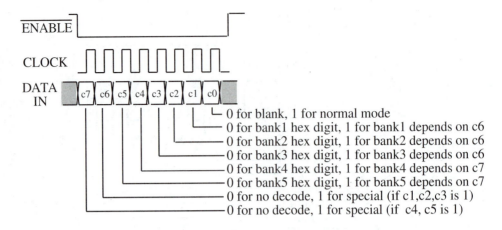

```
// MC68HC11A8                          // MC68HC812A4
// PD3/MOSI = MC14489 DATA IN          // PS5/MOSI = MC14489 DATA IN
// PD4/SCLK = MC14489 CLOCK IN         // PS6/SCLK = MC14489 CLOCK IN
// PD5 (simple output) = MC14489 ENABLE // PS7 (simple output) = MC14489 ENABLE
void ritual(void) {                    void ritual(void) {
    DDRD |= 0x38;  // outputs to MC14489    DDRS |= 0xE0;  // outputs to MC14489
    SPCR=0x50;                             SPOCR1=0x50;
// bit  meaning                        // bit  meaning
// 7 SPIE=0 no interrupts              // 7 SPIE=0 no interrupts
// 6 SPE=1  SPI enable                 // 6 SPE=1  SPI enable
// 5 DWOM=0 regular outputs            // 5 SWOM=0 regular outputs
// 4 MSTR=1 master                     // 4 MSTR=1 master
// 3 CPOL=0 match timing with MC14489  // 3 CPOL=0 match timing with MC14489
// 2 CPHA=0                            // 2 CPHA=0
// 1 SPR1=0 E/2 is 1Mhz SCLK           // 1 SSOE=1 PS7 is simple output
// 0 SPR0=0                            // 0 LSBF=0 MSB first
    PORTD|= 0x20; // ENABLE=1              SPOCR2=0x00;  // no pull-up, regular drive
    PORTD&= 0xDF; // ENABLE=0              SPOBR=0x02;   // 1Mhz SCLK
    SPDR=0x01;     // hex format           PORTS|= 0x80; // ENABLE=1
    while(SPSR&0x80)==0){};                PORTS&= 0x7F; // ENABLE=0
    PORTD|=0x20;} // ENABLE=1              SPODR= 0x01;  // hex format
void LEDout(unsigned char data[3]){       while(SPOSR&0x80)==0){};
// 24 bit packed BCD                       PORTS|=0x80;} // ENABLE=1
    PORTD &= 0xDF;    // ENABLE=0      void LEDout(unsigned char data[3]){ //packed
    SPDR = data[2];   // send MSbyte       PORTS &= 0x7F;    // ENABLE=0
    while(SPSR&0x80)==0){};                SPODR = data[2];   // send MSbyte
    SPDR = data[1];   // send middle byte  while(SPOSR&0x80)==0){};
    while(SPSR&0x80)==0){};                SPODR = data[1];   // send middle byte
    SPDR = data[0];   // send LSbyte       while(SPOSR&0x80)==0){};
    while(SPSR&0x80)==0){};                SPODR = data[0];   // send LSbyte
    PORTD |= 0x20;}   // ENABLE=1          while(SPOSR&0x80)==0){};
                                           PORTS |= 0x80;}   // ENABLE=1
```

Program 8.17 C software interface of an integrated LED display.

Maintenance Tip: It will be more reliable to design the system using a transmission
rate somewhat slower than the absolute maximum.

8.3 Liquid Crystal Displays

8.3.1
LCD Fundamentals

Liquid crystal displays are widely used in microcomputer systems (Figure 8.43). One advantage of LCDs over LEDs is their low power consumption. This allows the display, and perhaps the entire computer system, to be battery-operated. In addition, LCDs are more flexible in their sizes and shapes, permitting the combination of numbers, letters, words, and graphics to be driven with relatively simple interfaces. A LCD consists of a liquid-crystal material that behaves electrically as a capacitor. Whereas a LED converts electric power into emitted optical power, a LCD uses an alternating current (AC) voltage to change the light reflectivity (or sometimes transmittivity). The light energy is supplied by the room or a separate back light, and not by the electric power within the LCD (as with LEDs). The computer controls the display by altering the reflectivity of each segment. The disadvantage of LCDs is their slow response time. Fortunately, the bandwidth of most displays (both LEDs and LCDs) is limited by the human visual processing system (about 30 Hz).

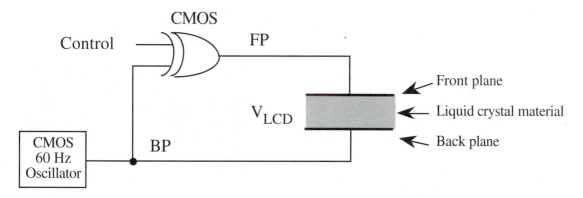

Figure 8.43 The basic idea of a liquid crystal interface.

It is important to provide the LCD with only AC and no DC signals. A DC signal above 50 mV will cause permanent damage to the LCD. Let V_{DD} be the supply voltage of the CMOS logic, and let V_{SS} be the ground voltage of the CMOS logic. The use of CMOS logic has two advantages. First, CMOS requires very little supply current. Second, CMOS logic has fairly stable high and low logic voltage levels. In other words, the V_{OH} of the CMOS NOR gate and CMOS 60-Hz oscillator will both be close to V_{DD} (hence close to each other). Similarly, the V_{OL} of the CMOS NOR gate and CMOS 60-Hz oscillator will also both be close to V_{SS} (hence close to each other). Therefore, for both **Control** high and low, V_{LCD} = **FP** – **BP** will contain no DC component. Obviously, it will be important for the oscillator to have a 50% duty cycle.

The oscillator output **BP** is a square wave (V_{SS} to V_{DD}) with a frequency of 60 Hz. When the **Control** signal is low, the front-plane voltage, **FP,** is in phase with the back-plane voltage, **BP.** Hence, V_{LCD} will be zero. In this state, the display does not reflect light, and the display is blank. When the **Control** signal is high, the front-plane voltage, **FP,** is out of phase with the back-plane voltage, **BP.** Hence, V_{LCD} will be an AC square wave ($-V_{DD}$ to $+V_{DD}$). In this state, the display reflects light, and the display is visible (Figure 8.44).

Observation: LCDs are controlled with waveforms in the 40- to 60-Hz range.

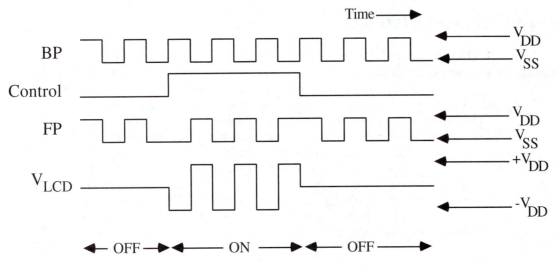

Figure 8.44
The basic timing of a liquid crystal interface.

One of the most important advantages of the LCD technology over lamps and LEDs is the flexibility in configuring the shapes and sizes of the segments. With LEDs the shapes of the segments are limited to simple regular shapes like circles and rectangles. Liquid crystal displays are created by sandwiching the liquid crystal material between a front and a back plane. The LCD segment is created in the overlap area of the front and back planes. Since the front and back planes are manufactured using techniques similar to PC board layout, there is a great flexibility in the sizes and shapes. In Figure 8.45 the entire phrase "**LCDs are**" represents a single segment, the overlap of FP1 and BP.

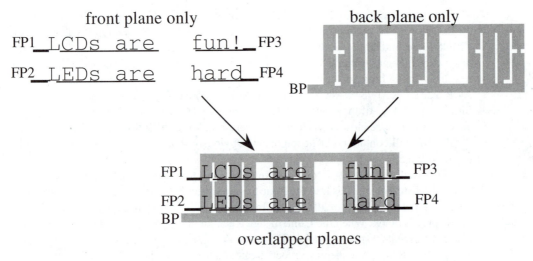

Figure 8.45
Example artwork for a LCD.

**8.3.2
Simple LCD
Interface with the
MC14543**
The LCD interface (hardware and software) is similar to that for LEDs. The direct-driven interface is simple but requires a large number of output bits and cable wires. The MC14543 is a CMOS BCD to seven-segment LCD driver. When **LD**=1 **BI**=0 and **Ph**=60 Hz square wave, one seven-segment LCD can be driven (Figure 8.46).

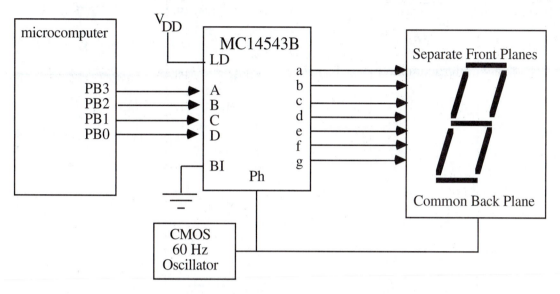

Figure 8.46 Direct interface of a LCD.

The ability to latch data into the driver allows multiple LCDs to be interfaced with a single microcomputer output port. To latch one digit, the software outputs the 4-bit BCD on PB3-0, then toggles one of PB7-4 high, then low, as, for example, shown in Program 8.18. To set all four digits, the software calculates the BCD from an unsigned input, as, for example, in Program 8.19.

```
void LCDOutDigit(unsigned char position, unsigned char data) {
// position is 0x80, 0x40, 0x20, or 0x10  and data is the BCD digit
     PORTB=0x0F&data;  // set BCD digit on the A-D inputs of the MC14543B
     PORTB|=position;  // toggle one of the LD inputs high
     PORTB=0x0F&data;} // LD=0, latch digit into MC14543B
```

Program 8.18 Helper function for a simple LCD display.

```
void LCDOutNum(unsigned int data){ unsigned int digit,num,i;
unsigned char pos;
     num=min(data,9999); // data should be unsigned from 0 to 9999
     pos=0x10;   // position of first digit (ones)
     for(i=0;i<4;i++){
         digit=num%10; num=num/10; // next BCD digit 0 to 9
         LCDOutDigit(pos,digit); pos=pos<<1;}}
```

Program 8.19 C software interface of a simple LCD display.

The 6812 requires a simple ritual that sets the direction register to outputs (e.g., DDRB=0×FF;). Port B on the 6811 is a fixed output port, so no ritual is required (Figure 8.47).

Figure 8.47 Latched interface of a LCD.

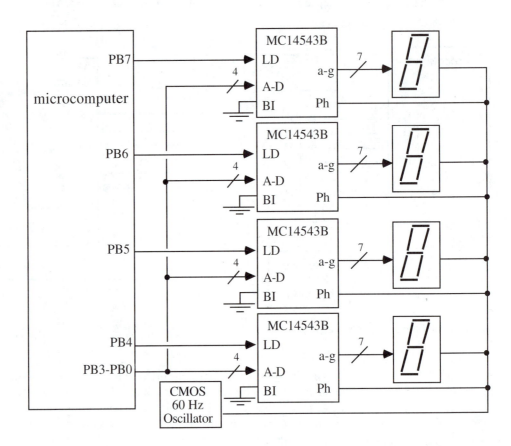

8.3.3
Scanned LCD
Interface with the
MC145000,
MC145001

Figure 8.48 shows a typical front-plane/back-plane configuration for an eight-segment LCD display. Different from LED displays, these LCD components do not have a common connection but rather utilize a 2 by 4 matrix. This configuration is compatible with LCD drivers like the MC145000 and MC145001.

Complex LCD artwork combines numbers, letters, words, and graphics on a single LCD. To simplify both the artwork and the interface, the front-plane and back-plane signals can be multiplexed. The concept is similar to the scanned LED displays described earlier. A four-digit display containing 32 segments can be multiplexed into four back-plane rows and eight front-plane columns (Figure 8.49). This reduces the number of cable wires from 32 (Figure 8.47) to 13 (Figure 8.49).

Consider the highlighted segment in Figure 8.50. To display this segment an AC voltage should be applied across BP3, FP5. To hide this segment, no AC voltage should be applied across BP3, FP5. It is impossible to control all 32 segments in this simple manner. Fortunately, Motorola has developed chips to simplify the interfacing of multiple LCDs. One MC145000 master and three MC145001 slaves accept serial input from the microcomputer and will directly drive 20 LCD digits or 180 segments. These

Figure 8.48
Artwork for an eight-segment liquid crystal digit.

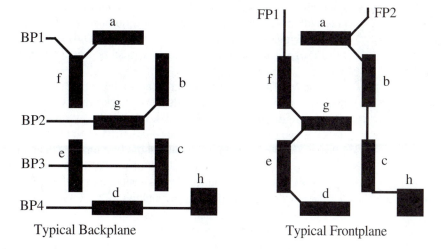

Typical Backplane   Typical Frontplane

Figure 8.49
Artwork for four eight-segment liquid crystal digits.

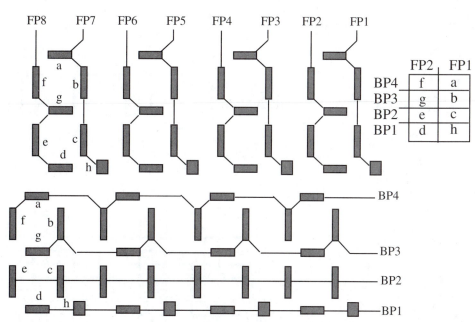

	FP2	FP1
BP4	f	a
BP3	g	b
BP2	e	c
BP1	d	h

Figure 8.50
Each segment has a unique front-plane/back-plane combination.

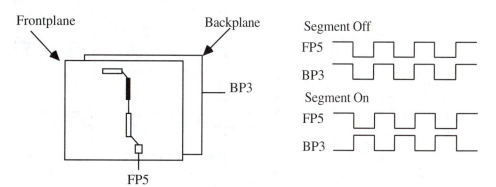

same chips (MC145000 and MC145001) could be used to drive a complex LCD display with numbers, letters, words, and graphics. The MC14500 master can control up to 48 segments organized in 4 back planes and 12 front planes. Each additional MC14501 slave can control up to 44 segments organized in 4 back planes and 11 front planes.

The MC145000 master creates the frame-sync pulses in Figure 8.51 each time the display is updated. The time-multiplexed back-plane voltages used to create the scanned display are divided into four phases. During the first phase, the BP1 signal goes +5 V, then 0. If you wish any of the front-plane/BP1 segments to be active, the corresponding FP signal will go 0, then +5 V, during the first phase. This will create an AC signal large enough to activate the segment. Each of the BP signals has 25% of the time (when BP goes +5 V, then 0) to produce either an activation with the BP/FP pair (by making FP go 0, then +5 V) or a deactivation (by making FP go 3.33 V, then 1.67 V). In both the on and off cases then DC component is zero. In the on situation, the BP to FP differential voltage is 5 to 0 V, then 0 to 5 V—that is, 5 V then −5 V. In the off situation, the BP to FP differential voltage is 5 to 3.33 V, then 0 to 1.67 V—that is, 1.67 V then −1.67 V. The ±5-V AC signal is large enough to active the LCD, but the ±1.67-V AC signal is not (Figures 8.52, 8.53).

Figure 8.51
Synchronization pulses
for the LCD display.

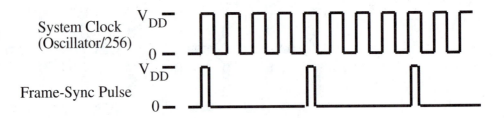

Figure 8.52
Fixed waveforms for the
four back-plane signals
for the LCD display.

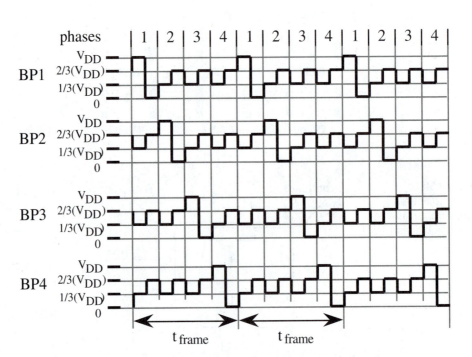

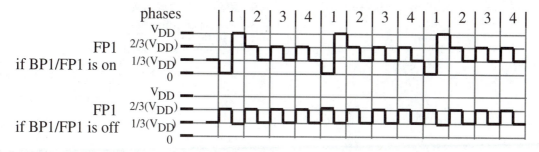

Figure 8.53 Variable waveforms for the front-plane signal for the LCD display.

The interface between the microcomputer and the LCD driver(s) utilizes a synchronized serial format (both serial data and serial clock). The MC68HC05C8, MC68HC708XL36, MC68HC11A8, MC68HC812A4, and MC68HC912B32 all have a SPI that automatically creates these signals. The computer interface is responsible for sending 48 bits in serial (one bit at a time). The SPI software simply outputs 6 bytes (most significant first), and the SPI hardware creates the MOSI data output and SCK clock. Once the information is loaded, the latch is activated, and the MC145000 LCD driver will maintain the display without software overhead (until the software wishes to change the display) (Figure 8.54).

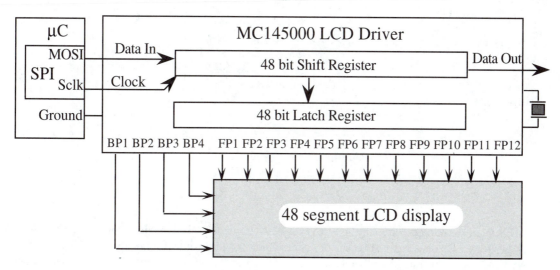

Figure 8.54 Interface of a 48-segment LCD display.

The details of the SPI port were discussed in Chapter 7. Even without a SPI port, the LCD interface would be possible with two ordinary output bits. Assume the MOSI, SCK outputs are implemented with a simple output port on PB0, PB1, respectively. For each bit, the software could output the data bit on the MOSI output, then toggle the SCK output. To fill the shift register and set all 48 segments, the software should output each bit one at a time, most significant bit first, as, for example, shown in Program 8.20. The SPI software to control this LCD interface is very similar to the LED interface using the MC14489 (Program 8.21).

```
void LCDOut (unsigned char *pt) {unsigned int i;  unsigned char mask;
    for(i=0;i<6;i++){
        for(mask=0x80;mask;mask=mask>>1){  // look at bits 7,6,5,4,3,2,1,0
            if((*pt)&mask) PORTB=1; else PORTB=0; // Serial data of the MC145000
            PORTB|=2;        // toggle the serial clock first high
            PORTB&=0xFD;}  // then low
        pt++; }}
```

Program 8.20 Bit-banged interface to a scanned LCD display.

```
// MC68HC11A8                              // MC68HC812A4
// PD3/MOSI = MC145000 DATA IN             // PS5/MOSI = MC145000 DATA IN
// PD4/SCLK = MC145000 CLOCK IN            // PS6/SCLK = MC145000 CLOCK IN
void ritual(void) {                        void ritual(void) {
    DDRD |= 0x18;  // outputs to MC145000       DDRS |= 0x60;  // outputs to MC145000
    SPCR=0x50; }                               SPOCR1=0x50;
// bit meaning                             // bit meaning
// 7 SPIE=0 no interrupts                  // 7 SPIE=0 no interrupts
// 6 SPE=1  SPI enable                     // 6 SPE=1  SPI enable
// 5 DWOM=0 regular outputs                // 5 SWOM=0 regular outputs
// 4 MSTR=1 master                         // 4 MSTR=1 master
// 3 CPOL=0 match timing with MC14489      // 3 CPOL=0 match timing with MC14489
// 2 CPHA=0                                // 2 CPHA=0
// 1 SPR1=0 E/2 is 1Mhz SCLK               // 1 SSOE=0 PS7 is simple output
// 0 SPR0=0                                // 0 LSBF=0 MSB first
void LCDout(unsigned char data[6]){            SPOCR2=0x00;  // no pull-up, regular drive
unsigned int j;                                SPOBR=0x02;} // 1Mhz SCLK
  for(j=5; j>=0 ; j--){                    void LCDout(unsigned char data[6]){
    SPDR = data[j];    // Msbyte first     unsigned int j;
    while(SPSR&0x80)==0){};}}                 for(j=5; j>=0 ; j--){
                                                SPODR = data[j];   // Msbyte first
                                                while(SPOSR&0x80)==0){};}}
```

Program 8.21 SPI interface to a scanned LCD display using a MC145000.

8.3.4 Parallel Port LCD Interface with the HD44780 Controller

Microprocessor-controlled LCD displays are widely used, having replaced most of their LED counterparts, because of their low power and flexible display graphics. This example will illustrate how a handshaked parallel port of the microcomputer will be used to output to the LCD display. The hardware for the display uses an industry standard HD44780 controller (Figure 8.55). The low-level software initializes and outputs to the HD44780 controller.

There are four types of access cycles to the HD44780, depending on RS and R/W (Table 8.8). Two types of synchronization can be used, blind cycle and gadfly. Most operations require 40 μs to complete, while some require 1.64 ms. This implementation uses OC5 to create the blind cycle wait. A gadfly interface would have provided feedback to detect a faulty interface but has the problem of creating a software crash if the LCD never finishes. The best interface utilizes both gadfly and blind cycle so that the software can return with an error code if a display operation does not finish on time (because of a broken wire

Figure 8.55
Interface of a HD44780
LCD controller.

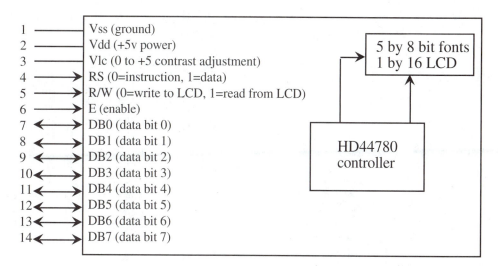

1	——	Vss (ground)
2	——	Vdd (+5v power)
3	——	Vlc (0 to +5 contrast adjustment)
4	——→	RS (0=instruction, 1=data)
5	——→	R/W (0=write to LCD, 1=read from LCD)
6	——→	E (enable)
7	←——→	DB0 (data bit 0)
8	←——→	DB1 (data bit 1)
9	←——→	DB2 (data bit 2)
10	←——→	DB3 (data bit 3)
11	←——→	DB4 (data bit 4)
12	←——→	DB5 (data bit 5)
13	←——→	DB6 (data bit 6)
14	←——→	DB7 (data bit 7)

Table 8.8
Two control signals
specify the type of
access to the HD44780.

RS	R/W	Cycle
0	0	Write to Instruction Register
0	1	Read Busy Flag (bit 7)
1	0	Write data from microprocessor to the HD44780
1	1	Read data from HD44780 to the microprocessor

or damaged display.) First we present a low-level private helper function (Program 8.22). This function would not have a prototype in the LCD.H file because it is private.

```
// MC68HC11A8                              // MC68HC812A4
// 1 by 16 char LCD Display (HD44780)      // 1 by 16 char LCD Display (HD44780)
//   ground = pin 1   Vss                  //   ground = pin 1   Vss
//   power  = pin 2   Vdd +5v              //   power  = pin 2   Vdd +5v
//   10Kpot = pin 3   Vlc contrast adjust  //   10Kpot = pin 3   Vlc contrast adjust
//   PB2    = pin 6   E  enable            //   PJ2    = pin 6   E   enable
//   PB1    = pin 5   R/W 1=read, 0=write  //   PJ1    = pin 5   R/W 1=read, 0=write
//   PB0    = pin 4   RS 1=data, 0=control //   PJ0    = pin 4   RS 1=data, 0=control
//   PC0-7  = pins7-14 DB0-7 8 bit data    //   PH0-7  = pins7-14 DB0-7 8 bit data
#define LCDdata    1 // PB0=RS=1            #define LCDdata    1 // PJ0=RS=1
#define LCDcsr     0 // PB0=RS=0            #define LCDcsr     0 // PJ0=RS=0
#define LCDread    2 // PB1=R/W=1           #define LCDread    2 // PJ1=R/W=1
#define LCDwrite   0 // PB1=R/W=0           #define LCDwrite   0 // PJ1=R/W=0
#define LCDenable  4 // PB2=E=1             #define LCDenable  4 // PJ2=E=1
#define LCDdisable 0 // PB2=E=0             #define LCDdisable 0 // PJ2=E=0
void LCDcycwait(unsigned short cycles){    void LCDcycwait(unsigned short cycles){
    TOC5=TCNT+cycles; // 500ns cycles          TC5=TCNT+cycles;  // 500ns cycles to wait
    TFLG1 = 0x08;  '  // clear C5F             TFLG1 = 0x20;     // clear C5F
    while((TFLG1&0x08)==0){};}                 while((TFLG1&0x20)==0){};}
```

Program 8.22 Private functions for an HD44780 controlled LCD display.

Next we show the high-level public members (Program 8.23). These functions would have prototypes in the LCD.H file.

```
// MC68HC11A8
void LCDputchar(unsigned short letter){
// letter is ASCII code
   PORTC=letter;
   PORTB=LCDdisable+LCDwrite+LCDdata;
   PORTB=LCDenable+LCDwrite+LCDdata;
// E goes 0,1
   PORTB=LCDdisable+LCDwrite+LCDdata;
// E goes 1,0
   LCDcycwait(80);} // 40 us wait
void LCDputcsr(unsigned short command){
   PORTC=command;
   PORTB=LCDdisable+LCDwrite+LCDcsr;
   PORTB=LCDenable+LCDwrite+LCDcsr;
// E goes 0,1
   PORTB=LCDdisable+LCDwrite+LCDcsr;
// E goes 1,0
   LCDcycwait(80);} // 40 us wait
void LCDclear(void){
   LCDputcsr(0x01);   // Clear Display
   LCDcycwait(3280);  // 1.64ms wait
   LCDputcsr(0x02);   // Cursor to home
   LCDcycwait(3280);} // 1.64ms wait
void LCDinit(void){
   DDRC=0xFF;
   LCDputcsr(0x06);
// I/D=1 Increment, S=0 nodisplayshift
   LCDputcsr(0x0C);   // D=1 displayon,
// C=0 cursoroff, B=0 blinkoff
   LCDputcsr(0x14);   /
/ S/C=0 cursormove, R/L=0 shiftright
   LCDputcsr(0x30);
// DL=1 8bit, N=0 1 line, F=0 5by7dots
   LCDclear();}    // clear display
```

```
// MC68HC812A4
void LCDputchar(unsigned short letter){
// letter is ASCII code
   PORTH=letter;
   PORTJ=LCDdisable+LCDwrite+LCDdata;
   PORTJ=LCDenable+LCDwrite+LCDdata;  // E=1
   PORTJ=LCDdisable+LCDwrite+LCDdata; // E=0
   LCDcycwait(80);} // 40 us wait
void LCDputcsr(unsigned short command){
   PORTH=command;
   PORTJ=LCDdisable+LCDwrite+LCDcsr;
   PORTJ=LCDenable+LCDwrite+LCDcsr;  // E=1
   PORTJ=LCDdisable+LCDwrite+LCDcsr; // E=0
   LCDcycwait(80);}    // 40 us wait
void LCDclear(void){
   LCDputcsr(0x01);   // Clear Display
   LCDcycwait(3280);  // 1.64ms wait
   LCDputcsr(0x02);   // Cursor to home
   LCDcycwait(3280);} // 1.64ms wait
void LCDinit(void){
   DDRH=0xFF; DDRJ=0xFF;
   TIOS |= 0x20;   // enable OC5
   TSCR |= 0x80;   // enable, no fast clear
   TMSK2=0xA2;     // 500 ns clock
   LCDputcsr(0x06);
// I/D=1 Increment, S=0 nodisplayshift
   LCDputcsr(0x0C);   // D=1 displayon,
// C=0 cursoroff, B=0 blinkoff
   LCDputcsr(0x14);
// S/C=0 cursormove, R/L=0 shiftright
   LCDputcsr(0x30);
// DL=1 8bit, N=0 1 line, F=0 5by7dots
   LCDclear(); }   // clear display
```

Program 8.23 Public functions for an HD44780 controlled LCD display.

8.4 Transistors Used for Computer-Controlled Current Switches

We can use individual transistors to source or sink current. In this chapter the transistors are used in saturated mode. This means that when the NPN transistors is on, current flows from the collector to the emitter. When the NPN transistor is off, no current flows from the collector to the emitter. Each transistor has an input and output impedance, h_{ie} and h_{oe}, respectively. The current gain is h_{fe}, or β. the model for the bipolar NPN transistor is shown in Figure 8.56.

Figure 8.56
NPN transistor model.

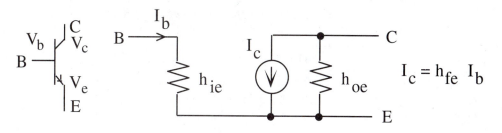

There are five basic design rules when using individual bipolar NPN transistors in saturated mode:

1. Normally $V_c > V_e$
2. Current can flow only in the following directions: from base to emitter (input current), from collector to emitter (output current), and from base to collector (doesn't usually happen but could if $V_b > V_c$)
3. Each transistor has maximum values for the following terms that should not be exceeded: I_b, I_c, V_{ce}, and $I_c \cdot V_{ce}$
4. The transistor acts like a current amplifier: $I_c = h_{fe} \cdot I_b$
5. The transistor will activate if $V_b > V_e + V_{be(SAT)}$, where $V_{be(SAT)}$ is typically above 0.6 V

The model for the bipolar PNP transistor is shown in Figure 8.57.

Figure 8.57
PNP transistor model.

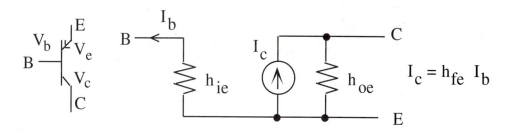

There are five basic design rules when using individual bipolar PNP transistors in saturated mode:

1. Normally $V_e > V_c$
2. Current can flow only in the following directions: from emitter to base (input current), from emitter to collector (output current), and from collector to base (doesn't usually happen but could if $V_c > V_b$)
3. Each transistor has maximum values for the following terms that should not be exceeded: I_b, I_c, V_{ce}, and $I_c \cdot V_{ce}$
4. The transistor acts like a current amplifier: $I_c = h_{fe} \cdot I_b$
5. The transistor will activate if $V_b < V_e - V_{be(SAT)}$, where $V_{be(SAT)}$ is typically above 0.6 V

Performance Tip: A good transistor is one in which the I/O response is independent of h_{fe}. We can design the interface so that I_b can be twice as large as needed to supply the necessary I_c.

Table 8.9 illustrates the wide range of bipolar transistors that we can use.

Type	NPN	PNP	Package	$V_{be(SAT)}$	$V_{ce(SAT)}$	h_{fe} min/max	I_c
General purpose	2N3904	2N3906	TO-92	0.85 V	0.2 V	100	10 mA
General purpose	PN2222	PN2907	TO-92	1.2 V	0.3 V	100	150 mA
General purpose	2N2222	2N2907	TO-18	1.2 V	0.3 V	100/300	150 mA
Power transistor	TIP29A	TIP30A	TO-220	1.3 V	0.7 V	15/75	1 A
Power transistor	TIP31A	TIP32A	TO-220	1.8 V	1.2 V	25/50	3 A
Power transistor	TIP41A	TIP42A	TO-220	2.0 V	1.5 V	15/75	3 A
Power Darlington	TIP120	TIP125	TO-220	2.5 V	2.0 V	1000 min	3 A

Table 8.9
Parameters of typical transistors used by microcomputer to source or sink current.

8.5 Computer-Controlled Relays, Solenoids, and DC Motors

Relays, solenoids, and pulse-width modulated DC motors are grouped together because their electric interfaces are similar. In each case, there is a coil, and the computer must drive (or not drive) current through the coil.

**8.5.1
Introduction to
Relays**

A relay is a device that responds to a small current or voltage change by activating switches or other devices in an electric circuit. It is used to remotely switch signals or power. The input control is usually electrically isolated from the output switch. The input signal determines whether the output switch is open or closed. Figure 8.58 shows typical circuit drawings of classic general-purpose electromagnetic relays. In each, the input current affects the position of the output switch.

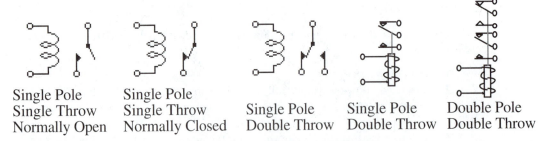

Single Pole   Single Pole
Single Throw   Single Throw   Single Pole   Single Pole   Double Pole
Normally Open   Normally Closed   Double Throw   Double Throw   Double Throw

Figure 8.58
Various types of relays.

The *American Heritage Dictionary* defines *relay* as "an act of passing something along from one person, group, or station to another." Electronic relays were originally used during the last century for extending the range of remote telegraph systems, *relaying* the signal from one station to another (Figure 8.59).

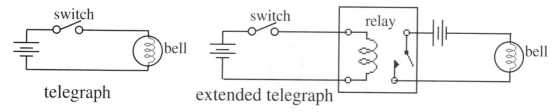

Figure 8.59
Original application of relays.

Relays are classified into four categories depending upon whether the output switches power (i.e., high currents through the switch) or electronic signals (i.e., low currents through the switch). Another difference is how the relay implements the switch. An electromagnetic (EM) relay uses a coil to apply EM force to a contact switch that physically opens and closes. The solid-state relay uses transistor switches made from solid-state components to electronically allow or prevent current flow across the switch. The four types (three shown in Figure 8.60) are:

- The classic general-purpose relay has an EM coil and can switch power
- The reed relay has an EM coil and can switch-low level DC electronic signals
- The solid-state relay (SSR) has an input-triggered semiconductor power switch
- The bilateral switch uses CMOS, FET, or biFET transistors

Actually, the bilateral switch is not a relay, but it is included in this discussion because it is a solid-state device that can be used to switch low-level signals.

Figure 8.60
Photo of an EM, solid-state, and reed relay.

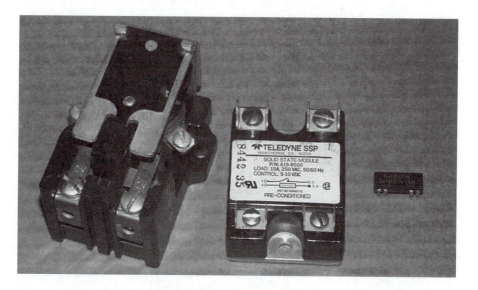

8.5.2
Electromagnetic
Relay Basics
Figure 8.61 illustrates a classical general-purpose EM relay. The input circuit is an EM coil with an iron core. The output switch includes two sets of silver or silver-alloy contacts (called *poles*). One set is fixed to the relay *frame,* and the other set is located at the end of leaf spring poles connected to the *armature.* The contacts are held in the "normally closed" position by the armature return spring. When the input circuit energizes the EM coil, a

Figure 8.61
Drawing of an EM relay.

Double Pole Double Throw   (DPDT)

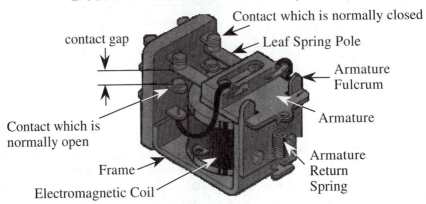

Contact which is normally closed

contact gap

Leaf Spring Pole

Armature Fulcrum

Contact which is
normally open

Armature

Frame

Armature
Return
Spring

Electromagnetic Coil

"pull-in" force is applied to the armature and the "normally closed" contacts are released (called *break*) and the "normally open" contacts are connected (called *made*). The armature pull-in can either energize or deenergize the output circuit, depending on how it is wired. The illustrated relay has its transparent protective polycarbonate cover removed. Relays are mounted in special sockets, or directly soldered onto a printed circuit board.

The number of poles (e.g., single-pole, double-pole, 3P, 4P) refers to the number of switches that are controlled by the input. The relay shown below is a double pole because it has two switches. *Single throw* means each switch has two contacts that can be open or closed. *Double throw* means each switch has three contacts. The common contact will be connected to one of the other two contacts (but not both at the same time).

The parameters of a relay are specified in terms of the input coil and output switch (Table 8.10). The input parameters include DC or AC excitation, coil resistance, pickup voltage, dropout voltage, and maximum coil power. The *pickup voltage* is the coil potential above in which activation is guaranteed. The *dropout voltage* is the coil potential below which deactivation is guaranteed. Typically the coil can handle sustained voltages a few volts above its nominal value before damage occurs. In microcomputer-based interfaces,

Table 8.10
Parameters of four
different types of
computer-controlled
switches.

Manufacturer Model number	Teledyne 712-5	Magnecraft W107DIP-2	Teledyne 611	PMI SW-01
Type	TO-5 relay	Reed	SSR	JFET
Coil resistance	50Ω	500Ω	500Ω	3-μA input
Pickup voltage	3.6 V	3.8 V	3.8 V	2 V
Dropout voltage		0.5 V	0.8 V	0.8 V
Contact load	AC/DC	DC	AC/DC	DC
Max contact power		10 W DC	2500 W AC	
Max contact voltage	250 V AC	100 V DC	250 V AC	+11 to −10 V
Max contact current	600 mA AC	0.5 A DC	10 A AC	5 mA DC
On resistance	0.2Ω	0.1Ω	0.15Ω	100Ω
Off resistance	Infinite	Infinite	28 KΩ	58 dB
Turn-on time	4 ms	600 μs	8.3 ms	400 ns
Turn-off time	3 ms	75 μs	16.6 ms	300 ns
Life expectancy	10^7	$25 \cdot 10^6$	Infinite	Infinite
Approximate cost	$2 to $4	$1 to $2	$10 to $20	$0.50–$2

DC coils are more convenient than AC coils. Unless there is an internal snubber diode, the polarity of the coil current does not matter.

The parameters of the output switch include maximum AC (or DC) power, maximum current, maximum voltage, on resistance, and off resistance. A DC signal welds the contacts together at a lower current value than an AC signal; therefore the maximum ratings for DC are considerably smaller than for AC. Other relay parameters include turn-on time, turn-off time, life expectancy, and I/O isolation. *Life expectancy* is measured in number of operations. Storage life for relays is usually quite long. Table 8.10 lists specifications for four devices.

Figure 8.62 illustrates the various configurations available. The sequence of operation is listed in Table 8.11.

Figure 8.62
Various relay configurations.

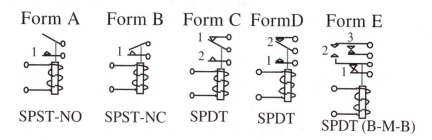

Table 8.11
Five relay configurations.

Form	Activation sequence	Deactivation sequence
A		Make 1 Break 1
B	Break 1	Make 1
C	Break 1, Make 2	Break 2, Make 1
D	Make 1, Break 2	Make 2, Break 1
E	Break 1, Make 2, Break 3	

Latching relays do not require continuous input current to remain in their present state. A short pulse on the two-input coil (1 to 50 ms) causes the switch to change state, after which it will remain in the new state (open or closed) indefinitely. These devices are suitable for low-power operation when the contact is infrequently switched. They are also convenient for low-noise applications, because they do not require continuous coil currents. Mercury-wetted relays provide for faster switching and eliminate contact bounce. Some devices have an internal snubber diode and/or electrostatic shielding. The shielding reduces the noise crosstalk from the input coil to the external circuits. A shielded relay should be used if the relay is switching simultaneously with critical low-noise functions.

8.5.3
Reed Relays

Reed relays (Figure 8.63) are often used in medical electronics, telecommunications, and automated test equipment (ATE) for signal-level switching (e.g., mode/gain/offset selection). The single-pole–single-throw (SPST) reed capsule has two contacts that are normally open. When the coil is activated, an EM force causes the contacts to close.

The coil is wound on a bobbin that surrounds the glass reed capsule. The reed capsule is hermetically sealed with inert gas. The contacts are typically made from precious metal like rhodium. Gold-cobalt contacts provide (at additional cost) reduced contact resistance (less than 0.05Ω) for applications that require higher accuracy. Reed relays are available as DIPs or single inline packages (SIPs). The relay can be purchased as SPST-NO (Form A), SPST-NC (Form B), DPST-NO (Form 2A), and SPDT (Form C).

Figure 8.63
Drawing of a reed relay.

Single Pole Single Throw (SPST) Reed Relay

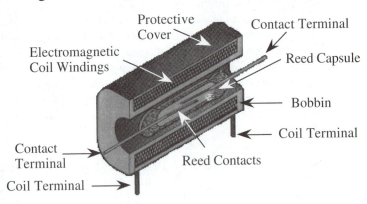

8.5.4
Solenoids

Solenoids are used in door locks, automatic disk/tape ejectors, and liquid/gas flow control valves (on/off type). Much like an EM relay, there is a frame that remains motionless and an armature that moves in a discrete fashion (on/off). A solenoid has an electromagnet (Figure 8.64 and Figure 8.65). When current flows through the coil, a magnetic force is created, causing a discrete motion of the armature. When the current is removed, the magnetic force stops, and the armature is free to move. The motion in the opposite direction can be produced by a spring, gravity, or by a second solenoid.

Figure 8.64
Mechanical drawing of a solenoid showing that the EM coil causes the armature to move.

Solenoid

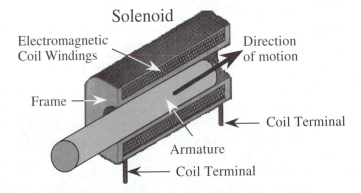

Figure 8.65
Photograph of two solenoids.

8.5.5
Pulse-Width
Modulated DC
Motors

Similar to the solenoid and EM relay, the DC motor has a frame that remains motionless and an armature that moves. In this case, the armature moves in a circular manner (shaft rotation). A DC motor has an electromagnet as well. When current flows through the coil, a magnetic force is created, causing a rotation of the shaft. Brushes positioned between the frame and armature are used to alternate the current direction through the coil so that a DC current generates a continuous rotation of the shaft. When the current is removed, the magnetic force stops and the shaft is free to rotate. In a pulse-width modulated DC motor, the computer activates the coil with a current of fixed magnitude but varies the duty cycle to control the motor speed. Software examples that generate variable duty-cycle square waves were presented in Section 6.2.5.

8.5.6
Interfacing EM
Relays, Solenoids,
and DC Motors

The interface circuit should provide sufficient current and voltage to activate the device. In the off state, the input current should be zero. Because of the inductive nature of the coil, huge back electromotive force (EMF) signals develop when the coil current is cut off. Because of the high-speed transistor switch, there is a large dI/dt when the computer deactivates the coil (some current to no current). There is also a dI/dt (but smaller because of the capacitances in the circuit) when the computer activates the coil (no current to some current). These voltages can range from 50 to 200 V, depending on the coil and driving circuit. To protect the driver electronics, a snubber diode is added to suppress the back EMF (Table 8.12). The choice of diode should consider the magnitude of the back EMF. For signals less than 75 V, the 1N914 is adequate. For larger and larger back EMFs the 1N4001 through 1N4007 can be used. All these diodes are fast enough to protect the driver electronics. The open-collector driver is used because the circuit requires two states: the need to sink current (during the activate state) and no current (during the off state). Because solenoids and DC motors (controlled with PWM) have coils similar to the EM relay, the interfaces shown in Figure 8.66 can be also used to control solenoids and DC motors.

Table 8.12
Maximum voltage parameter for typical snubber diodes.

Diode	Maximum voltage (V)
1N4001	50
1N914	75
1N4002	100
1N4003	200
1N4004	400
1N4005	600
1N4006	800
1N4007	1000

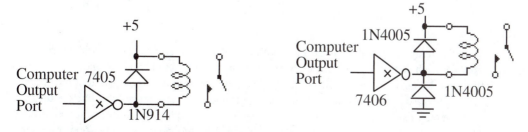

Figure 8.66 Two relay interfaces that sink current through the coil.

The output low current (I_{OL}) of the driver should be sufficient to activate the relay. Other open-collector drivers (e.g., 7406, 75492, 75451, ULN2074, IRF540) have larger output low currents. Some designers suggest using two diodes to suppress both positive and negative back EMF signals. Good engineering practice would be to measure the coil voltage and include both diodes if negative EMF signals are observed.

Transistors provide an alternative interfacing technique for relays. Choose a device that can source (PNP) or sink (NPN) the necessary coil current. Speed is not an issue in transistor selection (Figure 8.67).

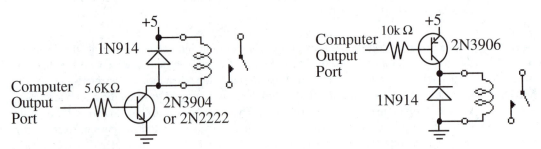

Figure 8.67   Two relay interfaces that source current through the coil.

For coils that require currents above 500 mA, a MOSFET can be used. The top NPN transistor sources the current to the gate (pin 1) to turn on the MOSFET. The bottom NPN transistor sinks the current from the gate to turn off the MOSFET. Although the MOSFET gate does not need a lot of current to maintain the on state, large currents are required to switch it from off to on and from on to off (Figure 8.68).

Figure 8.68
Motor interface using a
high-current MOSFET.

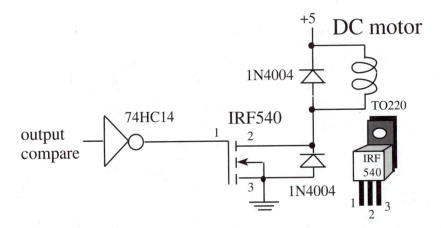

The relay by its nature isolates the computer and its driver electronics from the currents in the switch contacts. On the other hand, currents in the solenoid and DC motor must pass through the same ground as the computer. When these currents are large and noisy, we can decouple the coil currents from the computer ground using an optoisolator like the 6N139. Using electrical isolation can protect the computer electronics from surges in and around the motor. In the circuit of Figure 8.69, the computer (Vdd) and motor (Vpp) power are separate and their grounds are usually not connected. High-speed optoisolators, similar to

Figure 8.69
An isolated motor
interface using a 6N139.

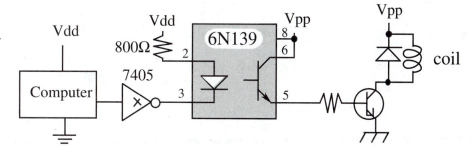

Figure 8.70
A high-current isolated
motor interface using a
6N139 and a MOSFET.

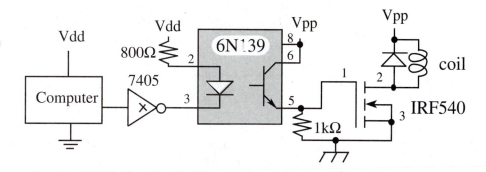

Figure 8.71
Another high-current
isolated motor interface
using a 6N139 and a
MOSFET.

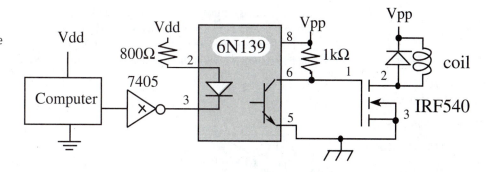

these, can be used in computer networks to protect one computer from another. For more current, we can use the IRF540 MOSFET (Figures 8.70 and 8.71).

In some applications we wish to drive the motor forward and backward. To achieve this result we must be able to drive current both forward and backward through the motor coil. One approach to this interface is called the H-bridge. It is important not to simultaneously drive both Q1, Q2 or both Q3, Q4. The basic approach to the H-bridge is illustrated in Figure 8.72. If Q2 and Q3 are on, then current flows right to left across the coil. If Q1 and Q4 are on, then current flows in the opposite direction. PNP transistors are used to source current into the coil, and NPN transistors are used to sink current out of the coil. Four diodes are required to prevent back EMF that occurs when the current is applied and removed.

The TIP125 PNP Darlington transistors can source up to 3 A. To activate the PNP transistor, the base voltage must be 0.6 V less than the emitter voltage (Vm). When light flows

Figure 8.72
An H-bridge is used to
drive current in both
directions.

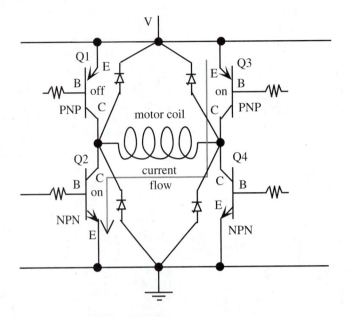

through its 6N139 isolation barrier, the 6N139 output goes to about 1 V, which drops the
TIP125 base voltage low enough and sinks enough base current to activate the PNP source
transistor. A low on the **Forward** signal will activate Q1 and Q4.

The TIP120 NPN Darlington transistors can sink up to 3 A. To activate the NPN transistor, the base voltage must be 0.6 V more than the emitter voltage (ground). When light
flows through its 6N139 isolation barrier, the 6N139 output goes to about Vm − 1 V, which
raises the TIP120 base voltage high enough and sources enough base current to activate the
NPN sink transistor. A low on the **Reverse** signal will activate Q2 and Q3 (Figure 8.73).

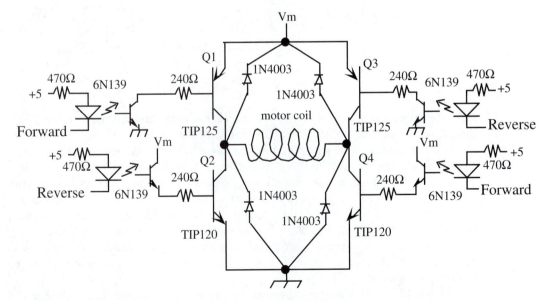

Figure 8.73   An isolated H-bridge can drive current in both directions.

Figure 8.74
A digital circuit used
with an H-bridge.

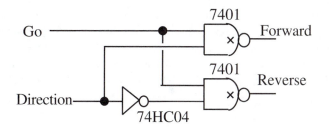

As mentioned earlier it is important not to activate both **Forward** and **Reverse** at the same time. To prevent this situation even when the software crashes, the digital interface shown in Figure 8.74 could be used. To prevent temporary current paths through Q1+Q2 or Q3+Q4, the software should change the **Direction** only while the motor has been stopped (**Go**=0) for enough time to allow all transistors to turn off.

8.5.7 Solid-State Relays

To solve the limited life expectancy and contact bounce problems, SSRs were developed. Figure 8.75 illustrates the major components of a SSR. Figure 8.76 is a photograph of two SSRs. The SSR has no moving parts. The optocoupler provides isolation between the input circuit (pseudocoil) and the triac (pseudocontact). The switch function is constructed from either two inverse-parallel SCRs or an electrically equivalent triac. In the next section we will use silicon devices for switching DC signals that use power bipolar transistors or MOSFETs.

The signal from the phototransistor triggers the output triac so that it switches the load current. The zero-voltage detector triggers the triac only when the AC voltage is zero, reducing the surge currents when the triac is switched. Surge currents can occur when

Figure 8.75
Internal components of
a SSR.

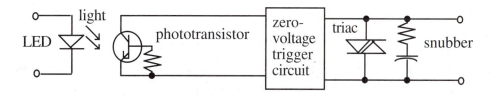

Figure 8.76
Photograph of two SSRs.

controlling capacitive loads like lamps. Once triggered the triac conducts until the next zero crossing. If the input signal continues, then the triac will continuously conduct. The RC circuit is called a snubber and is used to reduce the voltage transients that occur when switching inductive loads like motors and solenoids. SSRs cost five to ten times more than general-purpose relays but have the following advantages because there are no moving parts:

- Longer life
- Higher reliability
- Faster switching (although it occurs on the AC load zero crossing)
- Better mechanical stability
- Insensitive to vibrations and shock
- No contact bounce
- Reduces electromagnetic interference
- Quieter
- Eliminates contact arcing (can be used in the presence of explosive gases)

Interfacing the SSR to the microcomputer is identical to the LED interface (Figure 8.77).

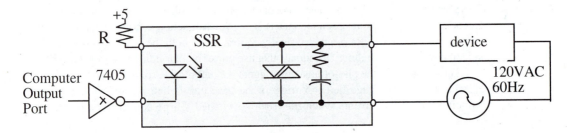

Figure 8.77   Interface of a SSR.

The value of the resistor, R, is chosen to select the V_d, I_d operating point of the LED.

$$R = \frac{+5 - V_d - V_{OL}}{I_d}$$

where V_{OL} is the output low voltage of the open-collector driver (7405). Again, the output low current (I_{OL}) of the driver should be sufficient to activate the LED (e.g., 7406, 75492, 75451, ULN2074).

8.6 Stepper Motors

Stepper motors are very popular for microcomputer-controlled machines because of their inherent digital interface. It is easy for a microcomputer to control both the position and velocity of a stepper motor in an open-loop fashion. Although the cost of a stepper motor is typically higher than an equivalent DC permanent magnetic field motor, the overall system cost is reduced because stepper motors may not require feedback sensors. For these reasons, they are used in many computer peripheral such as hard disk drives, floppy disk drives, and printers. Stepper motors can also be used as shaft encoders, measuring both position and speed (Figure 8.78).

Figure 8.78
Three stepper motors.

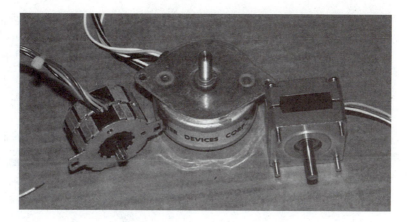

8.6.1
Stepper Motor
Example

In this section, we will introduce stepper motors by showing a simple example. The computer can make the motor spin by outputting the sequence . . . 10, 9, 5, 6, 10, 9, 5, 6 . . . over and over. For a motor with 200 steps per revolution each new output will cause the motor to rotate 1.8°. If the time between outputs is fixed at Δt s, then the shaft rotation speed will be $0.005/\Delta t$ in revolutions per second (rps). In each system, we will connect the stepper motor to the least significant bits of output PORTB. In this first hardware interface, we will connect a unipolar stepper with four 5-V coils, A, A', B, B'. Because the coil resistance is about 70Ω, it will require about 5 V/70Ω, or 70 mA, to activate. The output low current of the 75492 is sufficient to sink the 70 mA. The 1N914 diodes will protect the 75492 from the back EMF that will develop across the coil when the current is shut off. Figure 8.79 illustrates the state when 10 (binary 1010) is output to PORTB. Bits PB3 and PB1 are high, making their 75492

Figure 8.79 Simple stepper interface.

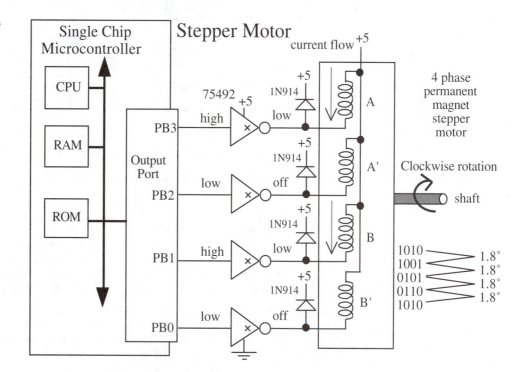

outputs low and driving current through coils A and B. Bits PB2 and PB0 are low, making their 75492 outputs off (HiZ), and driving no current through coils A' and B'.

We define the active state of the coil when current is flowing. The basic operation is summarized in Table 8.13.

Table 8.13
Stepper motor sequence.

PORTB output	A	A'	B	B'
10	Activate	deactivate	activate	deactivate
9	Activate	deactivate	deactivate	activate
5	Deactivate	activate	deactivate	activate
6	Deactivate	activate	activate	deactivate

We will implement a linked list approach to the stepper motor control software. This approach yields a solution that is easy to understand and change. If the computer outputs the sequence backward, then the motor will spin in the other direction. To ensure proper operation, this . . . 10, 9, 5, 6, 10, 9, 5, 6 . . . sequence must be followed. For example, assume the computer outputs . . . 9, 5, 6, 10, and 9. Now it wishes to reverse direction, since the output is already at 9, then it should begin at 10, and continue with 6, 5, 9 . . . . In other words, if the current output is "9," then the only two valid next outputs would be "5" if it wanted to spin clockwise or "10" if it wanted to spin counterclockwise. Maintaining this proper sequence will be simplified by implementing a double circular linked list. For each node in the linked list there are two valid next states, depending upon whether the computer wishes to spin clockwise or counterclockwise (Figure 8.80).

Linked List Data Structure

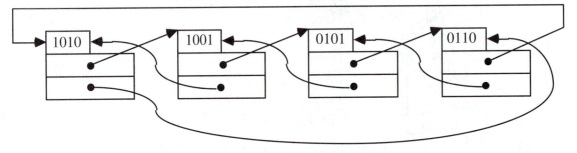

Figure 8.80   A double circular linked list used to control the stepper motor.

A slip is when the computer issues a sequence change, but the motor does not move. A slip can occur if the load on the shaft exceeds the available torque of the motor. A slip can also occur if the computer tries to change the outputs too fast. If the system knows the initial shaft angle, and the motor never slips, then the computer can control both the shaft speed and angle without a position sensor. The routines CW and CCW will step the motor once in the clockwise and counterclockwise directions, respectively. If every time the computer calls CW or CCW it were to wait for 5 ms, then the motor would spin at 1 rps. The linked list for the MC68HC705 uses 8-bit pointers (index from the start of the structure) because the computer does not support 16-bit registers. The MC68HC708, MC68HC11, and

```
;MC68HC705 or MC68HC708              ; MC68HC11A8 or MC68HC812A4
;Linked list stored in EEPROM       ;Linked list stored in EEPROM
LL:                                  S10:   dc.b 10    ;Output pattern
S10:   dc.b 10    ;Output pattern           dc.w S9    ;Next if CW
       dc.b S9-LL ;Next if CW               dc.w S6    ;Next if CCW
       dc.b S6-LL ;Next if CCW       S9:    dc.b 9
S9:    dc.b 9                               dc.w S10
       dc.b S10-LL                          dc.w S5
       dc.b S5-LL                    S5:    dc.b 5
S5:    dc.b 5                               dc.w S9
       dc.b S9-LL                           dc.w S6
       dc.b S6-LL                    S6:    dc.b 6
S6:    dc.b 6                               dc.w S5
       dc.b S5-LL                           dc.w S10
       dc.b S10-LL
;Global variables stored in RAM      ;Global variables stored in RAM
POS:   ds   1   ;0<=POS<=199         POS:   ds   1    ;0<=POS<=199
PT:    ds   1   ;to current state    PT:    ds   2    ;to current state
```

Program 8.24 A double circular linked list used to control the stepper motor.

MC68HC812 have similar solutions. In each case, the linked list data structure will be stored in EEPROM, and the variables are allocated into RAM and initialized at run time in the ritual (Program 8.24).

The programs that step the motor also maintain the position in the global, POS. If the motor slips, then the software variable will be in error. Also it is assumed the motor is initially in position 0 at the time of the ritual. Because the 6808 does have a 16-bit index register (HX), it is possible to implement the controller similar to the 6811/6812 example (Program 8.25).

```
;MC68HC705, or MC68HC708XL36         ; MC68HC11A8 or MC68HC812A4
;Move 1.8 degrees clockwise          ;Move 1.8 degrees clockwise
CW:   ldx PT       ;current state    CW:   ldx  PT     ;current state
      ldx LL+1,X ;Next clockwise           ldx  1,X    ;Next clockwise
      stx PT       ;Update pointer          stx  PT     ;Update pointer
      lda LL,X     ;Output pattern          ldaa ,X     ;Output pattern
      sta PORTB    ;Set phase control       staa PORTB  ;Set phase control
      lda POS      ;Update position         ldaa POS    ;Update position
      inca         ;clockwise               inca        ;clockwise
      cmp #200                              cmpa #200
      blo OK1      ;0<=POS<=199             blo  OK1    ;0<=POS<=199
      clra                                  clra
OK1:  sta POS                        OK1:  staa POS
      rts                                  rts
;Move 1.8 degrees counterclockwise   ;Move 1.8 degrees counterclockwise
CCW:  ldx PT       ;current state    CCW:  ldx  PT     ;current state
      ldx LL+2,X ;Next CCW                 ldx  3,X    ;Next CCW
      stx PT       ;Update pointer          stx  PT     ;Update pointer
      lda LL,X     ;Output pattern          ldaa ,X     ;Output pattern
```

continued on p. 480

Program 8.25 Helper functions used to control the stepper motor.

continued from p. 479

```
        sta PORTB   ;Set phase control
        lda POS     ;Update position
        deca        ;CCW direction
        cmp #255
        bne OK2     ;0<=POS<=199
        lda #199
OK2: sta POS
        rts
Init: lda #$FF
        sta DDRB    ;PA3-0 output
        clr POS
        ldx #S10-LL
        stx PT
        rts
```

```
        staa PORTB ;Set phase control
        ldaa POS    ;Update position
        deca         ;CCW direction
        cmpa #255
        bne  OK2     ;0<=POS<=199
        ldaa #199
OK2: staa POS
        rts
Init: clr POS
        ldx  #S10
        stx  PT
        movb #$FF,DDRB   ;6812 only
        rts
```

Program 8.25 Helper functions used to control the stepper motor.

The function in Program 8.26 will step the motor moving to the desired position. It will choose to go clockwise or counterclockwise, depending on which way is closer. In C, these routines could be written as shown in Program 8.27.

```
; MC68HC705 or MC68HC708
goal: ds 1      ;desired position
;Reg A=desired 0<=Reg A<=199
SEEK: sta goal   ;Save desired
        sub POS     ;Go CW or CCW?
        beq DONE    ;Skip if equal
        bhi HIGH    ;Desired>POS?
;Desired<POS
        nega        ;(POS-Desired)
        cmp #100
        blo GOCCW  ;Go CCW if
;Desired<POS and POS-Desired<100
GOCW: bsr CW       ;Reg A current
        cmp goal
        bne GOCW    ;POS=Desired?
        bra DONE
HIGH: cmp #100    ;(Desired-POS)
        blo GOCW    ;Go CW if
;Desired>POS and Desired-POS<100
GOCCW: bsr CCW    ;Reg A current
        cmp goal
        bne GOCCW ;POS=Desired?
DONE: rts         ;Return
```

```
; MC68HC11A8 or MC68HC812
;Reg B=desired 0<=Reg B<=199
SEEK: pshb         ;Save desired
        tsy
        subb POS    ;Go CW or CCW?
        beq  DONE   ;Skip if equal
        bhi  HIGH   ;Desired>POS?
;Desired<POS
        negb         ;(POS-Desired)
        cmpb #100
        blo  GOCCW ;Go CCW if
;Desired<POS and POS-Desired<100
GOCW: bsr  CW       ;Reg A current
        cmpa ,Y
        bne  GOCW   ;POS=Desired?
        bra  DONE
HIGH: cmpb #100   ;(Desired-POS)
        blo  GOCW   ;Go CW if
;Desired>POS and Desired-POS<100
GOCCW: bsr CCW    ;Reg A current
        cmpa ,Y
        bne  GOCCW ;POS=Desired?
DONE: pulb
        rts         ;Return
```

Program 8.26 High-level assembly function to control the stepper motor.

Program 8.27 C linked list and helper functions used to control the stepper motor.

```
const struct State
{   unsigned char Out;      /* Output for this state */
    const struct State *Next[2]; /* Next state CW or CCW motion */
};
#typedef struct State StateType;
#typedef StateType * StatePtr;
unsigned char POS;  /* between 0 and 199 representing shaft angle */
#define clockwise 0            /* Next index*/
#define counterclockwise 1     /* Next index*/
StateType fsm[4]={
    {10,{&fsm[1],&fsm[3]}},
    { 9,{&fsm[2],&fsm[0]}},
    { 5,{&fsm[3],&fsm[1]}},
    { 6,{&fsm[0],&fsm[2]}}
};
StatePtr Pt;  /* Current State */
void CW(void){
    Pt=Pt->Next[clockwise];     /* circulates around linked list */
    PORTB=Pt->Out;              /* step motor */
    if(POS++==200) POS=0;}      /* maintain shaft angle */
void CCW(void){
    Pt=Pt->Next[counterclockwise]; /* circulates around linked list*/
    PORTB=Pt->Out;                 /* step motor */
    if(POS==0)POS=199;
    else POS--; }                  /* maintain shaft angle */
void Init(void){
    DDRB=0xFF;   // 6812 only
    POS=0;
    Pt=&fsm[0];}
```

In the routine of Program 8.28 the computer will step the motor to the desired New position. We can use the current position, POS, to determine if it would be faster to go clockwise or counterclockwise. CWsteps is calculated as the number of steps from POS to New if the motor were to spin clockwise. If it is greater than 100, then it would be faster to get there by going counterclockwise.

Program 8.28 High-level C function to control the stepper motor.

```
void SEEK(unsigned char New){ int CWsteps,i;
    if((CWsteps=New-POS)<0)
        CWsteps+=200;
    if(CWsteps>100)
        for(i=CWsteps;i<200;i++)
            CCW();
    else
        for(i=0;i<CWsteps;i++)
            CW(); }
```

Figure 8.81
Simple stepper motor
with 20 steps per
revolution.

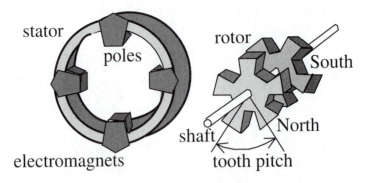

8.6.2
Basic Operation

Figure 8.81 shows a simplified stepper motor. The permanent magnet stepper has a rotor and a stator. The rotor is manufactured from a spool-shaped permanent magnet. This rotor has five teeth in both the north and south ends of the magnet. The teeth on the two magnetic poles are offset by half the tooth pitch. A stepper motor with 200 steps per revolution has 50 teeth on each end of the rotor. The stator consists of multiple iron-core electromagnets whose poles have the same width as the teeth on the rotor but are spaced slightly farther apart. The stator of this simple stepper motor has four electromagnets and four poles. The stator of a stepper motor with 200 steps per revolution has 8 electromagnets each with 5 teeth, making a total of 40 poles.

The operation of the stepper motor is illustrated in the next four figures. In a four-wire (or bipolar) stepper motor, the electromagnets are wired together, creating two phases. The five- and six-wire (or unipolar) stepper motors also have two phases, but each is center-tapped to simplify the drive circuitry.

Since there are five teeth on this simple motor, the shaft will rotate $360°/(4 \cdot 5)$, or $18°$ per step. Figure 8.82 shows the initial position of the rotor with respect to the fixed stator. The labels on the permanent magnet rotor are added to clarify the rotation steps. There is no net force on the left and right stator poles because the north and south teeth are equally covered. On the other hand, there is a strong attractive force on both the top and bottom,

Figure 8.82
Stable state 1.

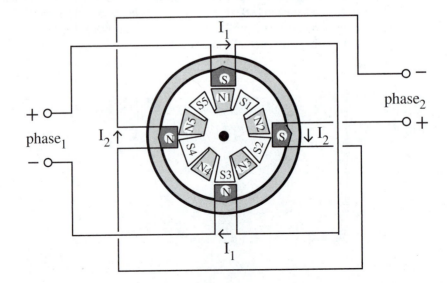

holding the rotor at this position. In fact, stepper motors are rated according to their holding torque. Typical holding torques range from 10 to 300 oz·in.

When the polarity of phase$_1$ is reversed, the rotor is in an unstable state, as shown in Figure 8.83. The rotor will move because there are strong repulsive forces on both the top and bottom. By observing the left and right poles, the closest stable state occurs if the rotor rotates clockwise, resulting in the stable state illustrated in Figure 8.84.

Figure 8.83
Unstable state as rotor goes from state 1 to state 2.

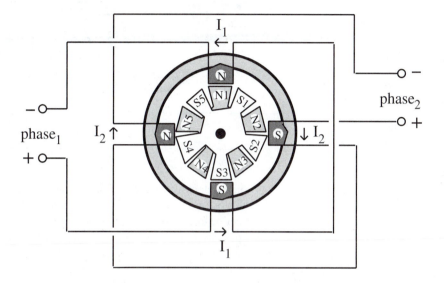

Figure 8.84
Stable state 2.

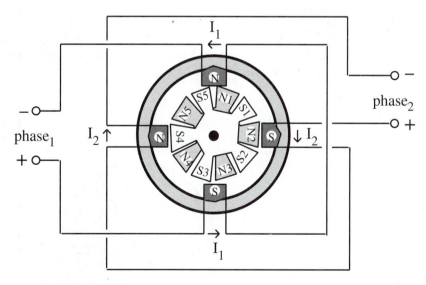

State 2 is exactly 18° clockwise from state 1. Now there is no net force on the top and bottom stator poles because the north and south teeth are equally covered. On the other hand, there are strong attractive forces on both the left and right, holding the rotor at this new position.

Next the polarity of phase$_2$ is reversed. Once again, the rotor is in an unstable state because there are strong repulsive forces on both the left and right. The closest stable state

Figure 8.85
Stable state 3.

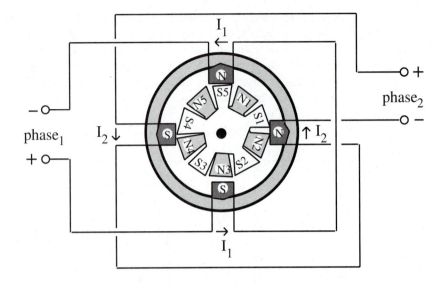

occurs if the rotor rotates clockwise, resulting in the stable state illustrated in Figure 8.85. State 3 is exactly 18° clockwise from state 2, with no net force on the left and right stator poles and strong attractive forces on both the top and bottom.

Once again the polarity of $phase_1$ is reversed, and the rotor rotates clockwise by 18°, resulting in the stable state shown in Figure 8.86. State 4 has no net force on the top and bottom stator poles and strong attractive forces on both the left and right.

When the polarity of $phase_2$ is reversed again, and the rotor rotates clockwise by 18°, it results in the stable state similar to Figure 8.82. These four states are cycled. The rotor will spin in a counterclockwise direction if the sequence is reversed.

The time between states determines the rotational speed of the motor. Let ΔT be the time between steps and θ the step angle; then the rotational velocity v is $\theta/\Delta T$. As long as the load on the shaft is below the holding torque of the motor, the position and speed can

Figure 8.86
Stable state 4.

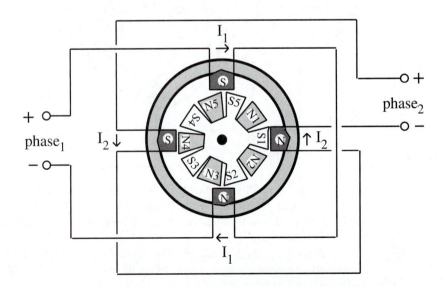

be reliably maintained with an open-loop software control algorithm. To prevent skips (digital commands that produce no rotor motion) it is important to limit the change in acceleration, or *jerk*. Let $\Delta T(n-2)$, $\Delta T(n-1)$, $\Delta T(n)$ be the discrete sequence of times between steps.

$$v(n) = \theta/\Delta T$$

The acceleration is given by

$$a(n) = (v(n) - v(n-1))/\Delta T(n) = (\theta/\Delta T(n)-\theta/\Delta T(n-1))/\Delta T(n) = \theta/\Delta T(n)^2 - \theta/(\Delta T(n-1)\Delta T(n))$$

The change in acceleration, or jerk, is given by

$$b(n) = (a(n) - a(n-1))/\Delta T(n)$$

For example, if the time between steps is to be increased from 1000 to 2000 µs, an ineffective approach (as shown in Table 8.14) would simply be to go directly from 1000 to 2000. This produces a very large jerk that may cause the motor to skip.

Table 8.15 shows that a more gradual change from 1000 to 2000 produces a ten times smaller jerk, reducing the possibility of skips. The optimal solution (the one with the smallest jerk) occurs when $v(t)$ has a quadratic shape. This will make $a(t)$ linear, and $b(t)$ a constant. Limiting the jerk is particularly important when starting to move a stopped motor.

Table 8.14
An ineffective approach to changing motor speed.

n	ΔT (µs)	$v(n)$ (°/s)	$a(n)$ (°/s*2)	$b(n)$ (°/s*3)
1	1000	1800		
2	1000	1800	0.00E+00	
3	2000	900	−2.50E+05	0.00E+00
4	2000	900	0.00E+00	−2.25E+08
5	2000	900	0.00E+00	2.25E+08
6	2000	900	0.00E+00	0.00E+00

Table 8.15
An effective approach to changing motor speed.

n	ΔT (µs)	$v(n)$ (°/s)	$a(n)$ (°/s**2)	$b(n)$ (°/s**3)
1	1000	1800		
2	1000	1800	0.00E+00	
3	1000	1800	0.00E+00	0.00E+00
4	1008	1786	−1.39E+04	−1.38E+07
5	1032	1744	−4.11E+04	−2.64E+07
6	1077	1671	−6.77E+04	−2.47E+07
7	1152	1563	−9.37E+04	−2.26E+07
8	1275	1411	−1.19E+05	−1.96E+07
9	1500	1200	−1.41E+05	−1.48E+07
10	1725	1044	−9.06E+04	2.91E+07
11	1848	974	−3.78E+04	2.86E+07
12	1923	936	−1.96E+04	9.44E+06
13	1968	915	−1.09E+04	4.43E+06
14	1992	904	−5.65E+03	2.63E+06
15	2000	900	−1.77E+03	1.94E+06
16	2000	900	0.00E+00	8.85E+05
17	2000	900	0.00E+00	0.00E+00

Figure 8.87
Bipolar stepper motor
interface.

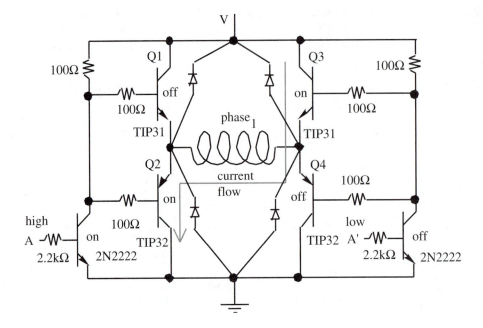

8.6.3
Stepper Motor
Hardware
Interface

The bipolar stepper motor can be controlled with a double-pole–double-throw driver, as shown in Figure 8.87. The resistors and transistors are chosen so that the appropriate current is driven across the phases. The choice of the 100Ω resistors is to provide sufficient base current to the TIP power transistors. The NPN and PNP power transistors should be selected to provide the necessary current to the motor coils. The diodes protect the transistors from "back EMF" caused by current changes in the inductive coils. Select a diode (1N914 or 1N4001 through 1N4007) that can handle the magnitude of the back EMF. Typical values of voltage and current are shown in Table 8.16. When the A input is high and the A' input is low, the NPN transistor Q3 and the PNP transistor Q2 conduct and current travels right to left across the coil. If the A input were low and the A' input high, the NPN transistor Q1 and the PNP transistor Q4 would conduct and current would travel left to right.

Table 8.16
Specifications of typical
stepper motors.

Manufacturer	Phase voltage (V)	Phase current (A)	Step angle	Holding torque (oz·in)	Type	No. of leads	Used cost
Eastern Air	2.7	1.9	1.8°	53	Unipolar	8	$ 8.95
Astro-Syn	5	0.45	1.8°	10	Bipolar	4	$ 4.95
Shinano Kenshi	6	1.2	1.8°	80	Unipolar	5	$12.95
AirPax	12	0.33	7.5°	10.5	Unipolar	6	$ 3.95
Oriental Vexta	12	0.68	1.8°	125	Unipolar	6	$24.50
Sigma	24	0.66	1.8°	325	Unipolar	5	$34.50

We use a similar circuit to coil the current through the other phase (Figure 8.88).
The unipolar relay architecture provides for bidirectional currents by using a center tap on each phase. The center tap is connected to the +V power source, and the four ends of

Figure 8.88
The other bipolar stepper motor interface.

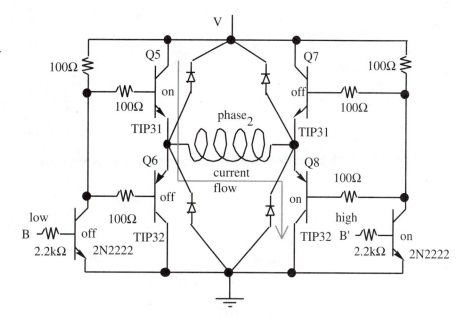

Figure 8.89
Unipolar stepper motor interface.

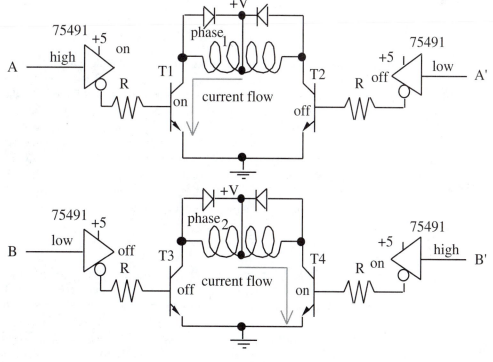

the phases are controlled with open-collector drivers, as shown in Figure 8.89. Only half the electromagnets are energized at one time. Again, the resistors and transistors are chosen so that the appropriate current is driven across the phases. The 75491 provides sufficient base current into the power transistors.

Figure 8.90
Stepper motors can also be used as shaft position sensors.

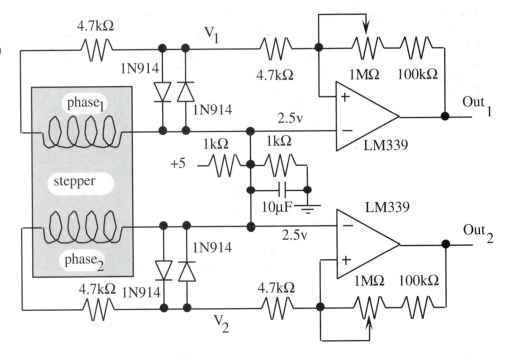

8.6.4 Stepper Motor Shaft Encoder

Stepper motors can also be used in a passive mode as shaft encoders, giving both speed and position. When the shaft of a stepper motor is rotated, a series of electric potentials is generated across its coils. The circuit shown in Figure 8.90 converts the AC voltage signals into two digital square waves. The frequency of the waves gives the rotational speed of the shaft, and the phase between the two waves gives the direction (clockwise or counterclockwise).

The 1N914 diodes produce signals that are ±1 V from the 2.5-V center. Four steps of the motor will produce one large cycle and one small cycle on V_1 and V_2. The distance traveled to produce these two cycles is the distance between two magnetic detents. The positive feedback hysteresis is adjusted so that the output signals (Out$_1$, Out$_2$) are sensitive to the large cycle but miss the small cycle. The shaft rotation speed, **r** in rps, is given by

$$r = \frac{4f}{N}$$

where **f** is the frequency of Out$_1$ in hertz and **N** is the number of steps in the motor. For example, a 200-step motor will yield 50 pulses per rotation. The microcomputer can measure the frequency of Out$_1$ to obtain shaft speed, or count pulses modulo 50 to establish position.

The direction of the rotation can be determined by the phase between Out$_1$ and Out$_2$. For a clockwise rotation, the value of Out$_2$ at the time of the fall of Out$_1$ is high (Figure 8.91).

For a counterclockwise rotation, the value of Out$_2$ at the time of the fall of Out$_1$ is low (Figure 8.92).

A unipolar stepper with six wires can be used exactly like a bipolar stepper by not connecting the common wires. A unipolar stepper with five wires is connected with the 2.5-V center at the common signal, then only half of each phase is used. This configuration will yield signals with half the amplitude.

Figure 8.91
Clockwise timing of a
stepper motor used as
shaft position sensor.

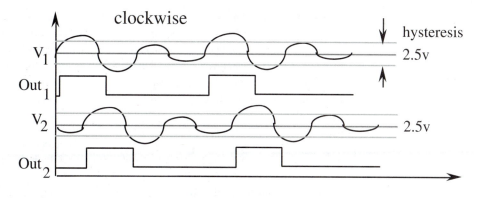

Figure 8.92
Counterclockwise timing
of a stepper motor used
as shaft position sensor.

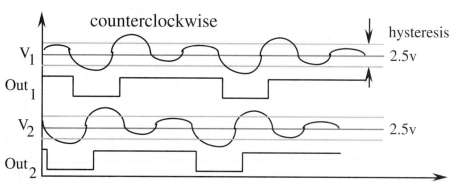

8.7 Glossary

Armature The moving structure in a relay, the part that moves when the relay is activated.

Bipolar stepper motor A stepper motor where the current flows in both directions (in/out) along the interface wires; a stepper with four interface wires.

Break before make In a double-throw relay or double-throw switch, there is one common contact and two separate contacts. Break before make means that as the common contact moves from one separate contact to another, it will break off (finish bouncing and no longer touch) the first contact before it makes (begins to bounce and starts to touch) the other contact. A *form C* relay has a *break before make* operation.

Common-anode LED display A display with multiple LEDs, configured with all the LED anodes connected together; there are separate connections to the cathodes (current flows in the common anode and out the individual cathodes).

Common-cathode LED display A display with multiple LEDs, configured with all the LED cathodes connected together; there are separate connections to the anodes (current flows in the individual anodes and out the common cathode).

Double-pole relay Two separate and complete relays, which are activated together (see single-pole).

Double-pole switch Two separate and complete switches, which are activated together (see single-pole).

Double-throw relay A relay with three contact connections, one common and two throws. The common will be connected to exactly one of the two throws (see single-throw).

Double-throw switch A switch with three contact connections, one common and two throws. The common will be connected to exactly one of the two throws (see single-throw).

Duty cycle When a LED display is scanned, it is the percentage of time each LED is active.

Frame The fixed structure in a relay, the part that doesn't move when the relay is activated.

I_{IH} Input current when the signal is high.

I_{IL} Input current when the signal is low.

I_{OH} Output current when the signal is high.

I_{OL} Output current when the signal is low.

Jerk The change in acceleration; the derivative of the acceleration.

LCD (Liquid-crystal display) The computer controls the reflectance or transmittance of the liquid crystal, characterized by its flexible display patterns, low power, and slow speed.

LED (Light-emitting diode) The computer controls the electrical power to the diode, characterized by its simple display patterns, medium power, and high speed.

Make before break In a double-throw relay or double-throw switch, there is one common contact and two separate contacts. Make before break means that as the common contact moves from one separate contact to another, it will make (finishing bouncing) the second contact before it breaks off (start bouncing) the first contact. A *form D* relay has a *make before break* operation.

Relay A switch that can be turned on and off by the computer.

Schmitt trigger A digital interface with hysteresis, making it less susceptible to noise.

Single-pole relay A simple relay with only one copy of the switch mechanism (see double-pole).

Single-pole switch A simple switch with only one copy of the switch mechanism (see double-pole).

Single-throw relay A relay with two contact connections. The two contacts may be connected or disconnected (see double-throw).

Single-throw switch A switch with two contact connections. The two contacts may be connected or disconnected (see double-throw).

Solenoid A discrete motion device (on/off) that can be controlled by the computer, usually by activating an electromagnet.

Stepper motor A motor that moves in discrete steps.

Three-pole relay Three separate and complete relays, which are activated together (see single-pole).

Three-pole switch Three separate and complete switches, which are activated together (see single-pole).

Two-pole relay Two separate and complete relays, which are activated together (same as double-pole).

Two-pole switch Two separate and complete switches, which are activated together (same as double-pole).

Unipolar stepper motor A stepper motor where the current flows in only one direction (on/off) along the interface wires; a stepper with five or six interface wires.

8.8 Exercises

8.1 Design a three-digit LED display interfaced only to the 6811/6812 PORTB (Figure 8.93). The software will utilize a 12-bit packed BCD global variable to contain the current value to be displayed. For example, if the value "456" is to be displayed, the main program will set the 16-bit global to $0456. Your interrupt software (part c) will read this global and output the appropriate signals to PORTB.

Figure 8.93
Three common-anode
LED digits.

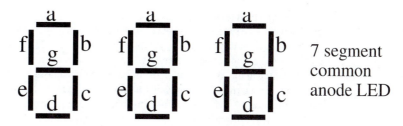

7 segment
common
anode LED

Available hardware devices include resistors, three common-anode seven-segment LED displays (Figure 8.94), any TTL chip (TTL, LSTTL, or HC) between 7400 and 7499, ULN2074s, 75491s, or 75492s.

a) Show the hardware interface between the display and the 6811 PORTB. Assume the desired LED operating point is 10 ma, 2 V.

b) Show the ritual that initializes the display software/hardware. *Full credit requires the appropriate usage of a circular statically allocated linked list data structure.*

c) Show the interrupt software that maintains the three-digit display. You do not write the `main()` program that sets the global variable. The main program is free to perform other unrelated tasks.

Figure 8.94
Connections for three
common-anode LED
digits.

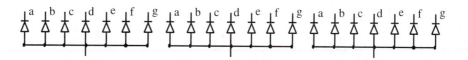

8.2 Design the interface between a computer output port, shown as B0 in Figure 8.95, and the EM relay. To activate the alarm the relay coil requires a voltage between +3.5 and 6.0 V and a current of 20 mA.

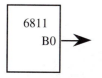

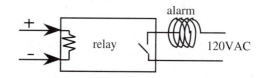

Figure 8.95 Relay interface.

8.3 Design a security code system. Assume there are four keys that have 10 ms of bounce. The keys are configured in a five-wire common line manner (Figure 8.96). Assume there is a global structure stored in EEPROM that contains the valid secret code

```
unsigned char secret[3];
```

Figure 8.96
Switch configuration.

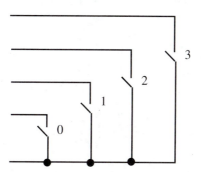

Don't worry about how this structure is initialized.

a) Show interface between the keypad and the microcomputer. Any interrupting mechanism can be used (input capture, key wake up, output compare, RTI, etc.).

b) Show the initialization software. Assume there is a main program (which you do not write that will call this ritual, then perform other unrelated tasks).

c) Show the interrupt handler(s) that performs the security functions in the background. Call the function `AlarmOff();` if the proper three-code sequence is typed. You will not write this `AlarmOff();` function.

8.4 The objective of this problem is to create a synchronized stepper motor network using single-chip 6811s (Figure 8.97). Each 6811 has an E clock of 2 MHz, 256 bytes of RAM, and 8 kbytes of ROM. Each 6811 is battery-operated and is separated by 100 m. Each system has a serial input, a serial output, and a 4-bit parallel output. Implement an 8-bit, no parity, one stop, 2400-baud serial communication. Upon receipt of a serial input, the 8-bit data are written to the stepper motor interface on Port B. After a 10-ms delay the 8-bit data are then transmitted in serial fashion to the next computer along the chain. At 2400 baud, the smallest interval between receive frames is about 4 ms (10 bits/2400 bits/s). This means it is possible for a second (and a third) input frame to arrive while one is waiting the 10 ms required before transmitting the first data. You will not design the master controller. Your software must be interrupt-driven. There is no BUF-FALO. You will specify all the software contained in each 6811 system. The +5-V stepper motor has a coil resistance of 500Ω.

Figure 8.97
Stepper motor
distributed network.

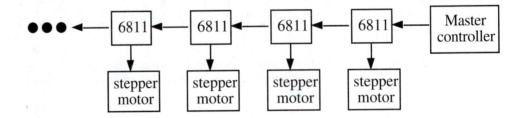

a) Show the hardware interface for one 6811 system. Please label chip numbers but not pin numbers. Choose the appropriate interface mechanism and be careful how the systems are grounded to each other. Clearly label all wires in the two 100-m cables (from the previous and to the next). The Stepper motor has +5-V power and the standard A, A', B, B' control.

b) Show all the information that will be located in the 256 byte RAM (0000 to $00FF). Use the assembly language directive `rmb`. Since the RAM contents are unspecified on power up, you cannot use directives `fcb`, `fdb`, or `fcc` in RAM. In the following example notice that the parallel output occurs right away, but the serial output is delayed by 10 ms. Also notice that the serial input may not (but could) occur at the full 2400 bits/s.

Time	Serial input	Parallel output	Serial output
0	6	6	None
4	9	9	None
9	11	11	None
10	None	None	6
13	2	2	None
14	None	None	9
19	None	None	11
23	None	None	2

c) Give *all* the software for each system. Use the `org` directive to place the programs in the ROM located from $E000 to $FFFF. The reset handler (the main program called on power up) will initialize all data structures, perform all the necessary 6811 I/O rituals, and then execute a *do-nothing* infinite loop. The interrupt processes will perform all the I/O operations. Give careful thought as to when and how to interrupt. Poll for 0s and 1s (jump to restart on error). *Good software has* no *backward jumps*. Use the `org` directive to set up the reset and interrupt vectors for your single-chip 6811. Use comments to explain your approach.

8.5 Design the hardware interface that allows the computer to control an EM solenoid (Figure 8.98). The computer controls the solenoid by driving current (activating the solenoid) or no current (deactivating it) through the coil, which has a resistance of 200 Ω. The solenoid activates when the coil voltage is above 4 V (i.e., a coil current above 20 mA). Limit the voltage across the coil to less than 6 V. There will be both positive and negative back EMF voltages, so protect the electronics.

Figure 8.98
Electromagnetic solenoid.

8.6 This question considers the concept of back EMF that can occur on a DC motor, stepper motor coil, relay coil, or solenoid coil when switched with a digital gate. A snubber diode is used to eliminate the back EMF. The choice of which diode to use depends on the magnitude of the back EMF voltage. The circuit configuration depends on the presence of either positive and/or negative amplitude back EMFs.

a) Aside from empirical (experimental) techniques, develop a procedure to select the proper diode in the circuit of Figure 8.99. We are looking for a different answer than the obvious method of measuring it in the lab with a scope (i.e., develop theoretical equations). Be as detailed as possible.

Figure 8.99
Relay interface with one diode.

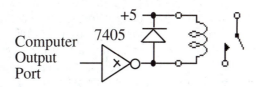

b) Why don't we usually add the second diode (Figure 8.100)? That is, what usually prevents negative-amplitude back EMF?

Figure 8.100
Relay interface with two diodes.

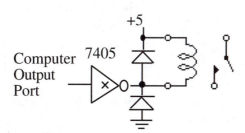

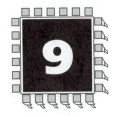

9 Memory Interfacing

Chapter 9 objectives are:

❏ Present the basic engineering design steps for interfacing memory to the computer
❏ Review the basic building blocks of computer architecture
❏ Develop descriptive language for defining time-dependent behavior of digital circuits
❏ Use timing diagrams to specify behavior of individual components as well as the entire system
❏ Compare and contrast synchronous, partially asynchronous, and fully asynchronous buses

For most applications, the embedded system uses a single-chip microcomputer with built-in memory devices. Therefore there is no need to attach additional memory. In other words, we select an available microcontroller with adequate RAM and ROM space to satisfy our constraints. Consequently most embedded system designers can skip this chapter. On the other hand, there are a few situations that might lead one to need to attach external devices to the address/data bus of the microcontroller. Memory interfacing and bus timing are general topics of computer architecture, and you may wish to study this chapter, not because you need to know how to attach memory for the embedded system but because you plan to apply this knowledge toward the design of general-purpose computers. In a similar fashion, understanding how to attach interface memory is an important step in the design of the single-chip microcontroller itself. The other situation that requires memory interfacing occurs when you need more memory than is available on the single-chip microcomputer. In this situation you can attach external memory to satisfy the constraint of the problem. For example, the MC68HC812A4 supports up to 4 Mbytes of external PROM and 1 Mbyte of external RAM.

9.1 Introduction

In this chapter, we will approach memory interfacing using both general concepts and specific examples. The processor is connected to the memory and I/O devices via the bus. The bus contains timing, command, address, and data signals. The timing signals, like the 6811/6812 E clock, determine when strategic events will occur during the transfer. The command signals, like the 6811/6812 R/W line, specify the bus cycle type. During a *read cycle* data flow from the memory or input device to the processor, where the address bus specifies the memory or input device location and the data bus contains the information. During a *write cycle* data are sent from the processor to the memory or output device, where the address bus specifies the memory or output device location and the data bus contains the information. During each particular cycle on most computers there is exactly one device sending the data and exactly one device receiving the data. Some buses like the IEEE488 and SCSI allow for broadcasting, where data is sent from one source to multiple destinations. In a computer system with memory-mapped I/O all slaves share the same bus (Figure 9.1).

In an Intel x86 computer system with isolated I/O, the slaves share the address and data buses but have separate control signals (Figure 9.2). From a programming perspective, processors with isolated I/O access their I/O devices and memory with separate instructions. On the Intel x86, there are four basic bus cycles: memory read, memory write, I/O read, and I/O write. The memory read/write cycles are similar to the memory-mapped cycles described above. During an *I/O read cycle* data flow from the input device to the

Figure 9.1
Architecture of a computer with memory-mapped I/O.

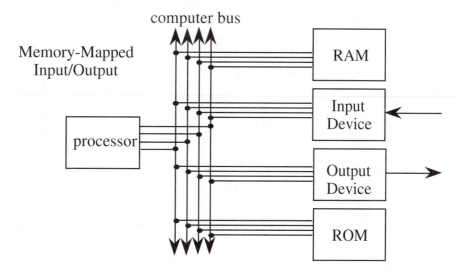

Figure 9.2
Architecture of a computer with isolated I/O.

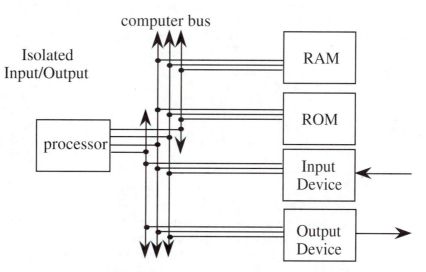

processor, where the address bus specifies the input device location and the data bus contains the information. During an *I/O write cycle* data are sent from the processor to the output device, where the address bus specifies the output device location and the data bus contains the information.

The interface to the slave devices, such as the RAM, ROM, and I/O devices, will consist of two parts. For each cycle, one packet of data is transmitted on the bus. We use the *command* part of the design to specify which slave will be active. The command portion of our slave design will specify whether or nor the current cycle is meant for our slave. If the slave can participate in both the read and write cycles, then the command part will also distinguish between the read and write functions. In a system without DMA, the processor will always be the master. The processor will determine the address and bus cycle type for each cycle. The 16 address lines and the R/W signal of the 6811/6812 specify the type of the current cycle (Figure 9.3). *There is no timing information in the address and R/W lines.* On computer systems that contain coprocessors or DMA controllers, there can be more than one bus master. For any given cycle there is only one bus master that will specify the slave address and bus cycle type. The processor, coprocessors, and DMA controllers vie for the use of the shared memory bus. More about DMA can be found in Chapter 10.

Figure 9.3
Both the 6811 and 6812 support external memory interfaces (expanded mode).

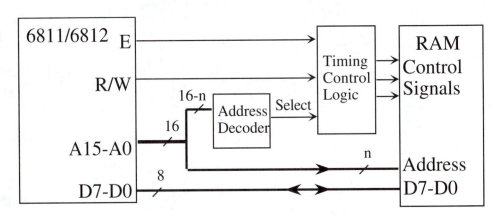

The first step in interfacing a device to the microcomputer is to construct an address decoder. Let SELECT be the positive logic output of the address decoder circuit. The command portion of our interface can be summarized by Table 9.1.

Table 9.1
The address decoder and R/W determine the type of bus activity for each cycle.

SELECT	R/W	Function	Rationale
0	0	Off	Because the address is incorrect
0	1	Off	Because the address is incorrect
1	0	Write	Data flow from 6811/6812 to our device
1	1	Read	Data flow from our device to the 6811/6812

The second part of the interface will be the *timing*. The 6811/6812 uses a *synchronous* bus, thus all timing signals will be synchronized to the E clock. It is important to differentiate timing signals from command signals within the interface. A timing signal is one that has rising and falling edges at guaranteed times during the memory access

Table 9.2
Available timing signals
and command signals
from the microcomputer.

Microcomputer	Timing signals	Command signals
MC68HC11A8	E AS	R/W A15-A8 AD7-AD0
MC68HC812A4	E CSP0 CSP1 CSD CS3 CS2 CS1 CS0	R/W A15-A0 D15-D0 LSTRB
MC68HC912B32	E DBE	R/W AD15-AD0 LSTRB

cycle. In contrast, command signals are generally valid during most of the cycle but do not have guaranteed times for their rising and falling edges. Table 9.2 divides the bus signals into the command or timing category.

The second objective of this section is to present a shorthand for drawing *timing diagrams*. The 6811 and 6812 bus timing will be presented. Various RAM and PROM memories will be interfaced.

Common Error: If control signals in our interface are derived only from command (type and address) signals, then we will have no guarantee when the rising and falling edges will occur.

Observation: We use the command signals to specify which cycles activate our slave, and we use the timing (clock) signals to specify when during the cycle to activate.

9.2 Address Decoding

The address decoder is usually located in each slave interface. The purpose of the address decoder is to monitor the 16 address lines from the bus master (the processor) and determine whether or not the slave has been selected to communicate in the current cycle. Each slave has its own address decoder, uniquely designed to select the addresses intended for that device. Care must be taken to avoid having two devices driving the data bus at the same time. We wish to select exactly one slave device during every cycle. The 6811 and MC68HC912B32 use a multiplexed address/data bus. In the 6811, the same eight wires, AD7–AD0, are used for low address (A7–A0) in the first half of each cycle and for the data (D7–D0) in the second half of each cycle. The 6811 address strobe output, AS, is usually used to capture the low address into an external latch (74HC573), as shown in Figure 9.4. In this way, the entire 16-bit address is available during the second half of the cycle.

Figure 9.4
The 6811 multiplexes
the low address on the
same pins as the data.

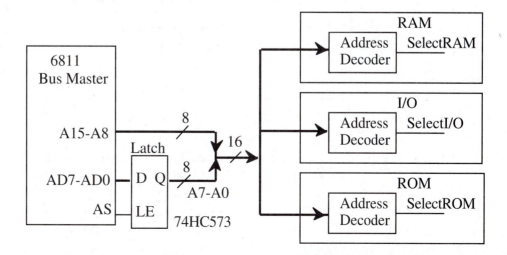

The MC68HC812A4 runs in nonmultiplexed mode, where the address pins are separate from the data pins. In this way, no external address latch is required (Figure 9.5).

Figure 9.5
The MC68HC812A4 supports a mode where the lines are dedicated to addresses (nonmultiplexed).

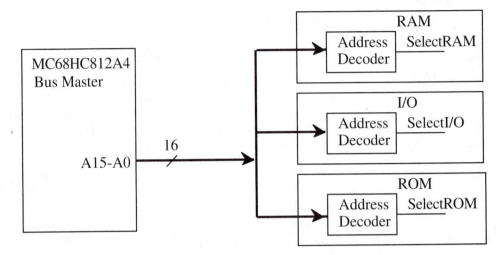

In the MC68HC912B32, the same 16 wires, AD15–AD0, are used for the address (A15–A0) in the first half of each cycle and for the data (D15–D0) in the second half of each cycle. The rising edge of the E clock is used to capture the address into an external latch (two 74HC374s), as shown in Figure 9.6. In this way, the entire 16-bit address is available during the second half of the cycle.

Figure 9.6
The MC68HC912B32 multiplexes the 16-bit address on the same pins as the 16-bit data.

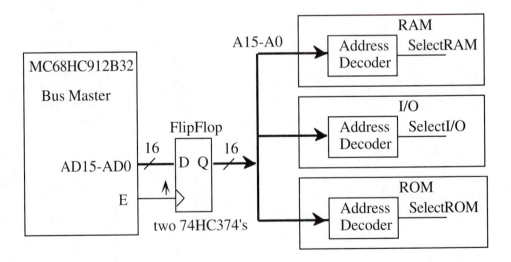

The address decoders should not select two devices simultaneously. In other words, no two select lines should be active at the same time. In this case,

$$SelectRAM \cdot SelectI/O = 0$$
$$SelectRAM \cdot SelectROM = 0$$
$$SelectI/O \cdot SelectROM = 0$$

Common Error: If two devices have address decoders with overlapping addresses, then a write cycle will store data at both devices, and during a read cycle the data from the two devices will collide, possibly causing damage to one or both devices.

9.2.1
Full-Address
Decoding

Full-address decoding is where the slave is selected if and only if the slave's address appears on the bus. In positive logic,

$$\text{Select} \quad \begin{matrix} = 1 \\ = 0 \end{matrix} \quad \begin{matrix} \text{if the slave address appears on the address bus} \\ \text{if the slave address does not appear on the address bus} \end{matrix}$$

Example. Design a fully decoded positive logic Select signal for a 1K RAM at $4000–$43FF.

Step 1. Write specified address using **0,1,X,** using the following rules:

a. There are 16 symbols, one for each of the address bits A15, A14 . . . , A0
b. **0** means the address bit must be 0 for this device
c. **1** means the address bit must be 1 for this device
d. **X** means the address bit can be 0 or 1 for this device
e. All the **X**s (if any) are located on the right-hand side of the expression
f. Let **n** be the number of **X**s. The size of the memory in bytes is 2^n
g. Let **I** be the unsigned binary integer formed from the $16-n$ **0s** and **1s**, then

$$\mathbf{I} = \frac{\text{beginning address}}{\text{memory size}}$$

In this example:

$$\text{address} = \mathbf{0100,00XX,XXXX,XXXX}$$
$$\text{beginning address} = \$4000$$
$$\mathbf{n} = 10 \quad \text{size} = 1024 \text{ bytes} = \$0400$$
$$\mathbf{I} = 010000_2 = 16_{10} = \frac{\$4000}{\$0400}$$

Step 2. Write the equation using all **0s** and **1s**. A **0** translates into the complement of the address bit, and a **1** translates directly into the address bit. For example,

$$\text{Select} \quad = \overline{\text{A15}} \cdot \text{A14} \cdot \overline{\text{A13}} \cdot \overline{\text{A12}} \cdot \overline{\text{A11}} \cdot \overline{\text{A10}}$$

Step 3. Build circuit using real TTL gates (Figure 9.7).

Figure 9.7
An address decoder identifies on which cycles to activate.

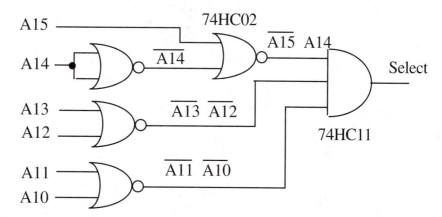

Observation: In each row of 0s, 1s, and Xs, the binary number formed by the 0s and 1s is equal to the beginning address divided by the memory size.

Observation: In each row of 0s, 1s, and Xs, 2 raised to the power of the number of Xs is equal to the memory size.

Observation: If we set all the Xs to 0 in the row of 0s, 1s, and Xs, we get the starting address.

Observation: If we set all the Xs to 1 in the row of 0s, 1s, and Xs, we get the ending address.

Common Error: If all the Xs are not in a group on the right-hand side, then the address range will become discontinuous.

Example. Design a fully decoded select signal for an I/O device at $5500 in negative logic.

Step 1. address = **0101,0101,0000,0000**
Step 2. The negative logic output can be created simply by inverting the output of a positive logic design. Recall that the address bus of the 6811/6812 is in positive logic.

$$\text{Select*} = \overline{\overline{A15} \cdot A14 \cdot \overline{A13} \cdot A12 \cdot \overline{A11} \cdot A10 \cdot \overline{A9} \cdot A8 \cdot \overline{A7} \cdot A6 \cdot A5 \cdot A4 \cdot A3 \cdot A2 \cdot A1 \cdot A0}$$

Step 3. Figure 9.8

Maintenance Tip: Use full-address decoding on systems where future expansion is likely.

Observation: We will use symbols ending in * to signify negative logic. We will also use a line over the symbol to signify negative logic.

Figure 9.8
An address decoder like this could be implemented with programmable logic (PAL).

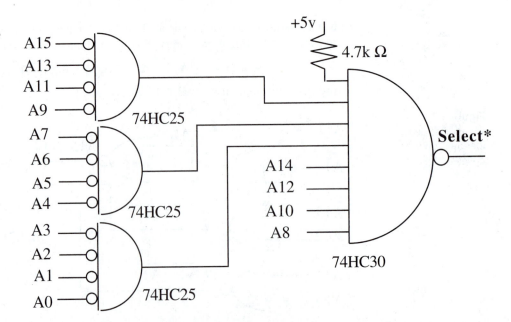

9.2.2
Minimal-Cost
Address Decoding

Minimal-cost address decoding is a mechanism to simplify the decoding logic by introducing don't care states for unspecified addresses. It does not guarantee lowest cost but often produces simpler solutions than fully decoded. Most embedded microcomputer systems do not use the entire 65,536 available addresses. An unspecified address is one at which there is no device. A reduction of logic may be realized if we introduce *don't care* outputs in the Karnaugh maps when the address equals one of the unspecified addresses. This shortcut should work because the software should never access an unspecified address. If a valid address is accessed, then one and only one device will be selected as expected. But, the address decoders may select zero, one, or more devices if an unspecified address were to be accessed. On most computers, the address bus is in positive logic, but we may construct the **Select** signals in either positive or negative logic, depending on which is more convenient. The formal definition of minimal cost is shown in Table 9.3.

Table 9.3
Minimal-cost decoding optimizes by taking advantage of the don't care states.

Address	Select
Matches our device	True
Matches other specified device	False
Unspecified address	Don't care

Example. Design four minimal-cost select signals in positive logic:

4K RAM	$0000 to $0FFF
Input	$5000
Output	$5001
16K ROM	$C000 to $FFFF

Step 1. Write out the addresses in binary for all devices. Include all specified addresses in the computer, not just the devices that we are designing.

RAM	0000,XXXX,XXXX,XXXX
Input	0101,0000,0000,0000
Output	0101,0000,0000,0001
ROM	11XX,XXXX,XXXX,XXXX

Step 2. Choose as many address lines that are required to differentiate between the devices. Consider the different devices in a pairwise fashion. If the decoder for only one device is being built, then only the addresses that differentiate that device from the others are required. In this example we choose A15, A14, A0. The address bits A15, A12, and A0 also could have been used to differentiate the devices.

Step 3. Draw a Karnaugh map for each device.

   a. Put a true for addresses specified by that device
   b. Put a false for other devices
   c. Put an "X" for unspecified addresses

Positive logic	true = 1 and false = 0
Negative logic	true = 0 and false = 1

Step 4. Minimize using Karnaugh maps and determine equations (Figure 9.9).

$$\text{RAMSelect} = \overline{\text{A14}}$$
$$\text{InputSelect} = \overline{\text{A15}} \cdot \text{A14} \cdot \overline{\text{A0}}$$

$$\text{ROMSelect} = \text{A15}$$
$$\text{OutputSelect} = \overline{\text{A15}} \cdot \text{A14} \cdot \text{A0}$$

Figure 9.9
Four Karnaugh maps.

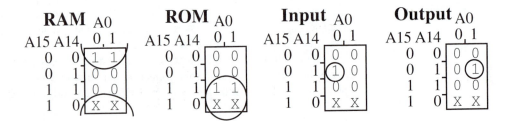

Step 5. Build circuit using real TTL gates (Figure 9.10).

Figure 9.10
Four address decoders like these could also be implemented with PAL.

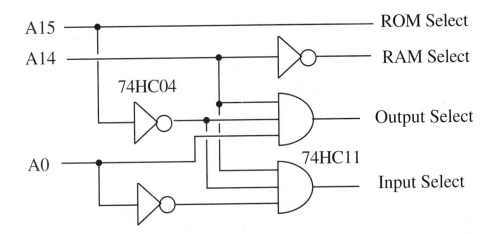

To implement **negative logic** either we can put true=0, false=1 into the Karnaugh map or we can build the decoders in positive logic and invert the outputs.

Performance Tip: Use minimal-cost decoding on systems where cost and speed are more important than future expansion.

Observation: If there are no unspecified addresses, then minimal-cost decoders will be the same as the full-address decoders.

Performance Tip: Address decoders for the entire computer system can sometimes be implemented with a single demultiplexer like the 74HC138.

Common Error: If one does not select enough address lines to differentiate between the devices, then some addresses may incorrectly select more than one device.

Observation: If one selects too many address lines, then the Karnaugh map will be harder to draw, but the resulting solution should be the same.

If all the slave select signals require the same address lines, then all the Karnaugh maps can be combined into a single map using slave names instead of 1s. We can use different colors for each slave. We then make the biggest circles that include each slave one by one. Added protection can be achieved by making none of the slave circles overlap. In this way, if the software inadvertently accesses the unspecified address, at most one slave will be activated. The "Cheaper" example in Figure 9.11 is the same solution as the one above. In this example, if the software read location is $8000, both the RAM and ROM would drive the data bus. If added protection is desired, we would eliminate the overlapping ROM and RAM circles. The "Safer" solution would activate only the ROM, if the unspecified address is accessed. If the ability to expand in the future is desired, we could leave holes.

Figure 9.11
A shorthand method for drawing multiple Karnaugh maps.

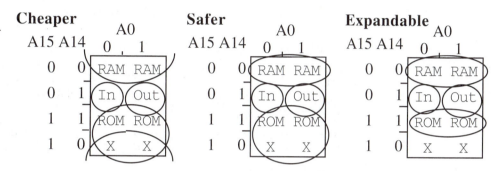

9.2.3
Special Cases
When Address
Decoding

If the size of the memory is not a power of 2 or if the memory size does not evenly divide the starting address, then the **0,1,X** address specification will require more than one line. To generate the multiline address specification, begin with the starting address and add **X**s to the right side as long as the address range does not exceed the ending address. Start over at the next address until the entire range is covered. The address decoder is the **OR** of the separate lines. Do not add **X**s to the left of a **0** or **1,** because doing so will make a discontinuous address range.

We define a *regular address range* as one with a size that is a power of 2, and one where the beginning address divided by the memory size is an integer. A regular address range can be expressed as a single line of 0s, 1s, and Xs. Two special cases with irregular address ranges are illustrated by the following examples. In the first case, the size of the memory is not a power of 2. In the second case, the beginning address divided by the memory size is not an integer. It requires more than one line of 0s, 1s, and Xs to specify the irregular address range. For irregular address ranges, we break the address range into multiple regular address ranges, solve each part separately, then combine the parts to form the decoder for the whole.

Example. Build a fully decoded positive logic address decoder for a 20K RAM with an address range from $0000 to $4FFF.
Start with

    0000,0000,0000,0000   Range $0000 to $0000

add Xs while the range is still within $0000 to $4FFF

    0000,0000,0000,000X   Range $0000 to $0001
    0000,0000,0000,00XX   Range $0000 to $0003
    0000,0000,0000,0XXX   Range $0000 to $0007
    0000,0000,0000,XXXX   Range $0000 to $000F

stop at

> **00XX,XXXX,XXXX,XXXX**   *Range $0000 to $3FFF*

Start over

> **0100,0000,0000,0000**   Range $4000 to $4000

add Xs while the range is still within $4000 to $4FFF

> **0100,0000,0000,000X**   Range $4000 to $4001
> **0100,0000,0000,00XX**   Range $4000 to $4003
> **0100,0000,0000,0XXX**   Range $4000 to $4007
> **0100,0000,0000,XXXX**   Range $4000 to $400F

stop at

> **0100,XXXX,XXXX,XXXX**   *Range $4000 to $4FFF*

$$\text{RAM SELECT} = \overline{A15} \cdot \overline{A14} + \overline{A15} \cdot A14 \cdot \overline{A13} \cdot \overline{A12}$$

Example. Build a fully decoded negative logic address decoder for a 32K RAM with an address range of $2000 to $9FFF.

Even though the memory size is a power of 2, the size 32,768 does not evenly divide the starting address 8192. We break the $2000 to $9FFF irregular address range into three regular address ranges.

> **001X,XXXX,XXXX,XXXX**   Range $2000 to $3FFF
> **01XX,XXXX,XXXX,XXXX**   Range $4000 to $7FFF
> **100X,XXXX,XXXX,XXXX**   Range $8000 to $9FFF

$$\text{SELECT*} = \overline{A15} \cdot \overline{A14} \cdot A13 + \overline{A15} \cdot A14 + A15 \cdot \overline{A14} \cdot \overline{A13}$$

Observation: If the beginning address divided by the memory size is not an integer, then the address range cannot be specified by a single row of 0s, 1s, and Xs.

Observation: If the memory size is not a power of 2, then the address range cannot be specified by a single row of 0s, 1s, and Xs.

9.2.4 Flexible Full-Address Decoder

We can design a flexible address decoder (Figure 9.12) with the following features:

It can decode any number of address bits by cascading multiple stages together; there is one stage for each address bit (e.g., the 6811/6812 would use 16 stages to decode all its address bits A15–A0)

It can be configured with DIP switches (there are two switches for each stage/address bit)

Figure 9.12
An approach for designing a programmable address decoder.

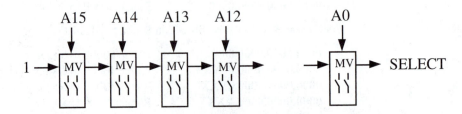

The M switch masks whether or not this address bit is to be considered—an open switch means consider, and a closed switch means ignore

If M is open, the V switch specifies whether this bit should be 1 or 0—an open switch means 1, and a closed switch means 0

Each stage has an address bit input, a cascade control input, and a cascade control output

The cascade control input, In, to the first stage is 1

The cascade control output, Out, of the last stage is the positive logic select

For example, assume the address range is \$4000 to \$5FFF:

	15	14	13	12	11	10	9	8	7	6	5	4	3	2	1	0
M switch	1	1	1	0	0	0	0	0	0	0	0	0	0	0	0	0
V switch	0	1	0													

For example, assume the address range is \$2010 to \$2011:

	15	14	13	12	11	10	9	8	7	6	5	4	3	2	1	0
M switch	1	1	1	1	1	1	1	1	1	1	1	1	1	1	1	0
V switch	0	0	1	0	0	0	0	0	0	0	0	1	0	0	0	

In both examples, 0 means closed, 1 means open, blank means either open or closed.

We will show the digital logic for one stage (Table 9.4 and Figure 9.13). We do not have to debounce the switches because they will not be toggled with the power on. There are four inputs and one output for each stage.

Table 9.4
Input/output function of one stage of a programmable address decoder.

In	Mn	An	Vn	Out	Meaning
0	X	X	X	0	There was an address mismatch in a previous stage
1	0	X	X	1	Ignore An, Vn and pass In to Out
1	1	0	0	1	OK because An matches Vn
1	1	0	1	0	Trouble because An does not match Vn
1	1	1	0	0	Trouble because An does not match Vn
1	1	1	1	1	OK because An matches Vn

Figure 9.13
Implementation of a programmable address decoder.

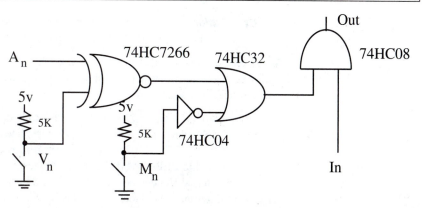

9.2.5
Integrated
Address Decoder
on the
MC68HC812A4

Some microcomputers like the MC68HC812A4 and the MC68HC340 have chip selects integrated into the microcomputer chip itself. In particular, the MC68HC812A4 has seven built-in address decoders that can be used as chip selects for external memory and I/O devices. The chip selects on the MC68HC340 behave similar to the example in the previous section, whereas the MC68HC812A4 negative chip selects have a simple dedicated use. The 5-bit **mmmmm** and **rrrrr** are the values in the INITRM and INITRG configurations, respectively. Notice that the CSD and CSP1 chip selects can be configured at different addresses. Each of these chip selects can be individually enabled or disabled by writing the CSCTL0 register.

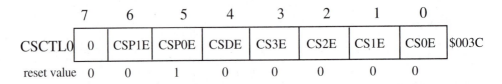

	7	6	5	4	3	2	1	0	
CSCTL0	0	CSP1E	CSP0E	CSDE	CS3E	CS2E	CS1E	CS0E	$003C
reset value	0	0	1	0	0	0	0	0	

In Table 9.5, the start and end addresses are calculated assuming the default values INITRG=0 and INITRM=$08.

Table 9.5
Built-in address decoder of the MC68HC812A4.

Port	Name	Range	Start	End	Size
PF3	CS3	mmmmm1xxxxxxxxxx	$0C00	$0FFF	1K (CS3EP=1)
PF3	CS3	rrrrr0101xxxxxxx	$0280	$02FF	128 (CS3EP=0)
PF2	CS2	rrrrr0111xxxxxxx	$0380	$03FF	128
PF1	CS1	rrrrr011xxxxxxxx	$0300	$03FF	256
PF0	CS0	rrrrr01xxxxxxxxx	$0200	$03FF	512
PF4	CSD	0111xxxxxxxxxxxx	$7000	$7FFF	4K (CSDFH=0)
PF4	CSD	0xxxxxxxxxxxxxxx	$0000	$7FFF	32K (CSDFH=0)
PF5	CSP0	1xxxxxxxxxxxxxxx	$8000	$FFFF	32K
PF6	CSP1	1xxxxxxxxxxxxxxx	$8000	$FFFF	32K (CSP1FL=0)
PF6	CSP1	xxxxxxxxxxxxxxxx	$0000	$FFFF	64K (CSP1FL=1)

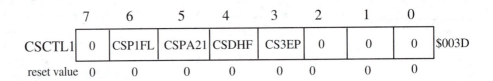

	7	6	5	4	3	2	1	0	
CSCTL1	0	CSP1FL	CSPA21	CSDHF	CS3EP	0	0	0	$003D
reset value	0	0	0	0	0	0	0	0	

Observation: Even though some of the built-in address decoders have overlapping ranges, at most one chip select will activate for each cycle.

When an overlapping address is accessed, the 6812 uses a priority system to determine which chip select is made active. Table 9.6 illustrates the procedure used by the 6812 to resolve overlapping addresses.

Highest   On-chip registers, 512 bytes at **rrrrr00xxxxxxxxx,** see the INITRG register
BDM space, 256 bytes at $FFxx when active
On-chip RAM, 1024 bytes at **mmmmm0xxxxxxxxxx,** see the INITRM register
On-chip EEPROM if enabled, 4 kbytes, see the INITEE register
External I/O devices, 128 bytes if CS3 is active
External I/O devices, 128 bytes if CS2 is active

Table 9.6
Size and range of CS0 as a function of whether or not CS1, CS2, CS3 are enabled.

CS1	CS2	CS3	CS0 size	CS0 range
Off	Off	Off	512	$0200 to $03FF
Off	Off	Enabled	384	$0200 to $027F and $0300 to $03FF
Off	Enabled	Off	384	$0200 to $037F
Off	Enabled	Enabled	256	$0200 to $027F and $0300 to $037F
Enabled	Off	Off	256	$0200 to $02FF
Enabled	Off	Enabled	128	$0200 to $027F
Enabled	Enabled	Off	256	$0200 to $027F
Enabled	Enabled	Enabled	128	$0200 to $027F

External I/O devices, 256 bytes if CS1 is active
External I/O devices, 512 bytes if CS0 is active
External Program, if CSP0 is active
External Data, 16 kbytes if CSD is active
External Program, if CSP1 is active
Lowest   Remaining external

Observation: The size of the CS0 chip select is affected by whether or not CS1, CS2, or CS3 is active.

This feature greatly simplifies the design of an expanded-mode computer system, because in many applications external memory and I/O devices can be connected directly to the MC68HC812A4 without additional interface logic (Figure 9.14).

Figure 9.14 The MC68HC812A4 has built-in address decoders.

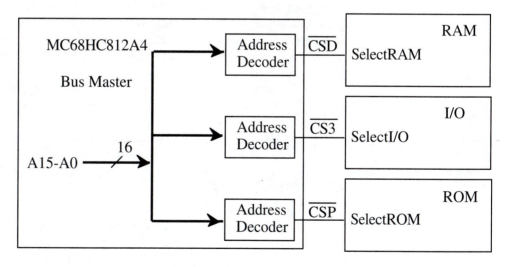

9.3 Timing Syntax

9.3.1 Available and Required Time Intervals

To specify an interval of time, we give its starting and stop times. These times may be relative numbers in nanoseconds referred to an edge of the system clock, signifying the start of the bus cycle. In the 6811 and 6812, we can refer times to the fall of the E clock because it marks the beginning of the current memory access cycle. For example,

(400, 520)

means the time interval begins at 400 ns after the fall of the E clock and ends at 520 ns after that same fall, 120 ns later. Sometimes we are not quite sure exactly when it starts or stops, but we can give upper and lower bounds. For example, assuming we know the interval starts somewhere between 400 and 430 ns after the E and stops somewhere between 520 and 530 ns, we would then write

$$([400, 430], [520, 530])$$

Sometimes we specify time intervals in terms of the rise and fall of control signals. For example, consider the simple circuit in Figure 9.15. We specify the time that the control signal **A** rises and falls as ↑**A** and ↓**A**, respectively. If we assume the propagation delay through the 74LS04 is exactly 10 ns, then the time interval when **Y** is high can be written

$$(\uparrow\mathbf{Y}, \downarrow\mathbf{Y}) = (\downarrow\mathbf{A}+10, \uparrow\mathbf{A}+10) = (\downarrow\mathbf{A}, \uparrow\mathbf{A}) + 10$$

Figure 9.15
A NOT gate.

If we assume the propagation delay through the 74LS04 can range from 0 to 20 ns, then the time interval when **Y** is high can be written

$$(\uparrow\mathbf{Y},\downarrow\mathbf{Y}) = ([\downarrow\mathbf{A},\downarrow\mathbf{A}+20], [\uparrow\mathbf{A},\uparrow\mathbf{A}+20]) = (\downarrow\mathbf{A}+[0,20], \uparrow\mathbf{A}+[0,20]) = (\downarrow\mathbf{A},\uparrow\mathbf{A}) + [0,20]$$

When data are transferred from one location (the source) and stored into another (the destination), there are two time intervals that will determine if the transfer is successful. The *data available* interval specifies when the data driven by the source are valid. The *data required* interval specifies when the data to be stored into the destination must be valid. For a successful transfer the data available interval must overlap (start before and end after) the data required interval (Figure 9.16). Let a, b, c, d be times relative to the same time reference, let the data available interval be (a, d), and let the data required interval be (b, c). The data will be successfully transferred if a is less than b and c is less than d.

Figure 9.16
The timing of data available should overlap data required.

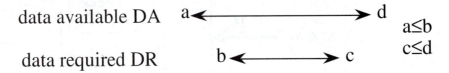

The following example illustrates the fundamental concepts. The objective is to transfer the data from the input **In** to the output **Out.** First, we assume the signal at the **A** input of the 74LS125 is always valid. When the tristate control **G*** is low, then the **A** is copied to the output **Y.** On the rising edge of **Clk,** the 74LS74 D flip-flop will copy its input **D** to the output **Q** (Figure 9.17).

Figure 9.17
Simple circuit to illustrate that the data available interval should overlap the data required interval.

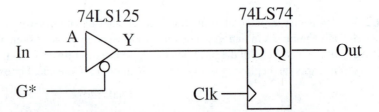

The data available interval is determined by the timing of the source (e.g., the 74LS125). The output of the 74LS125, **Y,** is valid between 10 and 20 ns after the fall of **G*.** It will remain valid until 0 to 15 ns after the rise of **G*.** The data available interval is

$$DA = (\downarrow G* + [10,20], \uparrow G* + [0,15])$$

The data required interval is determined by the timing of the destination (e.g., the 74LS74). The 74LS74 input **D** must be valid from 30 ns before the rise of **Clk** and remain valid until 0 after that same rise of **Clk.** The data required interval is

$$DR = (\uparrow \mathbf{Clk} - 30, \uparrow \mathbf{Clk} + 0)$$

Since the objective is to make the data available interval overlap the data required window, the worst case situation will be the shortest data available and the longest data required intervals. Without loss of information, we can write:

$$DA = (\downarrow G* + 20, \uparrow G*)$$

Thus the data will be properly transferred if $\downarrow G* + 20$ is less than or equal to $\uparrow \mathbf{Clk} - 30$ and if $\uparrow \mathbf{Clk}$ is less than or equal to $\uparrow G*$.

9.3.2 Timing Diagrams

We will use the timing symbols shown in Figure 9.18 to describe the causal relations in our interface. It is important to have it clear in our minds whether we are drawing an input or an output signal. To illustrate the graphic relationship of dynamic digital signals,

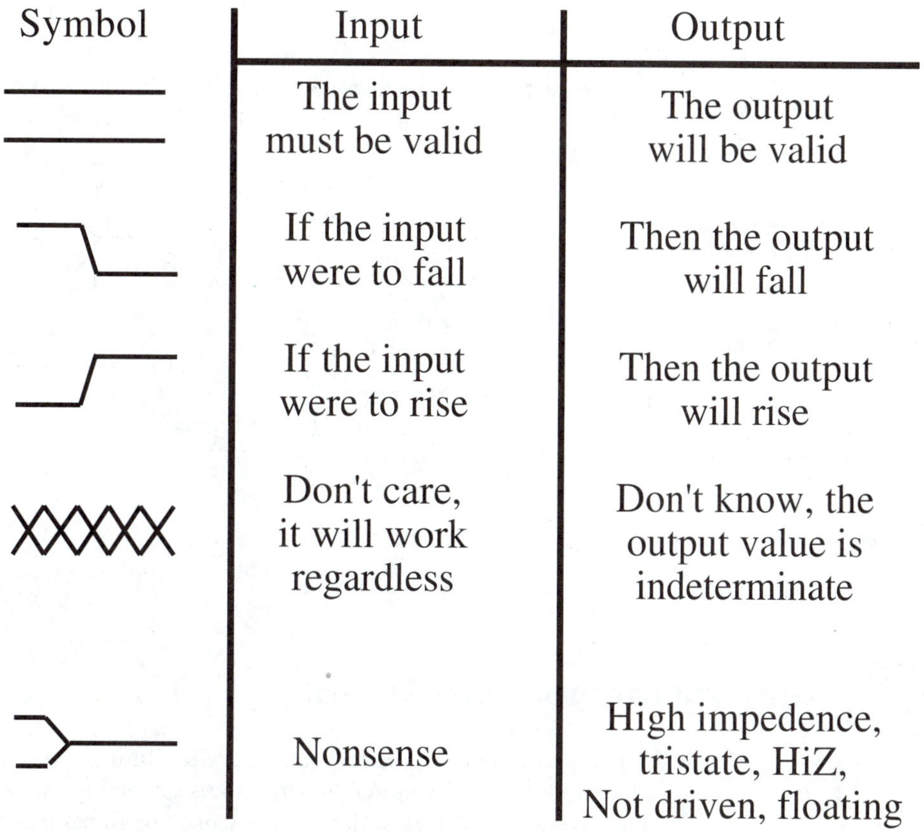

Figure 9.18
Nomenclature for drawing timing diagrams.

Symbol	Input	Output
———	The input must be valid	The output will be valid
⎍ (falling)	If the input were to fall	Then the output will fall
⎍ (rising)	If the input were to rise	Then the output will rise
XXXXX	Don't care, it will work regardless	Don't know, the output value is indeterminate
⟩—	Nonsense	High impedance, tristate, HiZ, Not driven, floating

we will present some simple examples of timing diagrams. The typical delay time for low-power Schottky logic is 10 ns. The subscript HL refers to the output changing from High to Low (Figure 9.19).

Figure 9.19
Timing of a NOT gate.

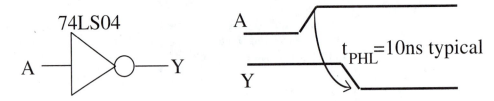

The shape of the tristate buffer timing is similar to the shape of the read cycle timing of the ROM and the RAM. Clearly, the 74LS244 is much faster than MOS or CMOS memories (Figure 9.20).

Figure 9.20
Timing of a tristate buffer gate.

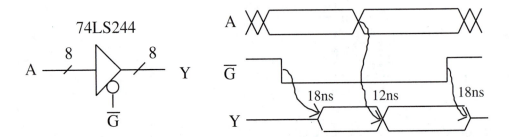

The shape of the timing for clocking the D flip-flop is similar to the shape of the write cycle timing of the RAM. Again, the 74LS374 is much faster than MOS or CMOS memories (Figure 9.21).

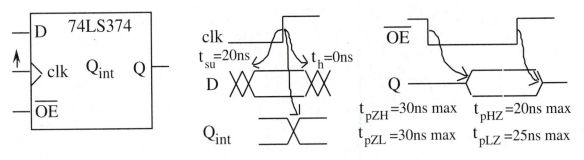

Figure 9.21 Timing of a D flip-flop with tristate outputs.

9.4 General Memory Bus Timing

There are two types of memory access cycles. During a *read cycle* data are passed from a slave device (memory or I/O) into the processor. During a *write cycle* data are passed from the processor to a slave device (memory or I/O). To participate in a read cycle, the slave

Figure 9.22
Simplified diagram showing circuits used during a read cycle.

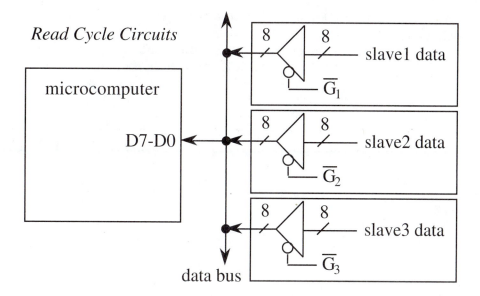

Figure 9.23
Simplified diagram showing circuits used during a write cycle.

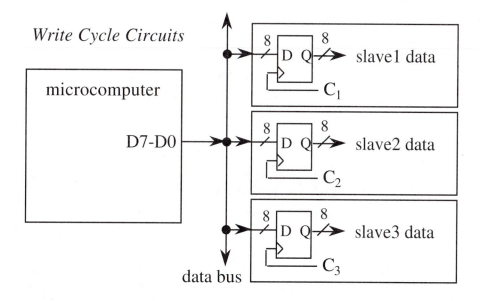

must have tristate logic so that it can drive data onto the data bus during read cycles involving that slave. In the first simplified diagram (Figure 9.22), we show the circuits to drive data on the bus during a read cycle for three slaves. In the second simplified diagram (Figure 9.23), we show the circuits to accept data from the bus during a write cycle for three slaves. Recall that the **command** portion of our slave design specified yes or no as to whether the current cycle is meant for our slave. In this general memory bus timing section, we will discuss three design approaches for generating the bus control signals (like $\overline{G1}$, $\overline{G2}$, $\overline{G3}$, C1, C2, C3). For the TIMING portion of our slave design we will specify when during a read cycle the data are driven onto the bus (e.g., $\overline{G1}$, $\overline{G2}$, $\overline{G3}$) and when during a write cycle data are clocked off the bus and into the slave (e.g., C1, C2, C3).

**9.4.1
Synchronous Bus
Timing**

The MC68HC11A8, MC68HC812A4, and MC68HC912B32 implement synchronous bus transfers. In a *synchronous bus* scheme the bus master (usually the processor) generates a clock (e.g., E clock), and all data transfer timings occur relative, or synchronous, to this clock. In our simplified circuit example above, a synchronous system has the control signals synchronized to the bus clock. In particular the read and write cycle timing will be as shown in Figure 9.24.

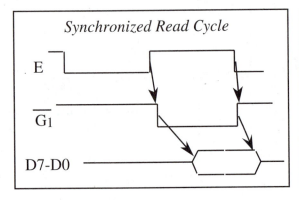 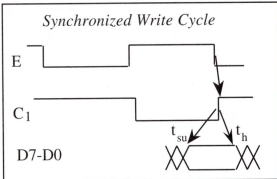

Figure 9.24   Synchronized bus timing.

Observation: In a synchronized bus interface the rising and falling edges of the control signals are synchronized (occur immediately) after an edge of the bus clock.

Observation: In a synchronized bus interface there is no feedback from the slave about whether or not the transfer was performed.

Observation: In a synchronized bus interface each bus cycle is exactly the same length, regardless of how fast the slave is.

Common error: If the slave cannot respond fast enough or if no slave exists at the address, data are not properly read or written, but no bus error signal is generated.

Common error: If one slave is replaced with an equivalent but faster device in a synchronized bus interface, the computer will not execute any faster.

**9.4.2
Partially
Asynchronous Bus
Timing**

As mentioned above, one of the limitations of a synchronous bus is that there is no feedback from the slave to the master. The two limiting consequences of having no feedback are inefficiency (requires all devices to operate at the speed of the slowest pair) and a lack of flexibility (all devices must be considered together when designed or redesigned.) We define a *partially asynchronous bus* as one that allows slaves to affect bus timing. For example, some computers (like the Motorola 6809 and 680x0 and the Intel x86) allow an external device to slow down the system clock, thus extending the data access time. This is called *cycle stretching*. In a 6809, the E and Q clocks operate normally when MRDY is high. The 6809 E, Q clocks will stop (or stretch) for an integer multiple of quarter bus cycles whenever MRDY is low. The 6809 MRDY must fall within 200 ns of when the end of the cycle would have been for the cycle to be stretched. In Figure 9.25, the cycle is stretched by half a bus cycle. If a slave can respond in the normal time for the cycle, it will without changing MRDY. In this way, a simple interface (one without the ability to drive MRDY low) can be designed using the regular rules of a synchronous interface.

Figure 9.25
Partially asynchronous
bus timing of a 6809.

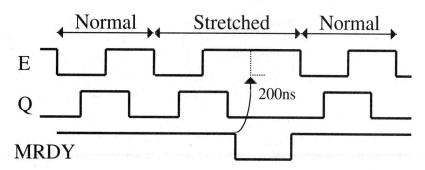

Typically a slow memory or I/O device pulls MRDY low until its read or write function is complete. It is much more efficient for the system to slow down only when accessing the slow devices, rather than changing the crystal frequency (which would slow down the system for every cycle). We will use this mechanism to interface dynamic RAMs and slow I/O chips. The 6811 and MC68HC912B32 does not allow cycle stretching.

The MC68HC812A4 implements cycle stretching on its built-in chip selects. This mechanism is different from other partially asynchronous interfaces in that there is no feedback from the slave. In a MC68HC812A4 system, the software must have prior knowledge of how fast each slave can respond. The amount of cycle stretching is predetermined by the initialization software using the CSSTR0 and CSSTR1 control registers:

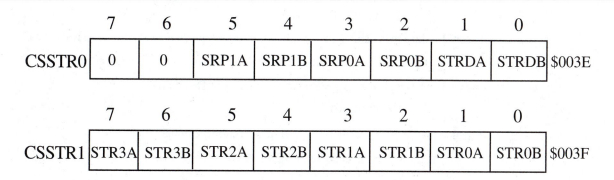

where the amount of stretch is fixed according to Table 9.7.

Table 9.7
MC68HC812A4 cycle
stretching.

STRxxA	STRxxB	Number of E clocks stretched
0	0	0
0	1	1
1	0	2
1	1	3 (default)

Observation: In a partially asynchronous bus, the bus cycle length can vary.

Common error: If the software uses a cycle-counting scheme to implement timing delays, then errors will occur if the programmer does not consider that memory accessed to slow devices will be stretched, thereby causing those cycles to be longer.

Observation: If a slow slave is replaced with an equivalent but faster device in a partially asynchronous bus interface, cycle stretching may be reduced and the computer will execute faster.

Common error: If the slave cannot respond fast enough or if no slave exists at the address, data are not properly read or written, but no bus error signal is generated.

9.4.3 Fully Asynchronous Bus Timing

The above two approaches cannot generate a hardware bus error signal when the slave does not or cannot respond. Another limitation of the previous methods is the difficulty in upgrading. To make a synchronous system run faster, each module (master and all slaves) must be redesigned. To upgrade a partially asynchronous system, the new cycle stretching amount must be determined and programmed into the software. In a *fully asynchronous bus* interface, there are control and acknowledge handshake signals that are generated for each bus cycle. Each phase in the following example is signified by the rise or fall of a control signal. The particular example is similar to the protocol used in the SCSI. The key to a fully asynchronous transfer (also called *handshaked* or *interlocked*) is that each phase can vary in length and will wait for the previous phase to occur.

Read Cycle Data Transferred from Slave to Master (Figure 9.26).

Phase 1. Master specifies the address and says, "Please give me data"

   I/O=0 and fall of REQ (SCSI)

Phase 2. Slave puts data on the bus and sends an acknowledge saying, "Here is the data"

   data are driven and fall of ACK (SCSI)

Phase 3. Master accepts the acknowledge and says "Thank you"

   rise of REQ (SCSI)

Phase 4. Slave makes its data HiZ and says "You're welcome"

   data are driven and rise of ACK (SCSI)

Figure 9.26
Fully asynchronous (interlocked or handshaked) read cycle bus timing.

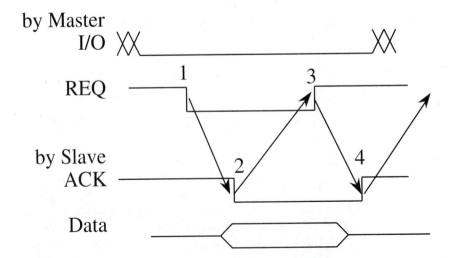

Figure 9.27
Fully asynchronous (interlocked or handshaked) write cycle bus timing.

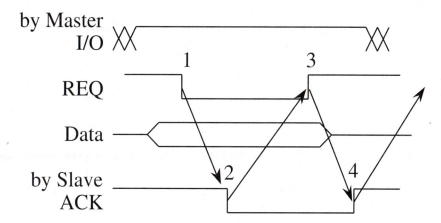

Write Cycle Data Transferred from Master to Slave (Figure 9.27).

Phase 1. Master puts data on the bus and says, "Please save these data"

         I/O=1 data driven by master and fall of REQ (SCSI)

Phase 2. Slave accepts the data and sends an acknowledge saying, "I saved the data"

         fall of ACK (SCSI)

Phase 3. Master makes its data HiZ and says "Thank you"

         rise of REQ (SCSI)

Phase 4. Slave says "You're welcome"

         rise of ACK (SCSI)

To pass each data correctly and to prevent the same data from being passed twice, there are four transitions in a fully interlocked protocol.

> ***Observation:*** If you place a time-out mechanism on the time the master waits for the slave response, then a hardware bus error can be generated on a broken or missing module.

> ***Observation:*** If you replace any module in a fully asynchronous bus system, then the data transfer automatically occurs at the fastest possible rate without changing the software configuration.

Most computer systems do not implement fully asynchronous bus transfers with their memory because the complexity makes the system too expensive and too slow. On the other hand, fully asynchronous transfers are used in many I/O bus protocols such as IEEE-488 and SCSI.

9.5   External Bus Timing

9.5.1
Synchronized
Versus Unsynch-
ronized Signals

When designing the control signals for our memory, we have two choices to make. The first choice is relatively easy: Do we use positive or negative logic? A positive logic control signal goes high when the memory is accessed, and a negative logic signal goes low during the cycle of interest. The other choice is whether or not to synchronize to the E clock. These

E	Command	Unsynchronized positive	Unsynchronized negative	Synchronized positive	Synchronized negative
0	0	0	1	0	1
1	0	0	1	0	1
0	1	1	0	0	1
1	1	1	0	1	0
		CS=command	CS=not(command)	CS=E·command	CS=not(E·command)

Table 9.8　A timing signal (like the E clock) can be combined with a positive logic command signal four ways.

choices are presented in Table 9.8, where **command** refers to a signal derived directly from A15–A0 or R/W. In this discussion, **command**=1 means activate this cycle. We will also use this table to design the digital logic to create the control signal **CS**.

The four possible shapes for the control signals are shown in Figure 9.28, where the "x" parameter refers to the gate delays through the digital logic required to implement the control signal. If we have a negative logic $\overline{command}$ signal, then the polarity of the command is reversed (Table 9.9).

Figure 9.28
The timing of four types of control signals.

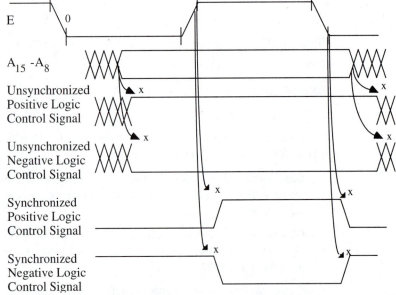

E	$\overline{command}$	Unsynchronized positive	Unsynchronized negative	Synchronized positive	Synchronized negative
0	1	0	1	0	1
1	1	0	1	0	1
0	0	1	0	0	1
1	0	1	0	1	0
		CS= not($\overline{command}$)	CS= $\overline{command}$	CS= E·not($\overline{command}$)	CS= not(E·not($\overline{command}$))

Table 9.9　A timing signal (like the E clock) can be combined with a negative logic command signal four ways.

**9.5.2
Motorola
MC68HC11A8
External Bus
Timing**

9.5.2.1
MC68HC11A8
Execution Modes

The 6811 can run in one of four modes. In *normal single chip* mode, Port B, Port C STRA, and STRB are used for I/O. In *expanded* mode, Port B contains the high address, Port C contains the low address/data, STRA is the address strobe AS, and STRB is the R/W line. *Special bootstrap* mode is a single-chip mode used to load data into the 256-byte RAM, and *special test* mode is used by Motorola for testing. The initial mode is determined by the values of the MODB, MODA pins at the time of the rise of RESET. Table 9.10 also shows the initial values for 4 bits of the HPRIO register.

Table 9.10
There are four execution
modes for the
MC68HC11A8.

MODB	MODA	Mode description	RBOOT	SMOD	MDA	IRV
1	0	Normal single chip	0	0	0	0
1	1	Normal expanded	0	0	1	0
0	0	Special bootstrap	1	1	0	1
0	1	Special test	0	1	1	1

Observation: To interface external memory devices to the 6811, we will utilize expanded mode.

If the computer is running in one of the special modes, the software can affect the mode by changing the SMOD and MDA bits in the HPRIO register.

	7	6	5	4	3	2	1	0	
HPRIO	RBOOT	SMOD	MDA	IRV	PSEL3	PSEL2	PSEL1	PSEL0	$103C

RBOOT: Read bootstrap ROM
   Can be written only while SMOD equals 1
   1 = Bootstrap ROM enabled at $BF40–$BFFF
   0 = Bootstrap ROM disabled and not present in memory map
The RBOOT control bit enables or disables the special bootstrap control ROM. This
   192-byte, mask-programmed ROM contains the firmware required to load a user's
   program through the SCI into the internal RAM and jump to the loaded program.
   In all modes other than the special bootstrap mode, this ROM is disabled and does
   not occupy any space in the 64-kbyte memory map. Although it is 0 when the
   MCU comes out of reset in test mode, the RBOOT bit may be written to 1 while
   in special test mode.
SMOD: Special mode
   May be written to 0 but not back to 1
   1 = Special mode variation in effect
   0 = Normal mode variation in effect
MDA: Mode A select
   Can be written only while SMOD equals 1
   1 = Normal expanded or special test mode in effect
   0 = Normal single-chip or special bootstrap mode in effect
IRV: Internal read visibility
   Can be written only while SMOD equals 1; forced to 0 if SMOD equals 0
   1 = Data driven onto external bus during internal reads
   0 = Data from internal reads not visible on expansion bus (levels on bus ignored)

9.5.2.2
MC68HC11A8 Timing

The E clock, AS strobe are timing signal outputs of the 6811. The clock frequency is determined by a crystal placed between the EXTAL and XTAL inputs (Figure 9.29). The 6811 system needs an external crystal that is four times the E clock period (t_1 in the timing diagram of Figure 9.31). The 2-MHz period is created using an 8-MHz crystal. The 25 pF includes all stray capacitances. To reduce noise, Motorola suggests placing a ground plane under the crystal (Figure 9.30).

Figure 9.29
6811 clock circuit.

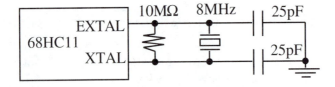

Figure 9.30
Layout for the 6811 clock crystal.

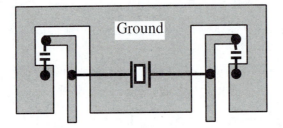

In Figure 9.31 the variables like t_1, t_2, t_3, refer to the timing parameters in Table 9.11. There are two equivalent ways to define the E clock period:

$$t_1 = t_2 + t_4 + t_3 + t_4$$

In some situations, the equivalent expressions might yield different results. This can lead to ambiguity when calculating data available and data required intervals. Another ambiguity results from the fact that a range of parameters may be given. In these situations it is best to design the interface considering the worst case.

Figure 9.31
Simplified bus timing for the 6811 in expanded mode.

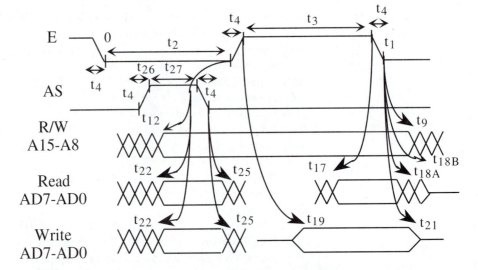

Table 9.11
Timing parameters for
the MC68HC11A8.

No.		Characteristic frequency at		
		1.0 MHz	**2.0 MHz**	**2.1 MHz**
t_1	Cycle time (ns)	1000	500	476
t_2	Pulse width E low (ns)	480	230	218
t_3	Pulse width E high (ns)	480	230	218
t_4	Rise/fall time (ns)	20	20	20
t_9	Address hold time (ns)	95.5 min	33 min	30 min
t_{12}	A15–A8, R/W valid time (ns)	281.5 min	94 min	85 min
t_{17}	Read data setup time (ns)	30 min	30 min	30 min
t_{18A}	Read data hold time (ns)	10 min	10 min	10 min
t_{18B}	Read data goes HiZ (ns)	145.5 max	83 max	80 max
t_{19}	Write data delay time (ns)	190.5 max	128 max	125 max
t_{21}	Write data hold time (ns)	95.5 min	33 min	30 min
t_{22}	A7–A0 valid time (ns)	271.5 min	84 min	75 min
t_{25}	A7–A0 hold time (ns)	95.5 min	33 min	30 min
t_{26}	E to AS rise time (ns)	115.5	53	50
t_{27}	AS pulse width (ns)	221	96	90

Performance Tip: It is proper design practice to consider the worse case—that is, the shortest data available interval and the longest data required interval.

For a 2-MHz 6811 the four potential control signals are plotted in Figure 9.32, where the "x" parameter refers to the gate delays through the digital logic required to implement the control signal.

Figure 9.32
Timing of four types of
control signals with the
6811 in expanded
mode.

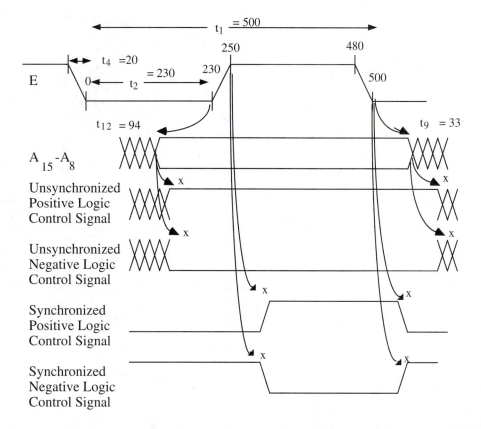

There are three important timing intervals we must determine from the microcomputer timing diagram. The first is the *address available,* AA, interval. This interval defines the time during the cycle when the address is valid.

$$AA = (AdV, AdN) = (\text{later } (t_2 - t_{12}, t_2 - t_{22} + 35 \text{ ns}), \text{earlier } (t_1 + t_{26}, t_1 + t_9)) = (181, 533)$$

where AdV is the time when the address lines A15–A0 are valid and AdN is the time when the address lines are no longer valid. The two terms of the later function represent A15–A8 and A7–A0, respectively. The 35 ns comes from the gate delay of the 74HC573 transparent latch. It will also be useful to know the address available for just A15–A8:

$$AA_{15-8} = (AdV_{15-8}, AdN_{15-8}) = (230 - 94, 500 + 33) = (136, 533)$$

The second important timing interval is *read data required,* RDR. During a read cycle the data are required by the 6811. Thus to determine the data required interval, we look in the 6811 data sheet. For data required, the worst case is the longest interval:

$$RDR = \text{read data required} = (t_1 - t_4 - t_{17}, t_1 + t_{18}) = (500 - 20 - 30, 500 + 10) = (450, 510)$$

The last important timing interval we get from the microcomputer is *write data available,* WDA. During a write cycle, the data are supplied by the 6811. Thus, to determine the data available interval, we look in the 6811 data sheet. We will specify the worst-case timing. For write data available, the worst case is the shortest interval:

$$WDA = \text{write data available} = (t_2 + t_4 + t_{19}, t_1 + t_{21}) = (230 + 20 + 128, 500 + 33) = (378, 533)$$

9.5.3 Motorola MC68HC812A4 External Bus Timing

9.5.3.1 MC68HC812A4 Execution Modes

The 68HC812 can run in one of eight modes (Table 9.12). Special modes are for testing and system development, while normal modes are intended for embedded products. In particular, special modes allow access to many mode control registers, like the MODE register shown below, that are inaccessible while in normal mode. In the two *single-chip* modes, Ports A–E are available for I/O. In all four expanded modes, Port E contains the bus control signals. In the two *expanded narrow* modes, Ports A, B implement the 16-bit address bus, and Port C implements the 8-bit data bus. In the two *expanded wide* modes, Ports A, B implement the 16-bit address bus, and Ports C, D implement the 16-bit data bus. *Special peripheral* mode is used by Motorola for testing. The initial mode is determined by the values of the BKGD, MODB, MODA pins at the time of the rise of RESET.

BKGD	MODB	MODA	Mode description	PortA	PortB	PortC	PortD
0	0	0	Special single chip	In/Out	In/Out	In/Out	In/Out
0	0	1	Special expanded narrow	A15–A8	A7–A0	D7–D0	In/Out
0	1	0	Special peripheral	A15–A8	A7–A0	D15–D8	D7–D0
0	1	1	Special expanded wide	A15–A8	A7–A0	D15–D8	D7–D0
1	0	0	Normal single chip	In/Out	In/Out	In/Out	In/Out
1	0	1	Normal expanded narrow	A15–A8	A7–A0	D7–D0	In/Out
1	1	0	Reserved	–	–	–	–
1	1	1	Normal expanded wide	A15–A8	A7–A0	D15–D8	D7–D0

Table 9.12 There are eight execution modes for the MC68HC812A4.

Observation: For a simple low-cost external memory interface, we will utilize expanded narrow mode.

Observation: For a high-speed external memory interface, we will utilize expanded wide mode.

When a 16-bit access is required in expanded narrow mode, it is performed by two sequential 8-bit accesses. If the computer is running in one of the special modes, the software can affect the mode by changing the SMOD and MDA bits in the HPRIO register.

Observation: In most situations a program running in expanded memory implemented in narrow mode runs twice as slow as the same program running in wide mode.

	7	6	5	4	3	2	1	0	
MODE	SMODN	MODB	MODA	ESTR	IVIS	0	EMD	EME	$000B

SMODN, MODB, MODA: Mode select special, B and A

These bits show the current operating mode and reflect the status of the BKGD, MODB, and MODA input pins at the rising edge of reset. Read anytime. SMODN may be written only if SMODN = 0 (in special modes), but the first write is ignored; MODB, MODA may be written once if SMODN = 1; anytime if SMODN = 0, except that special peripheral and reserved modes cannot be selected.

ESTR: E clock stretch enable

Determines if the E clock behaves as a simple free-running clock or as a bus control signal that is active only for external bus cycles. Normal modes: write once; special modes: write anytime, read anytime

0 = E never stretches (always free running)

1 = E stretches high during external access cycles and low during nonvisible internal accesses

IVIS: Internal visibility

This bit determines whether internal ADDR, DATA, R/W, and LSTRB signals can be seen on the external bus during accesses to internal locations. In special narrow mode if this bit is set and EMD = 1 when an internal access occurs, the data appear wide on Port C and Port D. This allows for emulation. Visibility is not available when the part is operating in a single-chip mode. Normal modes: write once; special modes: write anytime EXCEPT the first time. Read anytime

0 = No visibility of internal bus operations on external bus

1 = Internal bus operations are visible on external bus

EMD: Emulate Port D

This bit has meaning only in special expanded narrow mode. In expanded wide modes and special peripheral mode, PORTD, DDRD, KWIED, and KWIFD are removed from the memory map regardless of the state of this bit. In single-chip modes and normal expanded narrow mode, PORTD, DDRD, KWIED, and KWIFD are in the memory map regardless of the state of this bit. Removing the registers from the map allows the user to emulate the function of these registers externally. Normal modes: write once; special modes: write anytime EXCEPT the first time. Read anytime

 0 = PORTD, DDRD, KWIED, and KWIFD are in the memory map

 1 = If in special expanded narrow mode, PORTD, DDRD, KWIED, and KWIFD are removed

EME: Emulate Port E

In single-chip mode PORTE and DDRE are always in the map regardless of the state of this bit. Removing the registers from the map allows the user to emulate the function of these registers externally. Normal modes: write once; special modes: write anytime *except* the first time. Read anytime

 0 = PORTE and DDRE are in the memory map

 1 = If in an expanded mode, PORTE and DDRE are removed from the internal memory map

Another register that affects expanded mode interfacing is MISC:

	7	6	5	4	3	2	1	0	
MISC	EWDIR	NDRF	0	0	0	0	0	0	$0013

EWDIR: Extra window positioned in direct space

This bit is only valid in expanded modes. If the EWEN bit in the WINDEF register is cleared, then this bit has no meaning or effect

 0 = If EWEN is set, then a 0 in this bit places the EPAGE at $0400–$07FF

 1 = If EWEN is set, then a 1 in this bit places the EPAGE at $0000–$03FF

NDRC: Narrow data bus for Register Chip Select Space

This function requires at least one of the chip selects CS[3:0] to be enabled. It effects the (external) 512 byte memory space. If the narrow (8-bit) mode is being utilized, this bit has no effect

 0 = Makes the register-following chip select active space act as a full 16-bit data bus

 1 = Makes the register-following chip selects (2, 1, 0, and sometimes 3) active space [512-byte block] act the same as an 8-bit only external data bus (data goes through only port C externally). This allows 8-bit and 16-bit external memory devices to be mixed in a system.

9.5.3.2 MC68HC812A4 Timing

The E clock, and the six chip selects are timing signal outputs of the MC68HC812A4. The clock frequency is determined by a crystal placed between the EXTAL and XTAL inputs (Figure 9.33). The MC68HC812A4 system needs an external crystal that is two times the E clock period (t_1 in the timing diagram of Figure 9.34). The 8 MHz period is created using a 16 MHz crystal. The 22 pF includes all stray capacitances.

Figure 9.33
MC68HC812A4 clock circuit.

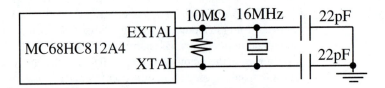

Figure 9.34
Simplified bus timing for the MC68HC812A4 in expanded mode.

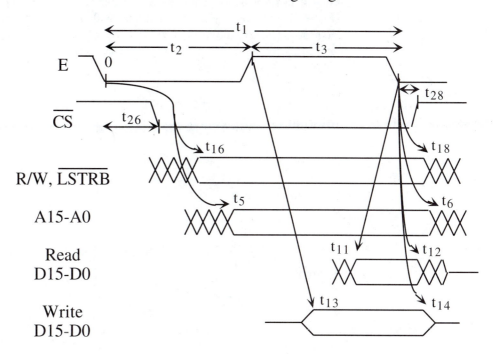

MC68HC812A4 Timing Diagram

In Figure 9.34 the symbols like t_1, t_2, t_3 refer to the timing parameters in Table 9.13. During a read cycle the MC68HC812A4 clocks the data in on the fall of E; t_{11} is the setup time, and t_{12} is the hold time for this transfer. During a write cycle the MC68HC812A4 drives the data bus.

No.		Characteristic frequency at 8 MHz for			
		0 stretch	**1 stretch**	**2 stretch**	**3 stretch**
t_1	Cycle time (ns)	125	250	375	500
t_2	Pulse width E low (ns)	60 min	60 min	60 min	60 min
t_3	Pulse width E high (ns)	60 min	185 min	310 min	435 min
t_5	A15–A0, R/W delay time (ns)	60 max	60 max	60 max	60 max
t_6	Address hold time (ns)	20 min	20 min	20 min	20 min
t_{11}	Read data setup time (ns)	30 min	30 min	30 min	30 min
t_{12}	Read data hold time (ns)	0 min	0 min	0 min	0 min
t_{13}	Write data delay time (ns)	46 max	46 max	46 max	46 max
t_{14}	Write data hold time (ns)	20 min	20 min	20 min	20 min
t_{16}	R/W delay time (ns)	49 max	49 max	49 max	49 max
t_{18}	R/W hold time (ns)	20 min	20 min	20 min	20 min
t_{26}	CS delay time (ns)	60 max	60 max	60 max	60 max
t_{28}	CS hold time	10 max	10 max	10 max	10 max

Table 9.13 Timing parameters for the MC68HC812A4 with an E clock of 8 MHz.

In expanded narrow mode, the data bus is only 8 bits. The $\overline{CS}$ signal shown in Figure 9.35 is one of the seven built-in chip selects available on Port F. With no cycle stretching, the external E clock follows the internal E clock. With one-cycle stretching, the access cycle is 250 ns long (Figure 9.36). With two-cycle stretching, the access cycle is 375 ns long (Figure 9.37). With three-cycle stretching, the access cycle is 500 ns long (Figure 9.38).

Figure 9.35
CS timing with no cycle stretching.

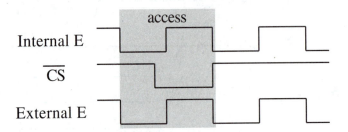

Figure 9.36
CS timing with one-cycle stretching.

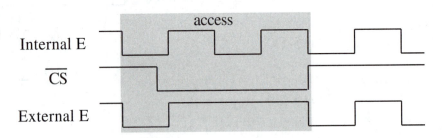

Figure 9.37
CS timing with two-cycle stretching.

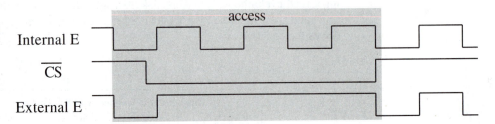

Figure 9.38
CS timing with three-cycle stretching.

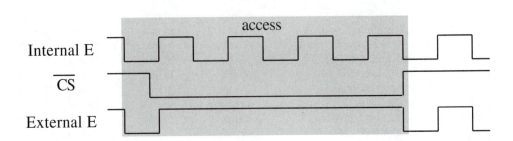

There are three important timing intervals we must determine from the microcomputer timing diagram. The first is the *address available,* AA, interval. This interval defines when during the cycle the address is valid:

$$AA = (AdV, AdN) = (t_5, t_1 + t_6)$$

where AdV is the time when the address lines A15–A0 are valid and AdN is the time when the address lines are no longer valid. For an 8-MHz MC68HC812A4 system, with no cycle stretching,

$$AA = (AdV, AdN) = (60, 145)$$

The second important timing interval is *read data required,* RDR. During a read cycle the data are required by the MC68HC812A4. Thus to determine the data required interval, we look in the MC68HC812A4 data sheet. For data required, the worst case is the longest interval:

$$RDR = \text{read data required} = (t_1 - t_{11}, t_1 + t_{12}) = (125 - 30, 125 + 0) = (95, 125)$$

The last important timing interval we get from the microcomputer is *write data available,* WDA. During a write cycle, the data are supplied by the MC68HC812A4. Thus, to determine the data available interval, we look in the MC68HC812A4 data sheet. We will specify the worst-case timing. For write data available, the worst case is the shortest interval:

$$WDA = \text{write data available} = (t_2 + t_{13}, t_1 + t_{14}) = (60 + 46, 125 + 20) = (106, 135)$$

**9.5.4
Motorola
MC68HC912B32
External Bus
Timing**

9.5.4.1
MC68HC912B32
Execution Modes

The 68HC912 can also run in one of eight modes (Table 9.14). Special modes are for testing and system development, while normal modes are intended for embedded products. In particular, special mode allows access to many mode control registers, like the MODE register shown below, that are inaccessible while in normal mode. In the two *single-chip* modes, Ports A, B, and E are available for I/O. In all four expanded modes, Port E contains the bus control signals. In the two *expanded narrow* modes, Port A implements the time-multiplexed A15–A8 address D7–D0 data bus, and Port B implements A7–A0, the low 8 bits of the address bus. In the two *expanded wide* modes, Ports A, B implement the time-multiplexed 16-bit address 16-bit data bus. *Special peripheral* mode is used by Motorola for testing. The initial mode is determined by the values of the BKGD, MODB, MODA pins at the time of the rise of RESET.

BKGD	MODB	MODA	Mode description	PortA	PortB
0	0	0	Special single chip	In/Out	In/Out
0	0	1	Special expanded narrow	A15–A8/D7–D0	A7–A0
0	1	0	Special peripheral	A15–A8/D15–D8	A7–A0/D7–D0
0	1	1	Special expanded wide	A15–A8/D15–D8	A7–A0/D7–D0
1	0	0	Normal single chip	In/Out	In/Out
1	0	1	Normal expanded narrow	A15–A8/D7–D0	A7–A0
1	1	0	Reserved	–	–
1	1	1	Normal expanded wide	A15–A8/D15–D8	A7–A0/D7–D0

Table 9.14 There are eight execution modes for the MC68HC912B32.

When a 16-bit access is required in expanded narrow mode, it is performed by two sequential 8-bit accesses. In most situations a program running in expanded memory implemented in narrow mode runs twice as slow as the same program running in wide mode. If the computer is running in one of the special modes, the software can affect the mode by changing the SMOD and MDA bits in the HPRIO register.

7	6	5	4	3	2	1	0	
SMODN	MODB	MODA	ESTR	IVIS	EBSWAI	0	EME	$000B

MODE (label to the left of the table)

SMODN, MODB, MODA: Mode select special, B and A

These bits show the current operating mode and reflect the status of the BKGD, MODB, and MODA input pins at the rising edge of reset. Read anytime. SMODN may be written only if SMODN = 0 (in special modes), but the first write is ignored; MODB, MODA may be written once if SMODN = 1; anytime if SMODN = 0, except that special peripheral and reserved modes cannot be selected

ESTR: E clock stretch enable

Determines if the E Clock behaves as a simple free-running clock or as a bus control signal that is active only for external bus cycles. Normal modes: write once; special modes: write anytime, read anytime

0 = E never stretches (always free running)

1 = E stretches high during external access cycles and low during nonvisible internal accesses

IVIS: Internal visibility

This bit determines whether internal ADDR, DATA, R/W, and LSTRB signals can be seen on the external bus during accesses to internal locations. In special expanded narrow mode, it is possible to configure the MCU to show internal accesses on an external 16-bit bus. The IVIS control bit must be set to 1. When the system is configured this way, visible internal accesses are shown as if the MCU was configured for expanded wide mode, but normal external accesses operate as if the bus was in narrow mode. In normal expanded narrow mode, internal visibility is not allowed and IVIS is ignored. Normal modes: write once; special modes: write anytime *except* the first time. Read anytime

0 = No visibility of internal bus operations on external bus

1 = Internal bus operations are visible on external bus

EBSWAI: External bus module stop in wait control

This bit controls access to the external bus interface when in wait mode. The module will delay before shutting down in wait mode to allow for final bus activity to complete

0 = External bus and registers continue functioning during wait mode

1 = External bus is shut down during wait mode

EME: Emulate Port E

In single-chip mode PORTE and DDRE are always in the map regardless of the state of this bit. Removing the registers from the map allows the user to emulate the function of these registers externally. Normal modes: write once; special modes: write anytime *except* the first time. Read anytime

0 = PORTE and DDRE are in the memory map

1 = If in an expanded mode, PORTE and DDRE are removed from the internal memory map

Another register that affects expanded mode interfacing is MISC. The MC68HC912B32 allows the software to select zero- to three-cycle stretching for Register-following and External spaces. When the MC68HC912B32 comes out of reset in any of the four expanded modes, the default cycle stretching is no cycle stretching.

	7	6	5	4	3	2	1	0	
MISC	0	NDRF	RFSTR1	RFSTR0	EXSTR1	EXSTR0	MAPROM	ROMON	$0013

NDRF: Narrow data bus for register-following map

This bit enables a narrow bus feature for the 512-byte register-following map. In expanded narrow (8 bit) modes, single-chip modes, and peripheral mode, NDRF has no effect. The register-following map always begins at the byte following the 512-byte register map. If the registers are moved, this space will also move

0 = Register-following map space acts as a full 16-bit data bus

1 = Register-following map space acts the same as an 8-bit external data bus

RFSTR1, RFSTR0: Register-following stretch bit 1 and bit 0 (Table 9.15)

These bits determine the amount of clock stretch on accesses to the 512-byte register-following map. It is valid regardless of the state of the NDRF bit. In single-chip and peripheral modes this bit has no meaning or effect.

Table 9.15
E clock stretching for the register-following space.

RFSTR1	RFSTR0	E clock stretch
0	0	None (default)
0	1	1
1	0	2
1	1	3

EXSTR1, EXSTR0: External Access Stretch Bit 1 and Bit 0 (Table 9.16)

These bits determine the amount of clock stretch on accesses to the external address space. In single-chip and peripheral modes this bit has no meaning or effect.

Table 9.16
E clock stretching for the external access space.

EXSTR1	EXSTR0	E clock stretch
0	0	None (default)
0	1	1
1	0	2
1	1	3

MAPROM: Map location of flash EEPROM

This bit determines the location of the on-chip flash EEPROM. In expanded modes it is reset to 0. In single-chip modes it is reset to 1. If ROMON is 0, this bit has no meaning or effect

0 = Flash EEPROM is located from $0000 to $7FFF

1 = Flash EEPROM is located from $8000 to $FFFF

ROMON: Enable flash EEPROM

In expanded modes ROMON is reset to 0. In single-chip modes it is reset to 1. If the internal RAM, registers, EEPROM, or BDM ROM (if active) are mapped to the same space as the flash EEPROM, they will have priority over the flash EEPROM

0 = Disables the Flash EEPROM in the memory map

1 = Enables the Flash EEPROM in the memory map

The E clock, and the six chip selects are timing signal outputs of the MC68HC912B32. The clock frequency is determined by a crystal placed between the EXTAL and XTAL inputs (Figure 9.39). The MC68HC912B32 system needs an external crystal that is two times the E clock period (t_1 in the timing diagram of Figure 9.40). The 8 MHz period is created using a 16-MHz crystal. The 22 pF includes all stray capacitances.

Figure 9.39
MC68HC812A4 clock
circuit.

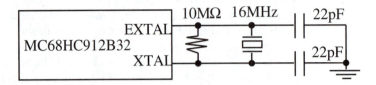

In Figure 9.40 the symbols like t_1, t_2, t_3 refer to the timing parameters in Table 9.17. In expanded narrow mode, the data bus is only 8 bits.

Figure 9.40
Simplified bus timing for
the MC68HC912B32 in
expanded mode.

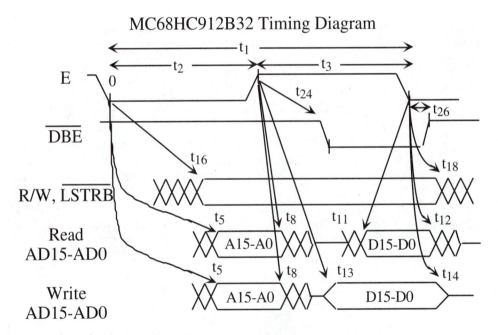

There are three important timing intervals we must determine from the microcomputer timing diagram. The first is the *address available,* AA, interval. This interval defines when during the cycle the address is valid. To calculate when the address is available, we first consider the 74HC374 octal D flips used to capture the address (Figure 9.41).

$$AA = (AdV, AdN) = (t_2 + 9, t_1 + t_2)$$

where AdV is the time when the address lines A15–A0 are valid and AdN is the time when the address lines are no longer valid. The +9 ns occurs because of the delay in the 74AHC374s. Because the address is clocked into the 16 D flip-flops on the rising edge of

No.		Characteristic frequency at 8 MHz for			
		0 stretch	1 stretch	2 stretch	3 stretch
t_1	Cycle time (ns)	125	250	375	500
t_2	Pulse width E low (ns)	60 min	60 min	60 min	60 min
t_3	Pulse width E high (ns)	60 min	185 min	310 min	435 min
t_5	A15–A0, R/W delay time (ns)	60 max	60 max	60 max	60 max
t_8	Address hold time (ns)	10 min	10 min	10 min	10 min
t_{11}	Read data setup time (ns)	30 min	30 min	30 min	30 min
t_{12}	Read data hold time (ns)	0 min	0 min	0 min	0 min
t_{13}	Write data delay time (ns)	46 max	46 max	46 max	46 max
t_{16}	R/W delay time (ns)	49 min	49 min	49 min	49 min
t_{18}	R/W hold time (ns)	20 max	20 max	20 max	20 max
t_{21}	Write data hold time (ns)	20 min	20 min	20 min	20 min
t_{24}	DBE delay time (ns)	37 max	37 max	37 max	37 max
t_{26}	DBE hold time (ns)	10 max	10 max	10 max	10 max

Table 9.17 Timing parameters for the MC68HC912B32 with an E clock of 8 MHz.

Figure 9.41
Address latch for the
MC68HC912B32 in
expanded mode.

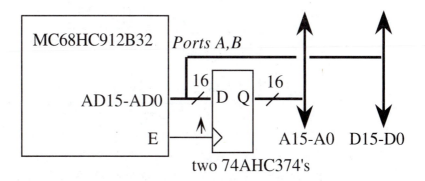

E, it is available during the second half of the cycle and continues to be valid until the next rising edge of E. For an 8-MHz MC68HC912B32 system, with no cycle stretching:

$$AA = (AdV, AdN) = (69, 185)$$

The second important timing interval is *read data required,* RDR. During a read cycle the data are required by the MC68HC912B32. Thus to determine the data required interval, we look in the MC68HC912B32 data sheet. For data required, the worst case is the longest interval.

$$RDR = \text{read data required} = (t_1 - t_{11}, t_1 + t_{12}) = (125 - 30, 125 + 0) = (95, 125)$$

The last important timing interval we get from the microcomputer is *write data available,* WDA. During a write cycle, the data are supplied by the MC68HC912B32. Thus, to determine the data available interval we look in the MC68HC912B32 data sheet. We will specify the worst-case timing. For write data available, the worst case is the shortest interval.

$$WDA = \text{write data available} = (t_2 + t_{13}, t_1 + t_{14}) = (60 + 46, 125 + 20) = (106, 135)$$

9.6   General Approach to Interfacing

The MC68HC11A1 contains the 6811 I/O ports and internal RAM, but it has no internal ROM. We connect MODA, MODB so that the computer executes in expanded mode. The general approach to interfacing a memory to the 6811 is shown in Figure 9.42.

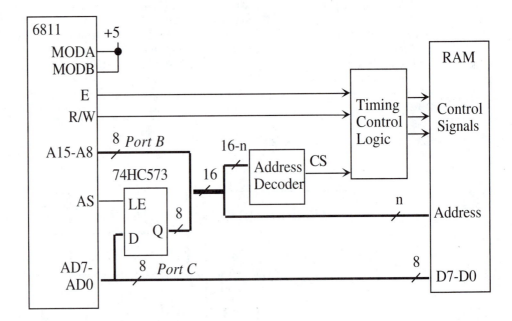

Figure 9.42 General approach to memory interfacing on a 6811 in expanded mode.

The 74HC573 *Latch* will capture the low address, A7–A0, from the AD7–AD0 signals on the fall of the AS so that the entire 16-bit 6811 address, A15–A0, is available for the duration of the cycle. The faster 74AHC573 can also be used. If the RAM contains 2^n bytes, then the low n bits of the address go directly to the RAM and specify which cell to access. The top $16 - n$ are used by the *Address Decoder* circuit to determine whether or not this RAM should activate during this cycle. If the **CS** signal is true, the *Timing Control Logic* will generate appropriate control signals for the memory. The control signals will activate a read operation (drive data out of memory onto the bus) if the **CS** is true and R/W is 1. Similarly, the control signals will activate a write operation (store data from the bus into the memory) if the **CS** is true and R/W is 0.

In expanded narrow mode, all external memory accesses utilize only 8 bits of data. We connect MODA, MODB so that the 6812 executes in expanded narrow mode. The 6812 will automatically divide 16-bit reads and writes into two separate 8-bit accesses. In many cases, we can interface external devices to the MC68HC812A4 in narrow mode without any additional digital logic. The MC68HC812A4 will use one of its built-in address decoders, shown as **CS** in Figure 9.43. The MC68HC912B32 interface will require an address decoder to generate the chip select, **CS** (Figure 9.44). A common reason for using narrow mode is that many I/O devices only support 8-bit accesses.

Figure 9.43

General approach to memory interfacing on a MC68HC812A4 in narrow expanded mode.

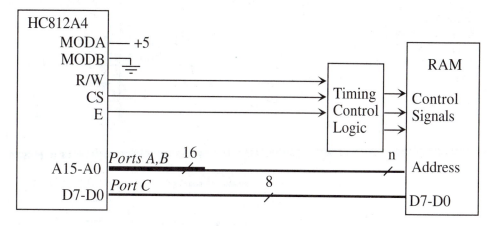

Figure 9.44

General approach to memory interfacing on a MC68HC912B32 in narrow expanded mode.

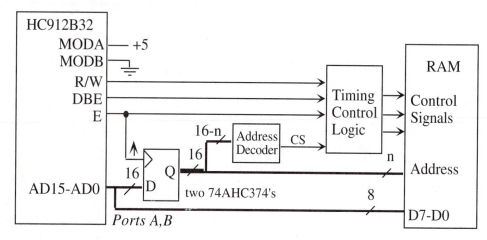

If the RAM contains 2^n bytes, then the low n bits of the address go directly to the RAM and specify which cell to access. One of the seven built-in negative logic chip selects will be used with the MC68HC812A4. If the **CS** signal is true, the *Timing Control Logic* will generate appropriate control signals for the memory. The control signals will activate a read operation (drive data out of memory onto the bus) if the **CS** is true and **R/W** is 1. Similarly, the control signals will activate a write operation (store data from the bus into the memory) if the **CS** is true and **R/W** is 0.

On both 6812 systems, the software must configure the cycle stretching, and enable the E clock and R/W signals as needed in the PEAR register. MC68HC812A4 initialization software should also activate the chip selects.

9.6.3
Interfacing to a 6812 in Expanded Wide Mode

In expanded wide mode, external memory accesses utilize 16 bits of data. We connect MODA, MODB so that the computer executes in expanded wide mode: 16-bit reads and writes to even addresses (aligned) will be performed in a single access; 16-bit reads and writes to odd addresses (misaligned) will be performed in two separate 8-bit accesses. The LSTRB (=0 when low byte data is valid) is used to implement 8-bit memory writes. The advantage of wide mode is execution speed, because opcode fetches will always occur as aligned 16-bit reads (Figures 9.45 and 9.46).

Figure 9.45
General approach to
memory interfacing on a
MC68HC812A4 in wide
expanded mode.

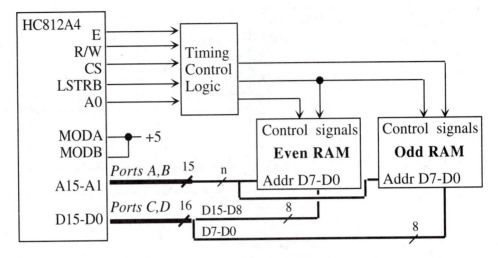

Figure 9.46
General approach to
memory interfacing on a
MC68HC912B32 in
wide expanded mode.

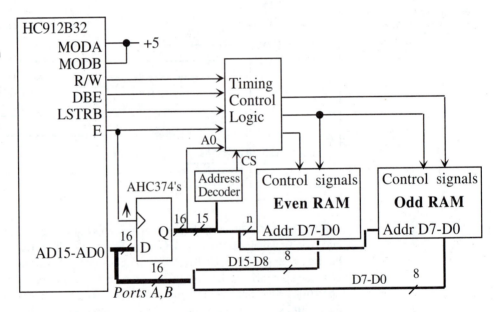

The typical approach to creating a 16-bit memory is to use two 8-bit memories called
even and *odd*. Assume the size of the memory is 2^n words (16 bits each). The two memory
chips share address lines (An–$A1$) and most control signals. One of the seven built-in negative logic chip selects will be used with the MC68HC812A4. The MC68HC912B32 will
require an address decoder that accepts the most significant $15 - n$ address lines and produces the chip select, **CS.** The 16-bit data bus is divided so that 8 bits go to each chip. The
"big endian" format places the most significant data, D15–D8, into the even address and the
least significant data, D7–D0, into the odd address. When the 6812 performs an 8-bit read,
there is usually no problem if the memory system responds with 16 bits of data (the processor simply takes the data it wants). On the other hand, if the 6812 performs an 8-bit write,
it would be a mistake to save all 16 bits D15–D0 into the memory system.

LSTRB	A0	R/W	Type of Access	Even RAM	Odd RAM
1	0	1	8-bit read of an even address	Activate	No action
0	1	1	8-bit read of an odd address	No action	Activate
1	0	0	8-bit write of an even address	Activate	No action
0	1	0	8-bit write of an odd address	No action	Activate
0	0	1	16-bit read of an even address	Activate	Activate
1	1	1	16-bit read of an odd address*	Not applicable	Not applicable
0	0	0	16-bit write to an even address	Activate	Activate
1	1	0	16-bit write to an odd address*	Not applicable	Not applicable
*Low/high data swapped.					

Table 9.18 LSTRB, A0, R/W specify when to activate even and odd RAM modules.

The external signals LSTRB, R/W, and A0 can be used to determine the type of bus access. Accesses to the internal RAM module are the only situation that requires the even and odd bytes to be swapped, LSTRB = A0 = 1 (Table 9.18). The internal RAM is specifically designed to allow misaligned 16-bit accesses in a single cycle. In a misaligned 16-bit access, the data for the address that was accessed are on the low half of the data bus and the data for address + 1 are on the high half of the data bus. We will use A0 and LSTRB to handle 8-bit data writes while running in 16-bit data mode.

Notice that we should activate the *even RAM* when $\overline{CS}$ =0 and A0=0, and we should activate the *odd RAM* when $\overline{CS}$ =0 and LSTRB=0. These digital logic functions can be implemented with one 74AHC32 gate. AHC or ACT logic is used because the propagation delay is less than 10 ns (Figure 9.47).

Figure 9.47
Circuit needed to handle even and odd 8-bit accesses.

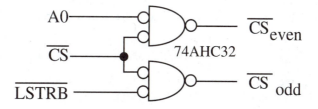

On both 6812 systems, the software must configure the cycle stretching and enable the E clock, LSTRB, and R/W signals as needed in the PEAR register. The NDRF bit in the MISC register can be set to configure the 512 bytes following the register space (typically $0200–$03FF) for narrow mode. When NDRF is set, all addresses except $0200–$03FF can utilize the 16-bit data bus. MC68HC812A4 initialization software should also activate the chip selects.

9.7 Memory Interface Examples

**9.7.1
32K PROM
Interface**

The objective of this example is to interface a 32K by 8 bit PROM to the microcomputer. We will place the 27256-25 PROM at locations $8000 to $FFFF. The **command portion** determines which CPU cycles will activate the PROM. The 27256-25 PROM will drive the data bus whenever both **CE*** and **OE*** are 0. The other three values of **CE*** and **OE*** will

Table 9.19

Function table of a typical PROM.

CE*	OE*	Function	I_{cc} (mA)
1	1	Off	40
1	0	Off	40
0	1	Off	100
0	0	Drive data bus	100

disable the PROM. The PROM chip requires less power (I_{cc} is the + 5 V supply current) whenever **CE*** is high (Table 9.19).

Using full decoding the address **SELECT** line in positive logic would be **SELECT**=A15. Because this chip is read only, the control lines will be activated only during read cycles (R/W=1) from this PROM (A15=1). To conserve power (lower I_{cc}), then the **CE***=1 state is used to deselect the PROM (Table 9.20). The **timing portion** of the interface determines the time during an active cycle when a control signal should rise and fall. The objective is to design the interface such that the data available interval overlaps the data required interval.

Table 9.20

Command table for the 27256 PROM interface.

SELECT	R/W	Function	CE*	OE*
0	0	Off	1	X
0	1	Off	1	X
1	0	Off	1	X
1	1	Drive data bus	0	0

During a read cycle, the data are supplied by the 27256-25. Thus, to determine the data available interval we look at the 27256-25 data sheet. We will specify the worst-case timing. For data available, the worst case is the shortest interval (Figure 9.48).

$$\text{Read data available} = (\text{later} (AdV + t_{ACC}, \downarrow CE^* + t_{CE}, \downarrow OE^* + t_{OE}),$$
$$\text{earlier} (AdN + t_{OH}, \uparrow CE^* + t_{OH}, \uparrow OE^* + t_{OH}))$$

To conserve power (minimize I_{cc}), we let CE* be 0 only during read cycles to \$8000 to \$FFFF. If we simply grounded CE*, the interface would be faster but would require more power. From the 27256-25 data sheet, t_{ACC}=250, t_{CE}=250 t_{OE}=100, t_{OH}=0.

Figure 9.48

Read timing for the 27256-25 32K PROM chip.

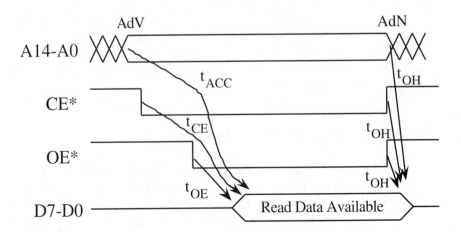

9.7.1.1
32K PROM Interface to a
MC68HC11A1

Entering the AdV, AdN times of a 2-MHz 6811, we calculate

$$\text{Read data available} = (\text{later} \,(181 + 250, \downarrow CE^* + 250, \downarrow OE^* + 100),$$
$$\text{earlier} \,(533 + 0, \uparrow CE^* + 0, \uparrow OE^* + 0))$$

During a read cycle the data are required by the 6811.

$$\text{Read data required} = (t_1 - t_4 - t_{17,} \, t_1 + t_{18}) = (450, 510)$$

We must choose the timing of the fall and rise of both CE* and OE* to make the data available interval overlap the data required window. Note that since CE* and OE* are active low, they will fall, then rise, during the active cycles. Thus,

$181 + 250$	≤ 450	and	533	≥ 510
$\downarrow CE^* + 250$	≤ 450	and	$\uparrow CE^*$	≥ 510
$\downarrow OE^* + 100$	≤ 450	and	$\uparrow OE^*$	≥ 510

or

$\downarrow CE^*$	≤ 200	and	$\uparrow CE^*$	≥ 510
$\downarrow OE^*$	≤ 350	and	$\uparrow OE^*$	≥ 510

The following signal is constructed inside the PROM chip:

$$ON = \overline{CE^*} \cdot \overline{OE^*}$$

The data bus is driven when ON equals 1. Since the 6811 is a synchronous computer, the operation of the ON signal must be synchronized to the 6811. There are three ways to synchronize the ON signal:

1. Synchronize only OE*
2. Synchronize only CE*
3. Synchronize both CE* and OE*

Performance Tip: In general, we will synchronize the control signal with the shortest access time.

It will be less expensive to synchronize just one of the signals. Because $t_{OE} < t_{CE,}$ it will be faster to synchronize OE*. This means that CE* will be controlled by bus signals A15 and R/W. The fall of CE* will be AdV plus the maximum LS gate delay (+10 ns). Similarly, the rise of CE* will be AdN plus the minimum LS gate delay (0). Assuming one 10-ns gate delay, we see that the CE* timing is satisfied,

$$\downarrow CE^* = \text{AdV}_{15-8} + 10 = 146 \le 200 \quad \text{and} \quad \uparrow CE^* = \text{AdN}_{15-8} \ge 510$$

Since the address bus is indeterminate during the first 181 ns of every cycle, we can never design a control signal that changes with the beginning of the cycle (at 0 ns). Thus, there is only one choice for the high-to-low and low-to-high transitions of OE*: OE* must fall at 260 and rise at 510 (Table 9.21). The beginning of RDA is determined by AdV (bits 7–0) and the end by $\uparrow OE^*$:

$$\text{Read data available} = (\text{AdV} + 250, \uparrow OE^* + 0) = (431, 510)$$

Table 9.21

Read timing table for the 27256 PROM interface.

E	OE*
0	1
1	0

Next we find the Boolean expressions that will generate CE* and OE* from E, AS, R/W, and SELECT. The fully decoded positive logic select signal is SELECT=A15. The command table is expanded to include the timing information (Table 9.22). Karnaugh maps are used to determine the logic equations. 74LSxx low-power Schottky logic has a good balance of speed, input loading current (I_{IL}), and power supply current. 74AHCxx advanced high-speed CMOS logic should be used for battery-operated systems because of its low power requirements.

Table 9.22
Combined timing table for the 27256 PROM interface.

E	R/W	A15	ON	CE*	OE*	
0	0	0	0	1	X	Address not on the chip
1	0	0	0	1	X	
0	1	0	0	1	X	Address not on the chip
1	1	0	0	1	X	
0	0	1	0	1	X	Correct address
1	0	1	0	1	X	But it is a write
0	1	1	0	0	1	
1	1	1	1	0	0	Enable ROM

Next, we build the interface (Figure 9.49). Finally, we draw the *Combined read cycle timing diagram* and use it to verify that read data available overlaps read data required (Figure 9.50).

Figure 9.49 Interface between a 6811 and a 27256-25 32K PROM chip.

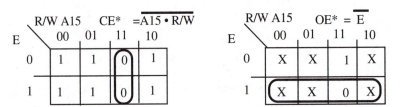

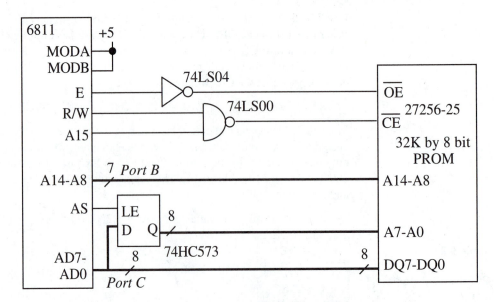

Figure 9.50
Timing of a 6811 and a
27256-25 32K PROM
chip.

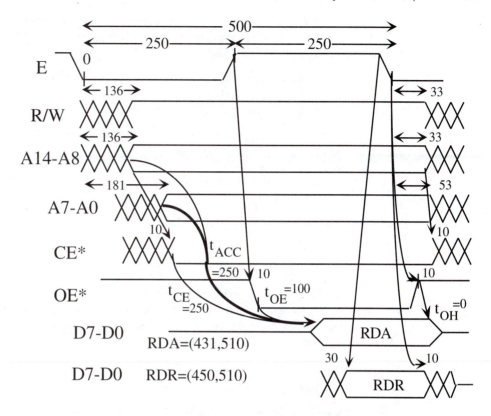

Two of the issues when interfacing to a 6812 is to determine data bus width (narrow or wide) and the number of cycles to stretch. For simplicity we start with an 8-bit narrow example. To stretch the E clock we will have to use one of the seven built-in chip selects (e.g., either CSP0 or CSP1). For now, we will develop the timing equations, and then we will use them to determine the cycle stretching requirements. The AdV time is 60 ns, independent of stretching, while AdN will be $t_1 + 20$ ns. We therefore calculate

Read data available = (later ($60 + t_{ACC}$, $\downarrow$CE*+ t_{CE}, $\downarrow$OE*+ t_{OE}), earlier (AdN + t_{OH}, $\uparrow$CE*+ t_{OH}, $\uparrow$OE*+ t_{OH}))
= (later (310, $\downarrow$CE* + 250, $\downarrow$OE* + 100), earlier ($t_1 + 20 + 0$, $\uparrow$CE* + 0, $\uparrow$OE* + 0))

During a read cycle the data are required by the 6812.

RDR = read data required = ($t_1 - t_{11}$, $t_1 + t_{12}$) = ($t_1 - 30$, t_1)

We must choose the timing of the fall and rise of both CE* and OE* to make the data available interval overlap the data required window. Note that since CE* and OE* are active low, they will fall, then rise, during the active cycles. Thus,

310	$\leq t_1 - 30$	and	$t_1 + 20$	$\geq t_1$
$\downarrow$CE*+ 250	$\leq t_1 - 30$	and	$\uparrow$CE*	$\geq t_1$
$\downarrow$OE*+ 100	$\leq t_1 - 30$	and	$\uparrow$OE*	$\geq t_1$

First line ($340 \leq t_1$) tells us that two extra cycles are needed to stretch the access time to 375 ns. The following signal is constructed inside the PROM chip:

$$ON = \overline{CE^*} \cdot \overline{OE^*}$$

The data bus is driven when ON equals 1. Since the 6812 is a synchronous computer, the operation of the ON signal must be synchronized to the 6812. There are three ways to synchronize the ON signal:

1. Synchronize only OE*
2. Synchronize only CE*
3. Synchronize both CE* and OE*

It will be less expensive to synchronize just one of the signals. Because $t_{OE} < t_{CE}$, it will be faster to synchronize OE*. On the other hand, it will be cheaper to synchronize CE* by connecting it directly to CSP0. This means CE* will be controlled by timing signal CSP0. The fall of CE* will be at 60 ns and, the rise of CE* will be at t_1 + 10 ns. The CE access time must be satisfied:

$$\downarrow CE* + 250 \leq t_1 - 30 \qquad \text{and} \qquad \uparrow CE* \geq t_1$$

or

$$60 + 250 \leq t_1 - 30 \qquad \text{and} \qquad t_1 + 10 \geq t_1$$

Again we need two extra cycles to stretch the access time to 375 ns. In this particular interface we will ground OE*, so the timing delays with OE* need not be calculated. Now that we have decided to use two-cycle stretching, we can calculate the actual RDA and RDR intervals.

Read data available = (310, 385)

Read data required = (345, 375)

Next we find the Boolean expressions that will generate CE* and OE* from E, CSP0, and R/W. In this simple case, we connected CE* to CSP0 and OE* to ground (Figure 9.51). The MC68HC812A4 will come out of reset in expanded mode, with CSP0 enabled having

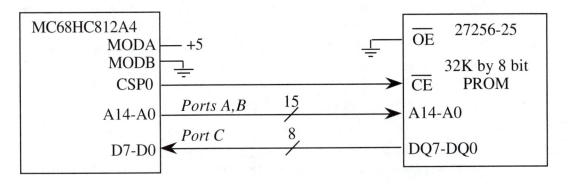

Figure 9.51 Interface between a MC68HC812A4 and a 27256-25 32K PROM chip.

three-cycle stretching. The reset software should enable any additional chip selects and change the cycle stretching on CSP0 to two cycles. Last, we draw the *Combined read cycle timing diagram* and use it to verify that read data available overlaps read data required (Figure 9.52).

Figure 9.52
Timing of a
MC68HC812A4 and a
27256-25 32K PROM
chip.

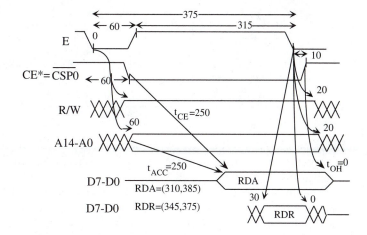

The software to activate this interface is shown in Program 9.1.

Program 9.1 Software to
configure the chip select
for the external PROM.

```
void PROMinit(void){
    CSSTR0=(CSSTR0&0xF3)|0x08;} // 2 cycle stretch on CSP0
```

9.7.2
8K RAM Interface

In this example we interface an 8K by 8-bit static RAM to the microcomputer using the Motorola MCM60L64. The memory will be placed at $6000 to $7FFF. The **command portion** determines which CPU cycles will activate the RAM. Using full decoding the address **Select*** line in negative logic is

$$\text{Select*} = \overline{\overline{A15} \cdot A14 \cdot A13}$$

A 74HC138 3 input to 8 output decoder could be used to generate the chip selects for multiple 8K devices in a single system. The Y3 output of the decoder implements the above address decoder. This RAM chip has both a positive logic (**E2**) and a negative logic ($\overline{E1}$) chip select. Negative logic is used because the output of the 74HC138 is in negative logic. With E2 tied to +5 V, the RAM has the functions shown in Table 9.23.

Table 9.23
Function table for the
MCM60L64 RAM.

$\overline{E1}$	$\overline{W}$	$\overline{G}$	Function
1	X	X	Disabled, low I_{cc} = 30 μA
0	1	1	Disabled, high I_{cc} = 3 mA
0	0	X	Write data into RAM
0	1	0	Read data out of RAM

To reduce power, we will let $\overline{E1} = 0$ only during accesses to this RAM. We will synchronize the read and write functions by synchronizing either $\overline{E1}$ or both $\overline{W}$ and $\overline{G}$. A read operation will occur when both $\overline{E1}$ and $\overline{G}$ are 0 and $\overline{W}$ is 1.

$$RD = \overline{\overline{E1} \cdot \overline{W} \cdot \overline{G}}$$

Similarly, a write operation will occur when both $\overline{E1}$ and $\overline{W}$ are zero.

$$WR = \overline{\overline{E1} \cdot \overline{W}}$$

Table 9.24
Command table for the
MCM60L64 RAM
interface.

Select	R/W	RD	WR	$\overline{E1}$	$\overline{W}$	$\overline{G}$	Function
0	0	0	0	1	X	X	Disable because address not on the chip
0	1	0	0	1	X	X	Disable because address not on the chip
1	0	0	1	0	0	X	Write cycle to this RAM
1	1	1	0	0	1	0	Read cycle from this RAM

The signals RD and WR are generated internal to the RAM. The command part of the design will have RD=1 only during read cycles from this RAM, and WR=1 only during write cycles to this RAM (Table 9.24).

The **timing portion** of the interface determines the timing of the rise and fall of the control signals. The objective is to design the interface such that the data available interval overlaps the data required. Since this is a RAM, we consider both the read and write cycles. Data are read from the memory when $\overline{E1}$ =0, $\overline{G}$ =0, and $\overline{W}$ =1. During a read cycle, the data are supplied by the 60L64. Thus, to determine the read data available interval we look in the Motorola MCM60L64 data sheet. We will specify the worst-case timing. For read data available, the worst case is the shortest interval (Figure 9.53).

$$\text{Read data available} = (\text{later } (\text{AdV} + t_{AVQV}, \downarrow \overline{E1} + t_{E1LQV}, \downarrow \overline{G} + t_{GLQV}),$$
$$\text{earlier } (\text{AdN} + t_{AXQX}, \uparrow \overline{E1} + t_{E1HQZ}, \uparrow \overline{G} + t_{GHQZ}))$$

Figure 9.53
Read timing for the
60L64 8K RAM chip.

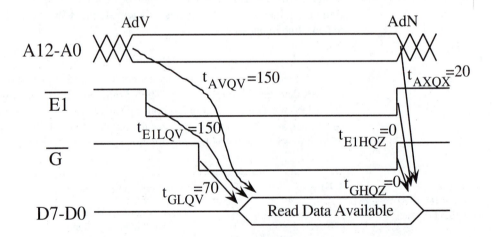

where AdV is the time when the address lines A12–A0 are valid and AdN is the time when the address lines are no longer valid. From the 60L64 data sheet t_{AVQV}=150, t_{E1LQV}=150, t_{GLQV}=70, t_{AXQX}=20, t_{E1HQZ}=0, and t_{GHQZ}=0.

During a write cycle the data are required by the 60L64. Thus to determine the write data required interval, we look in the 60L64 data sheet. Since the write operation occurs on the overlap of $\overline{E1}$ and $\overline{W}$, the first of these two signals to rise will cause data to be written into the memory. There are two possible timing diagrams for the write cycle. If $\overline{E1}$ is unsynchronized negative logic and $\overline{W}$ is synchronized negative logic, then it is the rise of $\overline{W}$ that stores data into the RAM (Figure 9.54).

$$\text{Write data required} = (\uparrow \overline{W} - t_{DVWH}, \uparrow \overline{W} + t_{WHDX}) = (\uparrow \overline{W} - 60, \uparrow \overline{W})$$

Figure 9.54
Write timing controlled by W for the 60L64 8K RAM chip.

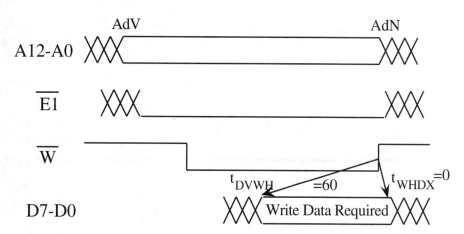

For write data required, the worst case is the longest interval. From the 60L64 data sheet, the setup time $t_{DVWH}=60$, and the hold time $t_{WHDX}=0$. Note also that the address must be valid whenever $\overline{E1}=0$. If $\overline{E1}$ is synchronized negative logic and $\overline{W}$ is unsynchronized negative logic, then it is the rise of $\overline{E1}$ that stores data into the RAM (Figure 9.55).

$$\text{Write data required} = (\uparrow\overline{E1} - t_{DVWH}, \uparrow\overline{E1} + t_{WHDX}) = (\uparrow\overline{E1} - 60, \uparrow\overline{E1})$$

Figure 9.55
Write timing controlled by W for the 60L64 8K RAM chip.

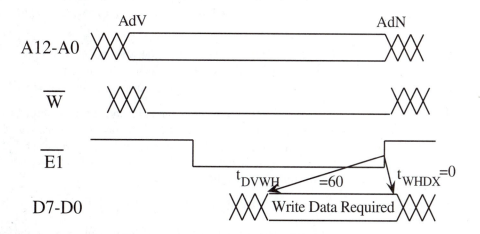

9.7.2.1
8K RAM Interface to a MC68HC11A1

To simplify the RAM interface to the 6811, we can let $\overline{E1} = $ **Select*** and make $\overline{G}$ and $\overline{W}$ synchronize the read and write cycles, respectively. Recall that for a 6811 running at 2 MHz:

$$AA = (AdV, AdN) = (181, 533)$$

$$\begin{aligned}\text{Read data available} &= (\text{later } (AdV + t_{AVQV}, \downarrow\overline{E1} + t_{E1LQV}, \downarrow\overline{G} + t_{GLQV}),\\ &\qquad \text{earlier } (AdN + t_{AXQX}, \uparrow\overline{E1} + t_{E1HQZ}, \uparrow\overline{G} + t_{GHQZ}))\\ &= (\text{later } (181+150, \downarrow\overline{E1} + 150, \downarrow\overline{G} + 70),\\ &\qquad \text{earlier } (533 + 20, \uparrow\overline{E1}, \uparrow\overline{G}))\end{aligned}$$

During a read cycle the data are required by the 6811. Thus to determine the data required interval, we look in the 6811 data sheet. For read data required, the worst case is the longest interval. Recall that

$$\text{Read data required} = (t_1 - t_4 - t_{17}, t_1 + t_{18}) = (450, 510)$$

We must choose the timing of the fall and rise of $\overline{E1}$ and $\overline{G}$ to make the read data available interval overlap the read data required interval. We must make $\overline{W} = 1$ during a read cycle. Note that since $\overline{G}$ is active low, it will fall, then rise, during the active cycles. During the inactive cycles, $\overline{G}$ will be high (with no transitions). Thus,

$181 + 150$	≤ 450	and	$533 + 20$	≥ 510
$\downarrow \overline{E1} + 150$	≤ 450	and	$\uparrow \overline{E1}$	≥ 510
$\downarrow \overline{G} + 70$	≤ 450	and	$\uparrow \overline{G}$	≥ 510

or

$\downarrow \overline{E1}$	≤ 300	and	$\uparrow \overline{E1}$	≥ 510
$\downarrow \overline{G}$	≤ 380	and	$\uparrow \overline{G}$	≥ 510

In our design, $\overline{E1}$ is derived only from the address signals A15, A14, and A13. Thus

$$\downarrow \overline{E1} = 230 - 94 + 36 = 172 \quad \text{and} \quad \uparrow \overline{E1} = 500 + 33 + 0 = 533$$

The 36 ns represents the maximum gate delay in the 74HC138 decoder, while the 0 ns represents the minimum gate delay in the 74HC138 decoder. Luckily, the actual timing of $\overline{E1}$ satisfies the constraints. If the actual $\downarrow \overline{E1}$ were greater than 300, then one would have to:

- Use a 74AHC138 advanced high-speed CMOS decoder
- Slow down the 6811 by increasing the E period
- Decrease t_{E1LQV} by spending more money on a faster RAM chip

Since the 6811 is a synchronous computer, the $\overline{G}$ control must be synchronized to the 6811. Since the 6811 has only the E clock, only two choices exist for the high-to-low and low-to-high transitions of $\overline{G}$. Assuming a gate delay between E and $\overline{G}$ of 10 ns, the choices are 260 and 510. $\overline{G}$ must fall at 260 and rise at 510. The beginning of RDA is determined by AdV (bits 7–0) and the end by $\uparrow \overline{G}$:

$$\text{Read data available} = (\text{AdV} + 150, \uparrow \overline{G} + 0) = (331, 510)$$

Observation: Synchronizing the memory read signals to the E clock on a 6811 interface prevents the memory from driving the data bus during the first half of the cycle, when the 6811 is outputting the low address on these same lines.

Common Error: If you do not synchronize driving the data bus on a multiplexed address/data computer like the 6811 and MC68HC912B32, then memory data might be driven onto the bus during the first half of the cycle and conflict with the address being driven at that time.

Data are written into the memory when $\overline{E1} = 0$, $\overline{G} = 1$, and $\overline{W} = 0$. During a write cycle, the data are supplied by the 6811. For write data available, the worst case is the shortest interval. Recall that

$$\text{Write data available} = (t_2 + t_4 + t_{19}, t_1 + t_{21}) = (378, 533)$$

We must choose the timing of the fall and rise of $\overline{W}$ to make the write data available interval overlap the write data required window. We will make $\overline{G} = 1$ during a write cycle (it will still write even if it were 0). Note that since $\overline{W}$ is active low, it will fall, then rise, during the active cycles. During the inactive cycles, $\overline{W}$ will be high (with no transitions).

$$378 \le \uparrow \overline{W} - 60 \quad \text{and} \quad \uparrow \overline{W} \le 533$$

Thus,

$$438 \le \uparrow \overline{W} \le 533$$

Since the 6811 is a synchronous computer, the $\overline{W}$ control line must be synchronized to the E clock. If we did not synchronize $\overline{W}$, then we could not guarantee that write data available will overlap write data required. In some situations a rising edge might not even occur at all. For example, consider the case where $\overline{W}$ is unsynchronized and equal to R/W. The STD $8000 instruction will produce successive write cycles to $8000 and $8001. In this case there may be only one rise of $\overline{W}$ and even that one rise may be too late to properly store data into the RAM (Figure 9.56).

Figure 9.56
Unsynchronized signals do not have guaranteed times of their rise and fall.

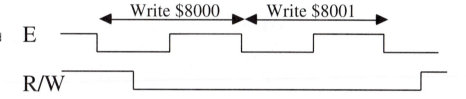

Common Error: If you do not synchronize clocking data into a memory during a write cycle, then data might not be stored properly.

There is only one choice that satisfies the design. The fall of $\overline{W}$ must be 260, and the rise of $\overline{W}$ must be 510. Since the rise of $\overline{W}$ is 510, the WDR becomes (450, 510). We combine the command, and timing aspects of the design to find the Boolean expression that will generate the control signals: $\overline{E1}$, $\overline{G}$, and $\overline{W}$ (Table 9.25).

$$\overline{E1} = \textbf{Select*}$$
$$\overline{G} = \overline{R/W \cdot E}$$
$$\overline{W} = \overline{R/W \cdot E}$$

Table 9.25
Combined timing table for the MCM60L64 RAM interface.

E	R/W	Select* = $\overline{E1}$	RD	WR	$\overline{G}$	$\overline{W}$	
0	0	1	0	0	X	X	Address not on the chip
1	0	1	0	0	X	X	
0	1	1	0	0	X	X	Address not on the chip
1	1	1	0	0	X	X	
0	0	0	0	0	1	1	Write cycle
1	0	0	0	1	1	0	
0	1	0	0	0	1	1	Read cycle
1	1	0	1	0	0	1	

The next step is to build the interface (Figure 9.57). To verify proper read cycle timing we draw the *Combined read cycle timing diagram* (Figure 9.58). Notice that read data available overlaps read data required.

Figure 9.57
Interface between the 6811 and the 60L64 8K RAM chip.

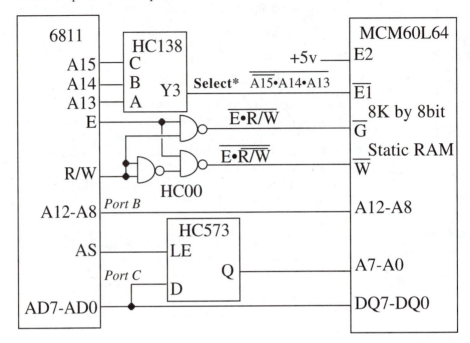

Figure 9.58
Read timing diagram of the 6811 and the 60L64 8K RAM chip.

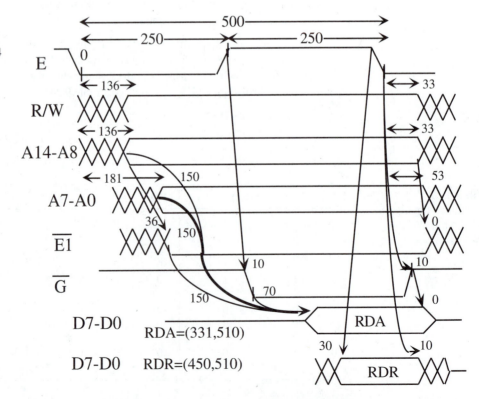

To verify proper write cycle timing we draw the *Combined write cycle timing diagram* (Figure 9.59). Notice that write data available overlaps write data required.

Figure 9.59
Write timing diagram of
the 6811 and the 60L64
8K RAM chip.

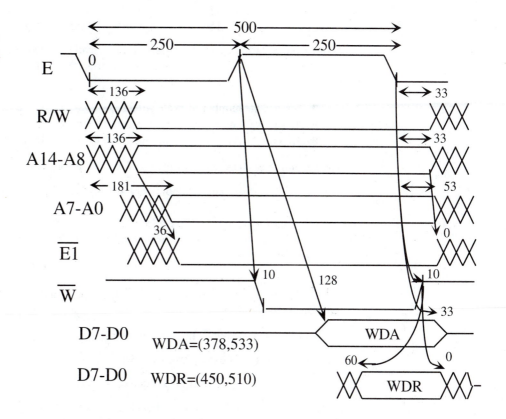

| 9.7.2.2 | To simplify the RAM interface to the MC68HC812A4, we will operate in expanded narrow |

9.7.2.2
8K RAM Interface to a
MC68HC812A4

To simplify the RAM interface to the MC68HC812A4, we will operate in expanded narrow mode and connect $\overline{E1}$ to the built-in CSD. We will have to configure CSD for the 32K bytes space $0000 to $7FFF. This will greatly simplify the interface, and we will still be able to add other external devices using other chip selects. Another simplification we will make is to ground $\overline{G}$. Because $\overline{E1}$ is synchronized to CSD, both the read and write operations will be synchronized. Just like the 32K PROM interface to the 6812 earlier, we must select the proper amount of cycle stretching. Recall that for a MC68HC812A4 running at 8 MHz,

$$AA = (AdV, AdN) = (t_5, t_1 + t_6) = (60, t_1 + 20)$$

Note that since $\overline{G}$ is grounded, we will neglect its timing constraints. Since $\overline{E1}$ is connected to CSD, $\downarrow\overline{E1}$ will occur at 60 ns, and $\uparrow\overline{E1}$ will occur at $t_1 + 10$. Entering this information into the memory timing,

$$\text{Read data available} = (\text{later} (AdV + t_{AVQV}, \downarrow\overline{E1} + t_{E1LQV}), \text{earlier} (AdN + t_{AXQX}, \uparrow\overline{E1} + t_{E1HQZ}))$$
$$= (\text{later} (60 + 150, 60 + 150), \text{earlier} (t_1 + 20 + 20, t_1 + 10))$$

During a read cycle the data are required by the 6812. Thus to determine the read data required interval, we look in the 6812 data sheet. For read data required, the worst case is the longest interval. Recall that

$$RDR = \text{read data required} = (t_1 - t_{11}, t_1 + t_{12}) = (t_1 - 30, t_1 + 0)$$

We must choose the number of cycle stretches to make the read data available interval overlap the read data required interval. We must make $\overline{W}$ =1 during a read cycle. Thus,

Address	$60 + 150$	$\leq t_1 - 30$	and	$t_1 + 40$	$\geq t_1$
E1	$60 + 150$	$\leq t_1 - 30$	and	$t_1 + 10$	$\geq t_1$

First column ($240 \leq t_1$) tells us that one extra cycle is needed to stretch the access time to 250 ns. If the read data available did not overlap the read data required, then one would have to:

- Increase the number of cycle stretches
- Slow down the 6812 by increasing the E period
- Decrease t_{E1LQV} by spending more money on a faster RAM chip

The beginning of RDA is determined by $\downarrow\overline{E1}$ and the end by $\uparrow\overline{E1}$:

$$\text{Read data available} = (\downarrow\overline{E1} + 150, \uparrow\overline{E1}) = (210, 260)$$

> **Observation:** You do not have to synchronize driving the data bus on a nonmultiplexed address/data computer like the MC68HC812A4, because the data bus is dedicated at all times.

Data are written into the memory when $\overline{E1}$ =0, and $\overline{W}$ =0. During a write cycle, the data are supplied by the 6812. For write data available, the worst case is the shortest interval. Recall that

$$\text{WDA} = \text{write data available} = (t_2 + t_{13}, t_1 + t_{14}) = (106, t_1 + 20)$$

Since we have chosen to synchronize $\overline{E1}$,

$$\text{Write data required} = (\uparrow\overline{E1} - 60, \uparrow\overline{E1})$$

where $\uparrow\overline{E1}$ is $t_1 + 10$. The number of cycle stretches will also affect whether or not the write data available interval overlaps the write data required interval.

$$106 \leq t_1 + 10 - 60 \qquad \text{and} \qquad t_1 + 10 \leq t_1 + 20$$

Thus,

$$156 \leq t_1$$

Therefore the one-cycle stretch needed for the read cycle also is sufficient for the write cycle.

> **Observation:** In most situations the read cycle timing is more critical than the write cycle timing.

> **Common Error:** If you do not synchronize clocking data into a memory during a write cycle, then data might not be properly stored.

We combine the command and timing aspects of the design to find the Boolean expression that will generate the control signals: $\overline{E1}$, $\overline{G}$, and $\overline{W}$ (Table 9.26).

$$\overline{E1} = \text{CSD}$$
$$\overline{G} = 0$$
$$\overline{W} = \text{R/W}$$

Table 9.26
Combined timing table for the MCM60L64 RAM interface.

E	R/W	CSD	RD	WR	$\overline{E1}$	$\overline{G}$	$\overline{W}$	
0	0	1	0	0	1	X	X	Address not on the chip
1	0	1	0	0	1	X	X	
0	1	1	0	0	1	X	X	Address not on the chip
1	1	1	0	0	1	X	X	
0	0	1	0	0	1	X	X	Write cycle
1	0	0	0	1	0	X	0	
0	1	1	0	0	1	X	1	Read cycle
1	1	0	1	0	0	0	1	

The next step is to build the interface (Figure 9.60). To verify proper read cycle timing we draw the *Combined read cycle timing diagram* (Figure 9.61). Notice that read data available overlaps read data required, although it is kind of close, and you may elect to be more

Figure 9.60
Interface between the MC68HC812A4 and the 60L64 8K RAM chip.

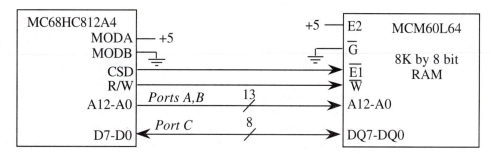

Figure 9.61
Read timing of the MC68HC812A4 and the 60L64 8K RAM chip.

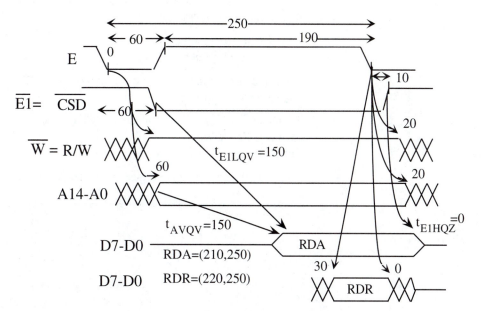

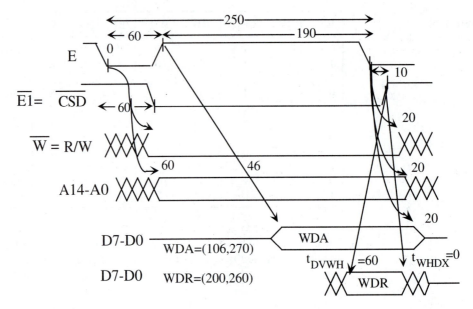

Figure 9.62
Write timing of the
MC68HC812A4 and the
60L64 8K RAM chip.

conservative and increase the cycle stretching to two cycles. To verify proper write cycle timing we draw the *Combined write cycle timing diagram* (Figure 9.62).

Notice that write data available overlaps write data required. One of the difficulties in interfacing additional RAM to a MC68HC812A4 is how to configure the system out of reset. If the above circuit is constructed, the system will come out of reset in expanded narrow mode (MODA=1, MODB=0). Unfortunately, the built-in 4K EEPROM is mapped to $1000 to $1FFF in expanded narrow mode. Therefore no memory exists at $FFFE, and the software cannot start. There are two solutions to this situation:

1. Include external PROM (Section 9.7.1.2) and come out of reset into expanded narrow mode and let the software:

   Enable CSD, R/W outputs
   Select one-cycle stretch on CSD

2. Come out of reset in special single-chip mode (BKGD = MODA = MODB = 0) and let the software:

   Switch over the special expanded narrow mode
   Enable CSD, R/W outputs
   Select one-cycle stretch on CSD

The software to activate this interface is shown in Program 9.2.

Program 9.2 Software
to configure the chip
select for the external
RAM.

```
void RAMinit(void){
        MODE=0x3B                          // special expanded narrow mode
        PEAR=0x2C;                         // enable E, R/W, LSTRB(not needed)
        WINDEF=WINDEF&0x7F;               // disable DPAGE
        CSCTL0=CSCTL0|0x10;               // enable CSD
        CSCTL1=CSCTL1|0x10;               // CSD $0000 to $7FFF
        CSSTR0=(CSSTR0&0xFC)|0x01;}       // 1 cycle stretch on CSD
```

9.7.2.3
8K RAM Interface to a
MC68HC912B32

To simplify the RAM interface to the MC68HC912B32, we will operate in expanded narrow mode. Since the B32 has no built-in chip selects, we will have to design our own for the 8K space $6000 to $7FFF. A 74AHC138 3 to 8 decoder could be used to generate the chip selects for multiple 8K devices in a single system. To prevent the memory from driving the data bus too early, we will connect $\overline{G}$ to DBE. In this way, the memory data output will not collide with the microcomputer address output during a read cycle. Just like the MC68HC812A4 interfaces earlier, we must select the proper amount of cycle stretching. Recall that for a MC68HC912B32 running at 8 MHz,

$$\text{AA} = (\text{AdV}, \text{AdN}) = (t_2 + t_9, t_1 + t_2) = (69, t_1 + 60)$$

All three control signals are negative logic, and we must synchronize either E2 or $\overline{\text{E1}}$ or both $\overline{G}$ and $\overline{W}$. We choose to make

E2	positive logic synchronized, connected directly to E
$\overline{\text{E1}}$	negative logic unsynchronized from the address lines A15, A14, A13
$\overline{G}$	negative logic synchronized, connected directly to DBE
$\overline{W}$	negative logic unsynchronized, connected directly to R/W

DBE, a synchronized control signal, falls at 97 ns and rises at $t_1 + 10$. Assuming 9-ns AHC gate delay max and 0 ns delay minimum, $\overline{\text{E1}}$ falls at 78 ns and rises at $t_1 + 60$. Entering this information into the memory timing:

Read data available $= (\text{later }(\text{AdV} + t_{AVQV}, \downarrow\overline{\text{E1}} + t_{E1LQV}, \uparrow\text{E2} + t_{E2HQV}, \downarrow\overline{G} + t_{GLQV}),$
earlier $(\text{AdN} + t_{AXQX}, \uparrow\overline{\text{E1}} + t_{E1HQZ}, \downarrow\text{E2} + t_{E2LQZ}, \uparrow\overline{G} + t_{GHQZ}))$

$= (\text{later }(69 + 150, 78 + 150, 60 + 150, 97 + 70), \text{ earlier }(t_1 + 60 + 20, t_1 + 60, t_1, t_1 + 10))$

During a read cycle the data are required by the 6812. Thus to determine the read data required interval, we look in the 6812 data sheet. For read data required, the worst case is the longest interval. Recall that

$$\text{RDR} = \text{read data required} = (t_1 - t_{11}, t_1 + t_{12}) = (t_1 - 30, t_1)$$

We must choose the number of cycle stretches to make the read data available interval overlap the read data required interval. We must make $\overline{W} = 1$ during a read cycle. Thus,

Address	$69 + 150$	$\leq t_1 - 30$	and	$t_1 + 80$	$\geq t_1$
$\overline{\text{E1}}$	$78 + 150$	$\leq t_1 - 30$	and	$t_1 + 60$	$\geq t_1$
E2	$60 + 150$	$\leq t_1 - 30$	and	t_1	$\geq t_1$
$\overline{G}$	$97 + 70$	$\leq t_1 - 30$	and	$t_1 + 10$	$\geq t_1$

The $\overline{\text{E1}}$ timing ($258 \leq t_1$) tells us that two extra cycles are needed to stretch the access time to 375 ns. If the read data available did not overlap the read data required, then one would have to:

- Increase the number of cycle stretches
- Slow down the 6812 by increasing the E period
- Decrease t_{E1LQV} by spending more money on a faster RAM chip

The beginning of RDA is determined by $\downarrow\overline{\text{E1}}$ and the end by $\downarrow\text{E2}$:

$$\text{Read data available} = (\downarrow\overline{\text{E1}} + 150, \downarrow\text{E2}) = (228, 375).$$

Common Error: If you do not synchronize driving the data bus on a multiplexed address/data computer like the 6811 and MC68HC912B32, then memory data might be driven onto the bus during the first half of the cycle and conflict with the address being driven at that time.

Data are written into the memory when $\overline{E1} = 0$, and $\overline{W} = 0$. During a write cycle, the data are supplied by the 6812. For write data available, the worst case is the shortest interval. Recall that

$$WDA = \text{write data available} = (t_2 + t_{13},\ t_1 + t_{14}) = (106,\ t_1 + 20)$$

Since we have chosen to synchronize E2 to the E clock,

$$\text{Write data required} = (\downarrow E2 - 60,\ \downarrow E2)$$

The number of cycle stretches will also affect whether or not the write data available interval overlaps the write data required interval. The worst-case delay selects the largest WDR interval:

$$106 \leq t_1 - 60 \qquad \text{and} \qquad t_1 \leq t_1 + 20$$

Thus,

$$166 \leq t_1$$

Therefore the two-cycle stretch needed for the read cycle also is sufficient for the write cycle. We cannot have a different number of stretches for the read and write cycles.

Common Error: If you do not synchronize clocking data into a memory during a write cycle, then data might not be properly stored.

We combine the command, and timing aspects of the design to find the Boolean expression that will generate the control signals: E2, $\overline{E1}$, $\overline{G}$, and $\overline{W}$ (Table 9.27).

$$
\begin{aligned}
E2 &= E \\
\overline{E1} &= \overline{\text{Select}} \\
\overline{G} &= DBE \\
\overline{W} &= R/W
\end{aligned}
$$

E	DBE	R/W	Select	RD	WR	E2	E1	G	W	
0	1	0	1	0	0	X	1	X	X	Address not on the chip
1	1	0	1	0	0	X	1	X	X	
0	1	1	1	0	0	X	1	X	X	Address not on the chip
1	0	1	1	0	0	X	1	X	X	
0	1	0	0	0	0	0	0	X	0	Write cycle
1	1	0	0	0	1	1	0	X	0	
0	1	1	0	0	0	0	0	1	1	Read cycle
1	0	1	0	1	0	1	0	0	1	

Table 9.27 Combined timing table for the MCM60L64 RAM interface.

The next step is to build the interface (Figure 9.63).

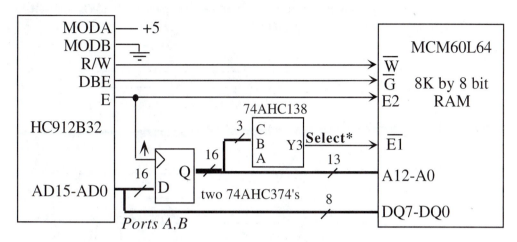

Figure 9.63   Interface between the MC68HC912B32 and the 60L64 8K RAM chip.

To verify proper read cycle timing we draw the *Combined read cycle timing diagram* (Figure 9.64). Notice that read data available overlaps read data required, although it is kind of close, and you may elect to be more conservative and increase the cycle stretching to two cycles.

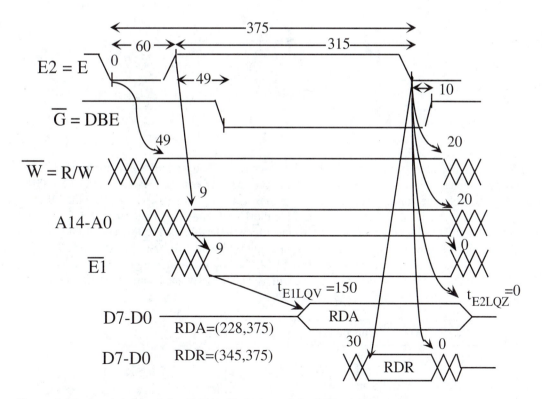

Figure 9.64   Read timing of the MC68HC912B32 and the 60L64 8K RAM chip.

To verify proper write cycle timing we draw the *Combined write cycle timing diagram* (Figure 9.65). Notice that write data available overlaps write data required. The software to activate this interface is shown in Program 9.3.

Figure 9.65
Write timing of the MC68HC912B32 and the 60L64 8K RAM chip.

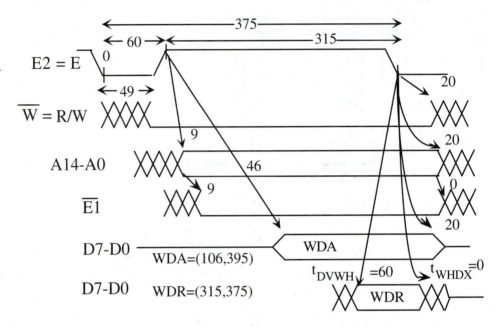

```
void RAMinit(void){
    MODE=0x39                    // special expanded narrow mode
    MISC=(MISC&0xF3)|0x08;       // 2 cycle stretch on external
    PEAR=0x2C;}                  // enable DBE, E, R/W, LSTRB(not needed)
```

Program 9.3 Software to configure the mode for the external RAM.

9.7.3
32K by 16-bit PROM Interface to a MC68HC812A4

The objective of this example is to interface two 16K by 8-bit PROMs to the 6812 in expanded wide mode. We will place the two 27C128-90 PROMs at locations $8000 to $FFFF. The **command portion** determines which CPU cycles will activate the PROMs. The two 27C128-90 PROMs will drive the 16-bit data bus whenever both $\overline{CE}$ and $\overline{OE}$ are 0. The other three values of $\overline{CE}$ and $\overline{OE}$ will disable the PROM. The PROM chip requires less power (I_{cc} is the +5 V supply current) whenever $\overline{CE}$ is high (Table 9.28).

Using full decoding the address **SELECT** line in positive logic would be **SELECT** = A15. To simplify the interface, we will neglect the R/W and assume the software will not

Table 9.28
Function table for the 27C128 PROM.

$\overline{CE}$	$\overline{OE}$	function	I_{cc}(mA)
1	1	Off	1
1	0	Off	1
0	1	Off	35
0	0	Drive data bus	35

attempt to write into this ROM. To conserve power (lower I_{cc}), then the $\overline{CE} = 1$ state is used to deselect the PROM (Table 9.29). The **timing portion** of the interface determines the time during an active cycle when a control signal should rise and fall. The objective is to design the interface such that the data available interval overlaps the data required interval (Figure 9.66).

Table 9.29
Command table for the 27C128 PROM interface.

SELECT	R/W	Function	$\overline{CE}$	$\overline{OE}$
0	0	Off	1	X
0	1	Off	1	X
1	0	Don't care	X	X
1	1	Drive data bus	0	0

Figure 9.66
Read timing for the 27C128-90 16K PROM chip.

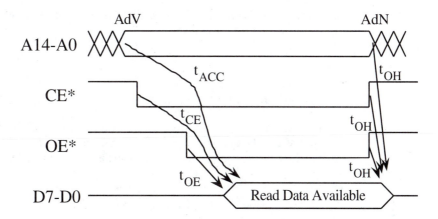

During a read cycle, the data are supplied by the 27C128-90. Thus, to determine the data available interval we look at the 27C128-90 data sheet. We will specify the worst-case timing. For data available, the worst case is the shortest interval.

$$\text{Read data available} = (\text{later } (\text{AdV} + t_{ACC}, \downarrow\overline{CE} + t_{CE}, \downarrow\overline{OE} + t_{OE}),$$
$$\text{earlier } (\text{AdN} + t_{OH}, \uparrow\overline{CE} + t_{OH}, \uparrow\overline{OE} + t_{OH}))$$

To conserve power (minimize I_{cc}), we let $\overline{CE}$ be 0 only during read cycles to \$8000 to \$FFFF. If we simply grounded $\overline{CE}$, the interface would be faster but would require more power. From the Fairchild 27C128-90 data sheet $t_{ACC}=90$, $t_{CE}=90$ $t_{OE}=50$, $t_{OH}=0$.

Two of the issues when interfacing to a 6812 are to determine data bus width (narrow or wide) and the number of cycles to stretch. For increased speed, we will utilize 16-bit wide mode. To stretch the E clock we will have to use one of the seven built-in chip selects (e.g., either CSP0 or CSP1). For now, we will develop the timing equations, and then we will use them to determine the cycle stretching requirements. The AdV time is 60 ns, independent of stretching, while AdN will be $t_1 + 20$ ns. We therefore calculate

$$\text{Read data available} = (\text{later } (60 + t_{ACC}, \downarrow\overline{CE} + t_{CE}, \downarrow\overline{OE} + t_{OE}),$$
$$\text{earlier } (\text{AdN} + t_{OH}, \uparrow\overline{CE} + t_{OH}, \uparrow\overline{OE} + t_{OH}))$$

$$= (\text{later } (150, \downarrow\overline{CE} + 90, \downarrow\overline{OE} + 50),$$
$$\text{earlier } (t_1 + 20 + 0, \uparrow\overline{CE} + 0, \uparrow\overline{OE} + 0))$$

During a read cycle the data are required by the 6812.

$$RDR = \text{read data required} = (t_1 - t_{11}, t_1 + t_{12}) = (t_1 - 30, t_1)$$

We must choose the timing of the fall and rise of both $\overline{CE}$ and $\overline{OE}$ to make the data available interval overlap the data required window. Note that since $\overline{CE}$ and $\overline{OE}$ are active low, they will fall, then rise, during the active cycles. Thus,

150	$\le t_1 - 30$	and	$t_1 + 20$	$\ge t_1$
$\downarrow \overline{CE} + 90$	$\le t_1 - 30$	and	$\uparrow \overline{CE}$	$\ge t_1$
$\downarrow \overline{OE} + 50$	$\le t_1 - 30$	and	$\uparrow \overline{OE}$	$\ge t_1$

First line ($180 \le t_1$) tells us that one extra cycle is needed to stretch the access time to 250 ns. The following signal is constructed inside the PROM chip. It will be cheaper to synchronize $\overline{CE}$ by connecting it directly to CSP0. This means $\overline{CE}$ will be controlled by timing signal CSP0. The fall of $\overline{CE}$ will be at 60 ns and, the rise of $\overline{CE}$ will be $t_1 + 10$ ns. The CE access time must be satisfied:

$$\downarrow \overline{CE} + 90 \le t_1 - 30 \qquad\qquad \text{and} \qquad\qquad \uparrow \overline{CE} \ge t_1$$

or

$$60 + 90 \le t_1 - 30 \qquad\qquad \text{and} \qquad\qquad t_1 + 10 \ge t_1$$

Again one extra cycle is needed to stretch the access time to 250 ns. In this interface we will ground $\overline{OE}$, so the timing delays with $\overline{OE}$ need not be calculated. Now that we have decided to use one-cycle stretching, we can calculate the actual RDA and RDR intervals.

$$\begin{aligned}\text{Read data available} &= (150, 260) \\ \text{Read data required} &= (220, 250)\end{aligned}$$

Next we find the Boolean expressions that will generate $\overline{CE}$ and $\overline{OE}$ from E, CSP0, and R/W. In this simple case, we connected $\overline{CE}$ to CSP0 and $\overline{OE}$ to ground (Figure 9.67). The MC68HC812A4 will come out of reset in expanded mode, with CSP0 enabled having three-cycle stretching. The reset software should enable any additional chip selects and change the cycle stretching on CSP0 to one cycle. Last, we draw the *Combined read cycle timing diagram* and use it to verify that read data available overlaps read data required (Figure 9.68). The software to activate this interface is shown in Program 9.4.

Figure 9.67 Interface between the MC68HC812A4 and the 27C128-90 16K PROM chip.

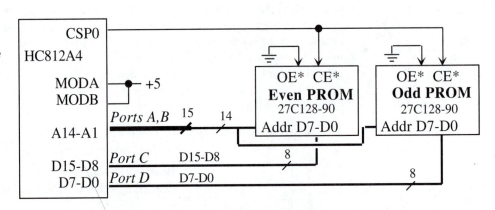

Figure 9.68
Read timing of the
MC68HC812A4 and the
27C128-90 16K PROM
chip.

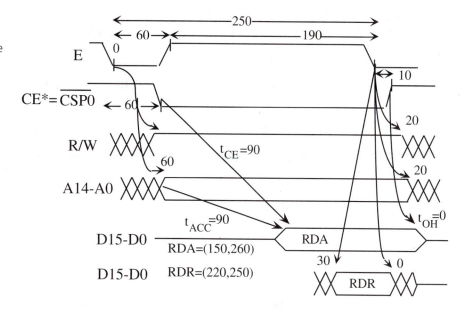

Program 9.4 Software to
configure the chip select
for the external PROM.

```
void PROMinit(void){
    CSSTR0=(CSSTR0&0xF3)|0x04;} // 1 cycle stretch on CSP0
```

9.7.4
8K by 16-bit RAM
Interface

In this example we interface an 8K by 16-bit static RAM to the 6812 using two Motorola MCM60L64s. To enhance the execution speed, we will operate in expanded wide mode. The memory will be placed at $4000 to $7FFF.

9.7.4.1
8K by 16-bit RAM
Interface to the
MC68HC812A4

The MC68HC812A4 interface will utilize the built-in address decoder CSD (0000 to $7FFF). The operation and timing of the MCM60L64 was presented earlier in Section 9.7.2. The solution uses 13 address lines (A13–A1) from the 6812. The 74AHC32 gates create chip selects that handle the situation where 8-bit data are written to this 16-bit RAM (Figure 9.69).

Figure 9.69
Interface between the
MC68HC812A4 and the
MCM60L64 RAM.

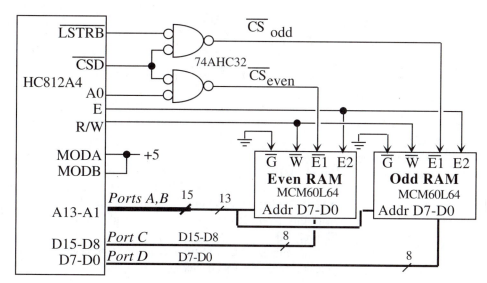

The timing considerations are almost identical to the 8-bit interface. For write data available, the worst case is the shortest interval. Recall that

$$WDA = \text{write data available} = (t_2 + t_{13,} \, t_1 + t_{14}) = (106, 270)$$

In the 8-bit interface, we connected $\overline{E1}$ directly to CSD, where the $\uparrow$CSD occurred at $t_1 + 10 = 260$ ns, making

$$\text{Write data required} = (\uparrow\overline{E1} - 60, \uparrow\overline{E1})$$

In this 16-bit interface, the 74AHC32 will delay the rising edge of $\overline{E1}$. If the gate delay is less than 10 ns, the write data available will overlap the write data required. Just to be safe, the E clock is added to the E2 RAM control signal, so the write data required interval (now triggered on the fall of E2) is not a function of the gate delay through the 74AHC32.

$$\text{Write data required} = (\downarrow E2 - 60 , \downarrow E2)$$

The addition of the E2 connection to the 6812 E clock does not affect the read cycle timing.

$$\text{Read data available} = (\text{later } (AdV + t_{AVQV}, \downarrow\overline{E1} + t_{E1LQV}, \uparrow E2 + t_{E2HQV}),$$
$$\text{earlier } (AdN + t_{AXQX}, \uparrow\overline{E1} + t_{E1HQZ}, \downarrow E2 + t_{E2LQZ}))$$

$$= (\text{later } (60 + 150 , 60 + 150, 60 + 150) ,$$
$$\text{earlier } (270 + 20, 250 + 0 , 250 + 0) = (210, 250))$$

With one-cycle stretching, recall that

$$RDR = \text{read data required} = (t_1 - t_{11,} \, t_1 + t_{12}) = (220, 250)$$

The initialization software must enable E, LSTRB, CSD, R/W outputs and select one-cycle stretch on CSD (Program 9.5).

Program 9.5 Software to configure the chip select for the external RAM.

```
void RAMinit(void){
    MODE=0x7B                     // special expanded wide mode
    PEAR=0x2C;                    // enable E, R/W, LSTRB
    WINDEF=WINDEF&0x7F;           // disable DPAGE
    CSCTL0=CSCTL0|0x10;           // enable CSD
    CSCTL1=CSCTL1|0x10;           // CSD $0000 to $7FFF
    CSSTR0=(CSSTR0&0xFC)|0x01;}   // 1 cycle stretch on CSD
```

**9.7.4.2
8K by 16-bit RAM
Interface to the
MC68HC912B32**

Using full decoding, the address **Select*** line in negative logic is

$$Select^* = \overline{A15} \cdot A14$$

The MC68HC912B32 interface will require this external address decoder. The operation and timing of the MCM60L64 was presented earlier in Section 9.7.2. The 74AHC32 gates create chip selects that handle the situation where 8-bit data are written to this 16-bit RAM. The timing considerations are identical to the 8-bit interface (Figure 9.70).

The initialization software must enable E, LSTRB, DBE, R/W outputs and select two-cycle stretch on external devices (Program 9.6).

Figure 9.70
Interface between the
MC68HC912B32 and
the MCM60L64 RAM.

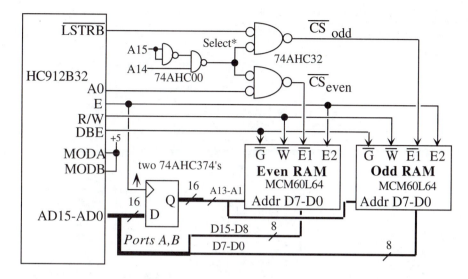

Program 9.6 Software
to configure the mode
for the external RAM.

```
void RAMinit(void){
    MODE=0x79                  // special expanded wide mode
    MISC=(MISC&0xF3)|0x08;     // 2 cycle stretch on external
    PEAR=0x2C;}                // enable DBE, E, R/W, LSTRB
```

9.7.5
Extended Address
Data Page
Interface to the
MC68HC812A4

One of the unique features of the MC68HC812A4 is its ability to interface large RAM and ROM using the data page and program page memory, respectively. Up to 1 Mbytes can be configured using the data page system. In this example, two 128K by 8-bit RAM chips are interfaced to the 6812 using 18 address pins (A17–A0). The 628128 is a 128K by 8-bit static RAM similar to the 60L64 static RAM presented earlier. When the paged memory is enabled, Port G will contain address lines A21–A16 (although the data page system can only use up to A19). The built-in address decoder CSD must be used with the data page system. The hardware circuit and timing equations are quite similar to the 8K by 16 bit RAM interfaced earlier. The only hardware difference is that 18 address lines (A17–A0) are needed instead of the 14 (A13–A0) (Figure 9.71).

Figure 9.71
Interface between the
MC68HC812A4 and the
628128 256K RAM.

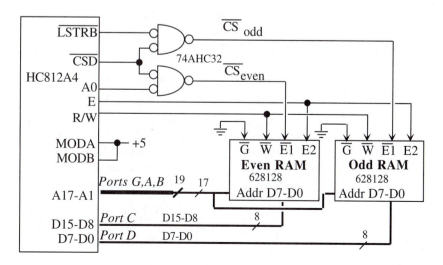

Program 9.7 Software
to configure the mode
for the extended RAM.

```
void RAMinit(void){
        MODE=0x7B                        // special expanded wide mode
        PEAR=0x2C;                       // enable E, R/W, LSTRB
        WINDEF=WINDEF|0x80;              // enable DPAGE
        MXAR=0x03;                       // enable A17, A16 on Port G
        CSCTL0=CSCTL0|0x10;              // enable CSD
        CSCTL1=CSCTL1&0xEF;              // CSD $7000 to $7FFF
        CSSTR0=(CSSTR0&0xFC)|0x01;}      // 1 cycle stretch on CSD
```

```
struct addr20
{   unsigned char msb;      // bits 19-12, only 17-12 used in this interface
    unsigned int lsw;       // bits 11-0
};
typedef struct addr20 addr20Type;
char ReadMem(addr20Type addr){ char *pt;
        DPAGE=addr.msb;                      // set address bits 19-12, only 17-12 used
        pt=(char *)(0x7000+addr.lsw); // set address bits 11-0
        return *pt;}                    // read access
void WriteMem(addr20Type addr, char data){ char *pt;
        DPAGE=addr.msb;                      // set address bits 19-12, only 17-12 used
        pt=(char *)(0x7000+addr.lsw); // set address bits 11-0
        *pt=data;}                          // write access
```

Program 9.8 Method for accessing extended RAM.

We divide the software into initialization (Program 9.7) and access (Program 9.8). The initialization software performs the usual steps:

Enable E, LSTRB, CSD, R/W outputs
Clear bit 4 in the CSCTL1 to set up CSD for the $7000–$7FFF range
Select one-cycle stretch on CSD

To enable the data page system we must also:

Set bit 7 in WINDEF to enable the Data Page Window
Set bits 1, 0 in MXAR to enable memory expansion pins A17–A16

Let A17–A0 be the desired 256K RAM location. To access that location requires two steps:

Set the most significant addresses A17–A12[1] into DPAGE register
Access $7000 to $7FFF, with the least significant addresses A11–A0

When MXAR is active, the MC68HC812A4 will convert all 16 addresses to the extended 22-bit addresses. When an access is outside the range of any active page window (EPAGE, DPAGE, or PPAGE), the upper 6 bits are 1. In Table 9.30 it is assumed that only the DPAGE is active, the DPAGE register contains DP7–DP0, and MXAR is 0x3F (activating all 22 address bits). From this table, we can see another trick when using paged

---

[1]With a 1 Mbyte RAM, you would set A19–A12 into DPAGE.

Internal	A21	A20	A19	A18	A17	A16	A15	A14	A13	A12	A11–A0
$0xxx	1	1	1	1	1	1	0	0	0	0	xxx
$1xxx	1	1	1	1	1	1	0	0	0	1	xxx
$2xxx	1	1	1	1	1	1	0	0	1	0	xxx
$3xxx	1	1	1	1	1	1	0	0	1	1	xxx
$4xxx	1	1	1	1	1	1	0	1	0	0	xxx
$5xxx	1	1	1	1	1	1	0	1	0	1	xxx
$6xxx	1	1	1	1	1	1	0	1	1	0	xxx
$7xxx	1	1	DP7	DP6	DP5	DP4	DP3	DP2	DP1	DP0	xxx

Table 9.30 Extended addressing using DPAGE.

memory. If we were to switch the initialization so that CSD activated on $0000–$7FFF, then seven 4K pages would overlap the regular address range $0000 to $6FFF. In this particular system, there are 64 data pages, numbered $00 to $3F. Notice that page numbers $30 through $36 overlap with regular data space $0000 to $6FFF. If we avoid using data pages $30–$36, we would have 28K of regular RAM from $0000 to $6FFF plus 57 4K windows of data paged memory space.

9.7.6
Extended Address
Program Page
Interface to the
MC68HC812A4

Up to 4 Mbytes can be configured using the program page system. In this example, two 1 Mbyte by 8-bit PROM chips are interfaced to the 6812 using 21 address pins (A20–A0). The 27C080-12 is a 1 Mbyte by 8-bit PROM similar to the 27C128-90 PROM presented earlier. Each 27C080-12 has 20 address lines, eight data lines, and a 120-ns access time. When the paged memory is enabled, Port G will contain address lines A21–A16 (although this program page system will only use up to A20). The built-in address decoders CSP0 or CSP1 must be used with the program page system. We will use CSP0 so that part of the PROM is available immediately after reset. The hardware circuit and timing equations are quite similar to the 32K by 16 RAM interfaced earlier. The only hardware difference is that 21 address lines (A20–A0) are needed instead of the 15 (A14–A0) (Figure 9.72).

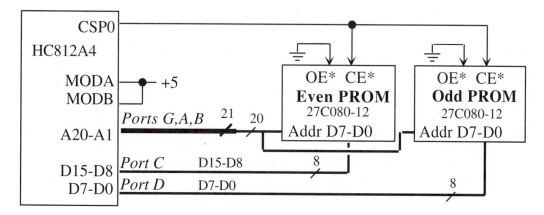

Figure 9.72   Interface between the MC68HC812A4 and the 27C080 1M PROM.

After a reset in expanded wide mode:

Ports A, B contain the address lines A15–A0
CSP0 is enabled for the $8000–$FFFF range (CSCTL0=$20)
There are three-cycle stretches on CSP0
The extended memory address lines are inactive (MXAR=0)
Port G is an input port (DDRG=0) but has pull-ups, so A21–A16 will be high

To enable the program page system we must also (Program 9.9):

Set bit 6 in WINDEF to enable the Program Page Window
Set bits 4–0 in MXAR to enable memory expansion pins A20–A16
Reduce cycle stretch to one because of the 120-ns access time

Program 9.9 Software to configure the mode for the extended PROM.

```
void PROMinit(void){
     MODE=0x7B                          // special expanded wide mode
     PEAR=0x2C;                         // enable E, R/W, LSTRB (none needed)
     WINDEF=WINDEF|0x40;               // enable PPAGE
     MXAR=0x01F;                        // enable A20-A16 on Port G
     CSSTR0=(CSSTR0&0xF3)|0x04;}       // 1 cycle stretch on CSP0
```

Let A20–A0 be the desired 2-Mbyte PROM location. To access that location requires two steps (Program 9.10):

Set the most significant addresses A20–A14[2] into PPAGE register
Access $8000 to $BFFF, with the least significant addresses A13–A0

```
struct addr22
{   unsigned char msb;     // bits 21-14, only 20-14 used in this interface
    unsigned int lsw;      // bits 13-0
};
typedef struct addr22 addr22Type;
char ReadPROM(addr22Type addr){ char *pt;
     PPAGE=addr.msb;                   // set address bits 21-14
     pt=(char *)(0x8000+addr.lsw);  // set address bits 13-0
     return *pt;}                      // read access
```

Program 9.10 Method for accessing extended PROM.

When MXAR is active, the MC68HC812A4 will convert all 16 addresses to the extended 22-bit addresses. When an access is outside the range of any active page window (EPAGE, DPAGE, or PPAGE), the upper 6 bits are 1. In Table 9.31 it is assumed that only the PPAGE is active, the PPAGE register contains PP7–PP0, and MXAR is 0x3F (activating all 22 address bits).

---

[2]With a 4 Mbyte PROM, you would set A21–A14 into PPAGE.

Internal	A21	A20	A19	A18	A17	A16	A15	A14	A13	A12	A11–A0
$8xxx	PP7	PP6	PP5	PP4	PP3	PP2	PP1	PP0	0	0	xxx
$9xxx	PP7	PP6	PP5	PP4	PP3	PP2	PP1	PP0	0	1	xxx
$Axxx	PP7	PP6	PP5	PP4	PP3	PP2	PP1	PP0	1	0	xxx
$Bxxx	PP7	PP6	PP5	PP4	PP3	PP2	PP1	PP0	1	1	xxx
$Cxxx	1	1	1	1	1	1	1	1	0	0	xxx
$Dxxx	1	1	1	1	1	1	1	1	0	1	xxx
$Exxx	1	1	1	1	1	1	1	1	1	0	xxx
$Fxxx	1	1	1	1	1	1	1	1	1	1	xxx

Table 9.31 Extended addressing using PPAGE.

From this table, we can see another trick when using paged memory. Notice that CSP0 activates on $8000–$FFFF. For the range $8000 to $BFFF, the program page system is activated, but for the range $C000-$FFFF, the CSP0 activates without utilizing the program page system. In this particular system, there are 128 program pages, numbered $00 to $7F. Notice that page number $7F overlaps with regular space $C000 to $FFFF. If we avoid using program page $7F, we would have 16K of regular PROM from $C000 to $FFFF plus 127 16K windows of paged program memory space. Also notice that this regular PROM is available immediately after reset and should contain the reset vectors and reset handler.

This extended memory system can be used to hold programs or fixed data. If it were constructed with a battery-backed nonvolatile static RAM (SRAM) like the following from Dallas Semiconductor,

DS1220Y-150	150 ns 16K
DS1225AD-150	150 ns 64K
DS1230Y-85	85 ns 256K

then the extended memory system could be used for programs and long-term (10-year) storage. When the software wishes to call a function that exists within the program paged system, it uses the `call` op code. This instruction pushes the old PPAGE and PC on the stack and loads a new PPAGE and PC as needed. The `rtc` instruction pulls the old PC and PPAGE from the stack. The effective address of the call instruction includes the most significant byte (8 bits), least significant word (14 bits) components similar to the above C programs. One of the most flexible ways to implement the program division is to use a jump table scheme and write relocatable code (Program 9.11). Notice that the software can use `call` to execute a function within the same page. Each of the following columns can be assembled separately, using the common `equ` definitions. For example,

```
fun1 equ 0
fun2 equ 3
fun3 equ 6
```

Observation: It is important to always use `call` to execute a function that ends with `rtc`.

Observation: It is important to always use `bsr` or `jsr` to execute a function that ends with `rts`.

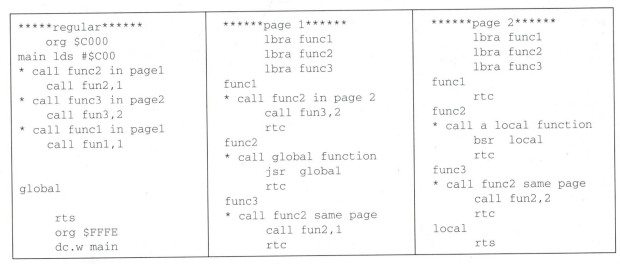

```
*****regular*****          *****page 1*****          *****page 2*****
    org $C000                  lbra func1                lbra func1
main lds #$C00                 lbra func2                lbra func2
* call func2 in page1          lbra func3                lbra func3
    call fun2,1            func1                     func1
* call func3 in page2     * call func2 in page 2        rtc
    call fun3,2               call fun3,2            func2
* call func1 in page1         rtc                   * call a local function
    call fun1,1           func2                         bsr  local
                          * call global function        rtc
                             jsr  global            func3
global                       rtc                    * call func2 same page
                          func3                         call fun2,2
    rts                   * call func2 same page         rtc
    org $FFFE                call fun2,1            local
    dc.w main                rtc                        rts
```

Program 9.11 Method for calling functions in extended PROM.

9.8   Dynamic RAM (DRAM)

Because of their high density and low cost DRAMs are used for situations requiring a large amount of storage. Consequently embedded systems will almost exclusively employ SRAMs for their RAM needs. As the complexity of embedded system software increases, the need for RAM space will grow as well. At some point, DRAMs may become cost-effective for embedded systems. The purpose of this section is to introduce the basic principles involved in DRAMs. The information in a DRAM cell is saved as a charge on a capacitor. The information in a SRAM cell is saved as the state of a set-reset flip-flop. This reduction is size comes at the expense of needing to refresh the capacitor charges. This refresh activity adds a fixed cost. Table 9.32 compares the two memory types.

Table 9.32
Comparison between DRAM and RAM.

DRAMs	SRAMs
High density	Low density
One transistor, one capacitor/bit	Three–four transistors/bit
Slower	Faster
High fixed cost (refresh)	Low fixed cost (address decoder)
Low incremental cost	Higher incremental cost
Address multiplexing	Direct addressing

The higher fixed cost and lower incremental cost of DRAM versus SRAM means for small memory applications (like embedded systems) the SRAMs will be cost-effective. At some point as the memory size gets large enough, the DRAMs will become cost-effective.

To interface a DRAM certain support functions are required. Because the DRAM devices have so many address lines, they employ a technique called address multiplexing. This means the computer first gives half the address and toggles a row address strobe, RAS.

Then the computer gives the second half of the address and issues a column address strobe, CAS. A refresh cycle occurs when a RAS is issued without a following CAS. This will refresh the entire row. The interface logic must handle the refresh process controlling RAS and CAS. In addition, the interface must arbitrate between normal read/write and refresh cycles. This arbitration is usually handled using the partially asynchronous bus protocol shown in Figure 9.25. Because the 6811 and 6812 do not support dynamic bus cycle stretching, DRAM interfacing to these computers is difficult.

9.9 Glossary

address decoder A digital circuit having the address lines as input and a select line as output (see select signal).

asynchronous bus A communication protocol without a central clock where the data are transferred using two or three control lines implementing a handshaked interaction between the memory and the computer.

command signals lines that specify general information about the current cycle; signals that specify whether or not to activate during this cycle; the specific times for the rise and fall edges are uncertain (see timing cycles).

cycle stretch The action where some memory cycles are longer, allowing time for communication with slower memories; sometimes the memory itself requests the additional time, and sometimes the computer has a preprogrammed cycle stretch for certain memory addresses.

direct memory access (DMA) The ability to transfer data between two modules on the bus; this transfer is usually initiated by the hardware (device needs service), and the software configures the communication, but the data are transferred without explicit software action for each individual piece of data.

dynamic RAM (DRAM) Volatile read/write storage built from a capacitor and a single transistor having a low cost, but requiring refresh (see static RAM).

extended address More than the usual 16-bit address. The MC68HC812A4 allows up to 22 address lines with its paged memory modes.

fan-out The number of inputs that one output can drive when connecting devices from the same logic family.

hold time The time after the active edge of the control signal that the data must continue to be valid when storing data into a device (e.g., into the computer during a read cycle and into a memory during a write cycle) (see also setup time).

isolated I/O Configuration where the I/O devices are interfaced to the computer in a manner different than the way memories are connected; from an interfacing perspective I/O devices and memory modules have separate bus signals, from a programmer's point of view the I/O devices have their own I/O address map separate from the memory map, and I/O device access requires the use of special I/O instructions.

Karnaugh map Tabular representation of the I/O relationship for a combinational digital function; the inputs possibilities are placed in the row and column labels, and the output values are placed inside the table.

memory-mapped I/O Configuration where the I/O devices are interfaced to the computer in a manner identical to the way memories are connected; from an interfacing perspective I/O devices and memory modules share the same bus signals, from a programmer's point of view the I/O devices exist as locations in the memory map, and I/O device access can be performed using any of the memory access instructions.

negative logic A signal where the true value has a lower voltage than the false value; in digital logic true is 0 and false is 1, in TTL logic true is less than 0.7 V and false is greater than 2 V, in RS232 protocol true is -12 V and false is $+12$ V (see also positive logic).

nonvolatile Storage that does not lose its information when power is removed.

nonvolatile RAM Read/write storage that achieves its long-term storage ability because it includes a battery.

paged memory A memory organization where logic pointers (used by software) have multiple and distinct components or fields. The least significant field defines the page size. The physical memory is usually continuous, having sequential addresses. There is a dynamic address translation (logic to physical).

partially asynchronous bus A communication protocol that has a central clock, but the memory module can extend the length of a bus cycle (cycle stretch) if it needs more time.

positive logic A signal where the true value has a higher voltage than the false value; in digital logic true is 1 and false is 0, in TTL logic true is greater than 2 V and false is less than 0.7 V, in RS232 protocol true is $+12$ V and false is -12 V (see also negative logic).

read cycle Data flows from the memory or input device to the processor; the address bus specifies the memory or input device location, and the data bus contains the information at that address.

read data available Time interval (start, end) during which the data will be valid during a read cycle, determined by the memory module.

read data required Time interval (start, end) during which the data should be valid during a read cycle, determined by the processor.

R/W signal A bus signal specifying whether it is a read (R/W = 1) or a write (R/W = 0) cycle.

select signal The output of the address decoder (each module on the bus has a separate address decoder); a Boolean (true/false) signal specifying whether or not the current address of the bus matches the device address.

setup time The time before the active edge of the control signal that the data must be valid when storing data into a device (e.g., into the computer during a read cycle and into a memory during a write cycle) (see also hold time).

static RAM (SRAM) Volatile read/write storage built from three transistors having fast speed and not requiring refresh (see dynamic RAM).

synchronous bus A communication protocol that has a central clock; there is no feedback from the memory to the processor, so every memory cycle takes exactly the same time; data transfers (put data on bus, take data off bus) are synchronized to the central clock.

timing signals Lines used to clock data on or off the bus; signals that specify when to activate during this cycle; the specific times for the rise and fall edges are synchronized to the E clock (see command signals).

volatile Storage that loses its information when power is removed.

write cycle Data are sent from the processor to the memory or output device; the address bus specifies the memory or output device location, and the data bus contains the information.

write data available Time interval (start, end) during which the data will be valid during a write cycle, determined by the processor.

write data required Time interval (start, end) during which the data should be valid during a write cycle, determined by the memory module.

9.10 Exercises

9.1 The objective is to interface an 8 K by 8-bit Motorola MCM60L64 RAM to a new (fictional) microprocessor. This new microprocessor is similar to a 6811 (16-bit address, 8-bit data, synchronous bus). In other ways this microprocessor is different from a 6811 (two control signals **R** and **W** instead of the **R/W** line and **E** clock, the address bus is not time-multiplexed). There are two types of bus cycles, *Read* and *Write*. One byte is transferred from the memory to the microprocessor during a read cycle, when **R**=1 and **W**=0. See part d for the specific timing of Read Data Required. One byte is transferred from the microprocessor to the memory during a write cycle, when **R**=0 and **W**=1. See part e for the specific timing of Write Data Available.

a) Show the minimal cost address decoder for your 8K RAM at $4000 to $5FFF. Other devices are at addresses $2000 to $27FF and $FF00 to $FFFF. Choose to build either a positive logic or negative logic decoder, depending on which would be most appropriate.

b) Create a combined table that has **Select, R, W** as inputs and $\overline{E1}$, $\overline{G}$, and $\overline{W}$ as outputs. Put a bar over the top of **Select,** if you decided to build a negative logic decoder in part a. Give the equations for $\overline{E1}$, $\overline{G}$, and $\overline{W}$ as a function of **Select, R, W.**

c) Show the digital interface between the microprocessor (R, W, A15-A0, D7-D0) and the 8K RAM. Label all chip numbers, but not pin numbers.

d) Finish the **Read Cycle Timing diagram** (Figure 9.73); use arrows to show causal relationships. Show $\overline{E1}$, $\overline{G}$, $\overline{W}$, and Read Data Available. Clearly label Read Data Available, showing that it does overlap Read Data Required.

Figure 9.73
Read timing.

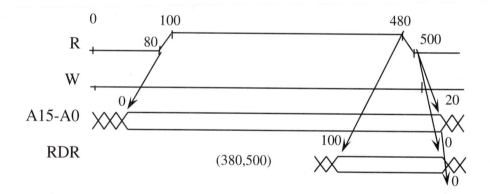

e) Finish the **Write Cycle Timing diagram,** (Figure 9.74); use arrows to show causal relationships. Show **E1, G, W,** and Write Data Required. Clearly label Write Data Required, showing that Write Data Available does overlap Write Data Required.

Figure 9.74
Write timing.

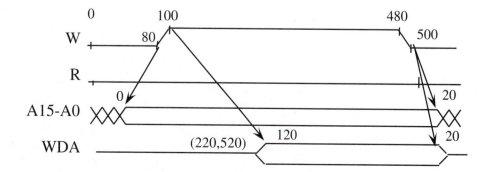

9.2 Interface a 16K by 8-bit RAM to a **6811** running at 2 MHz. The FULL address decoder is given, placing the RAM from $4000–$7FFF (SELECT= $\overline{A15}$ · A14). The address must be valid whenever the chip is enabled (**CS1**=1, $\overline{CS0}$ =0). Assume a 10-ns gate delay in each TTL chip and a 500-ns **E** clock period. The RAM has three control lines. A read (data out of the RAM) occurs on the overlap of **CS1**=1 $\overline{CS0}$ =0 and **WE** =0. The RAM read timing is shown in Figure 9.75. A write (data into the RAM) occurs on the overlap of **CS1** = 1, $\overline{CS0}$ =0 and **WE**=1. The earlier edge ↓**CS1** or ↑ $\overline{CS0}$ determines the RAM write timing. The RAM write timing is shown in Figure 9.76.

Figure 9.75
Read timing.

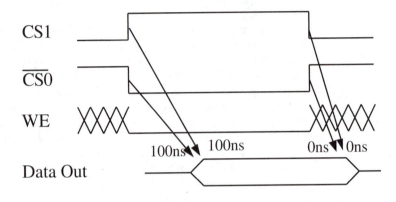

Figure 9.76
Write timing.

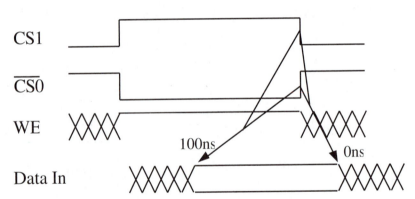

a) Either **CS1**=0 or $\overline{CS0}$ =1 will turn off the RAM. Create a status table with **Select** and **R/W** as inputs and **CS1**, $\overline{CS0}$, and **WE** as outputs.

b) What is the read data available interval? Express your answer as an equation using only the terms like AdV, AdN, ↓ $\overline{CS0}$, and ↑**WE**. Don't calculate (yet) the actual interval in nanoseconds.

c) What is the write data required interval? Express your answer as an equation using only the terms like AdV, AdN, ↓ $\overline{CS0}$, and ↑**WE**. Don't calculate (yet) the actual interval in nanoseconds.

d) Show the combined table with **Select, R/W,** and **E** as inputs and **CS1**, $\overline{CS0}$, and **WE** as outputs, and include digital equations for **CS1**, $\overline{CS0}$, and **WE**.

e) Show digital circuit for the interface between the 6811 and the RAM. Include the 74HC573 latch if and only if it is necessary for this interface. Please label TTL chip numbers but not pin numbers.

f) Show the combined **write cycle** timing diagram. All signals are outputs except Data Required. The A7–A0 signals are the ones connected to the RAM (not necessarily the AD7–AD0 of the 6811). Use arrows to signify causal relations.

9.3 Interface an AD557 8-bit DAC directly to the 6811/6812 bus (not to an output port). You will choose the fastest 6811 clock speed that allows this interface to operate properly. For the 6812, assume an 8 MHz E clock, use CS0, and choose the proper number of cycle stretches. Assume a 10-ns gate delay through each digital gate. The MINIMAL COST address decoder for the 6811 is given, placing the D/A at $7000 (SELECT= $\overline{A15}$ · A14 · A13 · A12). You *do not* design the address decoder. When both $\overline{CE}$ and $\overline{CS}$ are 0, the D/A output follows the 8-bit digital input (transparent mode). If $\overline{CE}$ rises while $\overline{CS}$ equals zero (or if $\overline{CS}$ rises while $\overline{CE}$ equals zero), the 8-bit digital inputs are latched into the chip. The setup time is 300 ns and the hold time is 10 ns. When either $\overline{CE}$ or $\overline{CS}$ is 1, the DAC ignores the input digital data and the analog output remains constant, depending on the digital input data at the time of the last rising edge (Figure 9.77).

Figure 9.77
Write timing.

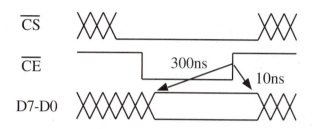

$\overline{CS}$

$\overline{CE}$ 300ns   10ns

D7-D0

a) Fill in the command table with Select and R/W as inputs and $\overline{CE}$ and $\overline{CS}$ as outputs.
b) What is the write data available interval? Express your answer as a function of the E clock period. Let t_{cyc} be the E clock period.
c) What is the write data required interval? Express your answer as an equation using only the terms like $\downarrow\overline{CS}$ and $\uparrow\overline{CE}$. Don't calculate (yet) the actual interval in nanoseconds.
d) Show the combined table with Select, R/W, and E as inputs and $\overline{CE}$ and $\overline{CS}$ as outputs. Include digital equations for $\overline{CS}$, and $\overline{CE}$.
e) Show a digital circuit for the interface between the expanded mode 6811/6812 and the AD557 8-bit DAC. Include the 74HC573 latch for the 6811 if and only if it is necessary for this interface. Please label TTL chip numbers but not pin numbers.
f) For the 6811, what is the smallest possible E clock period for this interface? For the 6812, what is the smallest number of cycle stretches? *Show your work.*

9.4 Interface a 4K by 8 bit RAM to a **6811** running at 2 MHz. The FULL address decoder is given, placing the RAM from $7000–$7FFF (SELECT= $\overline{A15}$ · **A14** · **A13** · **A12**). The RAM address must be valid whenever the chip is enabled $\overline{MEMR}$ =0 or $\overline{MEMW}$ =0. Assume a 10-ns gate delay in each TTL chip and a 500-ns **E** clock period. The RAM has two control lines. A read (data out of the RAM) occurs when $\overline{MEMR}$ =0. The RAM read timing is shown in Figure 9.78. A write (data into the RAM) occurs when $\overline{MEMW}$ =0. The RAM write timing is shown in Figure 9.79.

Figure 9.78
Read timing.

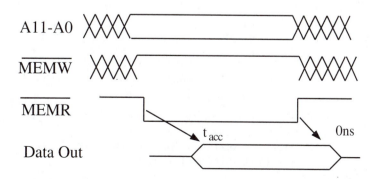

A11-A0

$\overline{MEMW}$

$\overline{MEMR}$

t_{acc} 0ns

Data Out

Figure 9.79
Write timing.

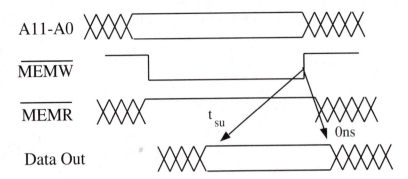

a) $\overline{\text{MEMR}}$ =1 and $\overline{\text{MEMW}}$ =1 turn off the RAM. Fill in the command table that has **Select** and **R/W** as inputs and $\overline{\text{MEMR}}$ and $\overline{\text{MEMW}}$ as outputs.

b) What is the read data available interval? Express your answer as an equation using only terms like $\downarrow\overline{\text{MEMR}}$ and $\uparrow\overline{\text{MEMW}}$. Don't calculate (yet) the actual interval in nanoseconds.

c) What is the write data required interval? Express your answer as an equation using only terms like $\downarrow\overline{\text{MEMR}}$ and $\uparrow\overline{\text{MEMW}}$. Don't calculate (yet) the actual interval in nanoseconds.

d) Show the combined table that has **Select, R/W,** and **E** as inputs and $\overline{\text{MEMR}}$ and $\overline{\text{MEMW}}$ as outputs. Include digital equations for $\overline{\text{MEMR}}$ and $\overline{\text{MEMW}}$.

e) Show a digital circuit for the interface between the 6811 and the RAM. Include the 74HC573 latch if and only if it is necessary for this interface. Please label TTL chip numbers but not pin numbers.

f) What is the allowable range (minimum and maximum) of values for t_{acc}?

g) What is the allowable range (minimum and maximum) of values for t_{su}?

9.5 The overall objective is to select the cheapest PROM that will operate with a **2.1 MHz** 6811. When PROM input **CS** is 1, the PROM **Data** outputs are driven with the 8-bit value stored at the specified **Address.** The address must be valid whenever **CS** is high and the **CS** access time is t_{CS} (Figure 9.80).

Figure 9.80
Read timing.

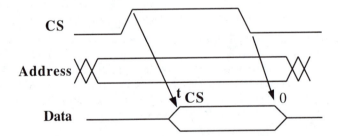

CS	Data
0	hiZ
1	Driven

a) The **CS** signal is positive logic, meaning it is normally low and goes high only during an access to this PROM. *Circle* the best shape for the PROM **CS** chip select control line:

Positive logic synchronized to the E clock
Positive logic synchronized to the AS 6811 signal
Positive logic unsynchronized

b) Why did you choose the above shape? That is, list the bad thing(s) that would happen with the other two choices.

c) What is the **Read Data Required** Interval for this 6811 running at **2.1 MHz**? Give your answer with exact explicit numbers.

d) Develop an equation for the **Read Data Available** Interval that is a function only of t_{CS}. All the other terms in the interval should be exact explicit numbers. Assume a 10-ns gate delay.

e) What is the maximum allowable value for t_{CS}?

9.6 Interface the following *Output Device* (Figure 9.81) to the address/data bus of an expanded mode 6811 running at 1 MHz. (*Hint: Think of this device as a memory that responds only to 6811 write cycles.* That is, this device ignores all read cycles.) The chip has two control inputs **C1, C2,** one address input **A0,** eight data inputs **D7–D0,** and two port outputs **Port0, Port1.** The minimal-cost positive logic address decoder is already completed, locating it at $4000 to $4001.

Figure 9.81
Output device.

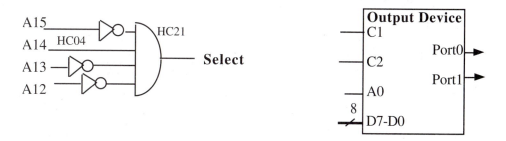

There are two different, but equally effective, timing situations. A write operation occurs on the overlap of **C1**=1 and **C2**=0. You will not be connecting to Port0, Port1 in this problem. The function table is:

C1	C2	A0	Function
0	X	X	Off low $I_{cc}=10\mu A$
1	1	X	Off high $I_{cc}=1mA$
1	0	0	Write data into Port0
1	0	1	Write data into Port1

Figure 9.82
Write timing.

The write timing when controlled by **C1** is shown on the left in Figure 9.82; the write timing when controlled by **C2** is shown on the right in Figure 9.82.

a) What is the **Write Data Available** Interval for this 6811 running at 1 MHz?

b) Show the combined table with **Select, R/W, E** as inputs and **C1** and **C2** as outputs. Include digital equations for **C1** and **C2.**

c) Show the digital circuit interface between the 6811 and this Output Device. Clearly label all the connections to the 6811. Include chip numbers but not pin numbers.

d) What is the allowable range (minimum and maximum) of values for t_{su}? Assume a 10ns gate delay.

e) Show the combined **write cycle** timing diagram. Show **E, AS, R/W, A15–A8, A7–A0, Select, C1, C2, Data Available,** and **Data Required.** All signals are outputs except Data Required. Use arrows to signify causal relations, with the timing delays as numbers in nanoseconds.

9.7 Interface a 16K by 8 bit PROM to a **6811** running at 2 MHz. Use FULL address decoding to place the PROM from $4000–$7FFF. The address must be valid whenever the chip is enabled (both **CS1**=1 and $\overline{\text{CS0}}$ =0). Assume a 10-ns gate delay in each digital chip and a 500-ns **E** clock period. The PROM has two control lines. A read (data out of the PROM) occurs on the overlap of **CS1**=1 and $\overline{\text{CS0}}$ =0. Notice that the **CS1** access time is larger than the $\overline{\text{CS0}}$ access time. The read timing is shown in Figure 9.83.

Figure 9.83
Read timing.

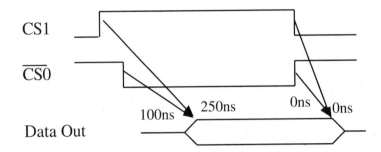

a) Either **CS1**=0 or $\overline{\text{CS0}}$ =1 will turn off the PROM. Fill in the command table that has **Select, R/W** as inputs and **CS1, $\overline{\text{CS0}}$** as outputs.

b) What is the read data available interval? Express your answer as an equation using only the terms like AdV, AdN, ↓ $\overline{\text{CS0}}$, and ↑**CS1** . Don't calculate (yet) the actual interval in nanoseconds.

c) Show the combined has **Select, R/W, E** as inputs and **CS1, $\overline{\text{CS0}}$** as outputs. Include digital equations for **CS1** and $\overline{\text{CS0}}$.

d) Show digital circuit for the interface between the 6811 and the PROM. The chip has two control inputs **CS1, $\overline{\text{CS0}}$** , 14 address inputs **A13–A0**, eight data outputs **D7–D0.** Include the 74HC573 latch if and only if it is necessary for this interface. Please label chip numbers but not pin numbers.

e) Show the combined **read cycle** timing diagram. All signals are outputs except Data Required. The A7–A0 signals are the ones connected to the PROM (not necessarily the AD7–AD0 of the 6811). Show **E, AS, R/W, A15–A8, A7–A0, Select, CS1, $\overline{\text{CS0}}$, Data Available,** and **Data Required.** Use arrows to signify causal relations.

9.8 Interface a Programmable Keyboard/Display Interface chip (Intel 8279-5) to a **6811.** Use minimal cost address decoding to place the I/O chip from $2018 to $2019. Assume a 10-ns TTL gate delay and a 640-ns **E** clock period. Your 6811 has the timing shown in Figure 9.84.

The Intel 8279-5 has the following function table:

$\overline{\text{CS}}$	$\overline{\text{RD}}$	$\overline{\text{WR}}$	A0	Function
0	0	1	0	Read data register
0	0	1	1	Read status register
0	1	0	0	Write data register
0	1	0	1	Write control register
0	0	0	X	Illegal (don't do this)
0	1	1	X	Off
1	X	X	X	Off

Figure 9.84
6811 timing.

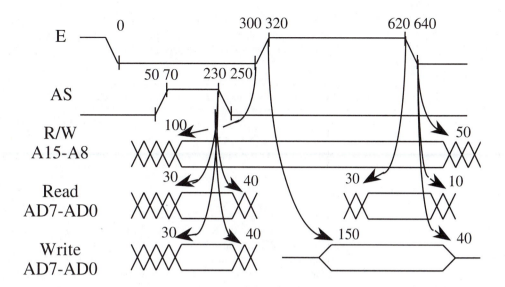

a) Design the minimal cost address decoder (your choice as to whether to make it positive or negative logic). Show digital equation and the TTL circuit, including chip numbers. If you choose positive logic, then label the output of the decoder **Select.** If you choose negative logic, then label the output of the decoder $\overline{\textbf{Select}}$. *Do not build all the decoders, just the one for this Intel 8279!*

b) Fill in the command table.

c) What is the read data available interval? Express your answer as an equation using only the terms like AdV, AdN, $\downarrow\overline{\textbf{RD}}$, and $\uparrow\overline{\textbf{WR}}$. The address access time is 250 ns, the $\overline{\textbf{CS}}$ access time is 250 ns, and the $\overline{\textbf{RD}}$ access time is 150 ns. Don't calculate (yet) the actual interval in nanoseconds.

d) What is the read data required interval? Express your answer in nanoseconds according to the time convention shown later in part i. Remember, this is a 6811 running with an E period of 640 ns. Give the actual numbers [e.g., RDR = (450, 510)].

e) What is the write data available interval? Express your answer in nanoseconds according to the time convention shown later in part i. Remember, this is a 6811 running with an E period of 640 ns. Give the actual numbers [e.g., WDA = (375, 533)].

f) What is the write data required interval? Express your answer as an equation using only the terms like $\downarrow\overline{\textbf{WR}}$ and $\uparrow\overline{\textbf{WR}}$. Don't calculate (yet) the actual interval in nanoseconds. The write occurs on the fall of $\overline{\textbf{WR}}$. The setup time is 150 ns, and the hold time is zero.

g) Show the combined table, and include digital equations for $\overline{\textbf{CS}}$, $\overline{\textbf{RD}}$, and $\overline{\textbf{WR}}$. You will build the circuit later in part h. The read and write functions must be synchronized to the **E** clock.

h) Show digital circuit for the interface between the 6811 and the I/O device. Please label TTL chip numbers but not pin numbers. You need not copy the address decoder circuit if shown earlier in part a, just use the **Select** label.

i) Show the combined **read cycle** timing diagram. All signals are outputs except Data Required. The A0 is the signal connected to the I/O device (not necessarily the AD7–AD0 of the 6811). Use arrows to signify causal relations.

9.9 The ultimate goal of this problem is to design a simple 14-pin LSI chip that supports XIRQ interrupts on an expanded mode 6811/6812. That is, you will be connecting up to the 6811/6812 address/data bus, not to the parallel ports. You will design the device using 74HC digital logic. We will use two 74HC74 D flip-flops to implement READY and ARM, and a 74HC125 tristate buffer to perform the read status function. Assume a 20-ns gate delay in each 74HC chip. Synchronize all read/write/clear operations to the **E** clock period. The device should have the following command table (others are no operation):

$\overline{CS3}$	$\overline{CS2}$	CS1	CS0	R/W	RS	Function
0	0	1	1	1	0	Read status (READY=bit 7)
0	0	1	1	0	0	Write status (READY=0, regardless of the data bus)
0	0	1	1	1	1	No operation (ARM is write-only)
0	0	1	1	0	1	Write mode (ARM=bit 7)

The READY signal is set on the rise of INPUT and cleared (acknowledged) on a write status. The software can observe the READY signal by performing a read status operation. The hardware can also observe READY because it is a direct output of your interface. The software can arm or disarm the interface by performing a write mode operation. An XIRQ interrupt is requested (open collector) if READY=1 and ARM=1. The device is connected to an expanded mode 6811 like the one shown in Figure 9.85.

Figure 9.85
External device.

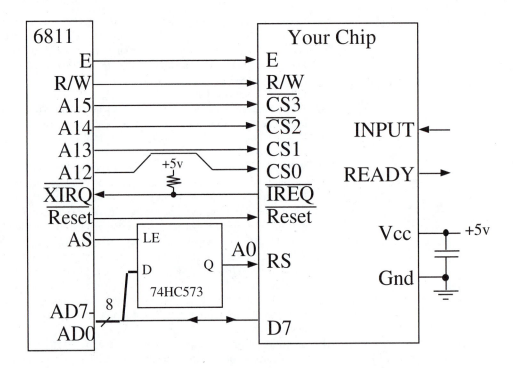

a) Complete the hardware design of your 14-pin chip. Do not change the interface to the 6811 as shown in Figure 9.85. Specify 74HC chip numbers but not pin numbers. Do not worry about simultaneous events: a rising edge of INPUT and a write status.

b) Show the ritual subroutine that arms and enables the XIRQ device. Use assembly language or C, but be consistent. Initialize an 8-bit unsigned global variable, CNT, to 0.

c) Show the XIRQ software that counts the number of rising edges of input (CNT=CNT+1). Execute SWI if an XIRQ interrupt occurs that is not from your chip.

9.10 Interface the following 32-kbyte RAM to a 2-MHz 6811 or an 8 MHz MC68HC812A4 using full-address decoding. The RAM will be located at $4000 to $BFFF. The RAM function table is:

CE	OE	Function
0	0	Off
0	1	Off
1	0	Write data into RAM
1	1	Read data out of RAM

The read timing is (with OE=1) shown on the left in Figure 9.86; the write timing is (with OE=0) shown on the right.

Figure 9.86
Read and write timing.

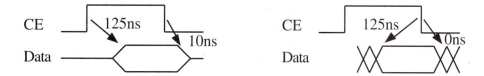

a) Show the interface between the 6811/6812 and this RAM. Clearly label all the connections to the 6811, especially A14.

b) Show the combined **write cycle** timing diagram. Show **E, AS, R/W, A15–A8, A7–A0, Select, CE, OE, WDA,** and **WDR.** All signals are outputs except Data Required. Use arrows to signify causal relations. Show the timing delays as arrows, with numbers in nanoseconds. Calculate the actual WDA and WDR intervals.

9.11 Interface a 27C128-90 16K by 8-bit PROM to the MC68HC812A4 address/data bus. The address will be $C000 to $FFFF. No other external devices exist. Assume the 6812 is running at 8 MHz in expanded narrow mode.

a) Show a digital circuit for the interface between the 6812 and the PROM. Please label TTL chip numbers but not pin numbers.

b) Develop equations for read data available and read data required, and use them to determine the amount of cycle stretching required to interface this 90-ns memory.

c) Show the software ritual required to initialize this memory interface.

d) Show the combined **read cycle** timing diagram. Assume a 10-ns gate delay through any digital gate. All signals are outputs except Data Required. Use arrows to signify causal relations.

9.12 Consider the 8K by 8-bit static RAM interface between the MC68HC812A4 and MCM60L64. Assume the MC68HC812A4 is running at 8 MHz in expanded narrow mode. Assume the same hardware presented in the MC68HC812A4/MCM60L64 example earlier in this chapter. The objective of this problem is to develop the specifications that allow this RAM interface to operate without cycle stretching. You may use timing equations or timing diagrams to justify your answers.

a) Develop the equations that specify the maximum allowable address access time t_{AVQV} and chip enable access time t_{E1LQV}. In other words, how fast would the RAM chip have to be to operate without cycle stretching?

b) Develop the equations that specify the maximum allowable t_{DVWH}.

9.13 Briefly explain the difference between single-chip, expanded narrow, and expanded wide mode on the MC68HC812A4 as they apply to memory interfacing. Give one sentence for each mode. In particular, explain the effect on Ports A, B, C, D, E.

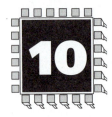

10 High-Speed I/O Interfacing

Chapter 10 objectives are to:

❏ Define the terms bandwidth, latency, and priority
❏ Introduce the concept of DMA synchronization
❏ Discuss the alternatives of hardware FIFOs, dual port memory, and bank-switch
   memory
❏ Use DMA to transfer memory blocks
❏ Present high-bandwidth/low-latency applications

Embedded system designers will not need DMA to solve most of their problems. But like the last chapter, there are two motivations for this chapter. The first motivation is basic knowledge, and the second is solving specific high-performance applications. DMA is an important yet complicated interfacing process. One of the advantages of learning DMA on a simple device like the MC68HC708XL36 is that it maintains all the fundamental concepts without most of the complexities found in larger computer systems. As the performance requirements of our embedded system grow, there comes a point when the simple methods of I/O interfacing are not adequate. This chapter introduces a number of techniques that produce high bandwidth and low latency.

10.1 The Need for Speed

Bandwidth, latency, and priority are quantitative parameters we use to evaluate the performance of an I/O interface. The basic function of an input interface is to transfer information about the external environment into the computer. In a similar way, the basic function of an output interface is to transfer information from the computer to the external environment. The bandwidth is the number of bytes transferred per second. The bandwidth can be expressed as a maximum or peak that involves short bursts of I/O communication. On the other hand, the overall performance can be represented as the average bandwidth. The latency of the hardware/software is the response time of the interface. It is measured in different ways, depending on the situation. For an input device, the *interface latency* is the time from when new input is available to when the data are transferred into memory. We can also define *device latency* as the response time of the external I/O device. For example, if we request that a certain sector be read from a disk, then the device latency is the time it takes to find the correct track and spin the disk (seek) so that the proper sector is positioned under the read head. For an output device, the interface latency is the time from when the output device is idle to when the interface writes new data. A *real-time* system is one that can guarantee a worst-case interface latency. Table 10.1 illustrates specific ways to calculate latency. In each case, however, latency is the time from when the need arises to when the need is satisfied.

If we consider the busy/done I/O states introduced in Chapters 3 and 4, the interface latency is the time from the *busy to done* state transition to the time of the *done to busy* state transition. Sometimes we are interested in the worst-case (maximum) latency and sometimes in the average. If we can put an upper bound on the latency, then we define the system as real time. A number of applications involve performing I/O functions on a fixed-interval basis. In a data acquisition system, the ADC is triggered (a new sample is requested) at the desired sampling rate.

Table 10.1
Interface latency is a measure of the response time of the computer to a hardware event.

The time a need arises	The time the need is satisfied
New input is available	Input data are read
New input is available	Input data are processed
Output device is idle	New output data are written
Sample time occurs	ADC is triggered, input data
Periodic time occurs	Output data, DAC is triggered
Control point occurs	Control system executed

10.2 High-Speed I/O Applications

Before introducing the various solutions to a high-speed I/O interface, we will begin by presenting some typical applications.

10.2.1 Mass Storage

The first application is mass storage, including floppy disk, hard disk, tape, and CD-ROM. Writing data to disk with these systems involves:

1. Establishing the physical location to write (position the record head at the proper block, sector, track, cylinder, etc.)
2. Specifying the block size
3. Waiting for the physical location to arrive under the record head
4. Transmitting the data

Reading data from disk with these systems is similar and involves:

1. Establishing the physical location to read (position the read head at the proper block, sector, track, cylinder, etc.)
2. Specifying the block size
3. Waiting for the physical location to arrive under the read head
4. Receiving the data

Figure 10.1
Data to/from the
read/write head of a
hard drive has a high
bandwidth.

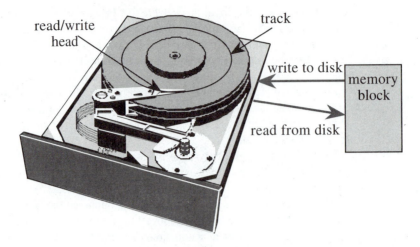

Under most situations the size of the data block transferred is fixed. The bandwidth depends on the rotation speed of the disk and the information density on the medium (Figure 10.1). A typical hard drive can sustain about 10 to 40 Mbytes/s. The time to locate the physical location is called the *seek time*. Although seek time has a significant impact on the disk performance, it does not affect the latency or bandwidth parameters. A 32X CD-ROM has a peak bandwidth of 4.8 Mbytes/s. There is a wide range of disk speeds, but it is important to note that for most situations, the disk bandwidth will be less than the computer bus bandwidth but greater than the maximum bandwidth that a software-controlled interface can achieve. If the disk interface is not buffered, then the interface must respond to each data byte at a rate as fast as the peak disk bandwidth. For example, in a disk read, once the data become available, the interface must capture and store them in memory, before the next data become available. If we do not meet the response time requirement in the disk interface, the rotation speed will have to be reduced. Notice that because of the seek time (time for the physical location to arrive under the head), the average and peak bandwidth will be quite different. Also notice that without buffering, the maximum interface latency will be inversely related to the peak bandwidth.

**10.2.2
High-Speed Data
Acquisition**

Examples of high-speed data acquisition are CD-quality sound recording (16-bit, two-channel, 44 kHz), real-time digital image recording, and digital scopes (8-bit 200 MHz). Sound recording actually has two high-speed data channels: one for recording into memory and a second for storing the memory data on hard disk or CD (Figure 10.2). A

Figure 10.2
Sound data from the
microphone are stored
in memory, processed,
then saved on CD.

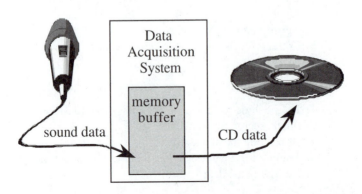

spectrum analyzer combines the high-speed data acquisition of a digital scope with the Discrete Fourier Transform (DFT) to visualize the collected data in the frequency domain. In the context of this chapter, we will define a high-speed data acquisition as one that samples faster than a software-controlled interface would allow.

10.2.3 Video Displays

Real-time generation of TV or video images requires an enormous data bandwidth. Consider the information bandwidth required to maintain an image on a graphics display (Figure 10.3). A video graphics array image is 256 colors (8-bit), 480 rows, 640 columns, and is refreshed at about 60 Hz. Calculating the bandwidth in bytes per second, we get $1 \cdot 480 \cdot 640 \cdot 60$, which is 18,432,000 bytes/s. Luckily, we don't have to communicate each pixel for each image but rather just transmit the changes from the previous image. To achieve the necessary bandwidth, video interface hardware will use a combination of DMA and dual port memories.

Figure 10.3
Image data from memory are displayed on the graphics screen.

10.2.4 High-Speed Signal Generation

Examples of high-speed signal generation are CD-quality sound playback (16-bit, two-channel, 44 kHz), and real-time waveform generation. Sound playback also has two high-speed data channels: one for loading sound data into memory from CD and a second for playing the memory data out to the speakers (Figure 10.4). In the context of this chapter, we will define a high-speed interface as one that samples faster than a software-controlled interface would allow.

Figure 10.4
Sound data from the CD are loaded in memory, processed, then output to the speakers.

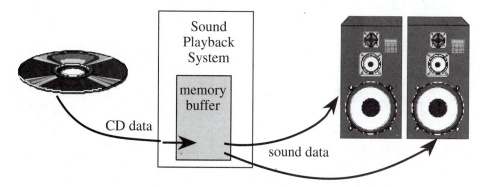

10.2.5
Network
Communications

For many networks the communication bandwidth of the physical channel will exceed the ability of the software to accept or transmit messages. For these high-speed applications, we will look for ways to decouple the software that creates outgoing messages and processes incoming messages from the hardware that is involved in the transmission and reception of individual bits (Figure 10.5). Because the network load will vary, the average bandwidth (determined by how fast the transmission software can create outgoing messages and the reception software can process incoming messages) will be slower than the peak/maximum bandwidth that is achieved by the network hardware during transmission. This mismatch allows one network to be shared among multiple potential nodes.

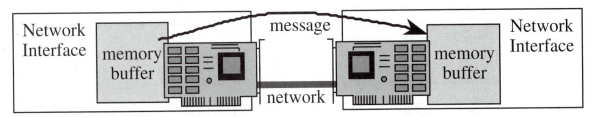

Figure 10.5   Message data are transmitted from the buffer of one computer to the buffer of another.

10.3   General Approaches to High-Speed Interfaces

10.3.1
Hardware FIFO

If the software-controlled interface can handle the average bandwidth but fails to satisfy the latency requirements, then a hardware FIFO can be placed between the I/O device and the computer. A common application of the hardware FIFO is the serial interface. Assume in this situation that the average serial bandwidth is low enough for the software to read the data from the serial port and write them to memory. We saw in Chapter 7 that the latency requirement of a SCI input port like the 6805/6808/6811/6812 is the time it takes to transmit one data frame. To reduce this interface latency requirement (without changing the average bandwidth requirement) we can add a hardware FIFO between the receive shift register and the receive data register, as illustrated in Figure 10.6.

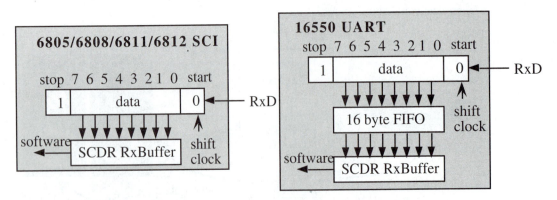

Figure 10.6   The 16550 UART employs a 16-byte FIFO to reduce the latency requirement of the interface.

Observation: With a serial port that has a shift register, a FIFO of size n, and one data register, the latency requirement of the input interface is the time it takes to transmit n+1 data frames.

A hardware FIFO, placed between the SCI data register and the transmit shift register, allows the software to write multiple bytes of data to the interface and then perform other tasks while the frames are being sent.

10.3.2
Dual Port Memory

One approach that allows a large amount of data to be transmitted from the software to the hardware is the dual port memory. A dual port memory allows shared access to the same memory between the software and hardware (Figure 10.7). For example, the software can create a graphics image in the dual port memory using standard memory write operations. At the same time the video graphics hardware can fetch information out of the same memory and display it on the computer monitor. In this way, the data need not be explicitly transmitted from the computer to the graphics display hardware. To implement a dual port memory, there must be a way to arbitrate the condition when both the software and hardware wish to access the device simultaneously. One mechanism to arbitrate simultaneous requests is to halt the processor using a MRDY signal so that the software temporarily waits while the video hardware fetches what it needs. Once the video hardware is done, the MRDY signal is released and the software resumes. None of the 6805/6808/6811/6812 expanded modes supports this sort of hardware-initiated cycle stretching. This hardware-controlled cycle stretching was defined in Chapter 9 as a partially asynchronous memory bus. Notice that except for the access conflict, both the software and graphics hardware can operate simultaneously at full speed.

Figure 10.7
A dual port memory can be accessed by two different modules.

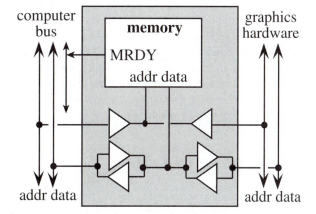

10.3.3
Bank-Switched Memory

Another approach similar to the dual port memory is the bank-switched memory. A bank-switched memory also allows shared access to the same memory between the software and hardware (Figure 10.8). The difference between bank-switched and dual port memory is that the bank-switched memory has two modes. In one mode (M=1), the computer has access to memory bank A, and the I/O hardware has access to memory bank B. In the other mode (M=0), the computer has access to memory bank B, and the I/O hardware has access to memory bank A. Because access is restricted in this way, there are no conflicts to resolve. The master controller determines the value of M.

Figure 10.8
A bank-switched memory can be accessed by two different modules, one at a time.

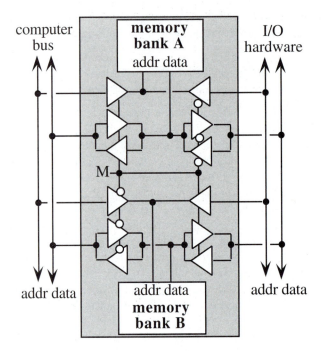

Observation: With a bank-switched memory, the latency requirement of the software is the time it takes the hardware to fill (or empty) one memory bank.

Many high-speed data processing systems employ this approach. The A/D hardware can write into one bank, while the computer software processes previously collected data in the other. When the A/D sampling hardware fills a bank, the memory mode is switched, and the software and hardware swap access rights to the memory banks. In a similar way, a real-time waveform generator or sound playback system can use the bank-switched approach. The software creates the data and stores them into one bank, while the hardware reads data from the other bank that was previously filled. Again, when the hardware is finished, the memory bank mode is switched.

10.4   Fundamental Approach to DMA

With a software-controlled interface (gadfly or interrupts) if we wish to transfer data from an input device into RAM, we must first transfer them from the input to the processor, then from the processor into RAM. To improve performance, we will transfer data directly from input to RAM or RAM to output using DMA. Because DMA bandwidth can be as high as the bus bandwidth, we will use this method to interface high-bandwidth devices like disks and networks. Similarly, because the latency of this type of interface depends only on hardware and is usually just a couple of bus cycles, we will use DMA for situations that require a very fast response. On the other hand, software-controlled interfaces have the potential to perform more complex operations than simply transferring the data to/from memory. For example, the software could perform error checking, convert from one format to another, implement compression/decompression, and detect events. These more complex I/O operations may preclude the usage of DMA.

10.4.1
DMA Cycles

During a *DMA read cycle,* the processor is halted and data are transferred from memory to output device (Figure 10.9). During a *DMA write cycle,* the processor is halted and data are transferred from input device to memory (Figure 10.10). In some DMA interfaces, two DMA cycles are required to transfer the data. The first DMA cycle brings data from the source into the DMA module, and the second DMA cycle sends the data to their destination.

Figure 10.9
A DMA read cycle copies data from RAM or ROM to an output device.

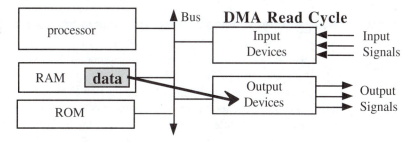

Figure 10.10
A DMA write cycle copies data from the input device into RAM.

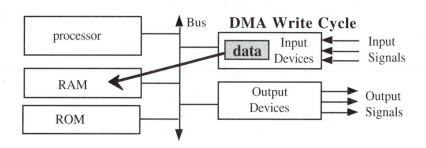

10.4.2
DMA Initiation

We can classify DMA operations according to the event that initiates the transfer. A *software-initiated* transfer begins with the program setting up and starting the DMA operation. Using DMA to transfer data from one memory block to another greatly speeds up the function. The efficiency of memory block transfers is very important in larger computer systems. Benchmarks on the Motorola 6808 show that even for block sizes as small as 4 bytes it is faster to initialize a DMA channel and perform the transfer in hardware. As the block size increases, the performance advantage of DMA hardware over traditional software becomes more dramatic.

Most DMA applications involve a *hardware-initiated* DMA transfer. For an input device, the DMA is triggered on new data available. For an output device, the DMA is triggered on output device idle. For periodic events, like data acquisition and signal generation, the DMA is triggered by a periodic timer. These are the exact issues involved in gadfly and interrupt synchronization discussed in Chapters 3 and 4. The difference with DMA is that the servicing of the I/O need will be performed by the DMA controller hardware without software explicitly having to transfer each byte.

10.4.3
Burst Versus Cycle
Steal DMA

When the desired I/O bandwidth matches the computer bus bandwidth, then the computer can be completely halted, while the block of data is transferred all at once (Figure 10.11). Once an input block is ready, a *burst mode DMA* is requested, the computer is halted, and the block is transferred into memory.

Figure 10.11
An input block is transferred all at once during burst mode DMA.

Figure 10.11 describes an input interface, but the same timing occurs on an output interface using burst mode DMA. For an output interface, the DMA is requested when the interface needs another block of data. During the burst mode DMA, the computer is halted, and an entire block is transferred from memory to the output device.

If the I/O bandwidth is less than the computer bus bandwidth, then the DMA hardware will steal cycles and transfer the data one DMA cycle at a time. In *cycle steal mode,* the software continues to run, although a little bit slower. In either case the processor is halted during the DMA cycles. Figure 10.12 describes an input interface, but the same timing occurs on an output interface using cycle steal mode DMA. For an output interface, the DMA is requested when the interface needs another byte of data. During the cycle steal DMA, one byte is transferred from memory to the output device.

Figure 10.12
Each time an input byte is ready, it is transferred to memory using cycle steal DMA.

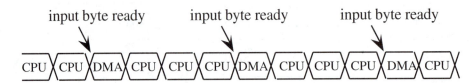

Observation: Some computers must finish the instruction before allowing a burst DMA. In this situation, the latency will be higher than cycle steal DMA, which does not need to finish the current instruction.

Since most I/O bandwidths are indeed less than the memory bandwidth, one technique to enhance speed is I/O buffering. In this approach a dedicated I/O memory buffer exists in the I/O interface hardware. This buffer is like the bank-switched memory discussed earlier. For example, on a hard disk read block operation, raw data comes off the disk and into the buffer. During this time the processor is not halted. When the buffer is full, burst DMA is used to transfer the data into the system memory. Similarly on a hard disk write block operation, the software initiates a burst DMA to transfer data from system memory into the I/O buffer. Once full, the I/O interface can write the data onto the disk.

**10.4.4
Single-Address
Versus Dual-
Address DMA**

Some computer systems allow the transfer of data between the memory and I/O interface to occur in one bus cycle, while others need two bus cycles to complete the transfer. In a *single-address DMA cycle,* the address and R/W line dictate the memory function to be performed, and the I/O interface is sophisticated enough to know it should participate in the transfer.

In this single-address example, the disk interface is reading bytes from a floppy disk (Figure 10.13). Cycle steal mode will be used because the floppy disk bandwidth is slower than the bus. When a new byte is available, **Request** will be asserted, and this will request a DMA cycle from the DMA controller. The DMA controller will temporarily suspend the processor and drive the address bus with the memory address and the R/W to write. During

Figure 10.13
Block diagram showing
the modules involved in
a floppy disk read.

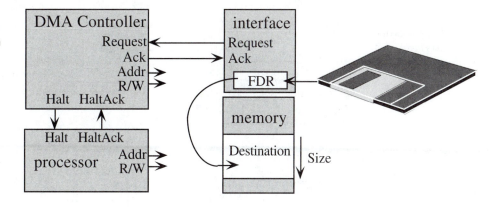

Figure 10.14
Timing diagram of a
single-address DMA-
controlled floppy disk
read.

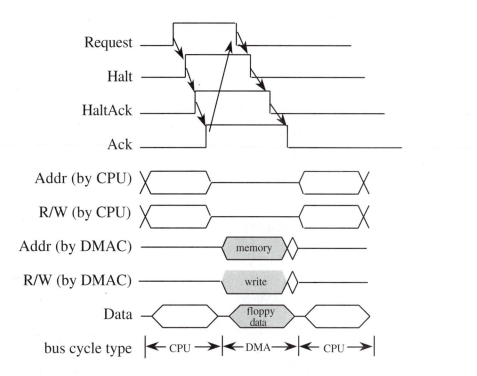

this cycle the DMA controller will respond to the floppy interface by asserting the **Ack.** The floppy uses the **Ack** (ignoring the address bus and R/W) to know when to drive its data on the bus (Figure 10.14).

    Observation: The MC68HC708XL36 does not support single-address DMA.

    In a *dual-address DMA cycle,* two bus cycles are required to achieve the transfer (Figure 10.15). In the first cycle the data are read from the source address and copied into the DMA controller. During the first cycle the address bus contains the source address, and R/W signifies read. The information from the data bus is saved in the **temp** register within the DMA controller. In the second cycle, the data are transferred to the destination address. During the second cycle, the address bus contains the destination address, the data bus has the **temp** data, and R/W signifies write.

Figure 10.15
Block diagram showing
the modules involved in
a SPI read.

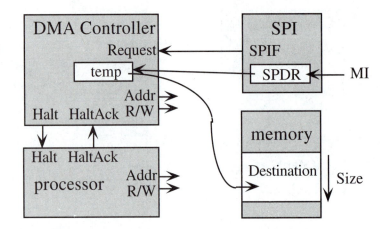

In this dual-address example, the SPI interface is receiving bytes from a synchronous serial network. Cycle steal mode will be used because the SPI bandwidth is slower than the bus. When a new byte is available, **Request** will be asserted, and this will request a DMA cycle from the DMA controller. The DMA controller will temporarily suspend the processor and first drive the address bus with the SPI data register address (R/W=read), then in the second cycle the DMA controller will drive the address bus with the memory address (R/W=write). The SPI knows it has been serviced, because its data register has been read (Figure 10.16).

Observation: Single-address DMA is twice as fast as dual-address DMA.

Figure 10.16
Timing diagram of a
dual-address DMA-
controlled SPI read.

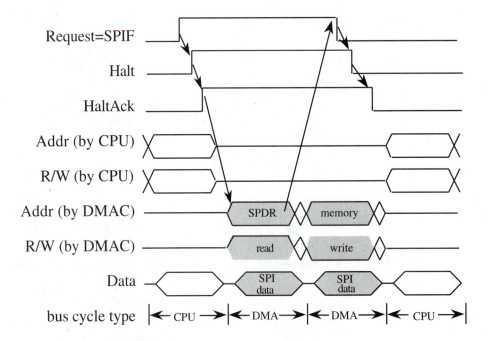

Observation: The dual-address DMA can be used with I/O devices not configured to support DMA.

10.4.5
DMA
Programming

Although DMA programming varies considerably from one system to another, there are a few initialization steps that most require. Table 10.2 lists the mode parameters that must be set to utilize DMA. There are two categories of DMA programming: initialization and completion. During initialization, the software sets the DMA parameters so that the DMA will begin.

Parameter	Possible choices
What initiates the DMA	Software trigger, input device, output device, periodic timer
Type	Burst versus cycle steal
Autoinitialization mode	Single event or continuous transfer
Precision	8-bit byte or 16-bit word
Mode	Single address or dual address
Priority	Should software completely halt? Should interrupts be serviced during DMA?
Synchronization	Set gadfly flag, or interrupt on block transfer complete

Table 10.2   DMA initialization usually involves specifying these parameters.

At the end of a block transfer, a done flag is set, and a number of additional actions may occur. If the system is armed, an interrupt can be generated. At the end of a block transfer in a continuous-transfer DMA, the parameters are autoinitialized so that the DMA process continues indefinitely. Table 10.3 lists additional parameters we will need to initialize.

Parameter	Definition
Source address	Address of the module (memory or input) that generates the data
Increment/decrement/static	Automatically $+1/-1/+0$* the source address after each transfer
Destination address	Address of the module (memory or output) that accepts the data
Increment/decrement/static	Automatically $+1/-1/+0$ the destination address after each transfer
Block size	Fixed number of bytes to be transferred

*Obviously the DMA controller will $+2/-2/+0$ when the precision is 16 bits.

Table 10.3   More DMA initialization parameters.

10.5   MC68HC708XL36 Examples

10.5.1
DMA I/O Registers

The MC68HC708XL36 has three DMA channels. The DMA register DC1 is used to specify bandwidth, enable DMA, and arm interrupts on DMA complete. The TEC2, TEC1, TEC0 bits enable DMA channels 2, 1, 0, respectively. If the corresponding IEC bit is set, then an interrupt will be requested when the entire block has been transmitted.

	7	6	5	4	3	2	1	0	
DC1	BB1	BB0	TEC2	IEC2	TEC1	IEC1	TEC0	IEC0	$004C

BB1 and BB0 specify the bus bandwidth during a burst mode DMA (Table 10.4).

BB1:BB0	DMA bus cycles	CPU bus cycles
00	2 (25%)	6 (75%)
01	2 (50%)	2 (50%)
10	2 (67%)	1 (33%)
11	All (100%)	0 (0%)

The DMA register DSC has both status bits (IFC2, IFC1, IFC0) and control bits. If DMAP is set, then CPU interrupts will be postponed during DMA operations. The L2, L1, L0 control bits enable autoinitialization on their respective DMA channels. If DMAWE is set, then DMA is enabled after a wait instruction.

	7	6	5	4	3	2	1	0	
DSC	DMAP	L2	L1	L0	DMAWE	IFC2	IFC1	IFC0	$004D

The IFC2, IFC1, IFC0 flags are set upon the completion of a block transfer. To clear a flag, the software must first read the register, then write 0s to them.[1] In looping mode, the flags are set at the end of each cycle.

The DMA resister DC2 is used by the software to trigger a DMA event. There are eight possible DMA trigger events, shown in the following table, then each event can be separately triggered. You set the bit to initiate a transfer and clear the bit to halt a transfer.

	7	6	5	4	3	2	1	0	
DC2	SWI7	SWI6	SWI5	SWI4	SWI3	SWI2	SWI1	SWI0	$004E

The DMA registers D0C, D1C, D2C specify source/destination address incrementing, precision, and DMA trigger event. When BWC is 1, 16-bit words are transferred.

	7	6	5	4	3	2	1	0	
D0C	SDC3	SDC2	SDC1	SDC0	BWC	DTS2	DTS1	DTS0	$0038
D1C	SDC3	SDC2	SDC1	SDC0	BWC	DTS2	DTS1	DTS0	$0040
D2C	SDC3	SDC2	SDC1	SDC0	BWC	DTS2	DTS1	DTS0	$0048

The SDC bits specify whether to increment, decrement, or maintain constant the source and destination addresses during a DMA transfer (Table 10.5). Static means the address is neither incremented nor decremented. When the precision is 16-bit words, then the

---

[1]Note that many other Motorola flags are cleared by writing 1s to the flag.

increment/decrement is obviously by 2. Bits SDC3, SDC2 affect the source, and bits SDC1, SDC0 affect the destination.

Table 10.5
Two control bits specify the effect on the address after each DMA cycle.

Bits	Address
00	Static
01	Decrement
11	Increment

The DTS bits specify the DMA trigger event (Table 10.6). In each case the trigger event corresponds to the event that would have normally caused an interrupt. We will see in the examples that the software arms the I/O device as if it were going to request an interrupt, but instead a DMA cycle occurs.

Table 10.6
Three control bits specify the DMA trigger event.

DTS[2:0]	Software trigger	DMA trigger event
000	SWI0	Timer channel 0
001	SWI1	Timer channel 1
010	SWI2	Timer channel 2
011	SWI3	Timer channel 3
100	SWI4	SPI receive
101	SWI5	SPI transmit
110	SWI6	SCI receive
111	SWI7	SCI transmit

The initialization software must set the source and destination addresses. D0S, D1S, D2S, D3S are the 16-bit source addresses for DMA channels 0, 1, 2, 3, respectively.

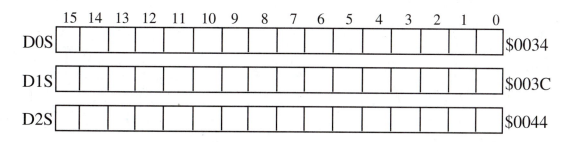

D0D, D1D, D2D, D3D are the 16-bit destination addresses for DMA channels 0, 1, 2, 3, respectively.

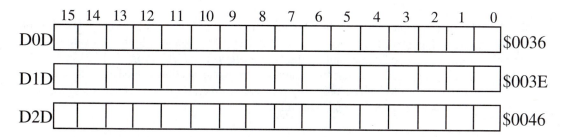

If programmed to increment or decrement, these registers will change during the DMA transfer. If autoinitialization is selected (looping), then the registers are reset back to the original values at the start of each loop. The block length registers, shown below, specify the number of bytes to transfer. Even if a 16-bit word mode is used, these registers still contain the number of bytes to be transferred.

	7	6	5	4	3	2	1	0	
D0BL									$0039
D1BL									$0041
D2BL									$0049

The number of bytes transferred in the current DMA transfer is contained in the Byte Count Registers, shown below:

	7	6	5	4	3	2	1	0	
D0BC									$003B
D1BC									$0043
D2BC									$004B

Writing to the source or destination address will clear the corresponding Byte Count Register. The software can poll these registers to check on the current progress of the DMA interfaces. The DMA registers used in the following examples are shown in Program 10.1.

Program 10.1 DMA register assignments for the MC68HC708XL36.

```
D0S    equ   $0034   ; DMA Channel 0 source address, 16-bit
D0D    equ   $0036   ; DMA Channel 0 destination address, 16-bit
D0C    equ   $0038   ; DMA Channel 0 control register, 8-bit
D0BL   equ   $0039   ; DMA Channel 0 block length, 8-bit
D0BC   equ   $003B   ; DMA Channel 0 byte count, 8-bit
DC1    equ   $004C   ; DMA Control 1 register, 8-bit
DSC    equ   $004D   ; DMA Status/control register, 8-bit
DC2    equ   $004E   ; DMA Control 2 register, 8-bit
```

**10.5.2
Memory-to-
Memory Block
Transfer**

This first example transfers a block of memory from the **Source** address to the **Destination** address (Figure 10.17). Since the DMA utilizes 100% bandwidth, the software will halt during transfer and will continue after the transfer is complete (Program 10.2).

In Figure 10.18, the time to transfer data from one location to another is plotted versus block size for a MC68HC708XL36. You can see that even for block sizes as small as 4 bytes, it is faster to activate a DMA transfer.

Figure 10.17
Block diagram showing
a memory-to-memory
block transfer.

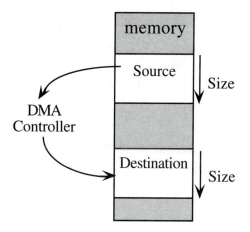

```
* Transfer Size bytes from Source to Destination
copy   ldhx  #Source       ; source address
       sthx  D0S
       ldhx  #Destination ; destination address
       sthx  D0D
       mov   #Size,D0BL    ; number of bytes to transfer (0 means 256)
       mov   #$A0,D0C      ; increment source, increment destination
       mov   #$01,DC2      ; software initiated
       mov   #$00,DSC      ; disable looping
       mov   #$C2,DC1      ; 100%, so software continues when done
       rts
```

Program 10.2 Function using burst DMA to copy memory blocks on the MC68HC708XL36.

Figure 10.18
Transfer speed versus
block size for a
MC68HC708XL36.

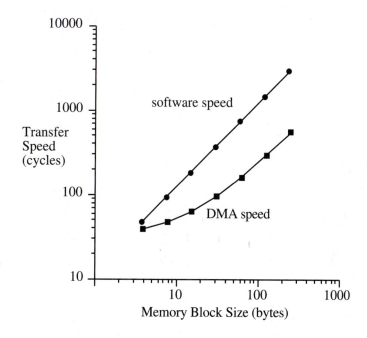

10.5.3
SCI Transmission

In this second example, DMA is used to transfer a message from memory to the SCI transmit hardware (Figure 10.19). DMA can be used because the software knows the message size. Every time the SCI transmitter is ready (SCTE[2] = 1), it will request a DMA cycle, and another byte will be transferred from the memory to the SCI transmit data register. The source address is incremented, but the destination address is static. Once the message is complete, the IFC0 flag will be set. Because the SCI bandwidth is much smaller than the bus bandwidth, it doesn't matter what values go in the BB1, BB0 field of the DC1 control register (Program 10.3).

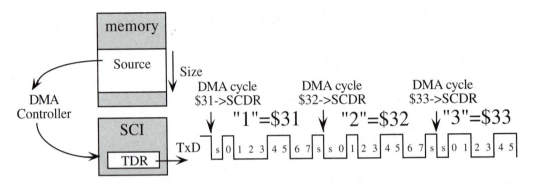

Figure 10.19 Block diagram showing a memory-to-SCI transfer.

```
* setup for transfer Size bytes from Source to the SCI transmit data register
init   mov  #$40,SCC1    ; 8 bits, no parity
       mov  #$10,SCC3    ; DMATE(Transmit DMA) enabled
       mov  #$03,SCBR    ; 9600 baud
       ldhx #SCD         ; destination address is SCI Transmit data reg
       sthx D0D
       mov  #$87,D0C     ; increment source, static destination, byte, SCI Tx
       mov  #$00,DSC     ; disable looping
       mov  #$0C,SCC2    ; enable SCI Tx and Rx
       bclr 0,DSC        ; acknowledge IFC0
       rts
* Transfer Size bytes from Source to the SCI transmit data register
send   ldhx #Source      ; source address
       sthx D0S
       mov  #Size,D0BL   ; number of bytes to transfer (0 means 256)
       mov  #$02,DC1     ; 25%, enable channel 0, no interrupts
       brclr 0,DSC,*     ; wait for IFC0 to be set
       bclr 0,DSC        ; acknowledge IFC0
       rts
```

Program 10.3 Functions using DMA to perform I/O transmission on the MC68HC708XL36.

Observation: This SCI transmit DMA example is analogous to writing data to a disk drive.

---

[2]Other Motorola SCI modules call this bit TDRE.

10.5.4
SPI Reception

In this third example, DMA is used to transfer a message from the SPI receiver hardware to a memory buffer (Figure 10.20). DMA can be used when receiving a message only if the software knows the message size. Every time the SPI receiver is ready (SPRF[3] = 1), it will request a DMA cycle and another byte will be transferred from the SPI data register to memory. The source address is static, but the destination address is incremented. Once the message is complete, the IFC0 flag will be set (Program 10.4).

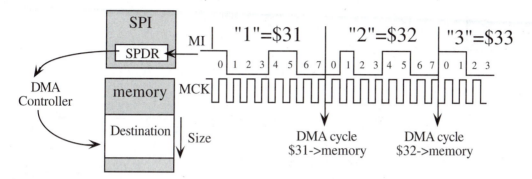

Figure 10.20 Block diagram showing a SPI-to-memory transfer.

```
* transfer Size bytes from the SPI to Destination
recv  mov   #$E2,SPCR     ; Receive DMA, master, SPI enabled
      mov   #$10,SPSCR    ; divide by 2 baud rate
      ldhx  #SPDR         ; source address is SPI data reg
      sthx  D0S
      ldhx  #Destination  ; destination address
      sthx  D0D
      mov   #Size,D0BL    ; number of bytes to transfer (0 means 256)
      mov   #$24,D0C      ; static source, increment destination, byte, SPI Rcv
      mov   #$00,DSC      ; disable looping
      mov   #$02,DC1      ; 25%, enable channel 0, no interrupts
      brclr 0,DSC,*       ; wait for IFC0 to be set
      bclr  0,DSC         ; acknowledge IFC0
      rts
```

Program 10.4 Functions using DMA to perform I/O reception on the MC68HC708XL36.

Observation: This SPI receive DMA example is analogous to reading data from a disk drive.

10.5.5
Simple Waveform
Generation

In this fourth example, DMA is used to transfer waveform data from memory to a DAC (Figure 10.21). Looping will be used, so the pattern is repeated continuously. The output compare timer is used to request a DMA cycle at 44 kHz. The output rate is set by the TMODH/TMODL register (e.g., 182 cycles). During each DMA cycle, another byte will

---

[3]Other Motorola SPI modules call this bit SPIF.

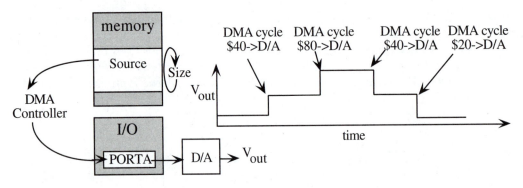

Figure 10.21 Block diagram showing a memory-to-DAC transfer.

```
* transfer Size bytes from Source (waveform) to an 8-bit D/A on port A
wave  mov   #$30,TSC     ; stop and reset timer
      mov   #$01,TDMA    ; service timer 0 with DMA
      ldhx  #181         ; timer DMA every 182 cycles
      sthx  TMODH        ; when the TCNT counts 0,1,2,...,181,0,1,2,...
      ldhx  #91
      sthx  TCH0H        ; DMA requested when TCNT = 91
      mov   #$5A,TSC0    ; squarewave on PTE4 (zero at 91, toggle at overflow)
      ldhx  #Source      ; source address is the waveform table
      sthx  D0S
      ldhx  #PORTA       ; destination address is the D/A
      sthx  D0D
      mov   #Size,D0BL   ; number of bytes to transfer (0 means 256)
      mov   #$80,D0C     ; increment source, static destination, byte, Timer0
      mov   #$10,DSC     ; DMA channel 0 looping
      mov   #$02,DC1     ; 25%, enable channel 0, no interrupts
      mov   #$00,TSC     ; start timer
      rts
```

Program 10.5 Function using DMA to perform periodic output on the MC68HC708XL36.

be transferred from the memory to the DAC. The source address is incremented, but the destination address is static. Once the message is complete, the looping feature will reset the counters and address registers, and the process will continue without software intervention (Program 10.5).

10.6 Glossary

autoinitialization The process of automatically reloading the address registers and block size counters at the end of a previous block transfer so that DMA transfer can occur indefinitely without software interaction.

bandwidth The number of bytes transferred per second.

bank-switched memory A memory module with two banks that interfaces to two separate address/data buses; at any given time one memory bank is attached to one address/data bus and the other bank is attached to the other bus, but this attachment can be switched.

burst DMA An I/O synchronization scheme that transfers an entire block of data all at once directly from an input device into memory or directly from memory to an output device.

bus bandwidth The number of bytes transferred per second between the processor and memory.

CPU cycle A memory bus cycle where the address and R/W are controlled by the processor.

cycle steal DMA An I/O synchronization scheme that transfers data one byte at a time directly from an input device into memory or directly from memory to an output device.

direct memory access (DMA) An I/O synchronization scheme that transfers data directly from an input device into memory or directly from memory to an output device.

DMA cycle A memory bus cycle where the address and R/W are controlled by the DMA controller.

dual-address DMA DMA that requires two bus cycles to transfer data from an input device into memory or from memory to an output device.

dual port memory A memory module that interfaces to two separate address/data buses and allows both systems read/write access to the data.

hardware FIFO High-speed memory that supports the basic FIFO operations of put and get; usually we can put data into one port of the chip while getting data from another port.

latency Time delay from hardware event requiring service (e.g., input device has new data, output device is idle and needs more data) and the transfer of data.

single-address DMA DMA that requires only one bus cycle to transfer data from an input device into memory or from memory to an output device.

10.7   Exercises

10.1 The objective of this exercise is to interface various devices to the computer using DMA synchronization. You may assume the bus bandwidth is at least 8 million bytes/s. For each device you are asked to select the most appropriate DMA mode. Assume the devices support single-address DMA. The 16-bit address of the memory buffer used in each case is 0x1234.

> **Write tape drive** Each tape block is 256 bytes. When a tape head is ready, the controller will signal that it is ready to accept all 256 bytes. At this time, the tape interface chip is ready to transfer as fast as possible all 256 bytes from the memory buffer at 0x1234 to the tape.
>
> **Sound input** The sound waveform buffer is located in memory at 0x1234. Your interface will read the 8-bit ADC 1024 times at 22 kHz and store the data in the buffer. Your software will be smart enough to create two 512-byte buffers out of the 1024 bytes (double buffer) so that it can process one buffer, while the A/D data are being stored automatically under DMA control into the other buffer. That is, when the 1024-byte wave buffer has been filled, the DMA system should repeat and fill it up again.
>
> **Read hard drive** There is a 256-byte buffer at 0x1234 that your DMA system will fill with data from the hard disk. When a hard drive read head is ready, the controller will signal that it has the next byte from the disk. It takes 10 ms for the read head to be ready, then the 256 bytes of data can be transferred from the disk to memory at 2 million bytes/s.

Fill in the following table that specifies the most appropriate mode for each device.

	Tape	Sound	Disk
Cycle steal or block transfer			
Read or write transfer			
Autoinitialization (yes or no)			
Address increment or decrement			
DMA address register value			
DMA count register value			

10.2 Write MC68HC708XL36 software that performs memory-to-memory copy. The three parameters are source address, destination address, and count. The example in the chapter restricted the count to values 1 to 256 bytes, but in this problem you will allow the count to be any 16-bit unsigned value greater than 0. You may assume the two buffers do not overlap.

10.3 Write MC68HC708XL36 software that performs memory-to-memory copy. The three parameters are source address, destination address, and count. You may assume the count ranges from 1 to 256 bytes (count=0 means 256 bytes). The example in the chapter assumed that the two buffers do not overlap. In this exercise you will remove the restriction of nonoverlapping buffers. For example, you should be able to handle these two examples. (*Hint:* The direction of transfer matters when the buffers overlap.)

move down   Source=$0100   Destination $0101   Count=128 bytes
move up   Source=$0101   Destination $0100   Count=128 bytes

10.4 When a 256-byte block is written to a floppy disk, there are 256 separate single-address DMA cycles in cycle steal mode. This question deals with just one of these DMA transfers. There are 14 events listed below. First you will eliminate the events that do not occur during the DMA cycle that saves one byte on the disk. In particular, list the events that will not occur. Second, you will list the events that do occur in the proper sequence.
a) An interrupt is requested.
b) Registers are pulled from the stack.
c) Registers are pushed on the stack.
d) The DMAC asks the processor to halt by activating its **Halt** signal.
e) The DMAC deactivates its **Halt** request to the processor.
f) The DMAC tells the FDC interface that a FMA cycle is occurring by activating its **Ack** signal; the DMA Controller drives the address bus with the FDC address; the DMAC drives the control bus to signify a write cycle (e.g., R/W=0); the memory drives the data bus; the FDC accepts the data.
g) The DMAC tells the FDC interface that a DMA cycle is occurring by activating its **Ack** signal; the DMAC drives the address bus with the memory address; the DMAC drives the control bus to signify a memory read cycle (e.g., R/W=1); the memory drives the data bus; the FDC accepts the data.
h) The FDC deactivates its DMA **Request** signal to the DMAC.
i) The FDC requests a DMA cycle to the DMAC by activating its **Request** signal.
j) The interrupt service routine is executed.
k) The write head is properly positioned over the place on the disk.
l) The processor address and control lines float; the processor responds to the DMAC that it is halted by activating its **HaltAck** signal.
m) The processor resumes software execution.
n) Wait until the current instruction is finished executing.

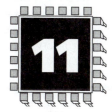

11 Analog Interfacing

Chapter 11 objectives are to

❑ Design analog amplifiers and filters
❑ Study building blocks for data acquisition, including sample and hold, multiplexers, DAC, and ADC
❑ Discuss the functionality of the built-in 8-bit ADCs

Most embedded systems include components that measure and/or control real-world parameters. These real-world parameters (like position, speed, temperature, and voltage) usually exist in a continuous (or analog) form. Therefore, the design of an embedded system involving these parameters rarely uses only binary (or digital) logic. Rather, we often will need to amplify, filter, and eventually convert these signals to digital form. In this chapter we will develop the analog circuit building blocks used in the design of data acquisition systems and control systems.

11.1 Resistors and Capacitors

11.1.1 Resistors

As engineers, we use resistors and capacitors for many purposes. The resistor or capacitor type is defined by the manufacturing process, the materials used, and the testing performed. The performance and cost of these devices varies significantly. For example, a 5% 0.25-W carbon resistor costs about 1 cent, while a 0.01% wire-wound resistor may cost $20. It is important to understand both our circuit requirements and the resistor parameters so that we match the correct resistor type to each application, yielding an acceptable cost/performance ratio. We must specify in our technical drawings the device type and tolerance (e.g., 1% metal film) so that our prototype can be effectively manufactured. The characteristics of various resistor types are shown in Table 11.1.

Table 11.1
General specification of various types of resistor components.

Type	Range	Tolerance	Temperature coef	Max power
Carbon composition	1Ω to $22M\Omega$	5 to 20 %	0.1 %/°C	2 W
Wire-wound	1Ω to $100k\Omega$	>0.0005 %	0.0005 %/°C	200 W
Metal film	$0.1\ \Omega$ to $10^{10}\Omega$	>0.005 %	0.0001 %/°C	1 W
Carbon film	10Ω to $100M\Omega$	>0.5 %	0.05 %/°C	2 W
Source: Wolf and Smith, *Student Reference Manual*, Prentice-Hall, p. 272, 1990.				

The most common type of resistor is carbon composition. It is manufactured with hot-pressed carbon granules. Various amounts of filler are added to achieve a wide range of resistance values. We add them to digital circuits as +5-V pull-ups. To improve the accuracy and stability of our precision analog circuits, we will use resistors with a lower tolerance and temperature coefficient. For most applications 1% metal film resistors will be sufficient to build our analog amplifier circuits. For very high precision analog circuits we could use wire-wound resistors. Wire-wound resistors are manufactured by twisting a very long, very thin wire like a spring. The wire is coiled up and down a shaft in such a way to try and cancel the inductance. Since some inductance remains, wire-wound resistors should not be used for high frequency (above 1 MHz) applications.

Observation: All resistors produce white (thermal) noise.

Observation: Metal film and wire-wound resistors do not generate 1/f noise, whereas carbon resistors do.

Common error: If you design an electronic circuit and neglect to specify explicitly the resistor types, then an incorrect substitution may occur in the layout/manufacturing stage of the project.

11.1.2
Capacitors

Similarly, capacitors come in a wide variety of sizes and tolerances. Polarized capacitors operate best when only positive voltages are applied. Nonpolarized or bipolar capacitors operate for both positive and negative voltages. We select a capacitor based on the following parameters: capacitance value, polarized/nonpolarized, voltage level, tolerance, leakage current (resistance), temperature coefficient, useful frequency response, and temperature range. Another parameter of capacitors is the maximum voltage rating. This parameter is important for high-voltage circuits but is of lesser importance for embedded microcomputer systems. Table 11.2 compares various capacitor types we could use in our circuit.

Type	Range	Tolerance	Temp coef	Leakage	Frequencies
Polystyrene	10 pF to 2.7 μF	±0.5%	Excellent	10 GΩ	0 to 10^{10} Hz
Polypropylene	100 pF to 50 μF	Excellent	Good	Excellent	
Teflon	1000 pF to 2 μF	Excellent	Best	Best	
Mica	1 pF to 0.1 μF	±1 to ±20%		1000 MΩ	10^3 to 10^{10} Hz
Ceramic	1 pF to 0.01 μF	±5 to ±20%	Poor	1000 MΩ	10^3 to 10^{10} Hz
Paper (oil-soaked)	1000 pF to 50 μF	±10 to ±20%		100 MΩ	100 to 10^8 Hz
Mylar (polyester)	5000 pF to 10 μF	±20%	Poor	10 GΩ	10^3 to 10^{10} Hz
Tantalum	0.1 to 220 μF	±10%	Poor		
Electrolytic	0.47 μF to 0.01 F	±20%	Ghastly	1 MΩ	10 to 10^4 Hz

Source: Modified from Wolf and Smith, *Student Reference Manual*, Prentice-Hall, p. 302, 1990; and Horowitz and Hill, *The Art of Electronics*, Cambridge University Press, p. 22, 1989.

Table 11.2   General specification of various types of capacitor components.

We will use capacitors for two purposes in our microcomputer-based embedded systems. First, we will place them on the DC power lines to filter the supply voltage to our circuits (Figure 11.1). A voltage supply typically will include ripple, which is added noise on top of the DC voltage level. There are two physical locations to place the supply filters. The

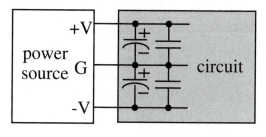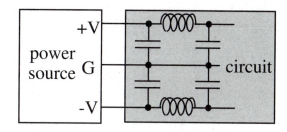

Figure 11.1 DC supply filters.

first location is at the entry point of the supply voltage onto the circuit board. There are two approaches to this filter. If the supply noise is mostly voltage ripple, then two capacitors in parallel can be used. The large-amplitude polarized capacitor (e.g., 1 to 47 μF electrolytic, tantalum) will remove low-frequency large-amplitude voltage noise, and the nonpolarized capacitor (e.g., 0.01 to 0.1 μF ceramic) will remove high-frequency noise. Two different types of capacitors are used because they are effective (i.e., behave like a capacitor) at different frequencies. The Π filter (CLC) is effective in situations where the supply is very noisy. The magnitudes of the parameters depend on the amplitude of the supply ripple.

In addition to the supply filter, we will add bypass capacitors at the supply pins of each chip. It will be important to place these capacitors as close to the pin as physically possible (Figure 11.2). A nonpolarized capacitor (e.g., 0.01 to 0.1 μF ceramic or polyester) will smooth the supply voltage as seen by the chip. Placing the capacitor close to the chip prevents current surges from one chip from affecting the voltage supply of another.

Figure 11.2

Printed circuit board layout positioning the bypass capacitors close to the chip.

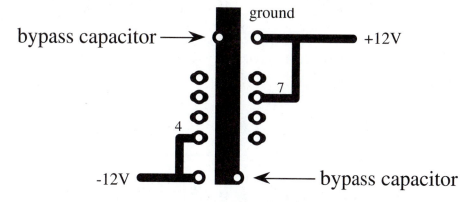

The second application of capacitors in our embedded systems will be in the linear analog circuits of the low-pass filter, the high-pass filter, the derivative circuit, and the integrator circuit. For these applications we will select a nonpolarized capacitor even if the signal amplitude is always positive. In addition, we usually want a low-tolerance, low-leakage capacitor to improve the accuracy of the linear analog circuit.

> **Performance Tip:** Teflon, polystyrene, and polypropylene capacitors are excellent choices when designing precision analog filters.

> **Observation:** The maximum voltage rating of a capacitor is limited by its size.

> **Common error:** Polarized capacitors often have a nonlinear capacitance versus frequency response, therefore using them in an analog filter will cause distortion.

11.2 **Operational Amplifiers (Op Amps)**

11.2.1
Transistor Models

Most of the components we use in the design in our embedded systems will be highly integrated, meaning the use of individual transistors is usually not required. Nevertheless, these integrated components indeed have transistors inside. Therefore the understanding of its behavior can be derived from the operation of its internal devices. For example, the speed and input current of a bipolar op amp is similar to individual bipolar transistors. In particular, bipolar transistors are fast yet require input base currents to activate the device. Therefore we will find that bipolar op amps will be fast but have a modest input impedance (Figure 11.3).

Figure 11.3
Bipolar transistor model.

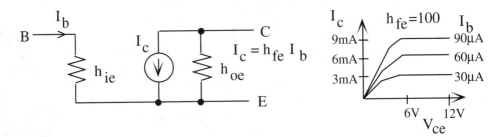

We begin with the DC analysis, where the capacitor acts as an open circuit (Figure 11.4). Assume h_{ie} is 1 kΩ and h_{fe} is 100. To make I_b-60 μA, we let R_1 be 12 V/60μA − h_{ie}, which equals 199 kΩ. Since I_c is h_{fe}· I_b, I_c is 6 mA, making V_{out} equal to 6 V. Next we consider the AC response, where the capacitor acts as a short circuit. The base current I_b will be V_{in}/h_{ie}, or $V_{in}/1000$. To prevent saturation, we will limit the amplitude of the AC input voltage to 60 mV. Again I_c is h_{fe}· I_b, which is $V_{in}/10$. The output current I_c is converted to an output voltage through R_2, thus V_{out} is − R_2 · I_c, or −100 · V_{in}. Putting the DC and AC responses together yields

$$V_{out} = 6 - 100 \cdot V_{in}$$

Figure 11.4
Bipolar transistor
amplifier.

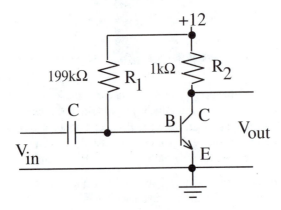

On the other hand, FET transistors are also fast but require input base currents only to turn the device on or off. Therefore we will find that FET op amps will be fast and have a large input impedance (Figure 11.5).

Figure 11.5
JFET/MOSFET transistor model.

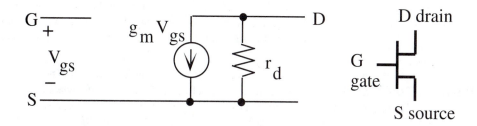

11.2.2
Ideal Op Amps

We will begin our discussion of op amps with the ideal op amp. For simple analog circuits using high-quality devices, the ideal model will be sufficient (Figure 11.6). The open-loop gain of op amps used in this book is shown in Table 11.3.

Figure 11.6
Ideal op amp.

$$V_{out} = K \cdot (V_y - V_x)$$

Table 11.3
Open-loop gain for various op amps used in this chapter.

Op amp	K
3140	100,000
OP07	200,000
OP15	100,000
TLC2272/TLC2274	135,000
TLC1078/TLC1079	525,000

Rule 1. Voltage ranges are bounded by the supply voltages Most op amps are powered with ± 12-V supplies. On the other hand, it will be convenient to power our embedded systems with a single voltage. Either way, we assume the input and output voltages will be between $+V_s$ and $-V_s$. A "rail-to-rail" op amp operates in its linear mode for output voltages all the way from $-V_s$ to $+V_s$. It will be convenient to power these analog components with the same +5-V supply as our digital devices.

Rule 2. Input currents are zero Because the input impedance of the op amp is large, we will assume that I_x and I_y are zero.

Rule 3. Negative feedback drives V_x to equal V_y If a feedback resistor is placed between the output and the negative terminal of the op amp, then this feedback will select an output such that V_x is very close to V_y. In the ideal model we let $V_x = V_y$. One way to justify this behavior is to recall that $V_{out} = K \cdot (V_y - V_x)$. Since K is very large, the only way for V_{out} to be between $-V_s$ and $+V_s$ is for V_x to be very close to V_y.

Rule 4. Positive feedback or no feedback drives V_{out} to equal $-V_s$ or $+V_s$ If a feedback resistor is placed between the output and the positive terminal of the op amp, then this feedback will saturate the output to either the positive or negative supply. With no feedback, the output will also saturate. In both cases the output will saturate to the positive supply if $V_y > V_x$ and to the negative supply if $V_x > V_y$. We will see later that positive feedback can be used to create hysteresis.

11.2.3
Realistic Op Amp
Models

Although the input impedance of an op amp is large, it is not infinite, and some current enters the input terminals. We define the *common-mode input impedance* as the common-mode voltage divided by the common-mode current. For most op amps this parameter is quite large (Figure 11.7). We define the *differential input impedance* as the differential voltage divided by the differential current. This parameter varies considerably from op amp to op amp (Figure 11.8). The input impedance of op amps used in this book is shown in Table 11.4.

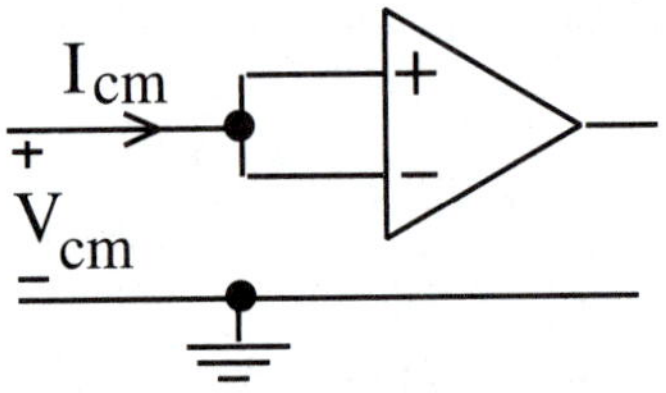

Figure 11.7
Definition of op amp common-mode input impedance.

$$R_{cm} = \frac{V_{cm}}{I_{cm}}$$

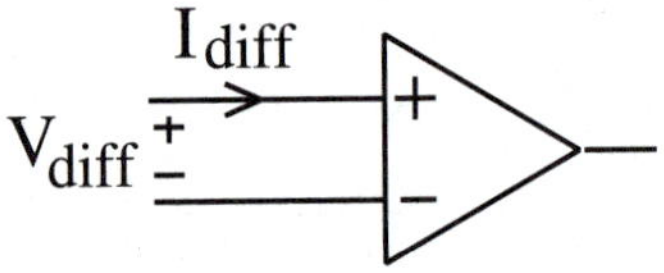

Figure 11.8
Definition of op amp differential input impedance.

$$R_{diff} = \frac{V_{diff}}{I_{diff}}$$

Table 11.4
Input impedances for various op amps used in this chapter.

Op amp	R_{cm}	R_{diff}
3140	1 TΩ	1 TΩ
OP07	200 GΩ	20 MΩ
OP15	1 TΩ	1 TΩ
TLC2272/TLC2274	1 TΩ	1 TΩ
TLC1078/TLC1079	1 TΩ	1 TΩ

The next realistic parameter we will define is *open-loop output impedance* (Figure 11.9). When the opamp is used without feedback, the opamp output impedance is defined as the open-circuit voltage divided by the short-circuit current. The output impedance is a measure of how much current the opamp can source or sink. The open-loop output impedance of the OP07, OP15, and 3140 is 60 Ω.

Figure 11.9
Definition of op amp open-loop output impedance.

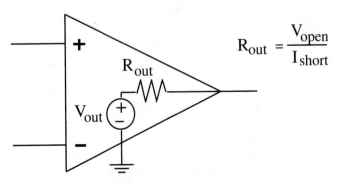

$$R_{out} = \frac{V_{open}}{I_{short}}$$

Observation: The input and output impedances of the op amp are not necessarily the same as the input and output impedances of the entire analog circuit.

The op amp *offset voltage* V_{os} is defined as the voltage difference between V_y and V_x, which yields an output of zero. The *offset current* I_{os} is defined as the current difference between V_y and V_x (Figure 11.10). If the gain of our circuit is large, these offsets can have a significant effect on our output (Table 11.5).

Figure 11.10
Definition of op amp offset voltage and offset current.

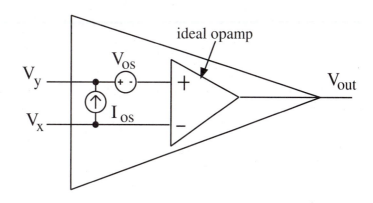

Table 11.5
Offset voltages and currents for various op amps used in this chapter.

Op amp	V_{os}	I_{os}
3140	5 mV	0.1 pA
OP07	75 µV	2.8 nA
OP15	0.5 mV	10 pA
LF351	10 mV	100 pA
TLC2272/TLC2274	2.5 mV	0.5 pA
TLC2272A/TLC2274A	950 µV	0.5 pA
TLC1078/TLC1079	850 µV	0.1 pA

We can eliminate these offsets either by using a more expensive lower-offset op amp, like the OP07, or using a null offset potentiometer (pot) (Figure 11.11). How to adjust this pot will be discussed later in Section 11.27.

Figure 11.11
Null offset pot to eliminate the offset voltage and offset current.

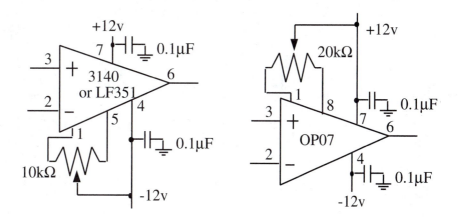

Observation: The use of a null offset pot increases manufacturing costs and incurs a labor cost to adjust it periodically. Therefore, the overall system cost may be reduced by using more expensive op amps that do not require a null offset pot.

Observation: While many op amps come in eight-pin packages with standard connections to pins 2, 3, 4, 6, 7, the pin connections, voltage levels, and pot values for the null offset pot are often different.

Observation: Some op amps, like the TLC227X and TLC107X, do not have a mechanism to attach a null offset pot.

Observation: If the gain and offset are small enough not to saturate the output, then the offset error can be corrected in software by adding/subtracting an appropriate constant.

The op amp *bias current* I_b is defined as the common current coming out of V_y and V_x (Figure 11.12 and Table 11.6).

Figure 11.12
Definition of op amp
bias current.

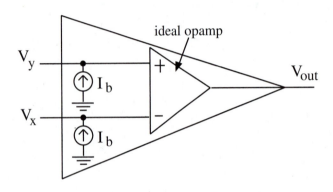

Table 11.6
Input bias currents for
various op amps used in
this chapter.

Op amp	I_b
OP07	3 nA
OP15	50 pA
TLC2272/TLC2274	1 pA
TLC1078/TLC1079	0.6 pA

We can usually reduce the effect of bias current by selecting resistors to equalize the effective impedance to ground from the two input terminals. Figure 11.13 shows two noninverting amplifier circuits. Both circuits have a gain of 5. The 8kΩ (10kΩ || 40kΩ) re-

Figure 11.13
Example approach to
reducing the effect of
bias current.

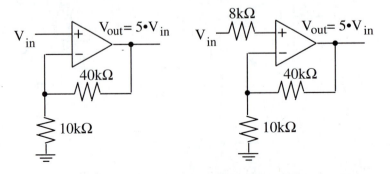

sistor was added to the circuit on the left to reduce the effect of bias current. The effective impedance to ground from both the plus and minus terminals is now 8 kΩ.

The *input voltage noise* V_n arises from the thermal noise generated in the resistive components within the op amp (Figure 11.14). Because of the white noise process, the magnitude of the noise is a function of the bandwidth (BW) of the analog circuit. This parameter varies quite a bit from op amp to op amp. To calculate the root-mean-square (RMS) amplitude of the voltage noise, we need to calculate $V_n = e_n \cdot \sqrt{BW}$. To reduce the effect of noise we can limit the BW of the analog system using an analog low-pass filter (Table 11.7).

Figure 11.14
Definition of op amp noise voltage and noise current.

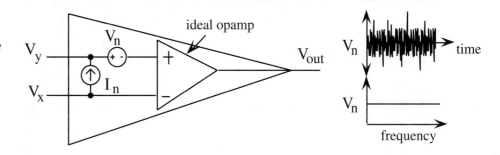

Table 11.7
Voltage noise densities for various op amps used in this chapter.

Op amp	e_n
OP07	13 nV/$\sqrt{Hz}$
OP15	20 nV/$\sqrt{Hz}$
TLC2272/TLC2274	50 nV/$\sqrt{Hz}$
TLC1078/TLC1079	68 nV/$\sqrt{Hz}$

There are two approaches to defining the transient response of our analog circuits. In the frequency domain we can specify the frequency and phase response (Figures 11.15 and 11.16). In the time domain, we can specify the step response. For most simple analog circuits designed with op amps, the frequency response depends on the op amp performance

Figure 11.15
Idealized gain versus frequency relationship of an OP07.

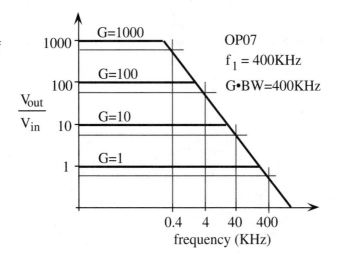

Figure 11.16
Actual gain and phase
versus frequency
relationship of a
TLC2272.

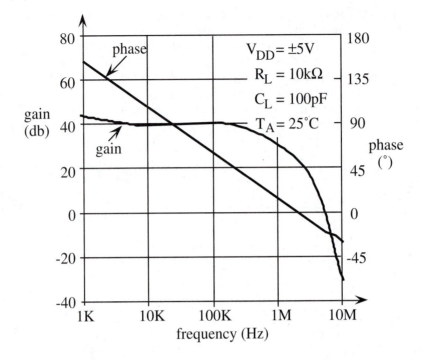

and the analog circuit gain. If the unity gain op amp frequency response is f_1, then the frequency response at gain G will be f_1/G, as illustrated in Figure 11.15. The BW is defined as the frequency at which the gain (V_{out}/V_{in}) drops to 0.707 of the original (Table 11.8).

The output of the op amp will be connected to other circuits. The input impedance and input capacitance of these circuits will load the op amp. Let R_L and C_L be the combined resistive and capacitive loads, respectively. The presence of R_L and C_L will reduce the gain and frequency response of the circuit (Figure 11.17).

Table 11.8
Closed-loop unity-gain
BWs (gain · BW product)
for various op amps
used in this chapter.

Op amp	f_1
OP07	400 kHz
3140	3.7 MHz
TLC2272/TLC2274	2.18 MHz
TLC1078/TLC1079	85 kHz

Figure 11.17
Resistive and capacitive
loads affect the gain and
frequency response.

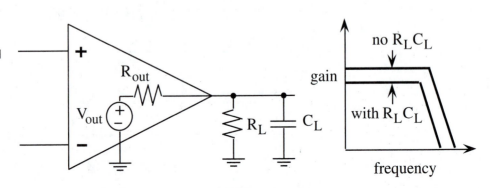

Figure 11.18
Definition of step response and settling time.

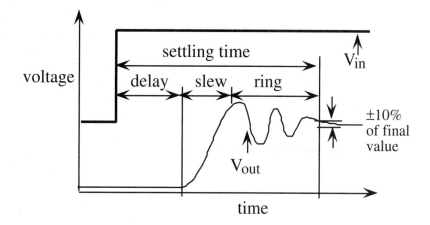

The *step response* of an analog circuit has three phases: delay, slewing, and ringing. The step response parameter can be specified in either the total settling time to 90% of the final value or in the slewing rate (V/s) (Figure 11.18).

When we consider the performance of a linear amplifier, normally we specify the voltage gain, input impedance, and output impedance. These three parameters can be lumped into a single parameter A_{db}, called the *power gain*. Let V_{in}, R_{in} be the inputs and V_{out}, R_{out} be the outputs of our amplifier. The input and output powers are $P_{in} = V^2_{in}/R_{in}$ and $P_{out} = V^2_{out}/R_{out}$, respectively. Then the power gain in decibels has voltage gain and impedance components.

$$A_{db} = 10 \log_{10} \frac{P_{out}}{P_{in}} = 20 \log_{10} \frac{V_{out}}{V_{in}} + 10 \log_{10} \frac{R_{in}}{R_{out}}$$

**11.2.4
Types of Op Amps**

Table 11.9 illustrates the wide range of available op amps. The bipolar, FET, and BiFET op amps have a good balance of features and will suffice for many applications. On the other hand, rail-to-rail micropower devices can be used in applications that run off a signal battery supply.

Table 11.9
Types of op amps.

Type	Description	Applications
Bipolar	General purpose	General purpose
FET	High input impedance	Instrumentation
CAZ	Bipolar plus auto-zero	Small signal applications
BiFET	Bipolar plus FET	General purpose
Superbeta	Bipolar with high input impedance	Weak signals
Micropower	CMOS, LinCMOS	Battery operation
Isolation	Transformer or optical	Medical and industrial
Chopper	DC to AC to DC	Low error
Varactor	Diode input, low bias current	Current amplifier, photomultiplier

**11.2.5
Saturation
Properties**

The open-loop or saturated mode performance of a 3140 op amp can be studied by looking at two very simple circuits. Saturated mode is used to create a threshold detector. When the input V_{in} in the first circuit is above the reference V_{ref}, then the output V_{out} saturates to $+V_s - 3$. Similarly, when V_{in} is below V_{ref}, then V_{out} saturates to $-V_s + 0.13$. The 3140 will operate on a supply range of ±2 to ±18 V. Notice that it will create a digital output when +V_s is +5 V and −V_s is ground. The 3140 offset voltage of 5 mV can be reduced with the null offset pot connected to pins 1, 5. The 3140 offset current (0.1 pA), and bias current (2 pA) are both very low. The input impedance is 1 TΩ. The input voltages on pins 2 and 3 of the 3140 must be between −V_s − 0.5 and +V_s + 8. The open-loop output impedance of the 3140 is 60 Ω. The 3140 can sink −1 mA output current and source up to +10 mA. The 3140 slewing rate is 7 V/μs, with a settling time of 80 ns (Figure 11.19). If the pin 2 and 3 inputs are reversed, the op amp can be used to create a negative-logic comparator (Figure 11.20).

Figure 11.19
A positive-logic voltage comparator using a 3140 op amp.

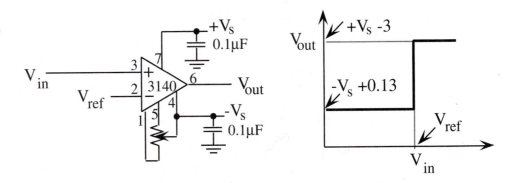

Figure 11.20
A negative-logic voltage comparator using a 3140 op amp.

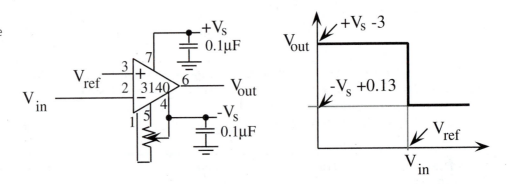

Observation: The 3140 will produce a digital output (high = 2 V, low = 0.13 V) when +V_s is +5 V and −V_s is ground.

Threshold detectors are very important in embedded systems. Recall that all through Chapter 6, a threshold detector was used to convert an external analog signal into a digital signal so that the period, pulse width, or frequency could be measured. An entire class of voltage comparators exist for this purpose. The LM311 is a typical voltage comparator with analog inputs (V_x, V_y), analog voltage supplies (+V_s, −V_s), digital output (D_{out}), and digital supply (pin 1 ground). The LM311 will operate on a wide range of analog supply voltages (+V_s to −V_s) from 0 to +5 V, all the way to −15 to +15 V. When the input V_y is above the input V_x, then the digital output D_{out} becomes high impedance, and the pull-up resistor creates a +5-V output. On the other hand, when V_y is below V_x, then D_{out} saturates to 0.4 V. The LM311 offset voltage of 7.5 mV can be reduced with the null offset pot connected to pins 5, 6. The LM311 offset current is 50 nA, and bias current is 250 nA. The input volt-

Figure 11.21
A voltage comparator using a LM311.

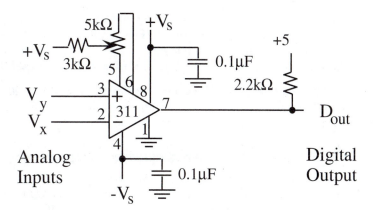

ages of the LM311 must be between $-V_s$ and $+V_s$. The LM311 can sink -8 mA output low current. Since the LM311 has an open-collector output, it cannot source any output high current. The LM311 settling time is 200 ns (Figure 11.21).

11.2.6
Simple Rules for Linear Op Amp Circuits

A wide range of analog circuits can be designed by following these simple design rules.

1. Choose quality components. It is important to use op amps with good enough parameters. Similarly, we should use low-tolerance resistors and capacitors. On the other hand, once the preliminary prototype has been built and tested, then we could create alternative designs with less expensive components. Because a working prototype exists, we can explore the cost/performance trade-off.

2. Negative feedback is required to create a linear mode circuit. As mentioned earlier, the negative feedback will produce a linear I/O response. In particular, we place a resistor between the negative input terminal and the output (Figure 11.22).

Figure 11.22
Negative feedback is created by placing a resistor between the negative terminal input and the output.

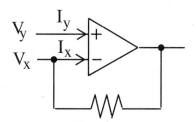

3. Assume no current flows into the op amp inputs. Since the input impedance of the op amp is large compared to the other resistances in the circuit, we can assume that $I_x = I_y = 0$.

4. Assume negative feedback equalizes the op amp input voltages. If the analog circuit is operating in linear mode with negative feedback, then we can assume $V_x = V_y$.

5. Choose resistor values in the 1 kΩ to 1 MΩ range. To have the resistors in the circuit be much larger than the output impedance of the op amp and much smaller than the input impedance of the op amp, we choose resistors in the 1 kΩ to 1 MΩ range. If we can, it is better to restrict to the 10 kΩ to 100 kΩ range. If we choose resistors below 1 kΩ, then currents will increase. If the currents get too large, the batteries will drain faster and the op amp may not be able to source or sink enough current. As the resistors go above 1 MΩ, the white noise increases, the current errors (I_{os}, I_b, I_n) become more significant. In addition, low-tolerance precision resistors are expensive in sizes above 2 MΩ.

6. The analog circuit BW depends on the gain and the op amp performance. Let the unity gain op amp frequency response be f_1 and the analog circuit gain be G. The frequency response or BW of the analog circuit will be f_1/G.

7. Equalize the effective resistance to ground at the two op amp input terminals. To study the bias currents, consider all other voltage sources as shorts to ground and all other current sources as open circuits. Adjust the resistance values in the circuit so that the impedance from the positive terminal to ground is the same as the impedance from the negative terminal to ground. In this way, the bias currents will create a common-mode voltage, which will not appear at the op amp output because of the common-mode rejection of the op amp (recall the op amp amplifies differential voltage inputs).

8. The input impedance of the analog circuit is the input voltage divided by the input current. If the analog circuit has a single input voltage, then the input impedance Z_{in} is simply the input voltage divided by the input current (Figure 11.23). If the input stage of the analog circuit is a differential amplifier with two input voltages, then we can specify the common-mode input impedance Z_{cm} and the differential mode input impedance Z_{diff} (Figure 11.24).

Figure 11.23
Definition of input impedance for an analog circuit with a single input.

$$Z_{in} = \frac{V_{in}}{I_{in}}$$

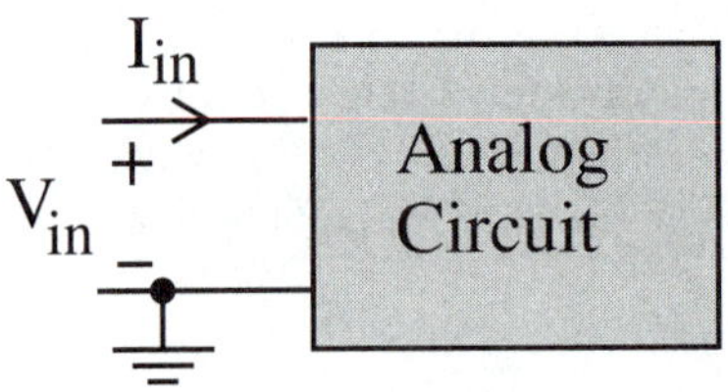

Figure 11.24
Definition of input impedance for an analog circuit with two inputs.

$$Z_{cm} = \frac{V_{cm}}{I_{cm}}$$

$$Z_{diff} = \frac{V_{diff}}{I_{diff}}$$

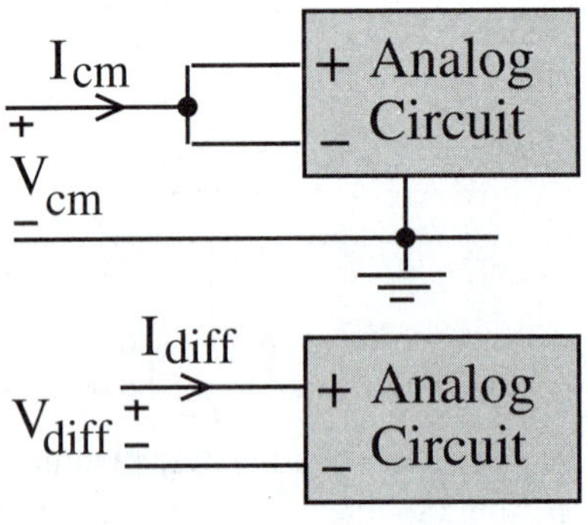

Observation: In most cases, the differential mode input impedance of the analog circuit will be the differential mode input impedance of the op amp.

9. **Match input impedances to improve common-mode rejection ratio (CMRR).** If the input stage of the analog circuit is a differential amplifier with two input voltages, a very important performance parameter is called CMRR. It is assumed that the signal of interest is the differential voltage, whereas common-mode voltages are considered noise. The CMRR is defined to be the ratio of the differential gain divided by the common-mode gain. In decibels, it is calculated as

$$CMRR = 20 \cdot \log_{10} \frac{G_{diff}}{G_{cm}}$$

Therefore a differential amplifier with a large CMRR will pass the signal and reject the noise (Figure 11.25).

Figure 11.25
Definition of common mode rejection ratio (CMRR).

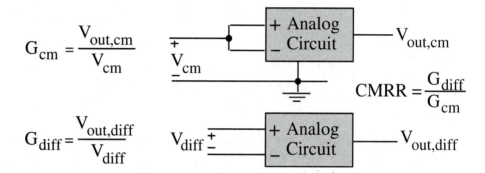

$$G_{cm} = \frac{V_{out,cm}}{V_{cm}}$$

$$G_{diff} = \frac{V_{out,diff}}{V_{diff}}$$

$$CMRR = \frac{G_{diff}}{G_{cm}}$$

There are two design sets of impedances we must match to achieve a good CMRR. Each amplifier input has a separate input impedance to ground, shown as Z_{in1} and Z_{in2} in Figure 11.26. To improve CMRR, we make Z_{in1} equal to Z_{in2}. Similarly, signal source has a separate output impedance to ground, Z_{out1} and Z_{out2}. Again, we try to make Z_{out1} equal to Z_{out2}. On the other hand, if Z_{in1} does not equal Z_{in2} or if Z_{out1} does not equal Z_{out2} then a common-mode signal (e.g., added noise in the cable) will appear as a differential signal to the analog circuit and thus be present in the output.

Figure 11.26
Circuit model for improving the CMRR.

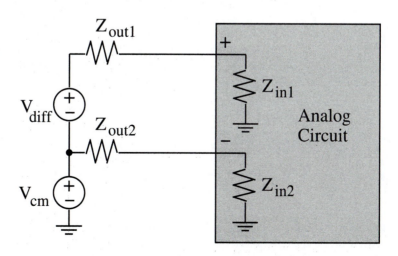

The first linear mode circuit we will study is the *inverting amplifier* (Figure 11.27). The gain is the R_2/R_1 ratio. Notice that the gain response is independent of R_3. We choose R_3 to be the parallel combination of $R_1\|R_2$ so that the effect of the bias currents is reduced. The bypass filtering and null offset pot circuit for the OP07 and 3140 was shown in Figure 11.11. Because I_y is zero, V_y is also zero. Because of negative feedback V_x equals V_y. Thus, V_x equals zero too. Because V_x is zero, I_{in} is V_{in}/R_1 and I_2 is $-V_{out}/R_2$. Because I_x is zero, I_{in} equals I_2. Setting I_{in} equal to I_2 yields

$$V_{out} = -\frac{R_2}{R_1} \cdot V_{in}$$

Figure 11.27
Inverting amplifier.

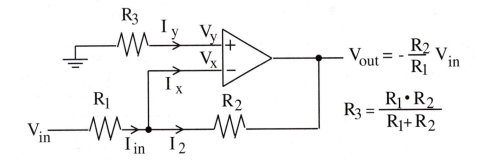

The input impedance Z_{in} of this circuit (defined as V_{in}/I_{in}) is R_1. If the circuit were built with an OP07 (which has a gain BW product of 400 kHz), the BW of this circuit will be 400 kHz divided by the gain.

Common error: This low input impedance of R_1 may cause loading on the previous analog stage.

Observation: The inverting amplifier input impedance is independent of the op amp input impedance.

To eliminate the offset voltage and offset current, connect the amplifier input to ground and adjust the pot until the output goes to zero (Figure 11.28).

Figure 11.28
Adjusting the null offset pot in an inverting amplifier built with an OP07.

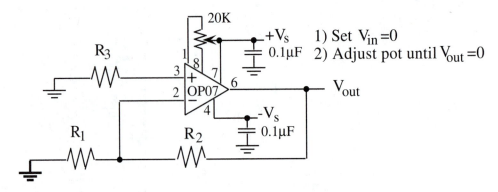

Common error: It is a mistake to ground the op amp input terminals when adjusting the null offset pot. This approach will only correct for offset voltage and not both offset voltage and offset current.

The negative feedback will reduce the output impedance of the amplifier, Z_{out}, to a value much less than the output impedance of the op amp itself, R_{out}. To calculate Z_{out}, we first determine the open-circuit voltage

$$V_{open} = -\frac{R_2}{R_1} \cdot V_{in}$$

Using Figure 11.29, we next determine the short-circuit current I_{short}. The amplifier output impedance is defined as V_{open}/I_{short}, which for this inverting amplifier is

$$Z_{out} = R_{out} \cdot \frac{R_2 + R_1}{K \cdot R_1}$$

Figure 11.29
Determination of the short-circuit current of an inverting amplifier.

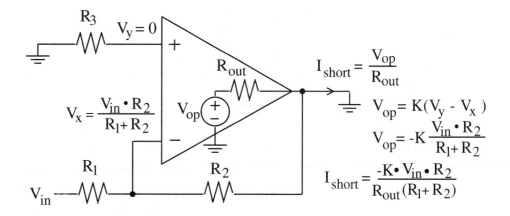

Observation: The output impedance of analog circuits using op amps with negative feedback is typically in the mΩ's.

11.2.7.2
Noninverting Amplifier

The second linear mode circuit we will study is the *noninverting amplifier* (Figure 11.30). The gain is $1 + R_2/R_1$. Notice that the noninverting amplifier cannot have a gain less than 1. Just like the inverting amp, the gain response is independent of R_3. So, we choose R_3 to be the parallel combination of $R_1 \| R_2$ so that the effect of the bias currents is reduced. Because I_y is zero, V_y equals V_{in}. Because of negative feedback V_x equals V_y. Thus, V_x equals V_{in} too. Calculating currents we get, I_1 is V_{in}/R_1 and I_2 is $(V_{out} - V_{in})/R_2$. Because I_x is zero, I_1 equals I_2. Setting I_1 equal to I_2 yields

$$V_{out} = \left(1 + \frac{R_2}{R_1}\right) \cdot V_{in}$$

Figure 11.30
Noninverting amplifier.

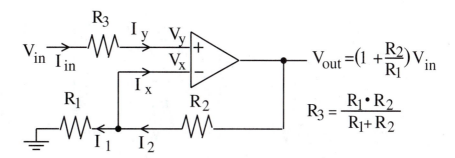

Using the simple op amp rules, I_y is zero, so the input impedance Z_{in} of this circuit (defined as V_{in}/I_{in}) would be infinite. In this situation we specify the amplifier input impedance to be the op amp input impedance. If the circuit were built with an OP07 (which has a gain BW product of 400 kHz), the BW of this circuit will be 400 kHz divided by the gain. To eliminate the offset voltage and offset current, connect the amplifier input to ground and adjust the pot until the output goes to zero (Figure 11.31).

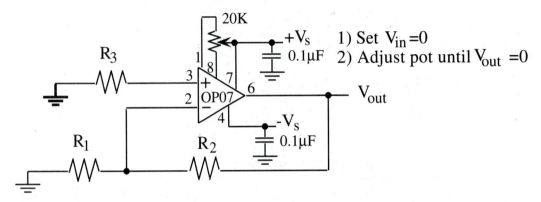

Figure 11.31   Adjusting the null offset pot in a noninverting amplifier built with an OP07.

To calculate Z_{out}, we first determine the open-circuit voltage

$$V_{open} = \left(1 + \frac{R_2}{R_1}\right) \cdot V_{open}$$

Using Figure 11.32, we next determine the short-circuit current I_{short}. The amplifier output impedance is defined as V_{open}/I_{short}, which for this noninverting amplifier is

$$Z_{out} = \frac{\left(1 + \dfrac{R_2}{R_1}\right) \cdot V_{in}}{\dfrac{K \cdot V_{in}}{R_{out}}} = R_{out} \cdot \frac{R_2 + R_1}{K \cdot R_1}$$

Figure 11.32
Determination of the
short-circuit current of a
noninverting amplifier.

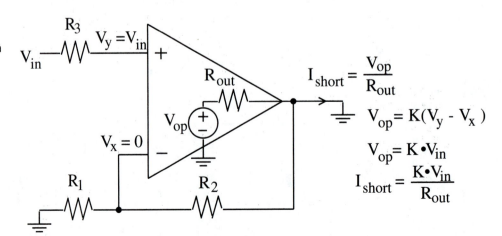

$$I_{short} = \frac{V_{op}}{R_{out}}$$

$$V_{op} = K(V_y - V_x)$$

$$V_{op} = K \cdot V_{in}$$

$$I_{short} = \frac{K \cdot V_{in}}{R_{out}}$$

<table>
<tr><td>11.2.7.3
Buffer Amplifier or
Voltage Follower</td><td>A variation of the noninverting amplifier is the buffer amplifier or voltage follower. The gain is 1. This analog circuit is used to separate an analog stage with a bad (high) output impedance from an analog stage with a bad (low) input impedance. It is useful because it has a very high Z_{in} and a very low Z_{out}. Any resistor, including none, could be used in place of the 10kΩ resistors in Figure 11.33.</td></tr>
</table>

Figure 11.33
Buffer amplifier or
voltage follower.

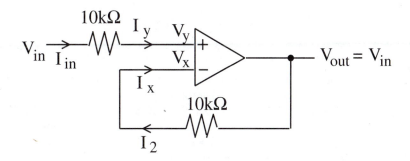

Because I_y is zero, V_y equals V_{in}. Because of negative feedback, V_x equals V_y. Thus, V_x equals V_{in} too. Because I_x is zero, I_2 is zero too, so this makes $V_{out} = V_{in}$. Just like the noninverting amplifier, when we use simple op amp rules I_y is zero, so the input impedance Z_{in} of this circuit (defined as V_{in}/I_{in}) would be infinite. Again we specify the amplifier input impedance to be the op amp input impedance. If the circuit were built with an OP15 (which has a gain BW product of 4 MHz), the BW of this circuit will be 4 MHz. To eliminate the offset voltage and offset current connect the amplifier input to ground and adjust the pot until the output goes to zero. To calculate Z_{out}, we first determine the open-circuit voltage $V_{open} = V_{in}$. Using Figure 11.34, we next determine the short-circuit current I_{short}. The amplifier output impedance is defined as V_{open}/I_{short}, which for this buffer amplifier is

$$Z_{out} = \frac{V_{in}}{\dfrac{K \cdot V_{in}}{R_{out}}} = \frac{R_{out}}{K}$$

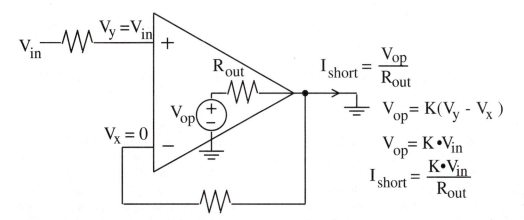

Figure 11.34 Determination of the short circuit current of a noninverting amplifier.

Figure 11.35
Inverting adder analog
circuit.

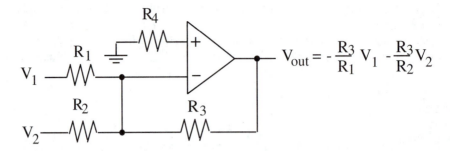

11.2.7.4   The next linear mode circuit is a variation of the inverting amplifier, called an *adder am-*
Inverting Adder   *plifier* (Figure 11.35). There are two gains, R_3/R_1 and R_3/R_2. Notice that the gain re-
Amplifier   sponse is independent of R_4. We choose R_4 to be the parallel combination of $R_1\|R_2\|R_3$
so that the effect of the bias currents is reduced. There are two input impedance's for this
circuit. The Z_{in} as seen by V_1 is R_1, and the Z_{in} as seen by V_2 is R_2. To eliminate the offset voltage and offset current connect both amplifier inputs to ground and adjust the pot
until the output goes to zero.

11.2.7.5   This *differential analog amplifier* has two input voltages, V_1 and V_2 (Figure 11.36). The ba-
Subtractor Amplifier   sic idea of this circuit is to create a voltage output that is the difference between the two
voltage inputs. The gain is determined by the R_2/R_1 ratio. Because of the negative feedback,

Figure 11.36
Differential analog
amplifier.

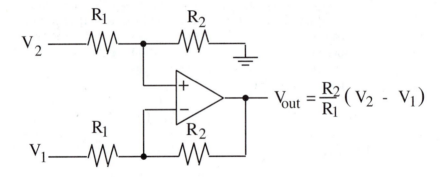

the two op amp input terminals will be at approximately the same voltage, in this case $V_2 \cdot$
$R_2/(R_2+R_1)$. The current through the bottom R_1 resistor is $[V_1-V_2 \cdot R_2/(R_2+R_1))/R_1$. The
current through the bottom R_2 resistor is $-[V_{out}-V_2 \cdot R_2/(R_2+R_1)]/R_2$. Since no current
enters the op amp terminal, these currents must will be equal. Solving for V_{out} we get

$$V_{out} = \frac{R_2}{R_1} \cdot (V_2 - V_1)$$

Since both the positive and negative op amp terminals have the same impedance to ground,
the op amp bias current will be subtracted.

11.2.7.6   The *instrumentation amp* will also amplify a differential voltage, shown in Figure 11.37 as
Instrumentation $V_2 - V_1$. We use instrumentation amplifiers in applications that require a large gain (above
Amplifier   100), a high input impedance, and a good CMRR. One can be built using three high-
quality op amps.

Figure 11.37
Instrumentation
amplifier.

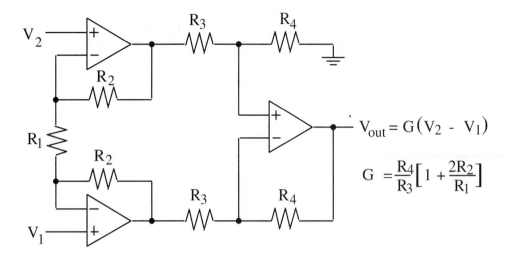

$$V_{out} = G(V_2 - V_1)$$

$$G = \frac{R_4}{R_3}\left[1 + \frac{2R_2}{R_1}\right]$$

Observation: To achieve quality performance with a three-op amp instrumentation amplifier circuit, we must use precision resistors and quality op amps.

Common Error: If you use a potentiometer in place of one of the gain resistors in the Figure 11.37 circuit, then fluctuations in the potentiometer resistance that can occur with temperature, vibration, and time will have a strong effect on the amplifier gain.

Because of the range of applications that require instrumentation amplifiers, chip manufacturers have developed a variety of integrated solutions. In many cases we can achieve higher performance at reduced cost by utilizing one of these ICs. The gain is selected by external jumpers or external resistors. The Analog Devices AD620 is a typical low-cost device ($6) (Figure 11.38).

Common Error: If you use a potentiometer as the R_G gain resistors in the Figure 11.38 circuit, then fluctuations in the potentiometer resistance that can occur with temperature, vibration, and time will have a strong effect on the amplifier gain.

Figure 11.38
Integrated
instrumentation
amplifier.

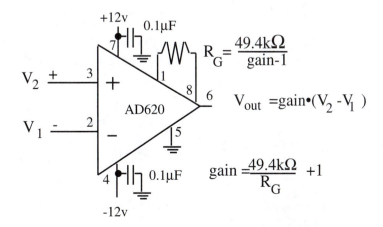

$$R_G = \frac{49.4k\Omega}{gain-1}$$

$$V_{out} = gain \cdot (V_2 - V_1)$$

$$gain = \frac{49.4k\Omega}{R_G} + 1$$

Figure 11.39
Calculating BW in a
multiple-stage analog
system.

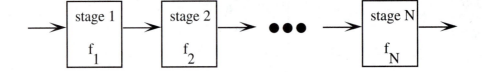

11.2.7.7 · When we combine multiple analog stages, then the overall BW of system (f_T) can be
Multiple-Stage Amplifier calculated from the BWs of the individual stages (f_n) (Figure 11.39).

$$f_T = \sqrt{\frac{1}{\left(\frac{1}{f_1}\right)^2 + \left(\frac{1}{f_2}\right)^2 + \cdots + \left(\frac{1}{f_N}\right)^2}}$$

11.2.8
Offset Adding Op
Amp Circuit

If we need to add constant offsets in our analog circuit, a precision voltage reference, like
ones shown in Table 11.10, can be used (Figure 11.40). In both chips, a +12-V supply
powers the chip and the output is a very stable +5.00 V. This approach is preferable to a
resistor divider from the +12-V supply.

Table 11.10
Parameters of various
precision reference
voltage chips.

Part	Voltage (V)	±Accuracy (mV)	Temperature stability (ppm/°C)
AD1580	1.215	1 to 10	100
AD589	1.235	5	10 to 100
REF191	2.048	2 to 10	5 to 25
AD580	2.5	10 to 75	1.75 to 15
AD680	2.5	5 to 10	20 to 30
AD1403	2.5	10 to 25	25 to 0
REF43	2.5	1.5 to 2.5	10 to 25
AD780	2.5/3.0	1 to 5	3 to 7
AD586	5.0	2.5 to 20	2 to 25
REF02	5.0	2.5 to 10	25 to 250
REF195	5.0	10	20
AD581	10.0	5 to 30	5 to 30
AD587	10.0	5 to 10	5 to 20
AD633	10.0	5	3
REF01	10.0	5 to 10	25 to 65
AD584	Selectable	3 to 7.5	3 to 30

Common Error: If you use a resistor divider from the +12-V supply to create a voltage constant, then the power supply ripple will be added directly to your analog
signal.

Figure 11.40
A precision reference
provides a very stable
voltage.

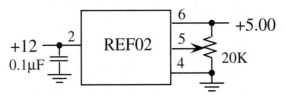

Common Error: Precision reference chips do not provide much output current and
should not be used to power other chips.

In this next circuit, we will study an application of adding offset to a circuit. Consider a system where the input voltage ranges from −5.0 to +5.0 V and we wish to map it into the 0 to +10 V range of our ADC. There are two possible conversion equations we could implement:

$$V_{out} = V_{in} + 5 \quad \text{or} \quad V_{out} = 5 - V_{in}$$

This second version is simpler to design. The basic idea is to start with the differential amplifier and connect the +5.00 reference voltage to one of the inputs (Figure 11.41). Using OP07 op amps the BW of each stage is 400 kHz. Therefore the total BW is 400 kHz/√2, or 283 kHz. The input impedance of the analog circuit is the input impedance of the op amp, 20 MΩ for the OP07.

Figure 11.41
Offset adding circuit.

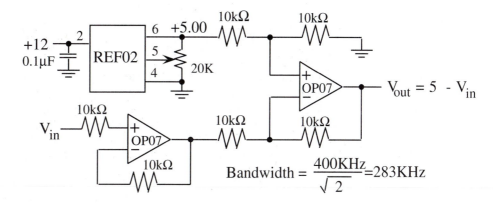

11.2.9
High-Gain Op
Amp Circuit

Amplifiers can be added in series to boost the gain. In this circuit, each stage has a gain of 50, for a total gain of 2500. Using OP07 op amps the BW of each stage is 400 kHz/50, or 8 kHz. Therefore the total BW is 8 kHz/√2, or 5.7 kHz. The input impedance of the analog circuit is 10 kΩ independent of the op amp parameters. In this first example, two inverting amplifiers are used (Figure 11.42).

Figure 11.42
High-gain amplifier
using two inverting op
amp stages.

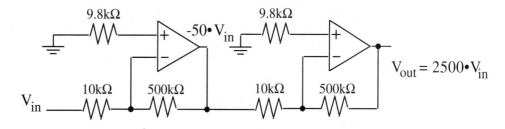

In the second example, two noninverting amplifiers are used. The input impedance of this analog circuit is the input impedance of the op amp, 20 MΩ for the OP07 (Figure 11.43).

Figure 11.43
High-gain amplifier
using two noninverting
op amp stages.

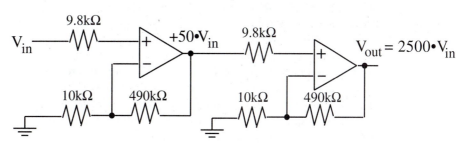

**11.2.10
Current-to-
Voltage Circuit**

In this circuit, the input is a current, I_{in}. Because no current enters the op amp terminal, this current will cross the feedback resistor R. Because of the negative feedback, the two op amp input terminals will be at approximately the same voltage, in this case zero. Therefore, the output voltage V_{out} will be $-I_{in} \cdot R$ (Figure 11.44).

Figure 11.44
Current-to-voltage
converter.

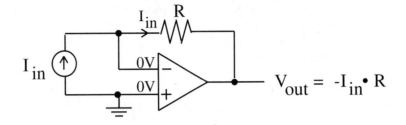

**11.2.11
Voltage-to-
Current Circuit**

In this circuit, the input is a voltage, V_{in}. The resistance R_1 is fixed and known (e.g., 10 kΩ), but the resistance R_L is external and unknown. The basic idea is that this circuit will deliver a constant current across R_L, independent of the value of R_L. Because of the negative feedback, the two op amp input terminals will be at approximately the same voltage, in this case V_{in}. The current across R_1 will be V_{in}/R_1. Since no current enters the op amp terminal, this same current must also pass through R_L (Figure 11.45).

> **Observation:** A precision reference voltage chip together with this voltage-to-current circuit creates a precision reference current system. In the application notes of the reference chips shown in Table 11.10 are many other circuits that create a precision reference current.

Figure 11.45
Voltage-to-current
converter.

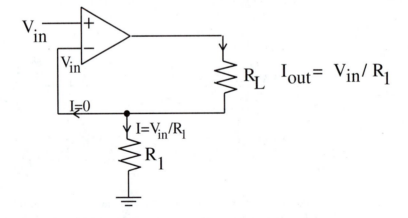

**11.2.12
Integrator Circuit**

In this circuit, the input is a voltage, V_{in}. The basic idea of this circuit is to create a voltage output that is the integral of the voltage input. Often a computer-controlled switch is added to short the voltage across the capacitor, initializing the output voltage to zero. Because of the negative feedback, the two op amp input terminals will be at approximately the same voltage, in this case zero. The current across the input R will be V_{in}/R. Since no current enters the op amp terminal, this same current must also pass through the capacitor. The second R on the positive terminal was added to subtract the bias current of the op amp (Figure 11.46). A low-offset op amp (e.g., OP07) should be used to reduce the error caused by the offset.

Figure 11.46
Analog integrator circuit.

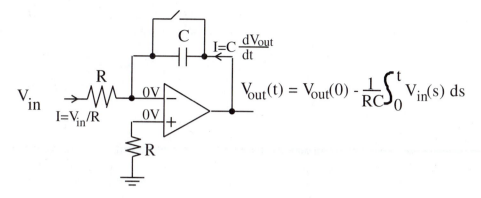

11.2.13
Derivative Circuit

In this circuit, the input is a voltage, V_{in}. The basic idea of this circuit is to create a voltage output that is the derivative of the voltage input. Because of the negative feedback, the two op amp input terminals will be at approximately the same voltage, in this case zero. The current across the input capacitor is related to the derivative of the input. Since no current enters the op amp terminal, this same current must also pass through the feedback resistor. The second R on the positive terminal was added to subtract the bias current of the op amp (Figure 11.47).

Figure 11.47
Analog derivative circuit.

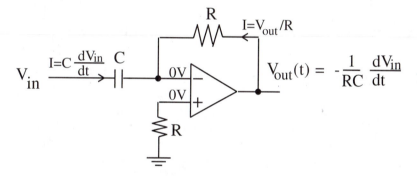

11.2.14
Voltage
Comparators with
Hysteresis

We can use a voltage comparator to detect events in an analog waveform. The range across the input voltage can vary and is usually determined by the analog supply voltages of the comparator. The output takes on two values, shown as V_h and V_l in Figure 11.48. A comparator with hysteresis has two thresholds, V_{t+} and V_{t-}. In both the positive- and negative-logic cases the threshold (V_{t+} or V_{t-}) depends on the present value of the output. Hysteresis prevents small noise spikes from creating a false trigger.

Figure 11.48
Input/output response of
voltage converters with
hysteresis.

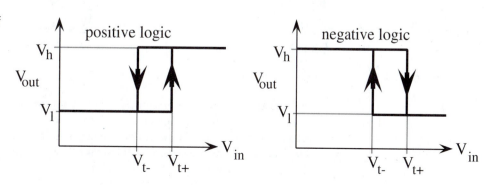

Performance Tip: To eliminate false triggering, we select a hysteresis level (V_{t+} − $V_{t−}$) greater than the noise level in the signal.

In this next circuit, a TLC2272 rail-to-rail op amp is used to design a voltage comparator (Figure 11.49). Since the output swings from 0.01 to 4.99 V, it can be connected directly to an input pin of the microcomputer. On the other hand, since +5 and 0 are used to power the op amp, the analog input must remain in the 0 to +5V range. The hysteresis level is determined by the R_2/R_1 ratio. If the output is at 0 V, then the input goes above +2.55 before the positive terminal of the op amp reaches 2.5 V. Similarly, if the output is at +5 V, then the input goes below +2.45 before the positive terminal of the op amp falls below 2.5 V. In linear mode circuits we should not use the supply voltage to create voltage references, but in a saturated mode circuit, power supply ripple will have little effect on the response.

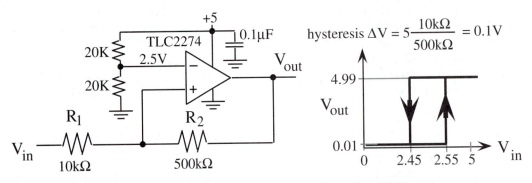

Figure 11.49   A voltage comparator with hysteresis using a rail-to-rail TLC2274.

In this second example, a 3140 op amp is used to design a voltage comparator (Figure 11.50). The output swings from −11.9 to +9 V. On the other hand, since −12 and +12 are used to power the op amp, the analog input can be any value in the −12 to +12 V range. The hysteresis level is again determined by the R_2/R_1 ratio.

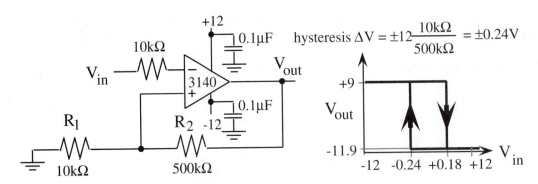

Figure 11.50   A voltage comparator with hysteresis using a 3140.

There exists a wide range of integrated voltage comparators (Figure 11.51). They vary in price, speed, voltage offset, output configuration, and power consumption. The positive feedback is added to this LM311 circuit to produce hysteresis. If the output is at 0 V, then V_y is 4.9 V and the input goes below +4.9 to make the output rise. Similarly, if the output is at +5 V, then V_y is 5 V and the input goes after 5V to make the output fall.

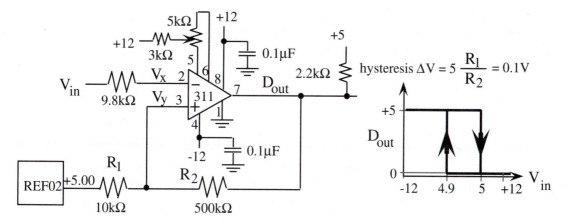

Figure 11.51 A voltage comparator with hysteresis using a LM311.

11.3   Analog Filters

**11.3.1
Simple Active
Filter**

We can add capacitors in parallel with the feedback resistor in the inverting amplifier to create a simple *one-pole low-pass filter* (Figure 11.52). The impedance of a resistor R2 in parallel with the capacitor C is a function of frequency.

$$Z = \frac{R_2}{1 + j\omega R_2 C}$$

Figure 11.52
Complex impedance
model of a resistor in
parallel with a capacitor.

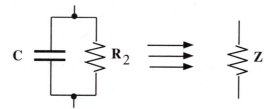

Therefore the gain of the circuit is $-Z/R_1$, which exhibits low-pass behavior. The *cutoff frequency* is defined to be the frequency at which the gain drops to 0.707 of its original value. In this simple low-pass filter, the cutoff frequency f_c is $1/(2\pi R_2 C)$ (Figure 11.53).

Figure 11.53
One-pole low-pass
analog filter.

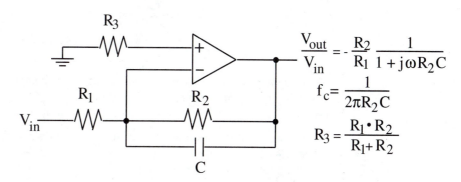

$$\frac{V_{out}}{V_{in}} = -\frac{R_2}{R_1}\frac{1}{1 + j\omega R_2 C}$$

$$f_c = \frac{1}{2\pi R_2 C}$$

$$R_3 = \frac{R_1 \cdot R_2}{R_1 + R_2}$$

Figure 11.54
Frequency response of a
one-pole low-pass
analog filter.

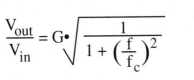

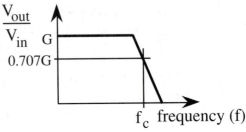

We classify this low-pass filter as one pole, because the transfer function has only one pair of poles in the s plane. One-pole low-pass filters have a gain versus frequency response as shown in Figure 11.54 (G is a constant).

**11.3.2
Butterworth
Filters**

Higher-order analog filters can be designed using multiple capacitors. One of the advantages of the *two-pole Butterworth analog filter* is that as long as the capacitors maintain the 2/1 ratio, the analog circuit will be a Butterworth filter. The design steps for the two-pole Butterworth low-pass filter are as follows (Figure 11.55):

1. Select the cutoff frequency f_c
2. Divide the two capacitors by $2\pi f_c$ (let C_{1A}, C_{2A} be the new capacitor values)

$$C_{1A} = \frac{141.4\ \mu F}{2\pi f_c} \qquad C_{2A} = \frac{70.7\ \mu F}{2\pi f_c}$$

3. Locate two standard-value capacitors (with the 2/1 ratio) with the same order of magnitude as the desired values; let C_{1B}, C_{2B} be these standard value capacitors and let x be this convenience factor

$$C_{1B} = \frac{C_{1A}}{x} \qquad C_{2B} = \frac{C_{2A}}{x}$$

4. Adjust the resistors to maintain the cutoff frequency

$$R = 10\ k\Omega \cdot x$$

Figure 11.55
Two-pole Butterworth
low-pass analog filter.

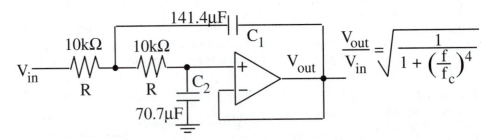

The analog filters in this section all require low-leakage, high-accuracy, and low-temperature coefficient capacitors like Teflon, polystyrene, or polypropylene. Lower-quality (lower cost) Mylar or ceramic capacitors could be used if filter accuracy were less important.

Performance Tip: If you choose standard-value resistors near the desired values, you will save money and the circuit will still be a Butterworth filter. The only difference is that the cutoff frequency will be slightly off from the original specification.

If a faster roll-off is required, a filter with more poles can be used. Higher-order filters require even higher-precision (expensive) capacitors to implement. The design steps for the three-pole Butterworth low-pass filter are as follows (Figure 11.56):

1. Select the cutoff frequency f_c
2. Divide the three capacitors by $2\pi f_c$ (let C_{1A}, C_{2A}, C_{3A} be the new capacitor values)

$$C_{1A} = \frac{354.6 \ \mu F}{2\pi f_c} \qquad C_{2A} = \frac{139.2 \ \mu F}{2\pi f_c} \qquad C_{3A} = \frac{20.2 \ \mu F}{2\pi f_c}$$

3. Locate capacitor values with the same order of magnitude as the desired values; let C_{1B}, C_{2B}, C_{3B} be the new values and let x be this convenience factor

$$C_{1B} = \frac{C_{1A}}{x} \qquad C_{2B} = \frac{C_{2A}}{x} \qquad C_{3B} = \frac{C_{3A}}{x}$$

4. Adjust the resistors to maintain the cutoff frequency

$$R = 10 \ k\Omega \cdot x$$

Figure 11.56
Three-pole Butterworth low-pass analog filter.

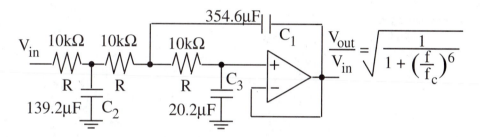

The design steps for the two-pole Butterworth high-pass filter are as follows (Figure 11.57):

1. Select the cutoff frequency f_c
2. Divide the two capacitors by $2\pi f_c$ (let C_A be the new capacitor value)

$$C_A = \frac{1 \ \mu F}{2\pi f_c}$$

3. Locate a standard value capacitor with the same order of magnitude as the desired value; let C_B be this standard value capacitor and let x be this convenience factor

$$C_B = \frac{C_A}{x}$$

4. Adjust the two resistors to maintain the cutoff frequency

$$R_1 = 707 \ k\Omega \cdot x \qquad R_2 = 1414 \ k\Omega \cdot x$$

Figure 11.57
Two-pole Butterworth high-pass analog filter.

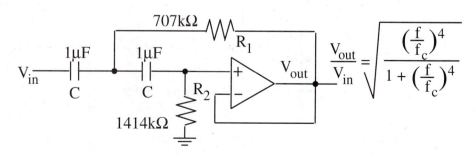

If a faster roll-off is required, a high-pass filter with more poles can be used. The design steps for the three-pole Butterworth high-pass filter are as follows (Figure 11.58):

1. Select the cutoff frequency f_c
2. Divide the three capacitors by $2\pi f_c$ (let C_A be the new capacitor value)

$$C_A = \frac{1\ \mu F}{2\pi f_c}$$

3. Locate a standard-value capacitor with the same order of magnitude as the desired value; let C_B be this standard value capacitors and let x be this convenience factor

$$C_B = \frac{C_A}{x}$$

4. Adjust the three resistors to maintain the cutoff frequency

$$R_1 = 282\ k\Omega \cdot x \qquad R_2 = 718\ k\Omega \cdot x \qquad R_3 = 4950\ k\Omega \cdot x$$

Figure 11.58
Three-pole Butterworth
high-pass analog filter.

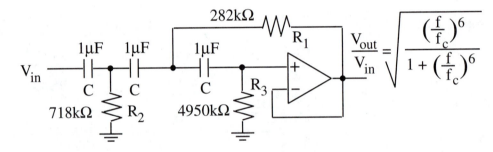

$$\frac{V_{out}}{V_{in}} = \sqrt{\frac{\left(\frac{f}{f_c}\right)^6}{1 + \left(\frac{f}{f_c}\right)^6}}$$

11.3.3
Bandpass and
Band-Reject Filters

Low-pass and high-pass filters can be cascaded to build a bandpass filter (Figure 11.59). Normally we put the high-pass filter first so that the low-pass filter can remove high-frequency noise generated in both stages.

This cascade filter should only be used for low-Q applications. Similarly, a band-reject filter can be constructed by running both filters in parallel and summing the output (Figure 11.60). The parallel band-reject filter also should be used only for low-Q applications. One implementation of a high-Q bandpass (frequency select) filter is called the multiple feedback bandpass filter (Figure 11.61). The Q is defined as the center frequency divided by the band-

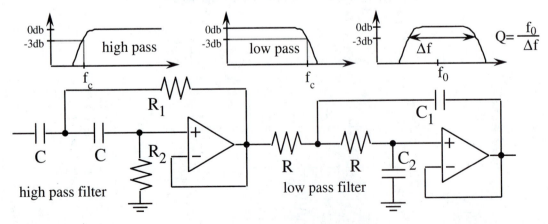

Figure 11.59   Cascade approach to bandpass analog filter design.

Figure 11.60
Parallel approach to bandreject analog filter design.

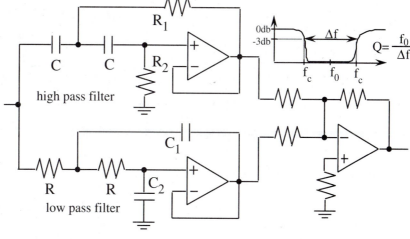

Figure 11.61
Multiple feedback bandpass filter.

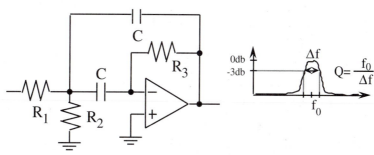

pass range, $f_0/\Delta f$. The two capacitors have the same value and are typically in the range from 0.001 to 10 μF. The design steps are:

1. Select a convenience capacitance value for the two capacitors
2. Calculate the three resistor values for $x = 1/(2\pi f_0 C)$

$$R_1 = Q \cdot x \qquad R_2 = x/(2Q - 1/Q) \qquad R_3 = 2 \cdot Q \cdot x$$

The resistors should be in the range from 5 kΩ to 5 MΩ. If not, then repeat the design steps with a different capacitance value. For a high-Q notch filter, the bootstrapped twin-T can be used (Figure 11.62). The notch frequency f_0 is determined by R_1 and C_1.

Figure 11.62
Bootstrapped twin-T band-reject analog filter.

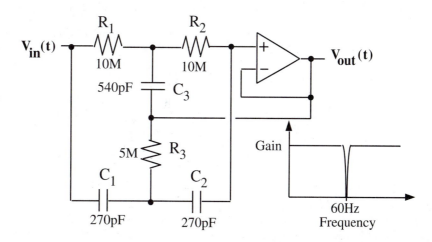

$$f_0 = \frac{1}{2\pi R_1 C_1} = 60 \text{ Hz}$$

where $R_1 = R_2 = 2 \cdot R_3$ and $C_1 = C_2 = 0.5 \cdot C_3$.

Common Error: If the capacitors are poor quality (leakage, tolerance, temperature coefficient), then the twin-T will not operate properly.

Observation: If the node between C3 and R3 is disconnected from the op amp output and then grounded, the filter becomes simply a twin-T. This filter has a lower Q and is easier to build with inexpensive capacitors.

11.4   Digital-to-Analog Converters

11.4.1 DAC Parameters

A DAC converts digital signals into analog form (Figure 11.63). A microcomputer output can be connected to a DAC. Many DACs can be interfaced to the SPI synchronous serial port. The D/A output can be current or voltage. Additional analog processing may be required to filter, amplify, or modulate the signal. We will also use DACs to design variable-gain or variable-offset analog circuits.

The D/A *precision* is the number of distinguishable D/A outputs (e.g., 256 alternatives, 8 bits). The D/A *range* is the maximum and minimum D/A output (volts, amperes). The

Figure 11.63
DACs provide analog output for our embedded microcomputer systems.

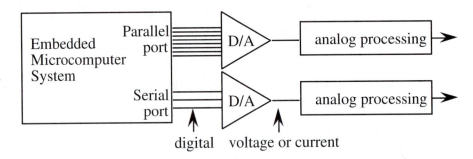

D/A *resolution* is the smallest distinguishable change in output. The units of resolution are in volts or amperes depending on whether the output is voltage or current. The resolution is the change that occurs when the digital input changes by 1.

Range (volts) = precision (alternatives) · Resolution (volts)

The D/A *accuracy* is (actual − ideal) / ideal, where ideal is referred to the National Bureau of Standards (NBS). There are two common 8-bit encoding schemes for a DAC:

$$\text{Unsigned:} V_{out} = V_{fs}\left(\frac{b_7}{2} + \frac{b_6}{4} + \frac{b_5}{8} + \frac{b_4}{16} + \frac{b_3}{32} + \frac{b_2}{64} + \frac{b_1}{128} + \frac{b_0}{256}\right) + V_{os}$$

$$\text{Signed 2s complement:} V_{out} = V_{fs}\left(-\frac{b_7}{2} + \frac{b_6}{4} + \frac{b_5}{8} + \frac{b_4}{16} + \frac{b_3}{32} + \frac{b_2}{64} + \frac{b_1}{128} + \frac{b_0}{256}\right) + V_{os}$$

where $b_7, b_6, b_5, b_4, b_3, b_2, b_1, b_0$ are the 8-bit digital inputs, V_{fs} is the full-scale voltage (typically +5 or +10), and V_{os} is the offset voltage (hopefully 0).

One can choose the full-scale range of the DAC to simplify the use of fixed-point mathematics. For example, if an 8-bit DAC had a full-scale range of 0 to 2.55 V, then the resolution would be exactly 10 mV. This means that if the D/A digital input were 123_{10}, then the D/A output voltage would be 1.23 V.

The following discussion will focus on a 3-bit DAC with a range of 0 to +7 V. This DAC can be used to generate a variable voltage V_{out}, a variable offset, or a variable gain V_{out}/V_{in} (Figure 11.64). A D/A gain error is a shift in the slope of the V_{out} versus V_{in} static response (Figure 11.65). A D/A offset error is a shift in the V_{out} versus V_{in} static response. The D/A transient response (Figure 11.66) has three components:

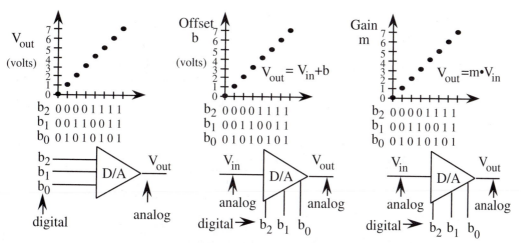

Figure 11.64   Three-bit DAC used to generate a variable output, a variable offset, or a variable gain.

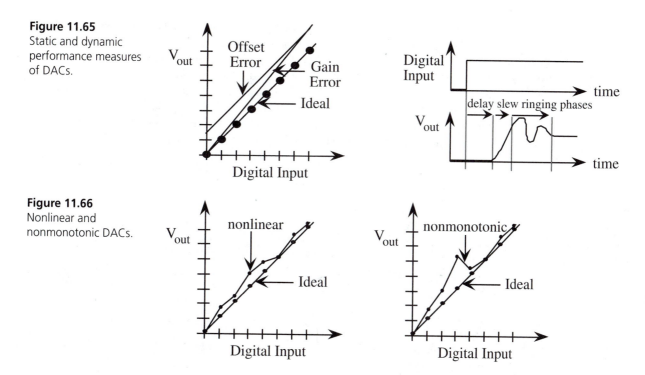

Figure 11.65
Static and dynamic performance measures of DACs.

Figure 11.66
Nonlinear and nonmonotonic DACs.

Delay phase
Slewing phase
Ringing phase

For purposes of linearity, let m, n be digital inputs, and let f(n) be the analog output of the D/A. Let Δ be the D/A resolution. The D/A is linear if

$$f(n + 1) - f(n) = f(m + 1) - f(m) = \Delta \qquad \text{for all n, m}$$

The D/A is monotonic Figure 11.64 if

$$\text{sign}[f(n + 1) - f(n)] = \text{sign}[f(m + 1) - f(m)] \qquad \text{for all n, m}$$

Conversely, the D/A is nonlinear if

$$f(n + 1) - f(n) \neq f(m + 1) - f(m) \qquad \text{for some m, n}$$

Practically speaking all DACs are nonlinear, but the worst nonlinearity is nonmonoticity. Table 11.11 lists sources of error. The DAC is nonmonotonic if

$$\text{sign}[f(n + 1) - f(n)] \neq \text{sign}[f(m + 1) - f(m)] \qquad \text{for some n, m}$$

Table 11.11
Sources and solutions to DAC errors.

Errors can be due to	Solutions
Incorrect resistor values	Precision resistors with low tolerances
Drift in resistor values	Precision resistors with good temperature coefficients
White noise	Reduce BW using a low-pass filter, reduce temperature
Op amp errors	Use more expensive devices with low noise and low drift
Interference from external fields	Shielding, ground planes

**11.4.2
DAC Using a
Summing
Amplifier**

We will need a −5.00-V reference, which we can create using an inverting op amp and voltage reference (Figure 11.67). We use the summing mode of another op amp to build the DAC. The exponential basis elements of the DAC are created by the resistance values 25 kΩ, 50, and 100 kΩ. This method can only be used to build low-precision DACs, because it is difficult to extend the exponential resistance network beyond 5 or 6 bits. In the D/A circuit of Figure 11.68, the SW02 switch is closed (75 Ω) when the digital input (IN) is high, and the SW02 switch is open when the digital input is low. The output is $V_{out} = 4b_2 + 2b_1 + b_0$.

The range of the DAC is 0 to +7 V with a resolution of 1 V.

$$V_{out} = \sum_{n=0}^{2} b_n 2^n$$

Figure 11.67
Negative reference that will be used in the DAC design.

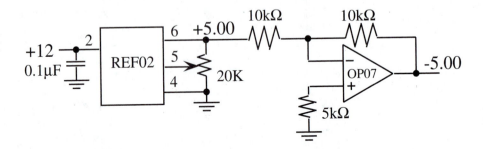

Figure 11.68
Three-bit unsigned
summing DAC.

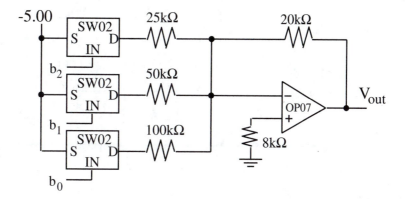

The operation of this 3-bit DAC can be summarized using the basis elements 1, 2, 4. The responses of these basis elements are presented in Table 11.12. To create a 2s complement signed DAC, we can implement the signed basis (Table 11.13). By comparing these two tables, we see that the signed DAC can be created by changing the response of the 1, 0, 0 input from +4 V to −4 V (Figure 11.69).

Table 11.12
Responses of the basis
elements for an
unsigned 3-bit DAC.

b_2	b_1	b_0	V_{out}
0	0	1	+1
0	1	0	+2
1	0	0	+4

Table 11.13
Responses of the basis
elements for a signed
3-bit DAC.

b_2	b_1	b_0	V_{out}
0	0	1	+1
0	1	0	+2
1	0	0	−4

Figure 11.69
Three-bit unsigned
summing DAC.

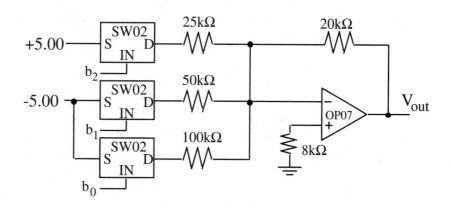

11.4.3
Three-Bit DAC
with an R-2R
Ladder

It is not feasible to construct a high-precision DAC using the summing op amp technique for two reasons. First, if one chooses the resistor values from the practical 10 kΩ to 1 MΩ range, then the maximum precision would be 1 MΩ/10 kΩ = 100, or about 7 bits. The second problem is that it would be difficult to avoid nonmonotonicity because a small

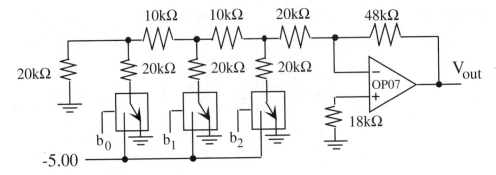

percentage change in the small resistor (e.g., the one causing the largest gain) would overwhelm the effects of the large resistor (e.g., the one causing the smallest gain). To address both these limitations, the R-2R ladder is used (Figure 11.70). It is practical to build resistor packages such that all the Rs are equal and all the 2Rs are two times the Rs. Resistance errors will change all resistors equally. This type of error affects the slope V_{fs} but not the linearity or the monotonicity.

To analyze this circuit we will consider the three basis elements (1, 2, 4). The unsigned basis elements were presented earlier in Table 11.12. If these three cases are demonstrated, then the law of superposition guarantees the other five will work. Notice that these analog switches are "three-pole" and are fundamentally different from the switches in the summing DAC. When one of the digital inputs is true then −5.00 V is connected to the R-2R ladder, and when the digital input is false, then the connection is grounded. In each of the three test cases, the current across the active switch is −1/6 mA. This current is divided by 2 at each branch point. Current injected away from the op amp will be divided more times. Since each stage divides by 2, the exponential behavior is produced. An actual DAC is implemented with a current switch rather than a voltage switch as shown in Figure 11.71. Nevertheless, this simple circuit illustrates the operation of the R-2R ladder function. The output voltage is a linear combination of the three digital inputs, $V_{out} = 4b_2 + 2b_1 + b_0$. To increase the precision one simply adds more stages to the R-2R ladder. One can purchase 16-bit DACs that utilize the R-2R ladder technique.

When the input is 001, −5.00 V is presented to the left. The effective impedance to ground is 3R (30 kΩ), so the current injected into the R-2R ladder is −1/6 mA. The current is divided three times, and −1/48 mA goes across the 48-kΩ resistor to create the +1-V output (Figure 11.71).

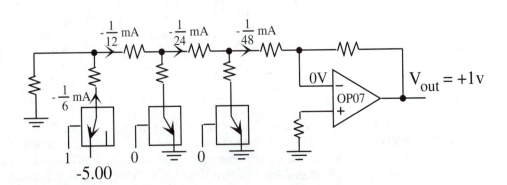

When the input is 010, -5.00 V is presented in the middle. The effective impedance to ground is again 3R (30 kΩ), so the current injected into the R-2R ladder is $-1/6$ mA. The current is divided twice, and $-1/24$ mA goes across the 48-kΩ resistor to create the +2-V output (Figure 11.72).

Figure 11.72
The DAC output is +2 V, then the digital input is 010.

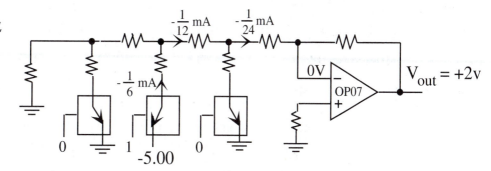

When the input is 100, -5.00 V is presented on the right. The effective impedance to ground is again 3R (30 kΩ), so the current injected into the R-2R ladder is $-1/6$ mA. This time the current is divided only once, and $-1/12$ mA goes across the 48-kΩ resistor to create the +4-V output (Figure 11.73).

Figure 11.73
The DAC output is +4 V, then the digital input is 100.

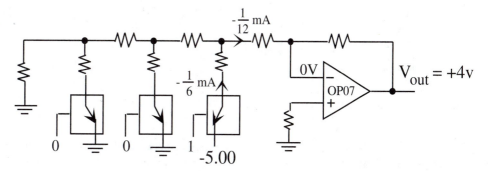

There are two mechanisms to implement a signed D/A. The first approach is to produce the signed basis directly (see Table 11.13). The b_2, b_1, b_0 = 1, 0, 0 output must be changed from +4 to -4 V. This can be accomplished by connecting the b_2 switch to +5.00 V instead of -5.00 V. The second approach is add an analog circuit at the output of an unsigned DAC. In this example, let V_{out} be the output of the 3-bit unsigned DAC. If one designs a simple subtraction ($V_{out} - 4$ V), the output becomes signed offset binary.

One of the reasons this simple 3-bit DAC is introduced is to illustrate the use of DACs in the design of variable-gain and variable-offset analog amplifiers. In this way, the software has control over the gain and offset of the analog circuit. The variable-offset approach simply connects the D/A output into one of the inputs of an analog adder or analog subtractor, as illustrated in Figure 11.74.

A multiplying DAC is one that allows a voltage input to be connected to the R-2R ladder instead of a voltage reference. If you connect a voltage reference to this pin, the DAC operates in the usual fashion. On the other hand, if this pin is an analog input, then the DAC

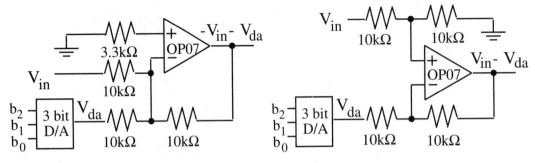

Figure 11.74   Variable-offset analog circuit using a 3-bit DAC.

is a variable-gain amplifier. Using the same 3-bit DAC circuit developed earlier, we can build a variable-gain amplifier. An inverting amplifier is added at the end to create positive-gain amplification. The 48-kΩ resistor is replaced with a 240-kΩ one so that the gain varies from 0 to +7 (Figure 11.75).

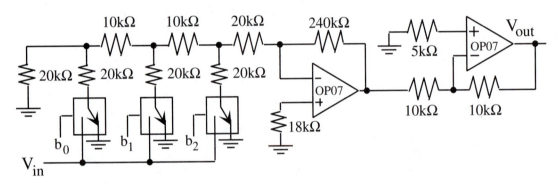

Figure 11.75   Variable-gain analog circuit using a 3-bit multiplying DAC.

11.4.4
Twelve-Bit DAC
with a DAC8043

The synchronous serial interface between the computer and the Analog Devices DAC8043 DAC was introduced previously in Section 7.7.6.1. Here in this section we will focus on the analog aspects of the circuit. If the DAC is used in the regular digital input/analog output mode, then Vin will be a reference voltage (e.g., +5.00 or +10.00 V). If the DAC is used in variable-gain mode (multiplying DAC), then Vin is the analog input (i.e., $-10 \le \text{Vin} \le +10$ V). The optional 100 Ω potentiometer is used to adjust the gain (Figure 11.76).

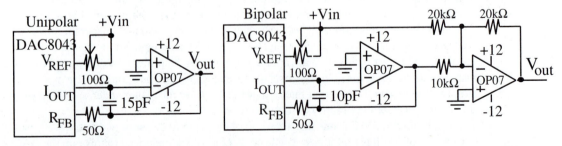

Figure 11.76   Unipolar and bipolar modes of the DAC8043 12-bit DAC.

Digital input	Unipolar Vout (V)	Bipolar Vout (V)	Unipolar gain	Bipolar gain
1111, 1111, 1111	−4.999	4.998	$-\dfrac{4095}{4096}$	$+\dfrac{2047}{2048}$
1000,0000,0001	−2.501	0.002	$-\dfrac{2049}{4096}$	$+\dfrac{1}{2048}$
1000,0000,0000	−2.500	0.000	$-\dfrac{2048}{4096}$	$+\dfrac{0}{2048}$
0111,1111,1111	−2.499	−0.002	$-\dfrac{2047}{4096}$	$-\dfrac{1}{2048}$
0000,0000,0001	−0.001	−4.998	$-\dfrac{1}{4096}$	$-\dfrac{2047}{2048}$
0000,0000,0000	0.000	−5.000	$-\dfrac{0}{4096}$	$-\dfrac{2048}{2048}$

Table 11.14   The 12-bit DAC8043 DAC can create an analog output or be used as a variable-gain amplifier.

The Vout in Table 11.14 is calculated for a Vin=+5.00 V reference voltage. The bipolar code is called *offset binary*. To convert to regular 2s complement we simply complement the most significant digital bit. When the unipolar circuit is used in variable-gain mode, it is called two-quadrant because the input (Vin) can be positive or negative, but the gain is always negative. When the bipolar circuit is used in variable-gain mode, it is called four-quadrant because the input (Vin) can be positive or negative, and the gain can be positive or negative.

**11.4.5
DAC Selection**

Many manufacturers, like Analog Devices, Burr Brown, Motorola, Sipex, and Maxim, produce DACs. These DACs have a wide range of performance parameters and come in many configurations. The following paragraphs discuss the various issues to consider when selecting a DAC. Although we assume the DAC is used to generate an analog waveform, these considerations will generally apply to most DAC applications.

Precision/range/resolution. These three parameters affect the quality of the signal that can be generated by the system. The more bits in the DAC, the finer the control the system has over the waveform it creates. As important as this parameter is, it is one of the more difficult specifications to establish a priori. A simple experimental procedure to address this question is to design a prototype system with a very high precision (e.g., 12, 14, or 16 bits). The software can be modified to use only some of the available precision. For example, the 12-bit DAC8043 hardware developed in Sections 7.7.6.1 and 11.4.4 can be reduced to 8 or 10 bits using the functions shown in Program 11.1. The bottom bits are set to zero, instead of shifting, so that the rest of the system will operate without change.

Program 11.1 Software used to test how many bits are really needed.

```
void DACout8(unsigned int code){
    DACout(code&0xFFF0);}   // ignore bottom 4 bits
void DACout10(unsigned int code){
    DACout(code&0xFFFC);}   // ignore bottom 2 bits
```

Figure 11.77
The waveform on the right was created by a DAC with one more bit than the left.

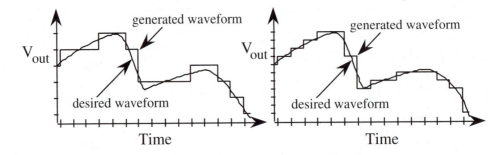

The three versions of the software (e.g., 8-, 10-, 12-bit DAC) are used to see experimentally the effect of DAC precision on the overall system performance. Figure 11.77 illustrates how DAC precision affects the quality of the generated waveform.

Channels. Even though multiple channels could be implemented using multiple DAC chips, it is usually more efficient to design a multiple-channel system using a multiple-channel DAC. Some advantages of using a DAC with more channels than originally conceived are future expansion, automated calibration, and automated testing.

Configuration. DACs can have voltage or current outputs. Current-output DACs can be used in a wide spectrum of applications (e.g., adding gain and filtering) but do require external components. DACs can have internal or external references. An internal-reference DAC is easier to use for standard digital input/analog output applications, but the external-reference DAC can be used often in variable-gain applications (multiplying DAC). As we saw with the DAC8043, sometimes the DAC generates a unipolar output, while other times the DAC produces bipolar outputs.

Speed. There are a couple of parameters manufacturers use to specify the dynamic behavior of the DAC. The most common is settling time; another is maximum output rate. When operating the DAC in variable-gain mode, we are also interested in the gain/BW product of the analog amplifier. When comparing specifications reported by different manufacturers, it is important to consider the exact situation used to collect the parameter. In other words, one manufacturer may define settling time as the time to reach 0.1% of the final output after a full-scale change in input given a certain load on the output, while another manufacturer may define settling time as the time to reach 1% of the final output after a 1-V change in input under a different load. The speed of the DAC together with the speed of the computer/software will determine the effective frequency components in the generated waveforms. Both the software (rate at which the software outputs new values to the DAC) and the DAC speed must be fast enough for the given application. In other words, if the software outputs new values to the DAC at a rate faster than the DAC can respond, then errors will occur. Figure 11.78 illustrates the effect of DAC output rate on the quality of

Figure 11.78
The waveform on the right was created by a system with twice the output rate than the left.

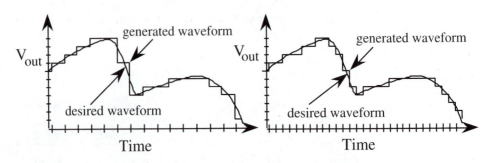

the generated waveform. We will learn in Chapter 12 that the Nyquist theorem states that to prevent error, the digital data rate must be greater than twice the maximum frequency component of the desired analog waveform.

Power. There are three power issues to consider. The first consideration is the type of power required. Some devices require three power voltages (e.g., +5, +12, and −12 V), while many of the newer devices will operate on a single voltage supply (e.g., +2.7, +3.3, +5, or +12 V). If a single supply can be used to power all the digital and analog components, then the overall system costs will be reduced. The second consideration is the amount of power required. Some devices can operate on less than 1 mW and are appropriate for battery-operated systems or for systems where excess heat is a problem. The last consideration is the need for a low-power sleep mode. Some battery-operated systems need the DAC only intermittently. In these applications, we wish to give a shutdown command to the DAC so that it won't draw much current when the DAC is not needed.

Interface. Three approaches exist for interfacing the DAC to the computer (Figure 11.79). In a digital logic interface, the individual data bits are connected to a dedicated computer output port. For example, a 12-bit DAC requires 12-bit output port bits to interface. The software simply writes to the parallel port(s) to change the DAC output. The second approach is called μP-bus or microprocessor-compatible. These devices are intended to be interfaced onto the address/data bus of an expanded-mode microcomputer. As we saw in Chapter 9, the MC68HC812A4 has built-in address decoders enabling most

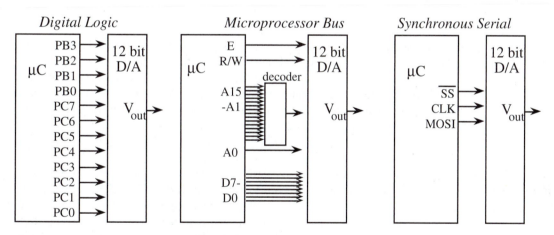

Figure 11.79   Three approaches to interfacing a 12-bit DAC to the microcomputer.

μP-bus DACs to be interfaced to the address/data bus without additional external logic. An example of this type of interface can be found in Exercise 9.5. The SPI/DAC8043 interface is an example of the last time of interface. This approach requires the fewest number of I/O pins. Even if the microcomputer does not support the SPI directly, these devices can be interfaced to regular I/O pins via the bit-banging software approach. The DS1620 interface in Section 4.4.9 is an example of attaching a synchronous serial I/O device to regular I/O pins.

Package. The standard DIP is convenient for creating and testing an original prototype. On the other hand, surface-mount packages like the SO and μMax require much less board

Figure 11.80
Integrated circuits come
in a variety of packages.

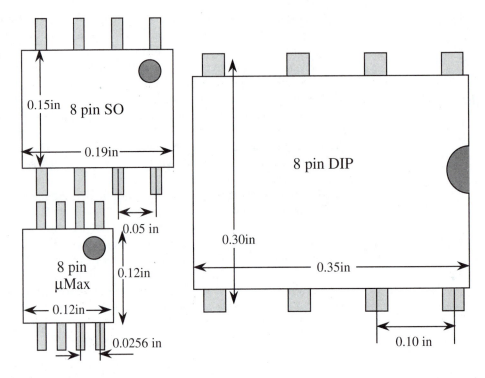

space. Because surface-mount packages do not require holes in the printed circuit board, circuits with these devices are easier/cheaper to manufacture (Figure 11.80).

Cost. Cost is always a factor in engineering design. Beside the direct costs of the individual components in the DAC interface, other considerations that affect cost include (1) power supply requirements, (2) manufacturing costs, (3) the labor involved in individual calibration if required, and (4) software development costs.

**11.4.6
DAC Waveform
Generation**

One application that requires a DAC is waveform generation. In this section, we will discuss various software methods for creating analog waveforms with a DAC. In each case, we will be using the DAC8043 hardware/software interface introduced in Sections 7.7.6.1 and 11.4.4. In addition, we will use an output capture interrupt for the timing so that the waveform generation occurs in the background. The rituals for initializing the periodic interrupt are shown in Section 6.2.3. To get a fair comparison between the various methods, each implementation will generate 32 interrupts per waveform.

In the first approach, we assume there exists a time-to-voltage function, called `wave()`, which we can call to determine the next DAC value to output. For example, the waveform in Figure 11.81 could be generated by Program 11.2. The simplest solution generates an output compare interrupt at a regular rate. The advantage of this approach is that complex waveforms can be encoded with a small amount of data. In this particular example, the entire waveform can be stored as five data points (2048.0, 1000.0, 31.25, −500.0, 125.0). The disadvantage of this technique is that not all waveforms have a simple function, and this software will run slower as compared to the other techniques (Program 11.3).

Figure 11.81
Generated waveform
using either the function
or the table look-up.

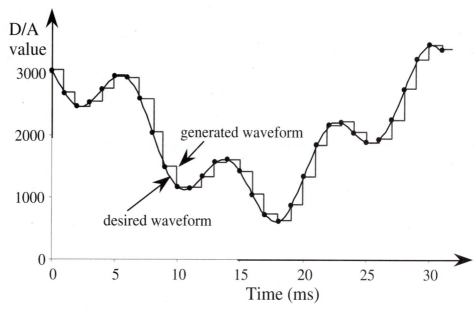

Program 11.2 Periodic
interrupt used to create
the analog output
waveform.

```
unsigned int wave(unsigned int t){
        float result,time;
        time=2*pi*((float)t)/1000.0;
// integer t in msec into floating point time in seconds
        result=2048.0+1000.0*cos(31.25*time)-500.0*sin(125.0*time);
        return (unsigned int) result;}
```

```
// MC68HC11A8                          // MC68HC812A4
#define Rate 2000                      #define Rate 2000
#define OC5  0x08                      #define OC5  0x20
unsigned int Time; // Inc every 1ms    unsigned int Time; // Inc every 1ms
#pragma interrupt_handler TOC5handler()  #pragma interrupt_handler TOC5handler()
void TOC5handler(void){                void TOC5handler(void){
   TFLG1=OC5;      // Ack interrupt       TFLG1=OC5;      // ack C5F
   TOC5=TOC5+Rate; // Executed every 1 ms  TC5=TC5+Rate;   // Executed every 1 ms
   Time++;                                Time++;
   DACout(wave(Time));}                   DACout(wave(Time));}
```

Program 11.3 Periodic interrupt used to create the analog output waveform.

In the second approach, we put the waveform information in a large statically allocated global variable. Every interrupt we fetch a new value out of the data structure and
output it to the DAC. In this case the output compare interrupt also occurs at a regular
rate. Assume the ritual initializes I=0. This waveform could be defined as shown in
Programs 11.4 and 11.5.

Program 11.4 Simple
data structure for the
waveform.

```
unsigned int I;  // incremented every 1ms
const unsigned int wave[32]= {
         3048,2675,2472,2526,2755,2957,2931,2597,
         2048,1499,1165,1139,1341,1570,1624,1421,
         1048,714,624,863,1341,1846,2165,2206,2048,
         1890,1931,2250,2755,3233,3472,3382};
```

```
// MC68HC11A8                              // MC68HC812A4
#define Rate 2000                          #define Rate 2000
#define OC5  0x08                          #define OC5  0x20
#pragma interrupt_handler TOC5handler()    #pragma interrupt_handler TOC5handler()
void TOC5handler(void){                    void TOC5handler(void){
   TFLG1=OC5;       // Ack interrupt           TFLG1=OC5;      // ack C5F
   TOC5=TOC5+Rate; // Executed every 1 ms      TC5=TC5+Rate;   // Executed every 1 ms
   if((++I)==32) I=0;                          if((++I)==32) I=0;
   DACout(wave[I]);}                           DACout(wave[I]);}
```

Program 11.5 Periodic interrupt used to create the analog output waveform.

Since the output rate is equal and fixed, these first two methods have the same performance as illustrated in Figure 11.81. The thick line is the desired waveform, and the thinner line is the actual generated curve. If the size of the table gets large, it is possible to store a smaller table in memory and use linear interpolation to recover the data points between the stored samples. Figure 11.82 shows the generated waveform derived from only 9 of the original 32 data points. To simplify the software, the first data point is repeated as the last data point. For each point we will need to save both the DAC value and time length of the current-line segment. For the 9 saved data points we simply output the data, but for the other points, we must perform a linear interpolation to get the value to output to the DAC.

Assume the ritual initializes I=J=0. Signed 16-bit numbers are used so that the subtractions operate properly. In other words, some of the intermediate calculations can be negative. This waveform could be defined as shown in Programs 11.6 and 11.7.

Figure 11.82
Generated waveform
using a short table with
linear interpolation.

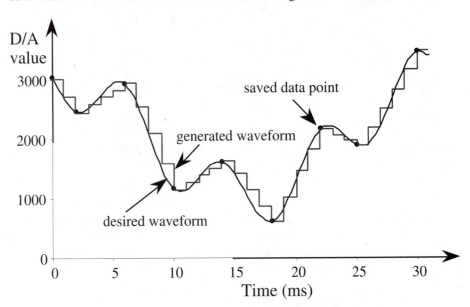

```
int I;  // incremented every 1ms
int J;  // index into these two tables
const int t[10]= {0,2,6,10,14,18,22,25,30,32};  // time in msec
const int wave[10]={3048,2472,2931,1165,1624,624,2165,1890,3472,3048}; //last=first
```

Program 11.6 Data structure with time and value for the waveform.

```
// MC68HC11A8                            // MC68HC812A4
#define Rate 2000                        #define Rate 2000
#define OC5  0x08                        #define OC5  0x20
#pragma interrupt_handler TOC5handler()  #pragma interrupt_handler TOC5handler()
void TOC5handler(void){                  void TOC5handler(void){
   TFLG1=OC5;       // Ack interrupt        TFLG1=OC5;       // ack C5F
   TOC5=TOC5+Rate; // Executed every 1 ms   TC5=TC5+Rate;    // Executed every 1 ms
   if((++I)==32) {I=0; J=0;}                if((++I)==32) {I=0; J=0;}
   if(I==t[J])                              if(I==t[J])
      DACout(wave[J]);                         DACout(wave[J]);
   else if (I==t[J+1]){                     else if (I==t[J+1]){
      J++;                                     J++;
      DACout(wave[J]);}                        DACout(wave[J]);}
   else                                     else
      DACout(wave[J]+((wave[J+1]-wave[J])      DACout(wave[J]+((wave[J+1]-wave[J])
        *(I-t[J]))/(t[J+1]-t[J]));}              *(I-t[J]))/(t[J+1]-t[J]));}
```

Program 11.7 Periodic interrupt used to create the analog output waveform.

The software in the previous techniques changes the DAC at a fixed rate. While this is adequate most of the time, there are some waveforms for which an uneven time between outputs seems appropriate. In our test signal, there are phases of the wave where the signal varies slowly while there are phases where the signal changes. Notice the data points in Figure 11.83 are placed at uneven time intervals to match the various phases of

Figure 11.83
Generated waveform using the uneven-time table look-up technique.

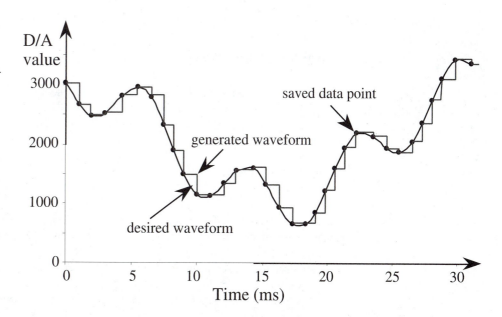

this signal. This generated waveform is still created with 32 points, but placing the points closer together during phases with large slopes improves the overall accuracy.

The table data structure will encode both the voltage (as a DAC value) and the time. The time parameter is stored as a Δt in cycles to simplify servicing the output compare interrupt. We will assume the TCNT is initialized to count every 500 ns, therefore the maximum Δt that can be generated is 32 ms. Assume the ritual initializes I=0 (Programs 11.8 and 11.9).

```
unsigned int I;  // incremented every sample
const unsigned int wave[32]= {
   3048,2675,2472,2526,2817,2981,2800,2337,1901,1499,1165,1341,1570,1597,1337, 952,
    662, 654, 863,1210,1605,1950,2202,2141,1955,1876,2057,2366,2755,3129,3442,3382};
const unsigned int dt[32]= { // time increment in 500 ns cycles
  2000,2000,2000,2500,2500,2000,2000,1500,1500,2000,4000,2000,2500,2000,2000,2000,
  2000,1500,1500,1500,1500,2000,2500,2000,2000,2000,1500,1500,1500,2000,2500,2000};
```

Program 11.8 Data structure with delta time and value for the waveform.

```
// MC68HC11A8                              // MC68HC812A4
#define OC5  0x08                          #define OC5  0x20
#pragma interrupt_handler TOC5handler()    #pragma interrupt_handler TOC5handler()
void TOC5handler(void){                    void TOC5handler(void){
   TFLG1=OC5;       // Ack interrupt           TFLG1=OC5;       // ack C5F
   if((++I)==32) I=0;                          if((++I)==32) I=0;
   TOC5=TOC5+dt[I]; // variable rate           TC5=TC5+dt[I];   // variable rate
   DACout(wave[I]);}                           DACout(wave[I]);}
```

Program 11.9 Periodic interrupt used to create the analog output waveform.

11.5 Analog-to-Digital Converters

11.5.1 ADC Parameters

An ADC converts an analog signal into digital form. The input signal is usually an analog voltage (V_{in}), and the output is a binary number. The ADC *precision* is the number of distinguishable ADC inputs (e.g., 256 alternatives, 8 bits). The ADC *range* is the maximum and minimum ADC input (volts, amperes). The ADC resolution is the smallest distinguishable change in input (volts, amperes). The *resolution* is the change in input that causes the digital output to change by 1.

$$\text{Range (volts)} = \text{precision (alternatives)} \cdot \text{Resolution (volts)}$$

Normally we don't specify accuracy for just the ADC, but rather we give the *accuracy* of the entire instrument (including transducer, analog circuit, ADC, and software). Therefore, accuracy will be described later in Chapter 12 (specifically, Section 12.1.1), part of the systems approach to data acquisition systems. An ADC is *monotonic* if it has no missing codes. This means that if the analog signal is a slow rising voltage, then the digital output will hit all values one at a time. The ADC is *linear* if the resolution is constant through the range.

The ADC *speed* is the time to convert, called t_c. The ADC cost is a function of the number and price of internal components. There are four common encoding schemes for an ADC. Tables 11.15 and 11.16 show the four encoding schemes for an 8-bit ADC.

Table 11.15
Unipolar codes for an 8-bit ADC.

Unipolar codes	Straight binary	Complementary binary
+5.00	1111, 1111	0000, 0000
+2.50	1000, 0000	0111, 1111
+0.02	0000, 0001	1111, 1110
+0.00	0000, 0000	1111, 1111

Table 11.16
Bipolar codes for an 8-bit ADC.

Bipolar codes	Offset binary	2s Complement binary
+5.00	1111, 1111	0111, 1111
+2.50	1100, 0000	0100, 0000
+0.04	1000, 0001	0000, 0001
+0.00	1000, 0000	0000, 0000
−2.50	0100, 0000	1100, 0000
−5.00	0000, 0000	1000, 0000

To convert between straight binary and complementary binary, we simply complement (change 0 to 1, change 1 to 0) all the bits. To convert between offset binary and 2s complement, we complement just the most significant bit. The exclusive-OR instruction can be used to complement a single bit.

Just like the DAC, one can choose the full-scale range to simplify the use of fixed-point mathematics. For example, if a 12-bit ADC had a full-scale range of 0 to 4.095 V, then the resolution would be exactly 1 mV. This means that if the ADC input voltage were 1.234 V, then the result would be 1234_{10}.

11.5.2
Two-Bit Flash ADC

The flash ADC can be used for high-speed low-precision conversions. Four equal-value resistors are used to create reference voltages 7.50, 5.00, and 2.50. The LM311 voltage comparators perform the ADC conversion. The digital circuit produces the binary code (Z1, Z0). The conversion speed is limited by the comparator delay, which can be as fast as 100 ns (Table 11.17, Figure 11.84).

Table 11.17
Output results for a 2-bit flash ADC.

V_{in}	X3	X2	X1	Z1	Z0
$2.5 > V_{in}$	0	0	0	0	0
$5.0 > V_{in} \cdot 2.5$	0	0	1	0	1
$7.5 > V_{in} \cdot 5.0$	0	1	1	1	0
$V_{in} \cdot 7.5$	1	1	1	1	1

Some flash converters include two more comparators. The underflow comparator is used to determine if $V_{in} < 0$. The overflow comparator is used to determine if $V_{in} > +10$ V.

A 3-bit flash ADC would place eight equal-value resistors in series to generate the reference voltages 1.25, 2.50, 3.75, 5.00, 6.25, 7.50, and 8.75. Seven comparators would

Figure 11.84
Two-bit flash ADC.

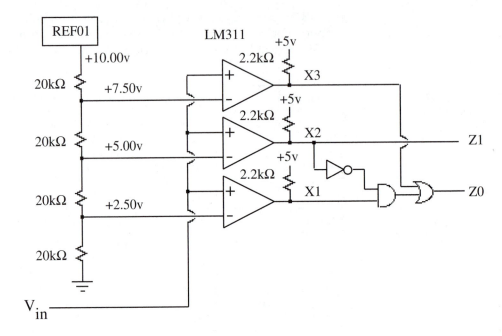

perform the ADC conversion. A 74LS148 8-to-3 encoder could be used to convert the X1,X2,X3,X4,X5,X6,X7 comparator outputs into the 3-bit binary code Z2,Z1,Z0. With a flash ADC there is an exponential relationship between the ADC precision and the number of components. In other words, an 8-bit flash requires 256 resistors, 255 comparators, and a 255 to 8-bit digital encoder.

Observation: The cost of a flash ADC relates linearly with its precision in alternatives.

There are two approaches to produce a signed flash ADC. The direct approach would be to place the equal resistors from a +10.00-V reference to a −10.00-V reference. The middle of the series resistors ("zero" reference) should be grounded. The digital circuit would then be modified to produce the desired digital code.

The other method would be to add analog preprocessing to convert the signed input to an unsigned range. This unsigned voltage is then converted with an unsigned ADC. The digital code for this approach would be offset binary. This approach will work to convert any unsigned ADC into one that operates on bipolar inputs.

**11.5.3
Eight-Bit Ramp
ADC**

The ramp ADC starts at the minimum voltage and counts up until the DAC output V_0 just exceeds the V_{in}. The AND gate is used to stop the counting when $V_{in}=V_0$. The full-scale range of V_0 should match the range of V_{in}. The conversion time of this ADC varies from 1 to 256 μs, depending on the input voltage. The maximum conversion time occurs when V_{in} at the maximum value of 10.0 V. If V_{in} is greater than 10.0 V, then the ADC never finishes (Figure 11.85).

The software issues the **Go** pulse, waits for **Busy** to fall, then reads the digital outputs from the counter (Figure 11.86).

There are two approaches to produce a signed ramp ADC. The direct approach would be to use a 2s complement signed DAC. The counter should be initialized to the minimum

Figure 11.85
Eight-bit ramp ADC.

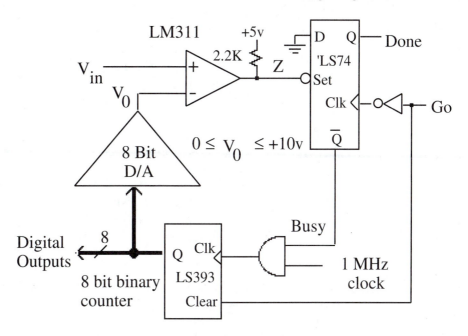

Figure 11.86
Operation of a ramp ADC.

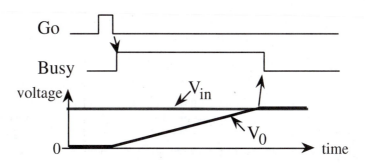

value. For the 8-bit ADC, the initial value should be $80. Again, the device counts up until the DAC output V_0 just exceeds the V_{in}. The output code would be 2s complement.

The other method would be to add analog preprocessing to convert the signed input to an unsigned range. This unsigned voltage is then converted with an unsigned ADC. The digital code for this approach would be offset binary.

**11.5.4
Successive
Approximation
ADC**

The most pervasive method for ADC conversion is the successive approximation technique. An ADC with a precision of n is clocked n times. At each clock another bit is determined, starting with the most significant bit. For each clock, the successive approximation hardware issues a new "guess" (V_0) by setting the bit under test to a 1. If V_0 is now higher than the unknown input V_{in}, then the bit under test is cleared. If V_0 is less than V_{in}, then the bit under test remains 1. The C program in Figure 11.87 illustrates the algorithm. In this program, n is the precision in bits, bit is an unsigned integer that specifies the bit under test (e.g., for an 8-bit ADC, bit goes 128, 64, 32, 16, . . .,1), Dout is the ADC digital output, and Z is the binary input that is true if V_0 is greater than V_{in}.

Figure 11.87
Successive
approximation ADC.

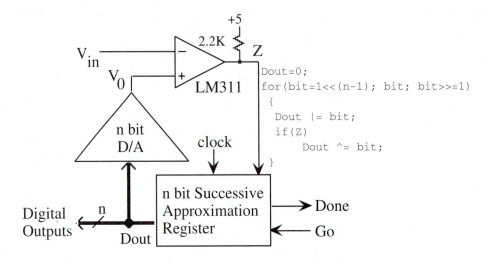

Most successive approximation ADCs use a current-output DAC and a current compara-tor, instead of the voltage DAC and voltage comparator shown above. In this case, each guess is converted to a current by the DAC. The input voltage is also converted to a current (notice the low-input impedance of this ADC). The current comparator determines if the guess is high or low. If the guess is low, the bit under test is left set to 1. If the guess is high, the bit under test is cleared to 0. This procedure starts with the most significant bit and stops with bit 0. An n-bit ADC requires exactly n cycles to convert, independent of the input voltage level.

Observation: The speed of a successive approximation ADC relates linearly with its precision in bits.

Figure 11.88 illustrates how one can configure a single device to operate in more than one voltage range. This type of ADC provides multiple input connections (e.g., In1, In2) that you can use to select the input voltage range. Table 11.18 illustrates how you use jumpers and input connections to specify the input voltage range. By connecting the posi-tive reference to the DAC, a fixed positive current is added to the current summing junction at the minus terminal of the comparator. This allows for the input voltage at In1 or In2 to be negative. The digital outputs for the bipolar ranges will be in offset binary. Since the in-put voltage is converted to a current using Ohm's law ($I = V/R$), when you jumper In1 to

Figure 11.88
Block diagram of an
ADC that supports
several analog input
ranges.

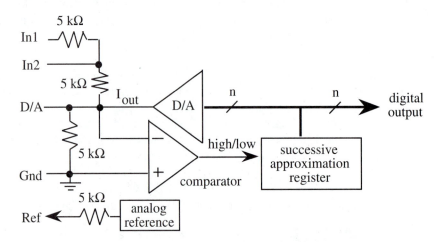

Analog voltage range	Jumper	Connect input to	Zin from input to ground
−10 to +10	Ref to DAC	In1	10 kΩ
−5 to +5	Ref to DAC	In2	5 kΩ
−2.5 to +2.5	Ref to DAC, In1 to Gnd	In2	2.5 kΩ
0 to +5	Ref to Gnd, In1 to DAC	In2	
0 to +10	Ref to Gnd	In2	

Table 11.18 Jumper settings to configure the input voltage range of an ADC.

ground, you reduce the input impedance at the In2 terminal by half, causing the voltage-to-current conversion to double.

Observation: Some ADCs have horribly low input impedance.

11.5.5 Software Schemes to Implement ADCs

The purpose of this section is to illustrate with software the operation of the ramp, tracking, and successive approximation ADC conversion scheme (Figure 11.89). If the range of the DAC does not match the desired range of the ADC (in this case, the ranges do match), then analog processing of either V_0 or V_{in} is required so that the two ranges at the 311 inputs match. Because the DAC8043 implements offset binary, the resulting ADC will also be offset binary. The 311 comparator output is connected to an input port.

Figure 11.89
Hardware interface to illustrate three software schemes for ADC conversion.

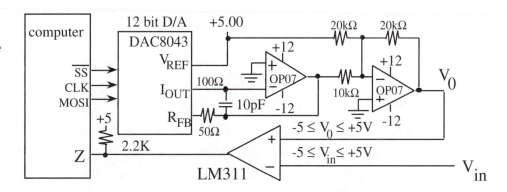

In each of these techniques, the software will adjust the DAC output until V_0 equals V_{in}. The ADC output result will equal the DAC digital input when V_0 equals V_{in}. Let the function Z() return a Boolean, true if V_0 is greater than V_{in} and false if V_0 is less than V_{in}.

In the ADC ramp technique, we start at the minimum and increment the DAC until the comparator changes. If the DAC output value reaches the maximum value, then the software will return with 4096 (Program 11.10). In the ADC tracking technique, there is no starting or stopping. It yields a continuous output that is maintained in a global variable. We

Program 11.10
Software implementation of a ramp ADC.

```
unsigned int rampAD(void){ unsigned int DOUT; // guess
    DOUT = 0;                // start at minimum, V0=-5.00 (offset binary)
    do {
       DACout(DOUT);         // set DAC output, V0
       DOUT++;               // ramp
    }
    while((!Z())&&(DOUT<4096));  // stop when V0>Vin or DOUT=4096
    return(DOUT);}
```

Program 11.11
Software implementation
of a tracking ADC.

```
int DOUT; // guess
void main(void){
    DOUT = 2048;          // start in the middle, V0=0.00 (offset binary)
    while(1) {
      DACout(DOUT);       // set DAC output, V0
      if(Z()) {           // check input
        if(DOUT>0)        // don't go below 0
          DOUT--;         // V0>Vin so decrement
      }
      else {
        if(DOUT<4095)
          DOUT++;         // V0<Vin so increment
      }
    }}
```

increment or decrement the DAC, depending on the comparator state. If the guess is too low, we increment; if it is too high, we decrement. This tracking algorithm is implemented in Program 11.11 as a main program, where the other software tasks must run in the background as interrupt handlers.

The tracking scheme could also be implemented in the background using output compare periodic interrupts. During each interrupt, one pass through the tracking algorithm is executed (Program 11.12).

In the ADC successive approximation technique, we start at the midpoint and use the comparator output to determine the ADC result from bit 11 down to bit 0. The algorithm is simpler because the DAC is implemented in offset binary. It is possible but more complicated to implement successive approximation with a 2s complement DAC (Program 11.13).

Program 11.12
Interrupting software
implementation of a
tracking ADC.

```
#pragma interrupt_handler TOC5handler()
void TOC5handler(void){
    TFLG1=OC5;            // ack C5F
    TC5=TC5+rate;
    DACout(DOUT);         // set D/A output, V0
    if(Z()) {             // check input
      if(DOUT>0)          // don't go below 0
        DOUT--;           // V0>Vin so decrement
    }
    else {
      if(DOUT<4095)
        DOUT++;           // V0<Vin so increment
    }
    DACout(DOUT);         // set DAC output, V0
}
```

Program 11.13
Software implementation
of a successive
approximation ADC.

```
unsigned int SuccAproxAD(void){ unsigned int DOUT,bit;
    DOUT = 0;
    for(bit=2048;bit;bit>>1){
      DOUT |= bit;        // try to turn on this bit
      DACout(DOUT);       // set DAC output, V0
      if(Z())
        DOUT ^| bit;      // too big, so remove this bit
    }
    return(DOUT);}
```

Figure 11.90
Hardware required to
implement the dual
slope ADC conversion
technique.

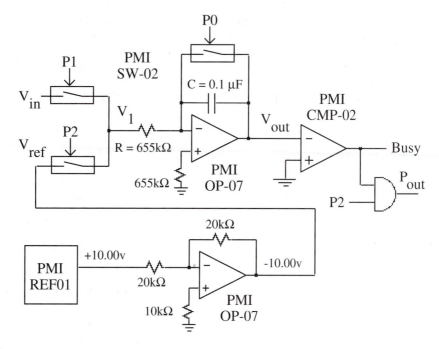

11.5.6
Sixteen-Bit Dual
Slope ADC

The dual slope technique can be used to construct an ADC with precision ranging from 16 to 20 bits (Figure 11.90). This 16-bit ADC will have a range of 0 to +10 V and a conversion time of 130 ms. There are three phases for the conversion. The three digital signals P0, P1, and P2 each control a SPST BiFET analog switch. Only one switch is activated at a time. Let the times t_0, t_1, and t_2 refer to the time at the end of each phase (Figure 11.91). During phase 0, P0 is active, which shorts the capacitor C. This initialization phase is used to make the output V_{out} equal to zero. Thus,

$$V_{out}(t_0) = 0$$

Figure 11.91
Digital and analog
waveforms during a dual
slope ADC conversion.

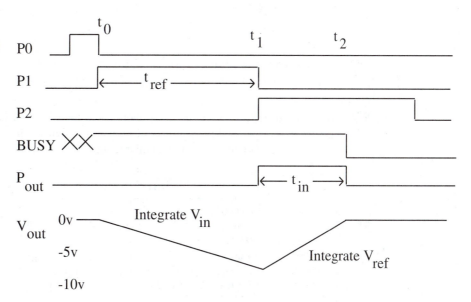

Phase 0 must be long enough to completely discharge the capacitor. During phase 1, P1 is active, which connects the unknown input voltage V_{in} to the integrator. The length of phase 1 is exactly 65,535 μs. We will call this length of phase 1 a time reference,

$$t_{ref} = t_1 - t_0 = 65,535 \text{ μs}$$

Since the input voltage is a constant and $V_{out}(t_0)$ is known, the integration can be simplified:

$$V_{out}(t_1) = V_{out}(t_1) - \frac{1}{RC} \int_{t_2}^{t_1} V_{in}(\uparrow) d\uparrow = -\frac{1}{RC} V_{in} \, t_{ref}$$

The largest negative voltage will occur when V_{in} equals +10 V. The values of R,C, and t_{ref} were chosen to prevent the integrating op amp from saturating. Output compare can be used to generate t_{ref}, and input capture can be used to measure t_{in}. During phase 2, P2 is active, which connects the reference voltage V_{ref} to the integrator. The negative voltage reference is created with the REF01 voltage reference and an inverting amplifier. During this phase V_{out} is integrated back to zero. The phase ends when $V_{out}=0$. Thus,

$$V_{out}(t_2) = 0$$

t_{in} in is the time spent in phase 2, $t_{in} = t_2 - t_1$. The pulse width of P_{out}, t_{in}, is measured with a precision timer. Again, since V_{ref} is constant, the integration can be simplified.

$$V_{out}(t_2) = V_{out}(t_1) - \frac{1}{RC} \int_{t_1}^{t_2} V_{ref}(\uparrow) d\uparrow = V_{out}(t_1) - \frac{1}{RC} V_{ref} \, t_{in} = 0$$

Substituting from above the value of $V_{out}(t_1)$

$$\frac{1}{RC} V_{in} \, t_{ref} - \frac{1}{RC} V_{ref} \, t_{in} = 0$$

The next algebraic step is critical to explain why this method can be used to produce high-precision ADCs. The RCs cancel!

$$V_{in} \, t_{ref} - V_{ref} \, t_{in} = 0$$

We can calculate the unknown voltage V_{in} from the voltage reference V_{ref}, the time reference t_{ref}, and the time measurement t_{in}.

$$V_{in} = \frac{V_{ref}}{t_{ref}} t_{in}$$

The 16-bit precision requires that the time t_{in} be measured with a resolution of 1 μs and a range of 0 to 65,535 μs. The operation can be summarized in Table 11.19.

**11.5.7
Sigma Delta ADC**

The sigma delta ADC is used in many audio applications (Figure 11.92). It is a cost-effective approach to 16-bit 44-kHz sampling (CD quality). Sigma delta converters have a DAC, a comparator, and digital processing similar to the successive approximation technique. While

Table 11.19
Output results of a dual slope ADC.

V_{in} (V)	$V_{out}(t_1)$ (V)	t_{in} (μs)
0.00000	0.00000	0
0.00015	−0.00015	1
0.00031	−0.00031	2
1.00000	−1.00000	6,554
5.00000	−5.00000	32,768
9.99985	−9.99985	65,535

Figure 11.92
Block diagram of the sigma delta ADC conversion technique.

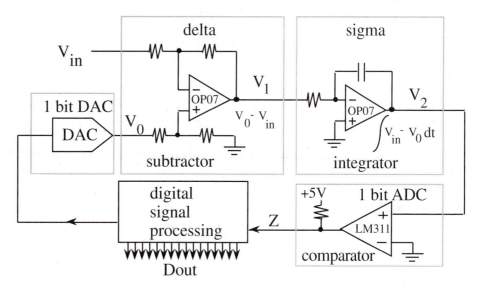

successive approximation converters have DACs with the same precision as the ADC, sigma delta converters use 1-bit DAC (which is simply a digital signal itself). The digital signal processing will run at a clock frequency faster than the overall ADC system (oversampled). It uses complex signal processing to drive the output voltage V_0 to equal the unknown input V_{in} in a time-averaged sense. The "delta" part of the sigma delta converter is the subtractor, where $V_1 = V_0 - V_{in}$. Next comes the "sigma" part that implements an analog integration. If V_0 to equal the unknown input V_{in} in a time-averaged sense, then V_2 will be zero. The comparator tests the V_2 signal. If V_2 is positive, then V_0 is made smaller. If V_2 is negative, then V_0 is made larger. This DAC-subtractor-integrator-comparator-digital loop is executed at a rate much faster than the eventual digital output rate.

A very simple software algorithm, shown in Program 11.14, is implemented as a periodic interrupt. This algorithm is much too simple to be appropriate in an actual converter, but it does illustrate the sigma delta approach. For an 8-bit conversion, the internal clock

```
unsigned char DOUT;  // 8-bit sample
unsigned char SUM;   // number of times Z=0 and V0=1
unsigned char CNT;   // 8-bit counter
#pragma interrupt_handler TOC5handler()
void TOC5handler(void){
   TFLG1=OC5;         // ack C5F
   TC5=TC5+rate;      // interrupt 256 times faster than the ADC output rate
   if(Z())            // check input
      DACout(0);      // too high, set DAC output, V0=0
   else {
      DACout(1);      // too low, set DAC output, V0=+5v
      SUM++;
   }
   if(++CNT==0){      // end of 256 loops?
      DOUT=SUM;       // new sample
      SUM=0;          // get ready for the next
   }
}
```

Program 11.14 Soft-ware implementation of a sigma delta ADC.

rate (interrupt rate) is 256 times the output rate of the 8-bit samples. We assume that the input voltage V_{in} is between 0 and +5 V. In this simple solution, the DAC is set to 1 ($V_0 = +5$) if Z is 0 ($V_2 < 0$). Conversely, the DAC is set to 0 ($V_0 = 0$) if Z is 1 ($V_2 > 0$). Each time the DAC is set to 1, the counter, SUM, is incremented. At the end of 256 passes, the value SUM is recorded as the ADC sample. Since there are 256 passes through the loop, the result will vary from 0 to 255. During each interrupt one pass through the sigma delta algorithm is executed. Similar to the other software ADC examples, Z() is a function that returns the value of the comparator output. Let DACout() be a function that takes a Boolean input parameter and sets the 1-bit DAC. Assume the ritual will clear all three global variables.

11.5.8
ADC Interface

Similar to the DAC interface, three approaches exist for interfacing the ADC to the computer (Figure 11.93). In a digital logic interface, the individual data bits are connected to a dedicated computer I/O ports. Usually the software toggles the **Start** ADC signal to begin an ADC conversion. The software can determine when the ADC is finished by reading the **Done** signal. For example, an 8-bit ADC requires one output port bit and 9-bit input port bits to interface. The software toggles the **Start** signal, waits for the **Done** to be true, then reads the ADC result. The second approach is called microprocessor-bus or microprocessor-compatible. These devices are intended to be interfaced onto the address/data bus of an expanded mode microcomputer. A microcomputer-bus ADC can be interfaced to a MC68HC812A4 without additional external logic. Discussion of this type of interface can be found in Chapter 9. The software interfaces to this type of ADC in a manner similar to the internal ADCs (discussed later in Section 11.10). The SPI/MAX1247 interface (discussed in Section 7.7.6.2) is an example of the last time of interface. This approach requires the fewest number of I/O pins. Even if the microcomputer does not support the SPI directly, these devices can be interfaced to regular I/O pins via the bit-banging software approach.

Figure 11.93
Three approaches to interfacing an ADC to the microcomputer.

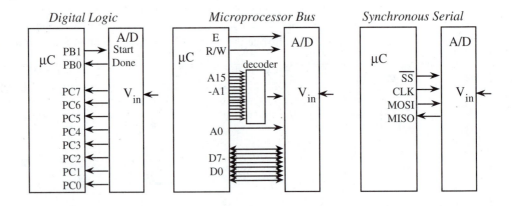

11.6   Sample and Hold

A sample and hold (S/H) is an analog latch (Figure 11.94). An alternative name for this analog component is track and hold. The purpose of the S/H is to hold the ADC analog input constant during the conversion. The digital input, **control,** determines the S/H mode. The

Figure 11.94
The S/H has a digital input, an analog input, and an analog output.

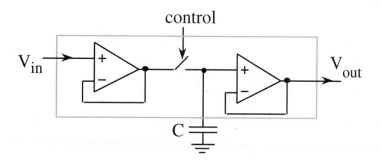

S/H is in *sample mode,* where V_{out} equals V_{in}, when the switch is closed. The S/H is in *hold mode,* where V_{out} is fixed, when the switch is open. The *acquisition time* is the time for the output to equal the input after the control is switched from hold to sample. This is the time to charge the capacitor C. The *aperture time* is the time for the output to stabilize after the control is switched from sample to hold. This is the time to open the switch that is usually quite fast. The *droop rate* is the output voltage slope (dV_{out}/dt) when **control** equals hold. Normally the gain K should be 1 and the offset V_{off} should be 0. The gain and offset error specify how close the V_{out} is to the desired V_{in} when **control** equals sample,

$$V_{out} = K V_{in} + V_{off}$$

where K should be one and V_{off} should be zero.
To choose the external capacitor C:

1. One should use a polystyrene capacitor because of its high insulation resistance and low dielectric absorption
2. A larger value of C will decrease (improve) the droop rate. If the droop current is I_{DR}, then the droop rate will be

$$\frac{dV_{out}}{dt} = \frac{I_{DR}}{C}$$

3. A smaller C will decrease (improve) the acquisition time

11.7 BiFET Analog Multiplexer

The analog multiplexer allows multiple analog signals to be sequentially connected to a single ADC. The operation of this type of computer-controlled switch was discussed previously in Section 8.5.8. Figure 11.95 illustrates the operation of a PMI MUX-08. This chip has four digital inputs, **Enable, A_2, A_1,** and **A_0.** It also has eight analog inputs (S_1–S_8) and one analog output, **Drain.** When the **Enable** is true, the three address bits determine which analog input is connected to the analog output.

R_{ON} is the on resistance of the switch (200 to 500Ω for the PMI MUX-08.) The "OFF" isolation, ISO_{OFF}, is a measure of how well the output is disconnected from its inputs when the device is shut off (60 dB for the PMI MUX-08.) To measure ISO_{OFF}, we place a 500-kHz +5-V sine wave on channel 8. We make all channels off (Enable=0) and measure the signal at the output with a R_L=1RΩ, CL=10pF. ISO_{OFF} is the gain V_{out}/V_{in} when all channels are off. The crosstalk (CT) is a measure of how much the input on one channel spills over into the output of another channel (70 dB for the PMI MUX-08.) To

Figure 11.95 An analog multiplexer allows the computer to select one of eight analog signals.

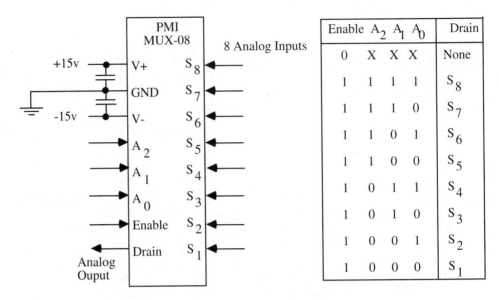

Enable	A_2	A_1	A_0	Drain
0	X	X	X	None
1	1	1	1	S_8
1	1	1	0	S_7
1	1	0	1	S_6
1	1	0	0	S_5
1	0	1	1	S_4
1	0	1	0	S_3
1	0	0	1	S_2
1	0	0	0	S_1

measure CT, we place a 500-kHz +5-V sine wave on channel 8. We activate channel 4 (Enable=1, $A_2A_1A_0$=011) and measure the signal at the output with a R_L=1 MΩ, C_L=10 pF. CT is the gain from V_{in}(channel 8) to V_{out} when channel 4 is on. The settling time t_S is the time it takes a 10-V step to reach within 0.02% of the final output (2.5 µs for the PMI MUX-08.)

A device similar to the multiplexer is the analog switch. The CD4066B CMOS switch was discussed in Section 8.5.8. The PMI SW02 is a SPST BiFET analog switch with

$$
\begin{aligned}
R_{on} &= 100\ \Omega\ \text{max} \\
ISO_{off} &= 58\ \text{dB typical} \\
CT &= 70\ \text{dB typical} \\
t_{on} &= 400\ \text{ns max} \\
t_{off} &= 300\ \text{ns max}
\end{aligned}
$$

Analog switches and/or multiplexers can be used to make variable-gain analog amplifiers. Because of the large (bad) on-resistance (ranges from 85 to 500 Ω), it is important to route the output signal from the switch input into a large input impedance. In this way, there will be a minimal voltage drop across the switch. The two circuits in Figures 11.96 and 11.97, show a good and bad approach to using the bilateral switch to implement a microprocessor-controlled analog amplifier gain. In this first circuit, the gain is either 1+R2B/ R1B or 1+R2A/R1A, independent of the bilateral switch resistance. Because of the large input impedance into the op amp input terminal, the voltage drop across the switch will be zero.

The problem with the circuit in Figure 11.97 is that the resistance of the switch affects the gain of the circuit. In this second circuit, the gain is either 1+(Ron+R2B)/R1 or 1+(Ron+R2A)/R1, where Ron is the bilateral switch resistance. Since Ron can drift with time and temperature, the gain of this circuit will vary as well.

The analog multiplexer in the circuit of Figure 11.98 is used to select the gain from 1, 2, 5, 10, 20, 50, 100, 200.

Figure 11.96
The proper way to use a bilateral switch to make a variable-gain amplifier.

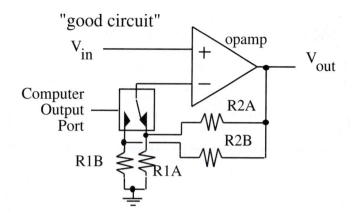

Figure 11.97
The improper way to use a bilateral switch to make a variable-gain amplifier.

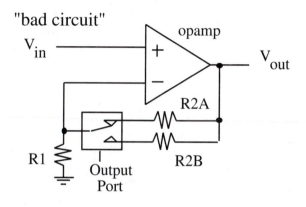

Figure 11.98
The use of an analog multiplexer to make a variable-gain amplifier.

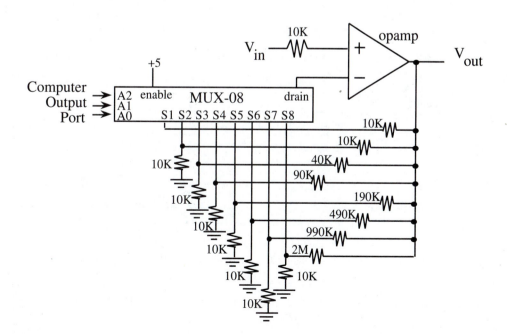

11.8   ADC System

11.8.1
ADC Block
Diagram
The block diagram in Figure 11.99 connects the devices introduced in this chapter to build a data acquisition system. The next chapter will discuss the overall design process, but here in this section we discuss the low-level interfacing issues. Recall that the Maxim MAX1147 ADC integrates the multiplexer, S/H, and ADC into a single package.

Figure 11.99
Together, the analog circuit, multiplexer, S/H, and ADC convert external signals to digital form.

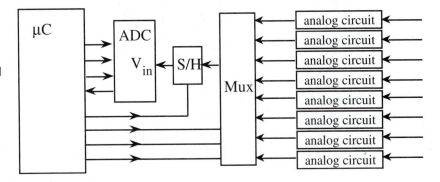

The flowcharts in Figures 11.100 and 11.101 show the tasks performed in the ISRs. It is assumed that the periodic interrupt occurs at the desired sampling rate. In the Figure 11.104, the system does not include a S/H. When only a single channel is being sampled, the multiplexer address is set once in the ritual. There are two approaches to sampling a single channel. If the ADC is started at the end of the interrupt handler, the interrupt software is faster and simpler. Unfortunately, this simple method introduces a delay in the data. In other words, data collected during one interrupt actually represent the signal at the time of the previous interrupt. Multiplexers are usually fast enough that normal software delays from one instruction to the next are sufficient to allow the multiplexer to settle, so no explicit software function need be added for the "Wait for Mux to settle" step.

Figure 11.100
Flowcharts for ADC interrupt software when no S/H is needed.

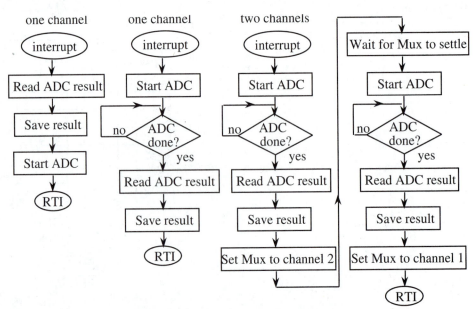

Figure 11.101
Flowcharts for ADC
interrupt software when
a S/H is needed.

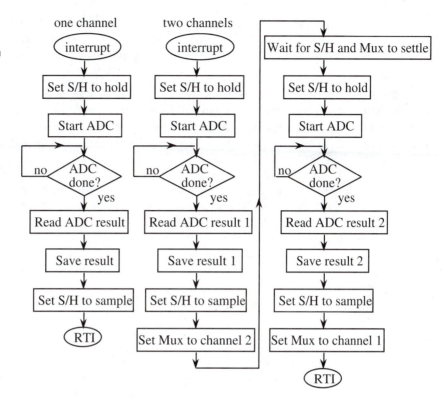

In Figure 11.105, the system does include a S/H. Just like the last example, if only a single channel is being sampled, the multiplexer address is set once in the ritual. Sample and hold modules are usually so slow that normal software delays from one instruction to the next are not sufficient for the S/H to settle, so some explicit software delay may be needed for the "Wait for S/H and Mux to settle" step.

11.8.2
Power and
Grounding for the
ADC System

The analog and digital grounds should be connected only at the ADC (Figure 11.102). Many ADC data sheets discuss layout techniques to minimize noise. A "star" pattern is used to decouple the devices. If the ground lines are in series, then a large current passing from an outer device could affect the ground voltage of an inner device. This can be a

Figure 11.102
The analog and digital
grounds are separate,
connecting only at the
ADC module.

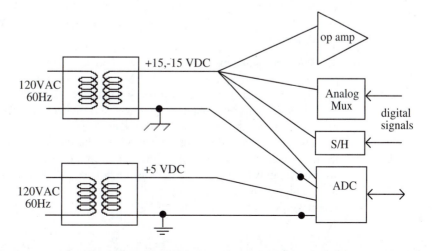

significant problem for high-speed devices that sink large transient currents during transitions. Because the S/H module requires careful circuit board layout for the capacitor, most manufacturers suggest PC board layout patterns for their S/H modules.

11.8.3
Input Protection for High-Speed CMOS Analog Inputs

It is necessary to protect CMOS inputs from large currents and out-of-range voltages. The microcomputer is particularly vulnerable to negative voltages. The information in this section was derived from the Motorola 6811 data sheets. Although no similar technical information on the 6812 is available, it is safe to assume this information applies to the 6812 as well. The equivalent circuit of the 6811 analog input is shown in Figure 11.103. Even though Port E on the 6811 is an input, part of an output channel provides some input protection against voltages below zero. The N-channel output transistor and the N-channel thick-field transistor will act like diodes if the input drops below −0.7 V. Similarly, the P-channel transistor on the input buffer will act like a diode if the input rises above 20 V.

Figure 11.103
Equivalent circuit of an analog input on the 6811.

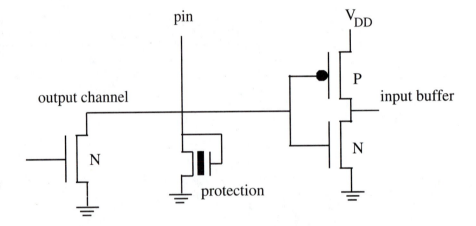

Common Error: It is a mistake to neglect the possibility of a broken transducer or disconnected cable.

Performance Tip: It is good engineering practice to design input protection based on reasonable expected failure modes. If you develop a collection of broken transducers (open and short) and cables, you could use these devices to design and test your systems.

When designing instrumentation that interfaces to the 6811 ADC, we must restrict input voltages to be between 0 and +5 V. Motorola advises against using external champing diodes to protect analog inputs from out-of-range voltages. Instead they suggest a 1- to 10-kΩ current-limiting resistor and a low-pass filter. Higher limiting resistor values provide for more protection but may cause conversion errors (Figure 11.104).

Instead of connecting protection diodes at the Port E inputs, it is possible to use protection diodes at an earlier stage of the analog circuit (Figure 11.105). For example, if you are using a unity gain low-pass filter as the last stage, then diodes can be used at the input of the low-pass filter. In this way the Port E input voltage could go out of range only if the low-pass filter were to malfunction.

Figure 11.104
Suggested interface
between analog circuits
and the analog input.

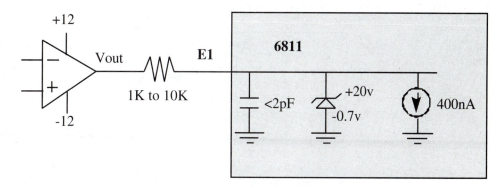

Figure 11.105
One possible approach
for protecting the
analog input.

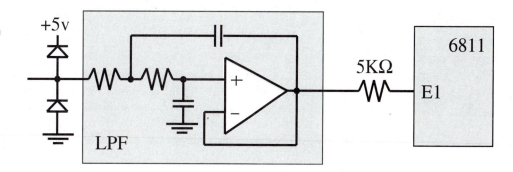

11.9 Multiple-Access Circular Queue

A *multiple-access circular queue* (MACQ) is used for data acquisition and control systems. A MACQ is a fixed-length order-preserving data structure. The source process (ADC sampling software) places information into the MACQ. Once initialized, the MACQ is always full. The oldest data are discarded when the newest data are **PUT** into a MACQ. The sink process can read any of the data from the MACQ. The **READ** function is nondestructive. This means that the MACQ is not changed by the READ operation. For example, consider

Figure 11.106
When data are put into
a MACQ, the oldest
data are lost.

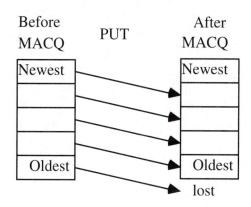

Figure 11.107
Application of the multiple access circular queue.

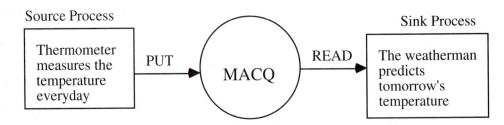

the problem of weather forecasting (Figure 11.107). The weatherman measures the temperature every day at 12 noon, and PUTs the temperature into the MACQ. To predict tomorrow's temperature, she looks at the trend over the last 3 days. Let T(0) be today's temperature, T(1) be yesterday's temperature, etc. Tomorrow's predicted temperature might be

$$U = \frac{170 \cdot T(0) + 60 \cdot T(1) + 36 \cdot T(2)}{256}$$

The MACQ is useful for implementing digital filters and linear control systems. One common application of the MACQ is the real-time calculation of derivative. Assume the function `Adin()` returns the current sample in millivolts. Also assume the function `ClkHandler()` is a periodic interrupt executed every 1 ms. x(n) will refer to the current sample, and x(n−1) will be the sample 1 ms ago (Δt = 1ms). There are a couple of ways to implement the discrete time derivative. The simple approach is

$$d(n) = \frac{x(n) - x(n - 1)}{\Delta t}$$

In practice, this first-order equation is quite susceptible to noise. In most practical control systems, the derivative is calculated using a higher-order equation like

$$d(n) = \frac{x(n) + 3x(n - 1) - 3x(n - 2) - x(n - 3)}{\Delta t}$$

The C implementation of this discrete derivative (Program 11.15) uses a MACQ. Since Δt is 1 ms, we simply consider the derivative to have units millivolts per millisecond and not actually execute the divide by Δt operation.

```
// MC68HC11A8
#define Rate 2000
#define OC5  0x08
unsigned int x[4];  // MACQ (mV)
unsigned int d;      // derivative (V/s)
#pragma interrupt_handler TOC5handler()
void TOC5handler(void){
   TFLG1=OC5;       // Ack interrupt
   TOC5=TOC5+Rate; // Executed every 1 ms
   x[3]=x[2];  // shift MACQ data
   x[2]=x[1];  // units of mV
   x[1]=x[0];
   x[0]=Adin(); // current data
   d=x[0]+3*x[1]-3*x[2]-x[3];} // mV/ms
```

```
// MC68HC812A4
#define Rate 2000
#define OC5  0x20
unsigned int x[4];  // MACQ (mV)
unsigned int d;      // derivative (V/s)
#pragma interrupt_handler TOC5handler()
void TOC5handler(void){
   TC5=TC5+Rate;    // Executed every 1 ms
   TFLG1=0x20;       // ack OC5F
   x[3]=x[2];  // shift MACQ data
   x[2]=x[1];  // units of mV
   x[1]=x[0];
   x[0]=Adin(); // current data
   d=x[0]+3*x[1]-3*x[2]-x[3];} // mV/ms
```

```
void ritual(void) {                         void ritual(void) {
asm(" sei");    // make atomic             asm(" sei");        // make atomic
  TMSK1|=OC5;   // Arm output compare 5       TIOS|=OC5;       // enable OC5
  TFLG1=OC5;    // Initially clear OC5F       TSCR|=0x80;      // enable
  TOC5=TCNT+Rate; // First one in 1 ms        TMSK2=0x32;      // 500 ns clock
asm(" cli"); }                                TMSK1|=OC5;      // Arm output compare 5
                                              TFLG1=OC5;       // Initially clear C5F
                                              TC5=TCNT+Rate; // First one in 1 ms
                                            asm(" cli"); }
```

Program 11.15 Software implementation of first derivative using a MACQ.

When the MACQ holds many data points, it can be implemented using a pointer or index to the newest data. In this way, the data need not be shifted each time a new sample is added. The disadvantage of this approach is that address calculation is required during the READ access.

11.10 Internal ADCs

The Motorola 6805/6808/6811/6812 microcomputers are available in versions that have built-in ADCs. We will begin by discussing the particular I/O registers used to interface analog signals to the individual microcomputers. The common features include:

- Eight-channel operation
- 8-bit resolution
- Successive approximation conversion technique
- A clock and charge pump is used to create higher voltages needed for the ADC
- Two operation modes: single sequence of conversions, then stop, and continuous conversion
- Two channel selection modes: multiple conversions of a single channel, (e.g., channel 1,1,1,1) and one conversion each on a group of channels (e.g., channels 0,1,2,3)
- External V_{RH}, V_{RL} analog high/low references

11.10.1
6805 ADC System

The B series of 6805 microcomputers includes an eight-channel 8-bit ADC. Some of the other 6805 series also have ADCs. In particular, some of the devices in the E, F, G, L, P, T, V, and X series have ADC capabilities. The discussions in this section and the next three sections cover the spectrum of ADC interfaces available for the Motorola 8-bit microcomputers.

This subsection focuses on the B series. The analog inputs are shared with Port D. The ADSTAT register is used to set the operation mode and start the conversion. This register contains both control and status information.

The 6805 ADC system is enabled by setting ADON equal to 1. The ADRC control bit selects the clock for the charge pump. ADRC should be 0 for normal E clock frequencies. For E clock frequencies below 1 MHz, ADRC should be 1. Each ADC conversion requires 32 external clock cycles (ADRC=1) or 32 μs (ADRC=0) to complete. We write to the ADSTAT with the channel number to start a new conversion. This register also contains a status bit, COCO, which we use to poll for the ADC conversion completion. The COCO flag is cleared by writing data into the ADSTAT (i.e., starting a new conversion).

	7	6	5	4	3	2	1	0	
ADSTAT	COCO	ADRC	ADON	0	CH3	CH2	CH1	CH0	$0009

COCO: Conversions complete flag (read only)
    0 = conversions in progress
    1 = conversion complete, ADDATA has valid data
ADRC: ADC RC oscillator control
    0 = use internal clock (conversion time is 32 μs)
    1 = use external RC clock (conversion time is 32 external clock periods)
ADON: Multiple channel/single channel control
    0 = disables the ADC saving power (Port D operates as a normal digital input port)
    1 = the ADC is enabled
CH3,CH2,CH1,CH0: Multiplexer control selects that channel to convert (Table 11.20)

Table 11.20
Multiplexer control for
the 6805B ADC.

CH3,CH2,CH1,CH0	Channel
0000	PD0 analog channel 0
0001	PD1 analog channel 1
0010	PD2 analog channel 2
0011	PD3 analog channel 3
0100	PD4 analog channel 4
0101	PD5 analog channel 5
0110	PD6 analog channel 6
0111	PD7 analog channel 7
1000	V_{RH} analog reference high
1001	$(V_{RH} + V_{RL})/2$
1010	V_{RL} analog reference low
1011	V_{RL} analog reference low
1100	V_{RL} analog reference low
1101	V_{RL} analog reference low
1110	V_{RL} analog reference low
1111	V_{RL} analog reference low

The ADC result register contains the digital output of the ADC conversion. For example, if we write $24 into ADSTAT, the 6805 will convert analog channel 4 and put the digital result in ADDATA. The result will vary from 0 to 255 as the input voltage varies from 0 to +5 V.

	7	6	5	4	3	2	1	0	
ADDATA	bit 7	bit 6	bit 5	bit 4	bit 3	bit 2	bit 1	bit 0	$0008

11.10.2
6808 ADC System

This subsection describes the ADC available on the AZ or AB series of the 6808 microcomputer. Many of the other Motorola 6808 microcomputers also have internal ADCs (e.g., AS, AT, LN, MP). On the MC68HC08AB and MC68HC08AZ the analog inputs are shared with Port B. The ADSCR register is used to set the operation mode and start the conversion. This register contains both control and status information. The 6808 ADC system is disabled by writing a $1F to the ADSCR register. Any other value written to the ADSCR register will enable the ADC. We write to the ADSCR with the channel number to start a new conversion. This register also contains a status bit, COCO, which we use to poll for the ADC conversion completion. The COCO flag is cleared by writing data into the ADSCR (i.e., starting a new conversion). When ADCO is 1, the ADC will continuously convert the selected channel and write each new sample into the ADR register, regardless of whether or not the previous sample has been read by the software.

	7	6	5	4	3	2	1	0	
ADSCR	COCO	AIEN	ADCO	CH4	CH3	CH2	CH1	CH0	$0038

COCO: Conversions complete flag (read only)
　0 = conversions in progress
　1 = conversion complete, ADR has valid data
AIEN: ADC interrupt enable
　0 = disables the ADC complete interrupt
　1 = enables the ADC complete interrupt
ADCO: ADC continuous mode
　0 = perform one ADC conversion and stop
　1 = continuous ADC sampling
CH4,CH3,CH2,CH1,CH0: Multiplexer control selects that channel to convert
　(Table 11.21)

Table 11.21
Clock control for the 6808AZ ADC.

CH4,CH3,CH2,CH1,CH0	Channel
00000	PTB0 analog channel 0
00001	PTB1 analog channel 1
00010	PTB2 analog channel 2
00011	PTB3 analog channel 3
00100	PTB4 analog channel 4
00101	PTB5 analog channel 5
00110	PTB6 analog Channel 6
00111	PTB7 analog Channel 7
other values	Unused
11111	ADC is turned off

For proper operation, we must configure the ADC conversion clock to be about 1 MHz. The ADICLK bit in the ADCLK register selects the clock for the charge pump. ADICLK should be 1 for E clock frequencies greater than or equal to 1 MHz. For example, if the bus clock is 8 MHz, set ADCLK to $70 (internal clock, divide by 8). For E clock frequencies below 1 MHz, ADICLK bit should be 0. For example, if the bus clock is slower than 1 MHz, set ADCLK to $00 (RC clock, divide by 1). Each ADC conversion requires 16 or 17 clock cycles, or 16 to 17 μs, to complete.

	7	6	5	4	3	2	1	0	
ADCLK	ADIV2	ADIV1	ADIV0	ADICLK	0	0	0	0	$003A

Table 11.22
Clock control for the
6808AZ ADC.

ADIV2, ADIV1, ADIV0	Divisor
000	÷1
001	÷2
010	÷4
011	÷8
1XX	÷16

ADICLK: ADC clock (Table 11.22)
    0 = External clock, cgmxclk
    1 = Internal bus clock

The ADC result register contains the digital output of the ADC conversion. For example, if we write $04 into ADSCR, the 6808 will convert analog channel 4 and put the 8-bit unsigned digital result in ADR.

	7	6	5	4	3	2	1	0	
ADR	bit 7	bit 6	bit 5	bit 4	bit 3	bit 2	bit 1	bit 0	$0039

11.10.3
6811 ADC System

The 6811 ADC system is enabled by setting ADPU equal to 1 in the OPTION register. When ADPU is 0, Port E operates as a normal digital input port. The CSEL control bit selects the clock for the charge pump. CSEL should be 0 for normal E clock frequencies. For E clock frequencies below 750 kHz, CSEL should be 1.

	7	6	5	4	3	2	1	0	
OPTION	ADPU	CSEL	IRQE	DLY	CME	0	CR1	CR0	$1039

The control part of the ADCTL register is used to set the operation mode and start the conversion. Each sequence of four conversions requires 128 E clock cycles to complete. For single-sequence mode, we write to the ADCTL with SCAN=0 to start a new sequence. This register also contains a status bit, CCF, which we use to poll for the ADC conversion completion. The CCF flag is cleared by writing data into the ADCTL (i.e., starting a new conversion). To start a continuous conversion, we write to the ADCTL with SCAN=1.

	7	6	5	4	3	2	1	0	
ADCTL	CCF	0	SCAN	MULT	CD	CC	CB	CA	$1030

CCF: Conversions complete flag (read only)
    0 = conversions in progress
    1 = conversion complete, ADR1, ADR2, ADR3, ADR4 have valid data
SCAN: Continuous scan control
    0 = single sequence of conversions, then stop
    1 = continuous conversion

MULT: Multiple channel/single channel control
    0 = sequence of conversions on a single channel
    1 = sequence of conversions on multiple channels (0,1,2,3) (4,5,6,7) (8,9,10,11)
        or (12,13,14,15)
        Multiplexer control bits CD and CC specify the group of channel
        Multiplexer control bits CB and CA are ignored
CD,CC,CB,CA: Multiplexer control selects that channel to convert (Table 11.23)

Table 11.23
Multiplexer control for the 6811A ADC.

CD,CC,CB,CA	Channel
0000	PE0 analog channel 0
0001	PE1 analog channel 1
0010	PE2 analog channel 2
0011	PE3 analog channel 3
0100	PE4 analog channel 4
0101	PE5 analog channel 5
0110	PE6 analog channel 6
0111	PE7 analog channel 7
1000	Reserved
1001	Reserved
1010	Reserved
1011	Reserved
1100	V_{RH} analog reference high
1101	V_{RL} analog reference low
1110	$(V_{RH} + V_{RL})/2$
1111	Reserved

The ADC result registers contain the digital outputs of the ADC conversions. In single-sequence mode (MULT=0), the specified single channel (CD,CC,CB,CA) is converted four times and the results placed in ADR1, ADR2, ADR3, and ADR4. In multiple-sequence mode (MULT=1), the specified group of four channels (CD,CC) is each converted once and the results placed in ADR1, ADR2, ADR3, and ADR4. For example, if we write $14 into ADCTL, the 6811 will convert analog channels 4, 5, 6, 7 and put the digital results in ADR1, ADR2, ADR3, ADR4, respectively. The 8-bit unsigned result will vary from 0 to 255 as the analog input varies from 0 to +5 V.

	7	6	5	4	3	2	1	0	
ADR1	bit 7	bit 6	bit 5	bit 4	bit 3	bit 2	bit 1	bit 0	$1031
ADR2	bit 7	bit 6	bit 5	bit 4	bit 3	bit 2	bit 1	bit 0	$1032
ADR3	bit 7	bit 6	bit 5	bit 4	bit 3	bit 2	bit 1	bit 0	$1033
ADR4	bit 7	bit 6	bit 5	bit 4	bit 3	bit 2	bit 1	bit 0	$1034

Performance Tip: If we are interested in a single conversion, we could start the ADC, wait 32 cycles, then read the result in ADR1.

Both the MC68HC812A4 and MC68HC912B32 have built in ADCs. The 6812 ADC system is enabled by setting ADPU equal to 1 in the ATDCTL2 register. With fast clear enabled, any access to the result registers will clear the associated flag. The ADC will request an interrupt if the arm bit ASCIE is set.

	7	6	5	4	3	2	1	0	
ATDCTL2	ADPU	AFFC	AWAI	0	0	0	ASCIE	ASCIF	$0062

ADPU: ADC power up
   0 = disables the ADC saving power
   1 = the ADC is enabled
AFFC: ADC fast clear
   0 = normal clearing of flags
      SCF is cleared by a write to ATDCTL5,
      CCFn are cleared by a read of the status register followed by a read of the
        result register
   1 = fast clear enabled
      SCF is cleared by a read from the first result register,
      CCFn are cleared by a read of the associated result register
AWAI: ADC wait mode
   0 = ADC operations continue while in wait mode
   1 = ADC stops while in wait mode
ASCIE: ADC sequence complete interrupt enable
   0 = disables the ADC sequence complete interrupt
   1 = enables the ADC sequence complete interrupt
ASCIF: ADC sequence complete interrupt flag
   0 = no interrupt has occurred
   1 = ADC sequence complete interrupt is pending

The freeze bits determine the ADC operation during background debug breakpoint (BDM mode) (Table 11.24).

	7	6	5	4	3	2	1	0	
ATDCTL3	0	0	0	0	0	0	FRZ1	FRZ0	$0063

Table 11.24
Freeze control for the
6812 ADC.

FRZ1	FRZ0	ADC response to BDM mode
0	0	Conversions continue in background mode
0	1	Reserved
1	0	Finish current conversion and stop
1	1	Freeze when BDM is active

The conversion time is determined by the ATDCTL4 register. The ATDCTL4 register selects the sample period and P-clock prescaler. The choice of these parameters depends on four factors. (1) Motorola recommends the ADC clock period be restricted to the 500 ns to

2 μs range. (2) For analog signals with the white noise, we can essentially add an analog low-pass filter by increasing the ADC sample time. (3) To increase conversion speed, we wish to select a fast clock and short sample period. For an 8-MHz 6812, the fastest conversion time of 1 μs per sample is selected by setting ATDCTL4 equal to $01. (4) The last factor to consider is the slewing rate of the input signal. For signals with a high slope, dV/dt, we need to select a faster conversion time (i.e., shorter sample time) (Table 11.25). More discussion about slewing rate will be made in Chapter 12 when considering the need for a S/H analog latch.

	7	6	5	4	3	2	1	0	
ATDCTL4	0	SMP1	SMP0	PRS4	PRS3	PRS2	PRS1	PRS0	$0064

Table 11.25
Sampling time control for the 6812 ADC.

SMP1	SMP0	Sample time
0	0	2 ADC clock periods
0	1	4 ADC clock periods
1	0	8 ADC clock periods
1	1	16 ADC clock periods

For an 8-MHz P clock, Table 11.26 shows the possible divide by factors for the ADC clock. Other choices are not recommended. The time to perform one ADC conversion is the product of the ADC period from Table 11.26 times the ADC sample time from Table 11.25.

Table 11.26
Clock control for the 6812 ADC.

PRS4,PRS3,PRS2,PRS1,PRS0	Divisor	ADC period
00001	÷4	500 ns
00010	÷6	750 ns
00011	÷8	1.00 μs
00100	÷10	1.25 μs
00101	÷12	1.50 μs
00110	÷14	1.75 μs
00111	÷16	2.00 μs

The ADCTL5 register is used to set the operation mode and start the conversion. S8CM determines if the sequence will be four or eight conversions. For single-sequence mode, we write to the ADCTL5 with SCAN=0 to start a new sequence. To start a continuous conversion, we write to the ADCTL5 with SCAN=1.

	7	6	5	4	3	2	1	0	
ADCTL5	0	S8CM	SCAN	MULT	CD	CC	CB	CA	$0065

S8CM: Select 8-channel mode
    0 = four-conversion sequence
    1 = eight-conversion sequence

SCAN: Continuous scan control
    0 = single sequence of conversions, then stop
    1 = continuous conversion
MULT: Multiple channel/single channel control
    0 = sequence of conversions on a single channel
    1 = sequence of conversions on multiple channels
        If S8CM = 1, then CD specifies the sequence 0, 1, 2, 3, 4, 5, 6, 7 or 8, 9, 10, 11, 12, 13, 14, 15
        If S8CM = 0, then CD,CC specifies the sequence 0, 1, 2, 3 or 4, 5, 6, 7 or 8, 9, 10, 11 or 12, 13, 14, 15
CD,CC,CB,CA: Multiplexer control selects which channel to convert (Table 11.27)

Table 11.27

Multiplexer control for the 6812 ADC.

CD,CC,CB,CA	Channel
0000	PAD0 analog channel 0
0001	PAD1 analog channel 1
0010	PAD2 analog channel 2
0011	PAD3 analog channel 3
0100	PAD4 analog channel 4
0101	PAD5 analog channel 5
0110	PAD6 analog channel 6
0111	PAD7 analog channel 7
1000	Reserved
1001	Reserved
1010	Reserved
1011	Reserved
1100	V_{RH} analog reference high
1101	V_{RL} analog reference low
1110	$(V_{RH}+V_{RL})/2$
1111	Reserved

The status register contains a status bit, SCF, which we use to poll for the ADC conversion completion. The SCF flag is cleared by writing data into the ADCTL5 (i.e., starting a new conversion). The CC2, CC1, CC0 bits are the sequence counter as the ADC steps through a conversion sequence. The CCFn bits are individual flags for each of the conversions. Recall that the number of conversions in a sequence is determined by S8CM.

	7	6	5	4	3	2	1	0	
ATDSTAT	SCF	0	0	0	0	CC2	CC1	CC0	$0066
ATDSTAT	CCF7	CCF6	CCF5	CCF4	CCF3	CCF2	CCF1	CCF0	$0067

SCF: Sequence complete flag (read only)
    0 = conversions in progress
    1 = conversion complete, result registers have valid data
        If AFFC = 0, then SCF is cleared by a write to ATDCTL5
        If AFFC = 1, then SCF is cleared by a read from the first result register,

CCFn: Conversion complete flag, n=0 to 7 (read only)
  0 = conversion not finished
  1 = conversion complete, associated result register has valid data
    If AFFC = 0, then CCFn are cleared by a read of the status register followed
      by a read of the result
    If AFFC = 1, then CCFn are cleared by a read of the associated result
      register

The ADC result registers contain the digital outputs of the ADC conversions. The analog input of 0 V maps into a digital result of 0. Similarly +5 V maps into 255. In single-sequence mode (MULT=0), the specified single channel (CD,CC,CB,CA) is converted multiple times (four times if S8CM=0 or eight times if S8CM=1) and the results placed in the ADRnH, registers. In multiple-sequence mode (MULT=1), the specified group of channels is each converted once and the results placed in the ADRnH, registers. For example if we write $14 into ADCTL5, the 6812 will convert analog channels 4, 5, 6, 7 and put the digital results in ADR0H, ADR1H, ADR2H, ADR3H, respectively. Similarly, if we write $50 into ADCTL5, the 6812 will convert analog channels 0 through 7 and put the digital results in ADR0H through ADR7H, respectively. If we add the SCAN bit and write $70 to ADCTL5, the 6812 will repeatedly convert analog channels 0 through 7 and put the digital results in ADR0H through ADR7H, respectively.

	7	6	5	4	3	2	1	0	
ADR0H	bit 7	bit 6	bit 5	bit 4	bit 3	bit 2	bit 1	bit 0	$0070
ADR1H	bit 7	bit 6	bit 5	bit 4	bit 3	bit 2	bit 1	bit 0	$0072
ADR2H	bit 7	bit 6	bit 5	bit 4	bit 3	bit 2	bit 1	bit 0	$0074
ADR3H	bit 7	bit 6	bit 5	bit 4	bit 3	bit 2	bit 1	bit 0	$0076
ADR4H	bit 7	bit 6	bit 5	bit 4	bit 3	bit 2	bit 1	bit 0	$0078
ADR5H	bit 7	bit 6	bit 5	bit 4	bit 3	bit 2	bit 1	bit 0	$007A
ADR6H	bit 7	bit 6	bit 5	bit 4	bit 3	bit 2	bit 1	bit 0	$007C
ADR7H	bit 7	bit 6	bit 5	bit 4	bit 3	bit 2	bit 1	bit 0	$007E

Performance Tip: If we are interested in a single conversion, we could start the ADC, wait for CCF0, then read the result in ADR0H.

11.10.5
ADC Software

In this example assume that an analog signal, ranging from 0 to +5 V, is connected to analog input channel 1 (Figure 11.108). The subroutine, Samp, will return an unsigned value from 0 to 255 that represents the analog input. Table 11.28 shows the straight binary format.

Figure 11.108
An analog signal is connected to a pin of the internal ADC.

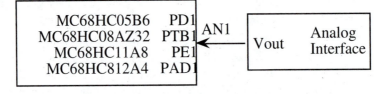

Table 11.28
Straight binary format used by the Motorola internal 8-bit ADCs.

Analog input voltage (V)	Digital output
0.00	%00000000 $00 0
0.02	%00000001 $01 1
2.50	%10000000 $80 128
3.75	%11000000 $C0 192
5.00	%11111111 $FF 255

The analog input range is 0 to +5 V, the analog input resolution is 5 V/256, which is about 20 mV, and the precision is 8 bits. The 6811 and 6812 ADCs will perform multiple conversions on the same analog signal. The Program 11.16 subroutine only returns the first one.

```
; MC68HC05B6
; Analog signal connected to D1
;ADSTAT Register ($0009)
;   7  COCO =0 on a write ADSTAT,
;           =1 on ADC done
;   6  0     internal clock
;   5  ADON =1   ADC enabled
;   4         not implemented
; 3-0 CH3-0 =0001 Mux addr D1
Init rts    ; no ritual required
Samp lda #$21
     sta ADSTAT  ;Start ADC
Loop lda ADSTAT
     and #$80    ;check COCO
     beq Loop
     lda ADDATA  ;result
     rts
```

```
; MC68HC08AZ32
; Analog signal connected to PTB1
;ADSCR Register ($0038)
;   7  COCO =0 on a write ADSCR,
;           =1 on ADC done
;   6  AIEN =0    no interrupts
;   5  ADCO =0   single conversion
; 4-0 CH4-0 =0001 Mux addr PTB1
Init lda #$70   ;divide by 8
     sta ADCLK  ;internal
     rts            ;E clk timing
Samp lda  #1
     sta  ADSCR ;Start ADC
Loop brclr 7,ADSCR,Loop
     lda  ADR  ;result
     rts
```

```
; MC68HC11A8
; Analog signal connected to E1
;ADCTL Register ($1030)
;   7  CCF  =0 on a write ADCTL,
;           =1 on ADC done
;   6  0     Not implemented
;   5  SCAN =0    one set of 4
;   4  MULT =0    single channel
```

```
; MC68HC812A4/MC68HC912B32
; Analog signal connected to PAD1
;ATDCTL5 Register ($0065)
;   7  0      Not implemented
;   6  S8CM=0     4 conversions
;   5  SCAN =0    one set of 4
;   4  MULT =0    single channel
; 3-0 CD-A =0001 Mux addr PAD1
```

```
;  3-0 CD-A =0001 Mux addr E1          Init ldaa  #$80    ;Turn on ADC
Init ldaa  #$80   ;Turn on ADC              staa  ATDCTL2 ;ADC power up
     staa  OPTION ;ADC power up            rts             ;E clk timing
     rts          ;E clk timing       Samp ldaa  #1
Samp ldx   #$1000                          staa  ATDCTL5 ;Start ADC
     ldaa  #1                         Loop brclr ATDSTAT,$80,Loop
     staa  $30,X  ;Start ADC               ldaa  ADR0H  ;first result
Loop brclr $30,X,$80,Loop                  rts
     ldaa  $31,X  ;first result
     rts
```

Program 11.16 Assembly software to sample data using the ADC (fixed channel).

It is easy to convert these subroutines to accept an input parameter in Register A (0 to 7) (Program 11.17).

```
; MC68HC05B6                            ; MC68HC08AZ32
; Analog signal to D0-7                 ; Analog signal connected to PTB0-7
;ADSTAT Register ($0009)               ;ADSCR Register ($0038)
;   7  COCO =0 on a write ADSTAT,       ;   7  COCO =0 on a write ADSCR,
;          =1 on ADC done               ;          =1 on ADC done
;   6  0      internal clock            ;   6  AIEN =0    no interrupts
;   5  ADON =1   ADC enabled            ;   5  ADCO =0    single conversion
;   4         not implemented           ; 4-0 CH4-0  Mux addr PTB0-7
; 3-0 CH3-0 Mux addr D0-7               Init lda #$70    ;divide by 8
Init rts    ; no ritual required            sta ADCLK   ;internal
;Reg A has channel 0 to 7                   rts             ;E clk timing
A2D  ora #$20                           ;Reg A has channel 0 to 7
     sta ADSTAT  ;Start ADC            A2D  sta  ADSCR ;Start ADC
Loop lda ADSTAT                         Loop brclr 7,ADSCR,Loop
     and #$80     ;check COCO                lda  ADR  ;result
     beq Loop                               rts
     lda ADDATA  ;result
     rts
```
```
; MC68HC11A8                            ; MC68HC812A4/MC68HC912B32
; Analog signal to E7-0                 ; Analog signal connected to PAD7-0
;ADCTL Register ($1030)                ;ATDCTL5 Register ($0065)
;   7  CCF  =0 on a write ADCTL,        ;   7  0      Not implemented
;          =1 on ADC done               ;   6  S8CM=0    4 conversions
;   6  0      Not implemented           ;   5  SCAN =0    one set of 4
;   5  SCAN =0    one set of 4          ;   4  MULT =0    single channel
;   4  MULT =0    single channel        ;  3-0 CD-A =0001 Mux addr PAD7-0
; 3-0 CD-A Mux addr E7-0                Init ldaa  #$80    ;Turn on ADC
Init ldaa  #$80   ;Turn on ADC              staa  ATDCTL2 ;ADC power up
     staa  OPTION ;ADC power up             rts            ;E clk timing
     rts          ;E clk timing        A2D  staa  ATDCTL5 ;Start ADC
Samp ldx   #$1000                       Loop brclr ATDSTAT,$80,Loop
     staa  $30,X  ;Start ADC                ldaa  ADR0H  ;first result
Loop brclr $30,X,$80,Loop                   rts
     ldaa  $31,X  ;first result
     rts
```

Program 11.17 Assembly software to sample data using the ADC (variable channel).

In the C versions of Program 11.18, the channel number (0 to 7) is passed in as an input parameter.

```
// MC68HC11A8
void Init(void){
   OPTION = 0x80;}  // Activate ADC
#define CCF 0x80
unsigned char A2D(unsigned char chan){
   ADCTL=chan;     // Start ADC
   while ((ADCTL & CCF) == 0){};
   return(ADR1); }
```

```
// MC68HC812A4/MC68HC912B32
void Init(void){
   ATDCTL2 = 0x80;}  // Activate ADC
#define SCF 0x8000
unsigned char A2D(unsigned char chan){
   ATDCTL5=chan;     // Start ADC
   while ((ATDSTAT & SCF) == 0){};
   return(ADROH); }
```

Program 11.18 Assembly software to sample data using the ADC.

11.11 Glossary

bandwidth The frequency at which the gain drops to 0.707 of the normal value. For a low-pass system, the frequency response ranges from 0 to a maximum value. For a high-pass system, the frequency response ranges from a minimum value to infinity. For a bandpass system, the frequency response ranges from a minimum to a maximum value. Same as frequency response.

bipolar Either a NPN or PNP transistor.

common-mode input impedance Common mode input voltage divided by common mode input current.

common-mode rejection ratio (CMRR) For a differential amplifier, CMRR is the ratio of the common-mode gain divided by the differential mode gain. A perfect CMRR would be zero.

differential mode input impedance Differential mode input voltage divided by differential mode input current.

FET Field-effect transistor, also junction FET (JFET).

follower An analog circuit with gain equal to 1, large input impedance, and small output impedance. Same as voltage follower.

frequency response The frequency at which the gain drops to 0.707 of the normal value. For a low-pass system, the frequency response ranges from 0 to a maximum value. For a high-pass system, the frequency response ranges from a minimum value to infinity. For a bandpass system, the frequency response ranges from a minimum to a maximum value. Same as bandwidth.

input bias current Difference between currents of the two op amp inputs.

input impedance Input voltage divided by input current.

input noise current Current noise refered to the op amp inputs.

input noise voltage Voltage noise refered to the op amp inputs.

input offset current Average current into the two op amp inputs.

input offset voltage Voltage difference between the two op amp inputs that makes the output zero.

instrumentation amplifier A differential amplifier analog circuit, which can have large gain, large input impedance, small output impedance, and a good CMRR.

MOSFET Metal oxide semiconductor field-effect transistor.

negative feedback An analog system with negative-gain feedback paths. These systems are often stable.

Op amp An integrated analog component with two inputs (V2, V1) and an output (Vout), where Vout=K•(V2 − V1). The output has a very large gain K.

open loop An analog system with no feedback paths.

output impedance Open-circuit output voltage divided by closed-circuit output current.

positive feedback An analog system with positive-gain feedback paths. These systems will saturate.

Q The Q of a bandpass filter (passes fmin to fmax) is the center pass frequency [fo=(fmax−fmin)/2] divided by the width of the pass region, Q=fo/(fmax−fmin). The Q of a band-reject filter (rejects fmin to fmax) is the center reject frequency [fo= (fmax−fmin)/2] divided by the width of the reject region, Q=fo/(fmax−fmin).

voltage follower An analog circuit with gain equal to 1, large input impedance, and small output impedance. Same as follower.

11.12 Exercises

11.1 Consider an ADC system.

a) Give a definition of the *ADC precision.* State the units.

b) Give a definition of the *ADC resolution.* State the units.

11.2 A 6811/6812 with *an 8-bit DAC* is used to create a software-controlled analog output. The 8-bit signed DAC (using offset binary format) is connected to the 6811/6812 Port B and has a −5 V to +5V range as shown in the following table. N is the Port B output (DAC input), and V is the DAC output

N=8-bit digital input	V=analog output (mvolts)
0 $00 %00000000	−5000
64 $40 %01000000	−2500
128 $80 %10000000	0
129 $81 %10000001	+ 39
192 $C0 %11000000	+2500
255 $FF %11111111	+4961

a) Derive the linear relationship between N and V. Show both the equation that expresses V in terms of N and the equation that expresses N in terms of V.

b) What is the *DAC precision?* State the units.

c) What is the *DAC resolution?* State the units.

d) Write a 6811/6812 assembly language function, DAout, which takes the desired voltage (in millivolts) and outputs the appropriate value to Port B. The 16-bit signed input parameter is passed by reference on the stack. Here are a couple of typical calling sequences:

```
     ldx #data pointer to value        ldy #neg2
     pshx       by reference           pshy         by reference
     jsr DAout                         jsr DAout
     pulx       discard parameter      puly
data fdb 2314   means 2.314 volts      neg2 fdb −2000 means −2 volts
```

11.3 What do ADC and ADC stand for?

11.4 What do DAC and DAC stand for?

11.5 Which port on the HC05B6, HC08AZ32, HC11A8, HC812A4, and HC912B32 is used by the ADC?

11.6 If a 10-bit ADC has a range of 0 to +10 V, what is the resolution?

11.7 Give three equivalent ways to specify the precision of a 12-bit ADC.

11.8 If an 8-bit ADC has a range of 5 V, what will be the output for an input voltage of 1 V, assuming straight binary encoding?

11.9 If an 8-bit ADC takes inputs ranging from −2.5 V to +2.5 V, what will the output corresponding to +1 V, assuming offset binary encoding?

Figure 11.109
Digital output versus analog input for Exercise 11.10.

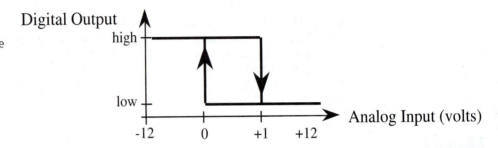

11.10 Design an ADC with the transfer function shown in Figure 11.109. Label all chip numbers (but not pin numbers). Specify the +12, −12, and +5 power supply connections, resistor values, and capacitor values. Offset null potentiometers are not required.

11.11 Design a variable-gain analog amplifier. The analog input is V_{in}, the 3-bit digital inputs (connected to a microcomputer output digital output) are B_2, B_1, B_0, and the analog output is V_{out}. Exactly one of the digital inputs will be 1, and the gain should be

B_2,B_1,B_0	Gain (V_{out}/V_{in})
001	1
010	10
100	100

11.12 In the EKG waveform shown in Figure 11.110, there are long periods of constant value together with short periods of rapidly changing values. Notice that the data points in this figure are placed at uneven time intervals to match the various phases of this signal. Implement software that generates this waveform using uneven time output compare interrupts. In particular, create ROM-based tables that store the nine or ten points and use the DAC8043 12-bit DAC to create the periodic waveform.

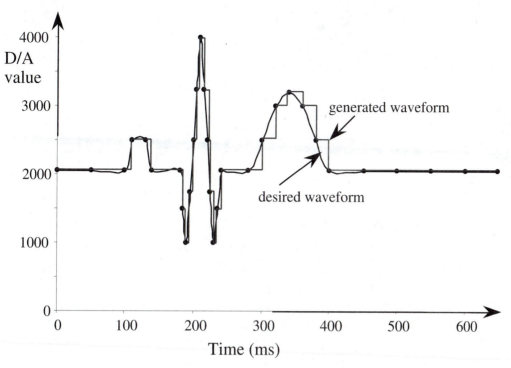

Figure 11.110
EKG output waveform for Exercise 11.12.

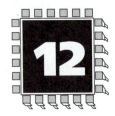

12 Data Acquisition Systems

Chapter 12 objectives are to:

❏ Define performance criteria to evaluate our overall data acquisition system
❏ Introduce specifications necessary to select the proper transducer for our data acquisition system
❏ Describe some typical transducers used in embedded systems
❏ Develop a methodology for designing data acquisition systems
❏ Analyze the sources of noise and suggest methods to reduce their effect
❏ Illustrate concepts of this chapter with case studies

Embedded systems are different from general-purpose computers in the sense that embedded systems have a dedicated purpose. As part of this purpose, many embedded systems are required to collect information about the environment. A system that collects information is called a data acquisition system (DAS). In this chapter, we will use the two terms, DAS and instrument interchangeably. Previous chapters presented the basic building blocks to acquire data into the computer, and in this chapter we will combine these blocks into DASs. Sometimes the acquisition of data is the fundamental purpose of the system, such as with a voltmeter, a thermometer, a tachometer, an accelerometer, an altimeter, a manometer, a barometer, an anemometer, an audio recorder, or a camera. At other times, the acquisition of data is an integral part of a larger system such as a control system or communication system. Control systems and communication systems will be discussed in Chapters 13 and 14.

12.1 Introduction

Figure 12.1 illustrates the integrated approach to instrument design. In this section, we begin with the clear understanding of the problem. We can use the definitions in this section to clarify the design parameters as well as to report the performance specifications. Next, in Section 12.2, we will define the parameters and discuss the selection of a suitable transducer. In Section 12.3, we put together the analog and digital components, introduced in

Figure 12.1 Individual components are integrated into a DAS.

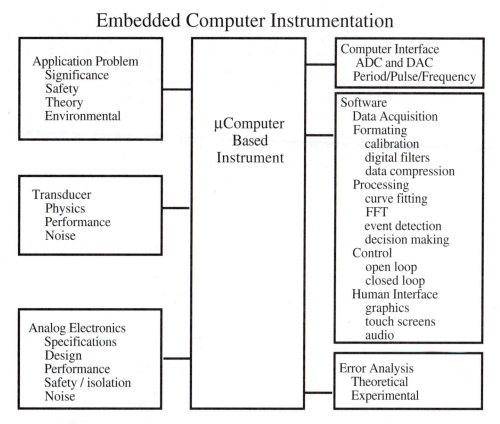

Embedded Computer Instrumentation

| Application Problem Significance Safety Theory Environmental | | Computer Interface ADC and DAC Period/Pulse/Frequency |

μComputer Based Instrument

Software
Data Acquisition
Formating
  calibration
  digital filters
  data compression
Processing
  curve fitting
  FFT
  event detection
  decision making
Control
  open loop
  closed loop
Human Interface
  graphics
  touch screens
  audio

Transducer
Physics
Performance
Noise

Analog Electronics
Specifications
Design
Performance
Safety / isolation
Noise

Error Analysis
Theoretical
Experimental

Chapter 11, to build DASs. The use of period/pulse/frequency as a means of collecting information was developed in Chapter 6. The design of digital filters will be developed later in Chapter 15. A closed-loop control system, as we will see in Chapter 13, includes a DAS. Noise can never be eliminated, but we will study techniques in Section 12.4 to reduce its effect on our system. The integrated approach to design will be illustrated using the case studies in Section 12.5.

The *measurand* is the physical quantity, property, or condition that the instrument measures. The measurand can be inherent to the object (like position, size, mass, or color), located on the surface of the object (like the human EKG or surface temperature), located within the object (e.g., fluid pressure or internal temperature), or separated from the object (like emitted radiation). In general, a *transducer* converts one energy level into another. In the context of this book, the transducer converts the measurand into an electric signal that can be processed by the microcomputer-based instrument. Typically, a transducer has a primary sensing element and a variable conversion element. The *primary sensing element* interfaces directly to the object and converts the measurand into a more convenient energy form. The output of the *variable conversion element* is an electric signal that hopefully depends on the measurand. For example, the primary sensing element of a pressure transducer is the diaphragm, which converts pressure into a displacement of a plunger. The variable conversion element is a strain gage that converts the plunger displacement into a change in electric resistance. If the strain gage is placed in a bridge circuit, the voltage output is directly proportional to the pressure. Some transducers perform a direct conversion without having a separate primary sensing element and variable conversion element. The instrumentation contains *signal processing*, which manipulates the transducer signal output to select, enhance, or translate the signal to perform the desired function, usually in the presence of disturbing factors. The signal processing can be divided into stages. The *analog signal processing* consists of instrumentation electronics, isolation amplifiers, amplifiers, analog filters, and analog calculations. The first analog processing involves calibration signals and preamplification. Calibration is necessary to produce accurate results. An example of a calibration signal is the reference junction of a thermocouple. The second stage of the analog signal processing includes filtering and range conversion. The analog signal range should match the ADC analog input range. Examples of analog calculations include RMS calculation, integration, differentiation, peak detection, threshold detection, PLLs, AM and FM modulation/demodulation, and the arithmetic calculations of addition, subtraction, multiplication, division, and square root. When period, pulse width, or frequency measurement is used, we typically use an analog comparator to create a digital logic signal to measure. Whereas Figure 12.1 outlined design components, Figure 12.2 shows the signal paths that exist in a data acquisition system or control system. The control system uses an actuator to drive a parameter in the real world to a desired value, while the DAS has no actuator because it simply measures the parameter in a noninstrusive manner.

The *data conversion element* performs the conversion between the analog and digital domains. This part of the instrument includes hardware and software computer interfaces, ADC, DAC, S/H, analog multiplexer, and calibration references. The *ADC* converts the ana-

Figure 12.2 Signal paths of a DAS.

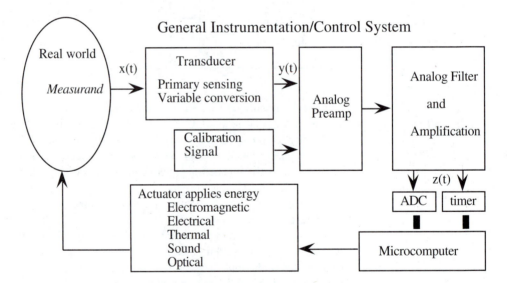

log signal into a digital number. In Chapter 6, we saw that the period, pulse width, and frequency measurement approach provides a low-cost high-precision alternative to the traditional ADC. The *digital signal processing* includes data acquisition (sampling the signal at a fixed rate), data formatting (scaling, calibration), data processing (filtering, curve fitting, Fast Fourier Transform, event detection, decision making, analysis), and control algorithms (open or closed loop). The *human interface* includes the input and output that is available to the human operator. The advantage of computer-based instrumentation is that sophisticated but easy to use and understand devices are possible. The *inputs* to the instrument can be audio (voice), visual (light pens, cameras), or tactile (keyboards, touch screens, buttons, switches, joysticks, roller balls). The *outputs* from the instrument can be numeric displays, cathode ray tube screens, graphs, buzzers, bells, lights, and voice. If the system can deliver energy to the real world, then it is classified as a control system. Control systems will be developed in Chapter 13. In this chapter, we focus on data acquisition. Table 12.1 lists the symbols used in the equations of this chapter.

Table 12.1
Nomenclature.

Symbol	Signal (units)
a	Thermistor radius (cm)
b	Offset of the analog signal processing (V)
C	Electric capacitance (F)
D	Thermistor dissipation constant (mW/°C)
e_n	Op amp noise voltage (V)
f_s	ADC sampling rate (Hz)
f_c	Analog system cutoff frequency (where gain = 0.707) (Hz)
f_{max}, f_{min}	Range of frequencies of interest (Hz)
G	Gain of the analog signal processing (V/V)
I	Electric current (A)
k	Boltzmann's constant = 1.38×10^{-23} J/K
K	Thermal conductivity [mW/(cm·°C)]
K_m	Thermal conductivity of water [6 mW/(cm· °C)]
m	Number of multiplexer signals
n	Index used in digital signal processing
n	Number of ADC bits = $\log_2 n_z$
n_x	Precision of x (alternatives)
n_y	Precision of y (alternatives)
n_z	Precision of z, the ADC analog input voltage (alternatives)
P	Electric power (W)
r_x	Range of $x = max_x - min_x$ (units of x)
r_z	Range of $z = max_z - min_z$ (V)
r_y	Range of $y = max_y - min_y$ (units of y)
R	Electric resistance (Ω)
R_0	Thermistor constant (Ω) $R = R_0 e^{\beta/T}$
t_{ap}	S/H aperture time (s)
t_{aq}	S/H acquisition time (s)
t_c	ADC conversion time (s)
t_{mux}	Settling time of the multiplexer (s)
T	Temperature (°C or K)
V	Voltage (V)
V_y	Transducer output (typically V)
V_z	ADC input voltage (V)
V_J	Johnson noise (alternatively thermal or white noise) = $\sqrt{4 k T R \Delta f}$
x	Real-world signal (cm, mmHg, °C, etc.)
$x(n)$	ADC output sequence, sampled at f_s
y	Transducer output (typically V)
$y(n)$	Digital filter output sequence
z	ADC input voltage (V)
z	Thermistor time constant (s)
Z	Electric impedance (Ω)
α	Thermistor thermal diffusivity (cm²/s)
β	Thermistor constant (K) $R = R_0 e^{\beta/T}$
Δ_x	Resolution of x (units of x)
Δ_y	Resolution of y (units of y)
Δ_z	Resolution of z, the ADC analog input voltage (V)

The rest of this section defines performance criteria we can use to characterize our instrument. Whenever reporting specifications of our instrument, it is important to give the definitions of each parameter, the magnitudes of each parameter, and the experimental conditions under which the parameter was measured, because engineers and scientists apply a wide range of interpretations for these terms.

12.1.1
Accuracy

The *instrument accuracy* is the absolute error referenced to the National Institute of Standards and Technology (NIST) of the entire system including transducer, electronics, and software. Let x_{mi} be the values as measured by the instrument, and let x_{ti} be the true values from NIST references. In some applications, the signal of interest is a relative quantity (like temperature or distance between objects). For relative signals, accuracy can be appropriately defined many ways:

$$\text{Average accuracy (with units of x)} = \frac{1}{n}\sum_{i=1}^{n}\left|x_{ti} - x_{mi}\right|$$

$$\text{Maximum error (with units of x)} = \max\left|x_{ti} - x_{mi}\right|$$

$$\text{Standard error (with units of x)} = \sqrt{\frac{1}{n-1}\sum_{i=1}^{n}[x_{ti} - x_{mi}]^2}$$

In other applications, the signal of interest is an absolute quantity. For these situations, we can specify errors as a percentage of reading or as a percentage of full scale:

$$\text{Average accuracy of reading (\%)} = \frac{100}{n}\sum_{i=1}^{n}\frac{\left|x_{ti} - x_{mi}\right|}{x_{ti}}$$

$$\text{Average accuracy or full scale (\%)} = \frac{100}{n}\sum_{i=1}^{n}\frac{\left|x_{ti} - x_{mi}\right|}{x_{tmax}}$$

$$\text{Maximum accuracy of reading (\%)} = 100\max\frac{\left|x_{ti} - x_{mi}\right|}{x_{ti}}$$

$$\text{Maximum accuracy or full scale (\%)} = 100\max\frac{\left|x_{ti} - x_{mi}\right|}{x_{tmax}}$$

Observation: The definitions of the terms accuracy, resolution, and precision vary considerably in the technical literature. It is good practice to include both the definitions of your terms and their values in your technical communication.

Since the Celsius and Fahrenheit temperature scales have arbitrary zeros (e.g., 0°C is the freezing point of water), it is inappropriate to specify temperature error as a percentage of reading or as a percentage of full scale when Celsius and Fahrenheit scales are used. When specifying temperature error, we should use average accuracy, maximum error, or standard error that have units of Celsius or Fahrenheit degrees.

12.1.2
Resolution

The *instrument resolution* is the smallest input signal difference, Δ, that can be detected by the entire system including transducer, electronics, and software. The resolution of the system is sometimes limited by noise processes in the transducer itself (e.g., thermal imaging) and sometimes limited by noise processes in the electronics (e.g., thermistors, resistance temperature devices [RTDs], and thermocouples).

The *spatial resolution* (or *spatial frequency response*) of the transducer is the smallest distance between two independent measurements. The size and mechanical properties of the transducer determine its spatial resolution. When measuring temperature, a metal probe will disturb the existing medium temperature field more than a glass probe. Hence, a glass probe has a smaller spatial resolution than a metal probe of the same size. Noninvasive imaging systems exhibit excellent spatial resolution because the instrument does not disturb the medium that is being measured. The spatial resolution of an imaging system is the medium surface area from which the radiation originates that is eventually focused onto the detector during the imaging of a single pixel, the so-called instantaneous field of view (IFOV). When measuring force, pressure, or flow, the spatial resolution is the effective area over which the measurement is obtained. Another way to illustrate spatial resolution is to attempt to collect a two- or three-dimensional image of the measurand. The spatial resolution is the distance between points in our image.

12.1.3
Precision

Precision is the number of distinguishable alternatives, n_x, from which the given result is selected. Precision can be expressed in bits or decimal digits. Consider a thermometer instrument with a temperature range of 0 to 100°C. The system displays the output using three digits (e.g., 12.3°C). In addition, the system can resolve each temperature T from the temperature $T + 0.1$°C. This system has 1001 distinguishable outputs and hence has a precision of 1001, or about 10 bits. For a linear system, there is a simple relationship between range (r_x), resolution (Δ_x), and precision (n_x). Range is equal to resolution times precision

$$r_x(100°C) = \Delta_x(0.1°C) \cdot n_x(1001 \text{ alternatives})$$

where "range" is the maximum minus minimum temperature and precision is specified in terms of number of alternatives. We will develop later in Section 12.3.3 a more complex relationship for nonlinear systems. Table 12.2 illustrates the approximate relationship between binary bits and decimal digits.

Table 12.2
Three approaches to specifying precision.

Alternatives	Binary bits	Decimal digits
1000	10	3
2000	11	3½
4000	12	3¾
10000	>13	4

Observation: A good rule of thumb to remember is $2^{10n} \approx 10^{3n}$.

12.1.4
Reproducibility or
Repeatability

Reproducibility (or repeatability) is a parameter that specifies whether the instrument has equal outputs given identical inputs over some period of time. This parameter can be expressed as the full range or standard deviation of output results given a fixed input, where the number of samples and time interval between samples are specified. One of the largest sources of this type of error comes from transducer drift. *Statistical control* is a similar parameter based on a probabilistic model that also defines the errors due to noise. The parameter includes the noise model (e.g., normal, chi-squared, uniform, salt and pepper[1]) and the parameters of the model (e.g., average, standard deviation).

---

[1] An example of salt and pepper noise is the white and black spots on a poor Xerox copy.

12.2   **Transducers**

In this section, we will start with quantitative performance measures for the transducer. Next, specific transducers will be introduced. Rather than give an exhaustive list of all available transducers, the intent in this section is to illustrate the range of possibilities and to provide specific devices to use in the design sections later in the chapter.

12.2.1
Static Transducer
Specifications

The input or measurand is **x**. The output is **y**. A transducer converts **x** into **y** (Figure 12.3). In this subsection, we assume that the input parameter **x** is constant or static.

The input **x** and the output **y** can be either absolute or differential. An absolute signal represents a parameter that exists in a single place at a single time. A differential signal is

Figure 12.3
Transducers in this book convert a physical signal into an electric signal.

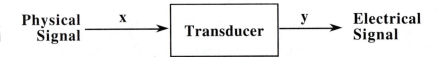

derived from the difference between two signals that exist at different places or at different times. Voltage is indeed defined as a potential difference, but when the voltage is referred to ground, we consider it an absolute quantity. On the other hand, if the signal is represented by the voltage difference between two points neither of which is ground, then we consider the signal as differential. Table 12.3 illustrates four types of transducers.

Type	Input → output	Example
Absolute→absolute	$x \rightarrow y$	Thermistor converts an absolute temperature to a resistance
Relative→absolute	$\Delta x \rightarrow y$	Mass balance converts a mass difference to an angle
Absolute→relative	$x \rightarrow \Delta y$	Strain gauge converts a displacement to a resistance difference
Relative→relative	$\Delta x \rightarrow \Delta y$	Thermocouple converts a temperature difference to voltage difference

Table 12.3   Four types of transducers.

12.2.1.1
Transducer Linearity

Let x_i, y_i be the I/O signals of the transducer as shown in Figure 12.4. The *linearity* is a measure of the straightness of the static calibration curve. Let $y_i = f(x_i)$ be the transfer function of the transducer. A linear transducer satisfies

$$f(ax_1 + bx_2) = af(x_1) + bf(x_2)$$

for any arbitrary choice of the constants a and b. Let $y_i = mx_i + b$ be the best-fit line through the transducer data. Linearity (or deviation from it) as a figure of merit can be expressed as percentage of reading or percentage of full scale. Let y_{max} be the largest transducer output.

$$\text{Average linearity of reading } (\%) = \frac{100}{n}\sum_{i=1}^{n}\frac{|y_i - mx_i - b|}{y_i}$$

$$\text{Average linearity of full scale } (\%) = \frac{100}{n}\sum_{i=1}^{n}\frac{|y_i - mx_i - b|}{y_{max}}$$

Figure 12.4
Input/output response of
a thermocouple
transducer.

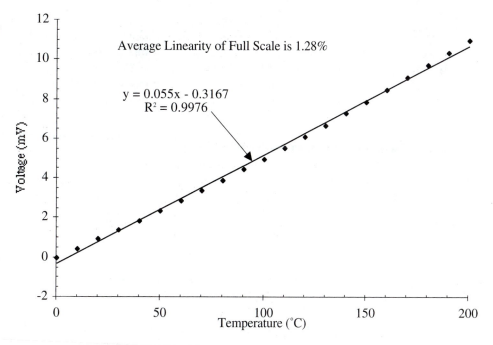

Figure 12.4 illustrates the concept of linearity. In particular, we should not assume this thermocouple to be linear over the 0 to 200°C temperature range.

12.2.1.2
Transducer Sensitivity

Two definitions for sensitivity are used for temperature transducers. The *static sensitivity* is

$$m = \frac{\Delta y}{\Delta x}$$

If the transducer is linear, then the static sensitivity is the slope m of the straight line through the static calibration curve that gives the minimum mean squared error. If x_i and y_i represent measured I/O responses of the transducer, then the least squares fit to $y_i = mx_i + b$ is

$$m = \frac{n\sum_{i=1}^{n}x_i y_i - \sum_{i=1}^{n}x_i \sum_{i=1}^{n}y_i}{n\sum_{i=1}^{n}x_i^2 - \left(\sum_{i=1}^{n}x_i\right)^2} \quad \text{and} \quad b = \frac{\sum_{i=1}^{n}y_i \sum_{i=1}^{n}x_i^2 - \sum_{i=1}^{n}x_i y_i \sum_{i=1}^{n}x_i}{n\sum_{i=1}^{n}x_i^2 - \left(\sum_{i=1}^{n}x_i\right)^2}$$

Thermistors can be manufactured to have a resistance value at 25°C ranging from 4 Ω to 20 MΩ. Because the interface electronics can just as easily convert any resistance into a voltage, a 20-MΩ thermistor is not more sensitive than a 30-Ω thermistor. In this situation, it makes more sense to define *fractional sensitivity* as

$$\alpha = \frac{1}{R} \cdot \frac{\partial R}{\partial T} \quad (1/°C)$$

12.2.1.3
Specificity

Unfortunately, transducers are often sensitive to factors other than the signal of interest. Environmental issues involve how the transducer interacts with its external surroundings (e.g., temperature, humidity, pressure, motion, acceleration, vibration, shock, radiation fields, electric fields, magnetic fields). *Specificity* is a measure of relative sensitivity of the transducer to the desired signal compared to the sensitivity of the transducers to these other unwanted influences. A transducer with a good specificity will respond only to the signal of interest and be independent of these disturbing factors. On the other hand, a transducer with a poor specificity will respond to the signal of interest as well as to some of these disturbing factors. If all these disturbing factors are grouped together as noise, then the *signal-to-noise (S/N) ratio* is a quantitative measure of the specificity of the transducer.

12.2.1.4
Transducer Impedance
The *input range* is the allowed range of input, **x**. The input impedance is the phasor equivalent of the steady-state sinusoidal effort (voltage, force, chemical concentration, pressure) input variable divided by the phasor equivalent of steady-state flow (current, velocity, molecular flux flow) input variable. The output signal strength of the transducer can be specified by the output resistance $\mathbf{R_{out}}$ and output capacitance $\mathbf{C_{out}}$ (Figure 12.5).

Figure 12.5
Output model of a transducer.

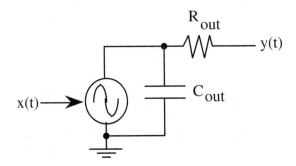

The input impedance of a thermal sensor is a measure of the thermal perturbation that occurs due to the presence of the probe itself in the medium. For example, a thermocouple needle inserted into a laser-irradiated medium will affect the medium temperature because heat will conduct down the stainless steel shaft. A thermocouple has a low input impedance (which is bad) because the transducer itself loads (reduces) the medium temperature. On the other hand, an infrared detector measures surface medium temperature without physical contact. Infrared detectors therefore have a very high input impedance (which is good) because the presence of the transducer has no effect on the temperature to be measured. In the case of temperature sensors, the driving force for heat transfer is the temperature difference ΔT. The resulting heat flow q can be expressed using Fourier's law of thermal conduction:

$$q = KA\frac{\partial T}{\partial x} \propto K4\pi a^2\frac{\Delta T}{a}$$

where K is the probe thermal conductivity, A is the probe surface area, and a is the radius of a spherical transducer. The steady-state input impedance of a spherical temperature probe can thus be approximated by

$$Z \equiv \frac{\Delta T}{q} \approx \frac{1}{4\pi a K}$$

Again, the approximation in the above equation assumes a spherical transducer. Similar discussions can be constructed for the sinusoidal input impedance. Since most thermal events can be classified as step events rather than as sinusoidally varying events, most researchers prefer the use of a time constant to describe the transient behavior of temperature transducers.

Some transducers are completely passive (e.g., thermocouple, EKG electrode), and others are active, requiring external power (e.g., ultrasonic crystals, strain gage, microphone, thermistors). Electric isolation is a critical factor in medical instrumentation. Some transducers are inherently isolated (e.g., thermistors, thermocouples, microphones), while others are not isolated (e.g., EKG electrodes, pacemakers, blood pressure catheters). Minimization of errors is important for all instruments. The sensitivity to disturbing factors (electric fields, magnetic fields, radiation, vibration, shock, acceleration, temperature, humidity) must be determined before a device can be used.

12.2.1.5 Transducer Drift

The *zero drift* is the change in the static sensitivity curve intercept **b** as a function of time or other factor (see Figure 12.6). The *sensitivity drift* is the change in the static sensitivity curve slope **m** as a function of time or some other factor. These drift factors determine how often the transducer must be calibrated. For example, thermistors have a drift much larger than that of RTDs or thermocouples. Transducers may be aged at high temperatures for long periods of time to improve their stability.

Figure 12.6
The two types of transducer drift: sensitivity drift and zero drift.

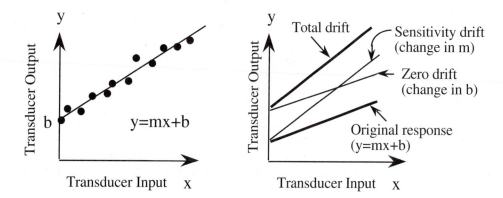

12.2.1.6 Manufacturing Issues

This subsection discusses issues involved in the manufacturing of the system. The transducer is often a critical device, affecting both the cost and performance of the entire system. A higher quality transducer may produce better signals but at an increased cost. An important manufacturing issue is the availability of components. The availability of a device may be enhanced by having a *second source* (more than one manufacturer that produces the device). The use of standard cables and connectors will simplify the construction of your system. The power requirements, size, and weight of the device are important in some systems and thus should be considered when selecting a transducer. Some transducers require calibration, which can greatly increase the cost of manufacturing.

12.2.2 Dynamic Transducer Specifications

The *transient response* is the combination of the *delay* phase, the *slewing* phase, and the *ringing* phase (Figure 12.7). The total transient response is the time for the output $y(t)$ to reach 99% of its final value after a step change in input, $x(t) = u_0(t)$.

The transient response of a temperature transducer to a sudden change in signal input can sometimes be approximated by an exponential equation (assuming first-order response):

$$y(t) = y_f + (y_0 - y_f) e^{-t/\tau}$$

Figure 12.7
The step response often
has delay, slewing, and
ringing phases.

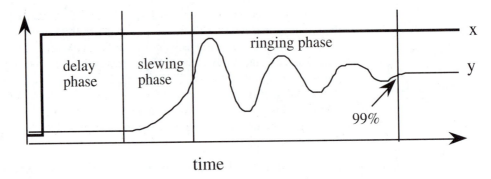

where y_0 and y_f are the initial and final transducer outputs, respectively. The *time constant* τ of a transducer is the time to reach 63.2% of the final output after the input is instantaneously increased. This time is dependent on both the transducer and the experimental setup. Manufacturers often specify the time constant of thermistors and thermocouples in well-stirred oil (fastest) or still air (slowest). In your applications, one must consider the situation. If the transducer is placed in a high-flow liquid like an artery or a water pipe, it may be reasonable to use the stirred-oil time constant. If the transducer is in air or embedded in a solid, then thermal conduction in the medium will determine the time constant almost independently of the transducer.

The *frequency response* is a standard technique to describe the dynamic behavior of linear systems. Let $y(t)$ be the system response to $x(t)$. Let

$$x(t) = A \sin(\omega t) \qquad y(t) = B \sin(\omega t + \phi) \qquad \omega = 2\pi f$$

The magnitude B/A and the phase ϕ responses are both dependent on frequency. Differential equations can be used to model linear transducers. Let $x(t)$ be the time domain input signal, $X(j\omega)$ be the frequency domain input signal, $y(t)$ be the time domain output signal, and $Y(j\omega)$ be the frequency domain output signal (Table 12.4).

Classification	Differential equation	Gain response	Phase response
Zero-order	$y(t) = m\,x(t)$	$Y/X = m$ = static sensitivity	
First-order	$y'(t) + a\,y(t) = b\,x(t)$	$YX = \dfrac{b}{\sqrt{a^2 + \omega^2}}$	Phase = arctan $(-\omega/a)$
Second-order	$y''(t) + a\,y'(t) + b\,y(t) = c\,x(t)$		
Time delay	$y(t) = x(t - T)$	$Y/X = \exp(-j\omega T)$	

Table 12.4   Classifications of simple linear systems.

**12.2.3
Nonlinear
Transducers**

Nonlinear characteristics include hysteresis, saturation, bang-bang, breakdown, and dead zone. *Hysteresis* is created when the transducer has memory. We can see in the Figure 12.8 that when the input was previously high it falls along the higher curve, and when the input was previously low it follows along the lower curve. Hysteresis will cause a measurement error, because for any given sensor output, y, there may be two possible measurand inputs. *Saturation* occurs when the input signal exceeds the useful range of the transducer. With saturation, the sensor does not respond to changes in input value when the input is either too high or too low. *Breakdown* describes a second possible result that may occur when in

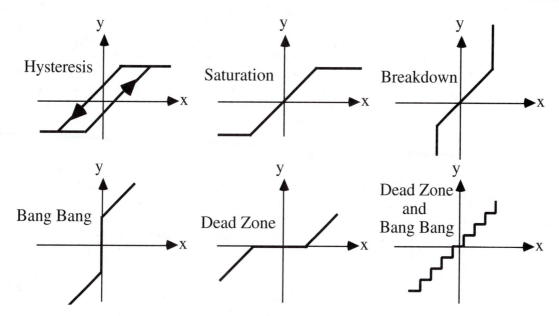

Figure 12.8 Nonlinear transducer responses.

the input exceeds the useful range of the transducer. With breakdown, the sensor output changes rapidly, usually the result of permanent damage to the transducer. Hysteresis, bang bang and dead zone all occur within the useful range of the transducer. *Bang bang* is a sudden large change in the output for a small change in the input. If the bang bang occurs predictably, then it can be corrected for in software. A *dead zone* is a condition where a large change in the input causes little or no change in the output. Dead zones can not be corrected for in software, thus if present will cause measurement errors.

There are many ways to model nonlinear transducers. A nonlinear transducer can be described as a piecewise-linear system. The first step is to divide the range of x into finite subregions, assuming the system is linear in each subregion. The second step is to solve the coupled linear systems so that the solution is continuous. Another method to model a nonlinear system is to use empirically determined nonlinear equations. The first step in this approach is to observe the transducer response experimentally. Given a table of x and y values, the second step is to fit the response to a nonlinear equation.

A third approach to model a nonlinear transducer uses a look-up table located in memory. This method is convenient and flexible. Let x be the measurand and y be the transducer output. The table contains x values, and the measured y value is used to index into the table. Sometimes a small table coupled with linear interpolation achieves equivalent results to a large table. There are two examples of this approach as part of the T*Ex*aS application called TBL.RTF and ETBL.RTF.

12.2.4
Position
Transducers

One of the simplest methods to convert position into an electric signal uses a position-sensitive potentiometer. These devices are inexpensive to build and are sensitive to small displacements. The transducer is constructed by wrapping a fine uninsulated wire around an insulating cylinder. The wires are close to one another, but do not touch. The total electric resistance of the transducer is given by

$$R = \rho\, L/A$$

where ρ is the resistivity of the wire, A is the cross-sectional area of the wire, and L is the length of the wire. A potentiometer is formed by placing a wiper blade that makes electric contact on the wires. The blade is free to move up and down along the axis of the cylinder. If the wire has been uniformly wrapped, then R_{out} will be linearly related to displacement x. The disadvantages of this transducer are its low-frequency response, its high mechanical input impedance, and its degeneration with time. Nevertheless, this type of transducer is adequate for many applications. This transducer will be interfaced two ways later in Section 12.5.5.

Given the fact that this transducer in actuality has multiple discrete outputs (i.e., R_{out} versus x is not linear but has multiple bang-bangs and dead zones), what is the maximum number of ADC bits that can be used? For a discrete wire potentiometer like the one shown in Figure 12.9, the maximum precision is determined by the number of times the wire is wrapped around the post. In other words, the resistance cannot change until the variable arm moves enough to jump to the next wire. For a continuous solid potentiometer there is no fundamental limit.

Figure 12.9
Potentiometer-based position sensor.

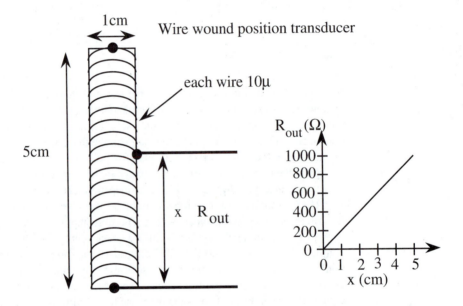

Similar transducers can be constructed using variable capacitors and inductors. The linear variable differential transducer (LVDT) uses a ferrite core to modify the mutual inductance between the single active primary coil and the two passive secondary coils (Figure 12.10). The primary coil is excited with an AC signal. The ferrite core is allowed to move up and down through the middle of the three inductors. If the ferrite core is midway between the two secondaries (x = 0), then the AC voltage across S1 will equal the AC voltage across S2, and the DC output V_{out} will be zero. If the ferrite core is closer to the secondary S1 (x > 0), then the AC voltage across S1 will be larger than the AC voltage across S2, and the DC output V_{out} will be positive. If the ferrite core is closer to the secondary S2 (x < 0), then the AC voltage across S2 will be larger than the AC voltage across S1, and the DC output V_{out} will be negative. An ADC converter is required to generate a digital signal proportional to displacement.

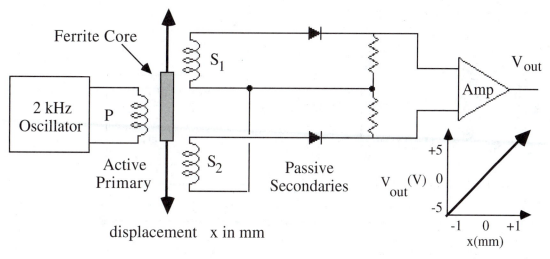

Figure 12.10 A LVDT measures displacement.

A method to measure the distance between two objects is to transmit a sound wave from one object at the other and listen for the reflection (Figure 12.11). The instrument must be able to generate the sound pulse, hear the echo, and measure the time, Δt, between pulse and echo. If the speed of sound, c, is known, then the distance, d, can be calculated. Our microcontrollers also have mechanisms to measure the pulse width Δt.

$$d = c\Delta t/2$$

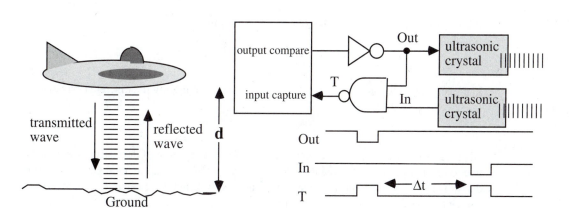

Figure 12.11 An ultrasonic pulse-echo transducer measures the distance to an object.

12.2.5
Velocity
Measurements

12.2.5.1
Velocity Transducers

One can use a LED-photosensor pair to measure the angular velocity of a rotating shaft (Figure 12.12). The circular disk is mounted on the rotating axle. The disk has eight equally spaced holes. The LED and sensor are positioned on opposite sides of the disk. The sensor output, Out, will be high when light passes through one of the holes in the disk. Out will be low when no light hits the sensor. If the shaft is rotating at a constant velocity, then Out will be a square wave. The rotations per minute (RPM) of the shaft can be calculated by

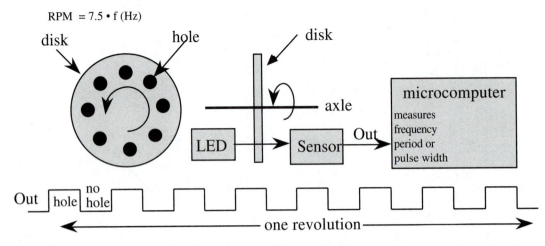

Figure 12.12   A LED-photosensor pair measures shaft rotation.

measuring the frequency of the square wave, f, in hertz. The frequency can also be indirectly calculated from measurement of period or pulse width.

$$RPM = 7.5 \cdot f \qquad (Hz)$$

12.2.5.2 Velocity Calculations   Given a method to measure position, we can use calculus to determine velocity and acceleration.

$$v(t) = \frac{dx(t)}{dt} \qquad a(t) = \frac{dv(t)}{dt}$$

The continuous derivative can be expressed as a limit.

$$v(t) = \frac{x(t + \Delta t) - x(t)}{\Delta t} \qquad \Delta t \to 0$$

The microcomputer can approximate the velocity by calculating the discrete derivative.

$$v(t) = \frac{x(t) - x(t - \Delta t)}{\Delta t}$$

where Δt is the time between velocity samples. This calculation of derivative is very sensitive to errors in the measurement of x(t). A more stable calculation averages two or more derivative terms taken over different time windows. In general we can define such a robust calculation as

$$v(t) = \frac{a}{a + b} \frac{x(t) - x(t - n\,\Delta t)}{n\,\Delta t} + \frac{b}{a + b} \frac{x(t - c\,\Delta t) - x(t - (c + m)\,\Delta t)}{m\,\Delta t}$$

If the integers n, m, and c are all positive, this calculation can be performed in real time. The first term is the derivative over the large time window of $n\Delta t$. The second window term has a smaller size of $m\Delta t$. It normally fits entire inside the first with $c > 0$ and $c - m < n$. The coefficients a, b create the weight for combining the short and long intervals. With $a = b = 1$, $n = 3$, $m = 1$, and $c = 1$, we get

$$v(t) = \frac{1}{2} \frac{x(t) - x(t - 3\Delta t)}{3\Delta t} + \frac{1}{2} \frac{x(t - \Delta t) - x(t - 2\Delta t)}{\Delta t} = \frac{x(t) + 3x(t - \Delta t) - 3x(t - 2\Delta t) - x(t - 3\Delta t)}{6\Delta t}$$

The acceleration can also be approximated by a discrete derivative.

$$a(t) = \frac{x(t) - 2x(t - \Delta t) + x(t - 2\Delta t)}{\Delta t^2}$$

To make a more stable calculation of second derivative, you could average two or more second-derivative terms taken over different time windows (Figure 12.13).

Figure 12.13
Robust calculation of acceleration.

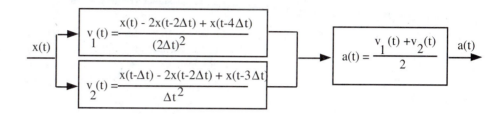

Another robust second-derivative calculation concatenates two robust first-derivative calculations (Figure 12.14).

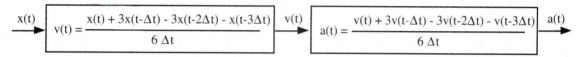

Figure 12.14   Sequential calculation of acceleration.

Observation: In the above calculations of derivative, a single error in one of the x(t) input terms will propagate to only a finite number of the output calculations.

Observation: Although the central difference calculation of $v(t)=[x(t+\Delta t)-x(t-\Delta t)]/2\Delta t$ is theoretically valid, we cannot use it for real-time applications, because it requires knowledge about the future, $x(t+\Delta t)$, which is unavailable at the time v(t) is being calculated.

Similarly, we can perform integration of velocity to determine position.

$$x(t) = \int_0^t v(s) \, ds$$

The microcomputer can perform a discrete integration by summation.

$$x(t) = x(0) + \sum_{n=0}^{t} v(n) \, \Delta t$$

There are two problems with this approach. The first difficulty is determining x(0). The second problem is the accumulation of errors. If one is calculating velocity from position and an error occurs in the measurement of x(t), then that error affects only two calculations of v(t). Unfortunately, if one is calculating position from velocity and an error occurs in the measurement of v(n), then that error will affect all subsequent calculations of x(t).

Observation: In the above calculation of integration, a single error in one of the x(t) input terms will propagate into all remaining output calculations.

The following function, which is quite similar to the derivative, is actually a low-pass digital filter. We will learn more about digital filters in Chapter 15.

$$y(t) = \frac{x(t) + x(t - \Delta t)}{2}$$

12.2.6 Force Transducers

The most common device to measure force is the *strain gage* (Figure 12.15). As a wire is stretched, its length increases and its cross-sectional area decreases. These geometric changes cause an increase in the electric resistance of the wire, R. The transducer is constructed with four sets of wires mounted between a stationary member (frame) and a moving member (armature). As the armature moves relative to the frame, two wires are stretched (increase in R1, R4), and two wires are compressed (decrease in R2, R3). The strain gage is a displacement transducer such that a change in the relative position between the armature and frame, Δx, causes a change in resistance, ΔR. The sensitivity of a strain gage is called its gage factor:

$$G = \frac{\Delta R/R}{\Delta x/x}$$

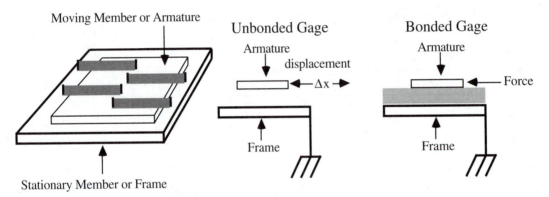

Figure 12.15   Strain gages used for displacement or force measurement.

The gage factor for an Advance strain gage is 2.1. The typical resistance R is 120 Ω. If the gage is bonded onto a material with a spring characteristic

$$F = -kx$$

then the transducer can be used to measure force. The wires each have a significant temperature drift. When the four wires are placed into a bridge configuration, the temperature dependence cancels (Figure 12.16). A high-gain high-input impedance high-CMRR differential amplifier is required.

Figure 12.16   Four strain gages are placed in a bridge configuration.

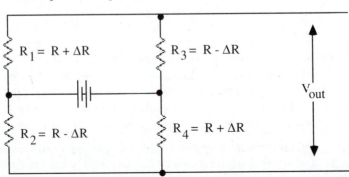

<table>
<tr><td>

12.2.7
Temperature
Transducers

12.2.7.1
Classification of
Temperature Transducers

</td><td>

An increase in thermal energy causes an increase in the average spacing between molecules. For an ideal gas the kinetic energy [$(\frac{1}{2}) mv^2$] equals the thermal energy [$(\frac{3}{2}) kT$]. For liquids and most gases, an increase in kinetic energy is also accompanied by an increase in potential energy. The mercury thermometer and bimetallic strips are common transducers that rely on thermal expansion. A RTD is constructed from long narrow-gage metal wires. Over a moderate range of temperature (e.g., 50°C) the wire resistance is approximately linearly related to its temperature.

</td></tr>
</table>

$$R = R_0 [1 + \gamma(T - T_0)]$$

The sensitivity γ for various metals is given in Table 12.5. R_0 is a function of the resistivity and geometry of the wire. Although many metals such as gold, nichrome (nickel-chromium), nickel, and silver can be used, platinum is typically selected because of its excellent stability.

Table 12.5

Electric properties of various materials.

Material	γ, fractional sensitivity (l/°C)	ρ, electric resistivity ($\Omega-cm$)
Gold	+0.004	$2.35 \cdot 10^{-6}$
Nickel	+0.0069	$6.84 \cdot 10^{-6}$
Platinum	+0.003927	10^{-5}
Copper	0.0068	$1.59 \cdot 10^{-6}$
Silver	+0.0041	$1.673 \cdot 10^{-6}$
NTC thermistor	−0.04	10_3
PTC thermistor	+0.1	

The rates of first-order processes are proportional to the Boltzmann factor, $e^{-E/kt}$. This strong temperature dependence can be exploited to make measurements. Rates that are governed by the Boltzmann factor include the evaporation of liquids and the population of charge carriers in a conduction band. The conductance (1/resistance) of a thermistor is proportional to the occupation of charge carriers in the conduction band:

$$G = G_0 e^{-E/kT}$$

The electric resistance is the reciprocal of the conductance:

$$R = R_0 e^{+E/kT} = R_0 e^{+\beta/T}$$

where β is E/k.

A thermocouple is constructed using two wires of different metals welded to form two junctions. One junction is placed at a known reference temperature (e.g., 0°C ice in thermal equilibrium with liquid and gaseous water at 1 atmosphere), and the other junction is used to measure the unknown temperature. The Seebeck effect involves thermal to electric energy conversion. In a closed-loop configuration, the current around the loop is proportional to the temperature difference between the two junctions. In the open-loop configuration, a voltage is generated proportional to the temperature difference:

$$V = a(T_2 - T_1) + b(T_2 - T_1)^2$$

Electromagnetic radiation is emitted by all materials above absolute zero, of which black or gray bodies are a special case. The spectral properties of a surface are a function of the material and its temperature. The total wide-band radiation power W is a function of the fourth power of surface temperature.

$$W = \sigma(T^4 - T_0^4)$$

The sudden transition of state that occurs for a particular substance can be used to measure temperature. In particular, this property is frequently exploited to produce temperature references. An ice bath at 0°C (water's triple point) is often used as the reference junction of a thermocouple. The melting point of gallium is conveniently 29.7714°C. A triple point occurs under pressure when a substance exists simultaneously in all three states: gas, liquid, and solid. The triple point of water is 0.01 ± 0.0005°C at 1 atmosphere, the triple point of rubidium is 39.265°C, and the triple point of succinonitrile is 58.0805°C.

12.2.7.2
Thermistors

Thermistors are a popular temperature transducer made from a ceramiclike semiconductor (Figure 12.17). A negative-temperature coefficient (NTC) thermistor is made from combinations of metal oxides of manganese, nickel, cobalt, copper, iron, and titanium. A mixture of milled semiconductor oxide powders and a binder is shaped into the desired geometry. The mixture is dried and sintered (under pressure) at an elevated temperature. The wire leads are attached and the combination is coated with glass or epoxy. By varying the mixture of oxides, a range of resistance values from 30 Ω to 20 MΩ (at 25°C) is possible. Table 12.6 lists the trade-offs between thermistors and thermocouples.

Figure 12.17
Thermistors come in many shapes and sizes.

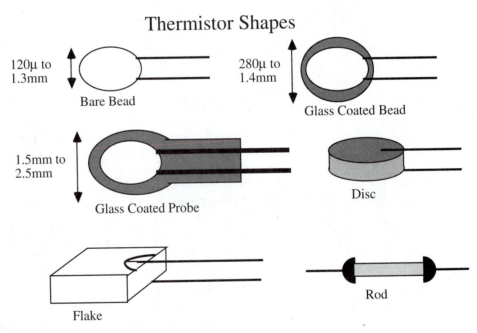

Thermistor Shapes

120μ to 1.3mm — Bare Bead

280μ to 1.4mm — Glass Coated Bead

1.5mm to 2.5mm — Glass Coated Probe

Disc

Flake

Rod

Table 12.6
Trade-offs between thermistors and thermocouples.

Thermistors	Thermocouples
More sensitive	More sturdy
Better temperature resolution	Faster response
Less susceptible to noise	Inert, interchangeable V versus T curves
Less thermal perturbation	Requires less frequent calibration
Does not require a reference	More linear

A precision thermometer, an ohmmeter, and a water bath are required to calibrate thermistor probes. The following empirical equation yields an accurate fit over a wide range of temperature:

$$T = \frac{1}{H_0 + H_1 \ln(R) + H_3 [\ln(R)]^3} - 273.15$$

where T is the temperature in °C and R is the thermistor resistance in Ω. The cubic term was added to the above equation to improve accuracy. It is preferable to use the ohmmeter function of the eventual instrument for calibration purposes so that influences of the resistance measurement hardware and software are incorporated into the calibration process.

12.2.7.3
Thermocouples

A *thermocouple* is constructed by spot welding two different metal wires together. Probe transducers include a protective casing that surrounds the thermocouple junction. Probes come in many shapes including round tips, conical needles, and hypodermic needles. Bare thermocouple junctions provide faster response but are more susceptible to damage and noise pickup. Ungrounded probes allow electric isolation but are not as responsive as grounded probes. Commercial thermocouples have been constructed in 16 to 30 gage hypodermic needles—a 30 gage needle has an outside diameter of above 0.03 cm. Bare thermocouples can be made from 30-μm wire, producing a tip with an 80-μm diameter. A spot weld is produced by passing a large current through the metal junction that fuses the two metals together.

If the wires form a loop, and the junctions are at different temperatures, then a current will flow in the loop. This thermal to electric energy conversion is called the Seebeck effect (see Figure 12.18). If the temperature difference is small, then the current I is linearly proportional to the temperature difference $T_1 - T_2$.

Figure 12.18
When the two thermocouple junctions are at different temperatures, current will flow.

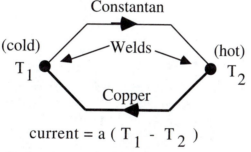

Constantan

(cold)   — Welds —   (hot)
T_1 T_2

Copper

current $= a \, (T_1 - T_2)$

Seebeck effect = thermal to electrical

If the loop is broken, and an electric voltage is applied, then heat will be absorbed at one junction and released at the other. This electric to thermal energy conversion is called the Peltier effect (see Figure 12.19). If the voltage is small, then the heat transferred is linearly proportional to the voltage V.

Figure 12.19
When voltage is applied to two thermocouple junctions, heat will flow.

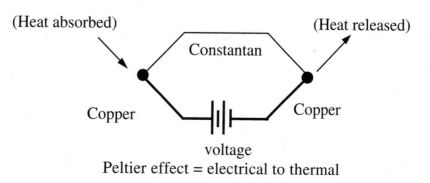

(Heat absorbed)   (Heat released)

Constantan

Copper   Copper

voltage

Peltier effect = electrical to thermal

If the loop is broken and the junctions are at different temperatures, then a voltage will develop because of the Seebeck effect. If the temperature difference is small, then the voltage V, is nearly linearly proportional to the temperature difference $T_1 - T_2$. Thermocouples are characterized by (1) low impedance (resistance of the wires), (2) low temperature sensitivity (45 μV/°C for copper/constantan), (3) low power dissipation, (4) fast response (because of the metal), (5) high stability (because of the purity of the metals), and (6) interchangeability (again because of the physics and the purity of the metals).

Figure 12.20 shows a simple approach to measure temperature using a thermocouple. Typically an ice bath ($T_{ref} = 0$°C) is used for the reference junction. A high-gain differential amplifier is used to convert the low-level voltage (e.g., 0 to 0.9 mV) into the range of the ADC (e.g., 0 to 5 V). Table 12.7 gives the sensitivities and temperature ranges of typi-

Figure 12.20
An instrumentation and low-pass filter are used to interface a thermocouple.

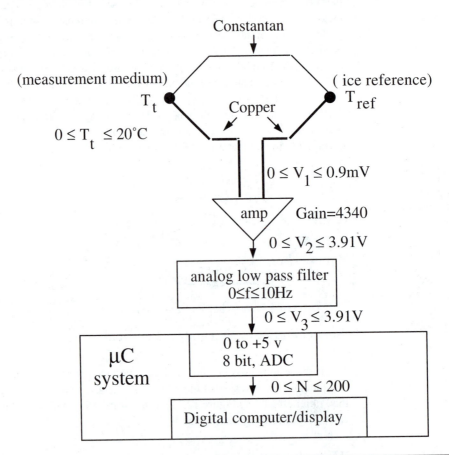

Type—Thermocouple	μV/°C at 20°C	Useful range, °C	Comments
T—Copper/constantan	45	−150 to +350	Moist environment
J—Iron/constantan	53	−150 to +1000	Reducing environment
K—Chromel/alumel	40	−200 to +1200	Oxidizing environment
E—Chromel/constantan	80	0 to +500	Most sensitive
R S—Platinum/platinum-rhodium	6.5	0 to +1500	Corrosive environment
C—Tungsten/rhenium	12	0 to 2000	High temperature

Table 12.7   Temperature sensitivity and range of various thermocouples.

cal thermocouple devices. The amplifier gain is chosen to provide a simple relationship between the A/D output N and the temperature of the medium, T_t. Assuming a linear relationship between T_t and V_1, $T_t = N/10$.

If the temperature range is less than 25°C, then the linear approximation can be used to measure temperature. Let N be the digital sample from the ADC for the unknown medium temperature T_1. A calibration is performed under the conditions of a constant reference temperature: typically, one uses the extremes of the temperature range (T_{min} and T_{max}). A precision thermometer system is used to measure the "truth" temperatures. Let N_{min} and N_{max} be the digital samples at T_{min} and T_{max}, respectively. Then the following equation can be used to calculate the unknown medium temperature from the measured digital sample:

$$T_1 = T_{min} + (N - N_{min}) \cdot \frac{T_{max} - T_{min}}{N_{max} - N_{min}}$$

Because the thermocouple response is not exactly linear, the errors in the above linear equation will increase as the temperature range increases. For instruments with a larger temperature range, a quadratic equation can be used,

$$T_1 = H_0 + H_1 \cdot N + H_2 \cdot N^2$$

where H_0, H_1, and H_2 are determined by calibration of the instrument over the range of interest. Linear regression can be used by letting $z = T_1$, $x = N$, and $y = N^2$.

12.3 DAS Design

12.3.1 Introduction and Definitions

Before designing a DAS we must have a clear understanding of the system goals. We can classify a system as a *quantitative DAS* if the specifications can be defined explicitly in terms of desired range (r_x), resolution (Δx), precision (n_x), and frequencies of interest (f_{min} to f_{max}). If the specifications are more loosely defined, we classify it as a *qualitative DAS*. Examples of qualitative DAS's include systems that mimic the human senses where the specifications are defined using terms like "sounds good," "looks pretty," and "feels right." Other qualitative DAS's involve the detection of events. In these systems, the specifications are expressed in terms of specificity and sensitivity. For binary detection systems like the presence/absence of a burglar or the presence/absence of cancer, we define a true positive (TP) when the condition exists (there is a burglar) and the system properly detects it (the alarm rings). We define a false positive (FP) when the condition does not exist (there is no burglar) but the system thinks there is (the alarm rings). A false negative (FN) occurs when the condition exists (there is a burglar) but the system does not think there is (the alarm is silent). *Sensitivity* is the fraction of properly detected events (a burglar comes and the alarm rings) over the total number of events (number of burglars). It is a measure of how well our system can detect an event. A sensitivity of 1 means you will not be robbed. *Specificity* is the fraction of properly detected events (a burglar comes and the alarm rings) over the total number of detections (number of alarms). It is a measure of how much we believe the system is correct when it says it has detected an event. A specificity of 1 means that when the alarm rings, the police will arrest a burglar when they get there.

$$\text{Sensitivity} = \frac{TP}{TP + FN} \qquad \text{Specificity} = \frac{TP}{TP + FP}$$

The *transducer* converts the physical signal into an electric signal. The *amplifier* converts the weak transducer electric signal into the range of the ADC (e.g., −10 to +10 V).

The *analog filter* removes unwanted frequency components within the signal. The analog filter is required to remove aliasing error caused by the ADC sampling. The *analog multiplexer* is used to select one signal from many sources. The *S/H* is an analog latch used to keep the ADC input voltage constant during the ADC conversion. The *clock* is used to control the sampling process. Inherent in digital signal processing is the requirement that the ADC be sampled on a fixed time basis. The *computer* is used to save and process the digital data. A *digital filter* may be used to amplify or reject certain frequency components of the digitized signal. The MACQ is a convenient data structure to use with the digital filter (Figure 12.21).

Figure 12.21
Block diagram showing how an analog multiplexer is used to sample multiple signals.

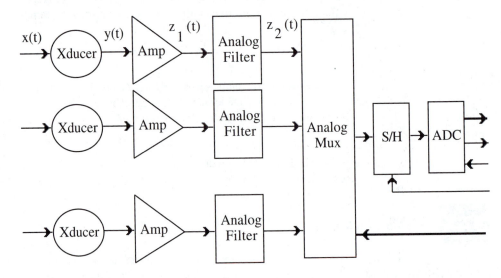

12.3.2 Using Nyquist Theory to Determine Sampling Rate

There are two errors introduced by the sampling process. *Voltage quantizing* is caused by the finite word size of the ADC. The *precision* is determined by the number of bits in the ADC. If the ADC has **n** bits, then the number of distinguishable alternatives is

$$n_z = 2^n$$

Time quantizing is caused by the finite discrete sampling interval. *Nyquist theory* states that if the signal is sampled at f_s, then the digital samples only contain frequency components from 0 to 0.5 f_s. Conversely, if the analog signal does contain frequency components larger than 0.5 f_s, then there will be an *aliasing* error. Aliasing is when the digital signal appears to have a different frequency than the original analog signal.

Simply put, if one samples a sine wave at a sampling rate of f_s,

$$V(t) = A \sin(2\pi ft + \phi)$$

is it possible to determine **A, f,** and ϕ from the digital samples? Nyquist theory says that if f_s is strictly greater than twice **f,** then one can determine **A, f,** and ϕ from the digital samples. In other words, the entire analog signal can be reconstructed from the digital samples. But if f_s less than or equal to twice **f,** then one cannot determine **A, f,** and ϕ. In this case, the apparent frequency, as predicted by analyzing the digital samples, will be shifted to a frequency between 0 and 0.5f_s.

In this first example the frequency of an input sine wave at 1000 Hz and the sampling rate f_s, is varied

$$V = \sin(2\pi \cdot 1000 \cdot t)$$

where **t** is in seconds. In this case the largest (and only) frequency component f_{max} is 1000 Hz. If the signal is sampled at 1333 Hz ($f_s \leq 2 f_{max}$), then an aliasing error will occur (Figure 12.22). This error occurs regardless of the phase between the signal and the ADC sampling. Notice that the apparent frequency of the digital samples (333 Hz) is different from the actual frequency of 1000 Hz.

Figure 12.22
Aliasing makes the 1000-Hz signal appear as a 333-Hz signal.

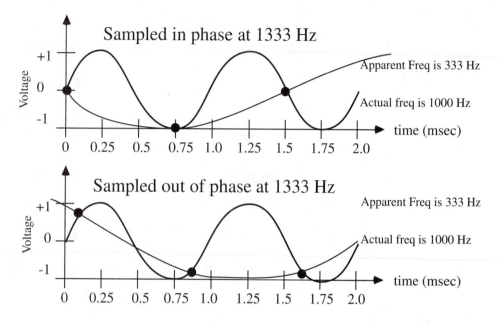

If the 1000-Hz sine wave is sampled in phase at 2000 Hz, then the digital samples appear as a constant (0 Hz). When the sampling frequency is exactly equal to twice the input frequency, the aliasing error is dependent on the phase between the signal and the ADC sampling (Figure 12.23).

Figure 12.23
Right at the Nyquist frequency aliasing may or may not occur.

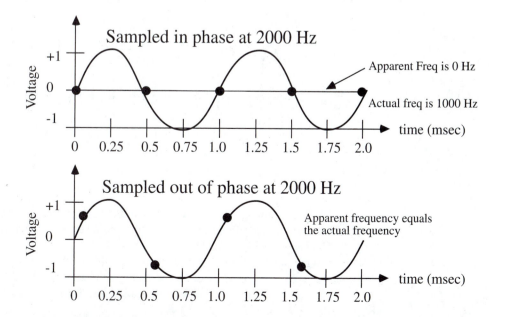

If the 1000-Hz sin wave is sampled above 2000 Hz, then the frequency, magnitude, and phase of the signal can be reconstructed from the digital samples. This reconstruction can be performed regardless of the phase between the signal and the ADC sampling (Figure 12.24).

Figure 12.24
Aliasing does not occur when the sampling rate is more than twice the signal frequency.

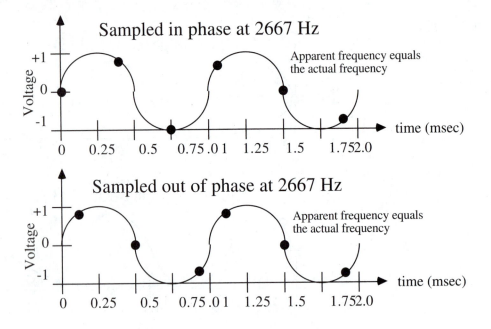

In this next example the sampling rate $\mathbf{f_s}$ will be fixed at 1000 Hz, and the frequency of the input signal will be varied. In the first two cases, the sampling rate is more than twice the input frequency, so the original signal can be properly reconstructed (Figure 12.25).

Figure 12.25
Aliasing does not occur when the sampling rate is more than twice the signal frequency.

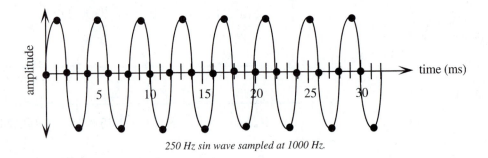

250 Hz sin wave sampled at 1000 Hz.

When sampling rate is exactly twice the input frequency, the original signal may or may not be properly reconstructed. In this specific case, it is frequency shifted (aliased) to (0 Hz) and lost (Figure 12.26).

When sampling rate is slower than twice the input frequency, the original signal cannot be properly reconstructed. It is frequency shifted (aliased) to a frequency between 0 and 0.5 $\mathbf{f_s}$. In this case the 533-Hz wave was aliased to 33 Hz (Figure 12.27).

Figure 12.26
Right at the Nyquist frequency aliasing may or may not occur.

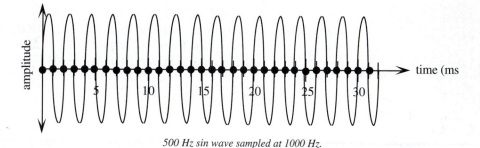

500 Hz sin wave sampled at 1000 Hz.

Figure 12.27
Aliasing makes the 533-Hz signal appear as a 33-Hz signal.

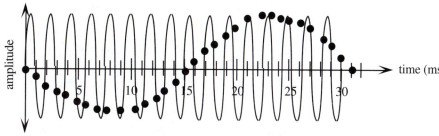

533 Hz sin wave sampled at 1000 Hz.

The choice of *sampling rate* f_s is determined by the maximum useful frequency contained in the signal. One must sample at least twice this maximum useful frequency. Faster sampling rates may be required to implement various digital filters and digital signal processing.

$$f_s > 2f_{max}$$

Even though the largest signal frequency of interest is $\mathbf{f_{max}}$, there may be significant signal magnitudes at frequencies above $\mathbf{f_{max}}$. These signals may arise from the input $\mathbf{x}$, from added noise in the transducer, or from added noise in the analog processing. Once the sampling rate is chosen at $\mathbf{f_s}$, then a low-pass analog filter may be required to remove frequency components above $0.5\mathbf{f_s}$. A digital filter cannot be used to remove aliasing.

**12.3.3
How Many Bits
Does One Need
for the ADC?**

The choice of the *ADC precision* is a compromise of various factors. The overall objective of the DAS will dictate the potential number of useful bits in the signal. If the transducer is nonlinear, then the ADC precision must be larger than the precision specified in the problem statement. For example, let $\mathbf{y}$ be the transducer output and let $\mathbf{x}$ be the real-world signal. Assume for now that the transducer output is connected to the ADC input. Let the range of $\mathbf{x}$ be $\mathbf{r_x}$, let the range of $\mathbf{y}$ be $\mathbf{r_y}$, and let the required precision of $\mathbf{x}$ be $\mathbf{n_x}$. The resolutions of $\mathbf{x}$ and $\mathbf{y}$ are $\mathbf{\Delta x}$ and $\mathbf{\Delta y}$, respectively. Let the following describe the nonlinear transducer.

$$\mathbf{y = f(x)}$$

The required ADC precision $\mathbf{n_y}$ (in alternatives) can be calculated by

$$\mathbf{\Delta_x = \frac{r_x}{n_x}}$$

$$\mathbf{\Delta y = min\ \{f(x + \Delta_x) - f(x)\}\ for\ all\ x\ in\ r_x}$$

$$\mathbf{n_y = \frac{r_y}{\Delta_y}}$$

For example, consider the nonlinear transducer $y = x^2$. The range of x is $0 \leq x \leq 1$. Thus, the range of y is also $0 \leq y \leq 1$. Let the desired resolution be $\Delta_x = 0.01$. $n_x = r_x/\Delta_x = 100$ alternatives, or about 7 bits. From the above equation, $\Delta y = \min\{(x+0.01)^2 - x^2\} = \min\{0.02x + 0.0001\} = 0.0001$. Thus, $n_y = r_y/\Delta y = 10000$ alternatives, or almost 15 bits.

12.3.4 Specifications for the Analog Signal Processing

If the analog signal processing is linear, then

$$z = Gy + b$$

where G is the gain and b is the offset. The resolution and range of z can be found from the previous section.

$$\Delta_z = G \Delta_y$$
$$r_z = G r_y$$

Thus, the precision at z equals the precision at y.

$$n_z = \frac{r_z}{\Delta_z} = \frac{Gr_y}{G\Delta_y} = \frac{r_y}{\Delta_y} = n_y$$

If the transducer and analog signal processing are both linear, then $n_z = n_y = n_x$. Another factor to consider in the choice of ADC word size is the electric noise in the signal. For example, if the signal ranges from 0 to $+8$ V and there is 1 mV of noise in the signal, then any ADC bits beyond 13 would be wasteful. Other factors include cost and convenience for digital processing.

An *analog low-pass filter* may be required to remove aliasing. The cutoff of this analog filter should be less than $0.5 f_s$. Some transducers automatically remove these unwanted frequency components. For example, a thermistor is inherently a low-pass device. Other types of filters (analog and digital) may be used to solve the DAS objective. One useful filter is a 60-Hz band-reject filter.

Let $X(s)$ be the Fourier transform of the physical signal, let $Y(s)$ be the Fourier transform of the transducer output, let $Z_2(s)$ be the Fourier transform of the ADC input, let $H_1(s)$ be the Fourier transfer function for the transducer, let $H_2(s)$ be the Fourier transfer function for the amplifier, let $H_3(s)$ be the Fourier transfer function for the analog filter, and let $f_{min} \leq f \leq f_{max}$ be the frequency range in the signal to be processed. The analog system (transducer, amplifier, filter) must pass frequencies f_{min} to f_{max}. To avoid aliasing, the analog system must reject frequencies above $0.5 f_s$ (Figure 12.28).

Let the gain of the analog filter be $G_3 = |H_3(s)|$. Then the system should pass, with no error as seen by the ADC, for signal frequencies between f_{min} and f_{max}. For example,

$$\frac{2^{n+1} - 1}{2^{n+1}} \leq G_3 \leq \frac{2^{n+1}+1}{2^{n+1}} \qquad \text{for } f_{min} \leq f \leq f_{max}$$

Figure 12.28
A DAS shown in the frequency domain.

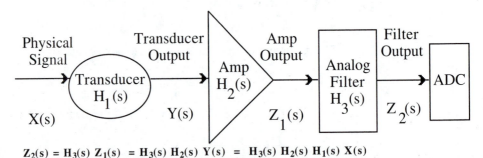

$$Z_2(s) = H_3(s) \, Z_1(s) = H_3(s) \, H_2(s) \, Y(s) = H_3(s) \, H_2(s) \, H_1(s) \, X(s)$$

Figure 12.29 Ideal and practical filter responses.

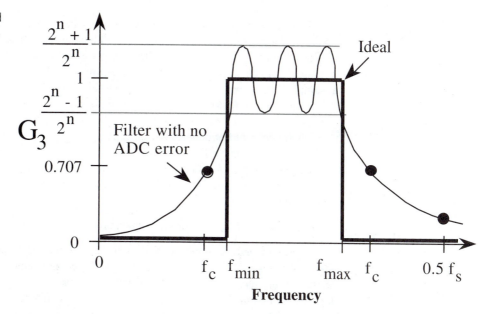

For example, consider a system that uses an 8-bit ADC. If the gain is lower than 255/256 or larger than 257/256, then an error will result in the ADC sample. Conversely, if the gain is between 255/256 and 257/256, then no error should occur in the ADC sample. We will add a safety factor of 1/2 and require that

$$\frac{511}{512} \leq G_3 \leq \frac{513}{512} \qquad \text{for } f_{min} \leq f \leq f_{max}$$

Observation: Most DASs do not need to pass "without ADC error" all frequencies from f_{min} to f_{max}. In this situation we simply place the filter cutoff frequencies at f_{min} and f_{max}.

To prevent aliasing, one must know the frequency spectrum of the ADC input voltage. This information can be measured with a spectrum analyzer. Typically, a spectrum analyzer samples the analog signal at a very high rate (>1 MHz), performs a discrete Fourier transform, and displays the signal magnitude versus frequency. Let $|Z_2|$ be the magnitude of the ADC input voltage as a function of frequency. For example a bandpass device might have a frequency spectrum like the following. There are three regions in the magnitude versus frequency graph shown in Figure 12.30. We will classify any signal with an amplitude less

Figure 12.30
To prevent aliasing there should be no measurable signal above $0.5f_s$.

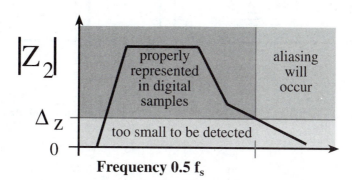

than the ADC resolution, $\mathbf{\Delta_z}$, to be undetectable. This region is labeled "too small to be detected." Undetectable signals cannot cause aliasing regardless of their frequency. We will classify any signal with an amplitude larger than the ADC resolution at frequencies less than $0.5f_s$ to be properly sampled. This region is labeled "properly represented in digital samples." It is information in this region that is available to the software for digital processing. The last region includes signals with an amplitude above the ADC resolution at frequencies greater than or equal to $0.5f_s$. Signals in this region will be aliased, which means their apparent frequencies will be shifted into the 0 to $0.5f_s$ range.

Aliasing will occur if $|\mathbf{Z_2}|$ is larger than the ADC resolution for any frequency larger than or equal to $0.5f_s$. In order to prevent aliasing, $|\mathbf{Z_2}|$ must be less than the ADC resolution. Our design constraint will again include a safety factor of 1/2. Thus, to prevent aliasing we will make

$$|Z_2| < 0.5\,\Delta_z \qquad \text{for all frequencies larger than or equal to } 0.5f_s$$

This condition usually can be satisfied by increasing the sampling rate or increasing the number of poles in the analog low-pass filter. We cannot remove aliasing with a digital low-pass filter, because once the high-frequency signals are shifted into the 0 to $0.5f_s$ range, we will be unable to separate the aliased signals from the regular ones.

There are errors caused by *impedance loading* between stages of the system. In general, one tries to maximize the input impedance and minimize the output impedance of the modules. A good rule is to design your system such that the error due to impedance loading causes no error in the ADC sample. The errors in the first stages must be multiplied by the amplifier gain so that they can be compared to the ADC resolution (Figure 12.31).

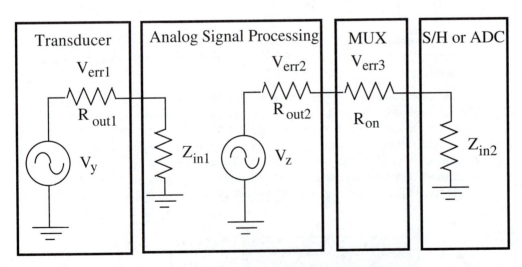

Figure 12.31 Block diagram for considering the problem of impedance loading.

In this example, assume the $\mathbf{V_y}$ and $\mathbf{V_z}$ are unipolar, and the analog signal processing is linear,

$$0 \le V_y \le r_y \qquad 0 \le V_z \le r_z$$

$$V_z = GV_y$$

Using the simple impedance model, the voltage errors are

$$V_{err1} = V_y \frac{R_{out1}}{R_{out1} + Z_{in1}} \approx V_y \frac{R_{out1}}{Z_{in1}}$$

$$V_{err2} = V_z \frac{R_{out2}}{R_{out2} + R_{on} + Z_{in2}} \approx V_z \frac{R_{out2}}{Z_{in2}}$$

$$V_{err3} = V_z \frac{R_{on}}{R_{out2} + R_{on} + Z_{in2}} \approx V_z \frac{R_{on}}{Z_{in2}}$$

The voltage error at the input increases by the gain, **G,** so the total error referred to the output is

$$G\,V_{err1} + V_{err2} + V_{err3} \approx G\,V_y \frac{R_{out1}}{Z_{in1}} + V_z \frac{R_{out2}}{Z_{in2}} + V_z \frac{R_{on}}{Z_{in2}} = V_z \left(\frac{R_{out1}}{Z_{in1}} + \frac{R_{out2} + R_{on}}{Z_{in2}} \right)$$

One wishes to keep the error below the ADC resolution, Δ_z. The factor 1/2 is again arbitrarily chosen. The largest error occurs when V_z equals r_z:

$$r_z \left(\frac{R_{out1}}{Z_{in1}} + \frac{R_{out2} + R_{on}}{Z_{in2}} \right) \approx 0.5\,\Delta_z$$

or

$$\left(\frac{R_{out1}}{Z_{in1}} + \frac{R_{out2} + R_{on}}{Z_{in2}} \right) \leq \frac{0.5}{n_z}$$

where the units of n_z are alternatives. In general, the ratio of the input impedance of each stage divided by the output impedance of the previous stage must exceed the precision of the ADC.

The voltage drop across the multiplexer can be a source of impedance-loading error (Figure 12.32). Consider the issue of impedance loading on the following two circuits. The input impedance of the ADC is 10 kΩ. The ADC is configured for −10 to +10 V. The max "On Resistance, at −25°C≤T_A≤85°C" of the PMI MUX08EP is 400 Ω. The output impedance of an inverting amplifier is the open-circuit voltage divided by the short-circuit current.

$$Zout \equiv \frac{V_{open}}{I_{short}} = \frac{R_{out}(R_1 + R_2)}{K\,R_1}$$

Figure 12.32
A DAS with an impedance-loading program.

$$Z_{out} \equiv \frac{R_{out}}{K}$$

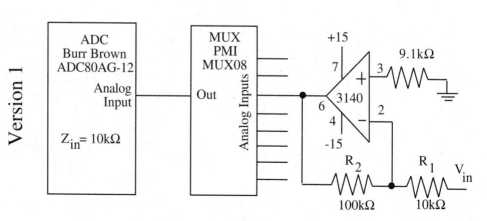

where R1 and R2 are as shown below. K, the op amp open-loop gain, is 100,000. Rout, the op amp output impedance, is 60Ω. The output impedance of a unity gain buffer is

$$Z_{out} \equiv \frac{R_{out}}{K}$$

Version 1 (incorrect): $V_{max} = +10 \text{ V}$

$$Z_{out}(\text{ of 3140 inverter}) = \frac{R_{out}(R_1 + R_2)}{K R_1}$$

$$= \frac{60(10 \text{ } k\Omega + 100 \text{ } k\Omega)}{100000 \times 10 \text{ } k\Omega} = 0.0066 \text{ } \Omega$$

$$R_{on} \text{ (of MUX08)} = 400 \text{ } \Omega$$

$$Z_{in} \text{ (of ADC)} = 10 \text{ k}\Omega$$

$$V_{err} = V_{max} \frac{R_{on} + Z_{out}}{Z_{in} + R_{on} + Z_{out}}$$

$$= 10 \text{ V} \frac{400 \text{ } \Omega + 0.0066 \text{ } \Omega}{10 \text{ } k\Omega + 500 \text{ } \Omega + 0.0066 \text{ } \Omega}$$

$$\approx 10 \text{ V} \frac{400 \text{ } \Omega}{10400 \text{ } \Omega}$$

$$= 0.4 \text{ V} \quad (80 \text{ times the ADC resolution of } 20/4096 = 0.005 \text{ V})$$

A unity gain amplifier is added between the MUX and the ADC (Figure 12.33). The purpose of this unity gain buffer amplifier in the second design is to prevent a significant voltage drop from occurring across the multiplexer. Ohm's law can be used to calculate the voltage drop across the MUX08 for both circuits.

Version 2 (correct):

Stage 1:

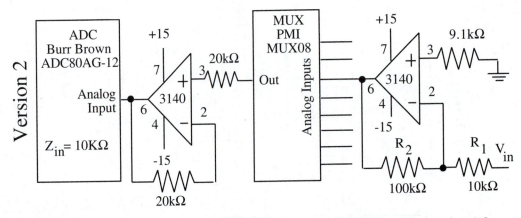

Figure 12.33   The impedance-loading program is solved with a voltage follower before the ADC.

$$Z_{out} \text{ (of 3140 inverter)} = \frac{R_{out}(R_1 + R_2)}{K\, R_1} = \frac{60\,(10\ k\Omega + 100\ k\Omega)}{100000 \times 10\ k\Omega} = 0.0066\ \Omega$$

$$R_{on} \text{ (of MV808)} = 400\ \Omega$$

$$Z_{in} \text{ (of 3140 buffer)} = 10^{12}\ \Omega$$

$$V_{err} = V_{max} \frac{R_{on} + Z_{out}}{Z_{in} + R_{on} + Z_{out}}$$

$$= 10\ V \frac{400\ \Omega + 0.0066\ \Omega}{10^{12}\ \Omega + 400\ \Omega + 0.0066\ \Omega} \bullet 10\ V \frac{400\ \Omega}{10^{12}\ \Omega}$$

$$= 4nv \quad (\textit{This is well below the ADC resolution} = 20/4096 = 5mv)$$

Stage 2:

$$Z_{out} \text{ (of 3140 buffer)} = \frac{R_{out}}{K} = \frac{60}{100000} = 0.0006\ \Omega$$

$$Z_{in} \text{ (of ADC)} = 10\ k\Omega$$

$$V_{err} = V_{max} \frac{Z_{out}}{Z_{in} + Z_{out}} = 10\ V \frac{0.0006\ \Omega}{10\ k\Omega + 0.0006\ \Omega} \approx 10\ V \frac{0.0006\ \Omega}{10\ k\Omega}$$

$$= 0.6\mu V \quad (\textit{This is well below the ADC resolution} = 20/4096 = 5mv)$$

12.3.5
How Fast Must the ADC Be?

The *ADC conversion time* must be smaller than the quotient of the sampling interval by the number of multiplexer signals. Let **m** be the number of multiplexer signals that must be sampled at a rate f_s, let t_{mux} be the settling time of the multiplexer, and let t_c be the ADC conversion time. Then without a S/H,

$$m \cdot (t_{mux} + t_c) < 1/f_s$$

With a S/H, one must include both the acquisition time, t_{aq} and the aperture time, t_{ap}:

$$m \cdot (t_{mux} + t_{aq} + t_{ap} + t_c) < 1/f_s$$

12.3.6
Specifications for the S/H

A S/H is required if the analog input changes more than one resolution during the conversion time. Let **dz/dt** be the maximum slope of the ADC input voltage, let Δ_z be the ADC resolution, and let t_c be the ADC conversion time. A S/H is required if

$$\frac{dz}{dt} \cdot tc > 0.5\, \Delta_z$$

If the transducer and analog signal processing are both linear, the determination of whether or not to use a S/H can be calculated from the input signals. A S/H is required if

$$\frac{dx}{dt} \cdot t_c > 0.5\, \Delta_x$$

Two S/Hs can be used to analyze the timing between two signals. For example, the circuit in Figure 12.34 can be used to measure the phase between two signals with one ADC. The two S/Hs are given the hold command simultaneously, then the signals are sequentially converted. The digital samples represent the two signals at the same time.

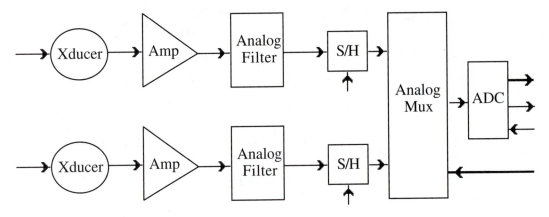

Figure 12.34   Multiple S/Hs can be used to implement synchronized sampling.

12.4   Analysis of Noise

The consideration of noise is critical for all instrumentation systems. The success of an instrument does depend on careful transducer design, precision analog electronics, and clever software algorithms. But any system will fail if the signal is overwhelmed by noise. Fundamental noise is defined as an inherent and nonremovable error. It exists because of fundamental physical or statistical uncertainties. We will consider four types of fundamental noise:

Thermal noise (also called white noise or Johnson noise)
Shot noise
1/f noise
Transducer limitations

Although fundamental noise cannot be eliminated, there are ways to reduce its effect on the measurement objective. In general, added noise includes the many disturbing external factors that interfere with or are added to the signal. We will consider seven types of added noise:

Magnetic induction
Displacement currents, or capacitive coupling
Impedance loading
CMRR
Frequency response
Motion artifact

12.4.1 Thermal Noise

Thermal fluctuations occur in all materials at temperatures above absolute zero. Brownian motion, the random vibration of particles, is a function of absolute temperature (Figure 12.35). As the particles vibrate, there is an uncertainty as to the position and velocity of the particles. This uncertainty is related to the thermal energy:

Absolute temperature T (K)
Boltzmann's constant $k = 1.67 \ 10^{-23}$ J/K
Uncertainty in thermal energy $\approx (1/2) \ kT$

Figure 12.35
Brownian motion of
individual particles.

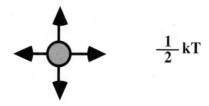

Brownian Motion $\frac{1}{2}kT$

Because the electric power of a resistor is dissipated as thermal power, the uncertainty in thermal energy produces an uncertainty in electric energy. The electric energy of a resistor depends on:

Resistance R (Ω)
Voltage V (V)
Time (s)
Electric power $= V^2/R$ (W)
Electric energy $= V^2 \cdot time/R$ (W $\cdot$ s)

By equating these two energies we can derive an equation for voltage noise similar to the empirical findings of J. B. Johnson. In 1928, he found that the open-circuit RMS voltage noise of a resistor was given by

$$V_J^2 = 4kTR\,\Delta\gamma$$

$$\Delta\gamma = f_{max} - f_{min}$$

where $f_{max} - f_{min}$ is the frequency interval, or bandwidth, over which the measurement was taken. For instance, if the system bandwidth is DC to 1000 Hz, then $\Delta\gamma$ is 1000 cycles/s. Similarly, if the system is a bandpass from 10 to 11 kHz, then $\Delta\gamma$ is also 1000 cycles/s. The term *white noise* comes from the fact that thermal noise contains the superposition of all frequencies and is independent of frequency. It is analogous to optics, where *white light* is the superposition of all wavelengths (Figure 12.36). Table 12.8 illustrates that white noise increases with resistance value and with system bandwidth.

Figure 12.36
White noise exists in all
resistors.

$$V_J = \sqrt{4kTR\,\Delta\gamma}$$

R

R = 1MΩ
T = 300 $^\circ$K
0<f<1000 Hz
$\Delta\gamma$ = 1000 cyc/sec
V$_J$ = 4.5 μv

Table 12.8
White noise for resistors
at 300K = 27°C.

	1Hz	10 Hz	100 Hz	1 kHz	10 kHz	100 kHz	1 MHz
10 kΩ	14 nV	45 nV	142 nV	448 nV	1.4 μV	4.5 μV	14 μV
100 kΩ	45 nV	142 nV	448 nV	1.4 μV	4.5 μV	14 μV	45 μV
1 MΩ	142 nV	448 nV	1.4 μV	4.5 μV	14 μV	45 μV	142 μV

Interestingly, only resistive, not capacitive or inductive, electric devices exhibit thermal noise. Thus a transducer that dissipates electric energy will have thermal noise, and a transducer that simply stores electric energy will not.

Figure 12.37 defines RMS as the square root of the time average of the voltage squared. RMS noise is proportional to noise power. The crest factor is the ratio of peak value divided by RMS. The peak value is half the peak-to-peak amplitude and can be measured easily from recorded data. From Table 12.9, we see that the crest factor is about 4. The crest factor can be defined for other types of noise.

Figure 12.37
RMS is a time average of the voltage squared.

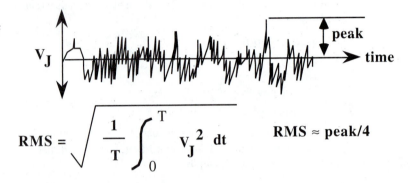

$$RMS = \sqrt{\frac{1}{T} \int_0^T V_J^2 \, dt} \qquad RMS \approx peak/4$$

Table 12.9
Crest factor for thermal noise.

Percent of the time the peak is exceeded	Crest factor (peak/RMS)
1.0	2.6
0.1	3.3
0.01	3.9
0.001	4.4
0.0001	4.9

Example analysis. The objective of the following DAS is to sample x(t) at 10 Hz and perform a software calculations based on the DC to 5-Hz components of x. The gain of this amplifier is −100. Assume the bandwidth of the amplifier without the capacitor C is 100 kHz. With C = 0.01 μF, the bandwidth of the amplifier is reduced to 100 Hz. This reduction does not affect the DC to 5-Hz components of x (Figure 12.38).

Figure 12.38
A DAS with a gain of 100 and a frequency range of 100 Hz.

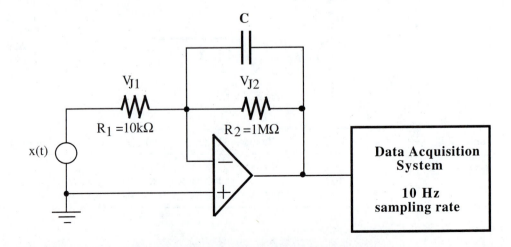

Let V_{J1} be the thermal noise of R_1 and V_{J2} be the thermal noise of R_2. Since the noise is related to the thermal power, the voltage amplitudes are combined at the power level. That is, if two devices generate voltage noises of V_1, and V_2 respectively, then together they will generate

$$V_{total} = \sqrt{V_1^2 + V_2^2}$$

When considering the overall effect of noise, V_{J1} is amplified by the op amp, whereas V_{J2} is not. The presence of the capacitor reduces the noise by 32 without degrading the system performance (Table 12.10). In addition to white noise in the resistor, there is white noise in the op amp.

Table 12.10
Reducing the system bandwidth reduces the thermal noise.

	Without C	With C
V_{J1}	4.5 µv	142 nv
V_{J2}	45 µv	1.4 µv
$\sqrt{[(100V_{J1})^2 + (V_{J2})^2]}$	452 µv	14.3 µv

12.4.2 Shot Noise

Shot noise arises from the statistical uncertainty of counting discrete events. Thermal cameras, radioactive detectors, photomultipliers, and O_2 electrodes count individual photons, gamma rays, electrons, and O_2 particles, respectively, as they impinge upon the transducer. Let dn/dt be the count rate of the transducer, and let Δt be the measurement interval or count time. The average count is

$$n = dn/dt\ \Delta t$$

On the other hand, the statistical uncertainty of counting random events is Δn. Thus the shot noise is

$$\text{Shot noise} = \sqrt{dn/dt\ \Delta t}$$

The S/N ratio is

$$S/N = \frac{n}{\sqrt{n}} = \sqrt{dn/dt\ \Delta t}$$

The solutions are to maximize the count rate (by moving closer or increasing the source) and to increase the count time. There is a clear trade-off between accuracy and measurement time.

12.4.3 1/f, or Pink Noise

Pink noise is somewhat mysterious. The origin of 1/f lacks rigorous theory. It is present in devices that have connections between conductors. 1/f noise results from a fluctuating conductivity. It is of particular interest to low-bandwidth applications because of the 1/f behavior. Wire-wound resistors do not have 1/f noise, but semiconductors do. One of the confusing aspects of 1/f noise is its behavior as the frequency approaches 0 Hz. The noise at DC is not infinite because although 1/f is infinite at DC, $\Delta\gamma$ is zero. Garrett gives an equation to calculate the 1/f noise of a carbon resistor (Table 12.11):

$$V_c = (10^{-6})\sqrt{1/f}\ R\ I\sqrt{\Delta\gamma}$$

Table 12.11
V_c versus frequency for R=10 kΩ, I = 1 mA, $\Delta\gamma$ = 1 kHz.

f(Hz)	V_c (µV)
1	316
10	100
100	32
1000	10

where V_c = 1/f voltage noise, V
        f = frequency, Hz
        R = resistance, Ω
        I = average direct current, A
        $\Delta\gamma$ = system bandwidth, Hz

12.4.4 Transducer Limitations

Certain transducers have intrinsic limitations due to their design or construction. For example, consider a position transducer constructed from a wire-wound potentiometer discussed earlier. Assume there are about 4000 turns of the wire around the 5-cm-long cylinder. Because of the construction of this transducer there is an inherent, nonremovable position uncertainty of 10 μm. Because there are 4000 distinguishable locations, the precision of the transducer is 4000.

12.4.5 Magnetic Field Induction

Magnetic fields can induce a voltage. Magnetic field induction is one of two sources of 60 Hz noise. The changing magnetic field must pass through a wire loop. This voltage noise (Figure 12.39) is proportional to the strength of the magnetic field, B (Wb/m^2), the area of the loop S (m^2), and a geometric factor K (V/Wb).

Figure 12.39
Magnetic fields are a significant source of instrumentation noise.

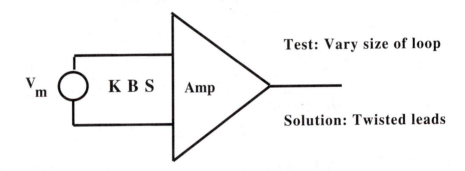

12.4.6 Displacement Currents or Capacitive (Electric Field) Coupling

Changing electric fields will capacitively couple into the lead wires. Displacement currents is the other source of 60 Hz noise. If the cable has twisted wires, then Id1 should equal Id2. The input impedance of the amplifier is usually much larger than the source impedance of the signal. Thus, V1 − V2 = Rc1 · Id1 − Rc2 · Id2. The ground shield should be connected only on the instrument side. If the cable is connected between two instruments, then connect the ground shield on only one instrument. If the ground shield is connected on both instruments, there could be significant currents in the shield, resulting in noise coupling into the signals (Figure 12.40).

12.4.7 Impedance Loading

Let V_s be the unipolar transducer output, and let R_s be the output resistance of the transducer. To reduce the voltage drop $V_{err} = V_s - V_{in}$, make $R_{diff} \ll R_s$. The details of this error were presented in Section 12.3.4 (Figure 12.41).

12.4.8 CMRR

CMRR is a critical factor for many instruments. Displacement currents into the object can create significant common-mode voltages. The simple approach is to use an electrode to ground the object and then design an amplifier with a high CMRR (Figure 12.42).

Figure 12.40
Electric fields are
another significant
source of
instrumentation noise.

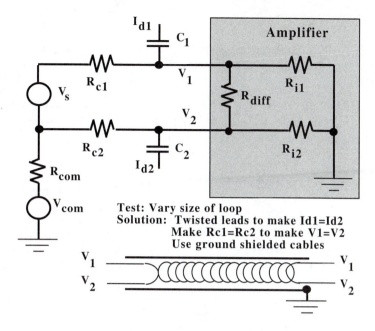

Test: Vary size of loop
Solution: Twisted leads to make Id1=Id2
Make Rc1=Rc2 to make V1=V2
Use ground shielded cables

Figure 12.41
If the input impedance is
not large enough, an
error will occur.

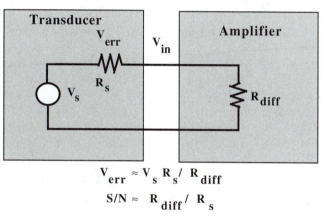

$$V_{err} \approx V_s R_s / R_{diff}$$
$$S/N \approx R_{diff} / R_s$$

Figure 12.42
If the CMRR is not large
enough, an error will
occur.

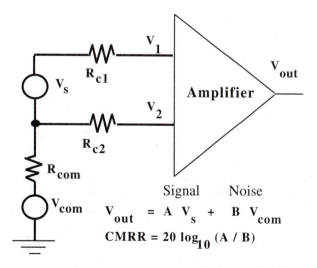

Signal   Noise
$$V_{out} = A V_s + B V_{com}$$
$$CMRR = 20 \log_{10} (A / B)$$

12.4.9
Frequency
Response

All instrumentation systems are band-limited. As the frequency increases, the gain decreases. This gain error usually affects the instrument objective.

12.4.10
Motion Artifact

Motion can introduce errors in many ways. Moving cables can induce currents. As the cables move, the connector impedance may change or disconnect. Acceleration of the transducer often affects its response.

12.5 DAS Case Studies

12.5.1
Temperature
Measurement
System

The objective is to measure temperature using a RTD (Figure 12.43). The range of temperature **T** is 0 to 50°C, and the desired resolution is 0.25°C. The frequency range of interest is 0 to 0.1 Hz. The transducer has a maximum slope of 10°C/s. The RTD resistance **R** in ohms is linearly proportional to temperature:

$$R = 100 + 0.4T$$

where **T** is in °C.

Figure 12.43
Block diagram of a temperature measurement system using a RTD.

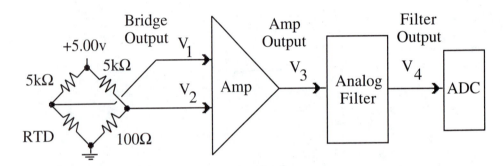

1. Since all devices are linear, the needed ADC precision is the range divided by resolution: 50°C / 0.25°C = 200. Thus, we will use an 8-bit ADC.

2. We will use a bridge circuit to convert the RTD resistance, **R**, into a voltage V_1. Note that $V_1 - V_2 = 0$ V when **T**=0°C.

3. The ADC range is 0 to +5 V. The range of $V_1 - V_2$ is 0 to 0.0191 V. Thus, we will use a differential amplifier with a gain of 261 to convert $V_1 - V_2$ into V_3.

4. Assume the ADC conversion time is 25 μs. No S/H is needed because the maximum slope of **T** multiplied by the ADC conversion time is less than the temperature resolution.

$$10°C/s \cdot 25 \text{ μs} = 0.00025°C \ll 0.25°C$$

5. To prevent noise in the ADC samples, the noise must be less than the resolution. The resolution of $V_1 - V_2$ is its range (0.0191 V) divided by its precision (256), which is 75 μV. Again a safety factor of 1/2 is included. Thus in the frequency range 0 to 1 Hz, the maximum allowable noise referred to the input of the differential amplifier should be

$$\text{Amplifier noise} \leq \frac{\text{resolution}}{2} = 37 \text{ μV}$$

6. A one-pole low-pass analog filter is used to:
    (a) Pass the temperature signal having frequencies from 0 to 0.1 Hz
    (b) Reject noise having frequencies above 0.1 Hz
    (c) Prevent aliasing

Figure 12.44
Block diagram of a temperature measurement system in the frequency domain.

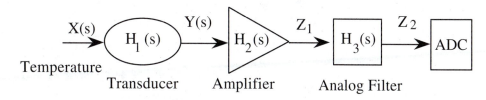

It is often illustrative to consider the DAS in the frequency domain (Figure 12.44). We design the filter so that the ADC error is less than half a resolution at 0.1 Hz (i.e., gain=511/512). The gain of the one-pole low-pass filter is

$$G_3 = |H_3| = \frac{1}{\sqrt{1 + (f/f_c)^2}}$$

where f_c is $1/(2\pi RC)$. f_c must be chosen large enough so the gain G is at least $(2^{n+1} - 1)/2^{n+1} = 511/512$ for frequencies 0 to 0.1 Hz. Thus f_c is

$$f_c = \frac{0.1\ \text{Hz}}{\sqrt{(512/511)^2 - 1}} = 1.6\ \text{Hz}$$

> ***Observation:*** In most situations, we simply place the low-pass filter cutoff frequency at the maximum frequency of interest.

7. To prevent aliasing, Z_2 must be less than the ADC resolution for all frequencies larger than or equal to $0.5f_s$. As an extra measure of safety, we make the amplitude less than $\mathbf{0.5\Delta_z}$ for frequencies above $0.5f_s$ (Figure 12.45). Thus,

$$|Z_2| < 0.5\Delta_z = \frac{5\ \text{V}}{512} \approx 0.01\ \text{V}$$

For discussion, assume the following criteria for frequencies above 1 Hz:

$$|X| = \frac{1°C}{\sqrt{1 + (f/1\ \text{Hz})^2}} \qquad |H_1| = \frac{382\ \mu V/°C}{\sqrt{1 + (f/10\ \text{Hz})^2}}$$

$$|H_2| = \frac{261\ \text{V}/\text{V}}{\sqrt{1 + (f/1533\ \text{Hz})^2}} \qquad |H_3| = \frac{1}{\sqrt{1 + (f/1.6\ \text{Hz})^2}}$$

Figure 12.45
To prevent aliasing there must be no measurable signal above $0.5f_s$.

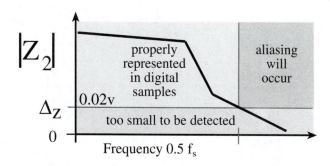

Thus

$$|Z_2| = \frac{0.1 \text{ V}}{\sqrt{[1 + (f/1 \text{ Hz})^2][1+(f/10 \text{ Hz})^2][1 + (f/1533 \text{ Hz})^2][1 + (f/1.6 \text{ Hz})^2]}}$$

We choose the sampling rate, f_s, to prevent aliasing:

$$\frac{0.1 \text{ V}}{\sqrt{[1 + (f_s/2)^2][1 + (f_s/20)^2][1 + (f_s/3066)^2][1 + (f_s/3.2)^2]}} < 0.01 \text{ V}$$

$$\sqrt{[1 + (f_s/2)^2][1 + (f_s/20)^2][1 + (f_s/3066)^2][1 + (f_s/3.2)^2]} > 10.24$$

$$[1 + (f_s/2)^2][1+(f_s/20)^2][1 + (f_s/3.2)^2] > 104.9$$

$$f_s > 7.4 \text{ Hz}$$

8. The effective output impedance of the bridge is 100 Ω. The input impedance of the differential amplifier must be high enough to not affect the ADC conversion.

$$Z_{in} > 100 \text{ Ω} \cdot 2^{n+1} = 51.2 \text{ kΩ}$$

Actually, because the bridge is used, we could get by with a smaller Z_{in} (Figure 12.46).

9. A table is often convenient, as in Table 12.12.

Figure 12.46
Amplifier and low-pass filter.

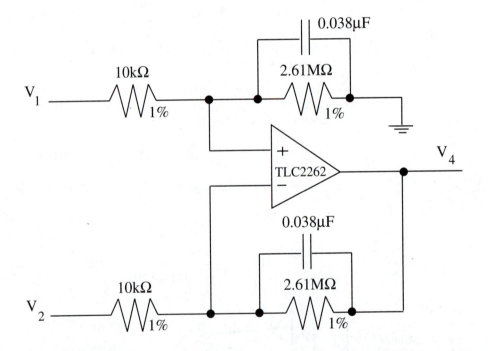

Table 12.12
Signals as they pass through the temperature DAS.

T(°C)	RTD (Ω)	V_1	V_2	$V_1 - V_2$	$V_3 = V_4$	ADC output
0.0	100.0	0.0980	0.0980	0.0000	0.0	0000,0000
0.25	100.1	0.0981	0.0980	0.0001	0.025	0000,0001
25.00	110.0	0.1076	0.0980	0.0096	2.500	1000,0000
50.0	120.0	0.1172	0.0980	0.0191	5.000	1111,1111

12.5.2 Force Measurement System

The objective of this system is to measure force, **x**. The useful range of **x** is from -1 to $+1$ N. The desired resolution, Δ_x, is 0.001 N. The frequencies of interest are from 2 to 100 Hz. Unfortunately, other frequency components exist in the signal, **x** (see Figure 12.47). Assume the transducer is linear and has a sensitivity of 0.1 V/N. The transducer output is a differential signal with an output impedance of 1 kΩ. The ADC range is from -10 to $+10$ V. Table 12.13 summarizes these parameters.

Figure 12.47
Specifications of a force measuring system.

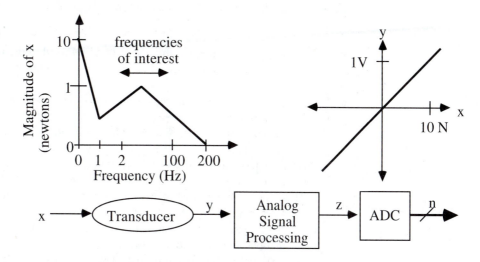

Table 12.13
Signals as they pass through the force DAS.

	x (N)	y (mV)	z (V)
Minimum	-1	-100	-10
Zero	0	0	0
Resolution	0.001	0.1	0.01
Maximum	1	100	10
Range	2	200	20
Precision	2000	2000	2000

1. A gain of 100 is needed to match the range of **y** into the ADC range. The purpose of the filter is to pass the frequencies of interest, remove the unwanted DC, and eliminate aliasing by removing high frequencies. The gain versus frequency is shown in Figure 12.48.

Figure 12.48
Ideal and realistic amplifier gain versus frequency response.

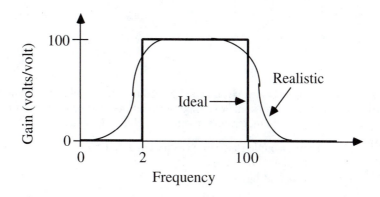

Observation: In most situations, we could simply place the filter cutoff frequencies at 2 and 100 Hz.

2. How many ADC bits are required (i.e., what is **n**)? The precision of **x** is 2 N/ 0.001 N = 2000. Since each component is linear, n_z, the required precision at **z** is also 2000. Thus 11 bits is required. We may have to purchase a 12-bit ADC because 11 bits is a nonstandard size.

3. What is the feasible range for the input impedance of the amplifier (Figure 12.49)?

$$\infty \geq \textbf{Zin} \geq 1 \text{ k}\Omega \cdot 4096 = 4 \text{ M}\Omega$$

4. What is the maximum allowable noise level referred to the input of the amplifier? The maximum allowable noise level is 0.5 Δ_y, the resolution at **y** = 200 mV/4096 = 0.05 mV.

5. Consider the factors that must be considered when choosing the ADC sampling rate.

  (a) You wish to correctly represent in memory the signal of interest. Thus, $f_s > 200$ Hz.

  (b) You do not wish to cause aliasing. Aliasing occurs when there is a significant component (larger than 0.01 V) of the signal at any frequency above $0.5f_s$. Thus, one must include a low-pass analog filter or increase the sampling rate, f_s, so that there is no measurable signal (as seen by the ADC) at any frequency above $0.5f_s$.

  (c) The computing power of the computer may limit the sampling rate. For example, the computer may only be able to process 1000 samples/s. In this case, $f_s \leq 1$ kHz.

  (d) As the sampling rate increases, more memory (or disk space) is required to store the information. Thus, the memory size may limit the sampling rate.

  (e) As the sampling rate increases, the ADC must have a smaller conversion time, $f_s \leq 1/t_c$. It costs more money to purchase fast ADCs.

  (f) For sampling rates above 50 kHz, DMA is required to input the data into memory. A DMA interface is complex and hence expensive.

Figure 12.49
Analog instrumentation for the force measuring system.

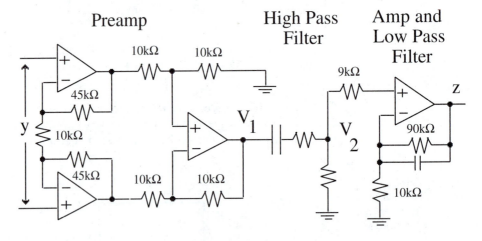

12.5.3 Thermocouple Interface Using an Electronic Reference

A thermistor bridge circuit can be used to add a voltage equal and opposite to voltage generated by the nonzero temperature reference. The objective of this example is to design a thermometer circuit with the following devices:

Thermocouple (used to measure temperature)
    Radius = 0.01 cm
    Iron/constantan

Sensitivity = 52 μV/°C
$\mathbf{V_c} = 52$ μV $(\mathbf{T_1} - \mathbf{T_2})$

Thermistor (used as the electronic reference)

Radius $\mathbf{a} = 0.2$ cm
$\mathbf{R_t} = 1.5 \times 10^{-6}$ exp $\{ 6154/\mathbf{T_2} \}$ ($\mathbf{R_t}$ in Ω, $\mathbf{T_2}$ in K)
Thermal conductivity = 0.6 mW/(cm·°C)

The specifications of the thermometer circuit (Figure 12.50) are:

Input temperature range 0 to 100 °C
Output voltage $\mathbf{V_{out}} = \mathbf{T_1}/100$ ($\mathbf{V_{out}}$ in volts, $\mathbf{T_1}$ in °C)
Output voltage range 0 to 1 V

Figure 12.50 Analog instrumentation for the thermocouple temperature acquisition system.

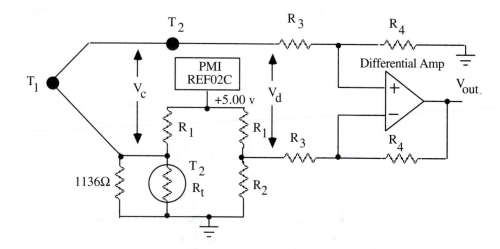

1. What is the maximum electric power that can be dissipated across the thermistor so that the self-heating error is less than 0.01°C? Assume the thermistor is in unstirred water. The thermal conductivity of water is 6.23 mW/(cm·°C).

$$D = \frac{4\pi a}{1/K_m + 0.2/K_b} = 5 \times 10^{-3} \text{ W}/°\text{C}$$

$$P_{max} = D \cdot T_{err} = 5 \times 10^{-3} \times 0.01 = 5 \times 10^{-5} \text{ W}$$

2. One thermocouple junction is at $\mathbf{T_1}$. The thermistor and the other thermocouple junction are at $\mathbf{T_2}$. Assume $0 \le \mathbf{T_1} \le 100°$C. Assume $20 \le \mathbf{T_2} \le 30°$C. A 1136-Ω parallel resistor is used to linearize the thermistor. Calculate the bridge output voltage for $\mathbf{T_2} = 20, 22, 24, 26, 28,$ and 30°C (Table 12.14).

$$R_t = 1.5 \times 10^{-6} \exp \frac{6154}{T_2}$$

$$R_{eqv} = R_t \| 1136 \text{ }\Omega = \frac{1136 \text{ } R_t}{1136 + R_t}$$

Table 12.14

Signals as they pass
through the temperature
DAS.

T_2 (°C)	R_t (Ω)	R_{eqv} (Ω)	V_b (mV)
20	1964	719.7	1.04
22	1703	681.5	1.14
24	1480	642.8	1.25
26	1289	603.8	1.35
28	1124	565.1	1.46
30	982.5	526.9	1.46

3. Find the best linear equation that expresses the bridge output voltage in terms of therm-istor temperature T_2(°C).

$$R_{eqv} \approx 1106 - 19.3T_2$$

$$V_b \approx -5\frac{R_{eqv} - R_2}{R_1} = -5\frac{1106 - 19.3\,T_2 - R_2}{R_1}$$

4. Choose the proper polarity of the +5-V bridge excitation voltage.

$$V_c = 52\ \mu V/°C \cdot (\,T_1 - T_2)$$

We want

$$V_d = 52\ \mu V/°C \cdot T_1$$

and

$$V_d = V_b + V_c$$

so that

$$V_b = 52\ \mu V/°C \cdot T_2$$

5. Choose values of R_1 and R_2 so that the thermistor bridge voltage offsets the potential due to the thermocouple T_2 junction. The idea is to make $V_d = 52\mu V \cdot T_1$, even though T_2 may vary between 20 and 30°C.

$$R_2 = 1106\ \Omega \text{ and } R_1 = 1.86\ M\Omega$$

$$\textbf{Actual power} = (5/1.86\ M\Omega)^2 \cdot 1000 \cdot = 7 \times 10^{-9}W$$

6. Choose values of R_3 and R_4 so that $0 \leq V_{out} \leq 1$. A low-pass filter should be added to prevent aliasing.

$$\frac{R_4}{R_3} = \frac{1\ V}{5.2\ mV} = 192$$

$$R_3 = 5\ k\Omega$$

$$R_4 = 960\ k\Omega \cdot$$

12.5.4
Heart Sounds

The objective of this system is to measure heart sounds, **x**. The useful range of **y** is from -10 to $+10$ mV. The desired resolution, Δ_y, is 0.1 mV. The frequencies of interest are from 5 to 2000 Hz. Below are the spectrum analyzer results obtained from the output of the microphone, **y**. The output of the microphone is a differential voltage. Unfortunately, other frequency components exist in the signal **y** (see Figure 12.51). The ADC range is from -10 to $+10$ V. The ADC input impedance is 5 kΩ. The maximum **dy/dt** is 2 V/s. There is no S/H in this system.

Figure 12.51 A heart sound measuring system.

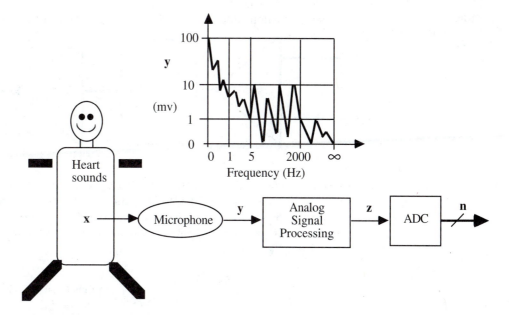

1. The first question we ask is how many ADC bits are required (i.e., what is $\mathbf{n_z}$)? Since the amplifier is linear,

$$\mathbf{n_z} = \mathbf{n_y} = \frac{\mathbf{r_y}}{\Delta_y} = \frac{20 \text{ mV}}{0.1 \text{ mV}} = 200 \text{ alternatives, or 8 bits}$$

2. Next we ask, what is the maximum allowable ADC conversion time so that the S/H is not needed?

$$\frac{dy}{dt} \cdot \mathbf{t_c} \leq 0.5 \, \Delta_y \qquad \mathbf{t_c} \approx \frac{0.5 \, \Delta_y}{dy/dt} = \frac{0.5 \cdot 0.0001 \text{ V}}{2 \text{ V/s}} = 25 \text{ μs}$$

or

$$\frac{dy}{dt} \cdot \mathbf{t_c} \leq \Delta_y \qquad \mathbf{t_c} \propto \frac{\Delta_y}{dy/dt} = \frac{0.0001 \text{ V}}{2 \text{ V/s}} = 50 \text{ μs}$$

3. Next we construct the analog circuit. We will pass with no ADC error all frequencies from 5 to 2000 Hz. We will select from the following analog building blocks presented in the last chapter.

Circuit	Equation	Constraints
Inverting op amp	$V_{out} = -GV_{in}$	$G > 0$
Noninverting op amp	$V_{out} = +GV_{in}$	$G \geq 1$
Subtractor	$V_{out} = +G(V_2 - V_1)$	$G > 0$, $Z_{in} = R_1 + R_2$
Instrumentation amplifier	$V_{out} = +G(V_2 - V_1)$	$G > 0$, $Z_{in} = 10^{12} \, \Omega$
Low-pass Butterworth	$V_{out} = \dfrac{V_{in}}{\sqrt{1 + (f/f_c)^4}}$	
High-pass Butterworth	$V_{out} = \dfrac{V_{in}}{\sqrt{1 + (f_c/f)^4}}$	

Figure 12.52
Heart sound measuring
system.

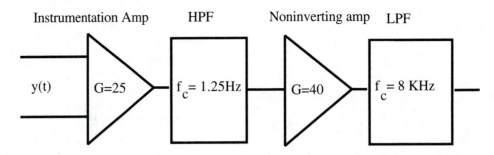

Instrumentation amplifier goes first to balance the microphone impedance and provide a small preamplifier (Figure 12.52). The gain of 25 will not saturate the output because 100 mV × 25 = 2.5 V. The preamplifier gain could have been as high as 100. The high-press filter removes large unwanted low frequencies. We want no ADC error in the pass-band of 5 to 2000 Hz, so

$$\frac{511}{512} = \sqrt{\frac{1}{1 + (f_c/5)^4}} \qquad \text{thus } \mathbf{f_c} = 1.25 \text{ Hz}$$

The remaining gain is provided by the noninverting amplifier. The low-pass filter will re-move aliasing. The cutoff for the low-pass filter is calculated so that no ADC error occurs in the passband:

$$\frac{511}{512} = \sqrt{\frac{1}{1 + (2000/f_c)^4}} \qquad \text{thus } \mathbf{f_c} = 8 \text{ kHz}$$

> **Observation:** In most situations, we could simply place the filter cutoff frequencies at 5 and 2000 Hz.

4. We choose the slowest ADC sampling rate that correctly samples the 5 to 2000 Hz and properly removes aliasing. From the picture, the largest available signal above 2 kHz is 0.001 V. The low-pass filter will drop it down, and $\mathbf{f_s}$ must be chosen large enough so that

$$0.01 \text{ V} \times 1000 \times \sqrt{\frac{1}{1 + \left[\dfrac{0.5f_s}{8000}\right]^4}} \leq 0.5\Delta_z = 0.5 \times 20/256 = 0.039\text{V}$$

thus $\mathbf{f_s} \geq 81$ kHz

5. Before we select the op amp devices, we need to know how much noise can be tolerated.

The maximum allowable noise referred to the input must be less than 0.5 $\mathbf{\Delta_y}$ = 39 μV or 50 μV

The maximum allowable noise referred to the output must be less than 0.5 $\mathbf{\Delta_z}$ = 39 mV or 50 mV

6. The output impedance of the amplifier must be small enough to prevent loading. If the ADC input impedance is 5 kΩ,

$$\frac{Z_{in} \text{ of ADC}}{Z_{out} \text{ of amp}} \geq 512$$

$$Z_{out} \leq 5 \text{ k}\Omega \cdot /512 = 9 \text{ }\Omega$$

12.5.5 Position Measurement System

In this section we will develop two interfaces for a class of position sensors based on the potentiometer. As we saw previously in Section 12.2.4, a potentiometer can be used to convert position into resistance. In this particular interface the full-scale range is 0 to 1000Ω.

$$R_{out} = 200 \cdot x$$

where the units of R_{out} and x are in Ω and centimeters, respectively. To interface this transducer to the microcomputer we drive the potentiometer with a stable DC voltage using a precision voltage reference (Figure 12.53). If we were to drive the bridge with the +5-V power, then any noise ripple in the power line would couple directly into the measurement. The ADC produces a digital output dependent on its analog input.

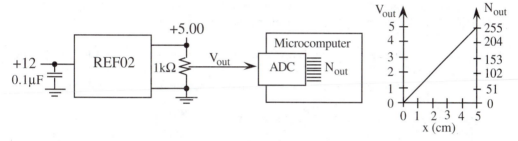

Figure 12.53 Potentiometer interface using an ADC.

Another approach to interface this transducer to the microcomputer would be to use an astable multivibrator. The period of a 555 timer is $0.693 \cdot C_T \cdot (R_A + 2R_B)$. In our circuit, R_A is R_{out}, R_B is 1 kΩ, and C_T is 2.2 μF. Given a fixed R_B, C_T, the period of the square wave, P_{out}, is a linear function of R_{out}. Our microcontrollers have a rich set of mechanisms to measure frequency, pulse width, or period. To change the slope and offset of the conversion between R_{out} and P_{out}, the fixed resistor and capacitor can be adjusted. Even though the period does not include zero, the precision of this measurement is over 3000 alternatives, or more than 11 bits. The precision can be improved by increasing the capacitor C_T, or decreasing the period on the measurement clock (Figure 12.54).

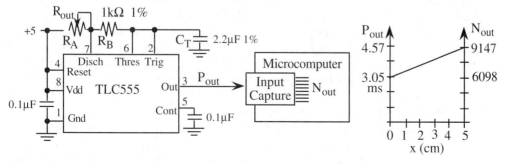

Figure 12.54 Potentiometer interfacing using input capture.

12.6 Glossary

1/f noise A fundamental noise in resistive devices arising from fluctuating conductivity. Same as pink noise.

60 Hz noise An added noise from electromagnetic fields caused by either magnetic field induction or capacitive coupling.

accuracy A measure of how close our instrument measures the desired parameter referred to the NIST.

ADC (Analog-to-digital converter) An electronic device that converts analog signals into digital form. Also written "A/D".

aliasing When digital values sampled at f_s contain frequency components above 0.5 f_s, then the apparent frequency of the data is shifted into the 0 to 0.5 f_s range. See Nyquist theory.

bang bang A sudden large change in the output for a small change in the input.

breakdown A device that stops functioning when its input goes above a maximum value or below a minimum value.

data acquisition system (DAS) A system that collects information, same as instrument.

dead zone A condition when a large change in the input causes little or no change in the output.

digital signal processing Processing of data with digital hardware or software after the signal has been sampled by the ADC, e.g., filters, detection and compression/decompression.

frequency response A measure of the system performance (e.g., gain) as a function of input frequency.

gage factor The sensitivity of a strain gage transducer, i.e., slope of the resistance versus displacement response.

hard real time A system that can guarantee that a process will complete a critical task within a certain specified range. In DASs, hard real time means there is an upper bound on the latency between when a sample is supposed to be taken (every 1/fs) and when the ADC is actually started. Hard real time also implies that no ADC samples are missed.

hysteresis A condition when the output of a system depends not only on the input, but also on the previous outputs, e.g., a transducer that follows a different response curve when the input is increasing than when the input is decreasing.

impedance The ratio of the effort (voltage, force, pressure) divided by the flow (current, velocity, flow).

impedance loading A condition when the input of stage n+1 of an analog system affects the output of stage n, because the input impedance of stage n+1 is too small and the output impedance of stage n is too large.

instrument A system that collects information, same as DAS.

Johnson noise A fundamental noise in resistive devices arising from the uncertainty about the position and velocity of individual molecules. Same as thermal noise and white noise.

latency The response time of the computer to external events. For a DAS the time between when the data should be sampled and the time when the ADC is started.

linear variable differential transformer (LVDT) A transducer that converts position into electric voltage.

measurand A signal measured by a DAS.

multiple-access circular queue (MACQ) A data structure used in DAS's to hold the current sample and a finite number of previous samples.

noninstrusive A characteristic when the presence of the DAS itself does not affect the parameters being measured.

Nyquist theory Digital values sampled at f_s only contain frequency components from 0 to $0.5 f_s$. See aliasing.

pink noise A fundamental noise in resistive devices arising from fluctuating conductivity. Same as 1/f noise.

precision The number of distinguishable inputs that can be measured reliably by the instrument. The units are in alternatives, bits, or decimal digits. When precision is in alternatives, range = precision × resolution.

qualitative DAS A DAS that collects information not in the form of numerical values but rather in the form of qualitative senses, e.g., sight, hearing, smell, taste, and touch. A qualitative DAS may also detect the presence or absence of conditions.

photosensor A transducer that converts reflected or transmitted light into electric current.

potentiometer A transducer that converts position into electric resistance.

range The difference between the largest and smallest input that can be measured by the instrument. The units are in the units of the measurand. When precision is in alternatives, range = precision × resolution.

reproducibility (or **repeatability**) A parameter specifying how consistent over time the measurement is when the input remains fixed.

resistance temperature device (RTD) A linear transducer that converts temperature into electric resistance.

resolution The smallest change in input that can be reliably detected by the instrument. The units are in the units of the measurand. When precision is in alternatives, range = precision × resolution.

sample and hold (S/H) A circuit used to latch a rapidly changing analog signal, capturing its input value and holding its output constant.

sampling rate The rate at which data is collected in a DAS.

saturation A device that is no longer sensitive to its inputs when its input goes above a maximum value or below a minimum value.

sensitivity The sensitivity of a transducer is the slope of the output versus input response. The sensitivity of a qualitative DAS that detects events is the percentage of actual events that are properly recognized by the system.

shot noise A fundamental noise that occurs in devices that count discrete events.

spatial resolution The volume over which the DAS collects information about the measurand.

specificity The specificity of a transducer is the relative sensitivity of the device to the signal of interest versus the sensitivity of the device to other unwanted signals. The sensitivity of a qualitative DAS that detects events is the percentage of events detected by the system that are actually true events.

strain gage A transducer that converts displacement into electric resistance. It can also be used to measure force or pressure.

thermal noise A fundamental noise in resistive devices arising from the uncertainty about the position and velocity of individual molecules. Same as Johnson noise and white noise.

thermistor A nonlinear transducer that converts temperature into electric resistance.

thermocouple A transducer that converts temperature into electric voltage.

time constant The time to reach 63.2% of the final output after the input is instantaneously increased.

transducer A device that converts one type of signal into another type.

ultrasound A sound with a frequency too high to be heard by humans, typically 40 kHz to 100 MHz.

white noise A fundamental noise in resistive devices arising from the uncertainty about the position and velocity of individual molecules. Same as Johnson noise and thermal noise.

Exercises

12.1 Design a computer-based DAS that measures pressure. The pressure transducer is built with four resistive strain gages placed in a DC bridge. When the pressure is zero, each gauge has a 120-Ω resistance making the bridge output y zero. When pressure is applied to the transducer, two gages are compressed (which lowers their resistance) and two are expanded (which increases their resistance). At full-scale pressure (p = 100 dynes/cm^2), the bridge output is y = 10 mV. The transducer/bridge output impedance is therefore 120 Ω. You may assume the transducer is linear. The desired pressure resolution is 1 dyne/cm^2. The frequencies of interest are 0 to 100 Hz, and the two-pole Butterworth analog low-pass filter will have a cutoff (gain = 0.707) frequency of 100 Hz (you will design it in part b). In terms of choosing a sampling rate, you may assume the low-pass filter removes all signals above 100 Hz.

a) Show the interface of the ADC to your computer. Justify your ADC precision.

b) Design the analog interface between the transducer/bridge and the your ADC. Use the full-scale ADC range even though it will complicate the conversion software. For example, if the ADC has a range of 0 to +5 V, then a pressure of p = 0 maps to a voltage at the ADC input of 0, and p = 100 dynes/cm^2 maps to a voltage at the ADC input of +5 V. Include the Butterworth low-pass filter. (Figure 12.55)

Figure 12.55
Transducers are placed in a bridge.

+V

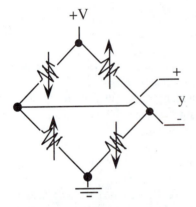

y

c) Show the ritual that initializes any global variables, the ADC, and an appropriate interrupt.

d) Show the interrupt handler(s) that samples the ADC, calculates pressure in dynes/cm^2, and stores the value in global variable `Pressure`. The software conversion maps a 0 ADC result into `Pressure=0`, and a 255 ADC result into `Pressure=100`. *Optimize the interrupt handler so that the number of execution cycles is minimized.*

12.2 The objective of this problem is to measure vibrations (displacement versus time) using a strain gage. The displacement range is -100 to $+100$ μm. The frequencies of interest are 1 to 200 Hz. The four resistors of the strain gage are placed in a bridge: two in compression, two in expansion. The bridge output voltage V_b has a sensitivity of 100 V/m. Each resistance R in the bridge is 100 Ω at a zero displacement. There is a lot of unwanted DC and 60-Hz noise in this system. You will remove the unwanted DC signal using analog signal processing and remove the unwanted 60-Hz signal using digital signal processing.

a) Show the interface between the transducer and the microcomputer. The S/H input impedance is 1 TΩ. The S/H aperture time is 10 μs, and the acquisition time is 100 μs. The ADC input voltage range is -5 to $+5$ V, the ADC digital output is 12-bit signed 2s complement, and the conversion time is 50 μs. (Figures 12.56 and 12.57)

b) What is the maximum allowable droop rate for the S/H? Show your work.

c) What is the displacement resolution? Give units and show your work.

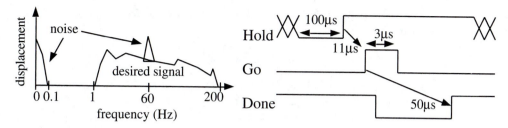

Figure 12.56   Vibration signals and ADC timing.

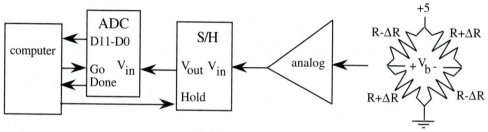

Figure 12.57   Vibration measuring system.

d) What is the slowest sampling frequency that is feasible for this system. Justify your answer. You will implement a sampling rate of 480 Hz and execute the following digital filter to remove the 60 Hz.

$$y(n) = (x(n) + x(n - 4))/2$$

e) Show the ritual subroutine that initializes the necessary microcomputer devices. Initialize all data structures including DONE. The output of your software is a global array, Y, containing the filtered displacement versus time. The data acquisition and digital filtering will be performed in real time under interrupt control. The main program will call this ritual and perform other unrelated tasks until DONE is −1.

```
int Y[1000];     // filtered displacement measurements
// The resolution and units are given in c), fs is 480 Hz
char DONE;       // initially zero, set to -1 when buffer is full
```

f) Show the interrupt software which samples the ADC, implements the digital filter in real time, and saves the filter outputs in the buffer Y. Set DONE to −1 when the buffer is full.

12.3 Design a remote-sensor temperature acquisition system. The goals are:

Range	$1 \le T \le 200$ °F
Resolution	$\Delta T = 1$ °F
Frequencies of interest	DC to 5 Hz

The system has the block diagram shown in Figure 12.58. The timer is Port A on the 6811 and Port T on the 6812.

Figure 12.58   Remote temperature measuring system.

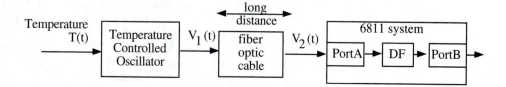

$V_1(t)$ is a low-power Schottky TTL square wave with a period $P(t)$ that is linearly related to the temperature $T(t)$.

$$P(t) = \frac{10\ \mu s}{°F} T(t)$$

Your interface should connect the fiber-optic cable so that $V_2(t)$ is also a TTL-level square wave with the same period. The 6811/6812 input capture/output compare system should be used to measure $T(t)$ at equal intervals. Let f_s be the fixed sampling rate, and $\Delta t = 1/f_s$. Your system should measure $T(t)$ exactly f_s times per second. Let $x(n)$ be the temperature sampled every Δt with a resolution of 1 °F. Your 6811/6812 system will implement the following low-pass Finite Impulse Response (FIR) digital filter:

$$y(n) = \frac{y(n-1) + x(n)}{2}$$

The output of this filter, $y(n)$, should be written to Port B every Δt s.

a) Show the interface between V_1 and the fiber-optic cable. The front end operates like a LED. Make $I_f = 20$ mA (Figure 12.59).

Figure 12.59
Fiber-optic cable transmitter interface.

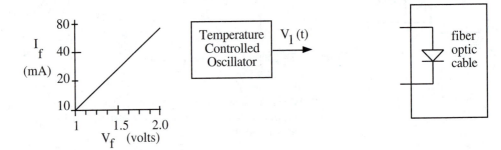

b) Show the interface between the fiber-optic cable and V_2 (Figure 12.60). The output of the fiber-optic cable is a photodiode detector. The detector conducts when light shines on it. It is open when dark. Provide for some noise rejection.

Figure 12.60
Fiber-optic cable receiver interface.

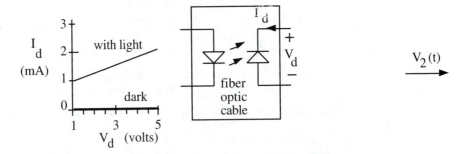

c) Choose the slowest possible sampling rate f_s. Justify your answer.
d) Show connections between the square wave $V_2(t)$ and the 6811/6812.
e) Show global data structure required to implement the digital filter. Full credit will be given to the best dynamic and static efficiency. This is not a question as to whether assembly is more efficient than C, but rather which data structure gets the job done in the simplest manner.
f) Show the ritual and interrupt handler(s). After calling the ritual, the main program (foreground) is free to execute other unrelated tasks. The interrupt handler(s) (background) will implement the fixed rate data acquisition. The output of the digital filter should be output to Port B f_s times per second. You may use either assembly or C.

12.4 The objectives of this problem are (1) to use a single-chip embedded 6811/6812 with its ADC to measure force at various locations on a robot and (2) to transmit the measurements across a master/slave distributed network. The robot system will have multiple slaves, each with its own force-measuring circuitry (Figure 12.61). You will design the slaves and specify/implement the communication channel to the master. The force measurement range is $0 \le F \le 200$ dynes. The desired force resolution is 1 dyne. The signals of interest are 0 to 10 Hz. Each slave has $+5$, $+12$ and -12 V power. The force transducers are placed in a resistance bridge powered by the $+5$-V supply (Figure 12.62). The bridge output has a sensitivity of 0.05 mV/dyne. The bridge output is *almost* zero when the force is zero. This transducer offset for each slave is a constant within the range from 0 to 0.2 mV. Each slave will require a no-force calibration that you will decide whether to handle in software or hardware.

Figure 12.61
Distributed force
measuring system.

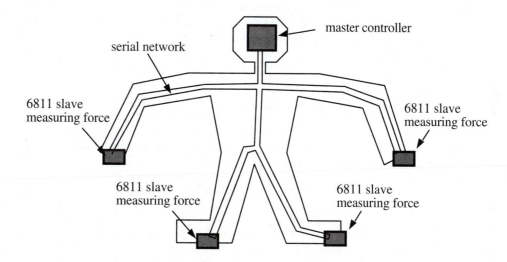

a) Show the analog interface between the bridge and the 6811/6812 ADC channel 1. Choose the appropriate op amp circuit, gain, offset, and analog filter. Include a mechanism to calibrate if you decide to handle the offset adjustment in hardware.

Figure 12.62
The force transducers
are placed in a bridge.

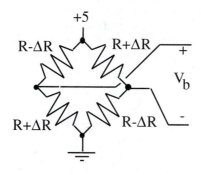

b) Design the distributed network between the slaves and their master. All slaves have identical network connections. To get a measurement from one of the slaves, the master will transmit the slave address that should activate exactly one slave. The slave will respond with the current force with a resolution of 1 dyne. The first priority of your network is to minimize cost (i.e., fewest wires and interface chips). Given the cheapest network, the second priority is to maximize bandwidth. Clearly show whether or not the grounds are connected between the multiple computers.

b1) Show the hardware interface between a single-chip 6811/6812 slave and the network. Remember that the hardware for all slaves will be identical. There is no noise interference.

b2) Show an example communication between the master and a single-chip 6811/6812 slave. Clearly identify the slave address, and force data components. Label the time axis.

c) Include all the software that will exist in each slave. Your solution will be segmented into three parts: RAM, EEPROM, and ROM. Specific software requirements include:

■ Other than a one-time initialization, there will be no foreground (main) program

■ You may use the 6811/6812 ADC continuous scan mode

■ Clearly show where the 6811/6812 is to begin execution on a power on reset

■ Include a mechanism to calibrate if you decide to handle the offset adjustment in software

c1) Show the software that goes in the RAM (uninitialized on power up).

c2) Show the software that goes in the EEPROM (nonvolatile, but can be different for each slave).

c3) Show the software that goes in the ROM (nonvolatile, and must be the same for each slave).

12.5 The objective of this problem is to design a DAS alarm. An alarm should sound when noise is present in the room. Assume the output of the microphone, x(t), is a ± 10-mV differential signal. The signals of interest are 100 to 500 Hz. You will sample the sound and calculate the sound energy once a second (sum of the signal squared). Be careful to subtract off the DC so that a quiet room (microphone = 0) results in an energy calculation of zero. If data is the 8-bit ADC, then x(n)=data−128 represents the current 8-bit signed sample. If you sample at 1000 Hz, sound the alarm if (x(0) + x(1)+ ... + x (999)) greater than 100,000. *32-bit-long integer operations are allowed.* The 6811/6812 system has a 2-MHz E clock. The 6812 system has an 8-MHz E clock. The alarm can be activated by driving 20 mA through a 5-V EM relay. The software should turn on the alarm if the energy level is above a threshold (simply pick any constant). Once on, the alarm should continue until the operator types the code "213" on the keypad. This 6811/6812 will have a secret code of 213, but allow each 6811/6812 to have a different three-number code and a different threshold. The computer is dedicated to this task.

a) Show the analog interface between the differential microphone output and the 6811/6812 ADC channel 2. Choose the appropriate op amp circuit, gain, offset, and analog filter. The Analog Devices AD680 is a three-wire low-cost 2.5-V reference (+5 V input, ground +2.500 V output). This is a *qualitative* and not a quantitative DAS.

b) Design the interface between the 6811/6812 signal B0 and the EM relay (Figure 12.63). To activate the alarm the relay coil requires a voltage between +3.5 and 6.0 V and a current of 20 mA.

Figure 12.63 The alarm is controlled by an EM relay.

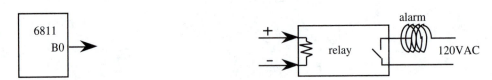

c) Show the interface between the keypad and the 6811/6812 (Figure 12.64). Each switch will bounce for about 10 ms, so you must implement a way to debounce. No other 6811/6812 connections may be used for this interface (although you may use the internal features of input capture/output compare). Your solution should minimize cost. *Periodic polling* should be used for this keypad because it is cheaper.

d) Include all the software that will exist in the system. Your solution will be segmented into three parts: RAM, EEPROM, and ROM. Specific software requirements include:

Figure 12.64
There are four switches in the keypad.

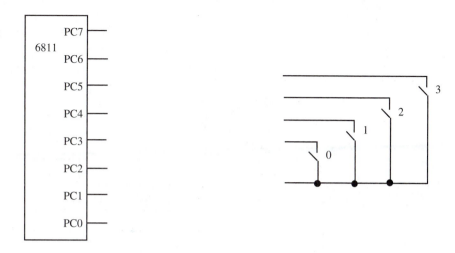

- Other than a one-time initialization, there will be no foreground (main) program
- Use the ADC continuous scan mode to simplify the software
- Clearly show where the 6811/6812 is to begin execution on a power on reset

d1) Show the software that goes in the RAM (uninitialized on power up).

d2) Show the software that goes in the EEPROM (nonvolatile, but can be different for each device).

d3) Show the initialization (ritual) software that goes in the ROM (nonvolatile, and must be the same for each device). This is where your system starts executing. Assume the two interrupt vectors are set by the compiler.

d4) Show the 1-kHz DAS interrupt handler that goes in the ROM (nonvolatile, and must be the same for each device). It is here that you will turn on the alarm if the sound energy goes above threshold.

d5) Show the keypad periodic polling interrupt handler that goes in the ROM (nonvolatile, and must be the same for each device). It is here that you will turn off the alarm if the types in the secret code.

12.6 This problem deals with the classification and reduction of noise.

a) Describe a *single experimental procedure* (measurement) that could *identify* (or differentiate) the *type*(s) of *noise* existing on the circuit.

b) For each of these three types of noise (white noise, 60-Hz noise, or 1/f noise), give a typical outcome of the experimental procedure.

c) Give one approach (other than analog or digital filtering) that will reduce white noise.

d) Give one approach (other than analog or digital filtering) that will reduce 60-Hz noise.

e) Give one approach (other than analog or digital filtering) that will reduce 1/f noise.

12.7 A temperature transducer has the relationship

$$R = 200 + 10T \qquad \text{where R is in } \Omega \text{ and T is in } °C.$$

The problem specifications are

Range is $30 \leq T \leq 50°C$
Resolution is $0.01°C$
Frequencies of interest are 0 to 100 Hz
Transducer dissipation constant is 20 mW/°C
ADC range is 0 to +5 V
Sampling rate is 1000 Hz

a) How many ADC bits are required?

b) What is the maximum allowable noise at the amplifier *output*?

c) Design the *analog amplifier/filter.*

12.8 Design an EKG amplifier interface to a 6811/6812 (Figure 12.65). The biopotential is a ± 5-mV signal measured between the left and right arms using silver–silver chloride electrodes. A third electrode is placed on the right leg and is used as a reference. To make the entire system battery-operated, low-power rail-to-rail TLC2274 op amps will be used, each powered with $+5$ and ground. Notice that the analog system uses the 2.5-V signal as its reference.

Figure 12.65
Block diagram of the
EKG instrumentation.

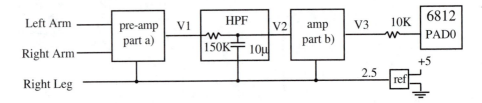

a) Design the preamplifier that has the following characteristics: differential input, gain$=50$, good CMRR, high Zin, low Zout, bandwidth > 200 Hz. Show resistor values, but not pin numbers.

b) Show the amplifier stage that has the following characteristics: single input, gain$=10$, high Zin, low Zout, bandwidth > 200 Hz. Show resistor values, but not pin numbers.

c) The signals of interest are 0.1 to 100 Hz. There is unwanted noise at 120, 240, 360, and 480 Hz. Discuss the trade-offs between the following two options:

   1. Add an active analog low-pass filter with a 100-Hz cutoff, and sample at 200 Hz.

   2. Sample at 1920 Hz and add a digital 120-, 240-, 360-, 480-Hz reject filter.

Which approach would you take and why? Under what conditions would the other choice be better?

12.9 List eight parameters that we can use to characterize a transducer.

12.10 Design a wind direction measurement instrument using the 8-bit ADC. You are given a transducer with a resistance that is linearly related to the wind direction. As the wind direction varies from 0 to 360 degrees, the transducer resistance varies from 0 to 1000 Ω. The frequencies of interest are 0 to 0.5 Hz, and the sampling rate will be 1 Hz. (See Exercise 6.6.)

a) Show the analog interface between the transducer and the ADC port channel 7. Only the $+5$-V supply can be used. Show how the analog components are powered. Give chip numbers, but not pin numbers. Specify the type and tolerance of resistors and capacitors.

b) Write the ritual and gadfly function/subroutine that measures the wind direction and returns a 16-bit unsigned result with units of degrees. That is, the value varies from 0 to 359. (You do not have to write software that samples at 1 Hz, simply a function that measures wind direction once.)

12.11 Design a position DAS using a sensor, analog electronics and a Motorola microcomputer with an 8-bit ADC. Let **x** be the position to be measured (Figure 12.66). The input range is -1 to $+1$ mm, and the signals of interest are 0 to 1 Hz. A LVDT will be used to convert position **x** into voltage **y.** When the input position is -1 mm, the LVDT output **y** is -100 mV. When the input position is zero, the LVDT output, **y,** is zero. When the input position is $+1$ mm, the LVDT output **y** is $+100$ mV. In between, the voltage output is linearly related to the position. A REF02 precision reference will provide the constant $+5.00$ V for your analog circuit and the microcomputer ADC.

Figure 12.66 Block diagram of the position measuring system.

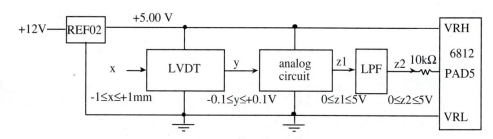

a) What is the transducer sensitivity? Give units.
b) What is the transfer relationship required to convert **y** into **z1**?
c) What is the maximum allowable noise for your analog circuit? Refer the noise to the amplifier input and give units.
d) Choose an appropriate sampling rate.
e) Design the preamplifier which has the following characteristics: single input (not differential), gain so that the 0 to +5 V ADC is used, high Zin, low Zout, bandwidth > 200 Hz. Show resistor values, but not pin numbers. You may use any analog op amps, but please include the chip number.
f) The signals of interest are 0 to 1 Hz. There is unwanted noise at 60 Hz. Add a two-pole low-pass filter with a cutoff of about 10 Hz.
g) What is the system resolution in millimeters?

12.12 Design an electronic scale using a Motorola microcomputer with an 8-bit ADC (Figure 12.67). Let **x** be in mass to be measured. The input range is 0 to 1 kg and the signals of interest are 0 to 10 Hz. A bonded strain gage bridge will be used to convert mass **x** into voltage, $V1 - V2$. When the input mass is zero, each arm of the bridge is 100 Ω, and the bridge output ($V1 - V2$) is zero. At full scale ($\mathbf{x} = 1$ kg), two resistors go to 99 Ω and the other two go to 101 Ω. In between 0 and 1 kg, the resistance change is linearly related to the mass:

$$\Delta R = x \qquad \text{where resistance is in ohms and the mass } \mathbf{x} \text{ is in kilograms}$$

Figure 12.67 Block diagram of the mass measuring system.

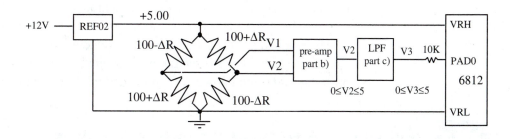

A REF02 precision reference will provide the constant +5.00 V for the bridge and the ADC.

a) What is the bridge output ($V1 - V2$) at full scale ($\mathbf{x} = 1$ kg)? What gain is required to match the full range of $0 \leq \mathbf{x} \leq 1$ kg to the 0 to +5 V range of the ADC?
b) Design the preamplifier that has the following characteristics: differential input, gain so that the 0 to +5 V ADC is used, good CMRR, high Zin, low Zout, bandwidth > 200 Hz. Show resistor values, but not pin numbers.
c) The signals of interest are 0 to 10 Hz. There is unwanted noise at 60 Hz. Add a two-pole low-pass filter with a cutoff of 20 Hz.
d) What is the system resolution? Give units.
e) What sampling rate would you choose? Explain your answer.

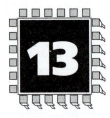

13 Microcomputer-Based Control Systems

Chapter 13 objectives are to:

❑ Introduce the general approach to digital control systems
❑ Design and implement some simple open-loop control systems
❑ Design and implement some simple closed-loop control systems
❑ Develop a methodology for designing PID control systems
❑ Present the terminology and give examples of fuzzy logic control systems

In the last chapter, we developed systems that collected information concerning the external environment. In this chapter we will use this information to control the external environment. To build this microcomputer-based control system we will need an output device that the computer can use to manipulate the external environment. If we review the list of applications introduced in Section 1.1.1, we find that many involve control systems. Control systems originally involved just analog electronic circuits and mechanical devices. With the advent of inexpensive yet powerful microcomputers, implementing the control algorithm in software provided a lower cost and a more powerful product. The goal of this chapter is to provide a brief introduction to this important application area. Control theory is a richly developed discipline, and most of the theory is beyond the scope of this book. Consequently, this chapter focuses more on implementing the control system with an embedded computer and less on the design of the control equations.

13.1 Introduction to Digital Control Systems

A *control system* is a collection of mechanical and electric devices connected for the purpose of commanding, directing, or regulating a *physical plant*. The *real state variables* are the properties of the physical plant that are to be controlled. The *sensor* and *state estimator* comprise a DAS, as discussed in Chapter 12. The goal of this DAS is to estimate the state variables. A *closed-loop* control system uses the output of the state estimator in a feedback loop to drive the system to a desired state. The control system compares these *estimated state variables* X'(t) to the *desired state variables* X*(t) to decide appropriate action U(t). The *actuator* is a transducer that converts the control system commands U(t) into driving forces V(t) that are applied to the physical plant. In general, the goal of the control system is to drive the real state variables to equal the desired state variables. In actuality though, the controller attempts to drive the estimated state variables to equal the desired state variables. It is important to have an accurate state estimator, because any differences between the estimated state variables and the real state variables will translate directly into controller errors. If we define the error as the difference between the desired and estimated state variables,

$$e(t) = X^*(t) - X'(t)$$

then the control system will attempt to drive e(t) to zero. In general control theory, X(t), X'(t), X*(t), U(t), V(t), and e(t) refer to vectors (e.g., position, velocity, acceleration), but the examples in this chapter control only a single parameter. The focus of this book is the microcomputer interfacing, and it should be straightforward to apply standard multivariate control theory to more complex problems. We usually evaluate the effectiveness of a control system by determining three properties: steady-state controller error, transient response, and stability. The steady-state controller error is the average value of e(t). The transient response is how long the system takes to reach 99% of the final output after X* is changed. A system is stable if a steady state (smooth constant output) is achieved. An unstable system oscillates (Figure 13.1).

Figure 13.1
Block diagram of a microcomputer-based closed-loop control system.

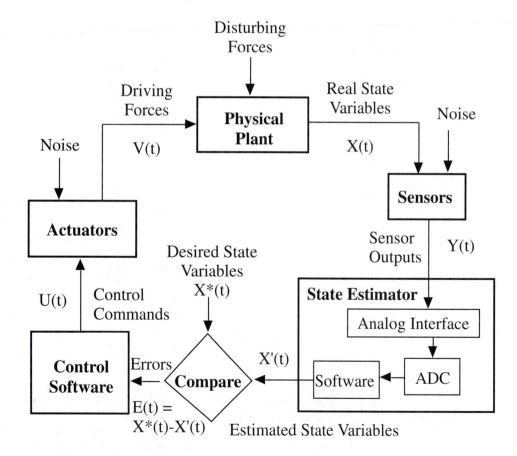

13.2 Open-Loop Control Systems

An open-loop control system does not include a state estimator. It is called open loop because there is no feedback path providing information about the state variable to the controller. It will be difficult to use open loop with the plant that is complex because the disturbing forces will have a significant effect on controller error. On the other hand, if the plant is well-defined and the disturbing forces have little effect, then an open-loop approach may be feasible. (Figure 13.2).

Figure 13.2
Block diagram of a microcomputer-based open-loop control system.

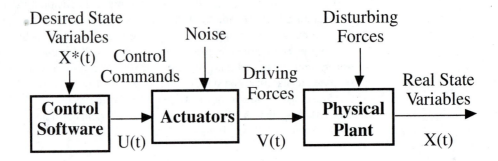

13.2.1
Open-Loop
Control of a
Toaster

This first example is a simple toaster (Figure 13.3). It is an open-loop control system because the control signal (heat applied to the bread) is independent of the state variable (taste of the toast). The controller simply applies heat for a fixed amount of time. If we provide an adjustable heat cycle to our toaster and allow the humans to adjust the heating cycle according to their taste, then the toast/humans combination becomes a closed-loop control system. Because an open-loop control system does not know the current values of the state variables, large errors can occur.

Figure 13.3
A simple open-loop control system.

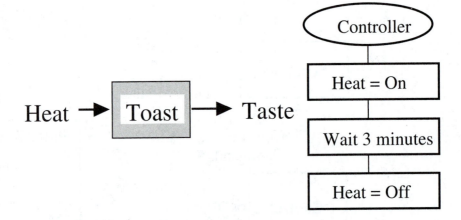

13.2.2
Open-Loop Traffic
Control

In a traffic control system the state variables are the positions and rates of cars as they travel along the streets. The usual goal of a traffic control system is to maximize traffic flow while minimizing waiting time and accidents. If the system cannot sense the presence of cars, then the control signals (lights) are independent of the state variables (traffic flow). In other words, without sensors, the traffic light is an open-loop control system. The control system

Figure 13.4
Interface for a simple open-loop traffic control system.

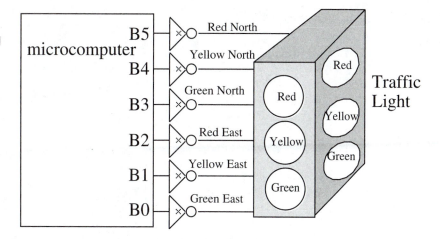

may attempt to reduce the worst-case waiting time, reduce the average waiting time, or maximize the overall traffic flow. Consider a simple intersection of two one-way streets that has a standard traffic signal (Figure 13.4). A simple (and frustrating to the driver) open-loop control system is to simply cycle through the four states. The yellow waiting time is adjusted to minimize accidents, and the green/red wait times are selected by considering the expected traffic flow (Figure 13.5).

Figure 13.5
Finite-state machine for a simple open-loop traffic control system.

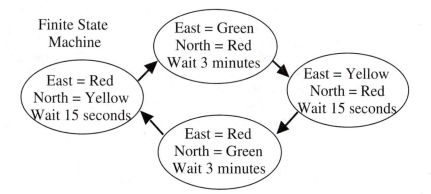

A circular linked list contains the four specific light patterns as well as the time to wait in each state (Program 13.1). This open-loop controller could be improved by adding sensors. If car sensors are available, then a Mealy or Moore FSM, like the FSM systems presented in Chapter 2, could be used to implement the improved control system.

Program 13.1 C implementation of a traffic light controller.

```
/* Port B bits 5-0 outputs that control the traffic signal */
const struct State {
   unsigned char Out;        /* Output to Port B */
   unsigned short Time;      /* Time in sec to wait */
   const struct State *Next; /* Next state */
};
typedef const struct State StateType;
#define NorthRed_EastGreen  &fsm[0]
#define NorthRed_EastYellow &fsm[1]
```

continued on p. 736

Program 13.1 C imple-
mentation of a traffic
light controller.

```
continued from p. 735
#define NorthGreen_EastRed  &fsm[2]
#define NorthYellow_EastRed &fsm[3]
StateType fsm[4]={
/* NorthRed_EastGreen, wait= 3 min, next state */
  {0x21, 180, NorthRed_EastYellow},
/* NorthRed_EastYellow, wait= 15 sec, next state */
  {0x22, 15, NorthGreen_EastRed},
/* NorthGreen_EastRed, wait= 3 min, next state */
  {0x0C, 180, NorthYellow_EastRed},
/* NorthYellow_EastRed, wait= 15 sec, next state */
  {0x14, 15, NorthRed_EastGreen}};
void main(void){ StatePtr *Pt;   /* Current State */
  Pt=NorthRed_EastGreen; /* Initial State */
  DDRB=0xFF;                /* Make Port B outputs  */
  while(1){
    PORTB=Pt->Out;    /* Perform output for this state */
    Wait(Pt->Time);   /* Time to wait in this state */
    Pt=Pt->Next;      /* Move to next state */
  }
};
```

13.2.3
Open-Loop
Stepper Controller

The next open-loop example is a stepper motor controller. The computer uses four digital signals to control the motor (Figure 13.6). If the software outputs the sequence 6,5,10,9,6,5,10,9 . . . , then the motor will spin. For each change in output (6 to 5, 5 to 10, 10 to 9 or 9 to 6), the motor turns 1.8 degrees. Therefore, it takes 200 outputs to turn the shaft one complete rotation. The goal of the system is to control the shaft position (angle, speed, acceleration, and jerk) of a stepper motor. Again without sensors to measure the shaft position this system is an open-loop controller. The control signals (power to the stepper coils) are independent of the state variables (shaft position). For low-torque applications (when the motor does not skip), the software can keep track of the shaft position without a sensor. For discussions of minimizing jerk in the stepper motor, see Chapter 8.

Figure 13.6
Interface for a simple
open-loop stepper
motor control system.

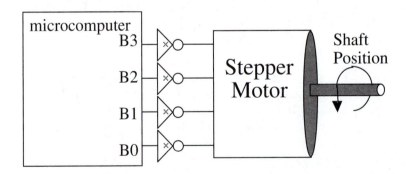

The software in Programs 13.2 and 13.3 will rotate the motor at a constant speed. The desired speed is given in the global variable, `Period`. It contains the number of E clock cycles (500 ns each) per step. For a 200 step/rotation motor, the rotation speed is

$$r(\text{rpm}) = \frac{2{,}000{,}000(\text{cycles/s}) \cdot 60(\text{s/min})}{200(\text{steps/rot}) \cdot \texttt{Period}(\text{cycles/step})} = \frac{600{,}000}{\texttt{Period}}$$

Program 13.2 Circular list used to spin a stepper motor.

```
/* Port B bits 3-0 outputs that control the stepper motor */
const struct State {
  unsigned char Out;        /* Output to Port B */
  const struct State *Next; /* Next state */
};
typedef const struct State StateType;
#define S6  &fsm[0]
#define S5  &fsm[1]
#define S9  &fsm[2]
#define S10 &fsm[3]
StateType fsm[4]={
  {0x06, S5},
  {0x05, S9},
  {0x09, S10},
  {0x0A, S6};
StatePtr *Pt;  /* Current State */
unsigned short Period;
```

```
// MC68HC05C8, Hiware compiler
interrupt 3 void TOChan(void){ char dummy;
   PORTB=Pt->Out; // output for this state
   Pt=Pt->Next;    // Move to next state
   dummy=TSR;      // part of acknowledge
   OCR=OCR+Speed;} // Executed every step
void ritual(void) { char dummy;
asm{ sei};        // make atomic
   TCR=0x40 ;      // Arm output compare
   DDRB=0xFF;      // Make Port B outputs
   Speed=10000;    // initial speed
   Pt=S6;          // initial state
   dummy=TSR;      // part of clear
   OCR=TCNT+2000;  // First one in 1 ms
asm{ cli}; }
```

```
// MC68HC708XL36, Hiware compiler
interrupt 8 void TOC3handler(void){
   PORTB=Pt->Out; // output for this state
   Pt=Pt->Next;    // Move to next state
   TSC3&=0x7F;     // acknowledge CH3F
   TCH3=TCH3+Speed;} // Executed every step
void ritual(void) {
asm{ sei};        // make atomic
   TSC=0x02;       // 500 ns clock
   TSC3=0x54;      // Arm output compare
   DDRB=0xFF;      // Make Port B outputs
   TSC3&=0x7F;     // Initially clear CH3F
   Speed=10000;    // initial speed
   Pt=S6;          // initial state
   TCH3=TCNT+2000; // First one in 1 ms
asm{ cli}; }
```

```
// MC68HC11A8, ICC11 compiler
#define OC5 0x08
#pragma interrupt_handler TOC5handler()
void TOC5handler(void){
   PORTB=Pt->Out; // output for this state
   Pt=Pt->Next;    // Move to next state
   TFLG1=OC5;      // Ack OC5F
   TOC5=TOC5+Speed; // Executed every step
   }

void ritual(void) {
asm(" sei");      // make atomic
   TMSK1|=OC5;     // Arm output compare 5
   TFLG1=OC5;      // Initially clear OC5F
   Speed=10000;    // initial speed
   Pt=S6;          // initial state
   TOC5=TCNT+2000; // First one in 1 ms
asm(" cli"); }
```

```
// MC68HC812A4, ICC12 compiler
#define OC5 0x20
#pragma interrupt_handler TC5handler()
void TC5handler(vOid){
   PORTB=Pt->Out; // output for this state
   Pt=Pt->Next;    // Move to next state
   TFLG1=OC5;      // ack C5F
   TC5=TC5+Speed;}  // Executed every step
void ritual(void) {
asm(" sei");      // make atomic
   TIOS|=OC5;      // enable OC5
   TSCR|=0x80;     // enable
   TMSK2=0x32;     // 500 ns clock
   TMSK1|=OC5;     // Arm output compare 5
   DDRB=0xFF;      // Make Port B outputs
   TFLG1=OC5;      // Initially clear C5F
   Speed=10000;    // initial speed
   Pt=S6;          // initial state
   TC5=TCNT+2000;  // First one in 1 ms
asm(" cli"); }
```

Program 13.3 C software to spin a stepper motor at a constant speed.

13.3   Simple Closed-Loop Control Systems

**13.3.1
Bang-Bang
Temperature
Control**

This digital control system applies heat to the room to maintain the temperature as close to T* (labeled `Tstar` in the software) as possible (Figure 13.7). This is a closed-loop control system because the control signals (heat) depend on the state variables (temperature). In this application, the actuator has only two states: *on,* which warms up the room, and *off,* which does not apply heat. For this application to function properly, there must be a passive heat loss that lowers the room temperature when the heater is turned off. A typical digital control algorithm for this type of actuator is *bang-bang* (Figure 13.8). Other names for bang-bang include *two-position, on-off,* or *binary* controller. A bang-bang controller turns on the power if the temperature is too low and turns off the power if the temperature is too high. To implement hysteresis, we need two set point temperatures, T_{HIGH} and T_{LOW}. The controller turns on the power (activates relay) if the temperature goes below T_{LOW} and turns off the power (deactivates relay) if the temperature goes above T_{HIGH}. The difference $T_{HIGH} - T_{LOW}$ is called hysteresis. The hysteresis extends the life of the relay by reducing the number of times the relay opens and closes.

Figure 13.7
Flowchart of a bang-bang temperature controller.

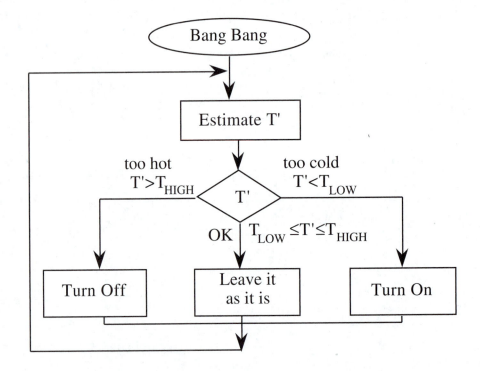

Figure 13.8
Algorithm and response of bang-Bang temperature controller.

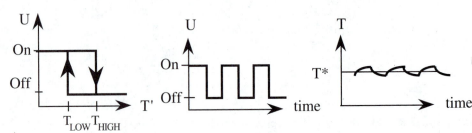

In this implementation, the RTD is made from a thin platinum wire and converts the room temperature into a resistance. The analog circuit matches the full-scale range of the room temperature to the 0 to +5 V ADC input voltage range. The software converts the ADC result into estimated temperature, T' (labeled T in the software) (Figure 13.9).

Figure 13.9
Interface of a simple bang-bang temperature controller.

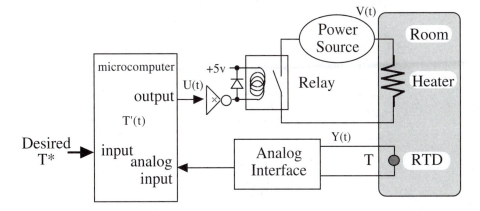

Assume that the function, SE, converts the ADC sample into temperature in degrees Celsius. The C code to implement the bang-bang controller uses a periodic interrupt so that the controller runs in the background (Program 13.4 and 13.5). The period interrupt rate is selected to be about ten times faster than the time constant of the physical plant.

Program 13.4 Port B is used to turn on the heater.

```
unsigned char Tlow, Thigh, T; int E;   // units in degrees C
void Actuator(unsigned char relay){
   PORTB=relay;}   // turns power on/off
```

```
// MC68HC05C8, Hiware compiler
interrupt 3 void TOChan(void){ char dummy;
   T=SE(); // estimated Temperature
   E=Tstar-T;              //error
   if(T<Tlow)
     Actuator(0);   // too cold->off
   else if (T>Thigh)
     Actuator(1);   // too hot->on
// leave as is if Tlow<T<Thigh
   dummy=TSR;        // part of acknowledge
   OCR=OCR+rate;} // periodic rate
```
```
// MC68HC708XL36, Hiware compiler
interrupt 8 void TOC3handler(void){
   T=SE(); // estimated Temperature
   E=Tstar-T;              //error
   if(T<Tlow)
     Actuator(0);   // too cold->off
   else if (T>Thigh)
     Actuator(1);   // too hot->on
// leave as is if Tlow<T<Thigh
   TSC3&=0x7F;     // acknowledge CH3F
   TCH3=TCH3+rate;} // periodic rate
```
```
//MC68HC11A8,  ICC11 compiler
#pragma interrupt_handler TOC5handler()
void TO5handler(void){
   T=SE(A2D(channel));   // estimatd T
   E=Tstar-T;           // error
   if(T<Tlow)
     Actuator(0);     // too cold so off
   else if (T>Thigh)
     Actuator(1);     // too hot so on
// leave as is if Tlow<T<Thigh
   TOC5=TOC5+rate;  // periodic rate
   TFLG1=0x08; }     // ack OC5F
```
```
// MC68HC812A4, ICC12 compiler
#pragma interrupt_handler TC5handler()
void TC5handler(void){
   T=SE(A2D(channel)); // estimated T
   E=Tstar-T;           // error
   if(T<Tlow)
     Actuator(0);      // too cold so off
   else if (T>Thigh)
     Actuator(1);      // too hot so on
// leave as is if Tlow<T<Thigh
   TC5=TC5+rate;     // periodic rate
   TFLG1=0x20; }      // ack C5F
```

Program 13.5 Bang-bang temperature control software.

Observation: Bang-bang control works well with a physical plant that has a very slow response.

Observation: The DS1620 interface presented first in Chapter 3 then again in Chapter 7 can be used as the SE function in this example.

13.3.2
Closed-Loop
Position Control
System Using
Incremental
Control

The objective of this *incremental control* system is to control the position of the robot arm, X (Figure 13.10). The control signals (power) are dependent on the state variables (position). The LVDT, which was described in Chapter 12, is used to sense the position of the robot arm.

Figure 13.10
Interface of a position controller.

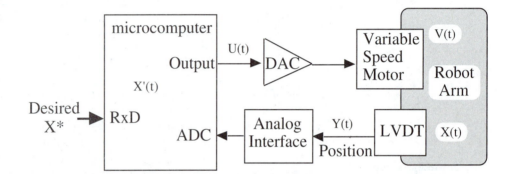

An incremental control algorithm simply adds or subtracts a constant from U depending on the sign of the error. In other words, if X is too small, then U is incremented, and if X is too large, then U is decremented. It is important to choose the proper rate at which the incremental control software is executed. If it is executed too many times per second, then the actuator will saturate, resulting in a bang-bang system. If it is not executed often enough, then the system will not respond quickly to changes in the physical plant or changes in X*. In this incremental controller we add or subtract "1" from the actuator, but a value larger than "1" would have a faster response at the expense of introducing oscillations (Figure 13.11).

Common error: An error will occur if the software does not check for overflow and underflow after U is changed.

Observation: If the incremental control algorithm is executed too frequently, then the resulting system behaves like a simple bang-bang controller.

Observation: Many control systems operate well when the control equations are executed about ten times faster than the step response time of the physical plant.

Assume the function, SE, converts the ADC sample into position in millimeters. The C code to implement the incremental controller (Program 13.6) uses a periodic interrupt so that the controller runs in the background. The period interrupt rate is selected to be about ten times faster than the time constant of the physical plant.

Observation: Incremental control will work moderately well (accurate and stable) for an extremely wide range of applications. Its only shortcoming is that the controller response time can be quite slow.

Figure 13.11
Flowchart of a position
controller implemented
using incremental
control.

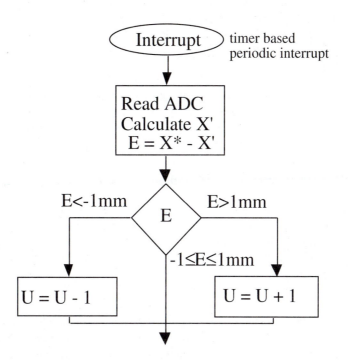

```
// MC68HC05C8, Hiware compiler
interrupt 3 void TOChan(void){ char dummy;
 int new;
   X=SE();  // estimated position (mm)
   E=Xstar-X;       // error (mm)
   new=PORTB;       // promote to 16 bits
   if(E<-1)     new--; // decrease
   else if (E>1) new++; // increase
// leave as is if -1<E<1
   if(new<0)    new=0;      // underflow
   if(new>255) new=255;  // overflow
   PORTB=new;   // output to actuator
   dummy=TSR;      // part of acknowledge
   OCR=OCR+rate;} // periodic rate
```

```
// MC68HC708XL36, Hiware compiler
interrupt 8 void TOC3handler(void){
int new;
   X=SE();  // estimated position (mm)
   E=Xstar-X;       // error (mm)
   new=PORTB;       // promote to 16 bits
   If(E< -1)    new--; // decrease
   else if (E>1) new++; // increase
// leave as is if -1<E<1
   if(new<0)    new=0;      // underflow
   if(new>255) new=255;  // overflow
   PORTB=new;   // output to actuator
   TSC3&=0x7F;     // acknowledge CH3F
   TCH3=TCH3+rate;} // periodic rate
```

```
// MC68HC11A8, ICC11 compiler
unsigned char Xstar,X; int E;   // in mm
#pragma interrupt_handler TOC5handler()
void TO5handler(void){ int new;
   X=SE(A2D(channel));  // estimated (mm)
   E=Xstar-X;       // error (mm)
   new=PORTB;       // promote to 16 bits
   if(E< -1)    new--; // decrease
   else if (E>1) new++; // increase
// leave as is if -1<E<1
   if(new<0)    new=0;      // underflow
   if(new>255) new=255;  // overflow
   PORTB=new;   // output to actuator
   TOC5=TOC5+rate; // establish periodic
   TFLG1=0x08; }        // ack OC5F
```

```
// MC68HC812A4, ICC12 compiler
#pragma interrupt_handler TC5handler()
void TC5handler(void){ int new;
   X=SE(A2D(channel));  // estimated (mm)
   E=Xstar-X;       // error (mm)
   new=PORTB;       // promote to 16 bits
   if(E< -1)    new--; // decrease
   else if (E>1) new++; // increase
// leave as is if -1<E<1
   if(new<0)    new=0;      // underflow
   if(new>255) new=255;  // overflow
   PORTB=new;   // output to actuator
   TC5=TC5+rate;        // periodic rate
   TFLG1=0x20; }        // ack C5F
```

Program 13.6 Incremental position control software.

13.4   PID Controllers

13.4.1
**General Approach
to a PID Controller**

The simple controllers presented in the last section are easy to implement, but they will have either large errors or very slow response times. To make a faster and more accurate system, we can use linear control theory to develop the digital controller. There are three components of a *PID controller.*

$$U(t) = Kp\, E(t) + K_I \int_o^t E(\tau)\, d\tau + K_D \frac{dE(t)}{dt}$$

We also can use just some of the terms. For example, a proportional/integral controller drops the derivative term. We will analyze the digital control system in the frequency domain. Let **X(s)** be the Laplace transform of the state variable **x(t),** let **X*(s)** be the Laplace transform of the desired state variable **x*(t),** and let **E(s)** be the Laplace transform of the error. Because the system is linear

$$\mathbf{E(s) = X^*(s) - X(s)}$$

Let **G(s)** be the transfer equation of the PID linear controller. PID controllers are unique in this aspect. In other words, we cannot write a transfer equation for a bang-bang, incremental, or fuzzy logic controller, but for a PID controller we have

$$\mathbf{G(s) = c\left(k_P + k_D s + \frac{k_I}{s}\right)}$$

Let **H(s)** be the transfer equation of the physical plant. If we assume the physical plant (e.g., a DC motor) has a simple single-pole behavior, then we can specify its response in the frequency domain with two parameters. **m** is the DC gain and τ is its time constant. The transfer function of a single-pole plant is

$$\mathbf{H(s) = \frac{m}{1 + \tau \cdot s}}$$

The overall gain of the control system (Figure 13.12) is

$$\mathbf{\frac{X(s)}{X^*(s)} = \frac{G(s)H(s)}{1 + G(s)H(s)}}$$

Theoretically we can choose controller constants $\mathbf{k_P}$, $\mathbf{k_I}$, and $\mathbf{k_D}$ to create the desired controller response. Unfortunately it can be difficult to estimate **c, m,** and τ. If the load to the motor varies, then **m** and τ will change.

Figure 13.12
Block diagram of a linear control system in the frequency domain.

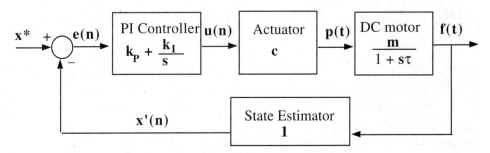

To simplify implementation of the PID controller, we break the controller equation into separate proportional, integral, and derivative terms. That is, let

$$U(t) = P(t) + I(t) + D(t)$$

where $U(t)$ is the actuator output and $P(t)$, $I(t)$, and $D(t)$ are the proportional, integral, and derivative components, respectively. The proportional term makes the actuator output linearly related to the error. Using a proportional term creates a control system that applies more energy to the plant when the error is large.

$$P(t) = K_p \cdot E(t)$$

To implement the proportional term we simply convert the above equation into discrete time.

$$P(n) = K_p \cdot E(n)$$

where the index n refers to the discrete time input of $E(n)$ and output of $P(n)$.

> *Observation:* To develop digital signal-processing equations, it is imperative that the control system be executed on a regular and periodic rate.

> *Common error:* If the sampling rate varies, then controller errors will occur.

The integral term makes the actuator output related to the integral of the error. Using an integral term often will improve the steady-state error of the control system. If a small error accumulates for a long time, this term can get large. Some control systems put upper and lower bounds on this term, called *anti-reset-windup,* to prevent it from dominating the other terms:

$$I(t) = K_I \cdot \int_0^t E(\tau)d\tau$$

The implementation of the integral term requires the use of a discrete integral or sum. If $I(n)$ is the current control output, and $I(n-1)$ is the previous calculation, the integral term is simply

$$I(n) = K_I \cdot \sum_1^n (E(n) \cdot \Delta t) = I(n-1) + K_I \cdot E(n) \cdot \Delta t$$

where Δt is the sampling rate of $E(n)$.

The derivative term makes the actuator output related to the derivative of the error. This term is usually combined with either the proportional and/or integral term to improve the transient response of the control system. The proper value of K_D will provide for a quick response to changes in either the set point or loads on the physical plant. An incorrect value may create an overdamped (very slow response) or an underdamped (unstable oscillations) response.

$$D(t) = K_D \cdot \frac{dE}{dt}$$

There are a couple of ways to implement the discrete time derivative. The simple approach is

$$D(n) = K_D \cdot \frac{E(n) - E(n-1)}{\Delta t}$$

In practice, this first-order equation is quite susceptible to noise. Figure 13.13 shows a sequence of $E(n)$ with some added noise. Notice that huge errors occur when the above equation is used to calculate derivative.

Figure 13.13
Illustration of the effect noise plays on the calculation of discrete derivative.

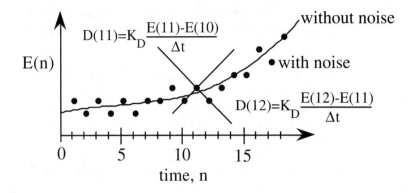

In most practical control systems, the derivative is calculated using the average of two derivatives calculated across different time spans. For example,

$$D(n) = K_D \cdot \left[\frac{1}{2} \frac{E(n) - E(n-3)}{3\Delta t} + \frac{1}{2} \frac{E(n-1) - E(n-2)}{\Delta t} \right]$$

which simplifies to

$$D(n) = K_D \cdot \frac{E(n) + 3E(n-1) - 3E(n-2) - E(n-3)}{6\Delta t}$$

13.4.2
Velocity PID
Controller

The objective of this example is to develop the fixed-point equations that implement a PID velocity controller (Figure 13.14).

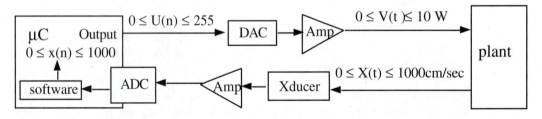

Figure 13.14   Interface of a PID velocity controller.

The maximum output of 255 maps linearly into a maximum applied power of 10 W to the physical plant. Similarly, a maximum speed of 1000 cm/s maps linearly into a maximum ADC result of 255. Also U=0 converts into V=0, and X=0 gives an ADC conversion of 0. Let `Xstar` be a 16-bit unsigned integer containing the desired speed or set point in centimeters per second. Let current error be defined as

$$e(t) = Xstar - X(t)$$

in centimeters per second. We will implement the following PID control equation

$$V(t) = 0.1e(t) + 0.5 \int_0^\tau e(\tau)\, d\tau + 0.005 \frac{de(t)}{dt}$$

where $e(t)$ is in centimeters per second and $V(t)$ is in watts. To simplify the problem, we will break the controller into separate proportional, integral, and derivative terms. That is, let

$$U(n) = P(n) + I(n) + D(n)$$

where $U(n)$ is the next DAC output and $P(n)$, $I(n)$, and $D(n)$ are the proportional, integral, and derivative components, respectively. Let data be the current 8-bit unsigned ADC sample. $x(n)$ will be the 16-bit estimated current speed in centimeters per second. $x(n)$ ranges from 0 to 1000 cm/s. The fixed-point calculation that converts data into the estimated speed $x(n)$ is

$$x(n) = \frac{1000 \cdot \text{data}}{256} = \frac{125 \cdot \text{data}}{32}$$

We define $e(n) = \text{Xstar} - x(n)$ as the 16-bit signed current error in centimeters per second. Let $e(n-1)$ be the previous calculation sampled at 100 Hz. Next, we develop fixed-point equations for $P(n)$, $I(n)$, and $D(n)$ in terms of the current and previous $e(n)$ calculations. The term $I(n-1)$ refers to the previous calculation of the integral.

$$e(n) = \text{Xstar} - x(n)$$

For the proportional term we have

$$V(t) = 0.1 \cdot e(t)$$

therefore

$$\frac{10 \cdot P(n)}{256} = 0.1 \cdot e(n)$$

Rearranging we get

$$P(n) = \frac{256 \cdot e(n)}{100}$$

For the integral term we have

$$V(t) = 0.5 \int_0^t e(\tau) d\tau = 0.5 \Delta t \sum_{i=0}^{n} e(n) = 0.005 \sum_{i=0}^{n} e(n)$$

therefore

$$\frac{10 \cdot I(n)}{256} = 0.005 \cdot \Sigma e(n)$$

Rearranging we get

$$I(n) = \frac{16 \cdot e(n)}{125} + I(n-1)$$

For the derivative term we have

$$V(t) = 0.005 \frac{de(t)}{dt} = 0.005 \frac{e(n) - e(n-1)}{\Delta t}$$

therefore

$$\frac{10 \cdot D(n)}{256} = 0.5 \cdot [e(n) - e(n-1)]$$

Rearranging, we get

$$D(n) = \frac{64 \cdot [e(n) - e(n-1)]}{5}$$

13.4.3
Integral Controller
with a PWM
Actuator

We will design a microcomputer-based integral control system. Let **X*** be the desired state variable. In this example, **X*** is fixed at +10 cm. Let **X'** be the estimated state variable that comes from the *state estimator* that encodes the current position as the period of a square wave (0 ≤ **X'** ≤ 100 cm). The period of this particular sensor is linearly related to the position **X**,

$$P = X + 10$$

where **P** is the period in milliseconds and **X** is the current position in centimeters. The state estimator measures the period in milliseconds and calculates **X'**

$$X' = P - 10$$

Let **U** be the actuator control variable (100 ≤ **U** ≤ 19900). It uses PWM with a 100-Hz square wave that applies energy to the physical plant. **U** will be the number of E clock cycles (out of 20000) that the output compare is high (Table 13.1, Figure 13.15).

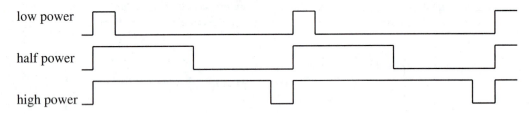

Figure 13.15   PWM actuator signals.

Table 13.1
Duty cycle affects
applied power to the
actuator.

U	Applied power
100	Low
10000	Half
19900	High

U is an output from the computer and an input to the actuator. We will execute the following integral control equation once a second:

$$U(n) = U(n-1) + 10 \cdot (X^* - X')$$

The controller rate is selected to be about ten times faster than the step response of the physical plant. In this way, the actuator output **U** does not saturate.

We use output compare to generate a 1-kHz real-time clock. The output compare handler will increment the current time in milliseconds. After 1000 output compare interrupts (1 s), the control algorithm is implemented (Program 13.7).

```
// MC68HC11A8, ICC11 compiler
unsigned int Time;   // Time in msec
unsigned int X:      // Est position in cm
unsigned int Xstar;  // Desired pos in cm
unsigned int U;      // Actuator duty cycle
unsigned int Cnt;    // once a sec
unsigned int Told;   // used to meas period
#pragma interrupt_handler TOC5handler()
void TOC5Handler(void){ int NewU;
```

```
// MC68HC812A4, ICC12 compiler
unsigned int Time;   // Time in msec
unsigned int X;      // Est position in cm
unsigned int Xstar;  // Desired pos in cm
unsigned int U;      // Actuator duty cycle
unsigned int Cnt;    // once a sec
unsigned int Told;   // used to meas period
#pragma interrupt_handler TC5handler()
void TC5handler(void){ int NewU;
```

```
  TFLG1=0x08;      // Ack OC5F
  TOC5=TOC5+2000; // every 1 ms
  Time++;      // used to measure period
  if((Cnt++)==1000){  // every 1 sec
     Cnt=0;
// 0<X<100, 0<Xstar<100, 100<U<19900, so
// Minimum occurs when U=100,Xstar=0,X=100
//    Minimum NewU = -900
// Max occurs when U=19900,Xstar=100,X=0
//    Maximum NewU = 20900
// so NewU will be a valid signed int
     NewU=U+10*(Xstar-X);
     if(newU<100) NewU=100;  // Constrain
     if(newU>19900) NewU=19900;
     U=NewU;  } }
```

```
  TFLG1=0x20;      // ack C5F
  TC5=TC5+2000;     // every 1 ms
  Time++;     // used to measure period
  if((Cnt++)==1000){  // every 1 sec
     CNT=0;
// 0<X<100, 0<Xstar<100, 100<U<19900, so
// Minimum occurs when U=100,Xstar=0,X=100
//    Minimum NewU = -900
// Max occurs when U=19900,Xstar=100,X=0
//    Maximum NewU = 20900
// so NewU will be a valid signed int
     NewU=U+10*(Xstar-X);
     if(NewU<100) NewU=100;  // Constrain
     if(NewU>19900) NewU=19000;
     U=NewU;  } }
```

Program 13.7 Integral position control software.

We use a second output compare to generate a 100-Hz real-time clock. This handler will be used to create the 100-Hz variable duty cycle square wave connected to the actuator. The output compare will be high for **U** cycles and low for 20000-**U** cycles (Program 13.8).

```
// MC68HC11A8, ICC11 compiler
#pragma interrupt_handler TOC4handler()
void TOC4Handler(void){
   TFLG1=OC4F; /* Ack */
   if (TCTL1&0x04)   /* OL4 bit */
/* OL4=1, High for the next U cycles */
      TOC4=TOC4+U;
   else
/* OL4=0, Low for the next 20000-U cyc */
      TOC4=TOC4+20000-U;
   TCTL1^=0x04; }     /* Toggle OL4 */
```

```
// MC68HC812A4, ICC12 compiler
#pragma interrupt_handler TC4handler()
void TC4handler(void){
   TFLG1=0x10;       // ack C4F
   if (TCTL1&0x01)   /* OL4 bit */
/* OL4=1, High for the next U cycles */
      TC4=TC4+U;
   else
/* OL4=0, Low for the next 20000-U cyc */
      TC4=TC4+20000-U;
   TCTL1^=0x01; }     /* Toggle OL4 */
```

Program 13.8 PWM actuator control software.

We use input capture to interrupt on each rise of the sensor square wave. The input capture interrupt handler will measure the period of the sensor square wave and estimate the current position (Programs 13.9 and 13.10, Figure 13.16).

```
// MC68HC11A8, ICC11 compiler
// Time is incremented every 1 ms, by OC5
// This handler is executed on rise
#pragma interrupt_handler TIC1handler()
void TIC1Handler(void){unsigned int p;
  TFLG1 = IC1F;  // Ack IC1F
  p = Time-Told; // period in msec
  X = p-10;    // estimated position (cm)
  Told=Time;
}
```

```
// MC68HC812A4, ICC12 compiler
// Time is incremented every 1 ms, by OC5
// This handler is executed on rise
#pragma interrupt_handler TC1handler()
void TC1Handler(void){unsigned int p;
  TFLG1 = 0x02;  // Ack C5F
  p = Time-Told; // period in msec
  X = p-10;    // estimated position (cm)
  Told=Time;
}
```

Program 13.9 Sensor measurement software.

```
// MC68HC11A8, ICC11 compiler
void ritual(void) {
  asm(" sei");       /* atomic */
  OC1M=0; OC1D=0;
  TCTL1=0x08; // Clear OC4
  TCTL2=0x10; // capture on rise of IC1
  TMSK1=OC5F+OC4F+IC1F; // Arm
  U=10000;        // Initial U , half power
  Time = 0; Told=0; Cnt=0;
  TFLG1=OC5F+OC4F+IC1F; // clear flags
  TOC5=TCNT+2000; // First OC5 in 1 ms
  TOC4=TCNT+100;  // First OC4 in 50us
  asm(" cli");}
```

```
// MC68HC812A4, ICC12 compiler
void ritual(void) { // Input capture IC1
  asm(" sei");        /* atomic */
  TIOS=0x30;  // output compare OC5, OC4
  TSCR|=0x80; // enable TCNT
  TMSK2=0x32; // 500 ns clock
  TCTL1=0x02; // Clear OC4
  TCTL4=0x08; // capture on rise of IC1
  TMSK1=0x32; // Arm OC5F+OC4F+IC1F
  U=10000;        // Initial U , half power
  Time = 0; Told=0; Cnt=0;
  TFLG1=0x32; // clear flags
  TC5=TCNT+2000; // First OC5 in 1 ms
  TC4=TCNT+100;  // First OC4 in 50us
  asm(" cli");}
```

Program 13.10 Initialization software.

Figure 13.16
Relative timing of the
IC1 and OC5 interrupts.

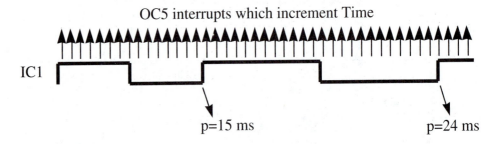

OC5 interrupts which increment Time

IC1

p=15 ms

p=24 ms

Observation: PID control will work extremely well (fast, accurate, and stable) if the physical plant can be described with a set of linear differential equations.

13.4.4 Empirical Method to Determine PID Controller Parameters

A simple empirical method can be used to determine the controller constants. It is described for a motor controller but will also apply to other PID implementations. This empirical approach starts with just a proportional term (k_P). This proportional controller will generate a smooth motor speed (actuator output achieves a constant value), but the speed will not be correct. It is important to choose the sign of the term k_P so that the system is stable. An incorrect sign will create an unstable controller. Try different k_P constants until the response times are fast enough. The response time is the delay after X^* is changed for the motor to reach a new constant speed. k_P is too big if the actuator saturates both at the maximum and minimum after X^* is changed. Steady-state controller accuracy is defined as the average difference between X^* and X'. The next step is to add some integral term (k_I) a little at a time to improve the steady-state controller accuracy without adversely affecting the response time. It is important to choose the sign of the term k_I so that the system is stable. An incorrect sign will create an unstable controller. *Overshoot* is defined as the maximum positive error that occurs when X^* is increased. Similarly, *undershoot* is defined as the maximum negative error that occurs when X^* is decreased. If the response time, overshoot, undershoot, and accuracy are within acceptable limits, then a proportional integral controller is adequate. On the other hand, if accuracy and response are okay but overshoot and undershoot are unacceptable, add a little derivative term ($\mathbf{k_D}$) to reduce the overshoots/undershoots in the step response. It is important to choose the sign of the term $\mathbf{k_D}$ so that the overshoots/undershoots are reduced.

13.5 Fuzzy Logic Control

There are a number of reasons to consider fuzzy logic approach to control. It requires less mathematics than PID systems. It will also require less memory and execute faster. In other words, an 8-bit fuzzy system may perform as well (same steady-state error and response time) as a 16-bit PID system. When complete knowledge about the physical plant is known, then a good PID controller can be developed. Since the fuzzy logic control is more robust (still works even if the parameter constants are not optimal), then the fuzzy logic approach can be used when complete knowledge about the plant is not known or can change dynamically. Choosing the proper PID parameters requires expert knowledge about the plant. The fuzzy logic approach is more intuitive, following more closely the way a human would control the system. It is easy to modify an existing fuzzy control system into a new problem. So if the framework exists, rapid prototyping is possible. The approach to fuzzy design can be summarized as:

- The *physical plant* has *real state variables* (like speed, position, temperature).
- The DAS monitors these signals, creating the *estimated state variables*.
- The *preprocessor* calculates relevant parameters called *crisp inputs*.
- *Fuzzification* will convert crisp inputs into *input fuzzy membership sets*.
- The *fuzzy rules* calculate *output fuzzy membership sets*.
- *Defuzzification* will convert output sets into *crisp outputs*.
- The *postprocessor* modifies crisp outputs into a more convenient format.
- The *actuator system* affects the physical plant based on these outputs.

13.5.1 DAC, ADC Fuzzy Controller

The objective of this example is to design a *fuzzy logic* microcomputer-based DC motor controller (Figure 13.17). The actuator is a DAC and linear power amplifier. The power to the motor is controlled by varying the 8-bit DAC output voltage. An amplifier provides power to the motor that is linearly related to the DAC output. The motor speed is estimated with a tachometer connected to an 8-bit ADC.

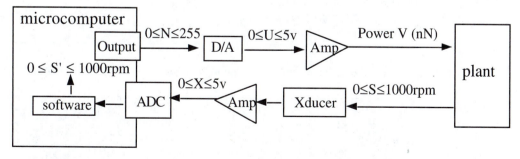

Figure 13.17   Interface of a motor controlled with fuzzy logic.

Our system has:

- Two control inputs
  - S*   desired motor speed, rpm
  - S'   current estimated motor speed, rpm
- One control output
  - N   digital value that we write to the DAC

To utilize 8-bit mathematics, we change the units of speed to 1000/256=3.90625 rpm.

T* =(256·S*)/1000   desired motor speed, 3.9 rpm
T' =(256·S')/1000   current estimated motor speed, 3.9 rpm

For example, if the desired speed is 500 rpm, then T* will be 128. Notice that the estimated speed T' is simply the ADC conversion value (i.e., just the ADR1 register without calculations). Inherent in most control systems is the concept of periodic execution. In other words, the control system functions (estimate state variables, control equation calculations, and actuator output) are performed on a regular and periodic basis, every Δt time units. This allows signal-processing techniques to be used. We will let T'(n) refer to the current measurement and T'(n−1) refer to the previous measurement (i.e., the one measured Δt time ago).

In the fuzzy logic approach, we begin by considering how a human would control the motor. Assume your hand were on a joystick (or your foot on a gas pedal) and consider how you would adjust the joystick to maintain a constant speed. We select crisp inputs and outputs to base our control system on. It is logical to look at the error and the change in speed when developing a control system. Our fuzzy logic system will have two crisp inputs

$E = T* - T'$ error in motor speed, 3.9 rpm
$D = T'(n) - T'(n-1)$ change in motor speed, 3.9 rpm/time

Notice that if we perform the calculations of D on periodic intervals, then D will represent the derivative of T, dT'/dt. T* and T' are 8-bit unsigned numbers, so the potential range of E varies from −255 to +255. Errors beyond ±127 will be adjusted to the extremes +127 or −128 without loss of information. Program 13.11 gives the calculations in C.

Program 13.11 Subtraction with overflow/underflow checking.

```
char Subtract(unsigned char N, unsigned char M){
/* returns N-M */
unsigned int N16,M16;
int Result16;
    N16=N;          /* Promote N,M */
    M16=M;
    Result16=N16-M16;    /* -255•Result16•+255 */
    if(Result16<-128) Result16 = -128;
    if(Result16>127)  Result16 = 127;
    return(Result16);}
```

Program 13.12 gives the global definitions of the input signals and fuzzy logic crisp input.

Program 13.12 Inputs and crisp inputs.

```
unsigned char Ts;     /* Desired Speed in 3.9 rpm units */
unsigned char T;      /* Current Speed in 3.9 rpm units */
unsigned char Told;   /* Previous Speed in 3.9 rpm units */
char D;               /* Change in Speed in 3.9 rpm/time units */
char E;               /* Error in Speed in 3.9 rpm units */
```

Common error: Neglecting overflow and underflow can cause significant errors.

The need for the special Subtract function can be demonstrated with the following example:

```
E=Ts-T; // if Ts=200 and T=50 then E will be -106!!
```

This function can be used to calculate both E and D (Program 13.13). Now, if Ts=200 and T=50, then E will be +127.

Program 13.13 Calculation of crisp inputs.

```
void CrispInput(void){
    E=Subtract(Ts,T);
    D=Subtract(T,Told);
    Told=T;}      /* Set up Told for next time */
```

To control the actuator, we could simply choose a new DAC value N as the crisp output. Instead, we will select ΔN, which is the change in N, rather than N itself because it better mimics how a human would control it. Again, think about how you control the speed of your car when driving. You do not adjust the gas pedal to a certain position but rather make small or large changes to its position to speed up or slow down. Similarly, when controlling the temperature of the water in the shower, you do not set the hot/cold controls to certain absolute positions. Again you make differential changes to affect the actuator in this control system. Our fuzzy logic system will have one crisp output:

ΔN change in output, $N=N+\Delta N$, in DAC units

Next we introduce fuzzy membership sets that define the current state of the crisp inputs and outputs. Fuzzy membership sets are variables that have true/false values. The value of a fuzzy membership set ranges from definitely true (255) to definitely false (0). For example, if a fuzzy membership set has a value of 128, you are stating the condition is halfway between true and false. For each membership set, it is important to assign a meaning or significance to it. The calculation of the input membership sets is called *fuzzification*. For this simple fuzzy controller, we will define six membership sets for the crisp inputs:

Slow	True if the motor is spinning too slow
OK	True if the motor is spinning at the proper speed
Fast	True if the motor is spinning too fast
Up	True if the motor speed is getting larger
Constant	True if the motor speed is remaining the same
Down	True if the motor speed is getting smaller

We will define three membership sets for the crisp output:

Decrease	True if the motor speed should be decreased
Same	True if the motor speed should remain the same
Increase	True if the motor speed should be increased

The fuzzy membership sets are usually defined graphically, but software must be written to actually calculate each. In this implementation, we will define three adjustable thresholds: TE, TD, and TN. These are software constants and provide some fine-tuning to the control system. We will set each threshold to 20. If you build one of these fuzzy system, try varying one threshold at a time and observe the system behavior (steady-state controller error and transient response). If the error E is −5 (3.9 rpm units), the fuzzy logic will say that *Fast* is 64 (25% true), *OK* is 192 (75% true), and *Slow* is 0 (definitely false).

If the error E is +21 (in 3.9 rpm units), the fuzzy logic will say that *Fast* is 0 (definitely false), *OK* is 0 (definitely false), and *Slow* is 255 (definitely true). TE is defined to be the error (e.g., 20 in 3.9 rpm units is 78 rpm) above which we will definitely consider the speed to be too fast. Similarly, if the error is less than −TE, then the speed is definitely too slow (Figure 13.18).

Figure 13.18
Fuzzification of the error input.

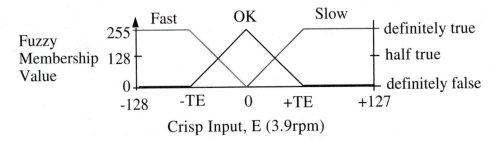

In this fuzzy system, the input membership sets are continuous piecewise-linear functions. Also, for each crisp input value (*Fast, OK, Slow*) sum to 255. In general, it is possible for the fuzzy membership sets to be nonlinear or discontinuous, and the membership values do not have to sum to 255. The other three input fuzzy membership sets depend on the crisp input D. TD is defined to be the change in speed (e.g., 20 in 3.9 rpm/time units is 78 rpm/time) above which we will definitely consider the speed to be going up. Similarly, if the change in speed is less than −TD, then the speed is definitely going down (Figure 13.19).

Figure 13.19
Fuzzification of the acceleration input.

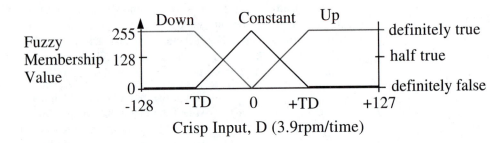

In C, we could define a fuzzy function that takes the crisp inputs and calculates the fuzzy membership set values. Again TE and TD are software constants that will affect the controller error and response time (Program 13.14).

```
#define TE 20
unsigned char Fast, OK, Slow, Down, Constant, Up;
#define TD 20
unsigned char Increase,Same,Decrease;
#define TN 20
void InputMembership(void){
     if(E <= -TE) {              /*    E≤-TE */
       Fast=255;
       OK=0;
       Slow=0;}
```

Program 13.14 Calculation of the fuzzy membership variables in C.

```
            else
                if (E < 0) {              /*  -TE<E<0    */
                    Fast=(255*(-E))/TE;
                    OK=255-Fast;
                    Slow=0;}
                else
                    if (E < TE) {         /*   0<E<TE    */
                        Fast=0;
                        Slow=(255*E)/TE;
                        OK=255-Slow;}
                    else {                /*  +TE≤E      */
                        Fast=0;
                        OK=0;
                        Slow=255;}
            if(D <= -TD) {                /*      D≤-TD  */
                Down=255;
                Constant=0;
                Up=0;}
            else
                if (D < 0) {              /*  -TD<D<0    */
                    Down=(255*(-D))/TD;
                    Constant=255-Down;
                    Up=0;}
                else
                    if (D < TD) {         /*   0<D<TD    */
                        Down=0;
                        Up=(255*D)/TD;
                        Constant=255-Up;}
                    else {                /*  +TD≤D      */
                        Down=0;
                        Constant=0;
                        Up=255;}}
```

The fuzzy rules specify the relationship between the input fuzzy membership sets and the output fuzzy membership values. It is in these rules that one builds the intuition of the controller. For example, if the error is within reasonable limits and the speed is constant, then the output should not be changed. In fuzzy logic we write:

If *OK* and *Constant* then *Same*

If the error is within reasonable limits and the speed is going up, then the output should be reduced to compensate for the increase in speed. That is,

If *OK* and *Up* then *Decrease*

If the motor is spinning too fast and the speed is constant, then the output should be reduced to compensate for the error. That is,

If *Fast* and *Constant* then *Decrease*

If the motor is spinning too fast and the speed is going up, then the output should be reduced to compensate for both the error and the increase in speed. That is,

If *Fast* and *Up* then *Decrease*

If the error is within reasonable limits and the speed is going down, then the output should be increased to compensate for the drop in speed. That is,

If *OK* and *Down* then *Increase*

If the motor is spinning too slow and the speed is constant, then the output should be increased to compensate for the error. That is,

If *Slow* and *Constant* then *Increase*

If the motor is spinning too slow and the speed is going down, then the output should be increased to compensate for both the error and the drop in speed. That is,

If *Slow* and *Down* then *Increase*

These seven rules can be illustrated in a table form (Figure 13.20).

Figure 13.20
Fuzzy logic rules shown in table form.

E \ D	Down	Constant	Up
Slow	Increase	Increase	
OK	Increase	Same	Decrease
Fast		Decrease	Decrease

It is not necessary to provide a rule for all situations. For example, we did not specify what to do for *Fast&Down* or for *Slow&Up*, although we could have added (but did not)

If *Fast* and *Down* then *Same*
If *Slow* and *Up*   then *Same*

When more than one rule applies to an output membership set, then we can combine the rules:

Same=(OKandConstant)
Decrease=(OKandUp)or(FastandConstant)or(FastandUp)
Increase=(OKandDown)or(SlowandConstant)or(SlowandDown)

In fuzzy logic, the *and* operation is performed by taking the minimum and the *or* operation is the maximum. Thus the C function that calculates the three output fuzzy membership sets is shown in Program 13.15.

Program 13.15 Calculation of the output fuzzy membership variables in C.

```
unsigned char min(unsigned char u1,unsigned char u2){
    if(u1>u2) return(u2);
    else return(u1);}
unsigned char max(unsigned char u1,unsigned char u2){
    if(u1<u2) return(u2);
    else return(u1);}
void OutputMembership(void){
    Same=min(OK,Constant);
    Decrease=min(OK,Up)
    Decrease=max(Decrease,min(Fast,Constant));
    Decrease=max(Decrease,min(Fast,Up));
    Increase=min(OK,Down)
    Increase=max(Increase,min(Slow,Constant));
    Increase=max(Increase,min(Slow,Down));}
```

The calculation of the crisp outputs is called *defuzzification*. The fuzzy membership sets for the output specifies the crisp output, ΔN, as a function of the membership value. For example, if the membership set *Decrease* were true (255) and the other two were false (0), then the change in output should be −TN (where TN is another software constant). If the membership set *Same* were true (255) and the other two were false (0), then the change in output should be 0. If the membership set *Increase* were true (255) and the other two were false (0), then the change in output should be +TN (Figure 13.21).

Figure 13.21
Defuzzification of the •N crisp output.

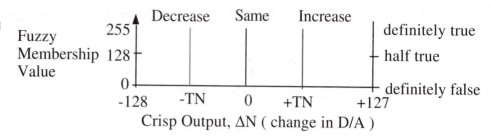

In general, we calculate the crisp output as the weighted average of the fuzzy membership sets:

$$\Delta N = (Decrease \cdot (-TN) + Same \cdot 0 + Increase \cdot TN)/(Decrease + Same + Increase)$$

The C compiler will promote the calculations to 16 bits and perform the calculation using 16-bit signed mathematics that will eliminate overflow on intermediate terms. The output dN will be bounded between −TN and +TN. Thus the C function that calculates the crisp output is shown in Program 13.16. When writing in assembly you will need to deal with converting *Increase, Same, Decrease* from 8-bit unsigned to 16-bit signed, and with overflow on intermediate terms. Just like in C, the output, ΔN, will be bounded between −TN and +TN. If the calculation is rearranged, you can still use the 8-bit unsigned multiply. The numerator is:

```
TN*Increase—TN*Decrease
```

Program 13.16 Calculation of the crisp output in C.

```c
char dN;
void CrispOutput(void){
       dN=(TN*(Increase-Decrease))/(Decrease+Same+Increase);
}
```

using 8-bit unsigned multiply and 16-bit subtract. The divide calculation for **dN** needs to be 16-bit signed (Program 13.17).

Program 13.17 Main program for fuzzy logic controller in C.

```c
unsigned int Time;
#define rate 2000   /* Rate must be less than 32767 E clocks */
void Initialize(void){
     OPTION=0x80;   /* Turn on A/D */
     PORTB=0;
     N=0;                /* Initial Actuator */
     Told=0;
     Ts=128; }        /* 500 rpm */
```

continued on p. 756

Program 13.17 Main program for fuzzy logic controller in C.

```
continued from p. 755
#define CCF 0x80
unsigned char Sample(unsigned char channel){
      ADCTL=channel;                 /* Start A/D */
      while((ADCTL&CCF)==0);   /* Wait for CCF */
      return(ADR1);}
void Main(void){ int dT;
      Initialize();                        /* Turn on A/D initialize globals */
      Time=TCNT+rate;                 /* TCNT value for first calculation */
      while(1){
         while((dT=Time-TCNT)>0){};
         Time=Time+rate;               /* TCNT value for next calculation */
         T=Sample(0);                    /* Sample A/D and set T */
         CrispInput();                    /* Calculate E,D and new Told */
         InputMembership();          /* Sets Fast,OK,Slow,Down,Constant,Up */
         OutputMembership();        /* Sets Increase,Same,Decrease */
         CrispOutput();                   /* Sets dN */
         N=max(0,min(N+dN,255));
         PORTB=N;}}                       /* Set Actuator */
```

13.5.2 PWM Fuzzy Controller

The objective of this section is to design a *fuzzy logic* microcomputer-based motor controller (Figure 13.22). The actuator is a PWM digital wave. The power to the motor is controlled by varying the duty cycle of the 200-Hz square wave. A switching power transistor provides current to the motor only when the digital output is high. The diodes protect the electronics from the back EMF (as high as 200 V) that occurs when the current is switched (large dI/dt) to the DC motor coil. The frequency of the square wave is chosen faster than the motor response so that the motor responds only to the duty cycle and not to the individual highs and lows of the square wave.

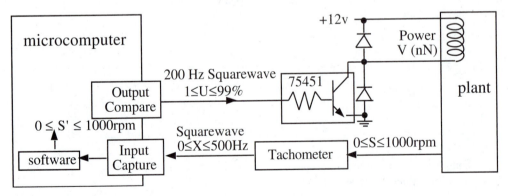

Figure 13.22   Interface of a motor controller with fuzzy logic.

The tachometer output is a frequency (in hertz) that is related to the motor speed (in rpm). The input capture system can be used to measure the period of this signal. Period measurement is faster than frequency measurement, so it is better suited for real-time control. The motor speed is a nonlinear function of the applied power. Because of friction and inertia, there is a range of power below which the motor will not spin. Below 40% duty cycle, the motor has a higher DC gain (Figure 13.23).

Figure 13.23
Steady-state response of
the physical plant.

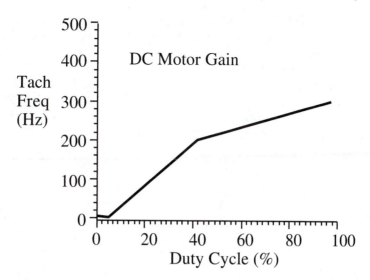

Our system has:
- Two control inputs

  S*   desired motor speed, rpm

  S'   current estimated motor speed, rpm
- One control output

  U   duty cycle, E clock cycles
- Two crisp inputs

  $E = S* - S'$ error in motor speed, rpm

  $D = dS'/dt = S'(n) - S'(n-1)$ change in motor speed, Hz/time or rpm/time
- One crisp output

  ΔU change in duty cycle, $U = U + \Delta U$, E clock cycles

Next we introduce fuzzy membership sets that define the current state of the crisp inputs
and outputs. The calculation of the input membership sets is called *fuzzification*. For this
simple fuzzy controller, we will define ten membership sets for the crisp inputs:

ENL	True if the motor is spinning much too slow
ENS	True if the motor is spinning a little bit too slow
EZE	True if the motor is spinning at the proper speed
EPS	True if the motor is spinning a little bit too fast
EPL	True if the motor is spinning much too fast
DNL	True if the motor speed is slowing down a lot
DNS	True if the motor speed is slowing down a little
DZE	True if the motor speed is remaining the same
DPS	True if the motor speed is speeding up a little
DPL	True if the motor speed is speeding up a lot

We will define five membership sets for the crisp output:

ONL	True if the motor speed should be decreased a lot
ONS	True if the motor speed should be decreased a little
OZE	True if the motor speed should remain the same
OPS	True if the motor speed should be increased a little
OPL	True if the motor speed should be increased a lot

Notice in the following rules that when the system is operating near the set point (ENS, EZE, EPS) with small changes in speed (DNS, DZE, DPS), then this fuzzy system is similar to the previous implementation. On the other hand, when the system has large errors and/or large acceleration, then it makes large changes in the output (OPL, ONL) in an attempt to reach the desired state faster. The 6811 implementation to this controller would be very similar to the previous example. On the other hand, the 6812 has a rich set of fuzzy logic assembly instructions that will be used to solve this problem. We begin with the fuzzy variables (Program 13.18).

Program 13.18 Global variables for fuzzy controller in 6812 assembly.

```
            org   $800
; crisp inputs
speed:    ds    1
acceleration: ds 1
; input membership variables
fuzvar:   ds    0     ; inputs
EPL:      ds    1     ; speed way too fast
EPS:      ds    1     ; speed too fast
EZE:      ds    1     ; speed OK
ENS:      ds    1     ; speed too slow
ENL:      ds    1     ; speed way too slow
DPL:      ds    1     ; speed decreasing a lot
DPS:      ds    1     ; speed decreasing
DZE:      ds    1     ; speed constant
DNS:      ds    1     ; speed increasing
DNL:      ds    1     ; speed increasing a lot

fuzout:   ds    0     ; outputs
OPL:      ds    1     ; add a lot of power to system
OPS:      ds    1     ; add some power to system
OZE:      ds    1     ; leave power as is
ONS:      ds    1     ; subtract some power from system
ONL:      ds    1     ; subtract a lot of power from system
BREAK:    ds    1     ; apply break?
; input membership variables relative offsets
epl:      equ   0     ; speed way too fast
eps:      equ   1     ; speed too fast
eze:      equ   2     ; speed ok
ens:      equ   3     ; speed too slow
enl:      equ   4     ; speed way too slow
dpl:      equ   5     ; speed decreasing a lot
dps:      equ   6     ; speed decreasing
dze:      equ   7     ; speed constant
dns:      equ   8     ; speed increasing
dnl:      equ   9     ; speed increasing a lot
;output membership variables
opl:      equ   10    ; add alot of power to system
ops:      equ   11    ; add some power to system
oze:      equ   12    ; leave power as is
ons:      equ   13    ; subtract some power from system
onl:      equ   14    ; subtract alot of power from system
break:    equ   15    ; apply break?
;crisp outputs
dpower:   ds    1
```

The error fuzzification is plotted in Figure 13.24 and defined in Program 13.19.

Figure 13.24
Fuzzification of the error input.

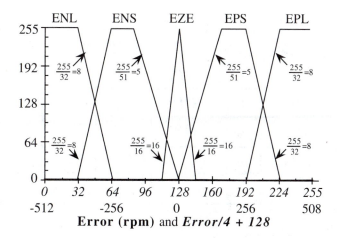

Program 13.19 Definition of the error input function in 6812 assembly.

```
        org   $F000
; format is Point1,Point2,Slope1,Slope2
s_tab:  dc.b  192,255,8,0      ;EPL
        dc.b  128,244,5,8      ;EPS
        dc.b  112,144,16,16    ;EZE
        dc.b  32,128,8,5       ;ENS
        dc.b  0,64,0,8         ;ENL
```

The acceleration fuzzification is plotted in Figure 13.25 and defined in Program 13.20.

Figure 13.25
Fuzzification of the acceleration input.

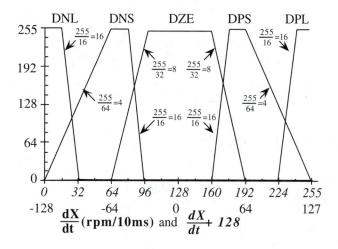

Program 13.20 Definition of the acceleration input function in 6812 assembly.

```
a_tab:  dc.b  244,255,16,0     ;DPL
        dc.b  160,255,16,4     ;DPS
        dc.b  64,192,8,8       ;DZE
        dc.b  0,96,4,16        ;DNS
        dc.b  0,32,0,16        ;DNL
```

The fuzzy roles are plotted in Figure 13.26 and defined in Program 13.21.

Figure 13.26
Fuzzy logic rules shown in table form.

$$D=X(n)-X(n-1)$$

$E=X^*-X$	DNL	DNS	DZE	DPS	DPL
ENL	OPL	OPL	OPL		
ENS	OPL	OPS	OPS		
EZE	OPL	OPS	OZE	ONS	ONL
EPS			ONS	ONS	ONL
EPL			ONL	ONL	ONL

```
rules:   dc.b   enl,dnl,$FE,opl,$FE           ; if ENL and DNL then OPL
         dc.b   ens,dnl,$FE,opl,$FE           ; if ENS and DNL then OPL
         dc.b   eze,dnl,$FE,opl,$FE           ; if EZE and DNL then OPL
         dc.b   enl,dns,$FE,opl,$FE           ; if ENL and DNS then OPL
         dc.b   ens,dns,$FE,ops,$FE           ; if ENS and DNS then OPS
         dc.b   eze,dns,$FE,ops,$FE           ; if EZE and DNS then OPS
         dc.b   enl,dze,$FE,opl,$FE           ; if ENL and DZE then OPL
         dc.b   ens,dze,$FE,ops,$FE           ; if ENS and DZE then OPS
         dc.b   eze,dze,$FE,oze,$FE           ; if EZE and DZE then OZE
         dc.b   eps,dze,$FE,ons,$FE           ; if EPS and DZE then ONS
         dc.b   epl,dze,$FE,onl,break,$FE ; if EPL and DZE then ONL
         dc.b   eze,dps,$FE,ons,$FE           ; if EZE and DPS then ONS
         dc.b   eps,dps,$FE,ons,$FE           ; if EPS and DPS then ONS
         dc.b   eps,dps,$FE,onl,break,$FE ; if EPL and DPS then ONL
         dc.b   eze,dpl,$FE,onl,break,$FE ; if EZE and DPL then ONL
         dc.b   eps,dpl,$FE,onl,break,$FE ; if EPS and DPL then ONL
         dc.b   epl,dpl,$FE,onl,break,$FE ; if EPL and DPL then ONL
         dc.b   $FF
```

Program 13.21 Definition of the fuzzy rules in 6812 assembly.

The defuzzification is shown in Table 13.2.

Table 13.2
Defuzzification converts the output memberships into a crisp output.

Output fuzzy set	Singleton value
ONL	−128
ONS	−10
OZE	0
OPS	10
OPL	127

Program 13.22 initializes the system and Program 13.23 implements the fuzzy logic controller.

Program 13.22 Ritual and main program for fuzzy controller in 6812 assembly.

```
addsingleton: dc.b 255,138,128,118,0
; 128 subtracted,   +127,10,0,-10,-128

ritual:   sei          ;make atomic
          bset #$20,TIOS  ;OC5
          movb #$80,TSCR  ;enable, no fast clr
          movb #$33,TMSK2 ;1us clk
          bset #$20,TMSK1 ;Arm OC5
          ldaa #$20       ;clear C5F
          staa TFLG1
          ldd  TCNT       ;current time
          addd #10000     ;first in 10 ms
          std  TC5
          cli             ;enable
          rts
main:     lds  #$0C00
          ldd  #100       ; initial duty cycle 1% is off
          std  dutycycle
          jsr  ritual     ; initialize OC interrupt
          bra  *
```

Program 13.23 Interrupt handler for fuzzy controller in 6812 assembly.

```
timehan: ldaa #$20  ;clear C5F
         staa TFLG1 ;Acknowledge
         ldd  TC5
         addd #10000 ;next in 10 ms
         std  TC5
         jsr  measurespeed ; crisp input speed
;reg A is speed 0 to 255
         ldx  #s_tab
         ldy  #fuzvar
         mem          ; calculate EPL
         mem          ; calculate EPS
         mem          ; calculate EZE
         mem          ; calculate ENS
         mem          ; calculate ENL
         jsr  measureacceleration ; crisp input acceleration
;reg A is acceleration 0 to 255
         ldx  #a_tab
         mem          ; calculate DPL
         mem          ; calculate DPS
         mem          ; calculate DZE
         mem          ; calculate DNS
         mem          ; calculate DNL
         ldab #6
cloop:   clr  1,y+  ; clear OPL,OPS,OZE,ONS,ONL,BREAK
         dbne b,cloop
         ldx  #rules
         ldy  #fuzvar
         ldaa #$FF
         rev
         ldy  #fuzout
```

continued on p. 762

Program 13.23 Inter-rupt handler for fuzzy controller in 6812 assembly.

continued from p. 762

```
                ldx   #addsingleton
                ldab  #5
                wav
                ediv
                tfr   y,d
                subb  #128
                stab  dpower
change          ldab  dpower
                sex   b,d
                addd  dutycycle
; 200 Hz squarewave is 10000 cycles
; correct range is 100 to 9600 cycles
                cpd   #100
                bhs   nolow
low:            ldd   #100    ; underflow
                bra   set
notlow:         cpd   #9600
                bls   set
high:           ldd   #9600   ; overflow
set:            std   dutycycle
                rti
```

13.5.3 Temperature Controller Using Fuzzy Logic

The objective of this section is to design a *fuzzy logic* microcomputer-based temperature controller. The desired temperature is **T***. The first step in designing a controller is choosing the input sensors and output actuators. There only one sensor input, **Temperature.** There are two actuator outputs. **Heat** is a variable output (0 to 255) that will apply heat to the physical plant. **Fan** is a variable output (0 to 255) that will force air across the physical plant. **Fan+Heat** will warm up a very cold environment, **Heat** alone will slowly warm up the environment, and **Fan** alone will cool down a very hot environment. To conserve power, the use of the fan will be restricted.

The second step is selecting the crisp inputs and crisp outputs. Knowledge of how this physical plant works is critical in this step. What we wish to do depends on both absolute temperature and temperature error, so two crisp inputs will be employed:

> **T** is the temperature (scaled to the range 0 to 255)
> **E** is the temperature error (also scaled to the range 0 to 255)

```
; crisp inputs
T: ds 1 ;temperature (units 0.5 F)
E: ds 1 ;temperature error (128 means no error) (units 0.125 F)
```

The crisp outputs will affect the two actuators:

> **Heat** is the heater actuator control (0 to 255)
> **Fan** is the fan actuator control (0 to 255)

```
;crisp outputs
Heat:  ds   1 ; 0 is off and 255 is maximum heat
Fan:   ds   1 ; 0 is off and 255 is maximum fan
```

The third step is choosing the fuzzy input and output membership sets. Here, we will divide each crisp input and output into three regions. Using five or seven regions would probably create a better controller, but for purposes of illustration, we will use only three. The three fuzzy membership inputs based on temperature are

Cold	means temperature of room is cold
Normal	means the temperature of room is a normal temperature
Hot	means temperature of room is hot

The three fuzzy membership inputs based on temperature error are

TooCold	means temperature of room is below the setpoint
OK	means temperature of room is correct
TooHot	means temperature of room is above the setpoint

The three fuzzy membership outputs based on the heater are

NoHeat	means heater should be off
SomeHeat	means heater should be on a little
MaxHeat	means heater should be on a lot

The three fuzzy membership outputs based on the fan are

NoFan	means fan should be off
SomeFan	means fan should be on a little
MaxFan	means fan should be on full speed

In the fourth step, we choose functions for the conversion of crisp inputs to input membership variables. The particular constants used in these functions will be used as a starting point. Once the system is built and tested, the values can be adjusted as needed (Figure 13.27 and Program 13.24).

Figure 13.27
Fuzzification of the temperature input.

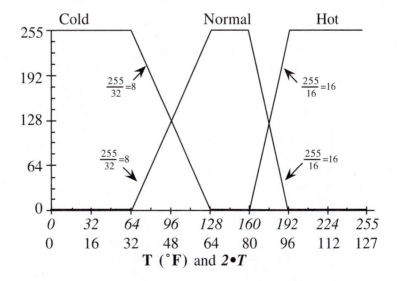

Program 13.24 Fuzzification function for the temperature input in 6812 assembly.

```
; format is Point1,Point2,Slope1,Slope2
T_tab:    dc.b  160,255,16,0      ;Hot
          dc.b  64,192,8,16        ;Normal
          dc.b  0,64,0,8           ;Cold
```

We will define the error as OK if it is within ±2·F of the setpoint (Figure 13.28 and Programs 13.25 and 13.26).

Figure 13.28
Fuzzification of the temperature error input.

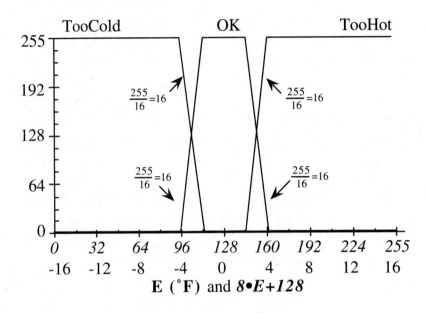

Program 13.25 Fuzzification function for the temperature error input in 6812 assembly.

```
E_tab:   dc.b  144,255,16,0    ;TooHot
         dc.b  64,192,8,16     ;OK
         dc.b  0,96,0,16       ;TooCold
```

Program 13.26 Global variables in 6812 assembly.

```
; input membership variables
fuzvar:   ds    0     ; inputs
Hot:      ds    1
Normal:   ds    1
Cold:     ds    1
TooHot:   ds    1
OK:       ds    1
TooCold:  ds    1
; output membership variables
fuzout:   ds    0     ; outputs
NoHeat:   ds    1     ; turn off heater
SomeHeat: ds    1     ; apply some heat
MaxHeat:  ds    1     ; turn on heater
NoFan:    ds    1     ; turn off fan
SomeFan:  ds    1     ; apply some fan
MaxFan:   ds    1     ; turn on fan
; input membership variables relative offsets
hot:      equ   0
normal:   equ   1
cold:     equ   2
toohot:   equ   3
ok:       equ   4
toocold:  equ   5
```

```
;output membership variables
noheat:   equ  6    ; turn off heater
someheat: equ 7     ; apply some heat
maxheat:  equ  8    ; turn on heater
nofan:    equ  9    ; turn off fan
somefan:  equ  10   ; apply some fan
maxfan:   equ  11   ; turn on fan
```

The fifth step is creating the fuzzy rules. Again, we use our intuition as a starting point. During the testing phase of the project, we develop a means to observe the values of the input membership variables and which rules apply in certain situations. The 6812 background debug module is a convenient nonintrusive debugging tool that we can use to observe memory locations while the program is running (Program 13.27).

```
rules:  dc.b hot,toohot,$FE,noheat,maxfan,$FE      ; use fan for max cooling
        dc.b hot,ok,$FE,noheat,somefan,$FE         ; use fan for some cooling
        dc.b normal,toohot,$FE,noheat,nofan,$FE    ; no fan for normal temps
        dc.b normal,ok,$FE,noheat,nofan,$FE        ; perfect
        dc.b normal,toocold,$FE,someheat,nofan,$FE ; a little heat
        dc.b cold,ok,$FE,noheat,nofan,$FE          ; cold but perfect
        dc.b cold,toocold,$FE,maxheat,maxfan,$FE   ; fast warmup
        dc.b $FF
```

Program 13.27 Fuzzy rules in 6812 assembly.

The last design step is defuzzification. It is here we convert the true/false fuzzy output variables (NoHeat, SomeHeat, MaxHeat, NoFan, SomeFan, MaxFan) to crisp outputs (Heat, Fan) (Program 13.28).

```
Heatsingleton: dc.b 0,50,255
Fansingleton: dc.b 0,128,255
timehan: ldaa #$20  ;clear C5F
         staa TFLG1 ;Acknowledge
         ldd  TC5
         addd #10000 ;next in 10 ms
         std  TC5
         jsr  MeasureTemperatire ; crisp input temperature, T
;reg A is temperature 0 to 255 (units 0.5•F)
         ldx  #T_tab
         ldy  #fuzvar
         mem         ; calculate Hot
         mem         ; calculate Normal
         mem         ; calculate Cold
         jsr  MeasureError ; crisp input error, E
;reg A is error 0 to 255  (128 means no error) (units 0.125•F)
         ldx  #E_tab
         mem         ; calculate TooHot
         mem         ; calculate OK
         mem         ; calculate TooCold
         ldab #6                          continued on p. 766
```

Program 13.28 Fuzzy controller in 6812 assembly.

continued on p. 766

continued from p. 765

```
cloop:    clr   1,y+   ; clear NoHeat, SomeHeat, MaxHeat, NoFan, SomeFan, MaxFan
          dbne  b,cloop
          ldx   #rules
          ldy   #fuzvar
          ldaa  #$FF
          rev
          ldy   #NoHeat    ; pointer to NoHeat, SomeHeat, MaxHeat
          ldx   #Heatsingleton
          ldab  #3
          wav
          ediv
          tfr   y,d
          stab  Heat
          ldy   #NoFan     ; pointer to NoFan, SomeFan, MaxFan
          ldx   #Fansingleton
          ldab  #3
          wav
          ediv
          tfr   y,d
          stab  Fan
          rti
```

Program 13.28 Fuzzy controller in 6812 assembly.

Output fuzzy set	Singleton value
NoHeat	0
SomeHeat	50
MaxHeat	255

Output fuzzy set	Singleton value
NoFan	0
SomeFan	128
MaxFan	255

Observation: Fuzzy logic control will work extremely well (fast, accurate, and stable) if the designer has expert knowledge (intuition) of how the physical plant behaves.

13.6   Glossary

actuator Electromechanical device that allows computer commands to affect the physical plant.

anti-reset-windup Establishing an upper bound on the magnitude of the integral term in a PID controller so that this term will not dominate when the errors are large.

bang-bang A control system where the actuator has only two states, and the system "bangs" all the way in one direction or "bangs" all the way in the other, same as binary controller.

binary controller Same as bang-bang.

closed-loop control system A control system that includes sensors to measure the current state variables; these inputs are used to drive the system to the desired state.

crisp input An input parameter to the fuzzy logic system, usually with units like centimeters, centimeters per second, degrees Celsius.

crisp output An output parameter from the fuzzy logic system, usually with units like dynes, watts.

defuzzification Conversion from the fuzzy logic output variables to the crisp outputs.

fuzzification Conversion from the crisp inputs to the fuzzy logic input variables.

fuzzy logic Boolean logic (true/false) that can take on a range of values from true (255) to false (0); fuzzy logic **and** is calculated as the minimum; fuzzy logic **or** is the maximum.

incremental control system A control system where the actuator has many possible states, and the system increments or decrements the actuator value depending on whether the error is positive or negative.

membership sets Fuzzy logic variables that can take on a range of values from true (255) to false (0).

open-loop control system A control system that does not include sensors to measure the current state variables.

physical plant The physical device being controlled.

PID controller A control system where the actuator output depends on a linear combination of the current error (P), the integral of the error (I) and the derivative of the error (D).

RTD (Resistance temperature device) A sensor used to measure temperature, usually made from platinum.

stepper motor A motor that moves in discrete finite steps.

13.7   Exercises

13.1 The objective of this exercise is to control the rotational speed of a DC motor. The +5 V DC motor has a resistance of 10 Ω. Your control system will apply a variable power to the motor by varying the duty cycle of a 50-Hz signal, as shown in Figure 13.29, for example.

Figure 13.29
PWM output to actuator.

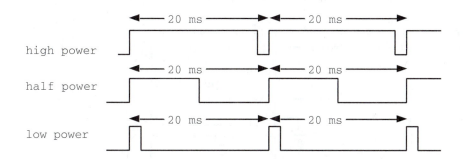

The rotation speed **R** of the motor is measured by a tachometer. You should measure **R** every 20 ms with a resolution of 1 rps. The rotational speed will vary from 0 to 250 rps ($0 \leq \mathbf{R} \leq 250$ rps). The output of the tachometer is an ugly-looking digital wave with a frequency 100 times the motor speed. Thus, $0 \leq \mathbf{f} \leq 25$ kHz (Figure 13.30).

Figure 13.30
Tachometer frequency is a function of the motor rotation speed.

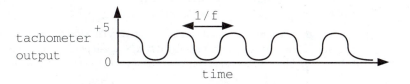

The desired rotational speed (**R*** also in rps) comes from an 8-bit parallel input port. A new value is available on the rise of Ready. The fall of Done is an acknowledge signal back to the input device signifying that the microcomputer no longer needs the data. The timing is shown in Figure 13.31.

Figure 13.31
Sensor timing.

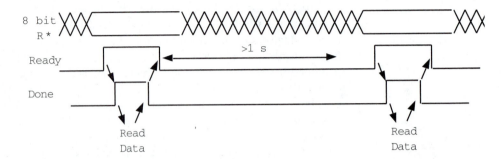

The interrupt-driven control algorithm should be implemented at 50 Hz (**U(t)** is the duty cycle in percent):

$$U(t) = \int_0^t (R^* - R(t))dt \qquad \text{where } R^* \text{ and } R(t) \text{ are in rps and } t \text{ is in seconds}$$

a) Let **u(n)** equal the cycle count (500 ns each) that controls the duty cycle of the output-compare variable-duty cycle 50-Hz square wave. If the range of duty cycle is $0 < U(t) < 100$, what is the relationship between **u(n)** and **U(t)**?

b) Let **R(n)** be the sampled sequence of measured rotational speed in rps. Convert the above integral control equation into discrete form. That is, determine the relationship that calculates **u(n)** from **u(n − 1)**, **u(n − 2)** . . . , **R(n)**, **R(n − 1)**, **R(n − 2)** . . . , and **R***.

c) Show the interface from the input device, the motor, and the tachometer to the microcomputer. Label chip numbers, resistors, and capacitor values (Figure 13.32).

Figure 13.32
Sensor and motor interface.

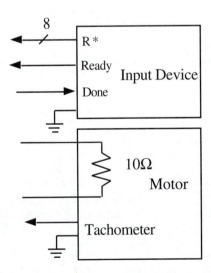

d) Show the ritual software including data structures. The main program executes the ritual, then performs other unrelated tasks (i.e., all processing occurs under interrupt control).

e) Show the interrupt software. You may poll any way you wish.

13.2 The objective of this exercise is to develop the fixed-point equations that implement a PID controller. You are to implement the following control system (Figure 13.33):

X(t)	is the state variable (V)
X*	is the desired state (V)
e(t)=X*-X(t)	is the error (V)
V(t)	is the actuator command (V)

$$V(t) = K_P e(t) + K_I \int_0^t e(\tau)\, d\tau + K_D \frac{de(t)}{dt}$$

where $K_P = 0.1$ (dimensionless)
$K_I = 10$ (1/s)
$K_D = 0.0001$ (s)

The state estimator and actuator output hardware/software interfaces are given. Your PID software is given a signed 16-bit decimal fixed-point input, $x(n)$, that represents the current state variable. Similarly, your PID software will calculate a signed 16-bit decimal fixed-point output $u(n)$ that will be fed to the actuator interface. For both cases, the fixed-point resolution, Δ, is 0.01 V.

Figure 13.33
Control system for Exercise 13.2.

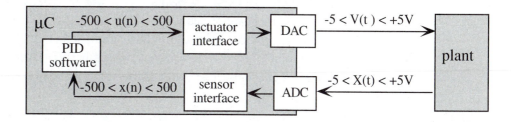

Assuming the digital controller is executed every 1 ms (1000 Hz), show the control equation to be executed in the periodic interrupt handler. In this process you will convert from floating to fixed-point numbers and convert from continuous to discrete time. No software is required; just show the equations. Explain how you would deal with overflow/underflow. You should assume some noise exists on the sensor input.

13.3 Briefly explain why it is important to choose the proper update rate for a fuzzy logic controller. In particular, explain what happens to a fuzzy logic controller if the controller is executed too infrequently. Similarly, explain what happens to a fuzzy logic controller if the controller is executed too frequently.

13.4 Write a 6812 assembly language program for the fuzzy logic controller described in Section 13.5.1 using the special fuzzy logic assembly instructions.

14 Simple Networks

Chapter 14 objectives are to:

❑ Introduce basic concepts of networks
❑ Design a master/slave network
❑ Show the fundamental issues in a packet-switching network like binary synchronous communication
❑ Define the protocols for the IEEE488 instrumentation bus and for the SCSI
❑ Define some of the terminology and approaches to modem communication
❑ Develop a building-wide control system using the X-10 communication protocol

The goal of this chapter is to provide a brief introduction to communication systems. Communication theory is a richly developed discipline, and most of the theory is beyond the scope of this book. Consequently, this chapter focuses on implementing simple communication systems appropriate for an embedded computer. Generally, requirements of an embedded system involve relatively low bandwidth, static configuration, and a low probability of corrupted data.

14.1 Introduction

In the serial interfacing chapter (Chapter 7), we considered the hardware interfaces between computers. In this chapter, we will build on those ideas and introduce the concepts of networks by investigating a couple of simple networks. A communication network includes both the physical channel (hardware) and the logical procedures (software) that allow users or software processes to communicate with each other. The network provides the transfer of information as well as the mechanisms for process synchronization. It is convenient to visualize the network in a hierarchical fashion. Figure 14.1 shows a three-layer communication system.

At the lowest level, frames are transferred between I/O ports of the two (or more) computers along the physical link or hardware channel. At the next logical level, the OS

Figure 14.1
A layered approach to
communication systems.

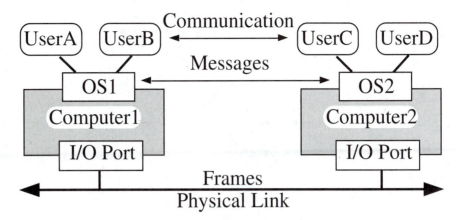

of one computer sends *messages* or *packets* to the OS on the other computer. The message protocol will specify the types and formats of these messages. Typically, error detection and correction is handled at this level. Messages typically contain four fields.

1. Address information field
   Physical address specifying the destination/source computers
   Logical address specifying the destination/source processes (e.g., users)
2. Synchronization or handshake field
   Physical synchronization like shared clock, start, and stop bits
   OS synchronization like request connection or acknowledge
   Process synchronization like semaphores
3. Data field
   ASCII text (raw or compressed)
   Binary (raw or compressed)
4. Error detection and correction field
   Vertical and horizontal parity
   Checksum
   Block correction codes (BCC)

Observation: Communication systems often specify bandwidth in total bits per second, but the important parameter is the data transfer rate.

Observation: Often the bandwidth is limited by the software and not the hardware channel.

At the highest level, we consider communication between users or processes. A *process* is a complete software task that has a well-defined goal. For example, when a file is to be printed on a network printer, the OS creates a process that

1. Establishes connection with the remote printer;
2. Reads blocks from the hard disk drive and sends the data to the printer:
   The OS printer driver may have to manipulate graphics/colors for the specific printer;
   The OS network driver will break the data into message packets;
3. Disconnects the printer.

Figure 14.2
A master/slave network
allows distributed data
acquisition and control.

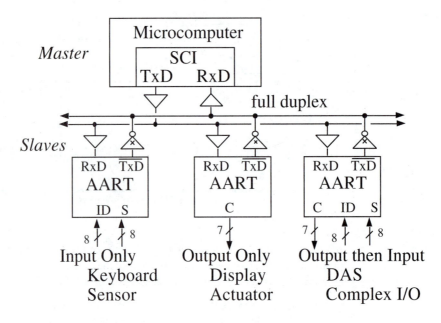

14.2 Serial Master/Slave Desktop Bus

This network allows the main computer to control multiple slave devices (Figure 14.2). The tabletop master/slave communication network is implemented with a microcomputer master and multiple slaves built with the Motorola MC14469 addressable asynchronous receiver/transmitter (AART). A typical application is a manufacturing process that requires sensors for temperature, pressure, etc., be placed at remote locations in an industrial plant. As the number of sensors grows, the cost and complexity of the wiring needed to interface the sensors to a central computer increases greatly. To alleviate this cost a common bus connecting both the processor and the monitored or controlled sensors is often used. A physical channel based on open-collector logic allows the TxDs of the multiple AART slaves to be wire-ANDed together. The other advantage of open collector is that it is easy to add or subtract AART devices without a major redesign. The nature of this particular master/slave communication protocol eliminates the possibility of a collision. In other words, two slaves will never attempt to communicate with the master at the same time.

As the microcomputer requires data from specific sensors, it "addresses" them to receive the data. The AART interprets its particular address and relays the information back to the processor via the serial data bus. The physical channel can be created in either full-duplex, as shown in Figure 14.2, or half-duplex, as shown in Figure 14.3. The advantage of half-duplex is that only two wires are required instead of three. In half-duplex mode, only 7 bits of the 8 bits in ID and S can be used to input data. You don't want an ID or S inadvertently activating another slave. The 6808, 6811, and 6812 can configure their asynchronous serial ports in open-drain mode. If the TxD output of the microcomputer can be made open collector (drives to zero, and float), then the 7407 driver is not needed.

There are two types of frames: address characters and command characters. An address character has bit 7=1. Command words contain 7 bits of data and have bit 7=0. Each AART has a 7-bit hardware-jumped physical address specified by the digital inputs A6–A0. This allows for up to 128 devices on the network (Figure 14.4). The AART mode is not

Figure 14.3
A master/slave network allows distributed data acquisition and control.

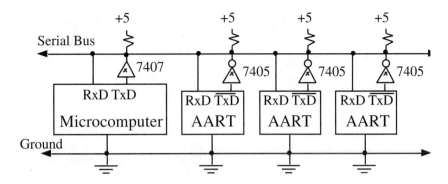

Figure 14.4
The MC14469 AART has an address determined by the A6–A0 inputs.

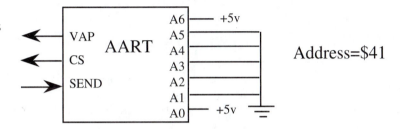

software-programmed but determined in hardware by how the external AART pins, VAP, CS, and SEND, are connected. There are three AART modes. It is okay to have AARTs on the network with different modes.

In *input-only mode (SEND=VAP),* the AART will turn on when

1. It is previously off, and
2. It receives an address frame that matches its address

In *input-only* mode, the AART will turn off automatically at the end of the AART transmission. If you are using a half-duplex channel, then you could connect the two pins ID7 and S7 to ground so that no other slave inadvertently gets an address match when the ID and S are transmitted (Figure 14.5).

Figure 14.5
The MC14469 AART in input-only mode.

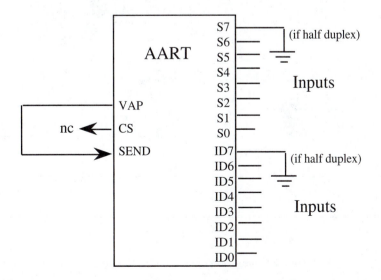

To input 16-bit (14 bits if using half-duplex) data from the slave device, the software in the master could

1. Output the slave address (bit $7=1$), which will cause a VAP pulse and activate the send operation: If half-duplex, the master must receive the echo of the address it just sent;
2. Receive the ID and S from the slave (the AART is automatically disabled after S is sent).

An example input-only communication in full-duplex mode is shown in Figure 14.6. In the C Programs 14.1 and 14.2 the gadfly routines InCh and OutCh shown earlier in Chapter 7 are assumed. If this input-only communication were in half-duplex mode, then TxD and RxD would be combined (Figure 14.7).

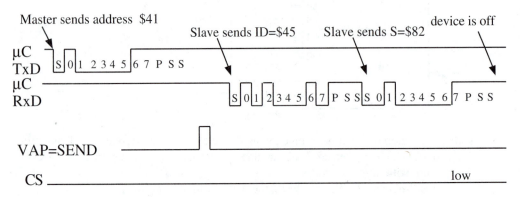

Figure 14.6   Full-duplex timing of the MC14469 AART in input-only mode.

Program 14.1 AART code to receive from an input-only full-duplex AART.

```
// AART full duplex Input only, returns ID:S as 16-bit
unsigned int GetAART(unsigned char addr){ // addr specifies which AART
    unsigned char ID,S;        // inputs from AART
    OutCh(addr|0x80);          // activate AART
    ID=InCh();                 // read ID
    S=InCh();                  // read status
    return((ID<<8)+S);}        // return ID:S as 16-bit unsigned int
```

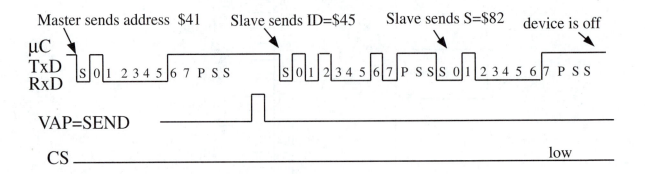

Figure 14.7   Half-duplex timing of the MC14469 AART in input-only mode.

In half-duplex each output to the network must be read back in to prevent overrun (Program 14.2).

Program 14.2 AART code to receive from an input-only half-duplex AART.

```
// AART half duplex Input only, returns ID:S as 16-bit
unsigned int GetAART(unsigned char addr){ // addr specifies which AART
      unsigned char ID,S;         // inputs from AART
      OutCh(addr|0x80);           // activate AART
      InCh();                     // read echo of previous addr output
      ID=InCh();                  // read ID
      S=InCh();                   // read status
      return((ID<<8)+S);}         // return ID:S as 16-bit unsigned int
```

In *output-only mode (SEND=0),* the AART will turn on when (Figure 14.8)

1. It is previously off;
2. It receives an address frame that matches its address.

In *output-only mode,* the AART will turn off when

1. It is previously on;
2. It receives any address frame including its own address.

Figure 14.8
The MC14469 AART in output-only mode.

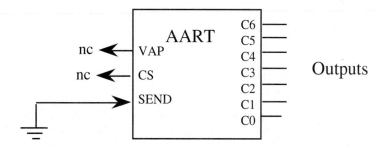

If using a half-duplex channel, the master must receive the echo of each frame it sends. To output 7 bits of data to the slave device, the software in the master could

1. Output the slave address (bit7=1), which causes a VAP pulse;
2. Output a data frame (bit7=0), which causes a CS pulse;
3. Output any address (bit7=1), which turns off the slave.

An example output-only communication in full-duplex mode is shown in Program 14.3. If this communication were in half-duplex mode (shown in Program 14.4), then the TxD and RxD in the Figure 14.9 timing diagram would be combined, and the echo of each output must be read to prevent overrun.

Program 14.3 AART code to transmit to an output-only half-duplex AART.

```
// AART full duplex output only
void PutAART(unsigned char addr,data){ // addr specifies which AART
      OutCh(addr|0x80);           // activate AART
      OutCh(data&0x7F);           // send data
      OutCh(addr|0x80);}          // deactivate AART
```

Program 14.4 AART code to transmit to an output-only half-duplex AART.

```
// AART half duplex output only
void PutAART(unsigned char addr,data){ // addr specifies which AART
            OutCh(addr|0x80);         // activate AART
            InCh();                   // read echo of addr output
            OutCh(data&0x7F);         // send data
            InCh();                   // read echo of data output
            OutCh(addr|0x80);         // deactivate AART
            InCh();}                  // read echo of previous output
```

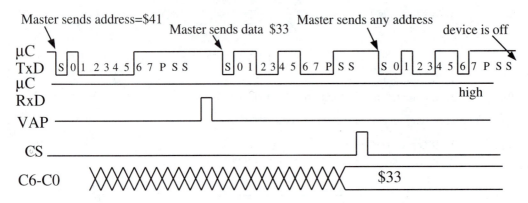

Figure 14.9 Full-duplex timing of the MC14469 AART in output-only mode.

To output multiple 7-bit data frames to the slave device, the software in the master could (Figure 14.10)

1. Output the slave address (bit7=1), which causes a VAP pulse;
2. Output the first data frame (bit7=0), which causes a CS pulse;
3. Output the second data frame (bit7=0), which causes a CS pulse;

• • •

n. Output the last data frame (bit7=0), which causes a CS pulse;
n+1. Output the slave address (bit7=1), which turns off the slave.

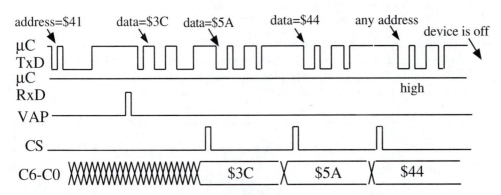

Figure 14.10 Multiple outputs to the MC14469 AART in output-only mode.

In *output-then-input mode (SEND=CS)*, the AART will turn on when

1. It is previously off;
2. It receives an address frame that matches its address.

Similar to the input-only mode, the AART will turn off at the end of the AART transmission. If you are using a half-duplex channel, then the master must receive the echo of each frame it sends (Figure 14.11).

Figure 14.11
The MC14469 AART in output-then-input mode.

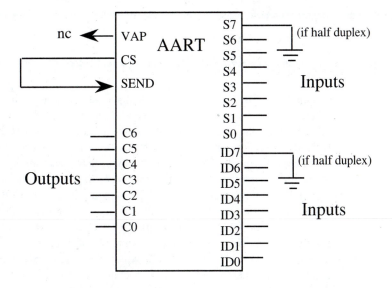

To output 7 bits of data to the slave device and input 16 bits, the software in the master could (Figure 14.12)

1. Output the slave address (bit7=1), which causes a VAP pulse;
2. Output a data frame (bit7=0), which causes a CS pulse initiating a send sequence;
3. Receive the ID and S from the slave (the AART is automatically disabled after S is sent) (Program 14.5).

Figure 14.12 Full-duplex timing of the MC14469 AART in output-then-input mode.

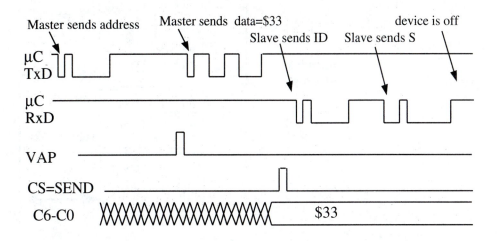

```
// AART full duplex Output then Input, returns ID:S as 16-bit
unsigned int PutGetAART(unsigned char addr,data){ // addr specifies which AART
      unsigned char ID,S;        // inputs from AART
      OutCh(addr|0x80);          // activate AART
      OutCh(data&0x7F);          // send data
      ID=InCh();                 // read ID
      S=InCh();                  // read status
      return((ID<<8)+S);}        // return ID:S as 16-bit unsigned int
```

Program 14.5 AART code to transmit-then-receive.

14.3   Parallel Bus Application Using Tristate Buffers

In this mode we can connect multiple devices that can potentially drive the I/O data pins. Example applications include IEEE488 and SCSI. We must control the system so that only one device drives the I/O bus at a time. We can use the 6811 Port C output handshake mode with DDRC=0 to implement the talk function of this interface (the device that drives the I/O data bus) (Figure 14.13). We can use input handshake mode, also with DDRC=0, to implement the listen function of this interface (the device that takes the data off the I/O data bus). With output handshake, STRB is an output ready. STRB is set to its active level (INVB) when the software writes to PORTCL. STRB returns to its inactive level either 2 E periods later (PLS=1) or on the active edge of the input STRA (PLS=0). This active edge of STRA will also set the STAF bit in the PIOC register. STAF is cleared by the two-step sequence: (1) read PIOC ($1002) with STAF=1; (2) write PORTCL. When STRA equals the complement of EGA, the Port C pins become outputs driven with the value of PORTCL ($1005) independent of the Port C direction register DDRC ($1007). When STRA equals EGA, the I/O direction of Port C pins are controlled by the Port C direction register. In this configuration DDRC=0, so the I/O bus pins will float (tristate) when STRA equals EGA. The role of the I/O bus controller (bus master) is to determine which device will talk and which device will listen.

Figure 14.13
Parallel I/O bus using the 6811 handshake modes.

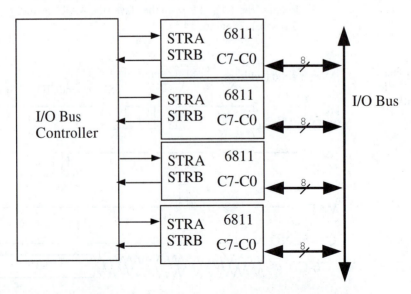

The sequence of operations that affect a 1-byte transfer are (Figure 14.14, Program 14.6)

1. The 6811 writes new data to PORTCL (but the Port C outputs are HiZ);
2. STRB goes to zero (its active level is INVB);
3. The external device (or bus controller) recognized the STRB request and makes STRA=0;
4. The 6811 automatically drives the data previously written into PORTCL on Port C outputs;
5. The external device makes STRA=1;
   The STAF flag is set;
   The Port C outputs are HiZ again;

6. The 6811 software gadfly loop finishes.

Figure 14.14
Timing of the parallel I/O bus using the 6811 Port C output handshake mode.

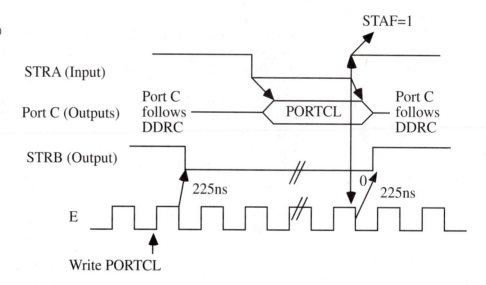

Program 14.6 6811 code to output on the parallel network.

```
/* Port C output handshake tristate output example    */
#define STAF 0x80
void main(void)
{   unsigned char Number;
    DDRC = 0x00;        /* Port C are outputs, only when STRA=0 */
/*  STAI=0   no interrupts,
    HNDS=1, OIN=1 output handshake mode
    CWOM=0   normal push/pull outputs on C
    PLS=0    makes STRB a level output
    EGA=1    means STRA active on the rise
    INVB=0   makes STRB=0 after write CL */
    PIOC = 0x1A;
    Number=100; OutC(Number);  }
void OutC(unsigned char data) {
    PORTCL=data;        /* set PortC flip flops, and make STRB=0  */
/* Port C data is driven only when STRA=0, otherwise Port C data=hiZ */
    while ((PIOC & STAF) == 0);  }  /* wait for rise of STRA */
```

14.4 Parallel Bus Application Using Open Collector Logic

Open-collector logic allows us to connect output signals together without damaging the circuit (Figure 14.15). In this application we will create a simple parallel bus network of microcomputers using the open-collector mode of Ports C and D.

> **Observation:** We can produce the same open-collector behavior of any I/O port that has a direction register. We initialize the port by writing a 0 to the data port. On subsequent accesses to the open-collector port, we write the complement to the direction register. That is, if we want the I/O port bit to drive low, we set the direction register bit to 1, and if we want the I/O port bit to float (open collector), we set the direction register bit to 0.

To transmit a byte to another microcomputer on the network, we

1. Make all bits of Port C open-collector outputs,
   Place the information on the 8-bit Port C open-collector outputs,
   Make bits 5, 3, 2, 1, 0 of Port D open-collector outputs with bit 5=1 (bit 4 is input);
2. Place the destination microcomputer address on Port D open-collector output bits 3, 2, 1, 0,
   Set Port D open-collector output bit 5=0 signifying data transfer;
3. Wait for Port D input bit 4 to be 0, signifying acknowledge;
4. Make Port D bit 5=1;
5. Wait for Port D input bit 4 to be 1, signifying transmission complete
   Initialize Ports C and D for receive mode.

To receive a byte from another microcomputer on the network, we

1. Make all bits of Port C inputs,
   Make bit 4 an open-collector output with bit 4=1, and the other bits input;

Figure 14.15
General parallel I/O bus network.

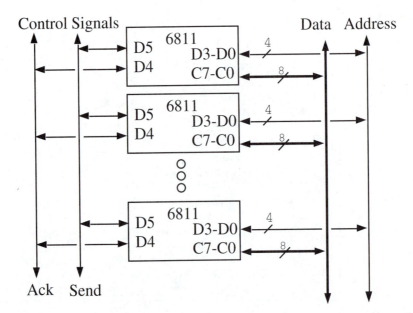

2. Wait for correct address on Port D inputs bits 3, 2, 1, 0 and with bit 5=0, read data from Port C inputs;
3. Set Port D open-collector output bit 4 to be 0, signifying acknowledge;
4. Wait for Port D bit 5=1;
5. Set for Port D open-collector output bit 4 to be 1, signifying transmission complete.

14.5 Binary Synchronous Communications

Binary synchronous communications (BSC) is a half-duplex protocol that transmits messages of character-based text. The discussion in this section is only a small subset of the BSC protocol. The purpose is to get an introduction to a higher-level communication protocol. This protocol includes *control/response* messages used for synchronization and *text/header* messages used to transmit ASCII character data. The control/response messages consist of either 3 or 4 bytes (Figure 14.16).

Table 14.1 lists some of the BSC control characters (*Ctrl* in Figure 14.16). An ETB or ETX at the end of a message requires a response (ACK0, ACK1, NAK, WACK, or RVI) (Figure 14.17).

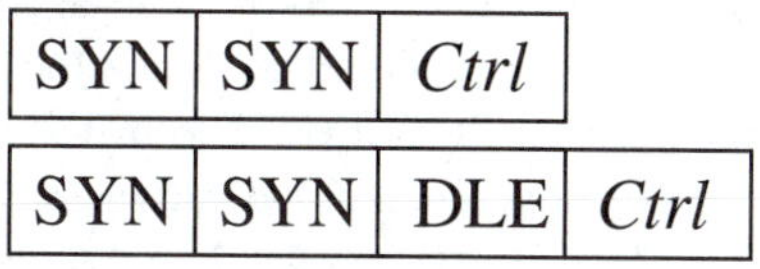

Figure 14.16
Two simple control messages in the BSC protocol.

Table 14.1 Some BSC control characters.

Symbol	Value	Name	Purpose
SYN	$16	Synchronous Idle	Establishes and maintains synchronization
SOH	$01	Start of Heading	Precedes header block
STX	$02	Start of Text	Precedes ASCII text block
ETB	$17	End of Transmission Block	End of block started with SOH or STX
ETX	$03	End of Text	End of block started with SOH or STX, no more blocks
EOT	$04	End of Transmission	End of message
ENQ	$05	Inquiry	Bid for line on a point to point link
ACK	$06	Acknowledge	Affirmative response
NAK	$15	Negative Acknowledge	Negative response
DLE	$10	Data Link Escape	Used to create WACK, ACK0, ACK1 and RVI

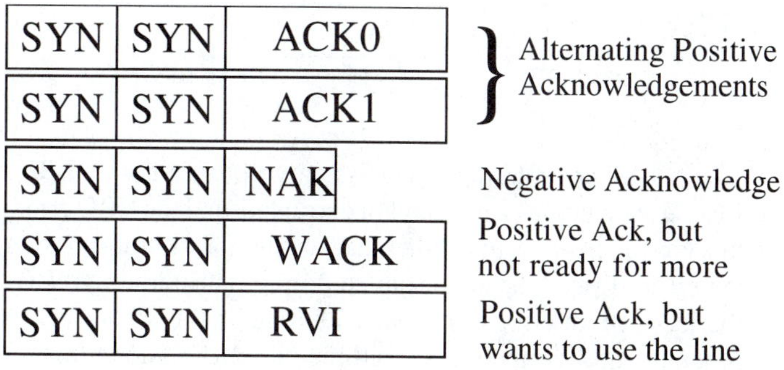

Figure 14.17
Five different control messages used to respond to an incoming message.

Immediately following an ETB or ETX is the block check character that can be either a 2-byte checksum or a longitudinal redundancy check (LRC). A 16-bit checksum is calculated by promoting each data byte in the message to an unsigned 16-bit number, adding the numbers with 16-bit precision ignoring overflow.

$$\text{checksum} = \Sigma_i \, \text{data}_i$$

A checksum detection scheme will detect all 1-bit errors and most multiple-bit errors. Unfortunately, it is possible to have certain multiple-bit transmission errors that cancel, causing the checksum to appear correct. The LRC or horizontal parity detection scheme appends 1 byte at the end of the message. The LRC is determined by calculating the *exclusive-OR* of each byte in the message.

$$C_j = \text{data}_{j1} \oplus \text{data}_{j2} \oplus \text{data}_{j3} \oplus \cdots \oplus \text{data}_{jn}$$

where $\oplus$ = exclusive-OR
$\quad\quad C_j$ = jth bit of the LRC
$\quad\text{data}_{ji}$ = jth bit of the ith byte in the n-byte message

The LRC is equivalent to an even parity calculated in the horizontal or longitudinal direction. The advantage of LRC is that if it is used with a regular even parity bit (vertical redundancy check [VRC]), it can detect all 1-, 2-, 3-bit errors and correct all 1-bit errors. The even parity bit is determined so that the total number of 1s in the data+parity is an even number. For example, consider the message "HELLO." The ASCII codes for these five characters are $48, $45, $4C, $4C, $4F. Table 14.2 shows the message in binary including both the VRC and LRC. Notice that the VRC as calculated on the LRC is that same value as the LRC calculated on the VRC. In other words, the bit in the upper right-hand corner of the figure is the exclusive-OR of all the other bits in the message. Since exclusive-OR is associative, it doesn't matter in what order the calculations are performed.

Table 14.2

Calculation of horizontal and vertical parity.

VRC	0	1	1	1	1	0
bit 7	0	0	0	0	0	0
bit 6	1	1	1	1	1	1
bit 5	0	0	0	0	0	0
bit 4	0	0	0	0	0	0
bit 3	1	0	1	1	1	0
bit 2	0	1	1	1	1	0
bit 1	0	0	0	0	1	1
bit 0	0	1	0	0	1	0
	'H'	'E'	'L'	'L'	'O'	LRC

A 1-bit error during transmission will cause both a LRC error and a VRC error. The LRC error specifies which bit got corrupted, and the VRC error identifies which byte. When this type of error occurs, rather than asking for a retransmission, the receiver can simply fix the error. If two bits are corrupted during transmission, the receiver will be able to detect the fact that an error has occurred but be unable to correct the error. More sophisticated schemes allow correction of multiple-bit transmission errors.

Observation: Noise in the communication channel is often correlated. In other words, the presence of one-bit reversal significantly affects the probability of errors in subsequent bits (e.g., the noise burst may affect many bits in a row).

Observation: Before designing an error detection/correction protocol, it is important to have a good understanding about the nature of the noise. A variety of schemes exist, some of which may be effective only for a narrow range of noise types.

Common error: An excellent error detection/correction scheme applied inappropriately to the wrong type of noise situation will yield ineffective performance.

DLE can also be used to create transparent messages, where each command code is preceded by a DLE. There are five types of *text/header* message formats (Figure 14.18).

Figure 14.18
Five different data messages used to transmit data.

SYN	SYN	SOH	*Header*	ETB	*bcc*		
SYN	SYN	SOH	*Header*	STX	*Text*	ETB	*bcc*
SYN	SYN	SOH	*Header*	STX	*Text*	ETX	*bcc*
SYN	SYN	STX	*Text*	ETB	*bcc*		
SYN	SYN	STX	*Text*	ETX	*bcc*		

One of the issues a network must handle is the retransmission of corrupted messages. Automatic repeat request (ARQ) includes a number of handshake techniques that handle message synchronization.

Stop-and-wait ARQ can be used on a half-duplex channel like BSC (Figure 14.19). The transmitting computer divides the overall message into smaller packets. The transmitter will

Figure 14.19
Flowcharts showing the transmit and receive operation of our simplified BSC protocol.

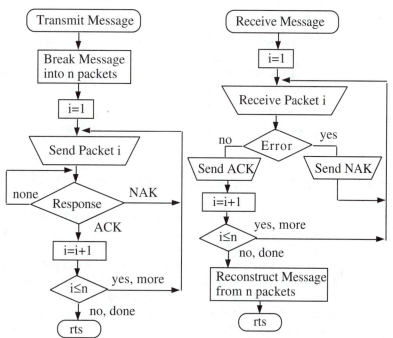

send the packets in sequence, waiting after each packet for an ACK or NAK. If an ACK is received, then the next message is sent. If a NAK is received, the packet is retransmitted. The receiver also accepts the packets in order. The block correction code is checked for each packet. ACK or NAK is sent back to the transmitter, signifying the status of the previous packet. The BSC protocol alternates ACK0 and ACK1 as a mechanism to handle either lost data or lost control packet.

14.6 IEEE488 General-Purpose Instrumentation Bus

Hewlett-Packard first defined this parallel communication standard in 1972. The IEEE488 bus is sometimes referred to as the Hewlett-Packard Instrumentation Bus (HPIB) or the General-Purpose Instrumentation Bus (GPIB.) Its purpose is to connect tabletop instruments, typically separated by less than 2 m. As is the goal of most networks, IEEE488 is intended to interconnect devices from different manufacturers. It implements an asynchronous handshake protocol; therefore it will operate with devices of varying speeds and will automatically improve performance as modules are upgraded with faster versions. IEEE488 is typically used for communications requiring bandwidths below 10 kbytes/s, but it will handle up to 1 Mbyte/s. It can handle 15 devices, but having eight or less will be more convenient. There are three types of modules, or capabilities, on the IEEE488 bus. A controller (usually attached to the computer) coordinates access and arbitrates conflicts. A talker is a device that can drive data onto the bus, and a listener is a device that can accept data from the bus. The bus supports a broadcast, which is when one talker sends information simultaneously to multiple listeners. The controller will select which device will talk and which device(s) will listen. The microcomputer usually includes all three (controller, talker, listener) capabilities, but the tabletop instruments may be

Talk only	can only transmit data
Listen only	can only receive data
Both talker and listener	can transmit or receive data

Observation: If a mixture of fast and slow modules exists on the same handshaked bus, then the speed of each data exchange will be determined by the speeds of the modules participating in the current data transfer and not affected by slower devices on the bus.

Observation: The performance of a handshaked interface will automatically improve when the speed of one or both modules is improved.

14.7 Small Computer Systems Interface

Another popular parallel communication standard is the SCSI. The initiator is the module that requests the data transfer (typically the I/O device), and the target is the module with which the initiator wishes to communicate (typically the computer). Figure 14.20 illustrates the handshake involved in the transfer of data from target to initiator. This is an example of an asynchronous handshake protocol. At time 1, the target drives the **C/D** line (command=1, data=0), the **I/O** line (1 means from target to initiator), and the data. Next (time 2), the target signifies that the **C/D, I/O,** and **Data** lines are valid by driving **REQ** low. Between times 2 and 3, the initiator captures the information from the **Data** lines. When the

Figure 14.20
SCSI data transfer from target to initiator.

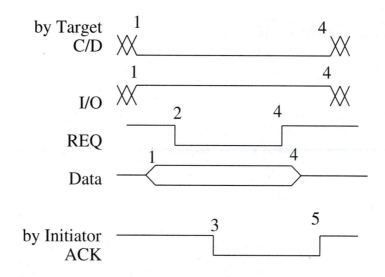

initiator no longer needs the data, it drives **ACK** low (time 3). Next (time 4), the target raises **REQ,** signifying that the **C/D, I/O,** and **Data** lines are no longer valid. The Initiator completes the transaction by raising **ACK** at time 5, which signifies it is ready for the next communication.

Figure 14.21 illustrates the handshake involved in the transfer of data from initiator to target. This is an example of an asynchronous handshake protocol. At time 1, the target drives the **C/D** line (command=1, data=0) and the **I/O** line (0 means from initiator to target). Next (time 2), the target signifies the **C/D** and **I/O** are valid by driving **REQ** low. Since **I/O** is low, it also means it wishes to receive data from the initiator. Between times 2 and 3, the initiator drives the information onto the **Data** lines. When the initiator knows the data are valid, it drives **ACK** low (time 3). Between times 3 and 4, the target captures the information from the data lines. When the target no longer needs the data, it drives **REQ** high (time 4). The rising of **REQ** also signifies that the **C/D** and **I/O** lines are no longer valid. The Initiator completes the transaction by raising **ACK** at time 5, which signifies it is ready for the next communication.

Figure 14.21
SCSI data transfer from initiator to target.

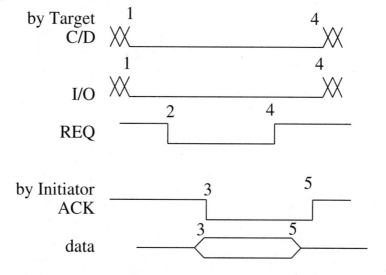

14.8   Modem Communications

14.8.1
FSK Modem

A *modem* is an electronic device that *MO*dulates and *DEM*odulates a communication signal. The transmitter of a FSK modem modulates the digital signals into frequency-encoded sine waves. The receiver demodulates the sine waves back into digital signals. The transmitting computer asynchronous serial port (SCI) converts the digital data into a RS232 timing serial signal. The Bell103 or V.21 standard achieves a data rate of 300 bits/s. The transmitting modem modulates the digital signal into 3.3-ms bursts of 1070- or 1270-Hz sounds. Figure 14.22 shows the encoded/decoded signals for the transmission of the ASCII character 'I', which is $49.

Figure 14.22 300
bits/sec FSK transmission
of the character 'I'
(originate to answer).

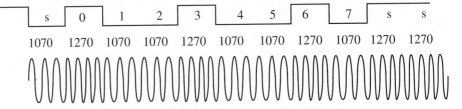

The transmitting bandpass filter guarantees there are no 2-kHz components in the output signal. The mixer adds the outgoing sound (1070 or 1270 Hz) with the incoming sounds on the telephone. The receiver bandpass filter will pass only the 1070/1270 Hz sounds to the demodulator. The receiver modem will demodulate the waveforms back into digital signals. If all goes well, the input to the receiving SCI, RxD, should be the same as the output from the transmitting SCI, TxD, only delayed in time. The software in the transmitting microcomputer should write data to the transmit serial data register (SCDR) when its TDRE=1. Eventually, that data will arrive in the receiver's receive serial data register (SCDR), setting the receiver's RDRF (Figure 14.23).

Figure 14.23
Block diagram of a FSK
modem communication
system, showing one
direction transfer.

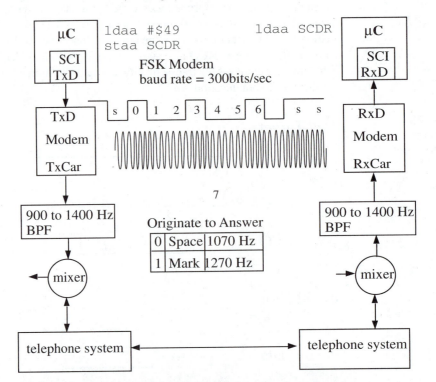

To implement a full-duplex channel, we use a different set of frequencies to simultaneously transmit data in the other direction. We define the *originate* modem as the device that places the telephone call. It dials the telephone, originating the communication. The *answer* modem is the device that answers the phone. The *carrier frequency* is defined as the average or midvalue frequency. In the FSK protocol, the two carrier frequencies are 1170 and 2125 Hz. By convention, the modem that initiates the connection uses the lower carrier frequency. Once a connection has occurred, full-duplex communication can occur. A logical true or Boolean 1 is encoded as a *mark,* while false (0) is defined as a *space.* Table 14.3 defines the full-duplex FSK protocol

Table 14.3
Frequencies used in a 300 bits/s FSK modem.

Direction	Space, false, /0	Mark, true, 1
originate to answer answer to originate	1070 Hz 2025 Hz	1270 Hz 2225 Hz

Figure 14.24 shows the encoded/decoded signals for the answer to originate transmission of a $49.

Figure 14.24
300 bits/s FSK transmission of the character 'I' (answer to originate).

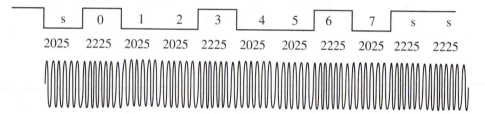

An important function of the bandpass filters is to separate the two pairs of frequencies so that transmission in one direction does not interfere with transmission in the other (Figure 14.25).

Figure 14.25
Block diagram of a FSK modem communication system, showing transfer in both directions.

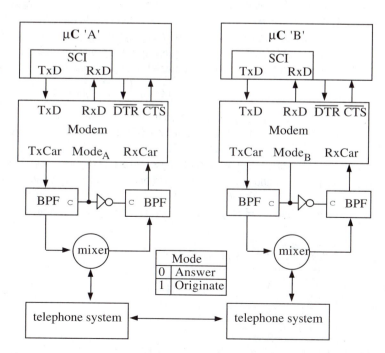

If we develop a modem interface that can both originate and answer telephone calls, then we will need bandpass filters that have digitally controlled center frequencies. We will build a bandpass filter that has the following digital control. c is a digital input from the interface to the bandpass filter (Table 14.4). If computer A originates the call, then $MODE_A = 1$, and $MODE_B = 0$. These values are reversed if computer B originates the communication.

Table 14.4

Filters used in a 300 bits/s FSK modem.

c	Low frequency	High frequency
0	1900	2400
1	900	1400

The bandwidth of the FSK protocol is limited by the frequency response of the telephone system. At 300 bits/s the four sounds have 3, 4, 6, 7 cycles per bit time. These differences allow for accurate demodulation. If we wished to increase the communication bandwidth by a factor of 10, we would have to increase the carrier frequencies by this factor as well. For example, at 3000 bits/s, we would need carrier frequencies of 11 and 21 kHz to maintain the desirable 3, 4, 6, 7 cycles per bit time property. Unfortunately, the telephone was designed so that humans could comprehend audio speech, and it will not pass 21-kHz sound waves.

14.8.2 Phase-Encoded Modems

There are three basic components of the wave that we can use to encode the information: frequency, phase, and amplitude. Each of these characteristics can be exploited to encode information. A *phase-shift keying* (PSK) protocol encodes the information as phase changes between the sounds.

A *baud* is defined as a pulse of sound. In the early days of modem communications, the FSK protocol encoded just 1 bit/baud. To increase the communication bandwidth without increasing the carrier frequency, multiple bits are encoded into each change in sound. The *baud rate* is the number of sounds (or sound changes) transmitted per second. With a FSK protocol, the baud rate equals the data rate. On the other hand, with the phase-encoded schemes, the transmission rate will usually be higher than the baud rate. This is a very confusing point because many people improperly interchange the terms baud rate, data rate, and bandwidth. If we encode n bits of data per baud, then

$$\text{data rate} = n \cdot \text{baud rate}$$

For example, the 2400 bits/s International Telegraph and Telephone Consultative Committee (CCITT) protocol uses an 1800-Hz carrier and a 1200 bits/s baud rate. It encodes 2 bits/baud as shown in Table 14.5.

Encoding is the process of transmitting one or more bits of information with each baud. To improve bandwidth we can also implement data compression. Most data contain repetitive or redundant information. With data compression, the information (the message, digitally encoded sound, or photograph) is reformatted so that it occupies fewer bytes. The compressed

Table 14.5

Phase shifts used in a 1200 bits/s PSK modem.

Data	CCITT phase
00	0°
01	90°
11	180°
10	270°

Figure 14.26
Compressed data allow information to be transferred with few data points.

data are transmitted across the channel and expanded on the other side. For example, a sampled waveform can be compressed into a sequence of linear line segments (Figure 14.26).

For modem communications, we use Huffman or Lempel-Ziv encoding. We can evaluate a compression/decompression algorithm by three factors. The *compression ratio* is defined as the ratio of the number of original bytes to the number of compressed bytes. If an algorithm achieves a compression ratio of 4:1, then a 1000-byte message can be compressed into 250 bytes. Since only the compressed message is transmitted, the effective bandwidth of the communication system is four times faster than the physical bandwidth of the channel. The second factor to consider is the speed of the compression/decompression algorithm. For most modem applications, we require the compression/decompression functions to proceed in real time so that the operation occurs transparent to the user. The last and most important factor is accuracy. In some compression/decompression applications, such as photography or sound recordings, we can tolerate some imperfections or errors. In most modem applications however, we expect the received message to exactly equal the message transmitted.

Consider the following example of a CCITT 2400 bits/s PSK transmission. To transmit the ASCII 'A' ($41=%01000001) with one start bit, 8-bit data, even parity, and two stop bits, we transmit the sequence start=0, data=10000010, parity=0, and stop=11. Remember the data bits are transmitted starting with the least significant bit. The sequence is divided into 2-bit pieces: 01, 00, 00, 01, 00, 11 and encoded as phase shifts: 90°, 0°, 0°, 90°, 0°, 180°. With a carrier frequency of 1800 Hz, we transmit 1.5 cycles and then phase-shift the signal. The baud rate is 1200 because we produce 1200 sound changes per second. The data rate is 2400 bits/s, twice the baud rate, because each sound change encodes 2 bits of data (Figure 14.27).

Figure 14.27
Waveform generated by a phase-encoded modem.

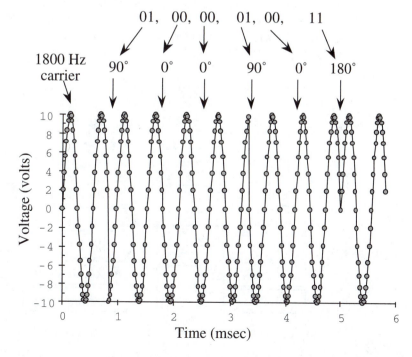

14.8.3 Quadrature Amplitude Modems

In a quadrature amplitude modem (QAM), we use both the phase and amplitude to encode up to 6 bits onto each baud. For standard QAM, only 4 bits is used to encode data. The *trellis-coded quadrature amplitude modem* (TCQAM or TCM) includes error-correcting signal processes so that all 6 bits can be reliably transmitted (Table 14.6).

Table 14.6
High-speed modems transmit many bits per sound.

Standard	Baud rate	bits/baud	Data rate (bit/s)	Modulation
V.21 Bell103	300	1	300	FSK
V.22 Bell212A	600	2	1,200	PSK
V.26	1200	2	2,400	PSK
V.27	1600	3	4,800	PSK
V.32	2400	4	9,600	QAM
V.33	2400	6	14,440	TCQAM

14.9   X-10 Protocol

X-10 protocol is a simple low-bandwidth communication protocol that can be used to control multiple devices within a single building. The Sears Home Control System and the Radio Shack Plug'n Power System first introduced the code format in 1978. Since then, many home automation and home security systems have been built around this standard. In this section we will develop X-10–based communication systems based on the PL513 and TW523 modules available from X-10 Powerhouse. An X-10 transmitter module sends an 11-bit packet to an X-10 receiver module over the existing power line wires. The key advantage of X-10 is its ability to use existing wires in the building and the wide range of available X-10 interfaces that can be installed. A communication system built with the PL513 allows the microcomputer to send commands that can control remote devices. With a TW523 interface the computer can both send and receive packets. So in addition to the ability to control remote devices, we can also receive information from remote locations. Figure 14.28 depicts a communication system with a single computer controller, but we can add multiple computer interfaces and use the X-10 network to communicate between them.

Figure 14.28
Communication systems built around the PL513 and TW523 modules.

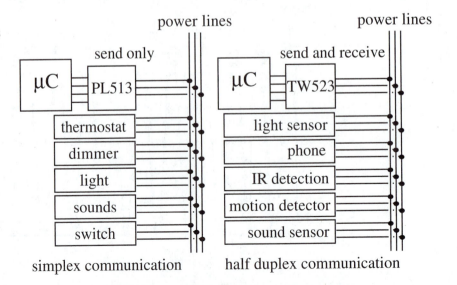

Since the protocol encodes the binary information as 1-ms-long 120-kHz bursts modulated at the zero crossings of the 60-Hz power line, the bandwidth will be less than 60 bits/s. For many applications, however, this is adequate. Figure 14.29 illustrates the transmission of a logic 1, which is signified by the presence of the 120-kHz pulses in the first half cycle of the 60-Hz power wave. The 120-kHz pulses are repeated at 2.778 and 5.556 ms so that communication can occur between modules that reside on any of the phases in a three-phase distribution system. A binary 0 is represented by the presence of the 120-kHz pulses in the second half cycle of the 60-Hz power wave. A special start code is used to synchronize the transmitter and receiver modules.

Figure 14.29

X-10 data is encoded as 120-kHz bursts on top of the 60-Hz power line.

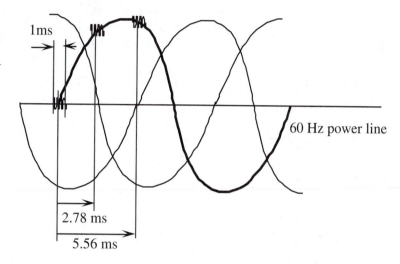

Each packet requires 11 power line cycles to transmit. The first two cycles are the Start Code (1110), the next four cycles represent the House Code, and the last five cycles include the Key Code. The House Code is used to address the device to which the packet is being sent. The X-10 protocol uses letters A through P to specify the House Codes (Table 14.7).

Table 14.7

X-10 house codes.

House code	H1 H2 H4 H8
A	0 1 1 0
B	1 1 1 0
C	0 0 1 0
D	1 0 1 0
E	0 0 0 1
F	1 0 0 1
G	0 1 0 1
H	1 1 0 1
I	0 1 1 1
J	1 1 1 1
K	0 0 1 1
L	1 0 1 1
M	0 0 0 0
N	1 0 0 0
O	0 1 0 0
P	1 1 0 0

The X-10 protocol allows for the transmission of a multiple-bit packet over a power grid. Each X-10 packet is defined as 22 bits long. The first 4 bits are a unique "start" code, represented as 1110. The next 8 bits in the X-10 packet are the 4-bit house code, sent with each bit followed by its complement. The last 10 bits in the X-10 packet are the 5-bit function or unit code, also sent with each bit followed by its complement. Thus the X-10 complimentary rule dictates that for both house code and unit/function code, a logic 1 is represented as 10, and a logic 0 is represented as 01. The X-10 protocol also insists that each packet must be transmitted twice, with each transmission separated by 3 (60-Hz) cycles of no-transmission. To activate a specific unit, an X-10 packet must first be sent that contains the house code and the unit code, followed by another X-10 packet that contains the same house code and the function code. Note that each X-10 packet must be sent twice to send house A, unit 1, function on. This complete package consists of a total of $22+3+22+3+22+3+22 = 97$ bits. The transmission looks like this:

Start	(4 bits)
House A	(8 bits)
Unit 1	(10 bits)
3-cycle no-transmission	(6 bits)
Start	(4 bits)
House A	(8 bits)
Unit 1	(10 bits)
3-cycle no-transmission	(6 bits)
Start	(4 bits)
House A	(8 bits)
Function on	(10 bits)
3-cycle no-transmission	(6 bits)
Start	(4 bits)
House A	(8 bits)
Function on	(10 bits)

Note that to perform the same function on multiple units in the same house code, the unit codes could be sent next to each other, followed by the function code. For example, to turn on both A1 and A2, send house A, unit 1, house A, unit 2, house A, function on. Also, functions that are addressed to all receivers, such as all units on, do not need the unit code to precede them. Finally, a special note on dim and bright functions: They do not need the 3-cycle delay. The X-10 protocol has 32 Key Codes to specify the data (Number Code) or command (Function Code) information (Table 14.8).

The hail request is sent to determine if there are any other X-10 transmitters on the system. This allows a transmitter at start-up to assign itself a unique house code. There are 5 bits or 32 levels specified in the preset dim command. In the preset dim command, the D8 bit represents the most significant bit, and H1, H2, H4, H8 are the least significant bits. For example, the H1,H2,H4,H8,D1,D2,D4,D8,D16 code of **011110111** represents the dim level **10111,** or 23/32. Normally there is a 3-cycle gap between packets, except for "dim" and "bright" that can be sent one right after another without gaps.

Except for the start code, the X-10 protocol uses complementary logic in the first and second halves of the 60-Hz cycles. Notice that the logic 1 is encoded by pulses followed by

Table 14.8
X-10 number and
function codes.

Number code	D1 D2 D4 D8 D16	Function code	D1 D2 D4 D8 D16
1	0 1 1 0 0	All lights off	0 0 0 0 1
2	1 1 1 0 0	All lights on	0 0 0 1 1
3	0 0 1 0 0	On	0 0 1 0 1
4	1 0 1 0 0	Off	0 0 1 1 1
5	0 0 0 1 0	Dim	0 1 0 0 1
6	1 0 0 1 0	Bright	0 1 0 1 1
7	0 1 0 1 0	All lights off	0 1 1 0 1
8	1 1 0 1 0	Extended code	0 1 1 1 1
9	0 1 1 1 0	Hail request	1 0 0 0 1
10	1 1 1 1 0	Hail acknowledge	1 0 0 1 1
11	0 0 1 1 0	Preset dim 0	1 0 1 0 1
12	1 0 1 1 0	Preset dim 1	1 0 1 1 1
13	0 0 0 0 0	Extended data	1 1 0 0 1
14	1 0 0 0 0	Status=on	1 1 0 1 1
15	0 1 0 0 0	Status=off	1 1 1 0 1
16	1 1 0 0 0	Status request	1 1 1 1 1

no pulses, and the logic 0 is encoded by no pulses followed by pulses. As an example, if the computer wished to turn on light "P," it would send House Code "P" (1100) with Function Code "on" (00101). Figure 14.30 illustrates this transmission.

Figure 14.30
X-10 transmission used
to turn on module "P".

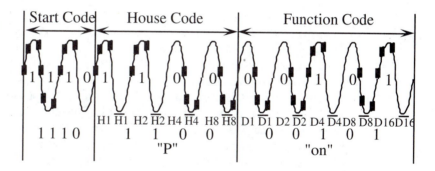

The extended code and extended data formats allow for packets to be longer than 9 bits. There are no standard protocols for these extended formats, except the bit stream must follow the first packet continuously without gaps. For example, we could send the first 8-bit byte as the data count, followed by the raw data. Unfortunately, the PL513 and TW523 interfaces can send extended formats, but neither one can receive extended formats.

When using the PL513 and/or TW523 interfaces, it is necessary to repeat all transmissions. To allow the receiver time to recover and reinitialize, most receivers require a 3 power cycle gap between packets. Figure 14.31 illustrates the waveforms used to first turn on module "P" then turn off module "N".

Figure 14.31
X-10 transmissions to
first turn on module "P"
then turn off module
"N".

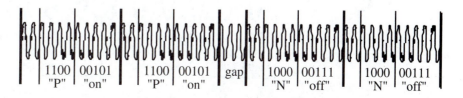

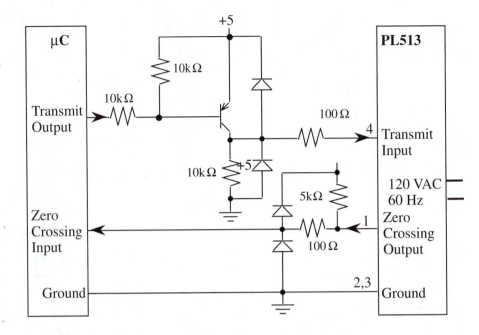

If all we require is the ability to transmit packets, then the PL513 interface can be used (Figure 14.32).

The zero-crossing signal from the PL513 is a square wave that the system uses to synchronize the 120-kHz bursts to the 60-Hz power wave. The microcomputer is responsible for placing a 1-ms (0.95 to +1.1 ms) positive logic within 50 µs of the rising and/or falling edge of the zero-crossing trigger. Figure 14.33 illustrates the signals that the microcomputer must generate to turn on module "P".

We will connect the zero-crossing input to a microcomputer pin that can generate interrupts on both the rising and falling edges. Either input capture or key wake-up mechanisms can be used. The **Transmit Output** signal can be connected to a simple output pin or to an output compare signal. To transmit a packet, the main program, foreground thread, will place 50 bits into a BitFifo and arm the zero-crossing interrupt signal so that the first interrupt occurs on the rising edge. At each zero crossing, the interrupt handler will get 1

Figure 14.33
Microcomputer signals
used to turn on module
"P".

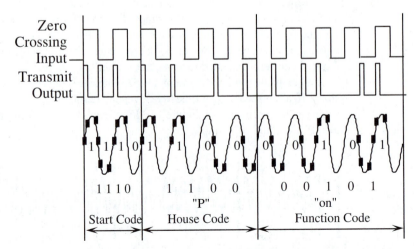

bit from the BitFifo and issue a 1-ms pulse if the bit is 1. When the BitFifo is empty, the interrupt handler can disarm itself. The C code in Program 14.7 can be used to implement the transmit BitFifo. This is a pointer implementation of the BitFifo. These routines can be used

```c
unsigned int TxFifo[FifoSize]; // storage for Bit Stream
struct BitPointer{
  unsigned int Mask;       // 0x8000, 0x4000,...,2,1
  unsigned int *WPt;};     // Pointer to word containing bit
typedef struct BitPointer BitPointerType;
BitPointerType PutTxPt;    // Pointer of where to put next
BitPointerType GetTxPt;    // Pointer of where to get next
/* BitFIFO is empty if PutTxPt==GetTxPt      */
/* BitFIFO is full  if PutTxPt+1==GeTxtPt after wrap  */
int SameBit(BitPointerType p1, BitPointerType p2){
  if((p1.WPt==p2.WPt)&&(p1.Mask==p2.Mask)) return(1); //yes
  return(0);} // no
//*****************InitTxBit*********************
// initializes the BitFifo to be empty
void InitTxBit(void) {
  PutTxPt.Mask=GetTxPt.Mask=0x8000;
  PutTxPt.WPt=GetTxPt.WPt=&TxFifo[0];}  /* Empty when PutTxPt==GetTxPt */
//*******************PutTxBit*********************
// returns TRUE=1 if successful, FALSE=0 if full and data not saved
// input is boolean FALSE if data==0
int PutTxBit (int data) {  BitPointerType TempPutPt;
  TempPutPt=PutTxPt;
  TempPutPt.Mask=TempPutPt.Mask>>1;
  if(TempPutPt.Mask==0) {
     TempPutPt.Mask=0x8000;
     ++TempPutPt.WPt;   // next word address
     if((TempPutPt.WPt)==&TxFifo[FifoSize]) TempPutPt.WPt=&TxFifo[0];} // wrap
  if (SameBit(TempPutPt,GetTxPt))
     return(0);         /* Failed, fifo was full */
  else {
     if(data)
       (*PutTxPt.WPt) |= PutTxPt.Mask;  // set bit
     else
       (*PutTxPt.WPt)&= ~PutTxPt.Mask;  // clear bit
     PutTxPt=TempPutPt;
     return(1);}}
//*******************GetTxBit*********************
// returns TRUE=1 if successful, FALSE=0 if empty and data not removed
// output is boolean 0 means FALSE, nonzero is true
int GetTxBit (unsigned int *datapt) {
   if (SameBit(PutTxPt,GeTxtPt))
      return(0);        /* Failed, fifo was empty */
   else {
      *datapt=(*GetTxPt.WPt)&GetTxPt.Mask;
      GetTxPt.Mask=GetTxPt.Mask>>1;
      if(GetTxPt.Mask==0) {
        GetTxPt.Mask=0x8000;
        ++GetTxPt.WPt;  // next word
        if((GetTxPt.WPt)==&TxFifo[FifoSize]) GetTxPt.WPt=&TxFifo[0];} // wrap
      return(1); }}
```

Program 14.7 Bit FIFO routines.

Figure 14.34
Input capture interrupt
occurs on both the
rising/falling edges of
the zero-crossing trigger.

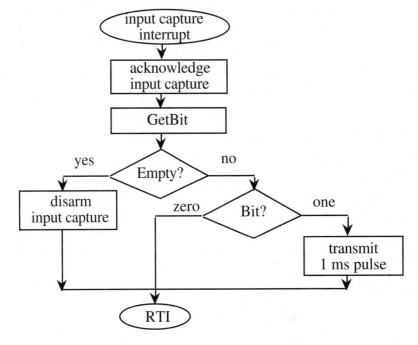

Figure 14.35
Half-duplex interface
between the
microcomputer and the
power system.

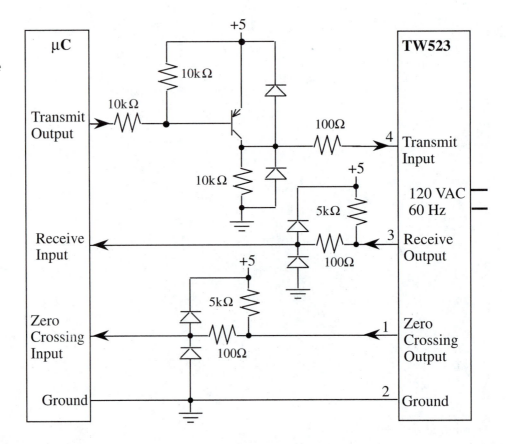

to save (PutTxBit) and recall (GetTxBit) binary data 1-bit at a time (bit streams). Information is saved/recalled in a FIFO manner. FifoSize is the number of 16-bit words in the FIFO. The FIFO is full when it has 16*FifoSize-1 bits.

To implement both transmission and receiving we will need two BitFifos. Assume we have a second BitFifo with functions GetRxBit and PutRxBit that manipulate the FIFO for the receive function (Figure 14.34). To both transmit and receive we will build the X-10 interface using the TW523 module (Figure 14.35). Again we will connect the zero-crossing input to a microcomputer pin that can generate interrupts on both the rising and falling edges. After each zero-crossing edge the software will interrogate the receive input to determine if a 120-kHz burst is present. Once the software detects a burst, it can store it in the receive BitFifo, then for the next 21 interrupts it will put a 1 if a burst is present or a 0 if no burst is detected. The main program can get bits from the receive BitFifo and decodes the transmission (Figure 14.36).

Observation: When a repeated packet is transmitted using the TW523 interface, the second packet is visible on the Receive Input line.

Observation: The receiver signals can be monitored during a TW523 transmission to detect noisy line conditions or the occurrence of a packet collision.

Figure 14.36
Flowchart for receiving X-10 communication.

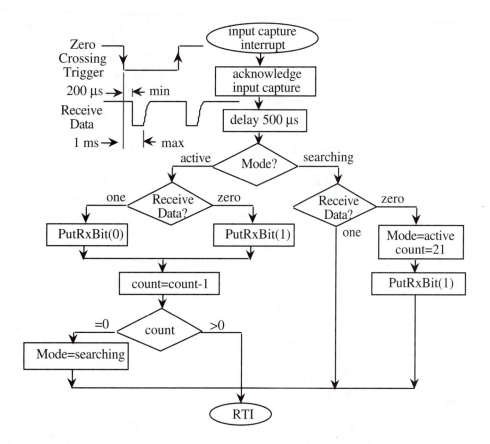

14.10   Glossary

answer modem The device that receives the telephone call.

bandwidth The number of information bits per time that are transmitted (same as bit rate and throughput).

baud rate In general the baud rate is the total number of bits (information, overhead, and idle) per time that are transmitted; in a modem application it is the total number of sounds per time are transmitted.

bit rate The number of information bits per time that are transmitted (same as bandwidth and throughput).

block correction code (BCC) A code (e.g., horizontal parity) attached to the end of a message used to detect and correct transmission errors.

carrier frequency The average or midvalue sound frequency in the modem.

channel The hardware that allows bits communication to occur.

checksum Simple sum of the data, usually in finite precision (8, 16, 24 bits).

compression ratio The ratio of the number of original bytes to the number of compressed bytes.

frame A complete and distinct packet of bits.

frequency shift keying (FSK) A modem that modulates the digital signals into frequency-encoded sine waves.

hold time When latching data into a register with a clock, it is the time after an edge that the data must continue to be valid (see setup time).

horizontal parity Parity calculated across the entire message on a bit-by-bit basis (e.g., the horizontal parity bit 0 is the parity calculated on all the bit 0s of the entire message, can be even or odd parity).

mark True, logic 1 (see space).

modem An electronic device that *MO*dulates and *DEM*odulates a communication signal.

originate modem The device that places the telephone call.

phase shift keying (PSK) A protocol that encodes the information as phase changes between the sounds.

quadrature amplitude modem (QAM) A protocol that uses both the phase and amplitude to encode up to 6 bits onto each baud.

setup time When latching data into a register with a clock, it is the time before an edge that the input must be valid (see hold time).

Small Computer Systems Interface (SCSI) A handshaking parallel data communication channel used to connect high-speed high-volume I/O devices to the computer.

space False, logic 0 (see mark).

throughput The number of information bits per time that are transmitted (same as bit rate and bandwidth).

vertical parity The normal parity bit calculated on each individual frame; can be even or odd parity.

14.11   Exercises

14.1 The objective of this exercise is to design a communication network using four single-chip microcomputers. The microcomputers are connected together by fiber-optic cables. The fiber-optic channel is virtually noise-free. The asynchronous simplex serial communication channels link the four computers in a circle. Each 8-bit message consists of a 2-bit source address, a 2-bit destination address, and 4 bits of data. The message data structure has three fields. In memory, messages are stored as 3 bytes, but when transmitted they are compressed into 1 byte.

```
struct message{
    unsigned char source;         / 0,1,2,3 computer that sent the message
    unsigned char destination;    / 0,1,2,3 computer that receives the message
    unsigned char information;}; / 0-15 data part of the message
typedef struct message messageType;
```

Seven steps occur as computer 0 sends "10" to computer 2:

The main program on computer 0 creates a message and passes it to its interrupt software—for example,

```
int SayHi(void){ messagetype Hello;
    Hello.destination=2;
    Hello.information=10;
    return(Send(&Hello)); } // Send knows the source is 0
```

The interrupt software on computer 0 transmits the message to computer 1

The interrupt software on computer 1 receives the message and notices the destination address does not match its computer number (DIP switch)

The interrupt software on computer 1 retransmits the message to computer 2

The interrupt software on computer 2 receives the message and notices the destination address does match its computer number (DIP switch)

The interrupt software on computer 2 passes the message to the main program on computer 2

The main program on computer 2 accepts the message—for example,

```
void WaitFor(void){ messagetype msg;
    while(Accept(&msg)); } // waits for message
    printf("%d %d\n",msg.source,msg.information);}
```

Each microcomputer has a 2-bit DIP switch specifying its computer number. You may assume that each computer has a unique 2-bit address. The electric switch being on (0Ω) signifies a zero address bit, and the electric switch being off (open) signifies a one address bit (e.g., both DIP switches in computer 0 are on (0Ω)) (Figure 14.37).

Figure 14.37
A ring network.

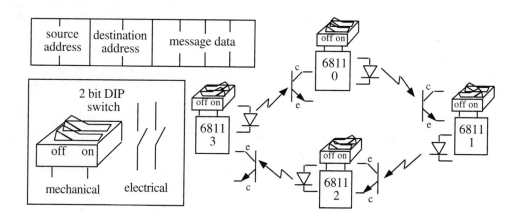

Each of the four fiber-optic cables contains a LED and a light-sensitive transistor. The LED operating point (shines light into the cable) is V = 2.5 V, I = 8 mA. When light is in the cable, the transistor voltage V_{ce} is less than 0.5 V for any I_{ce} less than 5 mA. If the LED current is zero, then the cable is dark and the transistor is off (I_{ce}=0). The fiber-optic cable can handle baud rates up to 2 Mb/s. Choose the fastest baud rate for this interrupt-driven asynchronous serial communication. You will not write the main program, but you will write subroutines executed by the main program to SEND and ACCEPT messages. Your network software must be interrupt-driven.

a) Show the hardware network interface for one single-chip microcomputer system. Except for the setting of the DIP switch, the network hardware/software is the same for all four computers. Please label chip numbers, but not pin numbers. Be careful how the systems are grounded to each other. If this is computer n, show the outgoing fiber-optic cable to computer n + 1 (diode) and the incoming fiber-optic cable from computer n − 1 (transistor).

b) Show all the information that will be located in the RAM. The globals in RAM are unspecified on power up, so you must initialize them in the ritual in part c. Give careful thought as to the most appropriate data structures.

c) Give the `Ritual`, `Send`, and `Accept` functions along with the interrupt handlers required. You may use FIFO queues by giving the global data structure definitions and the function prototypes without showing the FIFO function definitions. The same network communication software will be placed in each system (you can read the computer address from the DIP switches connected to an input port). The `Ritual` will initialize all data structures and configure the microcomputer. You will write two functions (`Send`, `Accept`) that will be called by the main program (you don't write the main program). The subroutine `Send` will initiate a send message communication. The prototype is shown above. `Send` returns right away, with the result equal to 0 if all is well so far but returns a 1 if the message cannot be sent (e.g., FIFO full). Use comments to explain how and why an error occurred. The subroutine `Accept` will check to see if a message has been received for this computer. The return value (RegD) is 0 if a message is ready and 1 if no incoming message is currently ready. If a message is available, then it returns by reference the source, destination, and information fields as illustrated in the above prototype. Whether or not an incoming message is available, `Accept` returns right away. The interrupt processes will perform the serial I/O operations. You need not poll for 0s and 1s. *Good interrupt software contains no gadfly loops.*

d) What is the bandwidth of your network? Specify units and show the equation that justifies your answer.

14.2 The objective of this exercise is to design a microcomputer-based IEEE488 to RS422 simplex converter. Your single-chip microcomputer system will perform handshaked parallel input from a IEEE488 output device, buffer it in an internal FIFO, then transmit the data on an asynchronous RS422 simplex serial channel. The serial protocol is 8-bit data, one stop, no parity, and 300 bits/s baud rate. The input bandwidth can vary from 0 to 1000 bytes/s, with an average of 10 bytes/s. The internal FIFO will allow temporarily the input bandwidth to exceed the output bandwidth. The IEEE488 sequence of events to transmit 1 byte is:

1. Your microcomputer signals it is ready for the next data by making RFD=1;
2. Eventually the IEEE488 output device will provide the 8-bit data and make $\overline{\text{DAV}}$ =0;
3. Your microcomputer makes RFD=0;
4. Your microcomputer should read the data, put it into the FIFO, then make DAC=1;
5. Eventually the IEEE488 output device will make $\overline{\text{DAV}}$ =1 and remove the data;
6. Your microcomputer makes DAC=0.

We can simplify this dedicated interface by connecting the $\overline{\text{DAV}}$ to RFD, skipping steps 1 and 3:

2. Eventually the IEEE488 output device will provide the 8-bit data and make $\overline{\text{DAV}}$ =0;
4. Your microcomputer should read the data, put it into the FIFO, then make DAC=1;
5. Eventually the IEEE488 output device will make $\overline{\text{DAV}}$ =1 and remove the data;
6. Your microcomputer makes DAC=0.

Figure 14.38 An IEEE488 to RS422 interface.

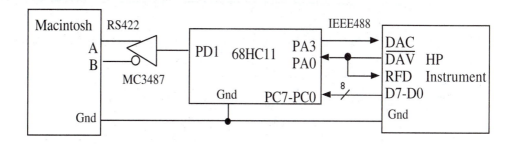

a) What should your microcomputer do when the FIFO is full? Why?
b) Show all the software including the interrupt handlers, rituals, main program, reset vector, interrupt vector, and ORG statements that place the code into the appropriate addresses on the *single-chip* microcomputer. You may use a FIFO without showing its implementation if you explain where in memory the data, pointers, and subroutines reside.

14.3 The objective of this exercise is to convert data from a parallel SCSI bus to serial RS232 using a single-chip 6811 microcomputer. The baud rate is 2400 bits/s. The RS232 protocol is 8-bit, no parity, and two stop bits. Assume a simple single-initiator/ single-target SCSI system and ignore initialization. The microcomputer acts as the only target device accepting 8-bit data from the Initiator. Your microcomputer will handshake (REQ, ACK) each data byte, buffer it for maximum bandwidth, and output data as fast as possible on the SCI serial output. The 6811 hardware is shown in Figure 14.39. For a 6812, you may use any available ports.

Figure 14.39 A SCSI to RS232 interface.

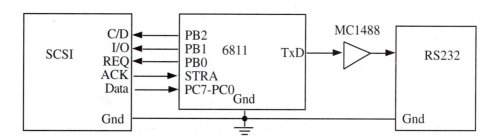

Show all the software, including the main program, interrupt vectors, reset vectors, data structures, interrupt handlers, and I/O. This is a single chip, so the dedicated software must be segmented into RAM (0 to $FF) and ROM ($E000-$FFFF). If you use C, explain the segmentation required to implement single-chip mode. Comments will be graded.

14.4 One of the problems with the X-10 interfaces shown in this chapter is that the digital ground is connected to the AC power line.

   a) Design the interface that uses two 6N139s to isolate the simplex PL513 interface. You will have to use a separate isolated power source different from the supply that powers the microcomputer.

   b) Design the interface that uses three 6N139s to isolate the half-duplex TW523 interface. You will have to use a separate isolated power source different from the supply that powers the microcomputer.

14.5 The objective of this exercise is to develop a message-passing facility that spans across two computers (Figure 14.40). A fully interlocked synchronization method will be implemented. The message is a simple 8-bit byte that is passed from the output port of computer 1 to the input port of computer 2. The following hardware connections are fixed. You may assume both computers initialize together, where computer 1 initializes its PT7 to 0 and computer 2 initializes its PT6 to 0. On the 6811 you may use the timer features of Port A and Port C for the data.

Figure 14.40
A handshaking parallel port interface.

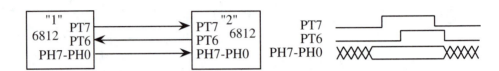

When computer 1 wishes to transmit to computer 2, it puts the data first on its Port H outputs. Next, computer 1 makes a rising edge on PT7, signifying that new data are available. Computer 1 will then wait for a rising edge on PT6, signifying that the data have been accepted. Next, computer 1 makes its PT7 low. Last, it waits for a low signal on its PT6 input line.

   a) Write the software for the transmitting computer 1. The transmission will occur in the background using input capture interrupts. You may use a FIFO queue without writing the three routines: `InitFifo`, `PutFifo`, and `GetFifo`. You may ignore FIFO full errors. There are three routines you will write.

    1. `void Ritual(void)` that clears the FIFO, makes the PT7 output low, and arms/enables the input capture interrupts as appropriate;

    2. `void PutMsg(unsigned char data)` function that is called by the main program (that you do not write); if a current message is in progress, then these data are entered in the FIFO; if there is no current message in progress, then one is started with the new data;

    3. `void IC6Handler(void)` interrupt handler called by the input capture on PT6.

When computer 2 wishes to receive from computer 1, it first waits for a rising edge on its PT7 input. Then it gets the data from its Port H inputs. Next, computer 2 makes a rising edge on its PT6 output, signifying that the data has been accepted. Computer 2 will then wait for a falling edge on its PT7 input. Last, it makes a low signal on its PT6 line.

   b) Write the software for the receiving computer 2. The reception will occur in the background using input capture interrupts. You may use a FIFO queue without writing the three routines: `InitFifo`, `PutFifo`, and `GetFifo`. You may ignore FIFO-full errors. There are three routines you will write

    1. `void Ritual(void)` that clears the FIFO, makes the PT6 output low, and arms/enables the input capture interrupts as appropriate;

    2. `unsigned char data GetMsg(void)` function that is called by the main program (that you do not write); if the FIFO is empty, then this routine will call the `get()` routine over and over until data are available;

    3. `void IC7Handler(void)` interrupt handler called by the input capture on PT7.

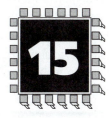

15 Digital Filters

Chapter 15 objectives are to:

❏ Introduce basic principles involved in digital filtering
❏ Define the Z transform and use it to analyze digital filters
❏ Design simple low-pass, frequency-reject, and high digital filters
❏ Discuss the effect of non-real-time sampling on digital filter error
❏ Develop digital filter implementations

The goal of this chapter is to provide a brief introduction to digital signal processing (DSP). DSP includes a wide range of operations such as digital filtering, event detection, frequency spectrum analysis, and signal compression/decompression. Similar to the goal of analog filtering, a digital filter will be used to improve the S/N ratio in our data. The difference is that a digital filter is performed in software on the digital data sampled by the ADC. The particular problem addressed in several ways in this chapter is removing 60-Hz noise from the signal. Like the control systems and communication systems discussed in the last two chapters, we will provide just a brief discussion to the richly developed discipline of DSP. Again, this chapter focuses mostly on the implementation on the embedded microcomputer. Event detection is the process of identifying the presence or absence of particular patterns in our data. Examples of this type of processing include optical character readers, waveform classification, sonar echo detection, infant apnea monitors, heart arrhythmia detectors, and burglar alarms. Frequency spectrum analysis requires the calculation of the discrete Fourier transform (DFT)[1]. Like the regular Fourier transform, the DFT converts a time-dependent signal into the frequency domain. The difference between a regular Fourier transform and the DFT is that the DFT performs the conversion on a finite number of discrete time digital samples to give a finite number of points at discrete frequencies. Data compression and decompression are important aspects in high-speed communication systems. Although we will not specifically address the problems of event detection, DFT, and compression/decompression in this book,

---

[1]A fast algorithm to calculate the DFT is called the fast Fourier transform (FFT).

these DSP operations are implemented using similar techniques as the digital filters that are presented in this chapter. Our goal for this chapter is to demonstrate that fairly powerful digital signal-processing techniques can be implemented on computers of modest performance like the 6811 and 6812. The limitation of single-chip computers like the 6811 and 6812 is not complexity but rather execution speed. If you have a DAS or control system with unwanted 60-Hz noise, then the techniques described in this chapter provide an effective means for removing this noise.

15.1   Basic Principles

The objective of this section is to introduce simple digital filters. Let $x_c(t)$ be the continuous analog signal to be digitized. $x_c(t)$ is the analog input to the ADC. If f_s is the sample rate, then the computer samples the ADC every T seconds ($T = 1/f_s$). Let . . . , $x(n)$ . . . , be the ADC output sequence, where

$$x(n) = x_c(nT) \qquad \text{with } -\infty < n < +\infty. \tag{1}$$

There are two types of approximations associated with the sampling process. Because of the finite precision of the ADC, amplitude errors occur when the continuous signal $x_c(t)$ is sampled to obtain the digital sequence $x(n)$. The second type of error occurs because of the finite sampling frequency. The Nyquist theorem states that the digital sequence $x(n)$ properly represents the DC to $(\frac{1}{2})f_s$ frequency components of the original signal $x_c(t)$. Two important assumptions are necessary to make when using digital signal processing:

1. We assume the signal has been sampled at a fixed and known rate f_s
2. We assume aliasing has not occurred

We can guarantee the first assumption by using a hardware clock to start the ADC at a fixed and known rate. A less expensive but not as reliable method is to implement the sampling routine as a high-priority periodic interrupt process. By establishing a high priority of the interrupt handler, we can place an upper bound on the interrupt latency, guaranteeing that ADC sampling is occurring at an almost fixed and known rate. We can observe the ADC input with a spectrum analyzer to prove there are no significant signal components above $(\frac{1}{2})f_s$. "No significant signal components" can be defined as either ADC input voltage $|Z|$ less than the ADC resolution Δz or less than $(\frac{1}{2})\Delta z$.

$$|Z|_f \geq (\frac{1}{2})_{f_s} < \Delta z \qquad \text{or} \qquad |Z|_f \geq (\frac{1}{2})_{f_s} < \frac{1}{2}\,\Delta z$$

A *causal* digital filter calculates $y(n)$ from $y(n-1)$, $y(n-2)$, . . . and $x(n)$, $x(n-1)$, $x(n-2)$ . . . . Simply put, a causal filter cannot have a nonzero output until it is given a nonzero input. The output of a causal filter, $y(n)$, cannot depend on future data [e.g., $y(n+1)$, $x(n+1)$].

A *linear* filter is constructed from a linear equation. A *nonlinear* filter is constructed from a nonlinear equation. One nonlinear filter is the median. To calculate the median of three numbers, one first sorts the numbers according to magnitude, then chooses the middle value.

A *finite-impulse response* (FIR) filter relates $y(n)$ only in terms of $x(n)$, $x(n-1)$, $x(n-2)$ . . . . In the next section we will determine the gain and phase response of four simple digital filters. These examples are presented in Equations 2 to 5. Equations 2 to 4 describe simple FIR filters:

$$y(n) = \frac{x(n) + x(n-1)}{2} \tag{2}$$

$$y(n) = \frac{x(n) + x(n-1) + x(n-2) + x(n-3) + x(n-4) + x(n-5)}{6} \tag{3}$$

$$y(n) = \frac{x(n) + x(n-3)}{2} \tag{4}$$

An **infinite-impulse response** (IIR) filter relates $y(n)$ in terms of both $x(n)$, $x(n-1)$, ..., and $y(n-1)$, $y(n-2)$ .... Equation 5 describes a simple IIR filter:

$$y(n) = \frac{y(n-1) + x(n)}{2} \tag{5}$$

One way to analyze linear filters is the Z transform. Equation 6 gives the definition of the Z transform:

$$X(z) = Z[x(n)] \equiv \sum_{n=-\infty}^{\infty} x(n)z^{-n} \tag{6}$$

The Z transform is similar to other transforms. In particular, consider the Laplace transform, which converts a continuous time-domain signal $x(t)$ into the frequency domain $X(s)$. In the same manner, the Z transform converts a discrete time sequence $x(n)$ into the frequency domain $X(z)$ (Figure 15.1).

Figure 15.1
A transform is used to study a signal in the frequency domain.

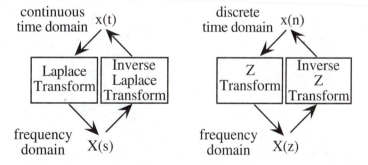

The input to both the Laplace and Z transforms are infinite time signals, having values at times from $-\infty$ to $+\infty$. The frequency parameters **s** and **z** are complex numbers, having a real and imaginary part. In both cases we apply the transform to study linear systems. In particular, we can describe the behavior (gain and phase) of an analog system using its transform, $H(s) = Y(s)/X(s)$. In this same way we will use the $H(z)$ transform of a digital filter to determine its gain and phase response (Figure 15.2).

Figure 15.2
A transform also can be used to study a system in the frequency domain.

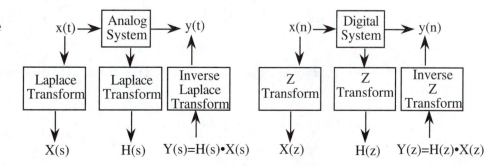

For an analog system we can calculate the gain by taking the magnitude of $\mathbf{H(s)}$ at $\mathbf{s} = \mathbf{j2\pi f}$, for all frequencies $\mathbf{f}$ from $-\infty$ to $+\infty$. The phase will be the angle of $\mathbf{H(s)}$ at $\mathbf{s} = \mathbf{j2\pi f}$. If we were to plot the $\mathbf{H(s)}$ in the $\mathbf{s}$ plane, the $\mathbf{s} = \mathbf{j2\pi f}$ curve is the entire y axis. For a digital system we will calculate the gain and phase by taking the magnitude and angle of $\mathbf{H(z)}$. Because of the finite sampling interval, we will only be able to study frequencies from DC to $(\frac{1}{2})\mathbf{f_s}$ in our digital systems. If we were to plot the $\mathbf{H(z)}$ in the $\mathbf{z}$ plane, the $\mathbf{z}$ curve representing the DC to $(\frac{1}{2})\mathbf{f_s}$ frequencies will be the unit circle.

We will begin by developing a simple, yet powerful rule that will allow us to derive the $\mathbf{H(z)}$ transforms of most digital filters. Let $\mathbf{m}$ be an integer constant. One can use the definition of the Z transform to prove that

$$\mathbf{Z[x(n\text{–}m)]} = \sum_{n=-\infty}^{\infty} \mathbf{z}^{-n}\mathbf{x(n-m)}$$

$$= \sum_{p+m=-\infty}^{p+m=+\infty} \mathbf{z}^{-p-m}\mathbf{x(p)} \qquad p=n-m,\ n=p+m$$

$$= \sum_{p=-\infty}^{p=+\infty} \mathbf{z}^{-p-m}\mathbf{x(p)} \qquad \mathbf{m} \text{ is a constant}$$

$$= \mathbf{z}^{-m}\sum_{p=-\infty}^{p=+\infty} \mathbf{z}^{-p}\mathbf{x(p)} \qquad \mathbf{m} \text{ is a constant}$$

$$= \mathbf{z}^{-m}\mathbf{Z[x(n)]} \tag{7}$$

For example if $\mathbf{X(z)}$ is the Z transform of $\mathbf{x(n)}$, then $\mathbf{z}^{-2} \cdot \mathbf{X(z)}$ is the Z transform of $\mathbf{x(n-2)}$. To find the Z transform of a digital filter, take the transform of both sides of the linear equation and solve for

$$\mathbf{H(z)} = \frac{\mathbf{Y(z)}}{\mathbf{X(z)}} \tag{8}$$

To find the response of the filter, let $\mathbf{z}$ be a complex number on the unit circle

$$\mathbf{z} \equiv \mathbf{e}^{\mathbf{j2\pi f/f_s}} \qquad \text{for } 0 \le \mathbf{f} < \frac{1}{2}\mathbf{f_s} \tag{9}$$

or

$$\mathbf{z} = \cos(2\pi \mathbf{f/f_s}) + \mathbf{j}\sin(2\pi \mathbf{f/f_s}) \tag{10}$$

Let

$$\mathbf{H(f)} = \mathbf{a} + \mathbf{bj} \qquad \text{where } \mathbf{a} \text{ and } \mathbf{b} \text{ are real numbers} \tag{11}$$

The gain of the filter is the complex magnitude of $\mathbf{H(z)}$ as $\mathbf{f}$ varies from 0 to $(\frac{1}{2})\mathbf{f_s}$.

$$\mathbf{Gain} \equiv \left|\mathbf{H(f)}\right| = \sqrt{\mathbf{a}^2 + \mathbf{b}^2} \tag{12}$$

The phase response of the filter is the angle of $\mathbf{H(z)}$ as $\mathbf{f}$ varies from 0 to $(\frac{1}{2})\mathbf{f_s}$.

$$\mathbf{Phase} \equiv \mathbf{angle[H(f)]} = \tan^{-1}\frac{\mathbf{b}}{\mathbf{a}} \tag{13}$$

15.2 Simple Digital Filter Examples

In this section, we perform the straightforward calculations to determine the filter response (gain and phase) given the filter equation. The gain and phase of these four filters are plotted in Figures 15.3 and 15.4. Even though the four examples are very simple, they can be used in actual applications. The first example, Equation 2, is a simple average of the current sample with the previous sample (Program 15.1).

$$y(n) = \frac{x(n) + x(n-1)}{2} \tag{2}$$

The first step is to take the Z transform of both sides of Equation 2. The Z transform of $y(n)$ is $Y(z)$, the Z transform of $x(n)$ is $X(z)$, and the Z transform of $x(n-1)$ is $z^{-1}X(z)$. Since the Z transform is a linear operator, we can write

$$Y(z) = \frac{X(z) + z^{-1}X(z)}{2} \tag{14}$$

The next step is to rewrite the equation in the form of $H(z) = Y(z)/X(z)$. We then can use Equations 9 to 13 to determine the gain and phase response of this filter.

$$H(z) \equiv \frac{Y(z)}{X(z)} = \frac{1 + z^{-1}}{2} \tag{15}$$

$$H(f) = \frac{1 + e^{-j2\pi f/fs}}{2} = \frac{1 + \cos(2\pi f/f_s) - j\sin(2\pi f/f_s)}{2} \tag{16}$$

```
// MC68HC11A8                          // MC68HC812A4
#define Rate 2000                      #define Rate 2000
#define OC5 0x08                       #define OC5  0x20
#pragma interrupt_handler TOC5handler()   #pragma interrupt_handler TOC5handler()
void TOC5handler(void){                void TOC5handler(void){
   TFLG1=OC5;      // Ack interrupt        TFLG1=OC5;      // ack OC5F
   TOC5=TOC5+Rate; // Executed every 1 ms   TC5=TC5+Rate;   // Executed every 1 ms
   x[1]=x[0];      // shift MACQ data       x[1]=x[0];      // shift MACQ data
   x[0] = A2D(channel);  // new data        x[0] = A2D(channel);  // new data
   y=(x[0]+x[1])>>1;}                       y=(x[0]+x[1])>>1;}

void ritual(void) {                     void ritual(void) {
asm(" sei");    // make atomic          asm(" sei");     // make atomic
   TMSK1|=OC5;  // Arm output compare 5     TIOS|=OC5;     // enable OC5
   x[0] = x[1] = 0;                         TSCR|=0x80;    // enable
   TFLG1=OC5;  // Initially clear OC5F      TMSK2=0x32;    // 500 ns clock
   TOC5=TCNT+Rate; // First one in 1 ms     TMSK1|=OC5;    // Arm output compare 5
asm(" cli"); }                              x[0] = x[1] = 0;
                                            TFLG1=OC5;     // Initially clear C5F
                                            TC5=TCNT+Rate; // First one in 1 ms
                                         asm(" cli"); }
```

Program 15.1 Real-time data acquisition with a simple digital filter, Equation 2.

$$\text{Gain} \equiv |H(f)| = 0.5\sqrt{\{1 + \cos(2\pi f/f_s)\}^2 + \{\sin(2\pi f/f_s)\}^2} \qquad (17)$$

$$\text{Phase} \equiv \text{angle}\,[H(f)\,] = \tan^{-1}\left[\frac{-\sin(2\pi f/f_s)}{1 + \cos(2\pi f/f_s)}\right] \qquad (18)$$

```
// C implementation, the TOC5handler interrupt occurs fs=1000 Hz
unsigned char x[2],y;
// x[0] is x(n) current sample
// x[1] is x(n-1) previous sample
// in general,        Rate=clockfrequency/fs, channel specifies the ADC channel
```

Observation: Equation 2 is a simple low-pass filter, passing DC and rejecting $0.5f_s$.

The second simple filter, Equation 3, performs the running average of the last six ADC samples (Program 15.2).

$$y(n) = \frac{x(n) + x(n-1) + x(n-2) + x(n-3) + x(n-4) + x(n-5)}{6} \qquad (3)$$

Just like the last example, first we take the Z transform of both sides of Equation 3:

$$Y(z) = \frac{X(z) + z^{-1}X(z) + z^{-2}X(z) + z^{-3}X(z) + z^{-4}X(z) + z^{-5}X(z)}{6} \qquad (19)$$

The next step is to rewrite the equation in the form of $H(z)=Y(z)/X(z)$.

$$H(z) \equiv \frac{Y(z)}{X(z)} = \frac{1 + z^{-1} + z^{-2} + z^{-3} + z^{-4} + z^{-5}}{6} \qquad (20)$$

We then can use Equations 9 to 13 to determine the gain and phase response of this filter.

$$H(f) = \frac{1 + e^{-j2\pi f/fs} + e^{-j4\pi f/fs} + e^{-j6\pi f/fs} + e^{-j8\pi f/fs} + e^{-j10\pi f/fs}}{6}$$

$$= [1 + \cos(2\pi f/f_s) + \cos(4\pi f/f_s) + \cos(6\pi f/f_s) + \cos(8\pi f/f_s) + \cos(10\pi f/f_s)$$

$$- j\{\sin(2\pi f/f_s) + \sin(4\pi f/f_s) + \sin(6\pi f/f_s) + \sin(8\pi f/f_s) + \sin(10\pi f/f_s)\}]/6 \qquad (21)$$

$$\text{Gain} \equiv |H(f)|$$
$$= 1/6 \cdot [\{1 + \cos(2\pi f/f_s) + \cos(4\pi f/f_s) + \cos(6\pi f/f_s) + \cos(8\pi f/f_s) + \cos(10\pi f/f_s)\}^2$$
$$+ \{\sin(2\pi f/f_s) + \sin(4\pi f/f_s) + \sin(6\pi f/f_s) + \sin(8\pi f/f_s) + \sin(10\pi f/f_s)\}^2] \qquad (22)$$

$$\text{Phase} \equiv \text{angle}[H(f)]$$
$$= \frac{\tan^{-1}[-\{\sin(2\pi f/f_s) + \sin(4\pi f/f_s) + \sin(6\pi f/f_s) + \sin(8\pi f/f_s) + \sin(10\pi f/f_s)\}}{\{1 + \cos(2\pi f/f_s) + \cos(4\pi f/f_s) + \cos(6\pi f/f_s) + \cos(8\pi f/f_s) + \cos(10\pi f/f_s)\}]} \qquad (23)$$

```
// C implementation, fs is 360 Hz, channel specifies the ADC channel
unsigned char x[6],y;
// x[0] is x(n) current sample
// x[1] is x(n-1) sample 1/360 sec ago
// x[2] is x(n-2) sample 2/360 sec ago
// x[3] is x(n-3) sample 3/360 sec ago
// x[4] is x(n-4) sample 4/360 sec ago
// x[5] is x(n-5) sample 5/360 sec ago
```

```
// MC68HC11A8
#define OC5  0x08
#pragma interrupt_handler TOC5handler()
void TOC5handler(void){ unsigned int i;
   TFLG1=OC5;        // Ack interrupt
   TOC5=TOC5+5556;    // fs=360Hz
   for(i=5;i>0;i++)
       x[i]=x[i-1];   // shift MACQ data
   x[0] = A2D(channel);  // new data
   y=(x[0]+x[1]+x[2]+x[3]+x[4]+x[5])/6;}

void ritual(void) {
asm(" sei");      // make atomic
   TMSK1|=OC5;  // Arm output compare 5
   x[0]=x[1]=x[2]=x[3]=x[4]=x[5]=0;
   TFLG1=OC5;  // Initially clear OC5F
   TOC5=TCNT+5556;
asm(" cli"); }
```

```
// MC68HC812A4
#define OC5   0x20
#pragma interrupt_handler TOC5handler()
void TOC5handler(void){ unsigned int i;
   TFLG1=OC5;         // ack OC5F
   TC5=TC5+5556;     // fs=360Hz
   for(i=5;i>0;i++)
       x[i]=x[i-1];        // shift MACQ data
   x[0] = A2D(channel);  // new data
   y=(x[0]+x[1]+x[2]+x[3]+x[4]+x[5])/6;}
void ritual(void) {
asm(" sei");       // make atomic
   TIOS|=OC5;        // enable OC5
   TSCR|=0x80;       // enable
   TMSK2=0x32;       // 500 ns clock
   TMSK1|=OC5;       // Arm output compare 5
   x[0]=x[1]=x[2]=x[3]=x[4]=x[5]=0;
   TFLG1=OC5;        // Initially clear C5F
   TC5=TCNT+5556;
asm(" cli"); }
```

Program 15.2 Real-time data acquisition with a simple digital filter, Equation 3.

Observation: Equation 3 is a multiple-notch filter rejecting $(\frac{1}{6})\mathbf{f_s}$, $(\frac{1}{3})\mathbf{f_s}$, and $(\frac{1}{2})\mathbf{f_s}$. In particular, if $\mathbf{f_s}$ is 360 Hz, then the digital filter in Equation 3 rejects 60, 120, and 180 Hz.

The third simple filter, Equation 4, performs the average of the current ADC sample with the sample collected three sampling periods ago. Although this filter appears to be simple, we can use it to implement a low-Q 60-Hz notch digital filter (Program 15.3).

$$y(n) = \frac{x(n) + x(n-3)}{2} \tag{4}$$

Again we take the Z transform of both sides of Equation 4:

$$Y(z) = \frac{X(z) + z^{-3}X(z)}{2} \tag{24}$$

Next we rewrite the equation in the form of $\mathbf{H(z)} = \mathbf{Y(z)/X(z)}$.

$$H(z) \equiv \frac{Y(z)}{X(z)} = \frac{1 + z^{-3}}{2} \tag{25}$$

We then can use Equations 9 to 13 to determine the gain and phase response of this filter.

$$H(f) = \frac{1 + e^{-j6\pi f/fs}}{2} = \frac{1 + \cos(6\pi f/f_s) - j\sin(6\pi f/f_s)}{2} \tag{26}$$

$$\mathbf{Gain} \equiv |\mathbf{H(f)}| = 0.5\sqrt{\{1 + \cos(6\pi f/f_s)\}^2 + \{\sin(6\pi f/f_s)\}^2} \tag{27}$$

$$\mathbf{Phase} \equiv \mathbf{angle[H(f)]} = \tan^{-1}\left[\frac{-\sin(6\pi f/f_s)}{1 + \cos(6\pi f/f_s)}\right] \tag{28}$$

```
// C implementation, fs is 360 Hz, channel specifies the ADC channel
unsigned char x[4],y;
// x[0] is x(n) current sample
// x[1] is x(n-1) sample 1/360 sec ago
// x[2] is x(n-2) sample 2/360 sec ago
// x[3] is x(n-3) sample 3/360 sec ago
```

```
// MC68HC11A8                              // MC68HC812A4
#define OC5  0x08                          #define OC5  0x20
#pragma interrupt_handler TOC5handler()    #pragma interrupt_handler TOC5handler()
void TOC5handler(void){                    void TOC5handler(void){
   TFLG1=OC5;      // Ack interrupt            TFLG1=OC5;     // ack OC5F
   TOC5=TOC5+5556;  // fs=360Hz               TC5=TC5+5556;   // fs=360Hz
   x[3]=x[2];      // shift MACQ data         x[3]=x[2];      // shift MACQ data
   x[2]=x[1];                                 x[2]=x[1];
   x[1]=x[0];                                 x[1]=x[0];
   x[0] = A2D(channel);  // new data          x[0] = A2D(channel);  // new data
   y=(x[0]+x[3])>>1;}                         y=(x[0]+x[3])>>1;}
                                           void ritual(void) {
void ritual(void) {                        asm(" sei");     // make atomic
asm(" sei");     // make atomic             TIOS|=OC5;      // enable OC5
   TMSK1|=OC5;  // Arm output compare 5      TSCR|=0x80;     // enable
   x[0]=x[1]=x[2]=x[3]=0;                    TMSK2=0x32;     // 500 ns clock
   TFLG1=OC5;  // Initially clear OC5F       TMSK1|=OC5;     // Arm output compare 5
   TOC5=TCNT+5556;                          x[0]=x[1]=x[2]=x[3]=0;
asm(" cli"); }                              TFLG1=OC5;      // Initially clear C5F
                                            TC5=TCNT+5556;
                                           asm(" cli"); }
```

Program 15.3 Real-time data acquisition with a simple digital filter, Equation 4.

Observation: Equation 4 is a double-notch filter rejecting $(\frac{1}{6})\mathbf{f_s}$ and $(\frac{1}{2})\mathbf{f_s}$. In particular, if $\mathbf{f_s}$ is 360 Hz, then the digital filter in Equation 4 rejects 60 and 180 Hz.

When writing digital filters it is important to consider the precision of the both the variables and the calculations. In particular, a serious error will occur if overflow or underflow occurs during the calculations. Consider the calculation x[0]+x[3] performed in Program 15.4. If simple 8-bit addition were to be used, then an error will occur when intermediate calculation x[0]+x[3] goes above 255. Observing the gain versus frequency plot in Figure 15.3, we are confident that the final calculation (x[0]+x[3])>>1 will fit into an 8-bit variable. To verify program correctness, we must look at the output of the compiler. The two listings in Program 15.4 were generated by the ImageCraft ICC11 and ICC12 compilers. These are proper implementations because the compiler first promotes the 8-bit data into 16 bits, performs the addition and shift using 16-bit operations, then demotes the result back to 8 bits when storing into y . The comments were added to clarify the operations.

The fourth simple filter, Equation 5, is an example of an infinite-impulse response. It performs the average of the current ADC sample with the last filter output. For this filter we

```
; MC68HC11A8, ICC11
; y=(x[0]+x[3])>>1;
   ldy #1
   pshy
   ldab  _x+3  ; 8-bit x[3]
   clra          ; promote into RegD
   pshb          ; save on stack
   psha
   ldab  _x    ; 8 bit x[0]
   clra          ; promote into RegD
   tsy
   addd 0,y    ; 16-bit x[0]+x[3]
   puly
   puly
   jsr  __asrd ; 16-bit shift
   stab _y     ; demote to 8 bit
```

```
; MC68HC812A4, ICC12
; y=(x[0]+x[3])>>1;
   ldab _x+3  ; 8-bit x[3]
   clra          ; promote into RegD
   std -10,x  ; save on stack
   ldab _x    ; 8 bit x[0]
   clra          ; promote into RegD
   addd -10,x ; 16-bit x[0]+x[3]
   asra
   rolb          ; 16-bit shift
   stab _y    ; demote to 8 bit
```

Program 15.4 Compiler listings for the filter implementation.

follow the same procedure to determine its gain and frequency response. Notice in Figure 15.3 that the gain is larger than 1 for some frequencies. Therefore it will be important when implementing this filter to use a number system with more bits than the ADC precision. In other words, even though the ADC in Program 15.5 has only 8 bits, we will perform the filter using 16-bit precision to avoid overflow.

$$y(n) = \frac{y(n-1) + x(n)}{2} \tag{5}$$

$$Y(z) = \frac{z^{-1}Y(z) + X(z)}{2} \tag{29}$$

$$H(z) \equiv \frac{Y(z)}{X(z)} = \frac{1}{2 - z^{-1}} \tag{30}$$

$$H(f) = \frac{1}{2 - e^{-j2\pi f/fs}} = \frac{1}{2 - \cos(2\pi f/f_s) + j\sin(2\pi f/f_s)}$$

$$= \frac{2 - \cos(2\pi f/f_s) - j\sin(2\pi f/f_s)}{[2 - \cos(2\pi f/f_s)]^2 - [\sin(2\pi f/f_s)]^2} \tag{31}$$

$$\text{Gain} \equiv |H(f)| = \left| \frac{2 - \cos(2\pi f/f_s) - j\sin(2\pi f/f_s)}{[2 - \cos(2\pi f/f_s)]^2 - [\sin(2\pi f/f_s)]^2} \right|$$

$$= \frac{\sqrt{\{2 - \cos(2\pi f/f_s)\}^2 + \{\sin(2\pi f/f_s)\}^2}}{[2 - \cos(2\pi f/f_s)]^2 - [\sin(2\pi f/f_s)]^2} \tag{32}$$

$$\text{Phase} \equiv \text{angle}[H(f)] = \tan^{-1}\left[\frac{-\sin(2\pi f/f_s)}{2 - \cos(2\pi f/f_s)} \right] \tag{33}$$

```
//C implementation, Rate should be 2,000,000/fs, channel specifies the ADC channel
unsigned short x,y[2];    // 16-bit precision used to prevent overflow
//y[0] is y(n) current filter output
//y[1] is y(n-1) filter output 1 sample ago
```

```
// MC68HC11A8                                    // MC68HC812A4
#define Rate 2000                                #define Rate 2000
#define OC5  0x08                                #define OC5  0x20
#pragma interrupt_handler TOC5handler()          #pragma interrupt_handler TOC5handler()
void TOC5handler(void){                          void TOC5handler(void){
    TFLG1=OC5;        // Ack interrupt               TFLG1=OC5;        // ack OC5F
    TOC5=TOC5+Rate;   // Executed at fs             TC5=TC5+Rate;     // Executed at fs
    y[1]=y[0];        // shift MACQ data            y[1]=y[0];        // shift MACQ data
    x = A2D(channel); // new data                   x = A2D(channel); // new data
    y[0]=(x+y[1])>>1;} // 16-bit                    y[0]=(x+y[1])>>1;} // 16-bit

void ritual(void) {                              void ritual(void) {
asm(" sei");       // make atomic                 asm(" sei");       // make atomic
    TMSK1|=OC5;    // Arm output compare 5           TIOS|=OC5;       // enable OC5
    y[0] = y[1] = 0;                                 TSCR|=0x80;      // enable
    TFLG1=OC5;     // Initially clear OC5F           TMSK2=0x32;      // 500 ns clock
    TOC5=TCNT+Rate; // First one in 1 ms            TMSK1|=OC5;      // Arm output compare 5
asm(" cli"); }                                       y[0] = y[1] = 0;
                                                     TFLG1=OC5;       // Initially clear C5F
                                                     TC5=TCNT+Rate;   // First one in 1 ms
                                                 asm(" cli"); }
```

Program 15.5 Real-time data acquisition with a simple digital filter, Equation 5.

The response of the four linear digital filters is plotted in Figure 15.3; the phase response of the four linear digital filters is plotted in Figure 15.4.

A linear phase versus frequency response is desirable because a linear phase causes minimal waveform distortion. Conversely, a nonlinear phase response will distort the shape or morphology of the signal.

Figure 15.3
Gain versus frequency response for four simple digital filters.

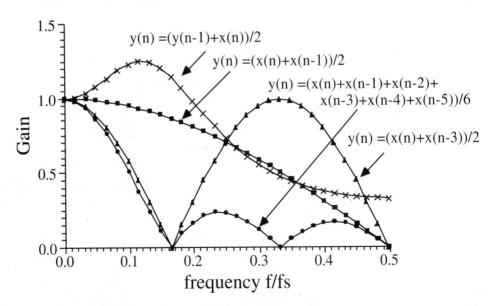

$y(n) = (y(n-1)+x(n))/2$

$y(n) = (x(n)+x(n-1))/2$

$y(n) = (x(n)+x(n-1)+x(n-2)+x(n-3)+x(n-4)+x(n-5))/6$

$y(n) = (x(n)+x(n-3))/2$

Figure 15.4
Phase versus frequency
response for four simple
digital filters.

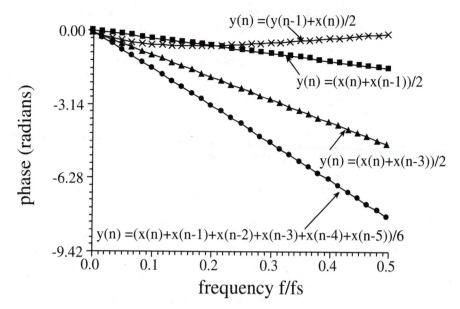

In general, if $\mathbf{f_s}$ is $2 \cdot \mathbf{k} \cdot \mathbf{f_c}$ Hz (where k is any integer k≥ 2), then the following is a $\mathbf{f_c}$ notch filter:

$$y(n) = \frac{x(n) + x(n-k)}{2} \tag{34}$$

Similarly, if $\mathbf{f_s}$ is $\mathbf{k} \cdot \mathbf{f_c}$ Hz (where **k** is any integer k≥ 2), then the following rejects $\mathbf{f_c}$ and its harmonics $2\mathbf{f_c}$, $3\mathbf{f_c}$ . . . :

$$y(n) = \frac{1}{k}\sum_{i=0}^{k-1} x(n-i) \tag{35}$$

15.3   Impulse Response

Another way to analyze digital filters is to consider the filter response to particular input sequences. Two typical sequences are the step and the impulse.

Step   . . . , **0, 0, 0, 1, 1, 1, 1** . . .
Impulse   . . . , **0, 0, 0, 1, 0, 0, 0** . . .

The step is defined as (Figure 15.5)

$$\mathbf{s(n)} \equiv \mathbf{0} \qquad \textbf{for n < 0}$$
$$\mathbf{1} \qquad \textbf{for n} \geq \mathbf{0}$$

Figure 15.5
Step input.

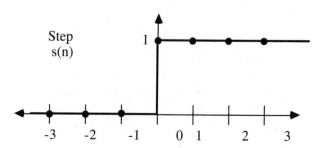

The step signal represents a sharp change (like an edge in a photograph).

Consider the following three digital filters:

FIR $$y(n) = \frac{x(n) + x(n-1)}{2}$$ (2)

IIR $$y(n) = \frac{y(n-1) + x(n)}{2}$$ (5)

Median $$y(n) = \text{median}\,[x(n), x(n-1), x(n-2)]$$ (36)

The median can be performed on any odd number of data points by sorting the data and selecting the middle value. The median filter can be performed recursively or nonrecursively. For a nonrecursive median filter, the original data points are not modified. That is, the median is calculated each time without regard to the previous filter outputs. For example, a five-wide nonrecursive median filter takes as the filter output the median of $[x(n), x(n-1), x(n-2), x(n-3), x(n-4)]$ On the other hand, a recursive median filter replaces the sample point with the filter output. For example, a five-wide recursive median filter takes as the filter output the median of $[x(n), y(n-1), y(n-2), y(n-3), y(n-4)]$, where $y(n-1)$, $y(n-2)$ ... are the previous filter outputs. A nonrecursive three-wide median filter is implemented in Program 15.6.

Program 15.6 The median filter is an example of a nonlinear filter.

```
unsigned char Median(unsigned char u1,unsigned char u2,unsigned char u3){
unsigned char result;

  if(u1>u2)
    if(u2>u3)    result=u2;    // u1>u2,u2>u3        u1>u2>u3
    else
       if(u1>u3) result=u3;    // u1>u2,u3>u2,u1>u3 u1>u3>u2
       else      result=u1;    // u1>u2,u3>u2,u3>u1 u3>u1>u2
  else
    if(u3>u2)    result=u2;    // u2>u1,u3>u2        u3>u2>u1
    else
       if(u1>u3) result=u3;    // u2>u1,u2>u3,u1>u3 u2>u1>u3
       else      result=u1;    // u2>u1,u2>u3,u3>u1 u2>u3>u1
  return(result);}
unsigned char x[3],y;
// x[0] is x(n) the current sample
// x[1] is x(n-1) the sample 1/fs ago
// x[2] is x(n-2) the sample 2/fs ago
void sample(void){
    x[2]=x[1];   // shift MACQ data
    x[1]=x[0];
    x[0] = A2D(0);  // new data from channel 0
    y=median(x[0],x[1],x[2]);}
```

Observation: A median filter can be applied in systems that have impulse or speckle noise. For example, if the noise occasionally causes one sample to be very different from the rest (like a speck on a piece of paper), then the median filter will completely eliminate the noise.

The step responses of the three filters are:

FIR	..., **0, 0, 0, 0.5, 1, 1, 1** ...
IIR	..., **0, 0, 0, 0.5, 0.75, 0.88, 0.94, 0.97, 0.98, 0.99** ...
Median	..., **0, 0, 0, 0, 1, 1, 1, 1, 1** ...

Except for the delay, the median filter passes a step without error (Figure 15.6).

Figure 15.6
Step response of four simple digital filters.

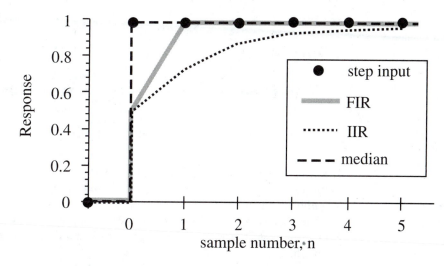

The impulse represents a noise spike (like spots on a Xerox copy). The definition of impulse is (Figure 15.7).

$$i(n) \equiv 0 \quad \text{for } n \neq 0$$
$$1 \quad \text{for } n = 0$$

The impulse response of a filter is defined as **h(n)**. The impulse responses of the three filters are:

FIR	..., **0, 0, 0, 0.5, 0.5, 0, 0, 0** ...
IIR	..., **0, 0, 0, 0.5, 0.25, 0.13, 0.06, 0.03, 0.02, 0.01** ...
Median	..., **0, 0, 0, 0, 0, 0, 0, 0** ...

Figure 15.7
Impulse input.

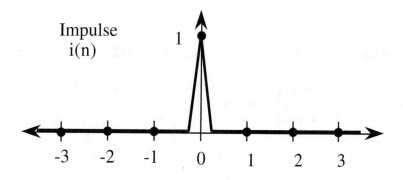

Figure 15.8
Impulse response of four
simple digital filters.

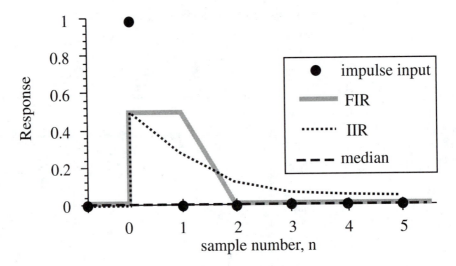

The median filter completely removes the impulse (Figure 15.8). Note that the median filter preserves the sharp edges and removes the spike or impulsive noise. The median filter is **nonlinear,** and hence **H(z)** and **h(n)** are not defined for this particular class of filters.

The impulse response **h(n)** can also be used as an alternative to the transfer function **H(z)**. **h(n)** is sometimes called the *direct form.* A causal filter has **h(n)** = 0 for **n**<0. For a casual filter.

$$H(z) = \sum_{n=0}^{\infty} h(n)\, z^{-n} \tag{37}$$

For a FIR filter, **h(n)** = 0 for **n**≥ **N** for some **N.** Thus,

$$H(z) = \sum_{n=0}^{N-1} h(n) z^{-n} \tag{38}$$

The output of a filter can be calculated by convolving the input sequence **x(n)** with **h(n).** For an IIR filter:

$$y(n) = \sum_{k=0}^{\infty} h(k)\ x\,(n-k) \tag{39}$$

For a FIR filter:

$$y(n) = \sum_{k=0}^{N-1} h(k)\ x(n-k) \tag{40}$$

15.4   High-Q 60-Hz Digital Notch Filter

There are two objectives for this example. The first goal is to show an example of a digital notch filter, and the second goal is to demonstrate the use of fixed-point mathematics. 60-Hz noise is a significant problem in most DASs. The 60-Hz noise reduction can be accomplished by:

1. Reducing the noise source (e.g., shut off large motors)

2. Shielding the transducer, cables, and instrument

3. Implementing a 60-Hz analog notch filter

4. Implementing a 60-Hz digital notch filter

Consider again the analogy between the Laplace and Z transforms. When the H(s) transform is plotted in the s plane, we look for peaks [places where the amplitude H(s) is high] and valleys (places where the amplitude is low). In particular, we usually can identify zeros [H(s)=0] and poles [H(s)=∞]. In the same way we can plot the H(z) in the z plane and identify the poles and zeros. The analogies in Table 15.1 apply:

Analog condition	Digital condition	Consequence
Zero near s=j2πf line	Zero near z=e$^{j2\pi f/fs}$ circle	Low gain at the f near the zero
Pole near s=j2πf line	Pole near z=e$^{j2\pi f/fs}$ circle	High gain at the f near the pole
Zeros in complex conjugate pairs	Zeros in complex conjugate pairs	Output y(t) is real
Poles in complex conjugate pairs	Poles in complex conjugate pairs	Output y(t) is real
Poles in left half plane	Poles inside unit circle	Stable system
Poles in right half plane	Poles outside unit circle	Unstable system
Pole near a zero	Pole near a zero	High-Q response

Table 15.1   Analogies between the analog and digital filter design rules.

It is the 60-Hz digital notch filter that will be implemented in this example. The signal is sampled at **f$_s$**=244.14 Hz. We wish to place the zeros (gain=0) at 60 Hz, thus (Figure 15.9):

$$\Theta = \pm 2\pi \cdot \frac{60}{f_s} = \pm 1.54416 \tag{41}$$

The zeros are located on the unit circle at 60 Hz:

$$z_1 = \cos(\theta) + j\sin(\theta) \qquad z_2 = \cos(\theta) - j\sin(\theta) \tag{42}$$

or

$$z_1 = 0.02663 + j\,0.99965 \qquad z_2 = 0.02663 - j\,0.99965 \tag{43}$$

Figure 15.9
Pole-zero plot of a 60-Hz digital notch filter.

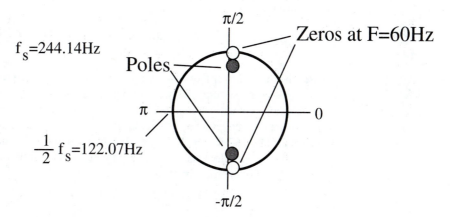

To implement a flat passband away from 60 Hz, the poles are placed next to the zeros, just inside the unit circle. Let α define the closeness of the poles, where $0 < \alpha < 1$.

$$p_1 = \alpha \, z_1 \qquad\qquad p_2 = \alpha \, z_2 \qquad\qquad (44)$$

for $\alpha = 0.95$

$$p_1 = 0.02530 + j\,0.94966 \qquad p_2 = 0.02530 - j\,0.94966 \qquad (45)$$

The transfer function is

$$H(z) = \prod_{i=1}^{k} \frac{z - z_i}{z - p_i} = \frac{(z - z_1)(z - z_2)}{(z - p_1)(z - p_2)} \qquad (46)$$

which can be put in standard form (i.e., with terms $1, z^{-1}, z^{-2} \ldots$)

$$H(z) = \frac{1 - 2\cos(\theta)z^{-1} + z^{-2}}{1 - 2\alpha\cos(\theta)z^{-1} + \alpha^2 z^{-2}} \qquad (47)$$

or

$$H(z) = \frac{1 - 0.0532672z^{-1} + z^{-2}}{1 - 0.0506038z^{-1} + 0.9025z^{-2}} \qquad (48)$$

The digital filter can be derived by taking the inverse Z transform of the H(z) equation

$$y(n) = x(n) - 2\cos(\theta)x(n-1) + x(n-2) + 2\alpha\cos(\theta)y(n-1) - \alpha^2 y(n-2) \qquad (49)$$

or

$$y_1(n) = x(n) - 0.0532672x(n-1) + x(n-2) + 0.0506038y_1(n-1) - 0.9025y_1(n-2) \qquad (50)$$

To implement this filter in real time without floating-point hardware we rewrite Equation 50 using fixed-point mathematics:

$$y_2(n) = x(n) + x(n-2) + \frac{-14x(n-1) - 13y_2(n-1) - 231y_2(n-2)}{256} \qquad (51)$$

It will be important to implement the software such that the intermediate calculations are performed in 16-bit arithmetic. Also, since this filter has a gain larger than 1 for some frequencies, the filter outputs must also be implemented with 16-bit variables.

Table 15.2 shows that the fixed-point calculations do not produce significant filter errors. There are two factors in choosing the fixed value (e.g., 256) in the fixed-point implementation. We choose a value large enough so that the fixed-point value closely approximates the floating-point value (e.g., $13/256 \approx 0.0506038$). The larger fixed-point value we choose, the better the approximation:

$$y_3(n) = x(n) + x(n-2) + \frac{-533x(n-1) + 506y_3(n-1) - 9025y_3(n-2)}{10000} \qquad (52)$$

On the other hand, larger fixed-point values will require higher-precision integer mathematics. For an 8-bit microcomputer we would prefer to do 8-bit by 8-bit into 16-bit multiplies, 16-bit addition/subtraction, and 16-bit divided by 8-bit divides. For a 16-bit microcomputer we would prefer to do 16-bit by 16-bit into 32-bit multiplies, 32-bit addition/subtraction, and 32-bit divided by 16-bit divides. Notice that the multiply by 2^n and divide by 2^n operations are very simple to implement using the arithmetic shift operations.

For $\alpha = 0.90$, the filter becomes

$$y_4(n) = x(n) + x(n-2) + \frac{-14x(n-1) + 12y_4(n-1) - 207y_4(n-2)}{256} \qquad (53)$$

Table 15.2
Gain versus frequency responses for the floating point and fixed implementations.

f (Hz)	$|Y_1(z)/X(z)|$	$|Y_2(z)/X(z)|$	$|Y_1(z)/X(z)| - |Y_2(z)/X(z)|$
0	1.052	1.049	0.003
10	1.060	1.060	0.000
20	1.048	1.048	0.000
30	1.052	1.052	0.000
40	1.048	1.048	0.000
50	1.040	1.036	0.004
55	0.984	0.980	0.004
56	0.944	0.944	0.000
57	0.888	0.888	0.000
58	0.744	0.740	0.004
59	0.464	0.464	0.000
60	0.012	0.016	−0.004
61	0.396	0.404	−0.008
62	0.752	0.752	0.000
63	0.880	0.880	0.000
64	0.944	0.944	0.000
65	0.980	0.980	0.000
70	1.036	1.036	0.000
80	1.048	1.048	0.000
90	1.052	1.052	0.000
100	1.056	1.060	−0.004
110	1.056	1.056	0.000
120	1.052	1.052	0.000

For $\alpha = 0.75$, the filter becomes

$$y_5(n) = x(n) + x(n-2) + \frac{-7x(n-1) + 5y_5(n-1) - 72y_5(n-2)}{128} \tag{54}$$

The gain of the three filters (Equations 51, 53, and 54) is plotted in Figure 15.10.

Figure 15.10
Gain versus frequency response of three 60-Hz digital notch filters.

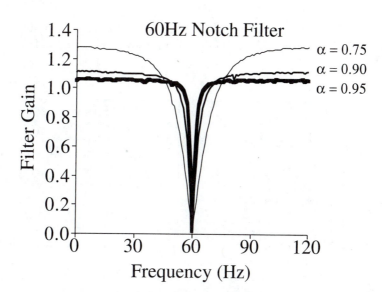

Sometimes we can choose $\mathbf{f_s}$ and/or α to simplify the digital filter equation. For example, if we choose $\mathbf{f_s} = 240$ Hz, then the $\cos(\theta)$ terms in Eq. (48) become zero. If we choose $\alpha = 7/8$ then the fixed-point digital filter becomes

$$y_6(n) = x(n) + x(n-2) - \frac{49 \cdot y_6(n-2)}{64} \tag{55}$$

Another consideration for this type of filter is the fact that the gain in the passbands is greater than 1. The DC gain can be determined two ways. The first method is to use the H(z) transfer equation and set z=1. The H(z) transfer equation for the filter in Eq. (55) is

$$H(z) = \frac{1 + z^{-2}}{1 + (49/64)z^{-2}} \tag{56}$$

At z=1 this reduces to

$$\text{DC gain} = \frac{2}{1 + 49/64} = \frac{128}{64 + 49} = \frac{128}{113} \tag{57}$$

The second method to calculate DC gain operates on the filter equation directly. In the first step, we set all the $x(n-k)$ terms in the filter to a single variable x and all the $y(n-k)$ terms in the filter to a single variable y.

Next we solve for the DC gain, which is y/x.

$$y = x + x - \frac{49y}{64} \tag{58}$$

This method also calculates the DC gain to be 128/113. We can adjust the digital filter so that the DC gain is exactly 1 by prescaling the input terms $[x(n), x(n-1), x(n-2)\ldots]$ by 113/128 (Program 15.7).

$$y_7(n) = \frac{113 \cdot x(n) + 113 \cdot x(n-2) - 98 \cdot y_7(n-2)}{128} \tag{59}$$

```
// y(n) = [113*x(n) + 113*x(n-2) - 98*y(n-2)]/128, channel specifies the ADC channel
unsigned char x[3],y[3];        // MACQs containing current and previous
```

Since the gain of this filter is always less than or equal to 1, the filter outputs will fit into 8-bit variables. The gain of this filter is shown in Figure 15.11.

```
// MC68HC11A8                              // MC68HC812A4
#define OC5  0x08                          #define OC5  0x20
#pragma interrupt_handler TOC5handler()    #pragma interrupt_handler TOC5handler()
void TOC5handler(void){                    void TOC5handler(void){
   TFLG1=OC5;       // Ack interrupt          TFLG1=OC5;       // ack OC5F
   TOC5=TOC5+8333;  // fs=240Hz              TC5=TC5+8333;    // fs=240Hz
   y[2]=y[1]; y[1]=y[0];  // shift MACQ       y[2]=y[1]; y[1]=y[0];  // shift MACQ
   x[2]=x[1]; x[1]=x[0];                      x[2]=x[1]; x[1]=x[0];
   x[0] = A2D(channel);  // new data          x[0] = A2D(channel);  // new data
   y[0]=(113*(x[0]+x[2])-98*y[2])>>7;}        y[0]=(113*(x[0]+x[2])-98*y[2])>>7;}
```

```
void ritual(void) {                          void ritual(void) {
asm(" sei");      // make atomic             asm(" sei");        // make atomic
   TMSK1|=OC5;  // Arm output compare 5         TIOS|=OC5;       // enable OC5
   y[0]=y[1]=y[2]=x[0]=x[1]=x[2]=0;             TSCR|=0x80;      // enable
   TFLG1=OC5;  // Initially clear OC5F          TMSK2=0x32;      // 500 ns clock
   TOC5=TCNT+8333;                              TMSK1|=OC5;      // Arm output compare 5
asm(" cli"); }                                  y[0]=y[1]=y[2]=x[0]=x[1]=x[2]=0;
                                                TFLG1=OC5;       // Initially clear C5F
                                                TC5=TCNT+8333
                                             asm(" cli"); }
```

Program 15.7 Real-time data acquisition with a 60-Hz notch digital filter, Equation 59.

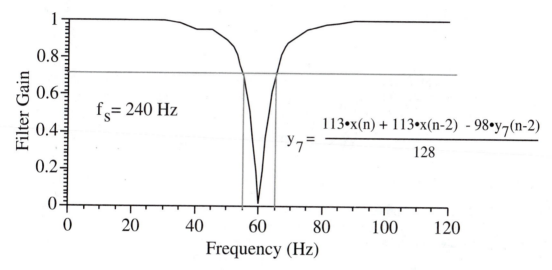

Figure 15.11 Gain versus frequency response of a 60-Hz digital notch filter scaled to have a DC gain of 1.

The Q of a digital notch filter is defined to be

$$Q \equiv \frac{f_c}{\Delta f} \tag{60}$$

where f_c is the center or notch frequency and Δf is the frequency range, where its gain is below 0.707 of the DC gain. For the filter in Equation (59) the gains at 55 and 65 Hz are about 0.707, so its Q is 6.

15.5 Effect of Latency on Digital Filters

The purpose of this analysis is to study the importance of "real time" in a real-time OS. We define a *hard real-time* system as one that can guarantee that a process will complete a critical task within a certain specified range. On the other hand, we define a *soft real-time* system as one that assigns higher priority to critical functions. You will, in this analysis, derive specific consequences that result when the OS delays service to a time-critical real-time process.

Consider a DAS that samples a signal at f_s and performs a 60-Hz digital notch filter to remove noise. For simplicity assume the actual signal is composed of a constant signal V_s and a 60-Hz noise signal V_n:

$$V(t) = V_s + V_n \sin(2\pi 60t) \tag{61}$$

The S/N ratio is V_s/V_n. Assuming the voltage range is between -5 to $+5$ V, the m-bit ADC will create an infinite digital sample sequence, $x(1)\ x(2), \ldots, x(n) \ldots$, with

$$x(n) = integer\left(\frac{2^{m-1}}{5} \cdot V(t_n)\right) \tag{62}$$

where t_n is the time the nth sample was taken. To perform DSP (filters, pattern recognition, and control) one usually samples the data at a regular interval. In our case, we want to sample at f_s. Ideally

$$t_n = \frac{n}{f_s} \tag{63}$$

In the DSP field, $x(n)$ refers to the current sample and $x(n-1)$ refers to the previous sample. If the sampling rate f_s is $120 \cdot k$ Hz, where k is an integer constant greater than 1, a simple 60-Hz digital notch filter is

$$y(n) = \frac{x(n) + x(n-k)}{2} \tag{34}$$

For large enough m, we can approximate $x(n)$ as $2^{m-1} \cdot V(t_n)/5$.

1. Consider a process that samples the ADC and calculates the digital filter, and displays the results. In a real-time OS this process will request service f_s times a second at exact intervals. Unfortunately, a time latency occurs between the process request and the process service. What happens to the filter performance if this delay is always exactly "d"? Assume the delay d is smaller than $1/f_s$ so that the sample is simply delayed and not skipped altogether. In this situation, we ask, how much noise passes the filter?

$$y(n) = \frac{x(n) + x(n-k)}{2} = \frac{(2^{m-1}/5) \cdot V(t_n + d) + (2^{m-1}/5) \cdot V[t_n - k/f_s + d]}{2}$$

$$= \frac{2^{m-1}}{10}\left[V\left(\frac{n}{120 \cdot k} + d\right) + V\left(\frac{n-k}{120 \cdot k} + d\right)\right]$$

$$= \frac{2^{m-1}}{10}\left\{V_s + V_n\sin\left[2\pi 60 \cdot \left(\frac{n}{120 \cdot k} + d\right)\right] + V_s + V_n\sin\left[2\pi 60 \cdot \left(\frac{n-k}{120 \cdot k} + d\right)\right]\right\}$$

$$= \frac{2^{m-1}}{10}\left[2V_s + V_n\sin\left(\frac{\pi n}{k} + 120\pi d\right) + V_n\sin\left(\frac{\pi(n-k)}{k} + 120\pi d\right)\right]$$

$$= \frac{2^{m-1}}{10}\{2V_s + V_n[\sin(a) + \sin(a - \pi)]\} \qquad \text{where } a = \frac{\pi n}{k} + 120\pi d$$

$$= \frac{2^{m-1}}{5}V_s \qquad \text{independent of d and } V_n \tag{64}$$

2. For the subsequent discussion we assume k=2, but the issues remain the same for any k. In this part, we will consider what happens if the request comes once at a time when the OS is performing a higher-priority job and the service is delayed. Let t_n be exactly n/240 for all n except at n=50, where $t_{50}=50/240+d$. Assume the delay d is smaller than 1/240 so that the sample is simply delayed and not skipped altogether. We can derive an expression for the error in y(50) as a function of the delay d, the ADC precision m, and the 60-Hz noise V_n.

$$y(n) = \frac{x(50) + x(48)}{2} = \frac{(2^{m-1}/5) \cdot V(t_{50} + d) + (2^{m-1}/5) \cdot V[t_{50} - 1/120]}{2}$$

$$= \frac{2^{m-1}}{10}\left[V\left(\frac{50}{240} + d\right) + V\left(\frac{48}{240}\right)\right]$$

$$= \frac{2^{m-1}}{10}\left[V_s + V_n\sin\left(\frac{2\pi60 \cdot 50}{240} + 120\pi d\right) + V_s + V_n\sin\left(\frac{2\pi60 \cdot 48}{240}\right)\right]$$

$$= \frac{2^{m-1}}{10}\left[2V_s + V_n\sin\left(25\pi + 120\pi d\right) + V_n\sin\left(24\pi\right)\right]$$

$$= \frac{2^{m-1}}{10}\left[2V_s + V_n\sin\left(\pi + 120\pi d\right)\right]$$

$$= \frac{2^{m-1}}{10}\left\{2V_s + V_n[\sin(\pi)\cos(120\pi d) + \cos(\pi)\sin(120\pi d)]\right\}$$

$$= \frac{2^{m-1}}{10}\left[2V_s - V_n\sin\left(120\pi d\right)\right] \tag{65}$$

$$\text{Error}(50) = -V_n\frac{2^{m-2}}{5}\sin(120\pi d) \approx -V_n\frac{2^{m-2}}{5}120\pi d \qquad \text{for } d << \frac{1}{60} \tag{66}$$

3. This single delay at n=50 will cause another error when y(52) is calculated.

$$\text{Error}(52) = V_n\frac{2^{m-2}}{5}\sin(120\pi d) \tag{66}$$

15.6 High-Q Digital High-Pass Filters

We will design a *two-pole* high-Q fixed-point digital high-pass filter (Figure 15.12). Let the sampling rate f_s be 100 Hz. The pole-zero plot of this type of filter is shown in Figure 15.13.

Figure 15.12
Gain versus frequency response of a high-pass filter.

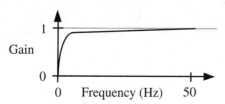

Figure 15.13
Pole-zero plot of a
high-pass filter.

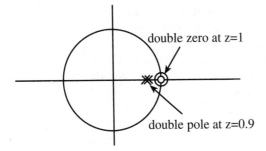
double zero at z=1

double pole at z=0.9

Next, we write the transform and convert it to standard form:

$$H(z) = \frac{(z-1)^2}{(z-0.9)^2} = \frac{z^2 - 2z + 1}{z^2 - 1.8z + 0.81} = \frac{1 - 2z^{-1} + z^{-2}}{1 - 1.8z^{-1} + 0.81z^{-2}}$$

From the transform, we can write the floating-point filter:

$$y(n) - 1.8y(n-1) + 0.81y(n-2) = x(n) - 2x(n-1) + x(n-2)$$

Next, we arrange the filter in standard form:

$$y(n) = x(n) - 2x(n-1) + x(n-2) + 1.8y(n-1) - 0.81y(n-2)$$

We determine the gain of this filter at 50 Hz by calculating the magnitude of H(z) at $z = -1$:

$$H(-1) = \frac{(-1-1)^2}{(-1-0.9)^2} = \frac{(-2)^2}{(-1.9)^2} = \frac{4}{3.61} \qquad 3.61/4 = 0.9025$$

So, to make the gain at 50 Hz equal to 1, we multiply all the **x** terms by 0.9025:

$$y(n) = 0.9025x(n) - 1.805x(n-1) + 0.9025x(n-2) + 1.8y(n-1) - 0.81y(n-2)$$

We can easily convert this filter to fixed-point without introducing any error:

$$y(n) = \frac{9025x(n) - 18050x(n-1) + 9025x(n-2) + 18000y(n-1) - 8100y(n-2)}{10000}$$

The following fixed-point approximation is easier to calculate:

$$y(n) = \frac{231x(n) - 462x(n-1) + 231x(n-2) + 461y(n-1) - 207y(n-2)}{256}$$

15.7   Low-Q Digital High-Pass Filters

Consider the following class of low-Q digital high-pass filters. Let f_s be the sampling rate. Let g0, g1, and g2 be the filter gains at DC, $0.25f_s$, and $(\frac{1}{2})f_s$, respectively. The parameter $0 < \alpha < 1$ determines the gain ratio g2/g0. The pole-zero plot is shown in Figure 15.14. First, we show the H(z) transform and derive the digital filter.

$$Hz = \frac{Y(z)}{X(z)} = \frac{z - \alpha}{z + \alpha} = \frac{1 - \alpha z^{-1}}{1 + \alpha z^{-1}}$$

$$y(n) = x(n) - \alpha x(n-1) - \alpha y(n-1)$$

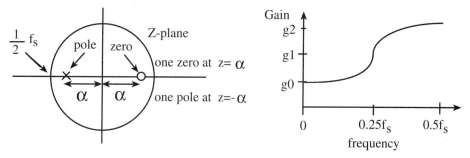

Figure 15.14   Pole-zero plot and gain versus frequency response of a low-Q high-pass filter.

To determine g0, the gain response at DC, we calculate the magnitude of H(z) at z=1. We solve both the general case of a and the specific case of $\alpha = 0.75$. We simplify the expression as much as possible.

$$g0 = |H(z)|_{at\ z=1} = \frac{1-\alpha}{1+\alpha} = \frac{0.25}{1.75} = \frac{1}{7}$$

To determine g1, the gain response at $0.25f_s$, we calculate the magnitude of H(z) at z=i. We notice the g1 is 1 for all α.

$$g1 = |H(z)|_{at\ z=1} = \left|\frac{i-\alpha}{i+\alpha}\right| = 1$$

To determine g2, the gain response at $(\frac{1}{2})f_s$, we calculate the magnitude of H(z) at z=-1. We solve both the general case of α and the specific case of $\alpha = 0.75$. We simplify the expression as much as possible.

$$g2 = \left|H(z)\right|_{at\ z=-1} = \frac{-1-\alpha}{-1+\alpha} = \frac{-1.75}{-0.25} = 7$$

For the specific case of $\alpha = 0.75$, we have the IIR

$$y(n) = x(n) - 0.75x(n-1) - 0.75y(n-1)$$

We can scale the filter so that the gain at $(\frac{1}{2})f_s$ is 1 by dividing the x(n) and x(n$-$1) terms by 7:

$$y(n) = \frac{4x(n) - 3x(n-1) - 21y(n-1)}{28}$$

15.8   Low-Q Digital Low-Pass Filters

Consider the following class of low-Q digital low-pass filters. Let f_s be the sampling rate. Let g0, g1, and g2 be the filter gains at DC, $0.25f_s$, and $(\frac{1}{2})f_s$ respectively. The parameter $0<\alpha<1$ determines the gain ratio g2/g0. The pole-zero plot is shown in Figure 15.15. First, we show the H(z) transform and derive the digital filter.

$$H(z) = \frac{Y(z)}{X(z)} = \frac{z+\alpha}{z-\alpha} = \frac{1+\alpha z^{-1}}{1-\alpha z^{-1}}$$

$$y(n) = x(n) + \alpha x(n-1) + \alpha y(n-1)$$

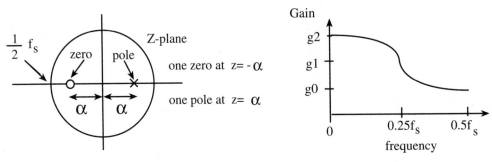

Figure 15.15   Pole-zero plot and gain versus frequency response of a low-Q low-pass filter.

To determine g0, the gain response at DC, we calculate the magnitude of H(z) at z=1. We solve both the general case of α and the specific case of $\alpha = 0.75$. We simplify the expression as much as possible.

$$g0 = |H(z)|_{at\ z=1} = \left| \frac{1 + \alpha}{1 - \alpha} \right| = \frac{1.75}{0.25} = 7$$

To determine g1, the gain response at $0.25f_s$, we calculate the magnitude of H(z) at z=i. We notice the g1 is 1 for all α.

$$g1 = |H(z)|_{at\ z=1} = \left| \frac{i + \alpha}{i - \alpha} \right| = 1$$

To determine g2, the gain response at $(\frac{1}{2})f_s$, we calculate the magnitude of H(z) at z=−1. We solve both the general case of α and the specific case of $\alpha = 0.75$. We simplify the expression as much as possible.

$$g2 = |H(z)|_{at\ z=-1} = \left| \frac{-1 + \alpha}{-1 - \alpha} \right| = \frac{-0.25}{-1.75} = \frac{1}{7}$$

For the specific case of $\alpha = 0.75$, we have the IIR

$$y(n) = x(n) + 0.75x(n-1) + 0.75y(n-1)$$

We can scale the filter so that the gain at DC is 1 by dividing the x(n) and x(n−1) terms by 7:

$$y(n) = \frac{4x(n) + 3x(n-1) - 21y(n-1)}{28}$$

15.9   Digital Low-Pass Filter

Another class of digital low-pass filters places the zeros equidistant along the unit circle. If k is a positive integer, then the equation

$$1 - z^{-k} = 0$$

has solutions evenly spaced along the unit circle at

$$z = e^{j\ 2\pi m/k} \qquad \text{for m} = 0,1 \ldots ,k-1$$

If the term $1 - z^{-k}$ appears in the numerator of the filter transform H(z), then zeros will occur at

$$f = \frac{m2\pi f_s}{k} \qquad \text{for } m = 0, 1 \dots, k-1$$

We can create a digital low-pass filter by placing a pole at z=1 (f=0) to cancel the zero at z=1 (Figure 15.16). The resulting transfer function for this class of low-pass filter is

$$H(z) = \frac{1 - z^{-k}}{1 - z^{-1}}$$

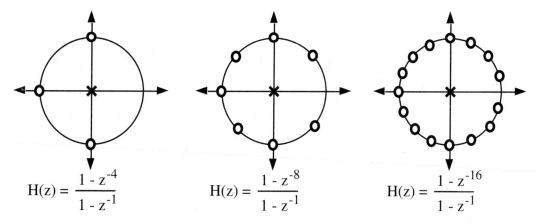

$$H(z) = \frac{1 - z^{-4}}{1 - z^{-1}} \qquad H(z) = \frac{1 - z^{-8}}{1 - z^{-1}} \qquad H(z) = \frac{1 - z^{-16}}{1 - z^{-1}}$$

Figure 15.16   Pole-zero plots of three low-pass filters.

We can use L'Hopital's rule to calculate the DC gain.

$$\lim_{z \to 1} H(z) = \lim_{z \to 1} \frac{kz^{-k-1}}{z^{-2}} = k$$

This class of digital low-pass filters can be implemented with a large MACQ and a simple calculation.

$$y(n) = \frac{x(n) - x(n-k)}{k} + y(n-1)$$

The gain of this class of filter is shown in Figure 15.17 for a sampling rate of 100 Hz. We can decrease the gain in the cutoff region by doubling all the poles and zeros.

$$H(z) = \left(\frac{1}{k}\frac{1 - z^{-k}}{1 - z^{-1}}\right)^2 = \frac{1}{k^2}\frac{1 - 2z^{-k} + z^{-2k}}{1 - 2z^{-1} + z^{-2}}$$

These double-pole–double-zero digital low-pass filters can also be implemented with a large MACQ (Figure 15.18),

$$y(n) = \frac{x(n) - 2x(n-k) + x(n-2k)}{k^2} + 2y(n-1) - y(n-2)$$

Figure 15.17
Gain versus frequency
plot of four low-pass
filters.

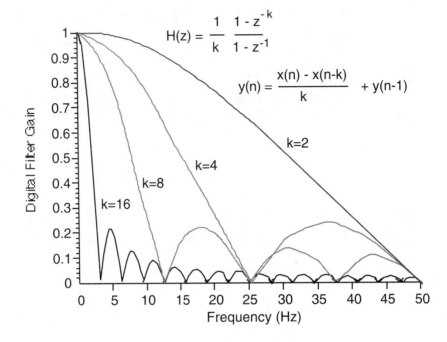

$$H(z) = \frac{1}{k} \frac{1 - z^{-k}}{1 - z^{-1}}$$

$$y(n) = \frac{x(n) - x(n-k)}{k} + y(n-1)$$

Figure 15.18   Gain
versus frequency plot of
four low-pass filters,
using double poles.

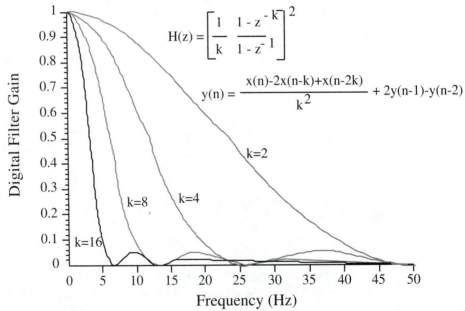

$$H(z) = \left[\frac{1}{k} \frac{1 - z^{-k}}{1 - z^{-1}} \right]^2$$

$$y(n) = \frac{x(n) - 2x(n-k) + x(n-2k)}{k^2} + 2y(n-1) - y(n-2)$$

15.10   Direct-Form Implementations

The general form for the transfer function for an IIR filter is

$$H(z) = \frac{Y(z)}{X(z)} = \frac{a_0 + a_1 z^{-1} + a_2 z^{-2} + \dots + a_M z^{-M}}{1 + b_1 z^{-1} + b_2 z^{-2} + \dots + b_N z^{-N}}$$

This converts to the standard difference equation

$$y(n) = a_0x(n) + a_1x(n-1) + a_2x(n-2) + \ldots + a_Mx(n-M) - b_1y(n-1) - b_2y(n-2) - \ldots - b_Ny(n-N)$$

The direct-form calculation of this filter requires two MACQs with lengths M and N. There are $(M+N-1)$ multiplies and $(M+N-2)$ additions. The data flow picture in Figure 15.19 illustrates the standard implementation. The z^{-1} boxes are time delays (MACQ). The other boxes are multiplications. The + circles are additions.

For the next implementation we specify the filter with N=M. We can do this without loss of generality by letting some of the coefficients be zero. An alternative implementation, called the *direct-form II realization*, requires only one MACQ of length N. There are still $(2N-1)$ multiplies and $(2N-2)$ additions. The data flow picture of Figure 15.20 illustrates the implementation.

Figure 15.19
General filter design using a direct-form calculation.

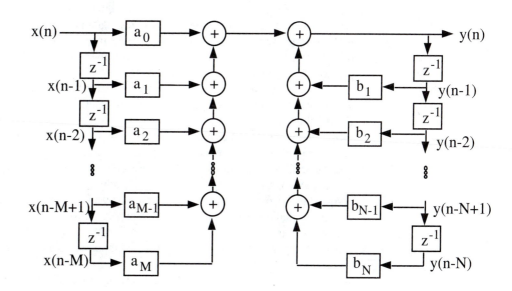

Figure 15.20
General filter design using a direct-form II calculation.

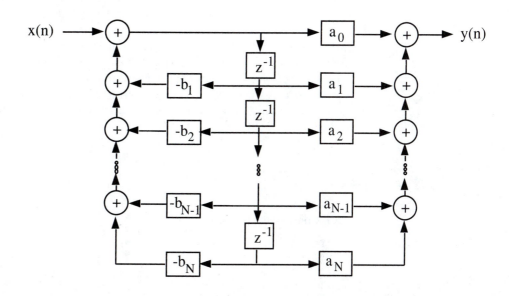

15.11 Glossary

ADC (Analog-to-digital converter) An electronic device that converts analog signals into digital form.

causal The system depends on the present and past values, but not on any future values.

finite-impulse response (FIR) filter A digital filter whose output is a function of a finite number of current and past data samples, but not a function of previous filter outputs.

fixed-point A technique where calculations involving nonintegers are performed using a sequence of integer operations [e.g., 0.123*x is performed in decimal fixed-point as (123*x)/1000 or in binary fixed-point as (126*x)>>10].

hard real-time system One that can guarantee that a process will complete a critical task within a certain specified range; in a DAS hard real time means there is an upper bound on the latency between when a sample is supposed to be taken (every 1/fs) and when the ADC is actually started; hard real time also implies that no ADC samples are missed.

infinite-impulse response (FIR) filter A digital filter whose output is a function of an infinite number of past data samples, usually by making the filter output a function of previous filter outputs.

latency The response time of the computer to external events; for a DAS, the time between when the data should be sampled and when the ADC is started.

linear filter The output is a linear combination of its inputs.

multiple access circular queue (MACQ) A data structure used in DASs to hold the current sample and a finite number of previous samples.

nonlinear filter The output is not a linear combination of its inputs (e.g., median, minimum, maximum are examples of nonlinear filters).

Nyquist theorem If a input signal is captured by an ADC at the regular rate of f_s samples/s, then the digital sequence can accurately represent the 0 to $(1/2)f_s$ frequency components of the original signal.

output compare A mechanism to cause an output pin to change when the TCNT matches a preset value.

pole A place in the frequency domain where the filter gain is infinite.

Q The Q of a bandpass filter (passes fmin to fmax) is the center pass frequency [fo=(fmax−fmin)/2] divided by the width of the pass region, Q=fo/(fmax−fmin); the Q of a band-reject filter (rejects fmin to fmax) is the center-reject frequency [fo=(fmax−fmin)/2] divided by the width of the reject region, Q=fo/(fmax-fmin).

sampling rate The rate at which data are collected in a DAS.

zero A place in the frequency domain where the filter gain is zero.

Z transform A transform equation converting a digital time-domain sequence into the frequency domain; in both the time and frequency domain it is assumed the signal is band-limited to 0 to $(1/2)f_s$.

15.12 Exercises

15.1 Consider the following digital filter:

$$y(n) = \frac{x(n) - x(n-2)}{2}$$

a) Using the Z transform derive general expressions for the gain and phase of the filter.

b) Using the general expressions from part a, calculate the gain and phase of the filter at DC and 60 Hz if the sampling rate is 240 Hz.

15.2. Design a *one-pole* high-Q fixed-point digital high-pass filter. Let $f_s = 100$ Hz, similar to Figure 15.13, but one zero and one pole.

a) Show the pole-zero plot of the filter.

b) Using the pole-zero plot derive the Z transform of the filter.

c) Using the Z transform of the filter, determine the digital filter. The filter gain at $f=0$ should be 0. The filter gain at $f=50$ Hz should be 1.

d) Show a fixed-point version of the filter.

15.3 Consider the use of the Z transform in the design and analysis of digital filters.

a) State the definition of the Z transform.

b) Why can't we use the Z transform on a median filter?

c) Use the Z transform to determine the DC gain and phase of the following digital filter:

$$y(n) = x(n) - x(n-2) + y(n-1)$$

15.4 Design a 10-Hz digital low-pass filter with a sampling rate of 1000 Hz. Make the gain at DC equal to 1, and the gain at 10 Hz equal to 0.707.

a) Show the pole-zero plot of your filter.

b) Show the H(z) transform.

c) Show the floating-point version of the digital filter.

d) Show the fixed-point version of the digital filter.

15.5 You will write the software that implements a real-time DAS with a sampling rate of 1000 Hz. You will also implement the following average filter [data x(n) are sampled at 1000 Hz, and the equation y(n) is also calculated 1000 times/s]:

$$y(n) = \frac{x(n) + x(n - 999)}{2}$$

Your software will also call the already-written display functions `InitDisplay` and `Display`. You do not write these functions but call them with the appropriate parameters at the appropriate times. Other than calls to `Display` your system provides no other operator I/O. The average execution time for calls to `Display` is 500 μs, but the worst case (occasionally) time is 1 s. The prototypes for the three additional functions are:

```
void Display(int);       // displays an integer value
void InitDisplay(void); // initializes the display (call this once)
int Sample(void);        // returns a 12 bit signed sample -2048 to +2047
```

The trick to this exercise is to implement a MACQ that does not require shifting the data on each sample.

15.6 You will write the software that implements a real-time DAS with a sampling rate of 1000 Hz. You will also implement the following sliding average filter [data x(n) is sampled at 1000 Hz, and the equation y(n) is also calculated 1000 times/s]:

$$y(n) = \frac{\sum_{i=0}^{999} x(n - i)}{1000}$$

Your software will also call the already-written display functions `InitDisplay` and `Display`. You do not write these functions but call them with the appropriate parameters at the appropriate times. Other than calls to `Display` your system provides no other operator I/O. The average execution time for calls to `Display` is 500 μs, but the worst case (occasionally) time is 1 s. The prototypes for the three additional functions are:

```
void Display(int);      // displays an integer value
void InitDisplay(void); // initializes the display (call this once)
int Sample(void);       // returns a 12 bit signed sample -2048 to +2047
```

The tricks to this problem are

1. To implement a MACQ that does not require shifting the data on each sample
2. To implement the sum without having to perform 1000 additions for each calculation (*Hint:* It can be done with one add, one subtract, and one divide)

15.7 The objective in this exercise is to design a modular and flexible software implementation of a high-Q digital notch IIR filter. In all cases, you may assume the data points are 16 bits integers sampled from a 12-bit ADC. Thus the input range is -2048 to $+2047$ inclusive. The sampling frequency and notch frequency will be set dynamically (at run time.) These parameters will be established when the user calls your `Filter()` function. You may implement any Q you wish, but it should be similar to the high-Q 60-Hz notch IIR filter developed in Section 15.4. Adjust the filter gain so that the gain at DC is 1.0. The data will be passed through the notch filter by calls to your `Filter()` function. You may assume without checking that the cutoff frequency fc is strictly greater than 0 and strictly less than half the sampling rate, $0 < fc < (1/2)fs$. Your software does not handle the establishment of the actual sampling rate. The following main program illustrates how a user might apply your filter software.

```
#include "FILTER.H"           // prototype file for your filter
void main(){ int x,y;         // filter input/output
    InitFilter(100,1000);     // notch at 100 Hz, sampling at 1000 Hz
    SetFs(1000);              // you do not write this
    do {
        Wait();               // you do not write this
        x=A2D();              // you do not write this
        y=Filter(x);          // call to your digital filter
        process(y);           // you do not write this
    } while (1);
}
#include "FILTER.C"   // implementation file for your filter
```

a) Show the prototype file `Filter.H`. Consider carefully what you put in the .H file and what you put in the .C file.

b) Show the prototype file `Filter.C`. In addition to other necessary software, please show the implementations of `InitFilter()` and `Filter()`. You may use floating-point within `InitFilter()`, but the `Filter()` function should use only fixed-point integer calculation.

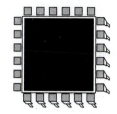

Index

Output parameters for various open collector gates

Family	Example	V_{OL} (V)	I_{OL} (A)
Standard TTL	7405	0.4	0.016
Schottky TTL	74W05	0.5	0.02
Low power Schottky TTL	74LS05	0.5	0.008
High speed CMOS	74HC05	0.33	0.004
High voltage output TTL	7406	0.7	0.040
High voltage output TTL	7407	0.7	0.04
Silicon monolithic IC	75492	0.9	0.25
Silicon monolithic IC	75451-75454	0.5	0.3
Darlington switch	ULN-2074	1.4	1.25
MOSFET	IRF-540	varies	28

Output parameters for various open emitter gates

Family	Example	V_{CE} (V)	I_{CE} (A)
Silicon monolithic IC	75491	0.9	0.05
Darlington switch	ULN-2074	1.4	1.25
MOSFET	IRF-540	varies	28

Parameters of typical transistors used by microcomputer to source or sink current

Type	NPN	PNP	Package	$V_{be(SAT)}$	$V_{ce(SAT)}$	h_{fe} min/max	I_c
General purpose	2N3904	2N3906	TO-92	0.85 V	0.2 V	100	10 mA
General purpose	PN2222	PN2907	TO-92	1.2 V	0.3 V	100	150 mA
General purpose	2N2222	2N2907	TO-18	1.2 V	0.3 V	100/300	150 mA
Power transistor	TIP29A	TIP30A	TO-220	1.3 V	0.7 V	15/75	1 A
Power transistor	TIP31A	TIP32A	TO-220	1.8 V	1.2 V	25/50	3 A
Power transistor	TIP41A	TIP42A	TO-220	2.0 V	1.5 V	15/75	3 A
Power Darlington	TIP120	TIP125	TO-220	2.4 V	2.0 V	1000 min	3 A

General specification of various types of resistor components[1]

Type	Range	Tolerance	Temperature coef	Max Power
Carbon composition	1 • to 22 M•	5 to 20 %	0.1 %/• C	2 W
Wirewound	1 • to 100 k•	>0.0005 %	0.0005 %/• C	200 W
Metal film	0.1 • to 10^{10}•	>0.005 %	0.0001 %/• C	1 W
Carbon film	10 • to 100 M•	>0.5 %	0.05 %/• C	2 W

[1] Wolf and Smith, *Student Reference Manual,* Prentice Hall, pg. 272, 1990.